Library of Congress Cataloging-in-Publication Data
Koerner, Robert M.,
Designing with geosynthetics / Robert M. Koerner. -- 4th ed.
p. cm.
Includes bibliographical references and index.
ISBN 0-13-726175-6 (hardcover)
1. Geosynthetics. I. Title.
TA455.G44K64 1997
624.1'5--dc21 97-30070
CIP

Acquisitions Editor: Bill Stenquist
Editor-In-Chief: Marcia Horton
Production Manager: Bayani Mendoza de Leon
Production Service: ETP Harrison
Director of Production and Manufacturing: David W. Riccardi
Creative Director: Paula Maylahn
Cover Designer: Patricia Woscayk
Full Service Coordinator/Buyer: Donna Sullivan
Editorial Assistant: Meg Weist
Composition: ETP Harrison

Simon & Schuster / A Viacom Company
Upper Saddle River, New Jersey 07458

Printed in the United States of America

10 9 8 7 6 5 4 3 2 1

ISBN 0-13-726175-6

Prentice-Hall International (UK) Limited, *London*
Prentice-Hall of Australia Pty. Limited, *Sydney*
Prentice-Hall Canada Inc., *Toronto*
Prentice-Hall Hispanoamericana, S.A., *Mexico*
Prentice-Hall of India Private Limited, *New Delhi*
Prentice-Hall of Japan, Inc., *Tokyo*
Simon & Schuster Asia Pte. Ltd., *Singapore*
Editora Prentice-Hall do Brasil, Ltda., *Rio de Janeiro*

Fourth Edition

Designing with Geosynthetics

Robert M. Koerner, Ph.D., P.E.

H. L. Bowman Professor of Civil Engineering,
Drexel University and Director, Geosynthetic
Research Institute

PRENTICE HALL
Upper Saddle River, New Jersey 07458

Dedication

For past and current support by the member organizations of the Geosynthetic Research Institute, and with an eye on an expanded and fruitful relationship within the Geosynthetic Institute, the author offers sincere appreciation.

Contents

Preface to the Fourth Edition

Since publication of the previous edition of *Designing with Geosynthetics*, the information database on the subject has grown enormously. New applications, design models, test methods, and materials have emerged showing the ongoing vibrant nature of geosynthetics. Every segment of the technology is involved: owners and regulators, engineers and consultants, testing and inspection firms, resin and additive producers, manufacturers, representatives and installers. Faculty are regularly teaching geosynthetics at many universities and colleges, and geosynthetic-related research involving graduate and undergraduate students is common. This involvement, and the interactions between the various groups, has sustained the continued strong use of geosynthetics in North America and Europe. New to the geosynthetics scene, however, is an exploding use of geosynthetics in Asia, Latin America, Africa, and Australia. Geosynthetics have truly positioned themselves as globally used construction materials.

With this global perspective in mind, a new expanded fourth edition was felt to be timely. It also offered the opportunity to incorporate SI units throughout the book (previous editions had been a mixture of standard, metric, and SI units).

The structure of the book remains as in the past, with the first chapter presenting an overview of geosynthetics and each subsequent chapter being a stand-alone unit on a particular geosynthetic material. Numeric examples, references, and problems accompany each chapter. Significant new material has been added in the geotextile chapter (walls, containers, and tubes); the geogrid chapter (reflective cracking, walls, and veneer reinforcement); the geomembrane chapter (slope stability, access ramps, waste stability, and heap leach pads); the geosynthetic clay liner chapter (new products and related performance); and the geocomposite chapter (new products, related behavior, and geofoam). Wherever possible, suggested areas for further research and development are mentioned.

Building upon the many meaningful experiences and interactions with past colleagues and students (mentioned in the prefaces of previous editions), this edition regularly co-opted my colleagues Drs. Y. (Grace) Hsuan, George R. Koerner, Te-Yang Soong, and Arthur E. Lord on many issues. My thanks to each of them. Graduate students during this interval have been William A. Harpur, Dhani B. Narejo, Wojciech (Alex) Gontar, Zeqing Guan, Rou-Chang (Danny) Liou, and Matthew A. Eberlé. Three individuals, however, have gone beyond being excellent in their cooperation in making this edition possible: Paula Koerner (in shielding me from many other duties), Marilyn Ashley (in typing draft after draft), and Te-Yang Soong (who was superb as a cooperative technical associate). A most appreciative thanks to all.

Robert M. Koerner
Philadelphia, Pennsylvania

1 An Overview of Geosynthetics

1.0 INTRODUCTION

An exciting new chapter in engineered materials has emerged for the civil engineering community—the rapidity at which the related products are being developed and used is nothing short of amazing. At no time in the author's experience have new materials within a specific area come on so strong. The reasons for this explosion of new products are numerous and include the following.

- They are quality-control manufactured in a factory environment.
- They can be installed rapidly.
- They generally replace raw material resources.
- They generally replace difficult designs using natural materials.
- Their timing is very appropriate.
- Their use is required by regulations in some cases.
- They are generally cost competitive against the natural soils that they replace.
- They make heretofore impossible designs and applications possible.
- They are actively marketed and widely available.

The professional groups most strongly influenced are geotechnical engineering, transportation engineering, environmental engineering and hydraulics engineering, although all soil-, rock-, and groundwater-related activities fall within the general scope of the various applications. This being the case, the term geosynthetics seems appropriate. *Geo,* of course, refers to earth. Acknowledgement that the materials are almost exclusively from human-made products gives the second part to the name—*synthetics.* The materials used in the manufacture of geosynthetics are almost entirely from the plastics industry; that is, they are primarily polymers, although fiberglass, rubber, and natural materials are sometimes used. This opening chapter presents an overview of the family of geosynthetic materials and serves to introduce the remainder of the book.

1.1 BASIC DESCRIPTION OF GEOSYNTHETICS

Lost to history are the initial attempts to reinforce soils; the adding of materials to enhance the properties of the soil was no doubt done long before our first historical records of it. It seems reasonable to assume that the first attempts were made to stabilize swamps and marshy soils using tree trunks, small bushes, and the like. These soft soils would accept the fibrous material until a mass was formed that had adequate properties for the intended purpose. It also seems reasonable to accept that either the continued use of such a facility was possible because of the properly stabilized nature of the now-reinforced soil (probably by trial and error), or was impossible due to a number of factors, among which were:

- insufficient reinforcement materials for the loads to be carried;
- pumping of the soft soil up through the reinforcement material; and
- degradation of the fibrous material over time, leading back to the original unsuitable conditions.

Such stabilization attempts were undoubtedly continued with the development of a more systematic approach in which timbers of nearly uniform size and length were lashed together to make a mattressed surface. Such split-log "corduroy" roads over peat bogs date back to 3000 B.C. [1]. This art progressed to the point where the ridged surface was filled in and smooth. Some of these systems were surfaced with a stabilized soil mixture or even paved with stone blocks. Here again, however, the deterioration of the timber and its lashings over time was an obvious problem.

The concept of reinforcing poor soils has continued until the present day. The first use of fabrics in reinforcing roads was attempted by the South Carolina Highway Department in 1926 [2]. A heavy cotton fabric was placed on a primed earth base, hot asphalt was applied to the fabric, and a thin layer of sand was put on the asphalt. The Department published the results of eight field experiments in 1935: Until the fabric deteriorated, the roads were in good condition and the fabric reduced cracking, raveling, and localized road failures. This project was the forerunner of the separation and reinforcement functions of geosynthetic materials as we know them today. The separation and reinforcement of soils is one major topic of this book.

A second major topic is the provision of an intermediate barrier between two dissimilar materials for the purpose of liquid (usually water) drainage and soil filtration. When requiring liquid flow across such a barrier it must, obviously, be porous, yet not so porous as to lose the retained soil—thus the necessity of using some sort of intermediate material. Again the historical development of filtration provides an important background for understanding the work that followed. Run-of-bank gravel, which was found to be naturally well graded, had been used as filter material since ancient times.

Purification of polluted water by running it through soil is an outmoded method, but is still apparently used on occasion. The idea of systematizing the filtration process seems to have been originated by K. Terzaghi and A. Casagrande in the 1930s and put to use by Bertram [3] shortly thereafter. This construction of soil filters, even multiple-graded soil filters, is a target area for the geosynthetic materials described in this book for reasons of quality control and cost effectiveness.

A third major topic of this book is the creation of a leak-proof barrier for preventing liquid movement from one point to another. Such containment liners have historically been made using poorly draining soils, notably low-permeability clay soils. The Roman aqueducts were lined in such a manner and the technology undoubtedly preceded them by many years [4]. Liners made from bitumen and various cements have been used since the 1900s, but the synthetic rubber developed in the 1940s was required before polymeric liners were developed. Today, such liners are regulatory-mandated for use in many environmental related applications. Interestingly, the newest barrier materials are combinations of both synthetic and soil materials used as a geocomposite material.

Thus geosynthetic materials perform five major functions: separation, reinforcement, filtration, drainage, and containment (as a liquid and/or gas barrier). The use of geosynthetics has two aims: to do the job better (e.g., with no deterioration of material or excessive leakage) and to do the job more economically (either through lower initial cost or through greater durability and longer life, thus reducing maintenance costs).

1.1.1 Families of Geosynthetics

The specific families of geosynthetics on which we focus are shown in Figure 1.1 and are discussed next.

Geotextiles. Geotextiles (see Section 1.3 and Chapter 2) form one of the two largest groups of geosynthetics described in this book. Their rise in growth during the past twenty years has been nothing short of awesome. They are indeed textiles in the traditional sense, but consist of synthetic fibers rather than natural ones such as cotton, wool, or silk. Thus biodegradation is not a problem. These synthetic fibers are made into flexible, porous fabrics by standard weaving machinery or are matted together in a random or nonwoven manner. Some are also knitted. The major point is that they are porous to liquid flow across their manufactured planes and also within their thickness, but to widely varying degrees. There are at least 100 specific application areas for geotextiles; however, the fabric always performs at least one of five discrete functions:

- Separation
- Reinforcement
- Filtration
- Drainage
- Containment (barrier, when impregnated)

Geogrids. Geogrids (see Section 1.4 and Chapter 3) represent a rapidly growing segment within geosynthetics. Rather than being a woven, nonwoven, or knitted

Figure 1.1 Typical geosynthetic materials.

textile fabric, geogrids are plastics formed into a very open, gridlike configuration; that is, they have large apertures. Geogrids are either stretched in one or two directions for improved physical properties or made on weaving machinery by unique methods. There are many application areas and they function almost exclusively as reinforcement materials.

Geonets. Geonets, called geospacers by some (see Section 1.5 and Chapter 4), constitute another specialized segment within geosynthetics. They are usually formed by a continuous extrusion of parallel sets of polymeric ribs at acute angles to one another. When the ribs are opened, relatively large apertures are formed into a netlike configuration. Their design function is completely within the drainage area where they have been used to convey liquids of all types.

Geomembranes. Geomembranes (see Section 1.6 and Chapter 5) represent the other largest group of geosynthetics described in this book, and in dollar volume their sales are probably greater than that of geotextiles. Their growth in the USA and Germany has been stimulated by governmental regulations originally enacted in the early 1980s. The materials themselves are "impervious" thin sheets of polymeric material used primarily for linings and for the covers of liquid- or solid-storage facilities. Thus the primary function in containment is as a liquid or vapor barrier. The range of applications, however, is very great, and in addition to the environmental area, applications are rapidly growing in geotechnical, transportation, and hydraulic engineering

Geosynthetic Clay Liners. Geosynthetic clay liners (or GCLs) (see Section 1.7 and Chapter 6) are the newest subset within the family of geosynthetic materials. They are rolls of factory-fabricated thin layers of bentonite clay sandwiched between two geotextiles or bonded to a geomembrane. Structural integrity is maintained by needle punching, stitching, or physical bonding. They are used as a composite component beneath a geomembrane or by themselves in environmental and containment applications, as well as in transportation, geotechnical, and hydraulic applications.

Geopipe (aka Buried Plastic Pipe). Buried plastic pipe (see Section 1.8 and Chapter 7) is perhaps the oldest geosynthetic material still available today. This orphan of the Civil Engineering curriculum was included due to an awareness that plastic pipe is being used in all aspects of geotechnical, transportation, environmental, and hydraulic engineering with little design and testing awareness, perhaps due to a general lack of formalized training. The critical nature of leachate collection pipes coupled with high compressive loads makes geopipe a bona fide member of the geosynthetics family.

Geocomposites. A geocomposite (see Section 1.9 and Chapter 8) consists of a combination of a geotextile and a geonet; a geotextile and a geogrid; a geogrid and a geomembrane; or a geotextile, a geonet, a geogrid, and a geomembrane; or any one of these four materials with another material (e.g., deformed plastic sheets, steel cables, or steel anchors). This exciting area brings out the best creative efforts of the engineer, manufacturer, and contractor. The application areas are numerous and growing steadily. The major functions encompass the entire range of functions listed for geosynthetics discussed previously: separation, reinforcement, filtration, drainage, and containment (barrier).

While not a geocomposite per se, the emerging material geofoam will be described in Chapter 8. This lightweight material offers many possibilities for the civil engineer.

Geo-Others. Geosynthetics have exhibited such innovation that many systems defy categorization. For want of a better phrase, *geo-others* describes items such as threaded soil masses, polymeric anchors, and encapsulated soil cells. As with geocomposites, their primary function is product-dependent and can be any of the five major functions of geosynthetics. These materials will be discussed throughout the book in the chapters most closely fitting the material.

1.1.2 Market Activity

To say that the market activity of geosynthetics in the geotechnical, transportation, and environmental areas is strong is decidedly an understatement. To obtain an insight into the vitality of geosynthetics note the curves in the graphs in Figure 1.2. The upper set of curves gives the estimated amount of geosynthetics used in North America over the years (geopipe is not shown), while the lower set gives the estimated in-place expenditures of these products. Used in the calculations were the data for 1995 (note that the values are in millions of square meters and millions of dollars):

geotextiles	500 Mm^2 @ \$0.90/m^2	=	\$450 M
geogrids	40 Mm^2 @ \$2.50/m^2	=	100 M
geonets	50 Mm^2 @ \$2.00/m^2	=	100 M
geomembranes	75 Mm^2 @ \$10.00/m^2	=	750 M
geosynthetic clay liners	50 Mm^2 @ \$2.50/m^2	=	125 M
geocomposite	25 Mm^2 @ \$5.00/m^2	=	125 M
geo-others	5 Mm^2 @ \$4.00/m^2	=	20 M
	total (1995)		\$1,670 M

The total represents an astounding \$1.67 billion for 1995! Never in the author's career has a construction material so enthralled and captivated the civil engineering community. The reasons for this activity and for why geosynthetics have only been recently accepted and used will be discussed next.

1.2 POLYMERIC MATERIALS

Well over 95% of the geosynthetics discussed in this book are made from synthetic polymers. Thus a brief discussion on the topics of polymer composition, structure, and identification is in order. This section is not meant to make a polymer engineer or a polymer scientist out of the reader, but only to afford an appreciation of the wealth of information that is available, the sophistication of the topic area, and the need for at least a rudimentary understanding of geosynthetics at the molecular level that will prove beneficial for the remainder of the book.

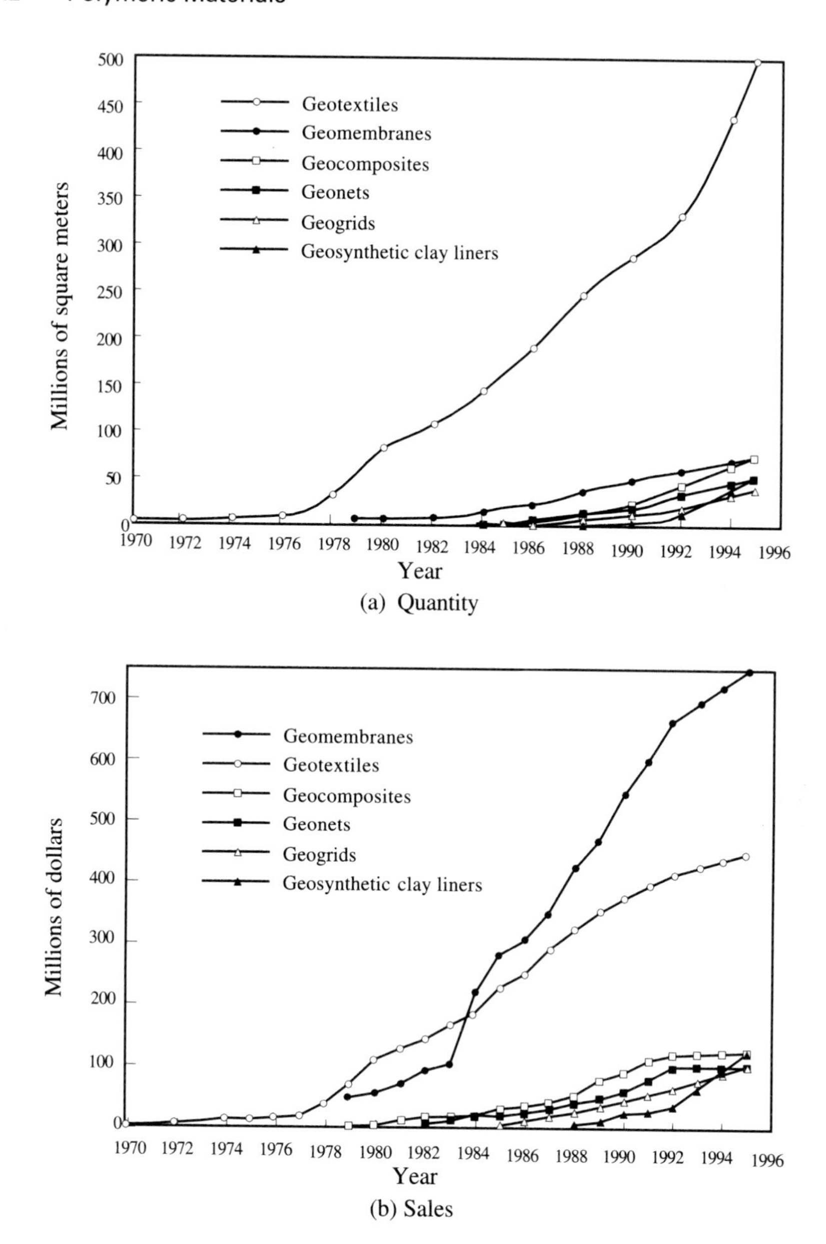

Figure 1.2 Estimated geosynthetic market in North America.

1.2.1 Brief Overview

The word *polymer* comes from the Greek *poly* meaning "many" and *meros* meaning "parts." Thus a polymeric material consists of many parts joined together to make the whole. Each part, or unit, is called a *monomer*. A monomer is a molecular compound used to produce the polymer. It should be recognized that the monomers and the repeating molecular units are different. This is due to the polymerization process. The functionality (i.e., the number of sites at which a monomeric molecule can link with other monomer molecules) determines the type and length of the chain.

The molecular weight of a polymer is the degree of polymerization (i.e., the number of times a repeating unit occurs) times the molecular weight of the repeating unit. The average molecular weight and its statistical distribution are very important in the resulting behavior of the polymer, since increasing *average* molecular weight results in:

- increased textile strength,
- increased elongation,
- increased impact strength,
- increased stress crack resistance,
- increased heat resistance,
- decreased flow behavior,
- decreased processability;

while narrowing the molecular weight *distribution* results in:

- increased impact strength,
- decreased stress crack resistance,
- decreased flow behavior,
- decreased processability.

While most of the polymers used in the manufacture of geosynthetics are from one type of monomer, thus called *homopolymers*, there are other possibilities. A polymer made from two repeating units in its chain is called a *copolymer*. Important here is the manner of linking or joining the repeating units, which can be random, alternating, block, or branch (graft). Such copolymerization greatly expands the structural properties of the resulting polymer. Furthermore, it is possible to have three repeating units in the chain; this is called a *terpolymer*. Easily seen is that the options are essentially limitless, which explains why there are approximately 50,000 commercialized polymers in existence. Table 1.1 gives sales of the more common synthetic polymers in the USA through 1990. It is an enormous and ever growing field. As the movie of the 1960s, *The Graduate*, predicted, "Get into polymers, young man [or woman]!"

Fortunately, there are only a few synthetic polymers that make up the vast majority of geosynthetic materials. See Table 1.2 for the repeating molecular units and the types of geosynthetics made from these specific polymers. Within the group shown, polyethylene and polypropylene are the most common, and collectively are called polyolefins.

Bonding between polymer molecules and their chains is critically important in understanding their behavior and performance. A number of excellent references are available for in-depth study [6, 7, 8, 9]; however, Moore and Kline [10] is particularly well-suited for an introduction to the subject. The bonds between polymer molecules,

TABLE 1.1 PRODUCTION OF PLASTICS BY RESIN TYPE IN THE U.S.

Resin	1986	1987	1988	1989	1990
HDPE	31,960	35,577	37,380	36,053	37,099
LDPE (Total)	39,618	42,715	46,266	43,142	49,608
LDPE	NA	NA	NA	29,258	32,284
LLDPE	NA	NA	NA	13,884	17,323
Nylon	2,069	2,256	2,528	2,532	2,483
Polypropylene	25,863	29,579	32,369	32,209	36,979
Polystyrene	19,891	21,271	23,082	22,712	22,343
PVC	32,289	35,470	37,157	37,727	40,481
Polyester	5,228	6,203	7,351	7,253	8,361
Resin	**1991**	**1992**	**1993**	**1994**	**1995**
HDPE	40,998	43,645	44,237	49,470	49,889
LDPE (Total)	49,609	53,030	53,698	56,070	57,342
LDPE	32,200	32,364	32,155	33,722	34,011
LLDPE	19,339	20,665	21,542	22,348	23,331
Nylon	2,563	2,972	3,417	4,196	4,539
Polypropylene	37,068	37,473	38,394	42,448	48,460
Polystyrene	22,045	22,677	23,949	26,023	25,169
PVC	40,779	44,451	45,643	52,118	54,712
Polyester	9,411	10,737	11,343	14,222	16,843

In millions of newtons.

Source: After SPI [5].

as just noted, are van der Waals forces, permanent dipoles, or hydrogen bonds. Between molecular chains, however, the bonds are usually much weaker and often must be supplemented by some form of cross-linking by means of covalent bonds or covalent bonding systems. Cross-links can be formed:

- by the use of monomers having a functionality greater than two,
- by the use of chemical agents (sometimes called *curing*), and
- by the use of nuclear radiation methods.

Cross-linking is an important conceptual consideration because it separates the two major types of polymeric materials, i.e., thermoplastic and thermoset. A *thermoplastic polymer* is one that can be repeatedly heated to its softening point, shaped or worked as desired, and then cooled to preserve that remolded shape. In a *thermoset polymer* the process cannot be repeated. Any additional heat after the first forming will only lead to charring and degradation of the material. The key to this behavior in thermoset materials is, of course, cross-linking, which does not exist in thermoplastic materials. Examples of thermoplastic materials are polyethylene (PE), polypropylene (PP), polyester (PET); examples of thermoset materials are nytrile, butyl, and EPDM. As seen in Table 1.2, geosynthetics consist entirely of thermoplastic materials; there are essentially no thermoset materials currently used in geosynthetic applications.

TABLE 1.2 REPEATING UNITS OF POLYMERS USED IN THE MANUFACTURE OF GEOSYNTHETICS

Polymer	Repeating Unit	Types of Geosynthetics
Polyethylene (PE)	$\left[-\mathrm{CH_2}-\mathrm{CH_2}-\right]_n$	Geotextiles, geomembranes, geogrids, geopipe, geonets, geocomposites
Polypropylene (PP)	$\left[-\mathrm{CH_2}-\mathrm{CH(CH_3)}-\right]_n$	Geotextiles, geomembranes, geogrids, geocomposites
Polyvinyl chloride (PVC)	$\left[-\mathrm{CH_2}-\mathrm{CHCl}-\right]_n$	Geomembranes, geocomposites, geopipe
Polyester (polyethylene terephthalate) (PET)	$\left[-\mathrm{O}-\mathrm{R}-\mathrm{O}-\mathrm{C(=O)}-\mathrm{R'}-\mathrm{C(=O)}-\right]_n$	Geotextiles, geogrids
Polyamide (PA) (nylon 6/6)	$\left[-\mathrm{NH}-(\mathrm{CH_2})_6-\mathrm{NH}-\mathrm{C(=O)}-(\mathrm{CH_2})_4-\mathrm{C(=O)}-\right]_n$	Geotextiles, geocomposites, geogrids
Polystyrene (PS)	$\left[-\mathrm{CH_2}-\mathrm{CH(C_6H_5)}-\right]_n$	Geocomposites, geofoam

Crystallinity can indeed exist in polymeric materials, but does so to widely varying degrees. In a rather difficult-to-visualize manner, the aligned portions of the polymer chain(s) in small regions form complex crystalline patterns called *crystallites*. The nonaligned regions are called *amorphous*. The crystalline patterns are very complex and are still being researched by scientists and engineers. For example, the aligned molecular chain can loop back on itself in a series of folds called *spherulites* and can form exotic patterns. Patterns such as "snowflakes" and "shish kebabs" are known configurations. The amount of crystallinity gives rise to a further polymer classification of noncrystalline (or amorphous) versus semicrystalline. (No polymer is completely crystalline.) Thus the three major classifications of polymers which can be used for geosynthetic materials are amorphous thermoplastic, semicrystalline thermoplastic, and thermoset (rarely used for geosynthetics).

Essentially all of the polymers used in the manufacture of geosynthetics are of the first two varieties. The amount of crystallinity varies from nil to 30% in some polyvinyl chlorides (PVCs), to as high as 65% in high-density polyethylene (HDPE). Crystallinity is significant, and in some instances critical, in the behavior of polymeric geosynthetics. Increasing crystallinity results in the following:

- increasing stiffness or hardness,
- increasing heat resistance,
- increasing tensile strength,
- increasing modulus,
- increasing chemical resistance,
- decreasing diffusive permeability (or vapor transmission),
- decreasing elongation or strain at failure,
- decreasing flexibility,
- decreasing impact strength, and
- decreasing stress crack resistance.

1.2.2 Polymer Identification

There are a number of possible ways to identify the specific polymer from which a material (in our case, a geosynthetic) is made. Table 1.3 gives a classification of polymer type on the basis of its characteristics while burning.

It must be recognized, however, that the burning tests described in Table 1.3 are very subjective. For a significantly more accurate identification of the particular type of any synthetic polymer, there are a number of *chemical analysis tests*. Such tests are finding a place in geosynthetic materials analysis for the following reasons:

- They are used in quality assurance and certification.
- They are used to evaluate the estimated lifetime of field-retrieval samples.
- They are used in laboratory investigations into degradation mechanisms and lifetime prediction.
- They aid in the forensic analysis of failures.

TABLE 1.3 BURNING CHARACTERISTICS OF POLYMERIC MATERIALS USED IN GEOSYNTHETICS

Polymer Type*	Behavior Beginning, During, and After Burning
Polyethylene (PE)	Before touching the flame the material shrinks, melts, and curls. Burns readily and is not self-extinguishing; burns rapidly when moved away from the flame; becomes clear when molten and tends to drip. When the flame is extinguished, the smell is of molten wax; the ash is soft and same color as material.
Polypropylene (PP)	Before touching the flame the material shrinks, melts, and curls. Burns in a manner similar to polyethylene and is not self-extinguishing; burns slowly when moved away from the flame; no clear blue color at the base of the flame, except with carbon black; there is a faint burning-asphalt-like odor. The ash is hard and light tan.
Polyvinyl chloride (PVC), unplasticized	Burns with difficulty and is self-extinguishing; flame is yellow, green at the bottom edges, with spurts of green and yellow; white smoke is given off; the material softens on ignition and has an unpleasant acidic smell.
Polyvinyl chloride (PVC), plasticized	Flammability behavior depends on the proportion of plasticizer present; most plasticizers burn readily with a yellow, smoky flame; black smoke is given off; the odors are mostly ester-like, but with an unpleasant acidic smell.
Polyester (PET)	Burns slowly with a yellow smoky flame; floral (ester) odor; flame jitters and dances; molten material drips.
Nylon (PA)	Difficult to ignite and is self-extinguishing. Slight green flame occurs due to chlorine; flame is blue with a yellow top; the material froths on ignition; odor is of burning wool or hair.
Polystyrene (PS)	Burns readily and is not self-extinguishing. The flame is orange-yellow, and black dense smoke containing soot is given off; flame jitters and dances somewhat; on ignition, the material softens and the odor is characteristic of benzene.
Chlorinated polyethylene (CPE)	Difficult to ignite and is self-extinguishing. Black smoke given off; green flame is noticeable; little odor; smells somewhat of wax; no drops; ash remains.
Chlorosuphonated polyethylene (CSPE) (Hypalon)	Difficult to ignite. White smoke while in flame; black smoke while burning; slight green flame; self-extinguishing; no drops; wax odor.
Most thermosets (e.g., EPDM)	Rapid burning, very intense; no drops; chars at edges; orange flame with black smoke. Char is tacky.

*Final check on a chlorine-containing polymer: Heat a copper wire in a flame and press the heated wire into the sample; then slowly put the wire back into the flame. A green flame points to a chlorine-containing polymer.

- They are used for research and development into new additive packages (stabilizers, antioxidants, and additives).
- They are used for new geosynthetic product development and application investigations.

A brief description of the most frequently used chemical analysis tests as they apply to the polymers used in the manufacture of geosynthetics follows. See Halse et al. [11] for additional insight into these methods.

Thermogravimetric Analysis (TGA). This is the first of a series of thermal methods in which a property of the polymer is tracked as a function of a controlled-temperature program (see Thomas and Verschoor [12] for a review). TGA follows mass as a function of temperature. The results of continuous weighing of the decreasing mass of a specimen being subjected to a constantly increasing temperature can be graphed as shown in Figure 1.3. The pronounced decreases in weight at specific temperatures signify specific component vaporization. For example, the plasticizer in the PVC is removed at about 300°C, while the resin is removed at between 450°C and 500°C. What remains above 500°C is probably carbon black and ash, since the tests were performed in a nitrogen atmosphere. The proportion of each component is readily obtained since the device automatically normalizes the vertical axis.

The technique can also be used to determine kinetic information concerning the stability of the polymer and the energy of activation for thermal decomposition. This latter piece of information can be used in an Arrhenius plot to predict the in-service lifetime at a specific temperature.

Differential Scanning Calorimetry (DSC). By maintaining a temperature balance between a reference cell and a test specimen cell, the heat flow into and out of a specimen can be monitored and plotted as a function of temperature. Figure 1.4 shows

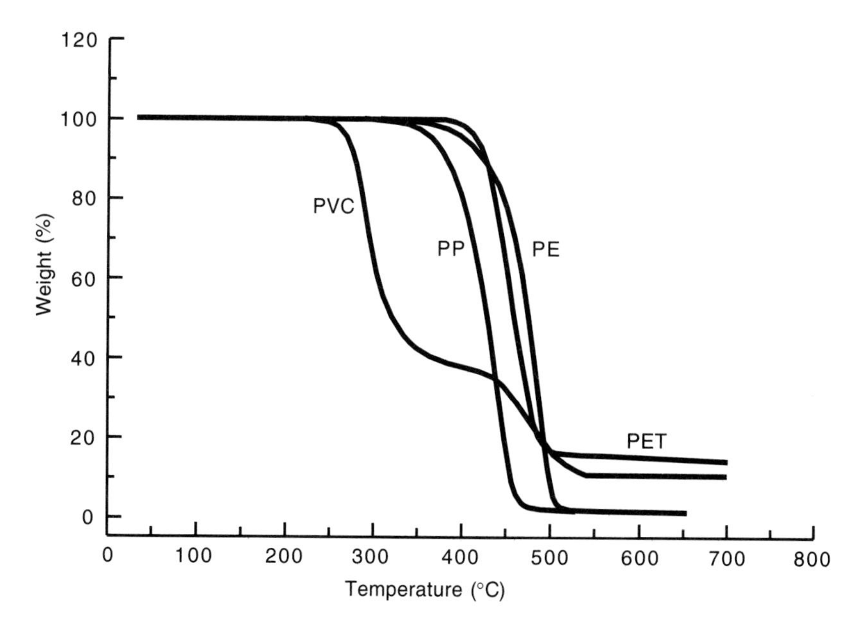

Figure 1.3 Thermogravimetric analysis curves of some common geosynthetic polymers. (After Thomas and Verschoor [12])

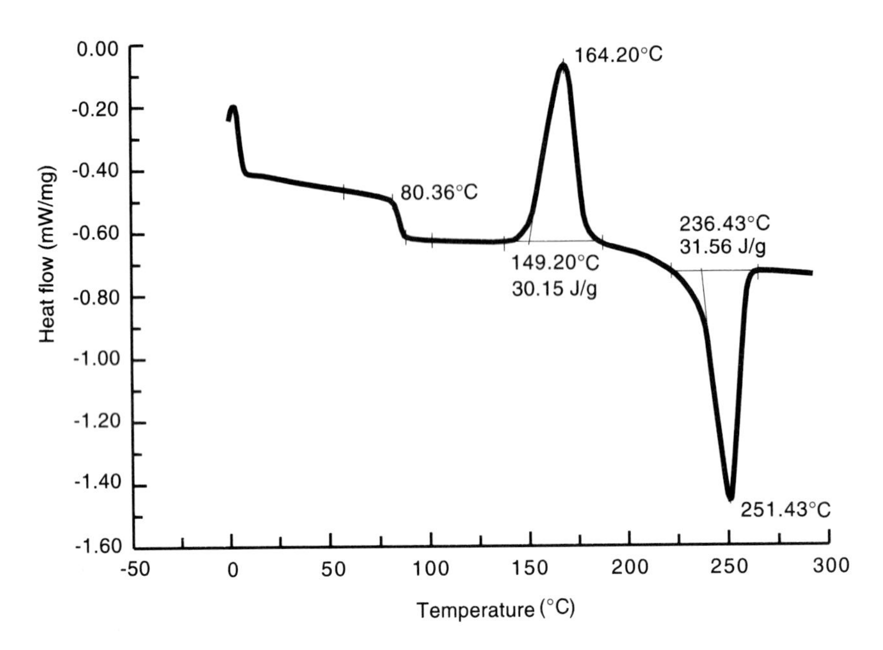

Figure 1.4 Differential scanning calorimeter curves for quenched polyester. (After Thomas and Verschoor [12])

such a trace for quenched polyester [12]. The glass transition temperature is at 80.36°C, the exothermic crystallization of the polymer backbone is at 164.20°C, and the endothermic melting of the crystallites is at 251.43°C. It is very important to recognize that the area under the curve of the crystallization melt, that is, the value of 31.56 J/g, is proportional to the percent crystallinity of the polymer. Reference standards are used as a calibration for obtaining the actual percent crystallinity value. Lastly, the beginning of melting of the crystalline portion of the polymer at 236.43°C is important, as it relates to proper seaming. This can be seen in the DSC curves of Figure 1.5 for different types of polyethylene [13]. Here the crystalline melting zone is clearly defined, and the higher the density of the PE, the narrower the temperature window in which melting takes place. In this regard, HDPE is a challenging material to seam properly.

Oxidative Induction Time (OIT). The OIT test uses a differential scanning calorimeter with a special testing cell capable of sustaining pressure. In the standard test (Std-OIT), per the American Society for Testing and Materials (ASTM) D3895, a 5 mg specimen is heated from room temperature to 200°C at a rate of 20°C/mm under a nitrogen atmosphere. Oxygen is then introduced and the test is terminated when an exothermal peak is reached (see Figure 1.6a) and the OIT time is obtained. This time is related to the quantity and type of antioxidants used in the polymer formulation. As seen in Figure 1.6 (lower), the OIT time is also related to incubation time of HDPE

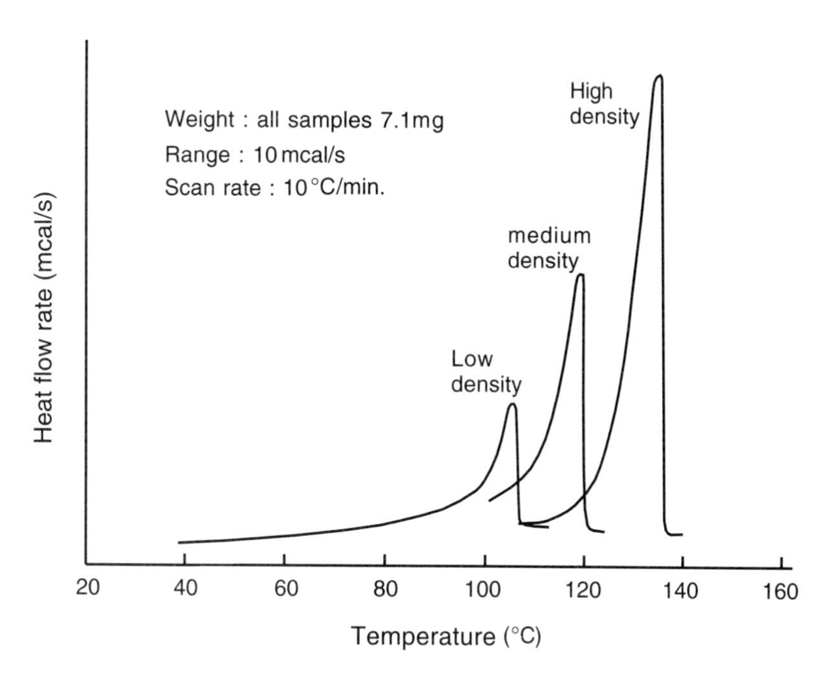

Figure 1.5 Differential scanning calorimeter curves for various densities of polyethylenes. (Compliments of Perkin-Elmer Instrument Co.)

geomembrane samples in forced-air ovens at elevated temperatures (Hsuan and Guan [14]). Data such as these can be used to predict antioxidant depletion lifetime at in situ (and lower) temperatures.

An alternative OIT test, the HP-OIT test, uses higher pressure and lower temperature than the standard test just described. It is designated ASTM D5885. Unless otherwise stated, the test is conducted at a pressure of 3.4 MPa and a temperature of 150°C. A response similar to Figure 1.6 (upper) is obtained. This test is considered to be more representative of the in situ behavior of low-temperature-functioning antioxidants insofar as prediction methods are concerned.

Thermomechanical Analysis (TMA). TMA measures a particular dimension of the polymer under controlled increase in temperature. A quartz probe rests on the test specimen and its displacement is precisely measured as the temperature increases (or decreases). The modes of operation are expansion, penetration, or shear flow of the polymer. The most straightforward property obtained from TMA is the coefficient of thermal expansion; it is simply the slope of the temperature-deformation curve. Figure 1.7 shows the results of such a test for PET, where the linear coefficient of thermal expansion is 76.4 μm/m°C for the initial (also called the glassy state) stage and 132 μm/m°C for the final (or rubbery state) stage. The transition value between the two stages clearly defines the glass-transition temperature, which is 80.55°C.

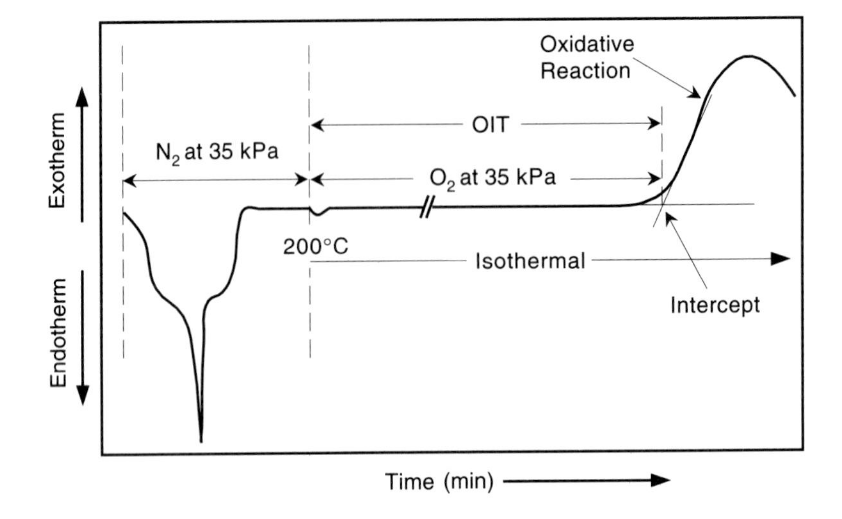

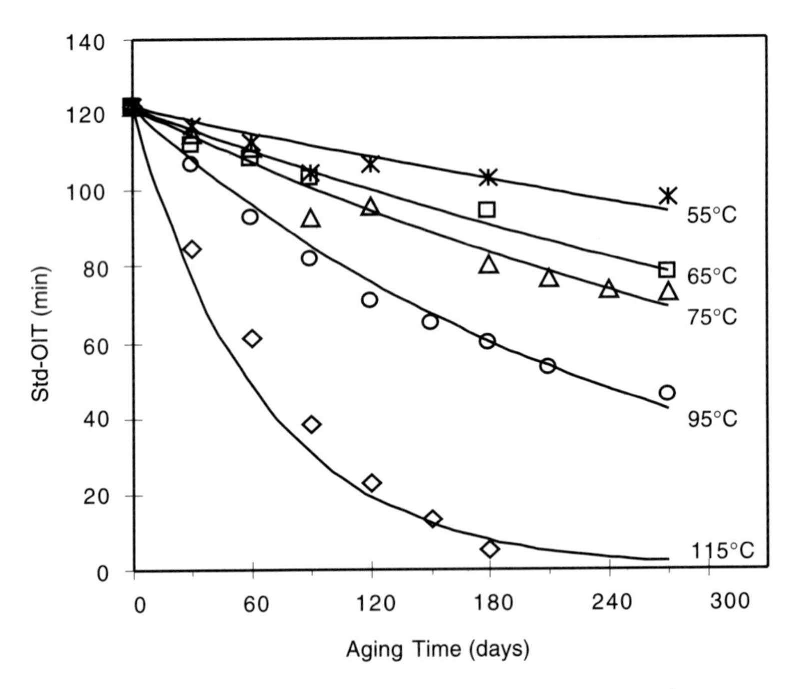

Figure 1.6 (Upper) Oxidative induction time test results and (lower) response after oven aging for a HDPE geomembrane test specimen. (After Hsuan and Guan [14])

Dynamic Mechanical Analysis (DMA). Another thermal technique measures the mechanical response of a polymer as it is deformed under a periodic stress in a controlled temperature environment. Thus the viscoelastic properties can be evaluated. The test measures the dynamic storage modulus, E' (a measure of stiffness), the dynamic

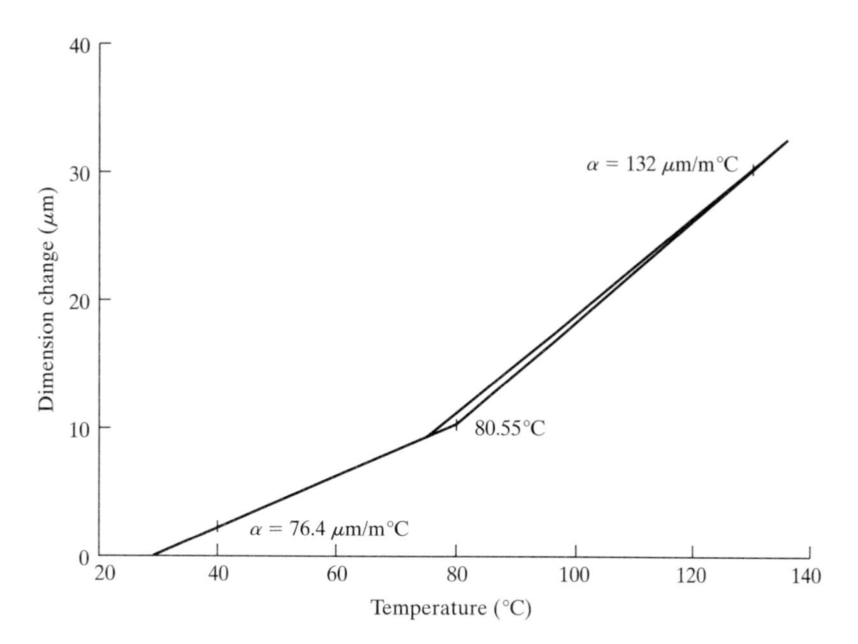

Figure 1.7 Thermomechanical analysis curves for PET under a temperature increase of 10°C/min. (After Thomas and Verschoor [12])

loss modulus, E'' (to measure glass transition and softening points) and the ratio of storage-to-loss moduli, which is called the *damping*, or *tan δ value*. Figure 1.8 gives the response of a HDPE geomembrane sample under the following conditions: at 1 Hz-fixed frequency, heated at 4°C per minute, under 5.7 J of energy. The amplitude of the frequency is 0.20 mm (peak-to-peak).

DMA devices are of additional interest in evaluating the viscoelastic engineering properties via creep and stress relaxation test methods. Since DMA units are computer-controlled, one can either preset the load and measure the deformation (i.e., the creep mode) or preset the deformation and measure the force (i.e., the stress relaxation mode). By repeating the particular test at a series of increasing (or decreasing) temperature increments, a family of curves results. Additionally, a master curve can be generated using the time-temperature superposition principle for use in long-term studies and lifetime prediction.

Infrared Spectroscopy (IR). Infrared spectroscopy, generally used as Fourier transform infrared spectroscopy (FTIR), is based upon the realization that the functional groups in molecules (such as the –CH– group in polyethylene) are always in motion. During an FTIR analysis, the polymer test specimen is subjected to radiation. The frequency of the incident radiation is in the infrared region. If the frequency matches a natural motion of the functional group, the polymer will absorb this energy and an absorption band will appear on the FTIR frequency sweep.

Figure 1.9 shows the spectrum of a polyethylene sample without compounding agents. Each peak in the spectrum represents the motion of a functional group either in bending or in stretching. For example, the strong peak at a frequency of 2850 cm^{-1} is the

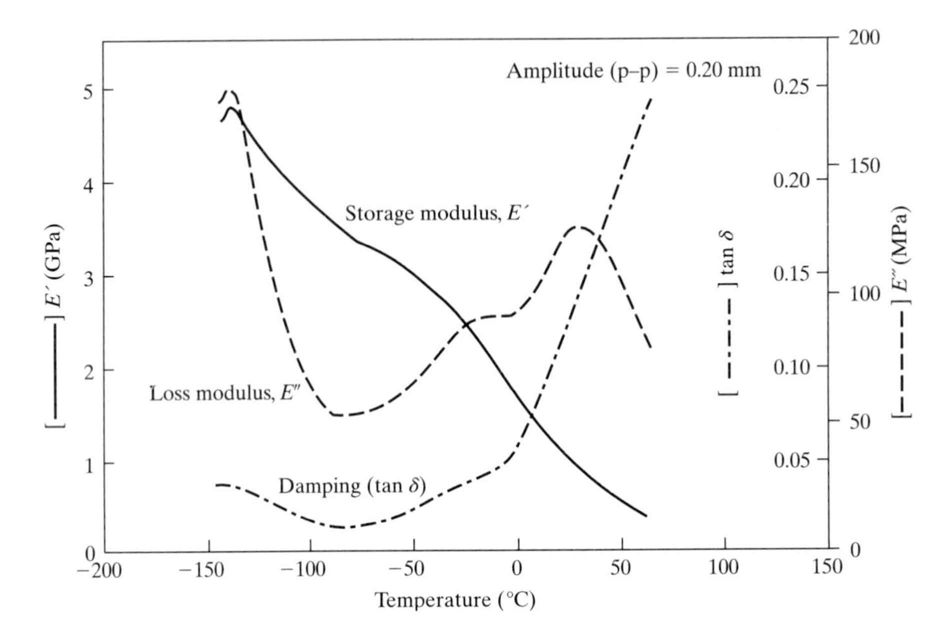

Figure 1.8 Dynamic mechanical analysis curves at fixed frequency for a HDPE geomembrane test specimen.

absorption peak due to C–H stretching motion. The area under the curve in this region is represented as I_{2850}.

For a majority of polymers, carbonyl groups (–C=O–) are produced after a certain amount of oxidation reaction (degradation) has occurred. The frequency corresponding to the motion of this molecular group is approximately 1715 cm^{-1}. (This frequency range does not appear in Figure 1.9 since it was a virgin nondegraded material.) The area of the peak (I_{1715}) is proportional to the amount of carbonyl groups formed in the polymer. Hence it can be used to monitor the progress of oxidation. Often the results are normalized with another peak area taken from the polymer spectrum. For ex-

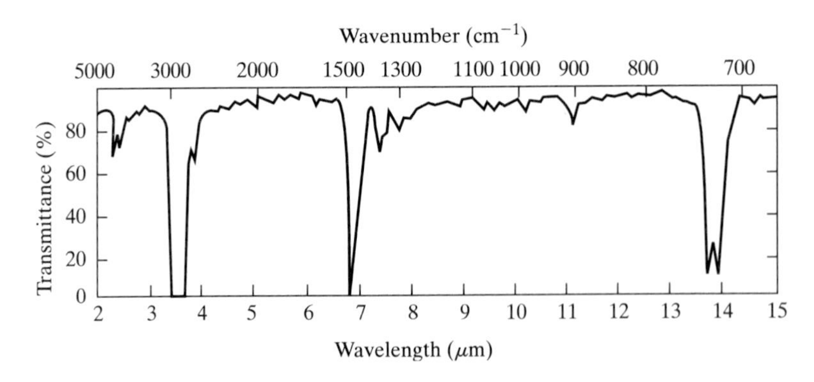

Figure 1.9 Infrared spectrum of an HDPE geomembrane test specimen. (After Halse et al. [11])

ample, using polyethylene, the I_{1715} value is normalized with the peak area at 2850 cm^{-1} (i.e., a ratio of I_{1715}/I_{2850}) would be obtained. It is then defined as the *carbonyl index*.

Chromatography (GC and HPLC). Chromatography is an analysis method that allows for the separation, isolation, and identification of complex mixtures. After the polymer is liquefied in a solvent carrier, the components of the mixture are carried through a stationary column and the migration rates indicate fundamental differences. The soluble mobile phase either dissolves, absorbs, or reacts with the stationary phase within the column.

In gas chromatography (GC) the mobile phase is a gas that is passed through the column, and a detector plots concentration versus time. The positions of the peaks serve to identify the components and the area under the peaks, the concentration. Plasticizers in PVC geomembranes are identified by this method.

In liquid chromatography (LC) the mobile phase is a liquid and the stationary phase is either liquid or solid. The separation process is very time-consuming, which has led to a high-pressure technique (called HPLC), which results in much improved flow rates. Additives in various polymers are identified and quantified by HPLC.

Molecular Weight Determination. There are four molecular weight averages in common use [9]: the number-average molecular weight, M_n; the weight-average molecular weight, M_w; the z-average molecular weight, M_z; and the viscosity-average molecular weight, M_v. These are defined below in terms of the number of molecules, N_i, having molecular weights M_i; or the weight of species, w_i, with molecular weights M_i.

$$M_n = \frac{\Sigma_i N_i M_i}{\Sigma_i N_i} = \frac{\Sigma_i w_i}{\Sigma_i (w_i / M_i)} \tag{1.1}$$

$$M_w = \frac{\Sigma_i N_i M_i^2}{\Sigma_i N_i M_i} = \frac{\Sigma_i w_i M_i}{\Sigma w_i} \tag{1.2}$$

$$M_z = \frac{\Sigma_i N_i M_i^3}{\Sigma N_i M_i^2} = \frac{\Sigma_i w_i M_i^2}{\Sigma w_i M_i} \tag{1.3}$$

$$M_v = \left[\frac{\Sigma_i N_i M_i^{1+a}}{\Sigma_i N_i M_i}\right]^{1/a} \tag{1.4}$$

These values are seen on the molecular weight distribution curves for a HDPE geomembrane in Figure 1.10, illustrating the noneffect of an additional heat cycle in degrading the molecular weight of the geomembrane [15]. M_n is seen to be close to the mean value (approximately 50,000 for this material), while M_w is near the upper inflection point (approximately 170,000). M_z is an indicator of the upper end of the curve (above M_w) and M_v is a zone varying between M_n and M_w. The curve (along with its various descriptors) is probably the most powerful indicator of any available method regarding the molecular structure of polymers and its subsequent degradation behavior.

To determine the entire distribution of the molecular weight, gel permeation chromatography (GPC) is sometimes performed. This is essentially a process for the fractionation of polymers according to their molecular size and, therefore, according to

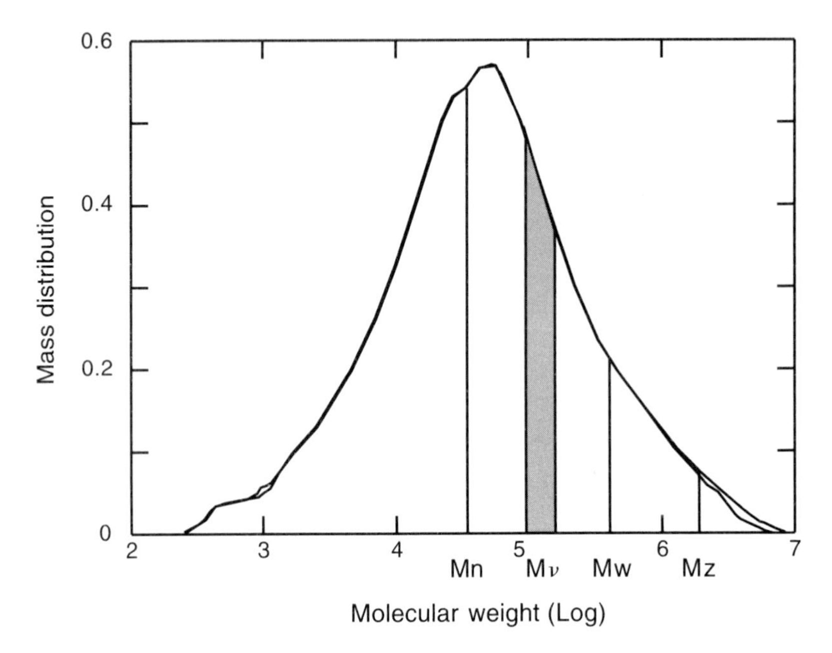

Figure 1.10 Gel permeation chromatography results of normalized molecular weight on HDPE geomembrane test specimens. There are two curves shown: one with one melt cycle, the other with two melt cycles. (After Struve [15])

their molecular weight. The molecular weight is determined indirectly by the calibration of the system in terms of the elution time expected for a particular polymer molecular-weight fraction with a particular piece of equipment. The column packages are made with microporous glass beads and powdered, swelled, and cross-linked polystyrene. It is a tedious test, requiring care and precision, but it is one that has had a dramatic effect on the procedures for polymer characterization and molecular weight determination [16].

It should be noted that a qualitative test to indirectly measure the molecular weight of a polymer is the *melt-flow index* test. ASTM D1238 is commonly used for geosynthetics. In this test, the polymer is heated until it melts. A constant load pushes the molten polymer through an orifice and the weight extruded in 10 min. is the melt index (MI) value. The lower the MI, other things being equal, the higher the molecular weight. By repeating the test at a different constant load, the two respective MI values can be made into a flow-rate ratio (FRR). High values of FRR, other things being equal, indicate broader molecular weight distributions.

These tests, particularly the MI test, are routinely used by the industry for quality control, conformance, and quality assurance testing.

Intrinsic Viscosity Determination. The molecular weight of the polymer from which geosynthetics are made can be indirectly determined using solution viscosity methods. The results are empirically related to the molecular weight in that higher viscosity values come about from higher molecular-weight resins, all other things being equal. The intrinsic viscosity value is particularly applicable to polyester (PET) resins used in the manufacture of reinforcement geotextiles and geogrids.

According to ASTM D4603, the inherent viscosity is obtained using a 0.50% concentration of PET resin in a 60/40 phenol/1,2,3,3-tetrachloroethane solution. The flow time of the solution is measured in a capillary viscometer at 30°C. The pure solvent is also measured under the identical test conditions and the ratio of the two values is the relative viscosity,

$$\eta_r = \frac{t}{t_o} \tag{1.5}$$

where

η_r = relative viscosity,
t = average solution flow time (s), and
t_o = average solvent flow time (s)

This value can then be used to calculate an inherent viscosity, if desired.

$$\eta_{\text{inh}} = \frac{\ln \eta_r}{C} \tag{1.6}$$

where

η_{inh} = inherent viscosity at 0.5% and 30°C, and
C = polymer solution concentration, g/dL

Usually one proceeds directly to the intrinsic viscosity.

$$[\eta] = \frac{0.25(\eta_r - 1 + 3 \ln \eta_r)}{C} \tag{1.7}$$

where $[\eta]$ = intrinsic viscosity

In its recent specification for geosynthetic reinforcement of walls and slopes, the American Association of State Highway and Transportation Officials (AASHTO) requires that the minimum number-average molecular weight (M_n) is 25,000 or higher. To obtain this value from intrinsic viscosity, the Mark-Honwock-Sakmada equation is used.

$$[\eta] = KM_n^a \tag{1.8a}$$

or

$$M_n = \exp\left[\frac{[\eta]}{K}\right]^{1/a} \tag{1.8b}$$

where K and a are constants for a particular solvent and temperature; for example, for the test performed at 35°C, $K = 0.175$ ml/g and $a = 0.65$.

Carboxyl End Group Analysis. A determination of carboxyl end groups for polyester (PET) fibers and yarns is an important indicator of the polymer's long-term durability. According to Pohl [17], a semimicroprocedure for the rapid determination of the number of carboxyl end groups of a polyester yarn entails dissolving the polymer in benzyl alcohol rapidly at a high temperature (e.g, 200°C), then quickly mixing the solution with chloroform, and titrating with sodium hydroxide and the aid of a phenol red indicator. The result of the procedure is expressed in equivalents per million grams. A maximum value of 30 is sometimes referenced for PET used in reinforcement geosynthetics (e.g., the AASHTO 1997 specifications).

Summary of Chemical Fingerprinting Tests. The previously described series of chemical analysis tests are sometimes referred to as "fingerprinting" tests. However, fingerprinting is perhaps too descriptive of a word, since no two products undergoing a single test will give identical response curves. Nevertheless, such response curves should be close enough to substantiate their equivalency or fundamental differences. Perhaps "signature" would be a better descriptive term; however, the term used generally is fingerprinting.

Taken collectively, the tests are very strong in their identification capability. Table 1.4 gives a summary of advantages and disadvantages of each method. While this collection of tests may initially be felt by the civil engineer to be excessive and unwarranted, there are numerous instances where the typical engineering tests (strength, elongation, puncture, etc.) are simply not sensitive enough to evaluate the situation under study. The fall-back position is invariably the use of these chemical analysis tests. They are continually important in many facets of geosynthetic engineering. See Halse et al. [11] for specifics on most of the tests.

1.2.3 Polymer Formulations

No geosynthetic material is 100% of the polymer resin associated with its name. In all cases, the primary resin is formulated with additives, fillers, extruders, and/or other agents for a variety of purposes. The total amount in a given formulation varies widely, from a minimum of 3% to as much as 65%. The additives, solids or liquids, are used as ultraviolet (UV) light absorbers, antioxidants, thermal stabilizers, plasticizers, biocides, flame retardants, lubricants, forming agents, or antistatic agents. The resulting mixture can be homogeneous or heterogeneous, depending upon the solubility parameters of the additives versus that of the primary resin polymer. Heterogeneous mixtures can also be particulate or fibrous [10].

Common particulate additives include: carbon blacks; various antioxidants; calcium carbonate; metallic powders and flakes; silicate minerals such as clay, talc, and mica; silica minerals such as quartz, diatomaceous earth, and novaculite; metallic oxides such as alumina; biocides; and other synthetic polymers. Common liquid additives include: plasticizers, fillers, and colorants. Common fibrous additives (although rarely used in geosynthetic materials) include: glass, carbon and graphite, cellulosics such as alpha cellulose, synthetic polymers such as nylon, metals such as steel strands, and boron. The formulation varies from product to product, but can be generalized for the most common polymers used to manufacture geosynthetics as shown in Table 1.5.

TABLE 1.4 CHEMICAL IDENTIFICATION METHODS

Method	Information Obtained	Advantages*	Disadvantages
TGA	polymer, additives ash content carbon black amount decomposition temperatures	straightforward measurements high accuracy all polymers	qualitative results high cost
DSC	melting point crystallinity oxidative induction time glass transition	straightforward measurement high accuracy all polymers	qualitative results limited to chlorinated polymers high cost
TMA	coefficient of linear thermal expansion softening point glass transition	straightforward measurement high accuracy all polymers	high cost
DMA	elastic constants loss modulus creep behavior stress relax. behavior	high accuracy versatile all polymers temp. controlled computer driven	high cost high maintenance complex unit
IR	identifies additives identifies fillers identifies plasticizers rate of oxidation reaction	all polymers	difficult specimen preparation no resin information high cost
GC and HPLC	identifies additives identifies plasticizers	straightforward measurement	difficult specimen preparation no resin information high cost
GPC	molecular weight distribution average molecular weight	accurate values only valid technique for molecular weight all polymers	not very common tedious test difficult specimen preparation very high cost
IV	intrinsic viscosity average molecular weight	straightforward measurement common test assesses hydrolysis	indirect measurements need correlations must be soluble
CEG	titration method carboxyl end group	assesses hydrolysis	tedious test not very common

*An advantage common to all methods is the extremely small sample size required for testing in comparison to traditional physical and mechanical test specimen sizes.

Source: Halse et al. [11].

Thus, an understanding of the formulation of a polymeric material is a complex and formidable task but a doable one. Unfortunately, it is rarely given a high priority in engineering curricula, the obvious exception being in a polymer (or materials) engineering program. Rarely (if at all) does a civil, mechanical, or industrial engineer have

TABLE 1.5 COMMONLY USED GEOSYNTHETIC POLYMERS AND THEIR APPROXIMATE FORMULATIONS

Polymer Type	Resin	Filler	Carbon Black or Pigment	Additives	Plasticizer
Polyethylene	97	0	2–3	0.5–1.0	0
Polypropylene	96	0	2–3	1–2	0
Polyvinyl chloride (unplasticized)	80	10	5–10	2–3	0
Polyvinyl chloride (plasticized)	35	25	5–10	2–3	30
Polyester	97	0	2–3	0.5–1.0	0
Nylon	97	0	2–3	0.5–1.0	0
Polystyrene	97	0	2–3	0.5–1.0	0
Chlorosuphonated polyethylene	45	20–25	20–25	5–7	0

any formal training in polymers, and even many chemical engineering programs are quite lean in this area. Future curricula must become more attuned to the necessities of modern material systems, in which polymers play a key and ever-increasing role.

1.3 OVERVIEW OF GEOTEXTILES

1.3.1 History

Geotextiles, as they are known and used today, were first used in erosion control applications and were intended to be an alternative to granular soil filters. Thus the original and still sometimes used term for geotextiles is *filter fabrics*. Barrett [18], in his now classic 1966 paper, tells of work originating in the late 1950s using geotextiles behind precast concrete seawalls, under precast concrete erosion control blocks, beneath large stone riprap, and in other erosion control situations. He used different styles of woven monofilament fabrics, all characterized by a relatively high percentage open area (varying from 6 to 30%). He discussed the need for both adequate permeability and soil retention, along with adequate fabric strength and proper elongation and set the tone for geotextile use in filtration situations. Note that an earlier paper by Agerschou [19] discussed applications along the same general lines.

In the late 1960s Rhone-Poulenc Textiles in France began working with nonwoven needle-punched fabrics for quite different applications. Here emphasis was on reinforcement for unpaved roads, beneath railroad ballast, within embankments and earth dams, and the like. The primary function in many of these applications was that of separation and/or reinforcement. Additionally, a quite different use for this particular style of fabric was also recognized, that is, that thick felt-like fabrics can also transmit water within the plane of their structure (i.e., they can act like drains). Such uses as dissipation of pore-water pressures and horizontal and vertical flow interception grew out of this particular fabric function. Today's use of the word *geotextiles* recognizes these many possible functions of fabric within a soil mass.

Credit for early work in the use of geotextiles should also be given to the Dutch (the Rijkswaterstaat was an early user) and the English. ICI Fibres was a major influence in the use of nonwoven, heat-bonded fabrics in a wide variety of uses. The first nonwovens used in the U.S. were imported from ICI Fibres by Mirafi, Inc., in the late 1970s. The worldwide movement can be seen in Figure 1.11. ICI Fibres provided early literature that was very significant in proper use of geotextiles in a variety of applications. Chemie Linz (now Polyfelt, Inc.) in Austria and du Pont in Europe and the U.S. (now BBA Nonwovens, Inc.) were also early leaders in the technology. These firms and many others have continued to introduce geotextiles worldwide. Today many manufacturers are involved in the production, sales, and distribution of geotextiles.

A number of early conferences were held exclusively on the subject of geotextiles. More recently, conferences have addressed the entire breadth of geosynthetics, the major ones being those held in Paris in 1977 [21], in Las Vegas in 1982 [22], in Vienna in 1986 [23], in The Hague in 1990 [24], in Singapore in 1994 [25], and in Atlanta in 1998. The original two books on the subject appeared almost simultaneously, Koerner and Welsh [26] in 1980 and Rankilor [20] in 1981. Today, additional books (e.g., van Zanten [27]; John [28]), along with thousands of separate papers and reports dealing with geotextiles (see Giroud [29]) are available and dedicated journals have been launched dealing with all types of geosynthetics [30, 31]. This massive generation and dissemination of information was led initially by geotextile manufacturers. Their influence in this market continues to be active, and indeed is very positive and welcome. It has been followed by the entire community of governmental, industrial, consulting, research, testing, and academic institutions.

Perhaps the culmination of this activity was the formation of the International Geosynthetics Society (IGS), which currently has seventeen national and regional chapters. All are active, with separate venues on a variety of geosynthetic-related topics and activities.

1.3.2 Manufacture

As noted, the role of fabric manufacturers in the stimulation and growth of the geotextile market has been both large and positive. Many fiber types and fabric styles have been developed both for general use and for specific applications. In fact, it seems that these two approaches to the marketing of geotextiles typify all geotextile manufacturers; manufacturers tend to target products either for the large, customary (or commodity) market or for the smaller, specialized (or engineered) market. Whatever the case may be, three points are important for manufacturing: the type of polymer, the type of fiber, and the fabric style. Each will be discussed separately.

Type of Polymer. The polymers used in the manufacture of geotextile fibers are made from the following polymeric materials, listed in order of decreasing use:

polypropylene ($\simeq$ 85%)
polyester ($\simeq$ 12%)
polyethylene ($\simeq$ 2%)
polyamide (nylon) ($\simeq$ 1%)

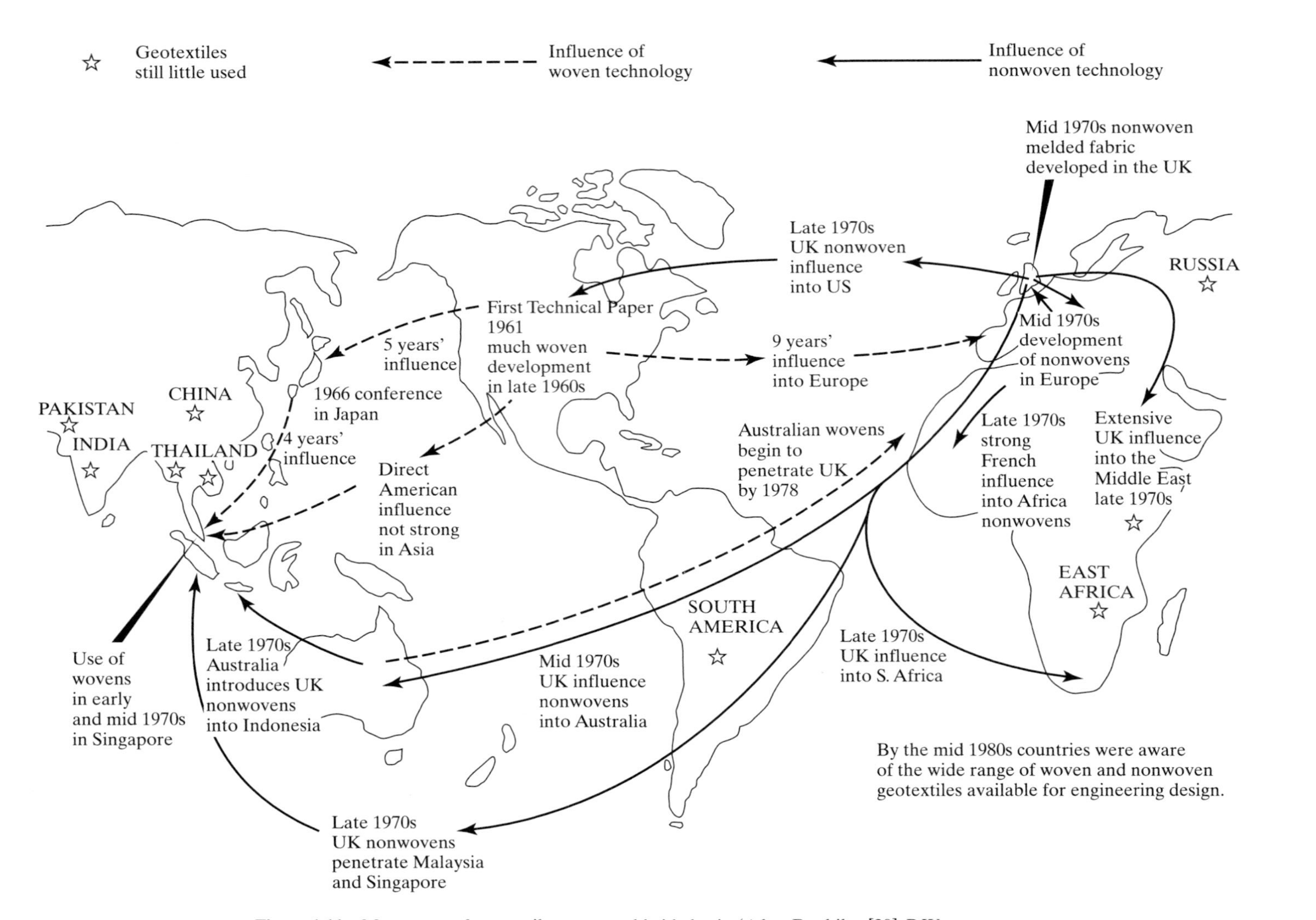

Figure 1.11 Movement of geotextiles on a worldwide basis. (After Rankilor [20]; P. W. Rankilor, *Membranes in Ground Engineering*, © 1981, John Wiley & Sons, New York)

Their respective repeating units are given in Table 1.2 and some of their relevant properties are given in Table 1.6. Note that moisture plays a relatively minor role in strength, that only polyolefins (polypropylene and polyethylene) are lighter than water, that polyester absorbs the least amount of water, and that all the polymeric materials have quite high melting points. While an extremely large database [32, 33] is available on these and other polymers, it is the final manufactured product that is of primary interest to the end user.

Type of Fiber. The properly formulated polymers are made into fibers (or yarns, where a yarn can consist of one or more fibers) by melting them and forcing them through a spinneret, similar in principle to a bathroom showerhead. The resulting fiber filaments are then hardened or solidified by one of three methods: wet, dry, or melt. Most geotextile fibers are made by the melt process; these include polyolefins, polyester, and nylon. Here hardening is by cooling, and simultaneously or subsequently they are stretched. Stretching reduces the fiber diameter and causes the molecules in the fibers to arrange themselves in an orderly fashion. In so doing the fiber's strength increases, its elongation at failure decreases, and its modulus increases. A wide range of stress versus strain responses can be achieved. These monofilaments can also be twisted together to form a multifilament yarn. Note that the diameter of the fiber is characterized by its *denier*. Denier is defined as the weight in grams of 9000 m of yarn. The related textile term *tex* is the weight in grams of 1000 m of yarn.

Staple fibers are somewhat different and are produced by continuous filaments of specific denier in a large ropelike bundle called a tow. A tow can contain thousands of continuous filaments. These bundles are then crimped and cut into short staple lengths of 25 to 100 mm. The short fibers, or staple, are then twisted or spun into long yarns for subsequent fabric manufacturing.

The last type of fiber to be mentioned is made completely differently from those discussed above. These fibers, called slit (or split) film or tapes, are made from a continuous sheet of polymer that is cut into fibers by knives or lanced by air jets. The resulting ribbonlike fibers are referred to as slit-film monofilament fibers. Obviously, these slit-film monofilaments can also be twisted together to make a slit-film multifilament.

Thus the principal fibers used in the construction of geotextiles are monofilament, multifilament, staple yarn, slit-film monofilament, and slit-film multifilament. See Figure 1.12.

Fabric Style. Once the *yarns*, as they are referred to in the textile industry, are made, they must be manufactured into fabrics. The basic manufacturing choices are woven, nonwoven, or knit (although knit fabrics are seldom used as geotextiles). Various woven and nonwoven types are shown in Figure 1.13. The woven fabrics are made on conventional textile-weaving machinery into a wide variety of fabric weaves. Kaswell [35] gives an excellent review of weaving technology in which each of the fabric weaves is clearly illustrated. Variations are many and most have a direct influence on the physical, mechanical, and hydraulic properties of the fabric.

For conventional industrial fabrics (of which geotextiles form a subset), the weaves are usually kept relatively simple. The particular pattern of the weave is determined by the sequence in which the warp yarns are threaded into the weaving loom and

TABLE 1.6 SOME PHYSICAL PROPERTIES OF SYNTHETIC FIBERS*

Fiber	Breaking Tenacity (g/denier)†		Specific Gravity	Standard Moisture Regain (%)	Coefficient of Thermal Expansion ($\times 10^{-5}$ per 1°C)	Effect of Heat
	Standard	Wet				
Polyethylene (high-density)	—	—	0.96	2.0	13	Melts between 110 and 140°C
Polypropylene (filament and staple)	4.8–7.0	4.8–7.0	0.91	3.0	6	Melts between 160 and 170°C
Polyester						
regular-tenacity filament	4.0–5.0	4.0–5.0	1.22 or 1.38	0.4 or 0.8	4 to 5	Melts between 250 and 290°C
high-tenacity filament	6.3–9.5	6.2–9.4	1.22 or 1.38	0.4 or 0.8	4 to 5	Melts between 250 and 290°C
regular-tenacity staple	2.5–5.0	2.5–5.0	1.22 or 1.38	0.4 or 0.8	4 to 5	Melts between 250 and 290°C
high-tenacity staple	5.0–6.5	5.0–6.4	1.22 or 1.38	0.4 or 0.8	4 to 5	Melts between 250 and 290°C
Nylon						
nylon 66 (regular-tenacity filament)	3.0–6.0	2.6–5.4	1.14	4.0–4.5	5.5	Sticks at 230°C; melts at about 260°C
nylon 66 (high-tenacity filament)	6.0–9.5	5.0–8.0	1.14	4.0–4.5	5.5	Same as above
nylon 66 (staple)	3.5–7.2	3.2–6.5	1.14	4.0–4.5	5.5	Same as above
nylon 6 (filament)	6.0–9.5	5.0–8.0	1.14	4.5	5.0	Melts between 210 and 220°C
nylon 6 (staple)	2.5	2.0	1.14	4.5	5.0	Melts between 160 and 220°C

*Standard laboratory conditions for fiber tests: 21°C and 65% relative humidity.

†Denier is the equivalent to the grams per 9000 meters in the thread used to make synthetic fabircs. The higher the denier, the heavier the fabric.

Source: Modified from Shreve and Brink [34].

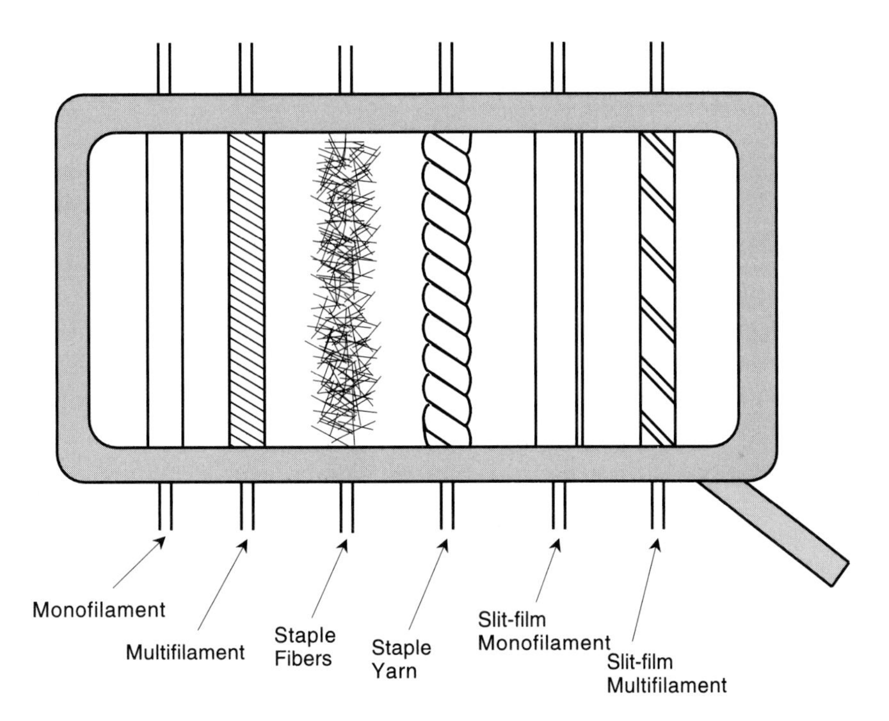

Figure 1.12 Types of polymeric fibers (or yarns) used in the manufacture of geotextiles.

the position of the warp harness for each filling pick. As shown in Figure 1.14a, reeds shed the warp yarns up, allowing a shuttle to insert the weft yarn. The reeds then shed downward (see Figure 1.14b), encapsulating the weft yarn and allowing the return of the shuttle in the opposite direction with another weft yarn. The reeds then shed back upward and the process continues as a cycle. This action gives rise to the nomenclature in woven fabrics—warp direction (the direction the fabric is being made or the long direction), weft or fill direction (the cross direction or short direction), and selvedge (the edges of the fabric where the weft yarns reverse direction and gather the outer warp yarns on each side of the fabric)—and to the various types of weaves common in the formation of fabrics for use as geotextiles.

Plain weave. This is the simplest and most common weave, and is also known as "one up and one down."

Basket weave. This weave uses two or more warp and/or filling yarns as one. For example, a "two-by-two basket weave" takes two warp and two weft yarns acting as individual units.

Twill weave. Here a diagonal or *twill* line moves across the fabric by moving yarn intersections one pick higher on successive warp yarns. Related patterns can also be formed, for example, steep twills and broken twills.

Satin weave. If the warp (or weft) yarn is carried over many weft (or warp) yarns, a smooth fabric surface will result. This is called a satin weave and the fabric is usually smooth and shiny. It is generally not used for geotextile fabrics.

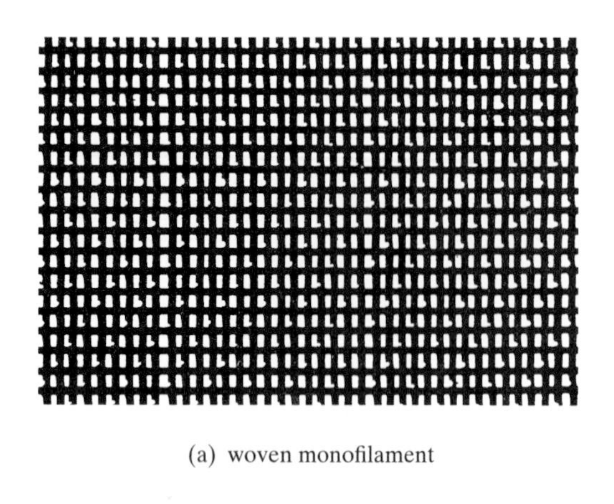

(a) woven monofilament

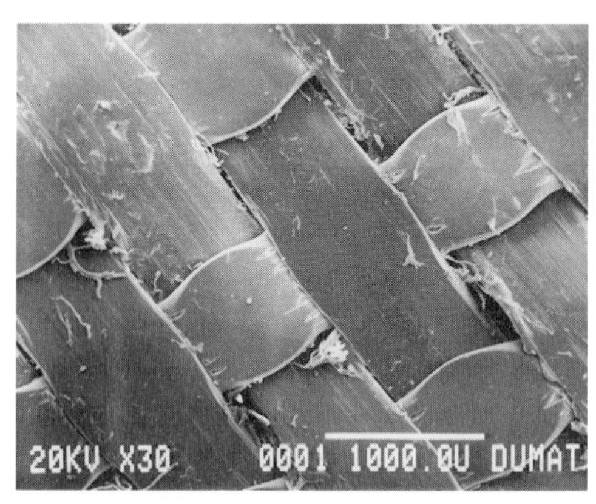

(b) woven monofilament, calendered

(c) woven multifilament

(d) woven slit (split) film

(e) nonwoven needle-punched

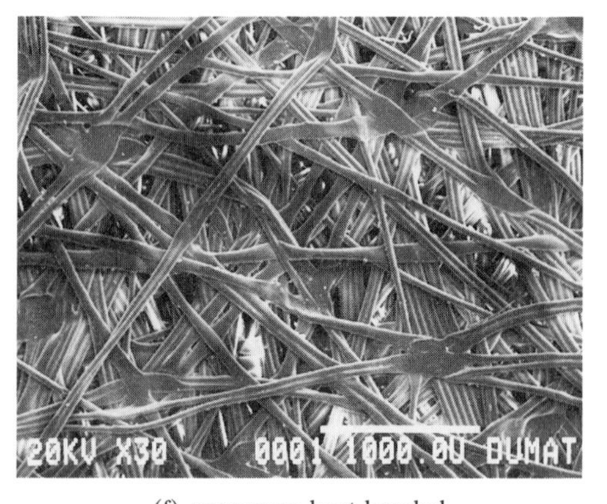

(f) nonwoven heat-bonded

Figure 1.13 Photomicrographs of various fabrics used as geotextiles. Magnification of (a) is × 5; all others are × 30.

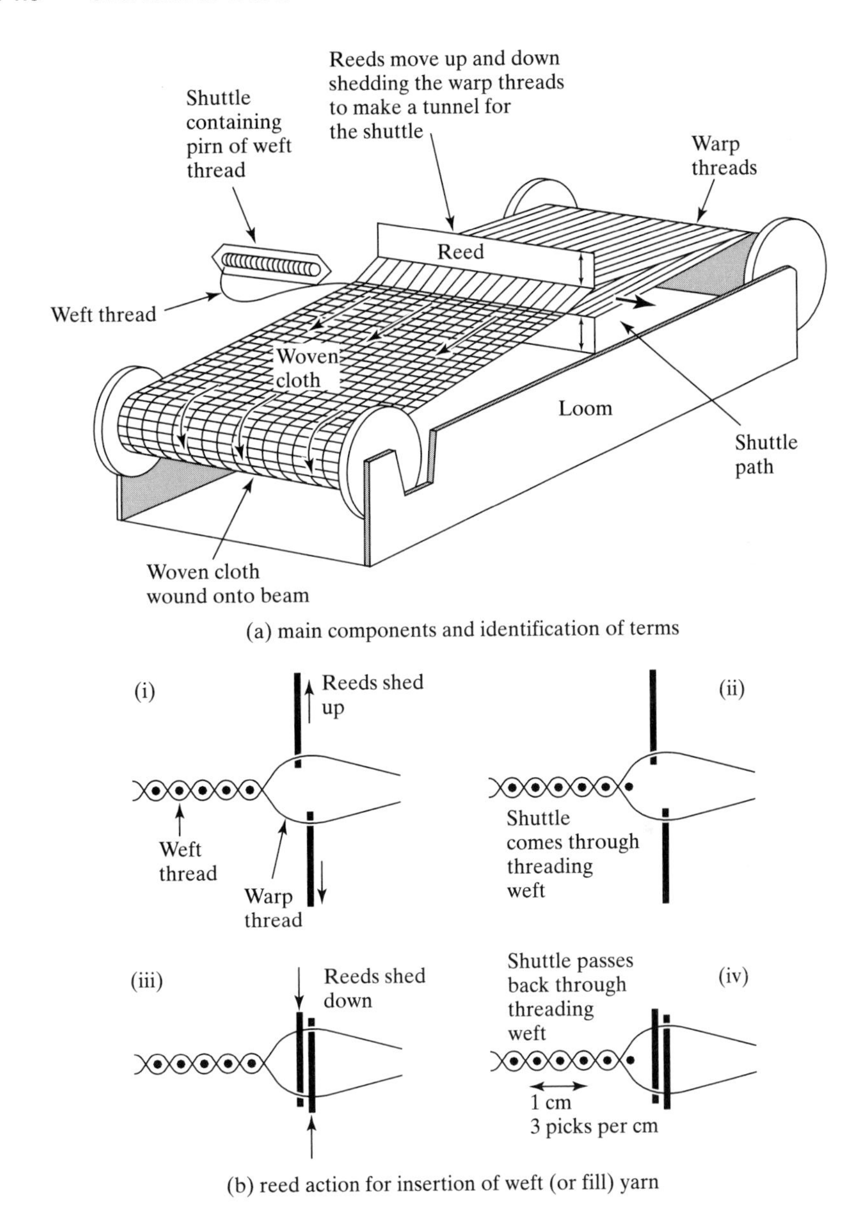

Figure 1.14 Basic functioning of a weaving loom. (After Rankilor [20]; P. W. Rankilor, *Membranes in Ground Engineering*, © 1981, John Wiley & Sons, New York)

Additional details on the weaving process using both natural and synthetic fibers are found in Kaswell [35].

The manufacture of nonwoven fabrics is very different from that of woven fabrics. Each nonwoven manufacturing system generally includes four basic steps: fiber preparation, web formation, web bonding, and post-treatment. Within each category there are many possibilities, so only those most common to current geotextiles will be described.

Of the four basic steps, fiber preparation has already been discussed. The process of *spun-bonding* encompasses the remaining three steps. Spun-bonding is a continuous process used to produce a finished fabric from a polymer. One or several polymers, such as polypropylene, polyester, polyamide, or polyethylene, are fed into an extruder. As it flows from the extruder, it is forced through a spinneret or a series of spinnerets. The fibers are usually stretched and after cooling are then laid on a moving conveyor belt to form a continuous web. In the lay-down process, the desired orientation of the fibers is achieved by various means, such as rotation of the spinneret, electrical charges, introduction of controlled air-streams, or by varying the speed of the conveyor belt. The mat of fabric is then bonded (i.e., the filaments are made to adhere to one another) by thermal, chemical, or mechanical treatment before being wound up into finished roll form (see Figure 1.15).

Alternatively, the web can be formed by starting the process with short chopped fibers of 50 to 100 mm in length. The fibers are made or purchased by the geotextile manufacturer in the form of bales, which are opened by forced air in what is referred to as a carding process. The fibers are then moved by conveyer in a lay-down process to form a web of desired width and mass per unit area. The process has enormous flexibility, particularly in the choice of initial fiber selection. Once the loose web is formed, the entanglement is similar to the spun-bonded process.

Usually, one of three processes is used to bond the filaments of the web together: heat-bonding, resin-bonding, or needle-punching. In the heat-bonding process (also called *melt-bonding* or *heat-setting*), the web, composed of continuous filaments or long

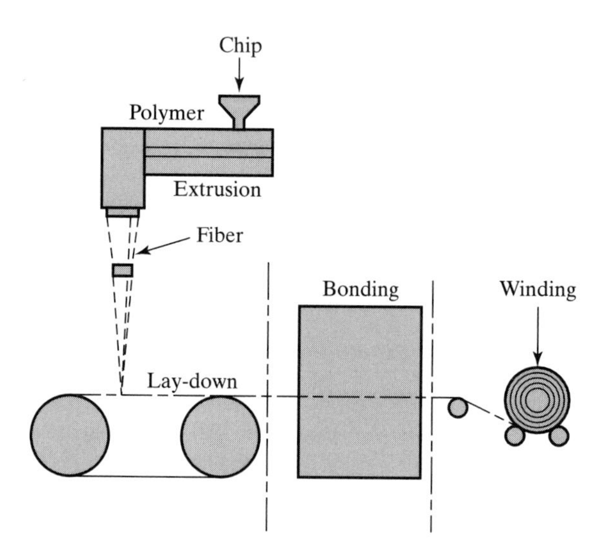

Figure 1.15 Diagram of the spun-bonding process to manufacture geotextiles. Note that bonding can be by needle-punching, heat-bonding, or resin-bonding. (Compliments of INDA)

staple fibers, is melted together at filament or fiber crossover points. The resultant fabrics are rather stiff in texture and feel. Higher product strength at lower fabric weights can be achieved with this type of manufacture than with other fabric styles, owing to the fiber bonding utilized in the process. The bonding operations differ among the commercially available fabrics, depending on the basic fiber characteristics. Thus, there are actually two kinds of heat-bonding: homofil and heterofil. In homofil bonding, all filaments are composed of a single polymer type, though some of the filaments have different melting characteristics. Bonding is achieved by a high-temperature calendering operation, accomplished by passing fabric between two counter-rotating hot rollers. In heterofil bonding, some of the filaments comprise two types of polymers of different melting points (heterofilaments), whereas the other filaments have only one polymer (homofilaments). The heterofilaments have different softening characteristics than the homofilaments; bonds can therefore be formed at the heterofilament crossover points by controlling the application of heat and pressure to fuse only the lower-melting-point polymer. In so doing, the lower-melting-point polymer forms a sheath, leaving the core and homofilaments unaffected.

In the second kind of nonwoven fabric-bonding process, resin-bonding, a fibrous web is either sprayed or impregnated with an acrylic resin. After curing and/or calendering, bonds are formed between filaments. Often a forced-air drying operation is used to reestablish the fabric's open-pore structure.

In the needle-punching process, the third and most common bonding, a fibrous web is introduced into a machine equipped with groups of specially designed needles. The needles are about 75 mm long and each have three or four barbs. While the web is trapped between a bed plate and a stripper plate, the needles punch through it and reorient the fibers so that mechanical bonding is achieved among the individual fibers. It is either on the downstroke or the upstroke where the entanglement process occurs. Often the batt of fibers is carried into the needle-punching section of the machine on a lightweight support material or substrate. This is done to improve finished fabric strength and integrity (see Figure 1.16). The needle-punching process is generally used to produce fabrics that have high mass per unit area yet retain considerable bulk. Fabric weights and thicknesses can be very large.

A major point to remember is that the textile industry is very mature and sophisticated and can produce a tremendously wide variety of fabrics. Indeed, tailoring a fabric for a specific purpose or property is well within the state-of-the-practice.

1.3.3 Current Uses

Although Chapter 2 deals with the uses of geotextiles more thoroughly, a few remarks can be made here. As already mentioned, the functions of geotextiles are separation, reinforcement, filtration, drainage and (when impregnated) as containment. Within these functions, however, there are a large number of applications, or use areas. These use areas are listed in Table 1.7 (obviously this is not an all-inclusive list, and it is constantly growing) to give an idea of the size and scope of the geotextile market. Actual sales by use area are given in Table 1.8.

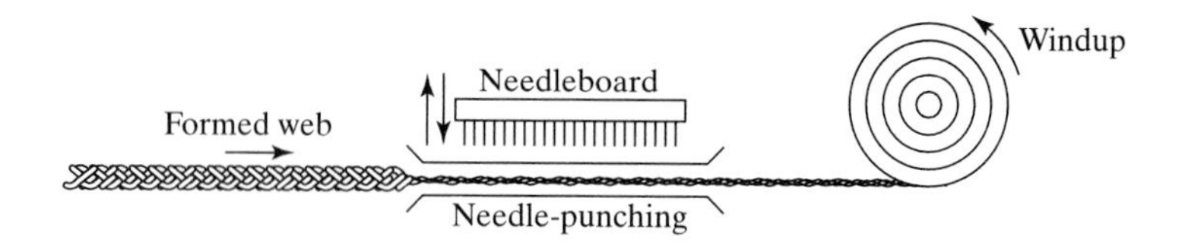

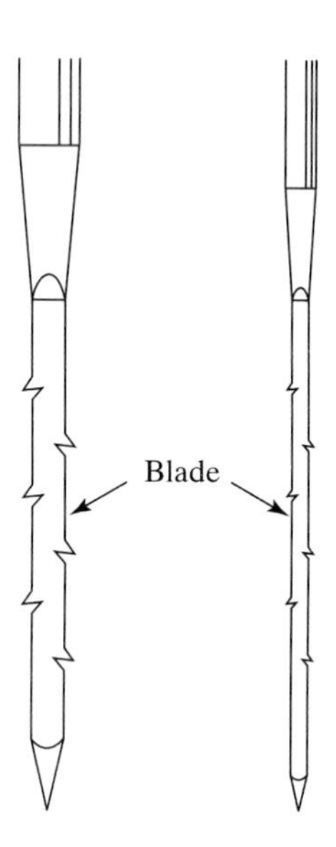

Figure 1.16 Diagram of needle-punched process and details of typical needles. (Compliments of INDA)

1.3.4 Sales

The use of geotextiles has experienced a growth that has probably never been equaled by any other system in civil engineering and heavy construction. Estimated sales in North America are given in Figure 1.2. Although no data are available to the author, European sales of geotextiles are thought to be equivalent to those in North America. The remainder of the world has perhaps 50% of the sales of North America or Europe. Within this total, the distribution on the basis of end use is approximately as follows.

The distribution of geotextiles to the ultimate user is handled (1) from the mill directly, (2) by means of commissioned agents, and (3) through individual distributors. Generally, but certainly not always, direct-mill sales efforts are focused on unusually large jobs where competition is very intense. Commissioned agents, who are often very well versed in geotextile uses, functions, properties, and design, work with professional engineers and consultants, and generally service the engineered job applications. Individual distributors service the standard applications and are often "wired-into" certain segments of the industry (e.g., road work or erosion control applications).

TABLE 1.7 MAJOR USES OF GEOTEXTILES

Separation of Dissimilar Materials

- Between subgrade and stone base in unpaved roads and airfields
- Between subgrade and stone base in paved roads and airfields
- Between subgrade and ballast in railroads
- Between landfills and stone base courses
- Between geomembranes and sand drainage layers
- Between foundation and embankment soils for surcharge loads
- Between foundation and embankment soils for roadway fills
- Between foundation and embankment soils for earth and rock dams
- Between foundation and encapsulated soil layers
- Between foundation soils and rigid retaining walls
- Between foundation soils and storage piles
- Between slopes and downstream stability berms
- Beneath sidewalk slabs
- Beneath curb areas
- Beneath parking lots
- Beneath sport and athletic fields
- Beneath precast blocks and panels for aesthetic paving
- Between drainage layers in poorly graded filter blankets
- Between various zones in earth dams
- Between old and new asphalt layers

Reinforcement of Weak Soils and Other Materials

- Over soft soils for unpaved roads
- Over soft soils for airfields
- Over soft soils for railroads
- Over soft soils for landfills
- Over soft soils in sport and athletic fields
- Over nonhomogeneous soils
- Over unstable landfills as closure systems
- For lateral containment of railroad ballast
- To wrap soils in encapsulated fabric systems
- To construct fabric-reinforced walls
- To reinforce embankments
- To aid in construction of steep slopes
- To reinforce earth and rock dams
- To stabilize slopes temporarily
- To halt or diminish creep in soil slopes
- To reinforce jointed flexible pavements
- As basal reinforcement over karst areas
- As basal reinforcement over thermokarst areas
- As basal reinforcement between pile foundation caps
- To bridge over cracked or jointed rock
- To hold graded-stone filter mattresses
- As substrate for articulated concrete blocks
- To stabilize unpaved storage yards and staging areas
- To anchor facing panels in reinforced earth walls
- To anchor concrete blocks in small retaining walls
- To prevent puncture of geomembranes by subsoils
- To prevent puncture of geomembranes by landfill materials or stone base
- To create more stable side slopes due to high frictional resistance
- To contain soft soils in earth dam construction
- For membrane-encapsulated soils
- For in-situ compaction and consolidation of marginal soils
- To bridge over uneven landfills during closure of the site
- To aid in bearing capacity of shallow foundations

Filtration (Cross-Plane Flow)

- In place of granular soil filters
- Beneath stone base for unpaved roads and airfields
- Beneath stone base for paved roads and airfields
- Beneath ballast under railroads
- Around crushed stone surrounding underdrains
- Around crushed stone without underdrains (i.e, French drains)
- Around perforated underdrain pipe
- Around stone and perforated pipe in tile fields
- Beneath landfills that generate leachate
- To filter hydraulic fills
- As a silt fence
- As a silt curtain
- As a snow fence
- As a flexible form for containing sand, grout, or concrete in erosion control systems
- As a flexible form for reconstructing deteriorated piles
- As a flexible form for restoring underground mine integrity
- As a flexible form for restoring scoured bridge pier bearing capacity
- To protect chimney drain material
- To protect drainage gallery material
- Between backfill soil and voids in retaining walls
- Between backfill soil and gabions
- Around molded cores in fin drains
- Around molded cores in strip drains
- Against geonets to prevent soil intrusion

(continued)

TABLE 1.7 *(continued)*

Filtration (Cross-Plane Flow) *(continued)*	
Against geocomposites to prevent soil intrusion	Around porous tips for piezometers
Around sand columns in sand drains	As a filter beneath stone riprap
Around porous tips for wells	As a filter beneath precast blocks
Drainage (In-Plane Flow)	
As a chimney drain in an earth dam	As a drain beneath sport and athletic fields
As a drainage gallery in an earth dam	As a drain for roof gardens
As a drainage interceptor for horizontal flow	As a pore water dissipator in earth fills
As a drainage blanket beneath a surcharge fill	As a replacement for sand drains
As a drain behind a retaining wall	As a capillary break in frost-sensitive areas
As a drain beneath railroad ballast	As a capillary break for salt migration in arid areas
As a water drain beneath geomembranes	To dissipate seepage water from exposed soil or rock surfaces
As an air drain beneath geomembranes	
Containment (When Impregnated)	
See list of applications under Geomembranes (Section 1.6).	

TABLE 1.8 UTILIZATION OF GEOTEXTILES IN NORTH AMERICA BY APPLICATION AREA

Applications	1987	1988	1989	1990	1991	1992	1995
Asphalt overlays	75	84	90	88	88	88	77
Separation/stabilization	65	73	81	85	85	87	115
Filtration/drainage	31	34	35	35	36	37	55
Protection for geomembranes	14	16	21	30	41	58	85
Erosion control	12	13	14	15	16	16	20
Silt fence	12	13	14	15	16	17	23
Reinforcement	12	13	15	16	17	18	25
Subgrade							
walls and slopes							
Total market	221	246	270	284	299	321	400

In millions of m^2.

Source: After Jagielski [37] and extended by author.

The sales of geotextiles are projected to grow at 5 to 10% per year, with one recent study (the most optimistic I have seen) showing sales at approximately 0.8 billion m^2 by 2000. This translates into an approximately 15% annual growth rate.

1.4 OVERVIEW OF GEOGRIDS

1.4.1 History

The development of methods of preparing high-modulus polymer materials by tensile drawing [38], in a sense "cold working," has raised the possibility that such materials could be used in the reinforcement of a number of construction materials, including soil.

Today, the major function of such geogrids is in the area of reinforcement. This area is very active, with a number of different styles, materials, connections, and so on, making up today's geogrid market. The key feature of geogrids is that the openings between the longitudinal and transverse ribs, called the *apertures*, are large enough to allow for soil strike-through from one side of the geogrid to the other. The ribs of geogrids are often quite stiff compared to the fibers of geotextiles. As we will discuss later, not only rib strength, but also junction strength is important. The reason for this is that the soil strike-through within the apertures bears against the transverse ribs, which transmits the load to the longitudinal ribs via the junctions. The junctions are, of course, where the longitudinal and transverse ribs meet and are connected.

The original geogrids were made in the United Kingdom by Netlon, Ltd., and were brought to the U.S. in 1982 by way of Canada by the Tensar Corp. Another type of drawn geogrid by the Tenax Corp. is also available. This product originates in Italy at RDB Plastotechnica SpA. More flexible textile-like geogrids, using polyester fibers as the reinforcing component, were developed by ICI in the United Kingdom around 1980. This led to the development of polyester geogrids made on textile weaving machinery. In this process hundreds of fibers are gathered together to form longitudinal and transverse ribs with large open spaces between. The cross-overs are joined by knitting or intertwining before the entire structural unit is protected by a separate coating. Geosynthetics within this group are manufactured by TC Mirafi, TC Nicolon, Huesker, Strata Systems, Akzo, and other companies having various trademarked products.

1.4.2 Manufacture

The polymer materials used in the manufacture of oriented geogrids are high-density polyethylene or polypropylene. The woven-type geogrids use polyester for the strength component and are coated with any of a number of materials (e.g., PVC, latex, bitumen). See Table 1.2 for the molecular structure of the repeating units of these polymers. The Tensar and Tenax products are manufactured in a similar manner. The process begins with heavy gauge sheets of polyethylene or polypropylene. Typical thicknesses are 4 to 6 mm. Holes are then punched into the sheeting on a regular pattern, and the sheet is then drawn uniaxially or biaxially (see Figure 1.17). Drawing is done under controlled temperatures and strain rates, so as to avoid fracture while allowing ductile flow of the molecules into an elongated condition. The key variable in the process is the draw ratio, but other variables, such as molecular weight, molecular weight distribution, and degree of branching or cross-linking, are also important [38, 39]. Aside from significant increases in modulus and strength, the creep sensitivity of the elongated ribs is greatly reduced by the drawing process. The resulting geogrids are referred to as homogeneous, utilized, or relatively stiff geogrids.

ICI's Paragrids are polyester fibers bundled together and enclosed within a polypropylene sheath. It is the polypropylene that is melt-bonded at rib intersections, thereby forming the junctions. Aperture sizes vary with the particular style of product being used.

TC-Mirafi's Miragrid, Huesker/Akzo/Wellman/ACFs Fortrac, TC-Mirafi's Matrex (marketed by Reco, Inc.), and Strata Systems Stratagrid are made from high-tenacity polyester yarns, woven into an open structure with the junctions being knitted together

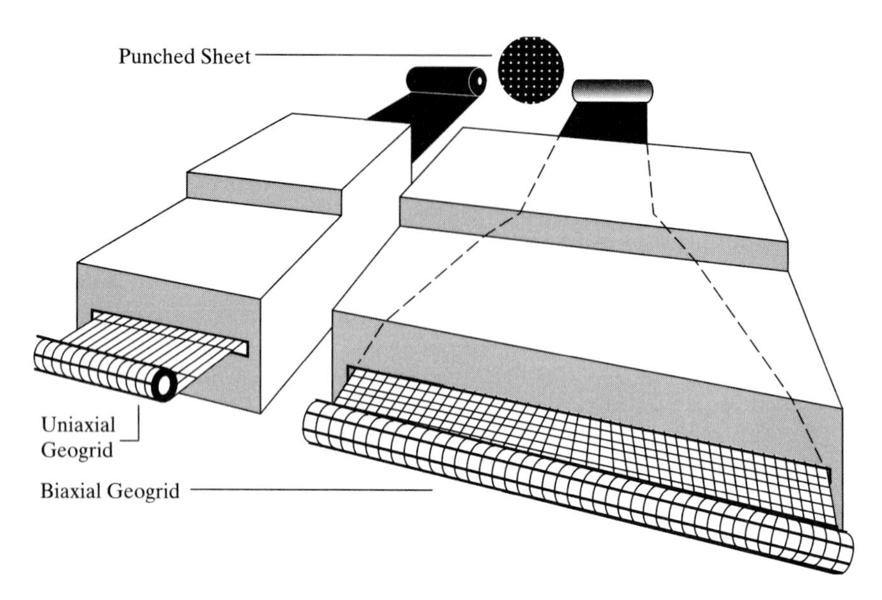

Figure 1.17 Method of manufacturing homogeneous, unitized geogrids. (After Netlon/Tensar [40])

or physically intertwined to link the transverse and longitudinal ribs. The entire geogrid is then coated with PVC, latex, or bitumen for dimensional stability and to provide protection for the ribs during installation. The resulting geogrids are referred to as textile-related or relatively flexible geogrids.

There are other types of prototype geogrids currently under development. Many are composite materials with intriguing junction assemblies, while some are continuous sheets of "super-tuff" polymeric materials with large holes punched in them. Others are made using fiberglass yarns, which are woven together forming a geogrid. This activity in the geogrid area stems from the excellent anchorage and pullout resistance afforded when placed in a soil system. As will be seen in the design portion of Chapter 3, the reinforcement function can profit handsomely from this type of geosynthetic material.

1.4.3 Current Uses

The geogrids that result from the process described above are relatively high-strength, high-modulus, low-creep-sensitive polymers with apertures varying from 10 to 100 mm in size. These holes are either elongated ellipses, near-squares with rounded corners, squares, or rectangles. Under some circumstances, separation may be a function, but only with very coarse gravels and large particle size materials. Invariably, geogrids are involved in some form of reinforcement. The following uses have been reported in the literature.

- Beneath aggregate in unpaved roads
- Beneath ballast in railroad construction
- Beneath surcharge fills or temporary construction sites

- To reinforce embankment fills and earth dams
- To repair slope failures and landslides
- As gabions for wall construction
- As gabions for erosion control structures
- As gabions for bridge abutments
- As basal reinforcement over soft soils
- As basal reinforcement over karst areas
- As basal reinforcement over thermokarst areas
- As basal reinforcement between pile foundation caps
- To bridge over cracked or jointed rock
- To construct mattresses for fills over soft soils
- To construct mattresses over peat, tundra, and muskeg
- As sheet anchors for retaining-wall facing panels
- As sheet anchors and facing panels to form an entire retaining wall
- As asphalt reinforcement in pavements
- As cement or concrete reinforcement in a wide variety of applications
- To reinforce disjointed rock sections
- To reinforcement disjointed concrete sections
- As inserts between geotextiles
- As inserts between geomembranes
- As inserts between a geotextile and a geomembrane
- To reinforce landfills to allow for vertical expansion
- To reinforce landfills to allow for lateral expansion
- To stabilize leachate collection stone as veneer reinforcement
- To stabilize landfill cover soil as veneer reinforcement
- As three-dimensional mattresses for landfill bearing capacity
- As three-dimensional mattresses for embankments over soft soils

1.4.4 Sales

Geogrids, in serving as a reinforcement material, compete directly against geotextiles in many of the above uses. Some of the manufacturers of geogrids also manufacture high strength geotextiles. As such, sales are difficult to separate out for the different products. For example, a 1987 survey of walls reinforced by geosynthetics shows that 50% of the geosynthetic-reinforced walls built in North America have been reinforced by geogrids. It is likely that an even greater proportion of walls and steep soil slopes are currently being reinforced with geogrids over that of geotextile reinforcement.

1.5 OVERVIEW OF GEONETS

1.5.1 History

Geonets owe their origin to Dr. B. Mercer, of Netlon, Ltd., in the U.K., who first developed and patented the machinery and processing methods for the lightweight plastic nets commonly seen in supermarkets for carrying produce, fruits and vegetables. Experimentation with gradually thicker ribs in various configurations led to drainage nets of the type used in civil engineering. The first known environmental application was in 1984 for leak detection in a double-lined hazardous liquid-waste impoundment in Hopewell, Virginia. Geonets are grid-like materials and could be included under geogrids. The reason for its separate treatment here lies not in the material or its configuration, but in its function. Geonets are used for their *in-plane drainage* capability, while geogrids (as discussed in Section 1.4) are used for *reinforcement*. It should be stated at the outset, however, that geonets are not weak, flimsy materials. They have considerable strength, but are used in drainage applications almost exclusively. Note that geonets are always used with a geotextile, geomembrane, or other material on their upper and lower surfaces to prevent soil intrusion into the apertures, which would block the in-plane flow function of the material. Hence, they are used as a composite and could equally be included in the chapter on geocomposites.

1.5.2 Manufacture

Almost all geonets are made of polyethylene. The specific gravity of geonets is in the range of 0.935 to 0.965; thus they are in the upper range of medium-density or lower range of high-density, depending on the classification system used. The division between medium- and high-density polyethylene established by the American Society for Testing and Materials (ASTM) is 0.940/0.941. The only additives in geonets are carbon black (1 to 2%) and a processing/antioxidant package (0.5 to 1.0%); thus the material is almost pure resin.

The ingredients are mixed and forced through an extruder, which ejects the melt into a die with slotted counter-rotating segments. This is called a *stenter* (see Figure 1.18). Here the polymer melt flows at angles, forming discrete ribs in two planes. As pressure from the ejected material forces the semisolid mass forward, it is pushed over an increasing diameter core (or mandrel), which separates the ribs and opens the net. Thus diamond-shaped apertures are formed that are typically 12 mm long by 8 mm wide. The resulting angles between sets of ribs are on the order of 70 and 110°. By the time the net has cooled completely, its full diameter is realized. After the full diameter is formed, the geonet is quenched in a water bath, cut along its manufactured axis, and formed into rolls for shipment. Final widths have been increasing with the development of newer production facilities so as to produce geonets up to 4.5 m wide. Because of this formation process, the intersecting ribs are generally not perpendicular to one another but are at slight angles. This is an important consideration when it comes to normal load carrying capability.

The above described nets are typically 5.0 to 7.0 mm in thickness. Since thickness is a key factor in determining in-plane drainage capability, it seems logical to try to in-

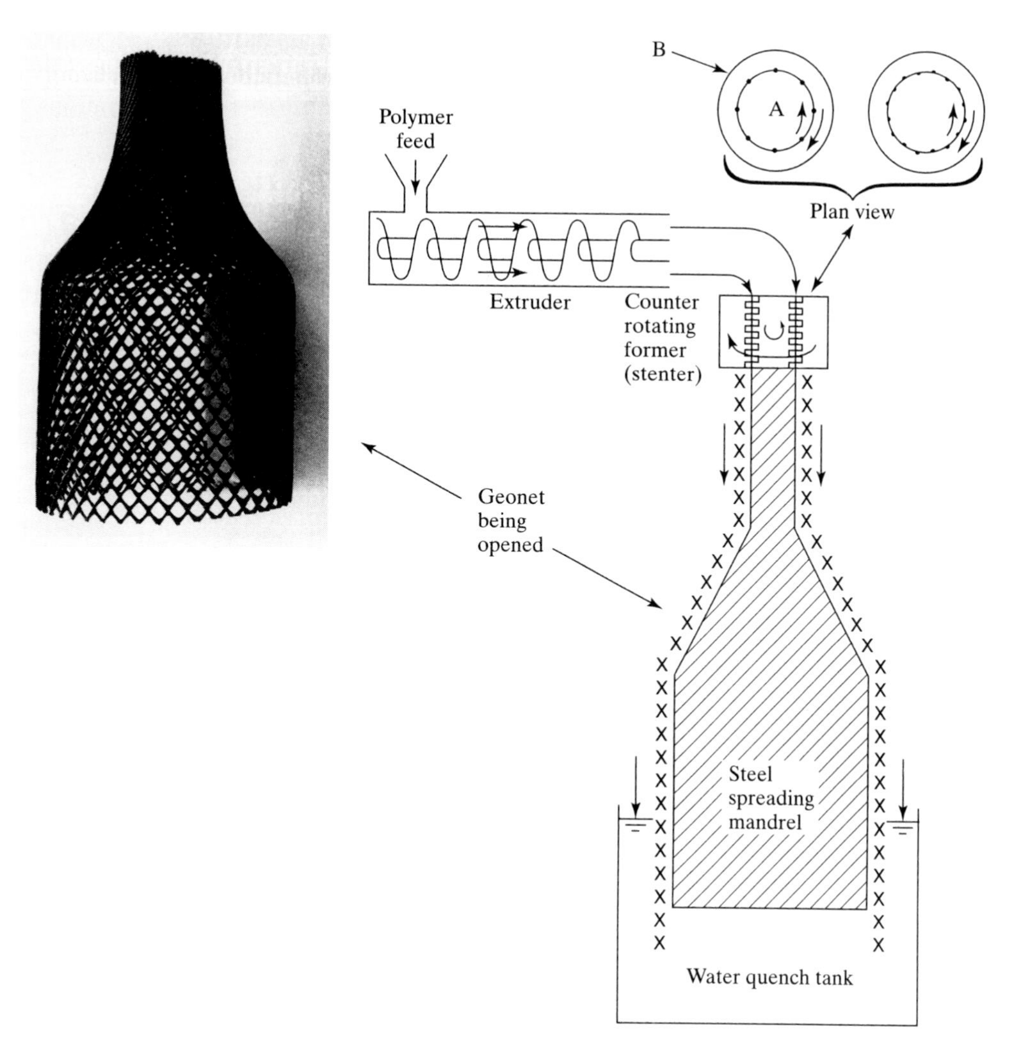

Figure 1.18 Diagram of geonet manufacturing process along with prototype shape as expanded over a steel spreading mandrel.

crease the above values. One way to do this is to add a foaming agent to the ingredients. The foaming agent reacts upon cooling to form microspheres in the solidified polymer. This has been done by a few manufacturers and results in nets up to 10 to 13 mm in thickness.

While the vast majority of the geonets in current use are formed as described above, there is no reason why other methods should be excluded. The above described extrusion process has recently been adapted to the manufacture of high-flow geonets with major ribs oriented parallel to the direction of flow (Austin [41]). Minor (stabilization) ribs are above and below the major ribs and are relatively closely spaced so as to minimize intrusion. The resulting geonets have flow rates 5 to 10 times higher in their manufactured direction than the conventional diamond shaped geonets. The cross machine direction has quite low flow rates; hence these geonets are used on side slopes where flow is predominately unidirectional. Prototype nets have also been formed by casting, injection molding, and other methods and can serve equally well, provided their mechanical and hydraulic properties are adequate. These considerations form the essence of Chapter 4, which is design-oriented.

1.5.3 Current Uses

Geonets are used almost exclusively for their drainage capability; they are single-function geosynthetics. The following uses have been documented in the literature.

- Water drainage behind retaining walls
- Water drainage of seeping rock slopes
- Water drainage of seeping soil slopes
- Water drainage beneath sport fields
- Water drainage of frost-susceptible soils
- Water drainage beneath building foundations
- Water drainage of plaza decks
- Polluted water drainage beneath highways
- Leachate collection in landfills and waste piles
- Leachate collection in heap leach pads
- Leak detection between double liners in landfills and surface impoundments
- Underdrain systems beneath landfills
- Surface water drains in landfills caps and closures
- Leak detection between two geomembranes in vertical containment walls
- Drainage blankets beneath a surcharge fill

1.5.4 Sales

Geonets as drainage materials are intermediate in their flow capability, between thick needle-punched, nonwoven geotextiles and numerous drainage geocomposites. As such, they compete with these materials at each end of their use spectrum. Yet their use has increased dramatically, from virtually nil in 1984 to an estimated 50 million square meters in 1995 in North America (see Figure 1.2). They are, indeed, viable geosynthetic materials in their own right.

1.6 OVERVIEW OF GEOMEMBRANES

1.6.1 History

In 1938, Goodyear cured (via vulcanization) natural rubber with sulfur, resulting in a synthetic rubber that is a thermoset polymer. The rubber industry was greatly stimulated by the cutoff of natural rubber supplies during World War II. Today, the production of various synthetic rubber materials is a major industry. The original geomembrane was a rubber product and was used as a potable water pond liner. It was butyl rubber, which is a copolymer of isobutylene with approximately 2% isoprene. Butyl rubber is quite impermeable and its major use is as inner tubes and as the liners of tubeless tires. Many other combinations and variants of rubber materials are possible, for example, nitrile and EPDM. Since the 1980s, however, the industry has shifted away from thermoset polymers to thermoplastic polymers. Thus, the geomembrane materials we will discuss fall into the category of polymers classified as thermoplastic materials. By definition, these are materials that when heated become soft and pliable without any substantial change in their inherent properties and when cooled revert back to their original properties. Thus they are readily seamed by heat, extrusion, or chemical methods.

Polyethylene is formed by the polymerization of compounds containing an unsaturated bond between two carbon atoms. Production in quantity began in 1943. Its main original uses were (and continue to be) in the packaging and molding industries. Polyethylene in its various densities is the most widely used polymer in the manufacturing of geomembranes. The development of crystallizing polypropylene is an outgrowth of low-pressure polymerization of ethylene and is the basic material from which many geotextiles are made. Polyvinyl chloride is another member of this group used to manufacture geomembranes. This product was developed in 1939 and has extensive uses. It ranks second in use to polyethylene. It is interesting to note that polyethylene geomembranes were first used in Europe and spread to North America and elsewhere, while polyvinyl chloride used for geomembranes had its roots in the U.S. and spread elsewhere. The global map of Figure 1.19 shows this movement, with polyethylene developing for use in geomembranes in Germany in the 1960s and spreading throughout Europe, to Africa, Australia, and North America. At roughly the same time, other types of geomembranes were being developed and used by the U.S. Bureau of Reclamation. These geomembranes served primarily as canal liners and their use spread to Canada, Hawaii, Russia, Taiwan, and Europe. Another early geomembrane, chlorosulfonated polyethylene (CSPE), resulting from the reaction of chlorine and sulfonyl chloride on polyethylene, was introduced for reservoir and landfill liners in the late 1960s and this geomembrane type was used in Europe shortly thereafter. Today's polymeric geomembranes are made from thermoplastic resins and are manufactured and distributed the world over, making all types of products readily available. However, what matters most to the owner/user/designer, and is the focus of this book, is to use the proper material for the particular project: *That* is the essence of the design-by-function concept.

1.6.2 Manufacture

The manufacturing of geomembranes begins with the production of the raw materials, which include the polymer resin itself; various additives, such as antioxidants, plasticizers, fillers, and carbon black; and lubricants (as a processing aid). Recall Table 1.5, which gives the amounts of the different materials used to make geosynthetic materials. These raw materials are then processed into geomembrane sheets of various widths and thicknesses in one of three ways (see Figure 1.20).

All polyethylene geomembranes [e.g., high-density (HDPE) and very flexible (VFPE) types [and some flexible polypropylenes (fPP)] are manufactured by the *extrusion* method. In this method the polymer resin, carbon black (usually in pellet form premixed with its carrier resin), and an additive package (antioxidant and lubricant) are

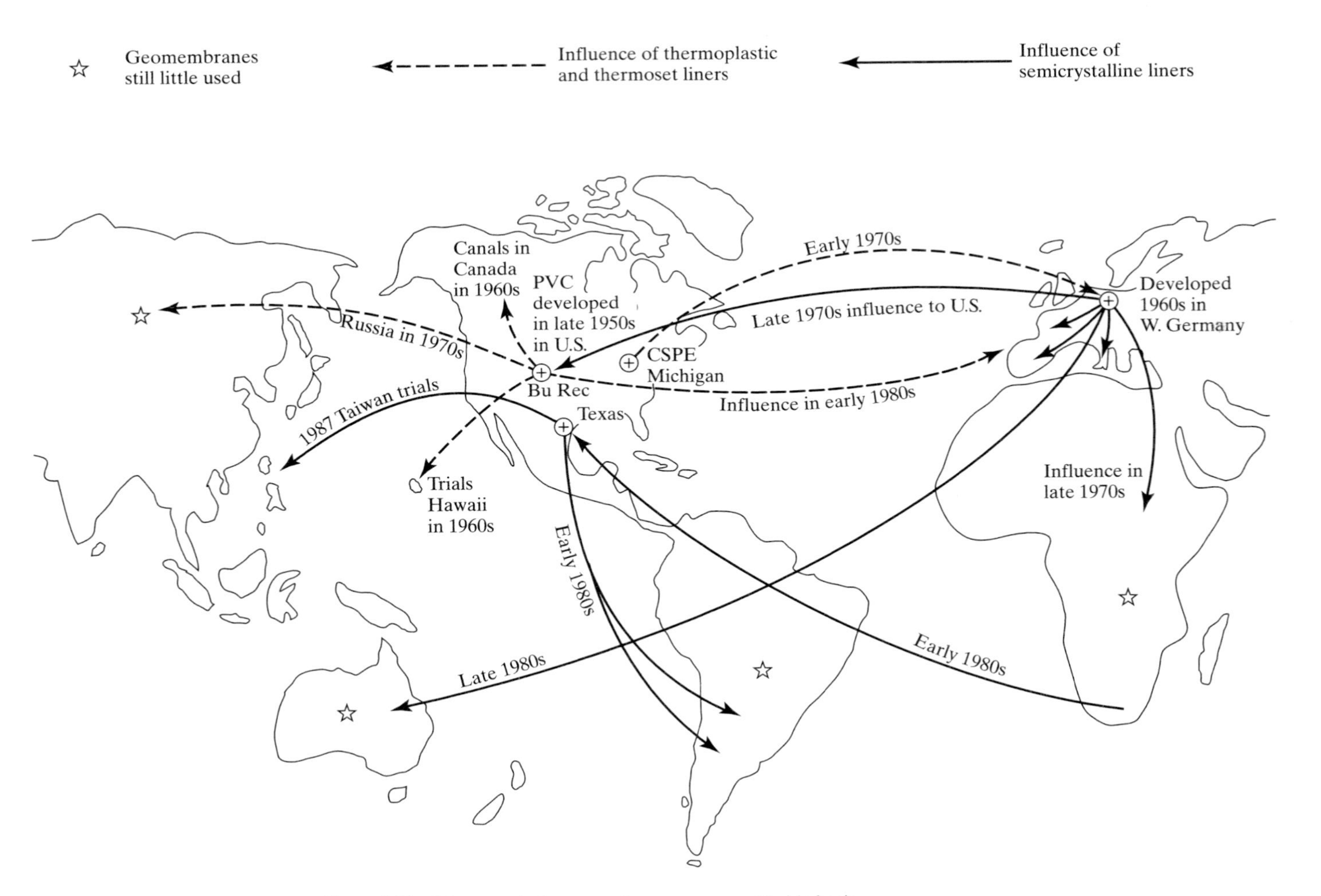

Figure 1.19 Movement of geomembranes on a worldwide basis.

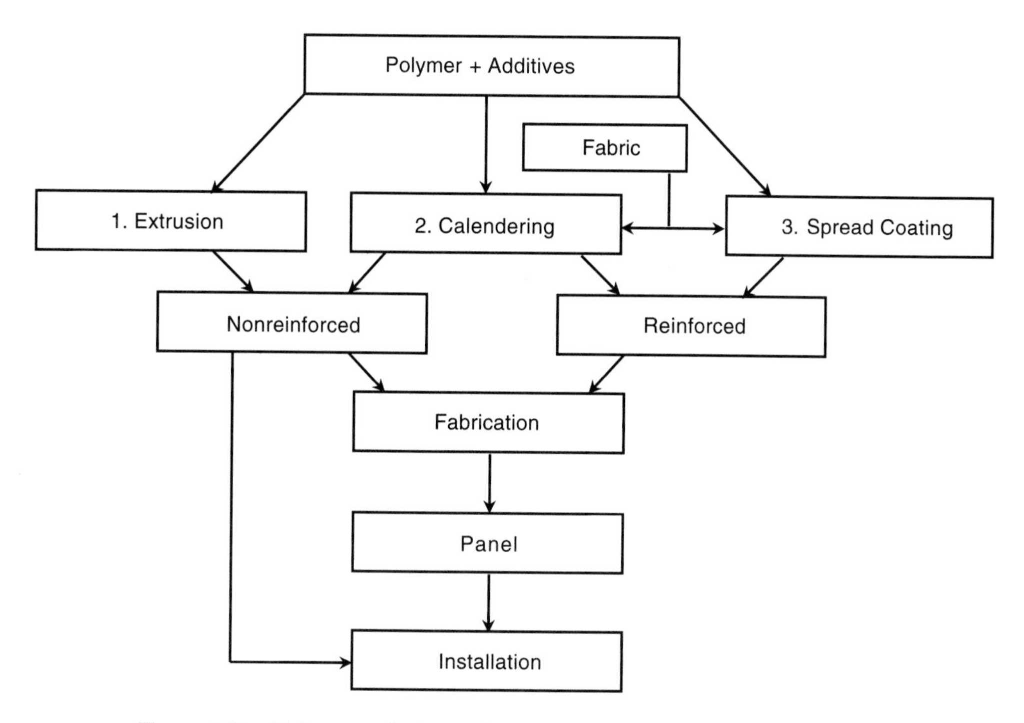

Figure 1.20 Various methods used to manufacture geomembranes. (After Haxo [42])

pneumatically loaded into the feed hopper of an extruder (see Figure 1.21). The extruder contains a rotating continuous flight screw. The formulation passes successively through a feed section, compression section, and metering section where it finally emerges as a filtered, mixed, and molten material into a die. Two variations of extrusion processing are then used to make geomembranes. One process uses a flat die (called *cast*

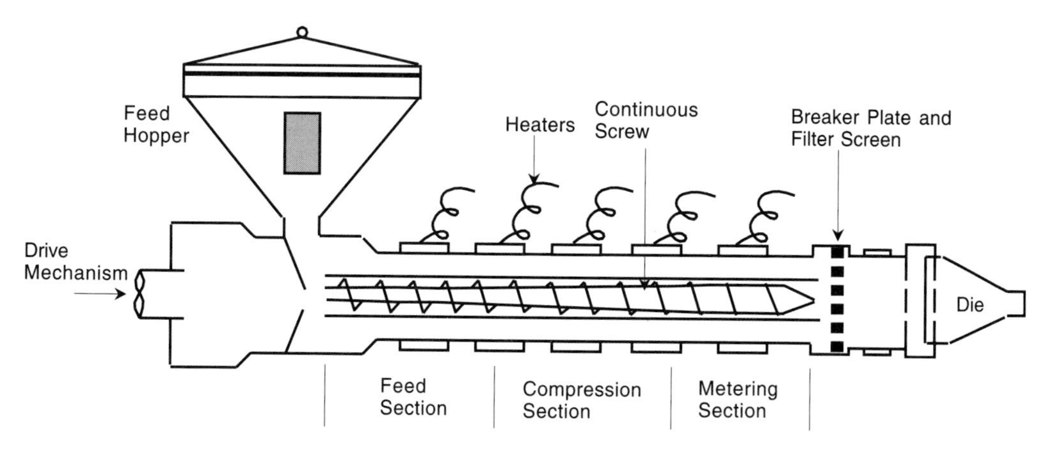

Figure 1.21 Cross-section diagram of a horizontal single-screw extruder for polymer processing.

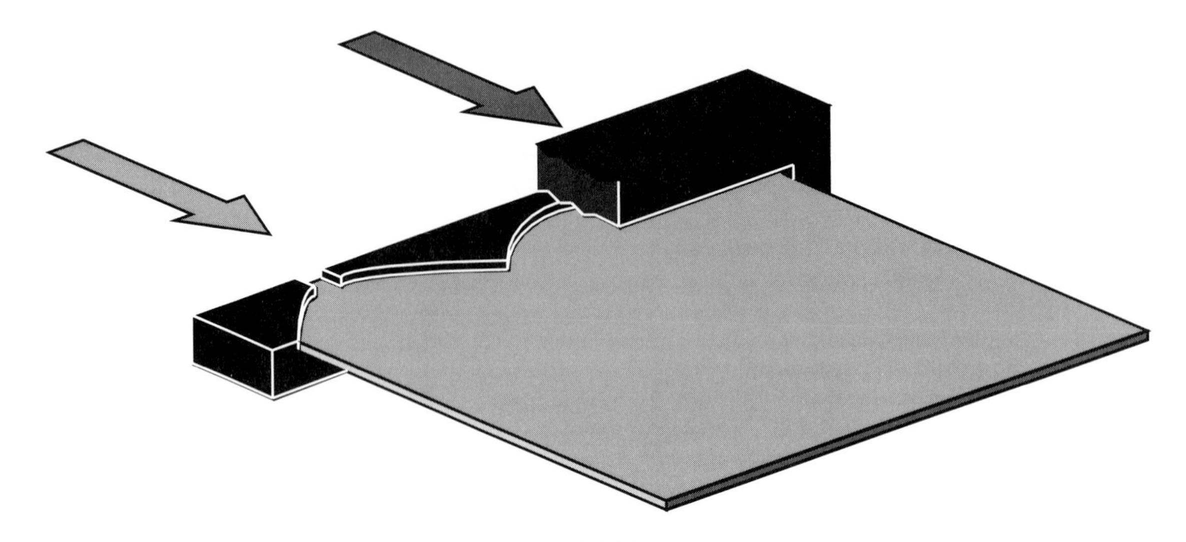

Figure 1.22 Processing of geomembranes by flat die extrusion using two extruders in parallel. (Compliments of National Seal Corp. and GSE Lining Technology, Inc.)

sheeting), which forces the polymer formulation between two horizontal die lips, in a coathanger-like manner, resulting in a sheet of closely controlled thickness from 0.75 to 3.0 mm. The sheet widths vary from 1.8 to 4.6 m. When two parallel extruders are used, the width can be increased to 9.5 m (see Figure 1.22). The second process uses a circular die (called *blown sheeting*), which forces the polymer formulation between two con-

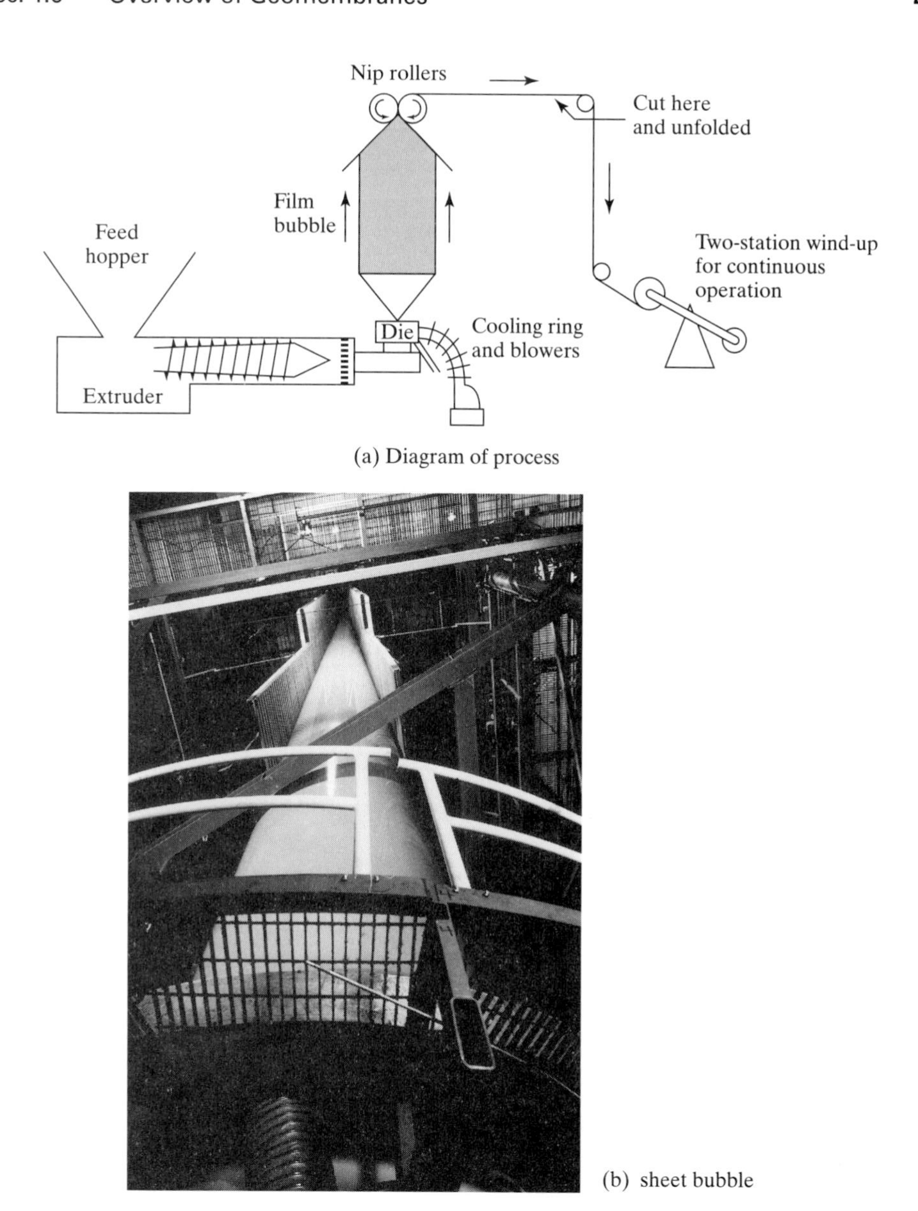

(a) Diagram of process

(b) sheet bubble

Figure 1.23 Processing of geomembranes by blown sheet extrusion. (Compliments of GSE Lining Technology, Inc.)

centric die lips oriented vertically. As seen in Figure 1.23a, the polymer formulation exits the die and is supported by a large circular internal mandrel as it eventually extends upward in an enormous cylinder (see Figure 1.23b). At the top of the system, two counter rotating rollers draw the cylinder upward and maintain stability. After passing over the rollers, the sheet is longitudinally cut, unfolded to its full width, and rolled onto a takeup

(c) cutting of sheet preceding wind-up

(d) final manufactured rolls with dedicated slings awaiting shipment to job site

Figure 1.23 (continued)

core (see Figure 1.23c). Geomembranes produced by both types of extrusion are shown in Figure 1.23c. They are transported to the job site and installed in the field as shown in Figure 1.24a, where they are field-seamed together into a completed liner system.

By creating a roughened surface on a smooth HDPE or VFPE sheet (or other polymer types), a process called "texturing" in this book, a high-friction surface can be

(a) roll of extruded polyethylene;

(b) accordion-folded pallet of calendered polyvinyl chloride

Figure 1.24 Geomembranes being deployed in the field.

created. There are currently four methods used to texturize geomembranes: coextrusion, impingement, lamination and structuring (see Figure 1.25).

The *coextrusion* method utilizes a blowing agent in the molten extrudate and delivers it from a small extruder immediately adjacent to the main extruder. When both sides of the sheet are to be textured, two small extruders (one internal and one external to the main extruder) are necessary. As the extrudate from these smaller extruders meets the cool air, the blowing agent expands, opens to the atmosphere, and creates the textured surface(s). A small width (approximately 300 mm) of the cylinder's circumference can be left smooth, which after central cutting becomes the two lengthwise edges of the roll for ease of seaming.

Impingement of hot HDPE particles against the finished polyethylene sheet is a second method of texturing. In this case, hot particles are actually projected onto the previously manufactured smooth sheet on one or both of its surfaces in a secondary operation. The adhesion of the hot particles to the cold surface(s) should be as great as or greater than the shear strength of the adjacent soil or other abutting material. The lengthwise edges of the sheets are left nontextured for approximately 150 mm, for ease of seaming.

The third method for texturing polyethylene sheet is by *lamination* of a foam on a previously manufactured smooth sheet in a secondary operation. In this method a

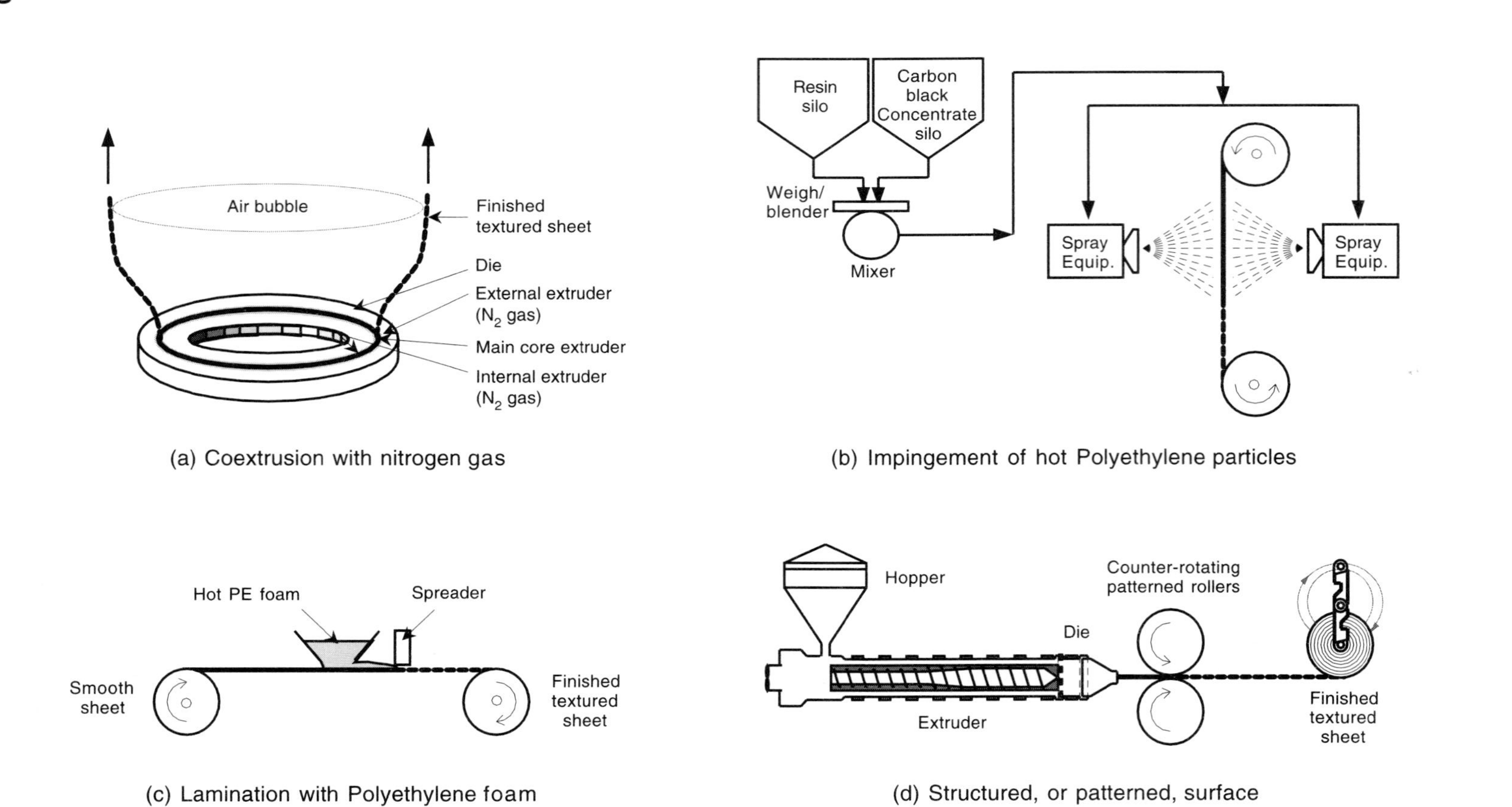

Figure 1.25 Various methods used to produce textured surfaces on HDPE, VFPE, and fPP geomembranes.

foaming agent contained within molten HDPE provides a froth that is adhered to the previously manufactured smooth sheet, providing a rough textured surface. The degree of adhesion is important with respect to the shear strength of the adjacent soil or other abutting material. If texturing on both sides of the geomembrane is necessary, the roll must go through another cycle on its opposite side. The lengthwise edges of the sheets are left nontextured for approximately 150 mm so that field-seaming can be readily accomplished.

The fourth method of increasing the friction of the surface(s) of a smooth geomembrane is called *structuring*. In this method a smooth sheet is made by the flat die method and immediately upon leaving the die lips, passes between two counter-rotating rollers. These rollers have patterned surface(s), allowing the still hot sheet (approximately at 120°C) to pass between and deform into the pattern(s). This gives a raised surface pattern(s) on the nonstructured sheet as it exits the rollers; this is typically a box and point pattern, but the variations are endless. The lengthwise edges of the sheets are left nonstructured for approximately 150 mm so that field-seaming can be readily accomplished.

All PVC and scrim-reinforced geomembranes like fPP-R and CSPE-R are manufactured by a *calendering* method. In this method the polymer resin, carbon black, filler, plasticizer (if any), and additive package are weighed (recall Table 1.5) and mixed in a (Banbury-type) batch or (Farrel-type) continuous mixer. During mixing, heat is added, which initiates a reaction between the components. The material exits the mixer and moves by conveyor to a roll mill where it is further blended and masticated. Now in a continuous roll, it is passed through a set of counter-rotating rollers (called a *calender*) to form the final sheet. The versatility of calenders is seen in Figure 1.26. This type of manufacturing gives rise to multiple plies of laminated geomembranes, sometimes with an open-weave fabric (called *scrim*) between the individual plies. The openings in the scrim must be large enough to allow the plies to adhere to one another (called *strike-through*) with adequate ply adhesion to prevent delamination. Geomembranes produced by calendering are available in widths up to 2.4 m. To produce much wider widths (called *panels*), the rolls are sent to a fabricator who factory-seams the rolls together and packages them in a double accordion-fold for shipment to the field (see Figure 1.24b).

Geomembranes can also be made by a manufacturing method called *spread coating*. In this method, the molten polymer (whatever its formulation) is spread in a relatively thin coating over a tightly woven fabric or even a nonwoven fabric. Factory- and field-manufactured bituminous geomembranes are of this type. Generally, the open pore spaces of the fabric are insufficient to allow for penetration to the opposite side, so if coating on both sides is required the material must be turned over and the process repeated. Research and some field trials are ongoing using a spray coating with various types of elastomers, for example, polyurea. The elastomer may also be applied directly to the subgrade soil as an in situ geomembrane. There are a large number of possibilities for the spread coating and/or spray coating manufacture of geomembranes.

In the descriptions of extrusion, calendering, and spread (or spray) coating manufacturing of geomembranes, it is important to appreciate that a complete production process involves several separate companies: the resin producer, the additive producers, the formulators, the manufacturers, the fabricators, and (eventually) the installers.

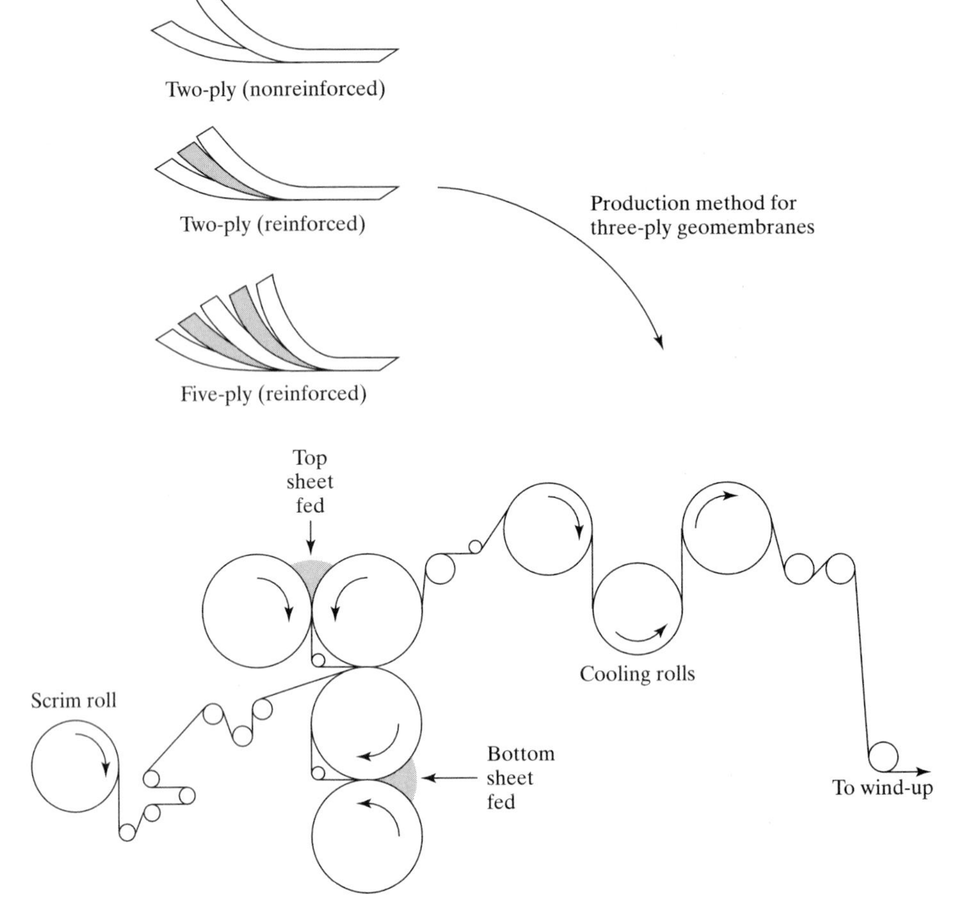

Figure 1.26 Processing of geomembranes by calendering, utilizing multiple plies of material. (Compliments of JPS Elastomerics, Inc.)

Communication between each party and proper liaison is critical in arriving at an acceptable and properly functioning installation. Problems and misunderstandings can arise because of the relative large number of parties involved. It is critical that proper manufacturing quality control (MQC) measures be taken by the manufacturer and fabricator when bringing to the job site the geomembrane that was designed, specified, and purchased. In this same light, manufacturing quality assurance (MQA), in seeing that the owner/operator of a facility has the proper geomembrane manufactured as per the project plans and specifications, is important and routinely practiced in the geomembrane industry.

1.6.3 Current Uses

A wide range of uses of geomembranes have been developed, all of which relate to the materials' primary function of being "impervious." Note at the outset that nothing is *strictly* impermeable in an absolute sense (this is treated further in Chapter 5). Here we are speaking of relative impermeability compared to that of competing materials. In the case of solid- or liquid-waste containment geomembranes, the competing material is often natural or amended clay, which usually has a targeted hydraulic conductivity (permeability) of approximately 1×10^{-9} m/s. By contrast, the equivalent diffusion permeability of a typical thermoplastic geomembrane will be 10^{-13} to 10^{-15} m/s. Thus we speak of geomembranes as being relatively impermeable.

Uses of geomembranes in environmental, geotechnical, hydraulic, and transportation activities are listed below.

- Liners for potable water
- Liners for reserve water (e.g., safe shutdown of nuclear facilities)
- Liners for waste liquids
- Liners for radioactive or hazardous waste liquid
- Liners for secondary containment of underground storage tanks
- Liners for solar ponds
- Liners for brine solutions
- Liners for water conveyance canals
- Liners for various waste conveyance canals
- Liners for solid-waste landfills and waste piles
- Liners for heap leach pads
- Covers (caps) for solid-waste landfills
- Liners for vertical walls: single or double with leak detection
- Cutoffs within zoned earth dams for seepage control
- Linings for emergency spillways
- To waterproof liners within tunnels
- To waterproof facing of earth and rockfill dams
- To waterproof facing for masonry dams
- Within cofferdams for seepage control
- As floating reservoirs for seepage control
- As floating reservoir covers for preventing pollution
- As a barrier to odors from landfills
- As a barrier to vapors (radon, hydrocarbons, etc.) beneath buildings
- To control expansive soils
- To control frost-susceptible soils
- To shield sinkhole-susceptible areas from flowing water
- To prevent infiltration of water in sensitive areas
- To form barrier tubes as dams
- To face structural supports as temporary cofferdams
- To conduct water flow into preferred paths
- Beneath highways to prevent pollution from deicing salts
- Beneath and adjacent to highways to capture hazardous liquid spills

- To act as containment structures for temporary surcharges
- To aid in establishing uniformity of subsurface compressibility and subsidence
- Beneath asphalt overlays as a waterproofing layer
- To correct seepage losses in existing above ground tanks
- As flexible forms where loss of material cannot be allowed

1.6.4 Sales

While there are some new resins entering the geomembrane market, it currently is divided among HDPE, VFPE, PVC, CSPE, fPP, and others such as ethylene interpolymer alloy (EIA) and ethylene propylene diene monomer (EPDM). Based on the 1995 estimated total of 75 Mm^2 in North America, the approximate proportions are as follows:

high-density polyethylene (HDPE)	$\simeq$ 40% or 30 Mm^2
very flexible polyethylene (VFPE)	$\simeq$ 25% or 19 Mm^2
polyvinyl chloride (PVC)	$\simeq$ 20% or 15 Mm^2
chlorosulphonated polyethylene (CSPE)	$\simeq$ 5% or 4 Mm^2
flexible polypropylene (fPP)	$\simeq$ 5% or 4 Mm^2
others	$\simeq$ 5% or 3 Mm^2

Note that within the VFPE category are a number of lower-density polyethylenes such as very low density (VLDPE), linear low density (LLDPE), and low density linear (LDLPE) polyethylenes.

Jagielski [43] estimates that landfill liners and covers represent 50% of the applications, while liquid impoundments represents another 25%. Thus environmental applications make up approximately 75% of the total geomembrane market. The balance is in mining, tunneling, geotechnical, and specialty applications. The European market is somewhat smaller than this, since regulatory use in the environmental field is not as stringent as in the U.S. (the exception being in Germany). The Asian market, while currently small, is rapidly growing, particularly in solid-waste and aquiculture liners. In South America, the main application for geomembranes is as liners for heap-leach mining.

Projections for geomembrane sales are extremely strong due to their becoming known only recently to many civil engineers; with familiarity comes additional use and branching into new areas. Thus estimates of 20 to 25% growth in the geomembrane field are readily understandable. From both sales and applications viewpoints, geomembranes are an exciting field of study.

1.7 OVERVIEW OF GEOSYNTHETIC CLAY LINERS

1.7.1 History

The use of geosynthetic clay liners (GCLs) as a separate category of geosynthetics is quite recent. It appears that the first use in the U.S. was in 1988 in solid-waste containment as a backup to a geomembrane; the product used was Claymax—bentonite clay mixed with an adhesive so as to bond clay between two geotextiles, one below (the *substrate*) and the other above (the *superstrate*). At about the same time a different product, named Bentofix, was manufactured in Germany by placing bentonite powder between two geotextiles and needle-punching the system together.

Other names for GCLs are "clay blankets," "bentonite blankets," "bentonite mats," and "prefabricated bentonite clay blankets." The engineering function of GCLs is as a hydraulic barrier to water, leachate, or other liquids. As such, they are used as replacements to either compacted clay liners or geomembranes, or they are used in a composite to augment the more traditional liner materials.

1.7.2 Manufacture

Today, many GCLs are available. In addition to the above-mentioned products there are: Bentomat, with two geotextiles needle-punched together containing bentonite powder; Gundseal, using an adhesive to bond bentonite powder onto a HDPE or VFPE geomembrane; NaBento, consisting of two geotextiles containing bentonite powder and stitch-bonded together; and others. Figure 1.27 shows a photograph of the resulting products with a small hydrated sample placed above the as-received materials. Figure 1.28 gives the production concept for GCLs, with the upper sketch depicting the two adhesive-bonded products and the lower sketch depicting the needle-punched and stitch-bonded products. All of the GCL products manufactured in North America use a sodium bentonite clay in the mass per unit area of 3.2 to 6.0 kg/m^2. The clay thickness varies in the range of 4.0 to 6.0 mm. The hydraulic conductivity (permeability) is typically in the range of 1×10^{-11} to 5×10^{-11} m/s. The various products come to the job site at a humidity-equilibrated moisture state that varies from 10 to 18%. This is sometimes referred to in the technical literature as the "dry" state. The types of geotextiles

Figure 1.27 Various types of geosynthetic clay liners currently available, showing the corresponding hydrated product directly on top of the as-received product.

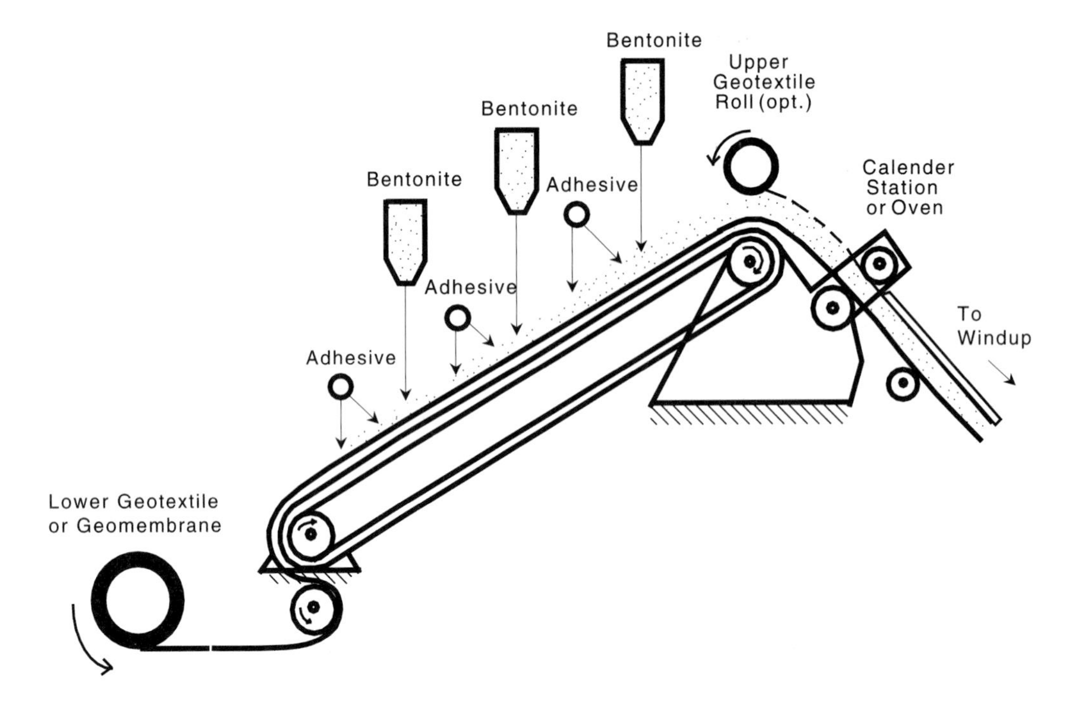

(a) Adhesive mixed with clay

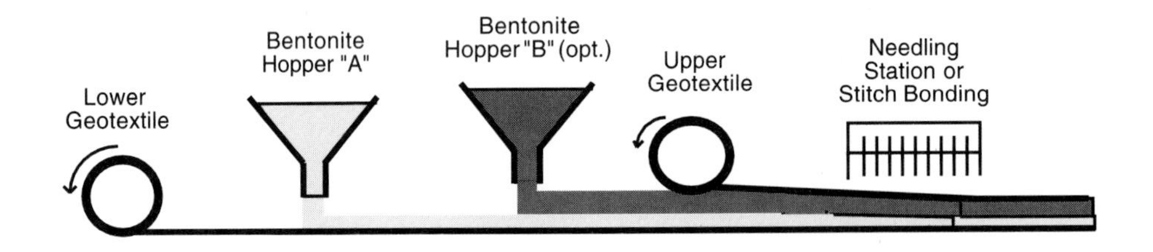

(b) Needle punched or stitch bonded through clay

Figure 1.28 Methods of manufacturing different types of geosynthetic clay liners.

used with the different products vary widely in their manufacturing (e.g., needle-punched nonwoven, woven-silt film, spunlaced, and composite) and in their mass per unit area (e.g., varying from 85 g/m^2 to 600 g/m^2). The particular product with a geomembrane backing can also vary in its type, thickness, and surface texture.

GCLs are factory-made in widths of 4.0 to 5.2 m and lengths of 30 to 60 m. Upon manufacturing they are rolled onto a core and are covered with a plastic film to prevent additional moisture absorption or wetting (hydration) during storage, transportation, and placement prior to their eventually being covered with an overlying layer.

1.7.3 Current Uses

GCLs are hydraulic barrier layers used to contain liquid movement and, as such, are competitive wherever geomembranes and compacted clay liners are used. However, GCLs have unique applications of their own.

- Beneath geomembranes in the primary liners of landfills
- Beneath geomembranes in the secondary liners of landfills
- Beneath geomembranes and above clay liners of landfills (i.e., three-component liners)
- Beneath geomembranes in the covers of landfills
- Adjacent to geomembranes in vertical cutoff walls
- Above geomembranes as puncture protection against coarse gravel
- As a portion of a compacted clay liner in primary composite liners
- As a portion of a compacted clay liner in secondary composite liners
- As secondary liners for underground storage tanks
- As single liners for surface impoundments
- Beneath geomembranes as composite liners for surface impoundments
- Beneath geomembranes as composite liners for heap leach ponds
- As liners for canals

1.7.4 Sales

The use of GCLs has moved from conception to application, perhaps more rapidly than any other geosynthetic material. It is estimated that approximately 50 M m^2 were installed in 1995 in North America. Recall Figure 1.2.

Within this total, the application breakdown is estimated as follows:

landfill liners (usually beneath a geomembrane)	$\simeq$ 40% or 20 Mm^2
landfill covers (usually beneath a geomembrane)	$\simeq$ 40% or 20 Mm^2
other environmental applications	$\simeq$ 10% or 5 Mm^2
other geotechnical, transportation, and hydraulic applications	$\simeq$ 10% or 5 Mm^2

1.8 OVERVIEW OF GEOPIPE (AKA PLASTIC PIPE)

1.8.1 History

The traditional materials used for the underground pipeline transmission of water, gas, oil, and various other liquids have been steel, cast iron, concrete, and clay. These pipe materials are classified as "rigid" and are strength-related as far as their material behavior is concerned. Plastics, however, are making significant inroads in these markets and they certainly deserve a separate chapter in a book devoted to polymeric-based geosynthetic materials. Pipes made from polymeric materials are classified as "flexible" and are deflection-governed as far as their material behavior is concerned. To be consistent with other geo-terms in this book, the category is titled *geopipe*, which refers to plastic pipe placed beneath the ground surface and subsequently backfilled.

Plastic pipe is one of the very first polymer materials fabricated and used to a wide extent. The area is very well-organized with a separately founded Plastic Pipe Institute and a world renown research arm, the Gas Research Institute. There are plastic pipe associations and institutes in most of the industrialized countries in the world. Annual plastic pipe conferences disseminate information and give great credibility to the manufacture, testing, design, and use of plastic pipe for all applications.

In the area of buried plastic pipe, over 95% of natural gas transmission lines are made of plastic pipe (generally HDPE). Additional major areas are in industrial, agricultural, transmission, and drainage applications. Geopipes' use in civil engineering applications is only a small fraction of this much larger and more mature market. Within the context of this book, geopipe has four areas of application: environmental, transportation, geotechnical, and (of course) hydraulic.

1.8.2 Manufacture

There are a number of polymer resins being used in the fabrication of plastic pipe. Currently they are polyvinyl chloride (PVC), high-density polyethylene (HDPE), polypropylene (PP), polybutylene (PB), acrylonitrile butadiene styrene (ABS), and cellulose acetate buytrate (CAB). They all entered the market as solid-wall, constant-thickness pipes of relatively small diameter. For example, every hardware store handles small diameter PVC pipe and fittings for household water and drainage systems for "do-it-yourselfers." Today's plastic pipes, however, can be very large in diameter and very thick in their wall dimensions.

Plastic pipe consists of a specific formulation (recall Table 1.5) that is fed into a extruder of the type shown in Figure 1.21. As the polymer melt exits the filter screen, however, the die is circular in its shape, with a concentric circular insert contained within it. The polymer thus has controlled outer and inner diameters. Obviously, the support system for the inner core must be appropriately designed, but plastic pipe manufacturing design is very developed in this regard; extremely close tolerances of diameter and thickness can be achieved.

In order to optimize behavior, economize on resin, and aid in installation, a number of differing types of wall sections consisting of ribs, cores, and corrugated profiles of a wide variety of cross-sectional shapes and sizes have appeared. These latter pipes are referred to as corrugated, or profiled, wall pipes. Additionally, many applications such as agriculture drains and leachate collection systems require holes, slots, or other types of perforations in the wall section to allow for the inflow of liquids. This has led to a designation system for corrugated-walled plastic pipe as follows.

CPP	corrugated (outside) plastic pipe
CPP-S	CPP pipe with smooth inside

Figure 1.29 Smooth and corrugated plastic pipe (geopipe) used in below-ground construction.

CPP-C	CPP pipe with corrugated inside
CPP-SP	CPP-S pipe with perforations
CPP-CP	CPP-C with perforations

Both solid-wall and profile-walled plastic pipe are shown in Figure 1.29.

1.8.3 Current Uses

Solid-wall and corrugated-wall plastic pipe are used in a wide variety of civil engineering applications. Some of these are:

- Highway, railway, and airfield edge drains
- Seepage drains in tunnels
- Pore water drains behind retaining walls
- Interceptor drains in seeping soil and rock slopes
- Interceptor drains for groundwater seepage
- Pipes used in dewatering projects
- Fluid transmission lines by gravity

- Force transmission lines under pressure
- Wastewater drainage systems
- Piping in leach fields of various types
- Chemical transmission pipelines
- Primary leachate removal systems in landfills and waste piles
- Secondary leachate removal systems in landfills and waste piles
- Pipe risers along landfill sidewalls
- Pipe manifold systems for landfill gas collection and removal
- Pipe manifold systems for leachate recycling into landfills
- Surface water removal systems in landfill covers
- Dredging pipelines

1.8.4 Sales

As noted in Figure 1.2, which gives the estimated geosynthetic market status, there is no data for geopipe. There are two reasons for this: the data cannot be placed on a unit area basis and there have been no market studies done specifically for geopipe to my knowledge. Additionally, the assessment would have had to be done on the basis of amount of polymer used, since heavy thick-walled pipes can use 10 times more polymer than corrugated-wall pipe. Suffice it to say that the plastic pipe industry is interested in geosynthetics and this interest should be reciprocated by the geosynthetics community.

1.9 OVERVIEW OF GEOCOMPOSITES

The basic philosophy behind geocomposite materials is to combine the best features of different materials in such a way that specific problems are solved in the optimal manner. In this book, these geocomposites will generally be geosynthetic materials but not always. In some cases it may be more advantageous to use a nonsynthetic material with a geosynthetic one for optimum performance and/or minimum cost. As you will see, the number of possibilities is almost infinite—the only limits are one's ingenuity and imagination.

In considering the following geocomposites, keep in mind that the basic functions we have to provide for are separation, reinforcement, filtration, drainage, and containment.

1.9.1 Geotextile-Geonet Composites

When a geotextile is used over or under a geonet, or makes a geotextile-geonet-geotextile sandwich, the separation and filtration functions are always satisfied, but the drainage function is vastly improved over the use of geotextiles by themselves. Placed horizontally, the composites make excellent drains to upwardly moving water in a capillary zone where frost heave or salt migration is a problem. When the water enters the sandwich, it travels horizontally within the geonet and away from where damage can occur. Such geocomposites have also been used in intercepting and conveying leachate in landfills and for conducting vapor or water beneath pond liners of various types.

1.9.2 Geotextile-Geomembrane Composites

Geotextiles are laminated on one or both sides of a geomembrane for a number of purposes. The geotextiles provide increased resistance to puncture, tear propagation, and friction related to sliding, as well as providing tensile strength in and of themselves. Quite often, however, the geotextiles are of the nonwoven needle-punched variety and are of relatively heavy weight. In such cases they act as drainage media, since their in-plane transmissivity feature can conduct water, leachate, or gases away from direct contact with the geomembrane.

1.9.3 Geomembrane-Geogrid Composites

Since some types of geomembranes and geogrids can be made from the same material (e.g., high-density polyethylene), they can be bonded together to form an impervious barrier with enhanced strength and friction capabilities.

1.9.4 Geotextile-Geogrid Composites

Those geotextiles with low modulus, low strength and/or high elongation at failure can be greatly improved by forming a composite material with a geogrid, or even with a woven-fabric scrim. The synergistic properties of all the components enhance the behavior of the final product.

1.9.5 Geotextile-Polymer Core Composites

A core in the form of a quasi-rigid plastic sheet can be extruded or deformed in such a way as to allow very large quantities of water to flow within its structure; it thus acts like a drainage core protected by a geotextile acting as a filter and separator on one or both sides. Many systems are available (e.g., wick drains fall into this group). The 100 mm wide by 5 mm thick polymer core is often fluted for ease of conducting water. A geotextile acting as a filter and separator is socked around the core. The emergence of such geotextile polymer-core composites has all but eliminated traditional sand drains as a rapid means of consolidating fine-grained saturated soils.

In the form of panels, the rigid polymer core can be nubbed, columned, or dimpled and with a geotextile on one side it makes an excellent drain on the backfilled side of retaining walls, basement walls, and plaza decks. These cores are often vacuum-formed as shown in Figure 1.30. As with wick drains, the geotextile is the filter/separator and the deformed polymer core is the drain. Many systems of this type are available, the latest addition having a thin pliable geomembrane on the side facing the wall, functioning as a vapor barrier.

Lastly within this area of drainage geocomposites comes the category of prefabricated edge drains. These materials, typically 500 mm high by 20 to 30 mm wide are

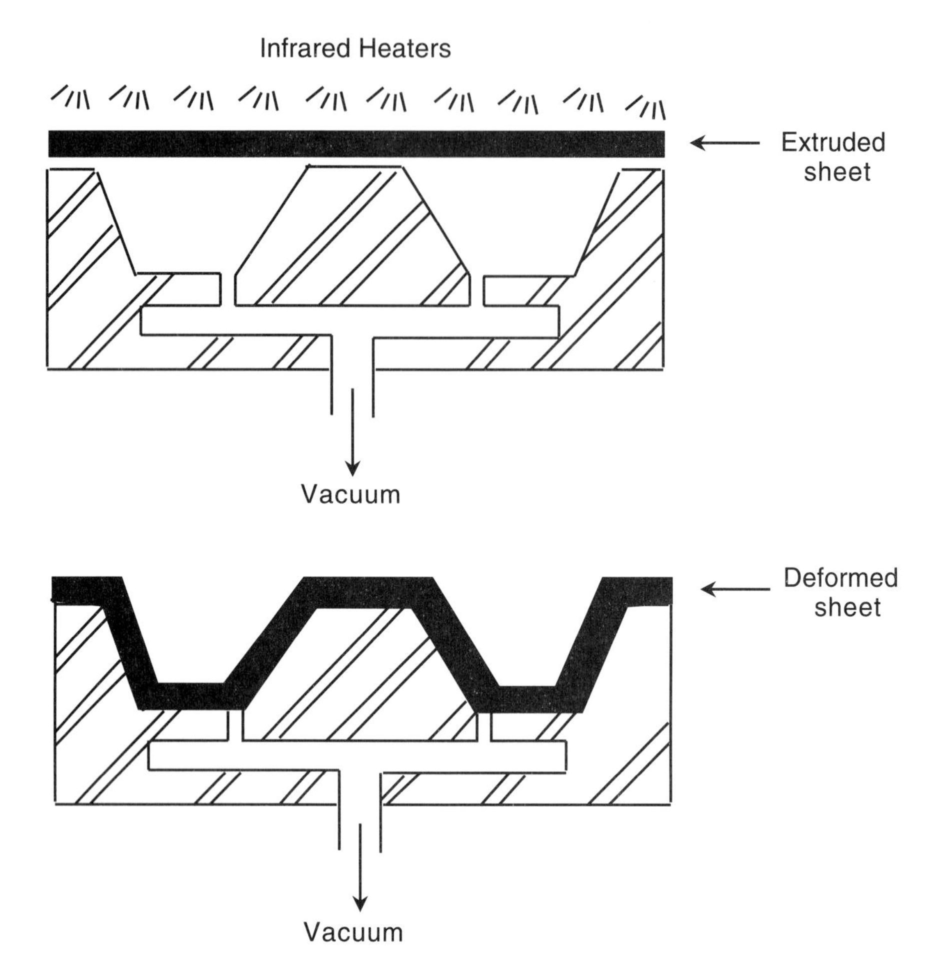

Figure 1.30 Method of vacuum-forming in the fabrication of cores for some types of drainage geocomposites.

placed adjacent to a highway, airfield pavement, or railroad right-of-way for lateral drainage out of and away from the pavement section. The installation of these systems are incredibly rapid and extremely cost effective.

1.9.6 Geosynthetic-Soil Composites

As typified by the geosynthetic clay liners described in Section 1.7, many other variations of geosynthetic products and soil can be developed. For example, geocells are rigid polymer strips and geotextiles that have been cleverly arranged vertically in a boxlike fashion, placed horizontally (standing upright), and filled with soil. Thus the material forms a cellular structure and, acting with the contained sand or gravel, makes an impressively strong and stable mattress. Sizable earth embankments have been built on such systems with the possibility of supporting structures over weak soils in the near future (i.e., an inexpensive mat foundation).

Another variation is to use continuous polymer fibers and sand to form a steep soil slopes with excellent strength properties. The fibers give the composite material a very pronounced apparent cohesion. One process called Texsol has seen worldwide use with impressive results.

1.9.7 Other Geocomposites

Weaving steel strands within a geotextile matrix can result in incredible composite material strengths. When it is used as a substrate, extremely large loads can be sustained and a measurable increase in bearing capacity for the support of buildings is also possible.

Geotextiles with prefabricated holes for the insertion of steel rod anchors have been used to stabilize slopes and as in situ compaction and consolidation systems. The rods act as anchors, stressing the geotextile against the soil, which is put into compression. The geotextile thus acts dually as a tensile-stressing mechanism and as a filter allowing the pore water to escape while retaining the individual soil particles.

Short fibers, grids, and nets to be placed in concrete or bitumen to form a high impact composite material can be added to this list.

1.9.8 Geofoam

The original use of lightweight expanded polystyrene (EPS) beads, bonded together and placed between two geotextiles, was to reduce lateral pressures behind basement walls and earth-sheltered homes. This North American product also acted as a partial drain and as thermal insulation. Parallel, and much more extensive, use of large blocks of EPS for lightweight fills in the Scandinavian countries of Europe were ongoing as early as 1980.

Currently the emerging category known as *geofoam* consists of expanded polystyrene (EPS) or extruded polystyrene (XPS) in below-grade applications. The area is experiencing tremendous growth as data about the properties and behavior are becoming more widely disseminated. Most notably, its lightweight characteristics, with a density between 10 and 20 kg/m^3, are very intriguing. Horvath (44) describes many applications, along with the requisite physical and mechanical properties and test methods required for an emerging material.

1.10 OUTLINE OF THE BOOK

To the author, and hopefully to many others, the area of geosynthetics is a vibrant, exciting and rapidly growing field within civil (i.e., geotechnical, environmental, transportation, and hydraulics) engineering. The information presented in Figure 1.2 (which is based on the author's best estimates) reflects this dynamic growth in each geosynthetic area. The data are approximate and only for North America, but are indicative of the general worldwide pattern. The field is in a constant state of flux and the designs presented here might well be superseded within the near future. This is to be expected. Nevertheless, the area demands a specific *design-by-function* methodology—an end to which I am committed. It is hoped that time will validate the effort.

The remainder of this book consists of seven chapters (one each on geotextiles, geogrids, geonets, geomembranes, geosynthetic clay liners, geopipe, and geocomposites), each structured after the pattern of this opening chapter. Each chapter is relatively self-contained and independent, although there is a logic and plan to the sequence in which the chapters appear. Design-by-function is emphasized throughout, with illustrative problems and homework problems in each chapter. Reference lists unique to each chapter are included.

REFERENCES

1. Dewar, S., "The Oldest Roads in Britain," *The Countryman,* Vol. 59, No. 3, 1962, pp. 547–555.
2. Beckham, W. K., and Mills, W. H., "Cotton-Fabric-Reinforced Roads," *Eng. News-Record,* Oct. 3, 1935, pp. 453–455.
3. Bertram, G. E., "An Experimental Investigation of Protective Filters," Graduate School of Engineering, Harvard University, Publ. 267, 1940.
4. Kays, W. B., *Construction of Linings for Reservoirs, Tanks and Pollution Control Facilities,* 2d ed. New York: John Wiley and Sons, 1982.
5. Society of Plastics Industries (SPI) Committee on Resin Statistics, compiled by Ernst & Young, LLP, 1995.
6. Mascia, L., *Thermoplastics: Engineering.* New York: Applied Science Publishers, 1982.
7. Rodriguez, F., *Principles of Polymer Systems,* 2d ed. New York: McGraw Hill, 1982.
8. Rosen, S. L., *Fundamental Principles of Polymeric Materials,* New York: John Wiley and Sons, 1982.
9. Sperling, L. H., *Introduction to Physical Polymer Science,* New York: John Wiley and Sons, 1986.
10. Moore, G. R., and Kline, D. E., *Properties and Processing of Polymers for Engineers,* Englewood Cliffs, NJ: Prentice-Hall, 1984.
11. Halse, Y., Wiertz, J., Rigo, J.-M., and Cazzuffi, D. A., "Chemical Indentification Methods Used to Characterize Polymeric Geomembranes," *Geomembranes: Identification and Performance Testing,* ed. A. Rollin and J-M Rigo, RILEM Report 4, London: Chapman and Hall, 1991, pp. 316–336.
12. Thomas, R. W., and Verchcoor, K. L., "Thermal Analysis of Geosynthetics," *Geotechnical Fabrics Rpt,* Vol. 6, No. 3, 1988, pp. 24–30.
13. Brennan, W. P., *Characterization of Polyethylene Films by Differential Scanning Calorimetry,* Norwalk, CT: Perkin-Elmer Instrument Div., 1978.
14. Hsuan, Y. G., and Guan, Z., "Evauation of Oxidation Behavior of Polyethylene Geomembranes Using Pressure Different Scanning Calorimetry," in *Oxidative Behavior of Materials by Thermal Analytical Techniques,* ASTM STP1326, ed. A. T. Riga and G. H. Patterson, West Conshohocken, PA: ASTM, 1997.
15. Struve, F., "Extrusion Fillet Welding of Geomembranes," *J. Geotextile and Geomembranes,* Vol. 9, Nos. 4–6, 1990, pp. 1–14.
16. Allcock, H., *Contemporary Polymer Chemistry,* Englewood Cliffs, NJ: Prentice-Hall, 1981.
17. Pohl, H. A., "Determination of Carboxyl End Groups in a Polyester (Polyethylene Terephthalate)," *J. of Analytical Chemistry,* Vol. 26, No. 10, 1954, pp. 1614–1616.
18. Barrett, R. J., "Use of Plastic Filters in Coastal Structures," *Proc. 16th Intl. Conf. Coastal Eng.,* 1966, pp. 1048–1067.
19. Agerschou, H. A., "Synthetic Material Filters in Coastal Protection," *J. Waterways Harbors Div.,* Vol. 87, No. WW1, 1961, pp. 111–124.

20. Rankilor, P. R., *Membranes in Ground Engineering,* New York: John Wiley and Sons, 1981.
21. *Proc. Intl. Conf. Use of Fabrics in Geotechnics,* Paris: Assoc. Amicale des Ingénieurs, 1977.
22. *Proc. 2d Intl. Conf. Geotextiles,* St. Paul, MN: Industrial Fabrics Assoc. International, 1982.
23. *Proc. 3d Intl. Conf. Geotextiles,* St. Paul, MN: Vienna Industrial Fabrics Assoc. International, 1986.
24. *Proc. 4th Intl. Conf. Geotextiles, Geomembranes and Related Products,* Rotterdam: A. A. Balkema, 1990.
25. *Proc. 5th Intl. Conf. Geotextiles, Geomembranes and Related Products,* Singapore: Southeast Asia Chapter, IGS, 1994.
26. Koerner, R. M., and Welsh, J. P., *Construction and Geotechnical Engineering Using Synthetic Fabrics,* New York: John Wiley and Sons, 1980.
27. van Zantan, R. V., ed., *Geotextiles and Geomembranes in Civil Engineering*, Rotterdam: A. A. Balkema, 1986.
28. John, M. W. M., *Geotextiles,* London: Chapman and Hall, 1987.
29. Giroud, J.-P., *Geosynthetics Bibliography. Vol. 1, Conferences; Vol. 2, Journals,* etc., St. Paul, MN: IFAI Publ., 1993.
30. *J. Geotextiles and Geomembranes* (Elsevier Applied Science Publishers, Ltd., U.K.).
31. *Geosynthetics International* (IFAI Publ., St. Paul, MN).
32. Corneliussen, R. D. (ed.), Maro Polymer Notes, Folcroft, PA: Craig Technologies, Inc.
33. Brostow, W., and Corneliussen, R. D., *Failure of Plastics,* New York: Macmillan, 1986.
34. Shreve, R. N., and Brink, J. A., Jr., *Chemical Process Industries,* 4th ed. New York: McGraw-Hill, 1977.
35. Kaswell, E. R., *Handbook of Industrial Textiles,* New York: West Point Pepperell, 1963.
36. INDA (Assoc. of Nonwoven Fabrics Industry, 10 East 40th Street, New York), 1962.
37. Jagielski, K., "Lining Systems Show Growth," *Geotechnical Fabrics Rept.,* Vol. 9, No. 6, 1991, pp. 26–28.
38. Capaccio, G. and Ward, I. M., "Properties of Ultra-high Modulus Linear Polypropylene," *Nature Phys. Sci.,* Vol. 243, 1974, pp. 130–143.
39. Ward, I. M., "The Orientation of Polymers to Produce High Performance Materials," *Proc. Symp. Polymer Grid Reinforcement in Civil Eng.,* London: ICI, 1984.
40. Netlon Ltd., United Kingdom and Tensar Inc., U.S., literature.
41. Austin, R. A., "The Manufacture of Geonets and Composite Products," *Proc. Geosynthetic Resins, Formulations and Manufacturing,* ed. Y. G. Hsuan and R. M. Koerner, St. Paul, MN: IFAI, 1995, pp. 127–138.
42. Haxo, H. E., Jr., "Quality Assurance of Geomembranes Used as Linings for Hazardous Waste Containment," *J. Geotextiles and Geomembranes,* Vol. 3, No. 4, 1986, pp. 225–248.
43. Jagielski, K., "Geomembrane Market Report: A Solution That Ensures Growth," *Geotechnical Fabrics Rept.,* Vol. 8, No. 6, 1990, pp. 27–29.
44. Horvath, J. S., *Conf. Proc. Geofoam as an Eng. Material,* St. Paul, MN: IFAI, 1995.

PROBLEMS

1.1. This book deals exclusively with geosynthetics, synthetic materials placed in the ground. What materials would *not* be considered in this category; that is, what would be some *non*synthetic geotextiles and *non*synthetic geomembranes?

1.2. Regarding the unit prices of geosynthetics given in Section 1.1.2, what factors do you think would influence these values?

1.3. Regarding the installation of geosynthetics, what labor group(s) would be involved in installing the following (considering both union and nonunion situations)?
- **(a)** geotextiles
- **(b)** geogrids
- **(c)** geonets

(d) geomembranes
(e) geosynthetic clay liners
(f) geopipes

1.4. What complications do you see arising in purchasing and providing quality assurance for geomembranes versus geotextiles? (*Hint:* Consider how many different firms are involved in the manufacture of each of the types mentioned.)

1.5. Name some major corporations that produce the following resins.
(a) polyethylene
(b) polypropylene
(c) polyester
(d) polyvinyl chloride

1.6. Identify the typical polymer used to manufacture the following commonly used products.
(a) milk jugs
(b) soft drink bottles
(c) disposable coffee containers
(d) household plumbing pipe
(e) automobile battery cases
(f) lightweight plastic canoes

1.7. The PVC curve in Figure 1.3 is described as consisting of plasticizer, resin, and carbon black/ash. What percentages of each are in this material?

1.8. In Figure 1.3, identify the components of PP, PE, and PET and their percentages.

1.9. The DSC traces in Figure 1.5 list LDPE, MDPE, and HDPE. Identify the temperature window at which each type begins to melt and at which the melting is complete.

1.10. Using the OIT data of Figure 1.6, estimate the time for depletion of all of the antioxidant in this particular geomembrane formulation at a service temperature of 20°C. Proceed using the following steps:
(a) Replot the data on a semi-log axis.
(b) Extrapolate each of the curves down to zero minutes.
(c) Plot these values on an arithmetic y-axis (in years) against inverse temperature on the x-axis.
(d) Determine the average slope of the resulting five points.
(e) Extrapolate this slope down to an estimated service-life temperature of 20°C to obtain the estimated time for antioxidant depletion.

1.11. Using the coefficients of thermal expansion shown in Figure 1.7:
(a) What change in length is involved in 10 m of PET yarn if there is a 20°C change in temperature below the glass-transition temperature (T_g)?
(b) Recalculate part (a) for temperature changes above T_g.

1.12. To perform a forensic analysis of a failure involving a geosynthetic product in which the polymeric material itself is suspect, which tests in Table 1.4 would you perform if
(a) the material is a polypropylene geotextile?
(b) the material is a polyester geogrid?
(c) the material is a polyethylene geonet?
(d) the material is a polyvinyl chloride geomembrane?

1.13. Name five major functions that geosynthetics perform and illustrate them by means of sketches.

1.14. In placing a geotextile beneath railroad ballast, the materials serve in four different functions simultaneously. Describe and illustrate these functions.

1.15. The first person to describe the general use of geotextiles was Barrett [18] in 1966. What major function did the geotextile serve? List the various uses he describes.

1.16. Plot the trend lines for the geotextile applications listed in Table 1.8. What is the fastest growing application area in current geotextile usage?

1.17. Permeability of geotextiles refers to liquid moving through the voids created by the fibers and yarns that make up the fabric. This permeability is called *Darcian permeability*.
- **(a)** Give the equation for Darcy's Law.
- **(b)** Identify the various terms.
- **(c)** What are the major variables involved in variations of the k-value?

1.18. Permeability of geomembranes refers to liquid vapor moving through the amorphous structure of the polymer. This permeability is called *diffusion permeability*.
- **(a)** Give the equation for Fick's Law.
- **(b)** Identify the terms.
- **(c)** What are the major variables involved in variations of the diffusion coefficient?

1.19. If a hole is created in a geomembrane, the flow through the hole is governed by Bernoulli's equation.
- **(a)** State Bernoulli's equation.
- **(b)** Identify the terms.
- **(c)** What kind of relationship do you think there would be between the diffusion permeability of an intact geomembrane and the Bernoulli flow through the hole(s) in a geomembrane?

1.20. What is a *corduroy* road, and how does it function?

1.21. Other than geosynthetics, what are some methods for strengthening soils (i.e., adding tensile strength to them)?

1.22. How would you estimate geosynthetic performance in severe climatic conditions? In arctic conditions? In desert conditions? (*Hint:* How do plastics respond to cold and hot temperatures in general?)

1.23. What two commonly used polymers in the manufacture of geosynthetic materials are in the polyolefin family?

1.24. Regarding the molecular structure of the polymers used to make geosynthetics:
- **(a)** What are typical lengths of the molecular chains?
- **(b)** What is meant by the "backbone" of the molecular chain?
- **(c)** Sketch the crystalline and amorphous phases of polyethylene and show how they are linked together.

1.25. The molecular structure of high-molecular-weight polymeric materials has been described as a "bowl of spaghetti." If this is the case:
- **(a)** What would be the length of a high-molecular-weight polymer like polyester if the diameter was typical of spaghetti (e.g., 1.5 mm)?
- **(b)** What happens to the "bowl of spaghetti" as the polymer structure is stressed?
- **(c)** What happens if it is stressed too high?

1.26. Degradation of polymeric materials involves *chain scission*.
- **(a)** Describe this process.
- **(b)** What mechanisms can bring it about?

1.27. In general, all polymeric materials are susceptible to UV attack. Considering their chemical structure, why is this the case?

1.28. The usual processing step taken to avoid UV-light degradation of polymers is the addition of carbon black. How does this material function? Are there other additives that can be used?

1.29. In the absence of ultraviolet-light degradation, what causes a polymer structure to age?

1.30. What are the major causes of degradation of the following polymers, considering that they have been "timely" covered and that no high-level radiation is in the vicinity?
- **(a)** polyethylene
- **(b)** polypropylene
- **(c)** polyester
- **(d)** polyvinyl chloride
- **(e)** polyamide

1.31. In considering the manufacturing of geomembranes as described in Section 1.6.2, do you think that residual stresses could exist in the as-manufactured sheet? If so, how would you measure the magnitude and orientation?

1.32. Regarding the production cost of geotextiles, rank the following styles on the basis of an equivalent mass per unit area (i.e., they are all the same in terms of g/m^2 weight and polymer type).
- **(a)** woven, monofilament
- **(b)** woven, slit-film
- **(c)** nonwoven, heat-bonded
- **(d)** nonwoven, resin-bonded
- **(e)** nonwoven, needle-punched

1.33. Regarding the material cost of geomembranes, rank the relative cost of the following styles on the basis of an equivalent thickness of material.
- **(a)** high-density polyethylene (HDPE)
- **(b)** very flexible polyethylene (VFPE)
- **(c)** polyvinyl chloride (PVC)
- **(d)** flexible polypropylene (fPP)
- **(e)** chlorosulphonated polyethylene scrim-reinforced (CSPE-R)

1.34. Regarding geosynthetic clay liners (GCLs):
- **(a)** List some advantages and disadvantages over a geomembrane (GM).
- **(b)** List some advantages and disadvantages over a compacted clay liner (CCL).

1.35. Regarding the use of a three-component liner consisting of a geomembrane over a GCL over a CCL:
- **(a)** List some advantages and disadvantages over a GM/CCL composite.
- **(b)** List some advantages and disadvantages over a GM/GCL composite.

1.36. Regarding soil-backfilled plastic pipe (or geopipe):
- **(a)** List some advantages and disadvantages of a polymer geopipe over concrete pipe.
- **(b)** List some advantages and disadvantages of a polymer geopipe over metal pipe.
- **(c)** List some advantages and disadvantages of HDPE polymer geopipe over PVC polymer geopipe.

1.37. If a new geoword "geospacer" were coined, what types of geosynthetic materials would you expect it to include?

1.38. List what you feel are the most relevant and unique properties of geofoam.

2 Designing with Geotextiles

2.0 INTRODUCTION

According to ASTM D4439, a geotextile is defined as

> **geotextile,** *n.* A permeable geosynthetic comprised solely of textiles. Geotextiles are used with foundation, soil, rock, earth, or any other geotechnical engineering-related material as an integral part of a human-made project, structure, or system.

The area of geotextiles is a rapidly growing and exciting field, with new uses being developed almost daily. As such, there are a number of possible applications and an even greater number of geotextiles to choose from. The vast majority of geotextiles are made from polypropylene or polyester polymers formed into fibers or yarns (the choices being monofilament, multifilament, staple yarn, slit-film monofilament, or slit-film multifilament) and finally into a woven or nonwoven fabric. When placed in the ground these fabrics are called geotextiles. In general, the words *fabric* and *geotextile* are used interchangeably. The choices of fabric styles are: woven monofilament, woven multifilament, woven slit-film monofilament, woven slit-film multifilament, nonwoven continuous filament heat-bonded, nonwoven continuous filament needle-punched, nonwoven staple needle-punched, nonwoven resin-bonded, other woven or nonwoven combinations, and (although very rarely) knitted. A complete description of the methods of manufacturing geotextiles is presented in Section 1.3.

Due to the very wide range of applications and the tremendous variety of available geotextiles having widely different properties, the selection of a particular design method or design philosophy is a critical decision. This decision must be made before the actual mechanics of the design process are initiated. These conceptual issues are followed by a detailed description of the various functions that geotextiles serve and then a lengthy description of geotextile properties and test methods. This material sets up the remainder of the chapter dealing with the specifics of designing-by-function using geotextiles.

2.1 DESIGN METHODS

While many possible design methods, or combinations of methods, are available to the geotextile designer, the ultimate decision for a particular application usually takes one of three directions: design-by-cost-and-availability, design-by-specification, and design-by-function.

2.1.1 Design-by-Cost-and-Availability

Geotextile *design-by-cost-and-availability* is quite simple. The funds available are divided by the area to be covered and a maximum available unit price that can be allocated for the geotextile is calculated. The geotextile with the best properties is then selected within this unit price limit and according to its availability. Intuition plays a critical role in the selection process. The method is obviously weak technically but is one that is still sometimes practiced. It perhaps typified the situation in the early days of geotextiles, but is outmoded by current standards of practice.

2.1.2 Design-by-Specification

Geotextile *design-by-specification* is very common and is used almost exclusively when dealing with public agencies. In this method several application categories are listed in association with various physical, mechanical and/or hydraulic properties. A specification of this type that is used by the Pennsylvania Department of Transportation is given in Table 2.1. While it is typical in its format (listing the various common applications against minimum or maximum property values), it is not typical insofar as the numeric values of the various properties. Different agencies have very different perspectives as to what properties are important and as to their method of obtaining the numeric values.

A federal agency that has formulated a unified approach in the U.S. is the American Association of State Highway and Transportation Officials (AASHTO). In its M288-96 geotextile specifications, AASHTO provides for three different classifications (see Table 2.2a). The classifications are essentially a list of strength properties meant to withstand varying degrees of installation survivability stresses.

Class 1. for severe or harsh survivability conditions where there is a greater potential for geotextile damage

Class 2. for typical survivability conditions; this is the default classification to be used in the absence of site specific information

Class 3. for mild survivability conditions

TABLE 2.1 PENNSYLVANIA DEPARTMENT OF TRANSPORTATION GEOTEXTILE REQUIREMENTS BY APPLICATION AREA

Fabric Property	Test Method	Application Area					
		Subsurface Filtration*	Erosion Control*		Sediment Control (Silt Fence)*		Separation*
			Type A	Type B	Type A	Type B	
Grab tensile strength, N	ASTM D 5034 MG-E	400	890	530	890	530	1200
Grab tensile elongation, %	ASTM D 5034 MG-E	20	15–45	15	15–45	15	15
Burst strength, MPa	ASTM D751 (Diaphragm Method)	1.03	2.80	1.40	2.80	1.40	2.96
Puncture, N	ASTM D3787 (8 mm flatrod)	280	550	280	550	280	760
Trapezoid tear strength, N	ASTM D1117	340	517	340	517	340	517
Apparent opening size (AOS) sieve no.	CW-02215§	$\leqslant$ 4.25 μm	425 μm–150 μm	$\leqslant$ 425 μm	—	—	$\leqslant$ 212 μm
Water permeability k, mm/s	PA Test Method No. 314	0.1‡	0.1‡	0.1‡	—	—	—
Abrasion resistance					—	—	
Slurry flow rate, l/sec-m	VTM-51**	—	—	—	0.06	0.06	—
Retention efficiency, %	VTM-51**	—	—	—	75	75	—
Percent open area	CW-02215§	—	4.0†	—	—	—	—
Sewn seam strength, N	ASTM D 5034 MG-E	—	1400	830	—	—	—
Ultraviolet resistance strength, retention percentage of grab	ASTM D 5034 MG-E, after 500 hr of Xenon or Atlas Twin Arc Weather-o-meter	—	—	—	70%	70%	

*The numerical values indicate the minimum average roll value or the minimum to maximum range. Minimum average roll values for Class 3 material is in the warp direction only.

†Design specified

‡$k_{fabric} > 10\ k_{soil}$

§U.S. Army Corps of Engineers Specification.

**Virginia DOT Test Method

TABLE 2.2a AASHTO M288-96 GEOTEXTILE STRENGTH PROPERTY REQUIREMENTS

			Geotextile Classification*,†					
			Class 1		Class 2		Class 3	
	Test Methods	Units	Elongation $< 50\%$	Elongation $\geqslant 50\%$	Elongation $< 50\%$	Elongation $\geqslant 50\%$	Elongation $< 50\%$	Elongation $\geqslant 50\%$
Grab strength	ASTM D4632	N	1400	900	1100	700	800	500
Sewn seam strength‡	ASTM D4632	N	1200	810	990	630	720	450
Tear strength	ASTM D4533	N	500	350	400§	250	300	180
Puncture strength	ASTM D4833	N	500	350	400	250	300	180
Burst strength	ASTM D3786	kPa	3500	1700	2700	1300	2100	950
Permittivity	ASTM D4491	s^{-1}	Minimum property requirements for permittivity, AOS and UV stability are based on geotextile application. Refer to Table 2.2b for subsurface filtration, Table 2.2c for separation, Table 2.2d for stabilization, and Table 2.2e for permanent erosion control.					
Apparent opening size	ASTM D4751	mm						
Ultraviolet stability	ASTM D4355	%						

*Required geotextile classification is designated in Tables 2.2b, c, d, e for the indicated application.

†As measured in accordance with ASTM D4632. Woven geotextiles fail at elongations (strains) < 50%, while nonwovens fail at elongation (strains) > 50%.

‡When sewn seams are required. Overlap seam requirements are application specific.

§The required MARV tear strength for woven monofilament geotextiles is 250 N.

TABLE 2.2b AASHTO M288-96 SUBSURFACE FILTRATION (CALLED "DRAINAGE" IN THE SPECIFICATION) GEOTEXTILE REQUIREMENTS

	Test Methods	Units	Requirements		
			Percent In Situ Soil Passing 0.075 mm		
			< 15	15 to 50	> 50
Geotextile Class				Class 2*	
Permittivity†,‡	ASTM D4491	s^{-1}	0.5	0.2	0.1
Apparent opening size†,‡	ASTM D4751	mm	0.43 max. avg. roll value	0.25 max. avg. roll value	0.22§ max. avg. roll value
Ultraviolet stability (retained strength)	ASTM D4355	%		50% after 500 hr of exposure	

*Default geotextile selection. The Engineer may specify a Class 3 geotextile from Table 2.2a if conditions are less severe.

†In addition to the default permittivity value, the Engineer may require geotextile permeability and/or performance testing in problematic soil environments.

‡Site specific geotextile design should be performed if unstable or highly erodable soils such as noncohesive silts; gap-graded soils; alternating sand/silt laminated soils; dispersive clays; and/or rock flour are encountered.

§For cohesive soils with a plasticity index greater than 7, geotextile Max ARV is 0.30 mm.

Table 2.2a is used in association with tables for various applications that are commonly encountered. They are as follows:

- Table 2.2b filtration applications (Class 2)
- Table 2.2c separation of soil subgrades (soaked CBR > 3; or shear strength > 90 kPa) (Class 2)
- Table 2.2d stabilization of soft subgrades (1 < CBR < 3; or shear strengths between 30 and 90 kPa) (Class 1)
- Table 2.2e erosion control, e.g., geotextiles beneath rock riprap (Classes 1 and 2)
- Table 2.2f temporary silt fences
- Table 2.2g prevention of reflective cracking

TABLE 2.2c AASHTO M288-96 SEPARATION GEOTEXTILE PROPERTY REQUIREMENTS

	Test Methods	Units	Requirements
Geotextile Class			Class 2*
Permittivity	ASTM D4491	s^{-1}	0.02†
Apparent opening size	ASTM D4751	mm	0.60 max. avg. roll value
Ultraviolet stability (retained strength)	ASTM D4355	%	50% after 500 hr of exposure

*Default geotextile selection. The Engineer may specify a Class 3 geotextile from Table 2.2a if conditions are less severe.

†Default value. Permittivity of the geotextile should be greater than that of the soil ($\Psi_g > \Psi_s$). The Engineer may also require the permeability of the geotextile to be greater than that of the soil ($k_g > k_s$).

TABLE 2.2d AASHTO M288-96 STABILIZATION GEOTEXTILE PROPERTY REQUIREMENTS

	Test Methods	Units	Requirements
Geotextile Class			Class 1*
Permittivity	ASTM D4491	s^{-1}	0.05†
Apparent opening size	ASTM D4751	mm	0.43 max. avg. roll value
Ultraviolet stability (retained strength)	ASTM D4355	%	50% after 500 hr of exposure

*Default geotextile selection. The Engineer may specify a Class 2 or 3 geotextile from Table 2.2a if conditions are less severe.

†Default value. Permittivity of the geotextile should be greater than that of the soil ($\Psi_g > \Psi_s$). The Engineer may also require the permeability of the geotextile to be greater than that of the soil ($k_g > k_s$).

Example 2.1 illustrates the use of these tables.

Example 2.1

Using the AASHTO M288-96 Specifications of Table 2.2, determine what geotextile properties are needed for the following applications:

(a) A nonwoven ($\varepsilon > 50\%$) pavement underdrain filter adjacent to soil with 60% passing the 0.075 mm sieve and under typical installation survivability conditions.

TABLE 2.2e AASHTO M288-96 PERMANENT EROSION CONTROL GEOTEXTILE REQUIREMENTS

			Requirements		
			Percent In Situ Soil Passing 0.075 mm		
	Test Methods	Units	< 15	15 to 50	> 50
Geotextile class					
woven monofilament geotextiles				Class 2*	
all other geotextiles				Class 1*,†	
Permittivity‡	ASTM D4491	s^{-1}	0.7	0.2	0.1
Apparent opening size§	ASTM D4751	mm	0.43 max. avg. roll value	0.25 max. avg. roll value	0.22** max. avg. roll value
Ultraviolet stability (retained strength)	ASTM D4355	%		50% after 500 hr of exposure	

*The default geotextile selection is appropriate for stone weights not to exceed 100 kg, stone drop height less than 1 m, and the geotextile protected by a 150 mm thick bedding layer. More severe applications require an assessment of geotextile survivability based on a field trial section and may require a geotextile with higher strength properties.

†The Engineer may specify a Class 2 geotextile from Table 2.2a if conditions are less severe.

‡The Engineer may require geotextile permeability and/or performance testing for erosion-control systems over problematic soil environments.

§Site-specific geotextile design should be performed especially if unstable or highly erodable soils such as noncohesive silts, gap-graded soils, alternating sand/silt laminated soils, dispersive clays, and/or rock flour are encountered.

**For cohesive soils with a plasticity index greater than 7, geotextile Max ARV for apparent opening size is 0.30 mm.

TABLE 2.2f AASHTO M288-96 TEMPORARY SILT FENCE PROPERTY REQUIREMENTS

	Test Methods	Units	Requirements: Supported Silt Fence*	Requirements: Unsupported Silt Fence, Geotextile Elongation $\geqslant 50\%$[†]	Requirements: Unsupported Silt Fence, Geotextile Elongation < 50[†]
Maximum post spacing			1.2 m	1.2 m	2.0 m
Grab strength	ASTM D4632	N			
machine direction			400	550	550
cross-machine direction			400	450	450
Permittivity	ASTM D4491	s^{-1}	0.05	0.05	0.05
Apparent opening size[‡]	ASTM D4751	mm	0.60 max. avg. roll value	0.60 max. avg. roll value	0.60 max. avg. roll value
Ultraviolet stability (retained strength)[†]	ASTM D4355	%	70% after 500 hr of exposure	70% after 500 hr of exposure	

*Silt fence support shall consist of 14 gage steel wire with a mesh spacing of 150 mm × 150 mm or prefabricated polymeric mesh of equivalent strength.

[†]As measured in accordance with ASTM D4632.

[‡]These default filtration property values are based on empirical evidence with a variety of sediments. For environmentally sensitive areas, a review of previous experience and/or site or regionally specific geotextile tests should be performed.

(b) A woven geotextile ($\varepsilon < 50\%$) pavement separator between firm soil subgrade (with CBR = 10) and stone base course, under harsh survivability conditions according to the design engineer.

Solution: Tables 2.2b to 2.2g are used for the appropriate application properties and then Table 2.2a is used for the required strength properties.

TABLE 2.2g AASHTO M288-96 PREVENTION OF REFLECTIVE CRACKING, i.e., PAVING FABRICS, PROPERTY REQUIREMENTS

	Test Methods	Units	Requirements
Grab strength	ASTM D4632	N	450
Mass per unit area	ASTM D3776	gm/m^2	140
Ultimate elongation	ASTM D4632	%	$\geqslant 50$
Asphalt retention*	Texas DOT Item 3099	l/m^2	*,†
Melting point	ASTM D276	°C	150

*Asphalt required to saturate paving fabric only. Asphalt retention must be provided in manufacturer certification. Value does not indicate the asphalt application rate required for construction. Refer to Appendix of AASHTO M288-96 titled Construction/Installation Guidelines for discussion of asphalt application rate.

[†]Product asphalt retention property must meet the MARV provided by the manufacturer's certification.

(a) From Table 2.2b and Table 2.2a, the required properties for the nonwoven geotextile filtration fabric are:

permittivity $\geqslant 0.1\ s^{-1}$
AOS $\leqslant$ 0.22 mm
grab strength $\geqslant$ 700 N
sewn seam strength $\geqslant$ 630 N
tear strength $\geqslant$ 250 N
puncture strength $\geqslant$ 250 N
burst strength $\geqslant$ 1300 kPa
UV stability $\geqslant$ 50% of 700 N $\geqslant$ 350 N after 500 hr

(b) From Table 2.2c and Table 2.2a, the required properties for the woven geotextile separation fabric are:

permittivity $\geqslant 0.02\ s^{-1}$
AOS $\leqslant$ 0.60 mm
grab strength $\geqslant$ 1400 N
sewn seam strength $\geqslant$ 1200 N
tear strength $\geqslant$ 500 N
puncture strength $\geqslant$ 500 N
burst strength $\geqslant$ 3500 kPa
UV stability $\geqslant$ 50% of 1400 N $\geqslant$ 700 N after 500 hr

It must be cautioned that when using a design-by-specification method, the specifications sometime list *minimum* required fabric properties, whereas current manufacturers' literature may list either *average-lot** or *minimum-average-roll* property values. By comparing such a specification value to the manufacturers' listed values, you may be comparing different sets of numbers. This is so because average lot value is the mean value for the particular property in question from all the tests made on that lot of fabrics. This may be the compilation of thousands of tests made over many months or even years of production of that particular geotextile style. Thus the average lot value is considerably higher than the minimum value (see Figure 2.1). An intermediate value between these two extremes is the *m*inimum *a*verage *r*oll *v*alue, or MARV. The minimum average roll value is the average of a representative number of tests made on selected rolls of the lot in question, which is limited in area to the particular site in question. This value is probably two standard deviations lower than the average lot, or mean, value. Thus it is seen that MARV is the minimum of a limited series of average roll values. These different values are shown schematically in Figure 2.1.

Note that in a true statistical sense about 16% of all values will be lower than $\overline{X} - S$; 2.5% will be lower than $\overline{X} - 2S$; and 0.15% will be lower than $\overline{X} - 3S$,

*In this regard, the word *lot* is defined as any unit of production taken for sampling or statistical testing, having one or more common properties and being readily separable from other similar units. Thus, a lot can be as large as an entire production run, or as small as a few rolls of fabric for a specific project. The point is that the unit must be agreed upon by the parties involved.

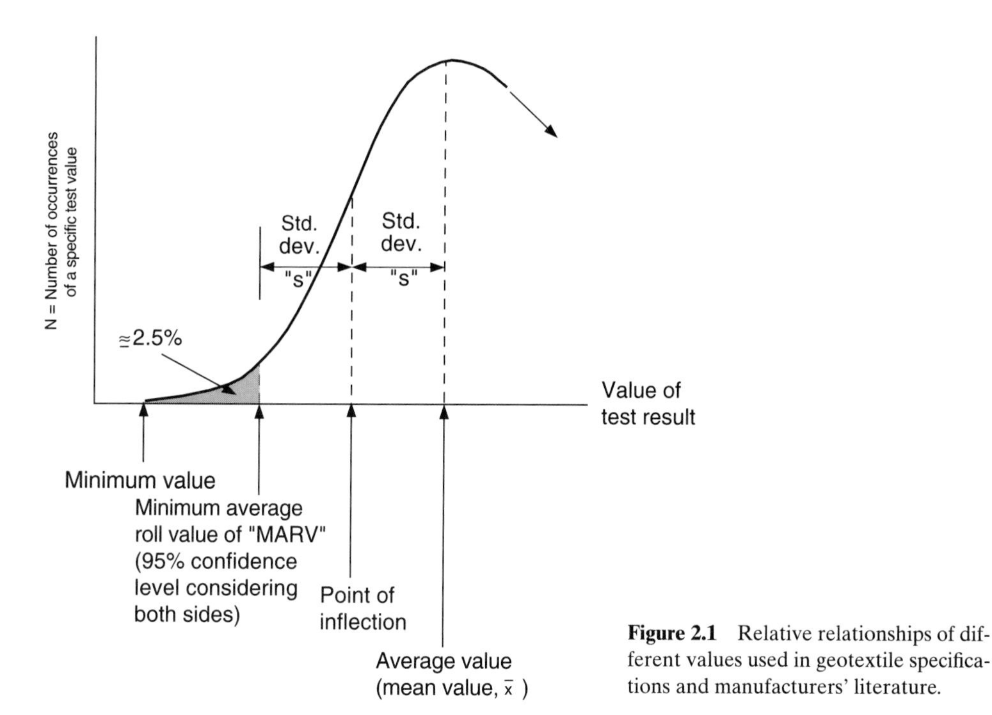

Figure 2.1 Relative relationships of different values used in geotextile specifications and manufacturers' literature.

where

$\overline{X}$ = mean value, and
S = standard deviation.

Furthermore, the minimum average roll value (MARV), with 2.5% of the values falling below $\overline{X} - 2S$ is also the 95% confidence level. (The other 2.5% is above $\overline{X} + 2S$, and is obviously not of concern since values are well in excess of that required). One other consideration should be mentioned—the case where a maximum value, for example, a maximum elongation, is targeted. Here you are considering the right side of the curve of Figure 2.1 and the comparable value to MARV would logically become Max ARV.

The mean value $\overline{X}$ is calculated using $\Sigma X_i/N$ and the standard deviation using

$$S = \left[\frac{(X_1 - \overline{X})^2 + (X_2 - \overline{X})^2 + \cdots + (X_N - \overline{X})^2}{N - 1}\right]^{1/2} \tag{2.1}$$

where

$\overline{X}$ = mean value,
X_i = measured value, and
N = number of measurements.

The coefficient of variation V, the variance, is calculated using $(S/\overline{X})(100)$. The variance should be as low as possible, thereby indicating good quality control. Most agencies, including AASHTO in its M288-96 specifications presented in Tables 2.2a to 2.2g, recommend the use of minimum average roll values in both the specification and the listing of manufacturers' product data. Example 2.2 illustrates the meaning of MARV insofar as field conformance testing is concerned.

Example 2.2

Consider a field construction site where 150 rolls of geotextile are delivered. This value then defines the *lot*. The quality control or quality assurance inspector would *sample** a representative number of these rolls to determine the MARV and see that it conforms with the value called for in the specification. Assume that the targeted value is grab-tensile strength. A geotextile sample, full-roll wide and 1.0 m long is taken from each of six randomly selected rolls in the lot. Note that according to ASTM D4354 on sampling technique, a lot consisting of from 126 to 216 rolls requires at least 6 rolls to be sampled. These samples are sent to an approved laboratory for testing. Within each sample, eight test specimens are taken and tested according to ASTM D4632, the grab tensile test method. Given the test data below, determine the MARV.

Solution: Assume that each of the six samples is cut into eight individual grab-tensile test specimens and are properly tested, and the following data set resulted in Newtons (N) at failure.

Test number	Roll number					
	1	2	3	4	5	6
1	643 N	627 N	637 N	642 N	652 N	637 N
2	627	615	643	646	641	624
3	652	621	628	658	639	631
4	629	616	662	641	657	620
5	632	619	646	635	642	618
6	641	621	633	642	651	633
7	662	622	619	658	641	641
8	635	628	636	662	645	625
Average =	640	621	638	648	646	629

From this data set, the MARV is 621 N, which must equal or exceed the MARV value of grab-tensile strength required by the specification. Note that there are five individual test values in the entire data set that are numerically less than 621. These represent the statistical 2.5% of the values less than MARV, as illustrated by the shaded portion in Figure 2.1.

In summary, the design-by-specification method compares like sets of numbers. If the intent of the specification is to list MARV (as it is in Tables 2.1 and 2.2a-g), then

*Throughout this book, a roll of geosynthetics is *sampled* by cutting a piece or swatch from it. This *sample* is then taken to a laboratory where *specimens* are cut to exact size for subsequent testing according to a particular test protocol. In some cases, the sample will be cut into smaller sections and incubated in an oven, in liquid, or under UV exposure, and so on. It is then called a *coupon*, and is subsequently cut into specimens for actual testing purposes. Thus, the order of hierarchy: a lot, roll, sample, coupon (sometimes), and specimen.

manufacturers' listed mean or average values must be decreased by two standard deviations (approximately 5 to 20%) if average lot values are given. Only if MARV are given by the manufacturer can they be directly compared to a MARV-based specification value on a like-set-of-numbers basis.

In closing, it is hoped that both specifications and manufacturers' literature come together with a common unit. It seems logical to the author that this coming together should center on MARV. It is a concept that everyone can live with, a value that can be field-verified, and a number that reflects the inherent variation in quality control of the manufacture of geotextiles.

2.1.3 Design by Function

Design-by-function consists of assessing the primary function that the geotextile will serve and then calculating the required numerical value of a particular property for that function. Dividing this value into the candidate geotextile's allowable property value gives a factor of safety (FS).

$$FS = \frac{\text{allowable (test) property}}{\text{required (design) property}} \tag{2.2a}$$

where

allowable property = a numeric value based on a laboratory test that models the actual situation
required property = a numeric value obtained from a design method that models the actual situation, and
FS = factor of safety against unknown loads and/or uncertainties in the analytic or testing process; sometimes called a global factor of safety.

If the factor of safety is sufficiently greater than 1.0, the candidate geotextile is acceptable. The above process can be repeated for a number of available geotextiles, and if others are acceptable then the final choice becomes one of availability and least cost. The individual steps in this process are as follows:

1. Assess the particular application, considering not only the candidate geotextile but the material system on both sides of it.
2. Depending on the criticality of the situation (i.e., "If it fails, what are the consequences?"), decide on a minimum factor of safety.
3. Decide on the geotextile's primary function.
4. Calculate numerically the required geotextile property value in question on the basis of its primary function.
5. Test for, or otherwise obtain, the candidate geotextile's allowable value of this particular property (recall the discussion in Section 2.1.2 on the recommended use of MARV values).

6. Calculate the factor of safety on the basis of the allowable property (Step 5) divided by required property (Step 4) per Equation 2.2a.
7. Compare this factor of safety to the required value decided upon in Step 2.
8. If it is not acceptable, check into geotextiles with more appropriate properties.
9. If it is acceptable, check whether any secondary function of the geotextile is more critical.
10. Repeat the process for other available geotextiles and if more than one satisfy the factor of safety requirement, select the geotextile on the basis of availability and least cost.

Note that the design-by-function process can also be used to solve for the required property value:

$$\text{required (design) property} = \frac{\text{allowable (test) property}}{\text{FS}} \tag{2.2b}$$

Both calculation procedures will be illustrated later.

Design-by-function will be used throughout this book. This method obviously necessitates identifying the primary function that the geotextile is to serve; thus this chapter (and the subsequent chapters) has been laid out accordingly. A brief treatment of the major functions that geotextiles serve is given in Section 2.2.

2.2 GEOTEXTILE FUNCTIONS AND MECHANISMS

Section 1.3.3 alluded to the many applications falling into categories vis-a-vis their major functions (recall Table 1.7). These categories—separation, reinforcement, filtration, drainage, and containment (*if* the geotextile is suitably impregnated)—when properly identified, lead to designing-by-function. The purpose of this section is to demonstrate technically what these functions mean with respect to geotextiles and to elaborate on the actual mechanisms embodied within each type of function.

2.2.1 Separation

The concept of separation can perhaps be illustrated by the engineering adage, "10 kilograms of stone placed on 10 kilograms of mud results in 20 kilograms of mud." A geotextile serving the function of separation can be defined as follows.

> **Geotextile separation:** the placement of a flexible porous textile between dissimilar materials so that the integrity and functioning of both materials can remain intact or be improved.

When placing stone aggregate on fine-grained soils there are two simultaneous mechanisms that tend to occur over time. One is that the fine soils attempt to enter into the voids of the stone aggregate, thereby ruining its drainage capability; the other is that the stone aggregate attempts to intrude into the fine soil, thereby ruining the stone ag-

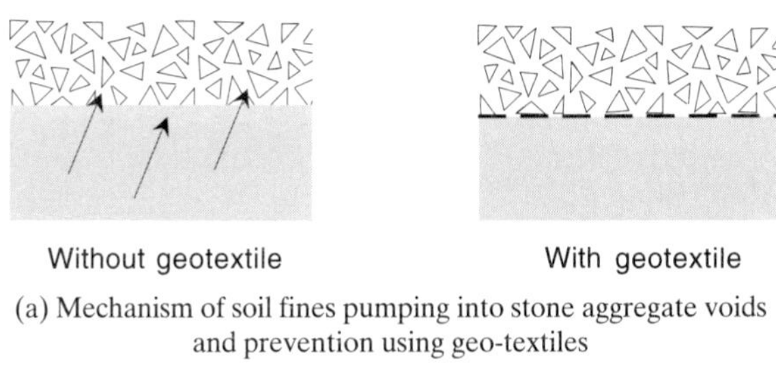

(a) Mechanism of soil fines pumping into stone aggregate voids and prevention using geo-textiles

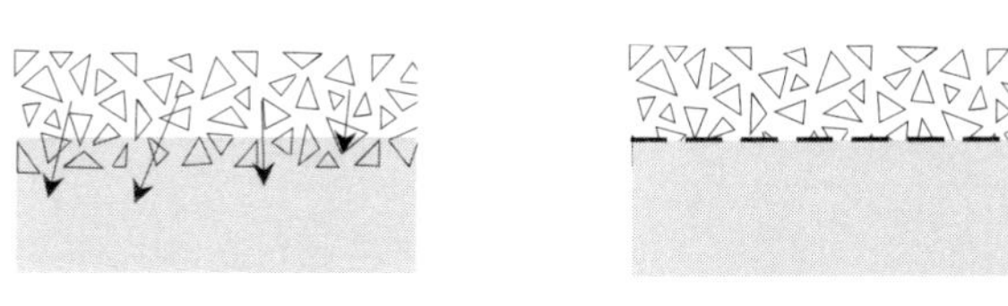

(b) Mechanism of stone aggregate intrusion into soil subgrade and prevention using geotextiles

Figure 2.2 Different mechanisms involved in the use of geotextiles involved in the separation function.

gregate's strength. When this occurs we have a situation that has been called *sacrificial aggregate*, which occurs all too often without the use of a proper separating geotextile. The two mechanisms are shown schematically in Figure 2.2.

2.2.2 Reinforcement

Geotextiles, as a material possessing tensile strength, can nicely complement materials that are good in compression but weak in tension. Thus low-strength fine-grained silt and clay soils are prime candidates for geotextile reinforcement. A convenient definition is the following.

> **Geotextile reinforcement:** the synergistic improvement of a total system's strength created by the introduction of a geotextile (that is good in tension) into a soil (that is good in compression but poor in tension) or into other disjointed and separated material.

Improvement in strength can be evaluated in a number of ways. The triaxial tests conducted by Broms [1] illustrate the beneficial effects of a geotextile when properly placed. Figure 2.3 shows two sets of triaxial tests on dense sand samples at confining pressures of 21 kPa and 210 kPa for different soil and geotextile configurations. Curves 1 represent the base-line shear strength data of the sand by itself. Curves 2 have geotextiles on the top and bottom of the soil and do not show improved shear strength behavior. Since the geotextile locations are in the nonacting dead zones in conventional triaxial tests, this behavior is both logical and instructive, for if the geotextile is placed at the wrong location, it will have no beneficial effect. Upon placing the geotextile in

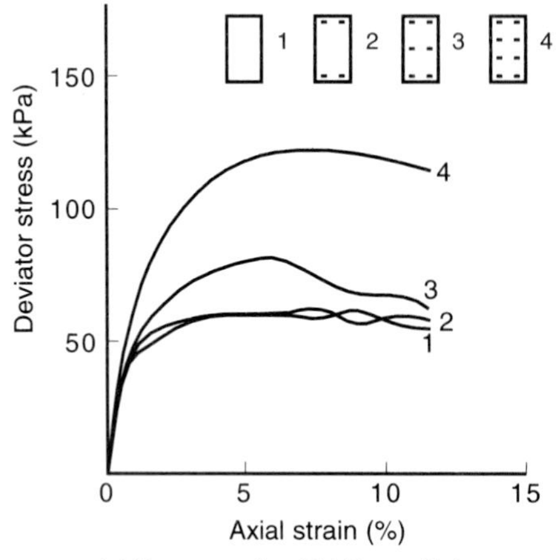

(a) Dense sand at 21 kPa confining pressure

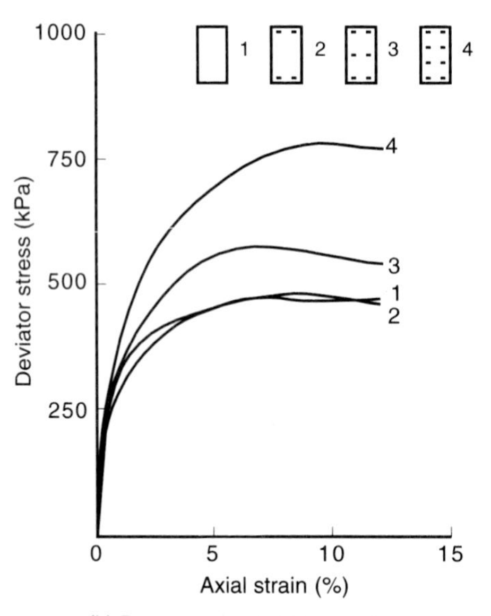

(b) Dense sand at 210 kPa confining pressure

Figure 2.3 Triaxial test results showing influence of geotextiles placed at various locations within soil specimen. (After Broms [1])

the center of the sample as with Curve 3 or at the one-third points as with Curve 4, however, the beneficial effects are easily seen. Here the geotextile interrupts potential shear planes and has the influence of increasing the overall shear strength of the now-reinforced soil. As expected, the double layers placed at the one-third points (Curves 4) are more beneficial than the single layer placed at the center of the sample (Curves 3).

Within the general function of geotextile reinforcement of soils there are three reinforcement mechanisms: membrane type, shear type, and anchorage type.

Membrane Type. Membrane reinforcement occurs when a vertical force is applied to a geotextile that has been placed on a deformable soil. Depending on the depth at which the geotextile is placed from the force application, it is well established [2] that

$$\sigma_h = \frac{P}{2\pi z^2}\left[3 \sin^2 \theta \cos^3 \theta - \frac{(1 - 2\mu) \cos^2 \theta}{1 + \cos \theta}\right] \tag{2.3}$$

where

σ_h = horizontal stress at depth z and angle θ,
P = applied vertical force,
z = depth beneath surface where σ_h is being calculated,
μ = Poisson's ratio, and
θ = angle from the vertical beneath the surface load P.

Note that directly beneath the load, where $\theta = 0°$,

$$\sigma_h = -\frac{P}{\pi z^2}\left(\frac{1}{2} - \mu\right) \tag{2.4}$$

Since μ is less than 0.5, σ_h is negative (=tension); that is, the applied vertical downward force produces tension on a horizontal plane beneath it. Thus tension results in the geotextile, which is precisely the objective in placing it there. As seen in the Equation (2.4), the larger the magnitude of P, the higher the tensile stress and the higher the requirement of tensile strength of the geotextile. Also, the closer the geotextile is to the force (i.e., low values of z), the higher will be the geotextile's stress. Many situations in which geotextiles are placed on soft soils or in a yielding situation use this particular reinforcement mechanism.

Shear Type. Shear reinforcement is illustrated by the triaxial tests of Figure 2.3, but can be better visualized by means of direct shear tests. Here a geotextile placed on a soil is compressed in a normal direction, and then the two materials are sheared at their interface. The resulting shear-strength parameters (adhesion and geotextile-to-soil friction angle) can be obtained as described in a traditional geotechnical manner using an adapted form of the Mohr-Coulomb failure criterion,

$$\tau = c_a + \sigma'_n \tan \delta \tag{2.5a}$$

where

τ = shear strength (between the geotextile and soil),
σ'_n = effective normal stress on the shear plane,
c_a = adhesion (of the geotextile to the soil), and
δ = friction angle (between the geotextile and soil).

The shear-strength parameters c_a and δ can be compared to the shear-strength parameters of the soil by itself (i.e., soil against soil) as follows:

$$\tau = c + \sigma'_n \tan \phi \tag{2.5a}$$

where

c = cohesion (of soil-to-soil), and
ϕ = friction angle (of soil-to-soil).

Furthermore,

$$E_c = (c_a/c) \times 100 \tag{2.6}$$

$$E_\phi = (\tan \delta / \tan \phi) \times 100 \tag{2.7}$$

where

E_c = efficiency of cohesion mobilization, and
E_ϕ = efficiency of friction angle mobilization.

These ratios, generally called *efficiencies*, have limiting values of zero to unity. While a value higher than unity is possible, such values cannot be mobilized since the failure plane would simply move into the soil itself and the situation would revert from Eq. (2.5a) to Eq. (2.5b).

Anchorage Type. Anchorage reinforcement is similar to the shear type just described, but now the soil acts on both sides of the geotextile as a tensile force tends to pull the geotextile out of the soil. The laboratory modeling of this type of mechanism is similar to direct shear except that now the upper and lower soil is stationary in both halves of the test device and the geotextile extends out of the device at its center. It is gripped externally and pulled, while normal compressive stresses act on the soil and geotextile within the test box setup. The situation is readily described in terms of shear-strength parameters by themselves and efficiencies as just discussed. Another approach could be to express the efficiency as a function of the amount of mobilized geotextile strength. Wide-width tensile values should be used in this case. Here anchorage efficiencies greater than unity can occur but are usually limited by the tensile strength of the geotextile. As with the other types of mechanisms of geotextile reinforcement, this category of geotextile anchorage is used quite often. The applications mentioned in Section 1.3.3 illustrate the point.

For calculations, we will use the shear strength mobilized by the geotextile with the soil above and with the soil below and arithmetically sum the two values as the limiting anchorage value. In the absence of anchorage tests, we will use direct-shear-generated values for this purpose.

2.2.3 Filtration

The geotextile function of filtration involves the movement of liquid through the geotextile itself, that is, across its manufactured plane. At the same time, the geotextile serves the purpose of retaining the soil on its upstream side. Both adequate permeability, requiring an open fabric structure, and soil retention, requiring a tight fabric structure, are required simultaneously. A third factor is also involved, that being a long-term soil-to-geotextile flow compatibility that will not excessively clog during the lifetime of the system. Thus a definition of filtration is:

> **Filtration:** the equilibrium soil-to-geotextile system that allows for adequate liquid flow with limited soil loss across the plane of the geotextile over a service lifetime compatible with the application under consideration.

This function of filtration is a major one for the geotextile industry (recall the application areas presented in Section 1.3.3). Geotextiles, when properly designed and constructed, offer a practical remedy to many problems involving the flow of liquids.

Permeability. This particular discussion of geotextile permeability refers to cross-plane permeability when liquid flow is perpendicular to the plane of the fabric. Some of the geotextiles used for this purpose are relatively thick and compressible. For this reason the thickness is included in the permeability coefficient and is used as a *permittivity*, which is defined as

$$\Psi = \frac{k_n}{t} \tag{2.8}$$

where

Ψ = permittivity,
k_n = cross-plane permeability coefficient (the subscript n is often omitted), and
t = thickness at a specified normal pressure.

The testing for geotextile permittivity follows lines similar to those used for testing soil permeability. It should be noted that some designers prefer to work directly with permeability and require the geotextile's permeability to be some multiple of the adjacent soil's permeability, that is, 0.1, 1.0, or 10.0 (see Christopher and Fisher [3]).

Soil Retention. For a greater flow of liquid to be allowed through a geotextile, the void spaces in it must be made larger. There is, however, a limit—that being when the upstream soil particles start to pass through the geotextile voids along with the flowing liquid. This can lead to an unacceptable situation called *soil piping*, in which the finer soil particles are carried through the geotextile, leaving larger soil voids behind. The velocity of the liquid then increases, accelerating the process, until the soil structure begins to collapse. This collapse often leads to minute sinkhole-type patterns that grow larger with time.

This process is prevented by making the geotextile voids small enough to retain the soil on the upstream side of the fabric. It is the coarser soil fraction that must be initially retained and that is the targeted soil size in the design process. These coarser-sized particles eventually block the finer-sized particles and build up a stable upstream soil structure. Fortunately, filtration concepts are well established in the design of soil filters, and those same ideas will be used to design an adequate geotextile filter.

There are many formulae available for soil-retention design, most of which use the soil particle size characteristics and compare them to the 95%-opening size of the geotextile, defined as 0_{95} of the geotextile. The test method used in the U.S. to determine this value is called the apparent opening size (AOS) and it is a dry-sieving method. In Europe and Canada, the test method is called filtration opening size (FOS) and it is accomplished by wet and hydrodynamic sieving, respectively. Both of these latter methods are preferable to the dry-sieving method used in the U.S.

The simplest of the design methods examines the percentage of soil passing a No. 200 sieve, with openings of 0.075 mm. According to Task Force #25, the following is recommended [4]:

- For soil $\leqslant$ 50% passing the No. 200 sieve:

$$0_{95} < 0.60 \text{ mm}, \qquad \text{that is, AOS of the fabric} \geqslant \text{No. 30 sieve}$$

- For soil > 50% passing the No. 200 sieve:

$$0_{95} < 0.30 \text{ mm} \qquad \text{that is, AOS of the fabric} \geqslant \text{No. 50 sieve}$$

Beginning in 1972, a series of direct comparisons of geotextile-opening size (0_{95}, 0_{50}, or 0_{15}) was made in ratio form to some soil particle size to be retained (d_{90}, d_{85}, d_{50}, or d_{15}) (see Christopher and Fischer [3]). The numeric value of the ratio depends upon the geotextile type, the soil type, the flow regime, and so on. For example, Carroll [5] recommends the following:

$$0_{95} < (2 \text{ or } 3)\, d_{85} \tag{2.9}$$

where d_{85} is the soil particle size in mm, for which 85% of the total soil is finer.

In contrast to the simplified methods above, a more comprehensive approach to soil retention criteria is given in Figures 2.4a and 2.4b, for steady-state and dynamic flow conditions, respectively (Luettich et al. [6]). To utilize the figures we must first completely characterize the upstream soil. A grain-size distribution, along with Atterberg limits and dispersivity characteristics for the fine fraction, are necessary. Examples 2.3 and 2.4 illustrate the use of the figures.

Example 2.3

What is the appropriate formula to obtain the required 0_{95} of a geotextile filter under steady-state flow conditions if the upstream soil is 25% less than 0.002 mm and the fine fraction is nondispersive?

Solution: From Figure 2.4a, $0_{95} < 0.21$ mm, which is equivalent to a No. 70 sieve or tighter.

Example 2.4

What is the appropriate formula to obtain the required 0_{95} of a geotextile filter under dynamic flow conditions if the upstream soil is less than 50% fines and less than 90% gravel and the situation is one of severe wave attack.

Solution: From Figure 2.4b, $0_{95} < d_{50}$, where d_{50} is the median particle size of the upstream soil.

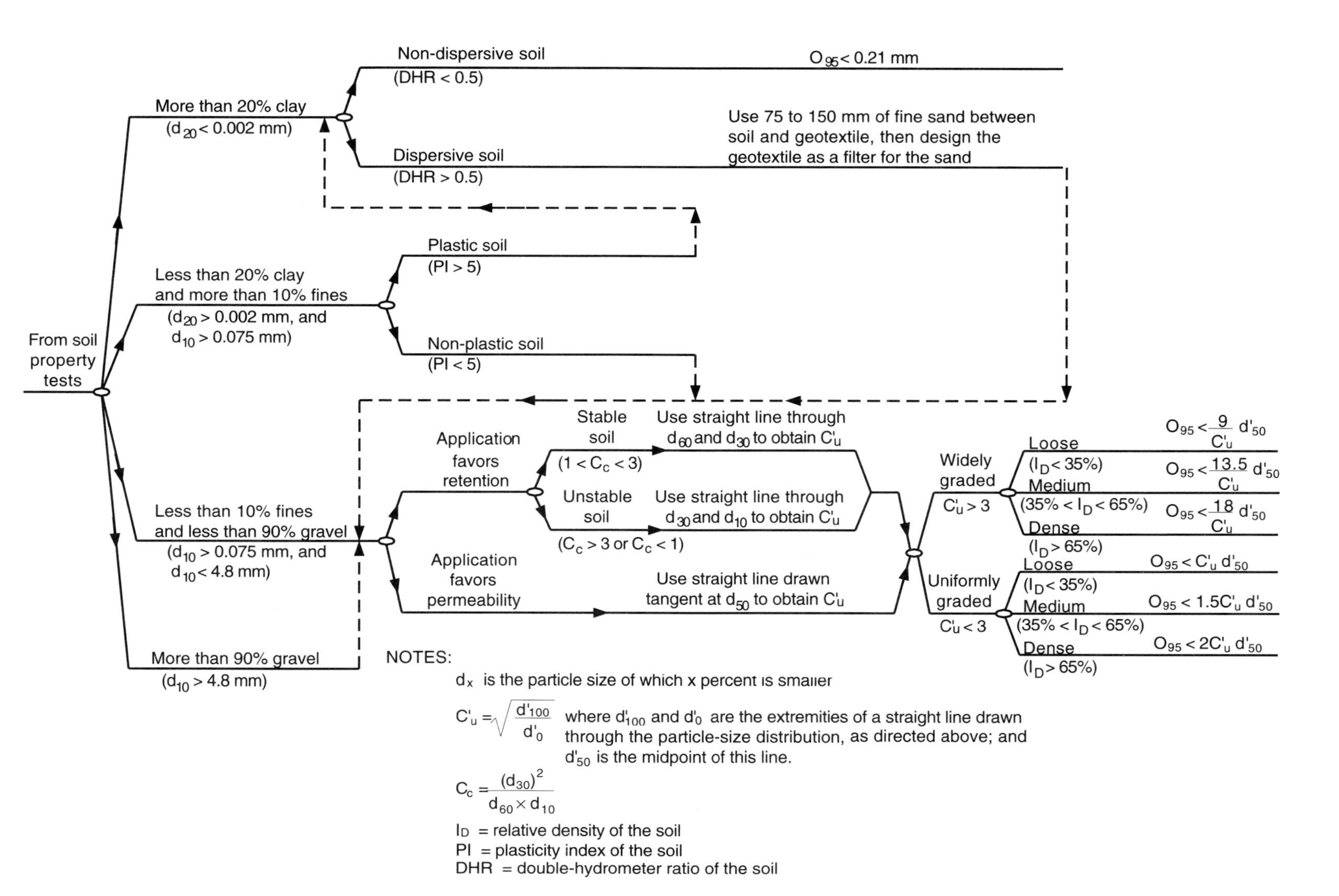

Figure 2.4 (a) Soil retention criteria for geotextile filter design using steady-state flow conditions. (After Luettich et al. [6])

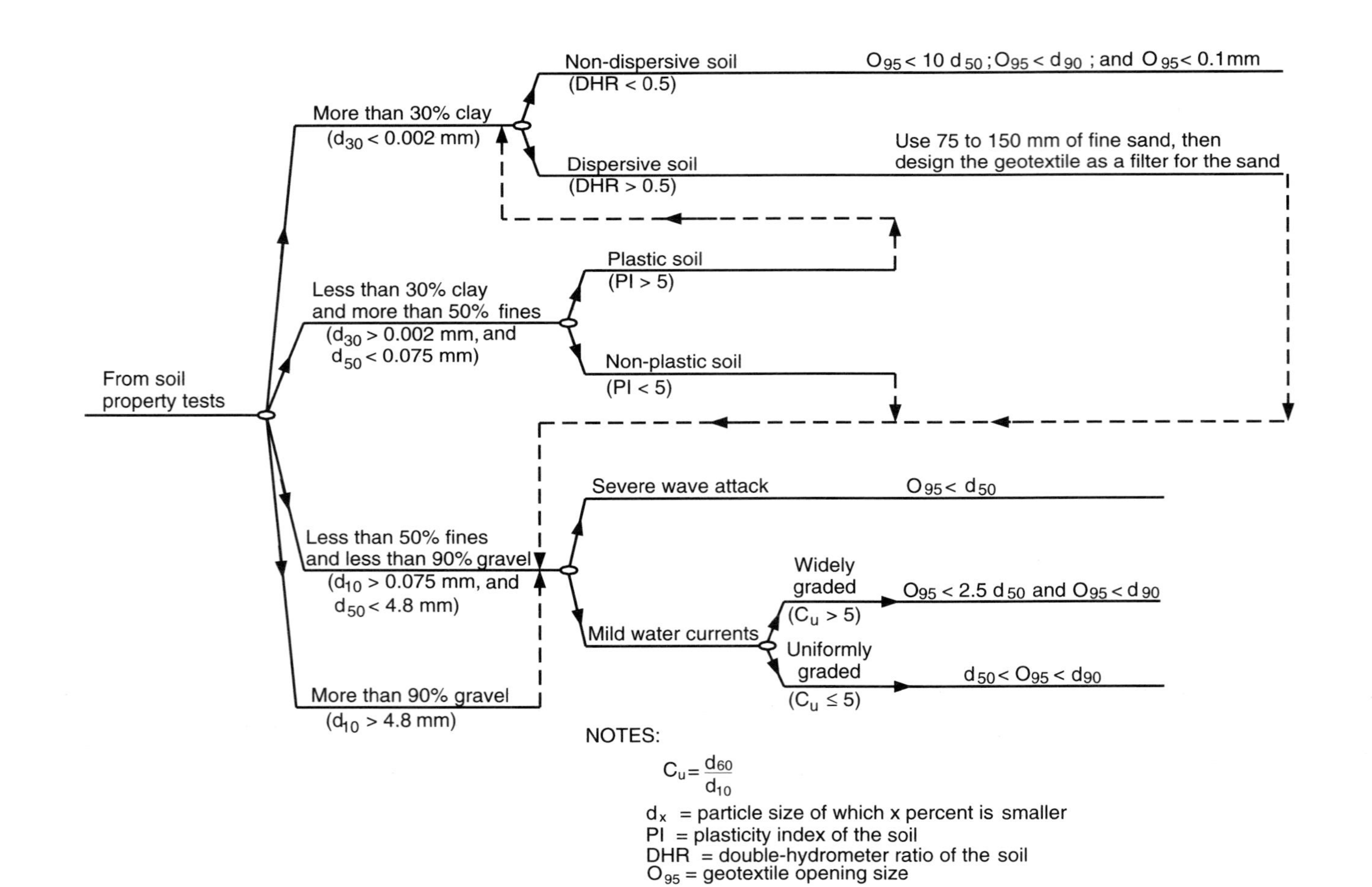

Figure 2.4 (b) Soil retention criteria for geotextile filter design under dynamic flow conditions. (After Luettich et al. [6])

Long-Term Flow Compatibility. Perhaps the most asked question regarding the use of geotextiles in hydraulic-related systems is, "Will it eventually clog?" Obviously, some soil particles will embed themselves on or within the geotextile structure and an understandable reduction in permeability or permittivity will occur. This type of *partial clogging* is expected. A more perceptive question is whether the geotextile will *excessively clog,* such that the flow of liquid through it will be decreased to the point where the system will not adequately perform its function. There are guidelines available for noncritical, nonsevere cases [3], but the question can be answered directly by taking a soil sample and the candidate geotextile(s) and testing them in the laboratory. Either gradient ratio (GR) test [7] to see that the GR ≤ 3.0, long-term flow (LTF) tests [8] to see that the terminal slope of the flow rate versus time curve is adequate for site specific conditions, or a hydraulic conductivity ratio (HCR) test [9] with resulting HCR values between 0.7 and 0.3 should be performed. These tests will be described later.

A different approach to the clogging question is simply to avoid situations that have been known to lead to excessive clogging problems. From experience it has been shown that the following conditions give rise to concerns about geotextile filter applications:

- Poorly graded (i.e., all uniform size) fine, cohesionless soils such as loess, rock flour, and quarry stone fines
- Cohesionless soils consisting of gap-graded particle size distributions functioning under high hydraulic gradients
- High alkalinity groundwater where the slowing of the liquid as it flows through the geotextile causes deposits of calcium, sodium, or magnesium precipitates
- High-suspended solids in the permeating liquid as in turbid river water that can build up on or within the geotextile
- High-suspended solids coupled with high-microorganism content, as in landfill leachates [10], that can combine to build up on or within the geotextile.

For all of these cases we could use a relatively open geotextile and allow fine particles, sediment, or microorganisms to pass through into the downstream drain. In such cases we would generally consider woven geotextiles with open area $\geq 10\%$, or nonwoven geotextiles with porosity $\geq 50\%$ (under site-specific normal stress conditions). Recognize, however, that whatever the downstream drain is (gravel, drainage core, perforated pipe, etc.), it must be designed to adequately accept and transport the particulate matter without itself clogging excessively.

This discussion of soil-to-geotextile compatibility assumes the establishment of a set of mechanisms that are in equilibrium with the flow regime being imposed on the system. Numerous attempts at insight into these phenomena have been attempted; see McGown [11], Heerten [12], and Giroud [13]. There exists a number of possibilities, including upstream soil-filter formation, blocking, arching, partial clogging, and depth filtration. These are shown schematically in Figure 2.5. With respect to how these mechanisms interact, it has been suggested that the geotextile serves as a catalyst to promote

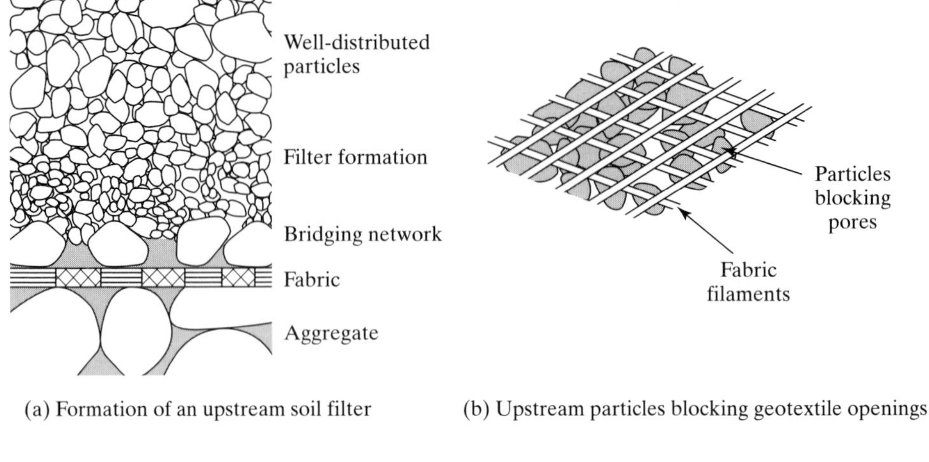

(a) Formation of an upstream soil filter

(b) Upstream particles blocking geotextile openings

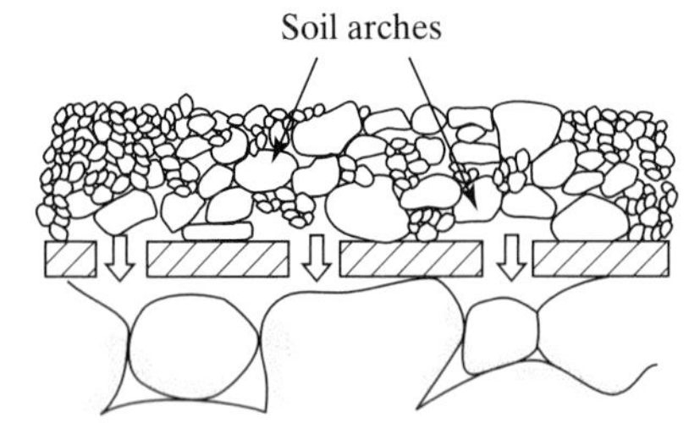

(c) Upstream particles arching over geotextile openings

(d) Soil particles clogged within geotextile structure

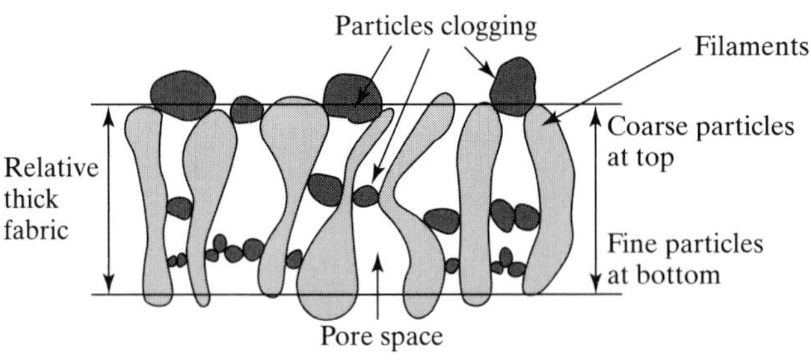

(e) Depth filtration concept using thick geotextiles

Figure 2.5 Various hypothetical mechanisms involved in long-term soil-to-fabric flow compatibility. (Parts a, b, c, d after McGown [11]; part e after Heerten [12])

the upstream soil, and the now soil-modified geotextile, to generate its own internal filter system. Obviously, a number of phenomena are working together simultaneously, and just what mechanism dominates under what conditions of soil type, geotextile type, and flow regime is still an issue that is being investigated.

2.2.4 Drainage

Fabrics placed in such a way as to transmit liquid within the plane of their structure provide a drainage function. Thus a definition of geotextile drainage is:

> **Drainage:** the equilibrium soil-to-geotextile system that allows for adequate liquid flow with limited soil loss within the plane of the geotextile over a service lifetime compatible with the application under consideration.

All geotextiles provide such a function, but to widely varying degrees [14]. For example, thin woven geotextiles, by virtue of their fibers crossing over and under one another, can transmit liquid within the spaces created at these crossover points but to an extremely low degree. Conversely, thick, nonwoven needle-punched geotextiles have considerable void space in their structure, and this space is available for liquid transmission. Furthermore (to preview the discussion in Chapters 4 and 8), geonets and drainage geocomposites can transmit much more liquid than geotextiles, even thick, bulky ones, can. Obviously, proper design will dictate just what type of geosynthetic drainage material is necessary.

Note that this discussion on drainage overlaps considerably the preceding section on filtration. For these two functions (except for the consideration of flow direction), the soil retention and long-term compatibility concepts are the same.

Permeability. Referring now to in-plane permeability for the drainage function, we must recognize that the geotextile's thickness will decrease with increasing normal stress on it. For this reason we will define a term called *transmissivity* as follows:

$$\theta = k_p t \tag{2.10}$$

where

θ = transmissivity,
k_p = in-plane permeability coefficient (the subscript p is often omitted), and
t = thickness at a specified normal pressure.

The testing method for geotextile transmissivity will be covered later.

Soil Retention. The criteria used to design the opening spaces of a geotextile so that it retains the adjacent soil were covered in Section 2.2.3. The concepts and design guides are precisely the same for the drainage function as they were for filtration.

Long-Term Flow Compatibility. As with the filtration function, we must ensure the compatibility of the soil with the geotextile over the lifetime of the system being built. The criteria discussed in Section 2.2.3 hold for drainage function the same as they do for filtration.

2.2.5 Liquid Containment (Barrier)

With regard to the geotextiles being discussed in this chapter, a liquid barrier is created by rendering the geotextile relatively impermeable to both cross-plane and in-plane flow. This in essence creates a geomembrane, albeit one having a fibrous structure rather than being a solid sheet of polymeric material. The liquid barrier referred to here is generally obtained by spraying bitumen, rubber-bitumen, elastomers, or other polymeric mixes onto a properly deployed geotextile thus creating an in situ liquid barrier rather than a factory-fabricated one. While the permeability created is obviously not zero (on an absolute basis, no geosynthetic has zero permeability), it is very low compared to that of the original geotextile. Quite possibly its permeability would now be in the range 1×10^{-7} to 1×10^{-9} m/s. This is comparable to the hydraulic conductivity of many fine-grained clay soil liners.

Within the function of moisture barrier, we are obviously referring to the impedance of the flow of liquid. However, we are also referring to the movement of *vapor* across the barrier. There are numerous situations in which the system on one side of the barrier must be kept free of liquid *and* vapor; for example, a barrier placed in the cover of a landfill to keep methane gas from escaping.

2.2.6 Combined Functions

The introduction to this chapter described design-by-function. The procedure as outlined identified the geotextile's primary function and set the design accordingly. Where geotextiles are used for a single function, this can indeed be done. However, geotextiles often serve multiple or combined functions. Some examples are:

- For prevention of crack reflection in asphalt pavement overlays, where both reinforcement and moisture-barrier functions are involved.
- Beneath railroad ballast, where separation, reinforcement, filtration, and drainage are involved.
- In flexible-forming systems to contain concrete, grout or soil, where separation, reinforcement, and filtration are involved.

In these situations the primary, secondary, tertiary, and so on, functions must all be evaluated. They must all satisfy the required factor of safety. If the situation is properly assessed, the calculated factors of safety should increase progressively as we proceed through the primary, secondary, tertiary, and so on, functions. If not (i.e., if the factors of safety jump around as we proceed through the calculations), it means that the critical functions were not properly assessed to begin with. Thus the minimum factor of safety will always indicate the primary function, the next highest value of factor of safety will indicate the secondary function, and so on. This approach, of course, assumes that a reasonably accurate quantitative analysis can be developed for each of the functions described (see Sections 2.5 to 2.9). Before we discuss this, however, we will treat a very important aspect of specific geotextile properties and how they are obtained: the quantification of geotextile properties via their current test methods and procedures.

2.3 GEOTEXTILE PROPERTIES AND TEST METHODS

This section presents the necessary test methods, relevant details, and selected data for the design-by-function procedure to be developed later in the chapter. The reader should refer back to this section continuously as the various design methods are developed, since from these test methods come the numerator of the design-by-function equation; recall Eqs. (2.2a) and (2.2b).

2.3.1 General Comments

In a growing area such as geotextiles, it should come as no surprise that a completely unified set of worldwide standards and test methods is currently not available. Yet the activity toward such an ultimate goal is very intense. Organizations that are involved in this activity are spread across the entire spectrum of potential users: raw material suppliers, manufacturers, manufacturers representatives, contractors and installers, testing organizations, design engineering firms, research organizations, owners, and regulators. Within these groups, there is often reference to either *index* or *performance* tests. This terminology is somewhat unfortunate, since what is an index test to one group might be (and usually is) very much a performance test to another group. For example, a geotextile puncture test using a steel probe may be an index test to a geotechnical engineer, but to the manufacturer it is instead a measure of the quality control performance of the particular manufacturing process. Thus, this book does not make continual reference to a particular test method as being either index- or performance-related, but when it does so, the test method is labeled from the perspective of the design engineer.

In any review of available geotextile test methods to follow, it should be recognized that many of the test methods are not fully standardized as far as their test procedures are concerned. Since this is a known industry-wide problem, standards groups all over the world are actively involved in proposing, evaluating, modifying, and finalizing geotextile standards, on their own and cooperatively. If such a process were done completely from first principles, it would take a prohibitively long time; however, there are tests that can be borrowed in whole or in part from other areas. For geotextile standards, there is an obvious overlap with textile products (clothing textiles, industrial textiles, etc.) where many standards already exist. The overall textile industry is, of course, a large and mature one that has a direct interest and involvement in geotextiles. In fact, the manufacturing methods are identical; only the applications are different. Thus it should come as no surprise that many physical and mechanical test methods for geotextiles are partially or completely taken from existing textile standards [15, 16, 17, 18]. The tests that differ for textiles and geotextiles are those that involve hydraulic, endurance, and environmental properties; these are generally new tests oriented completely toward geotextiles.

In the U.S., the ASTM has a standards committee specifically organized for geosynthetics (D35); however, the activity is actually worldwide, as the following list indicates.

- International Organization for Standards (ISO) (international)
- Permanent International Association of Road Congresses (PIARC) (international)

- European Disposables and Nonwovens Associated (EDANA) (international)
- Permanent International Association of Navigation Congresses (international)
- Geosynthetic Research Institute (GRI) of Drexel University (international)
- European Community Normalization (CEN) (Europe)
- Ministry of Public Works (Belgium)
- Canadian General Specification Board (CGSB) (Canada)
- British Standards Institution (U.K.)
- The Institution of Civil Engineers (U.K.)
- Transport and Road Research Laboratory (TRRL) (U.K.)
- Technical Research Center of Finland (Finland)
- Comité Français des Géotextiles et Géomembranes (France)
- L'Association Français de Normalisation (AFNOR) (France)
- Réunion Internationale des Laboratoires d'Essais et de Recherche sur les Matériaux et les Construction (RILEM) (France)
- German Standards Committee for Geotextiles (DIN) (Germany)
- Franzius Institut (für Wasserbau und Küstingenieurswesen der Universität Hannover) (Germany)
- Bundesanstalt für Wasserbau (Germany)
- Deutsche Gesellschaft für Erd- und Grundbau (Germany)
- Nederlands Normalisatie Institut (NNI) (The Netherlands)
- Consiglio Nazionale Richerche (CNR) (Italy)
- ENEL/CRIS (Italy)
- Unitex (Italy)
- Korea Highway Corp. (South Korea)
- Norwegian Road Research Laboratory (NRRL) (Norway)
- João de Matos Rosa (Portugal)
- The Swedish National Road Administration (Sweden)
- Swedish Geotechnical Institute (Sweden)
- Schweizerisches Verband der Geotextilfachleute (Switzerland)
- South Africa Bureau of Standards (South Africa)
- The American Society for Testing and Materials (ASTM) (U.S.)
- U.S. Army Corps of Engineers, Waterways Experiment Station (U.S.)
- American Association of State Highway and Transportation Officials (AASHTO) (U.S.)

One goal of the Standards and Specifications Committee of the International Geosynthetics Society (IGS) is to collect, compare, and provide information on specific standards for geosynthetics, including geotextiles; see [19] for a preliminary collection.

At this point in time, however, a unified document on standards or test methods is not available. Thus this section provides insights into test methods in use or at least currently favored for the testing of geotextiles. Reference will be made, whenever possible, to the existing ASTM standard or to the proposed test method being considered by Committee D35 on Geosynthetics. Cross-references to the most recent ISO standards will also be made. The section is subdivided into the following categories: physical properties, mechanical properties, hydraulic properties, endurance properties, degradation properties.

2.3.2 Physical Properties

The properties discussed in this subsection all refer to the geotextile in its manufactured or as-received condition. These tests are usually referred to as index tests.

Specific Gravity. The specific gravity of the fibers from which geotextiles are made is actually the specific gravity of the polymeric feedstock (see ASTM D792 or D1505). As customary, specific gravity is defined as the ratio of the material's unit-volume weight (without any voids) to that of distilled, de-aired water at 4°C. Typical values for the specific gravity of commonly used polymeric materials that are made into geotextiles are listed below (steel, soil, glass and cotton have been added for comparison).

Steel = 7.87
Soil/rock = 2.9 to 2.4
Glass = 2.54
Polyvinyl chloride = 1.69
Cotton = 1.55
Polyester = 1.38 to 1.22
Nylon = 1.14 to 1.05
Polyethylene = 0.96 to 0.90
Polypropylene = 0.91

Note that the specific gravity of some of the polymers (e.g., the polyolefins) is less than 1.0, which must be considered when working under water (i.e., some of them will float).

Mass per Unit Area (Weight). *Mass per unit area* is the proper term for what most people mean when they state or ask for the weight of a geotextile. It is also sometimes called *basis weight*, but this is equally incorrect since neither weight nor basis weight explicitly considers area. Geotextile mass per unit area is given in grams per square meter (g/m^2). Unfortunately, other values are listed in the literature, such as grams per linear meter for a geotextile of given width. Sometimes the latter value is given inversely as linear meters per kilogram. The point here is that we must clearly state which value is being communicated. Methods for the test are ASTM D5261 and ISO 9864.

Testwise, the mass (or weight) should be measured to the nearest 0.01% of the total specimen mass; length and width should be measured under zero geotextile tension. The range of typical values for most geotextiles is from 150 to 750 g/m^2. Since fabric cost (and, in general, mechanical properties) is directly related to mass per unit area, it is an important property.

Thickness. Thickness of geotextiles is sometimes mentioned in specifications, but this is really more of a descriptive property than design-oriented property. It is measured as the distance between the upper and lower surfaces of the fabric, measured at a specified pressure. ASTM D5199 stipulates that the thickness of a geotextile be measured to an accuracy of at least 0.02 mm under a pressure of 2.0 kPa. The ISO 9863 test method allows the specifier to select the pressure. The thicknesses of commonly used geotextiles range from 0.25 to 7.5 mm.

Stiffness. Stiffness, or flexibility, of a geotextile should not be confused with its modulus (which is determined as the initial portion of the stress-versus-strain curve). In this test, stiffness is a measure of the interaction between the geotextile weight and its bending stiffness, as shown by the manner in which the geotextile gravitationally bends under its own weight; the test method is designated ASTM D1388. It is more appropriately called *flex stiffness*. The method takes a 25 mm wide strip of geotextile specimen and slides it out lengthwise over the edge of a horizontal surface. The length of overhang is measured when the tip of the geotextile bends under its own weight and just touches an inclined plane making an angle of 41.5° with the horizontal. One-half of this length is the bending length of the specimen. The cube of this quantity multiplied by the mass per unit area of the geotextile is its flexural rigidity or stiffness. The value is expressed in mg-cm.

This property is indicative of the geotextile's inherent capability of providing a suitable working surface for installation. In placing a geotextile on extremely soft soils, a high geotextile stiffness is very desirable. Haliburton et al. [20] have related this property to various soil subgrade strength values as given in Table 2.3.

TABLE 2.3 RECOMMENDED GEOTEXTILE STIFFNESS VALUES FOR VARYING DEGREES OF REQUIRED WORKABILITY

Subgrade CBR* (%)	Workability Benefit of Vegetative Cover†	Field Workability Requirements	Minimum Fabric Stiffness‡ (mg · cm)
CBR ≤ 0.5	Poor	Very high	25,000
	Good	High	15,000
0.5 < CBR ≤ 1.0	Poor	High	15,000
	Good	Moderate	10,000
1.0 < CBR ≤ 2.0	Poor	Moderate	10,000
	Good	Low	5,000
CBR > 2.0	Poor	None	1,000
	Good	None	1,000

*CBR refers to soaked California Bearing Ratio, which is a test routinely used in geotechnical engineering to evaluate soil subgrade strength. It is standardized as ASTM D1883.

†Medium to dense root system will probably exhibit some inherent workability benefits, whereas little to no root system will be of no benefit.

‡Test conforms to ASTM D1388, except uses 300 mm long by 50 mm wide test specimens.

Source: After Haliburton et al. [20].

2.3.3 Mechanical Properties

The mechanical properties discussed here indicate the geotextile's resistance to tensile stresses mobilized from applied loads and/or installation conditions. Some are performed with the geotextile by itself (called *index* or *in-isolation* tests) while others are associated with a standard soil or with the site-specific soil (often called *performance* tests).

Compressibility. The compressibility of a geotextile is its thickness at varying applied normal stresses. For most geotextiles the compressibility is relatively low and of little direct consequence as far as design is concerned (e.g., with woven fabrics and with nonwoven heat-bonded and heavily calendered geotextiles). For nonwoven needle-punched or bulky resin-bonded geotextiles, however, compressibility is important. This is because such geotextiles are often used to convey liquid within the plane of their structure. The more a fabric compresses under load, the lower its transmissivity. Figure 2.6 illustrates the compressibility of several geotextile types where the influence of normal stress on thickness is clearly seen. The nonwoven needle-punched geotextiles are the most compressible, and this in turn is directly related to their mass per unit area.

Tensile Strength. Perhaps the single most important property of a geotextile is its tensile strength. Invariably all geotextile applications rely on this property, either as the primary function (as in reinforcement applications) or as a secondary function (as

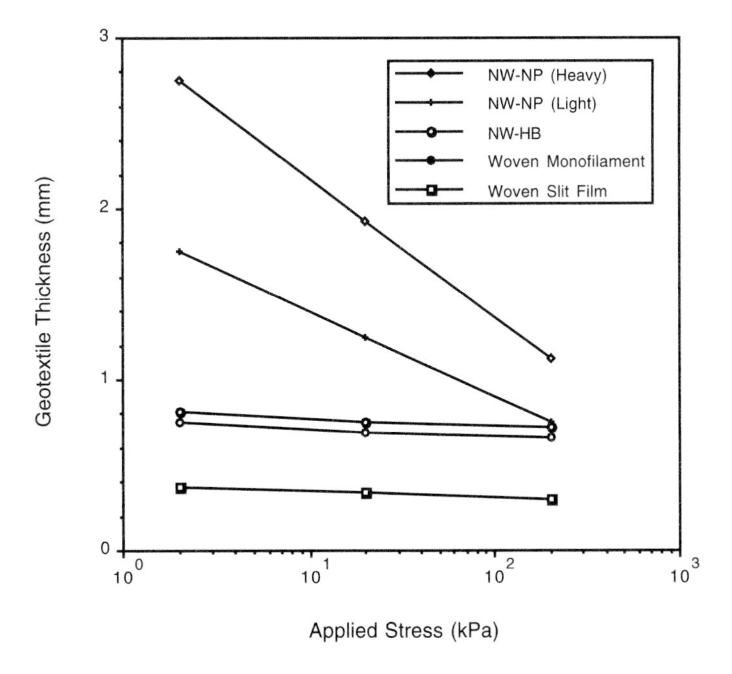

Figure 2.6 Compressibility of different types of geotextiles, where NW-NP = nonwoven needle-punched, and NW-HB = nonwoven heat-bonded.

in separation, filtration, drainage, or containment). The basic test is to place the geotextile within a set of clamps or jaws, place this assembly in a mechanical testing machine, and stretch the geotextile in tension until failure occurs. Fabric failure is generally easy to identify and often it is even audible. During the extension process, it is customary to measure both load and deformation in such a way that a stress-versus-strain curve can be generated. (Stress here is usually given as load per unit width.) From the stress-versus-strain curve (strain calculated as deformation divided by the original specimen length), four values are obtained:

1. Maximum tensile stress (referred to as the geotextile's strength)
2. Strain at failure (generally referred to as *maximum elongation* or simply *elongation*)
3. Toughness (work done per unit volume before failure, usually taken as the area under the stress-strain curve)
4. Modulus of elasticity (the slope of the initial portion of the stress-versus-strain curve)

Typical responses of geotextiles made from different manufacturing processes are given in Figure 2.7. Note that the vertical axis is in force per unit width of fabric (i.e., kN/m), which is not a bona fide stress unit. To obtain stress units, this value is divided by the geotextile's thickness, but this is not conventionally done, since the thickness varies greatly under load and during the extension process. This, of course, has implications for the toughness and modulus values, as well, since they too would have to be divided by thickness to obtain conventional engineering units. Example 2.5 illustrates these features.

Example 2.5

For the nonwoven heat-bonded geotextile illustrated in Figure 2.7 (Curve D), determine the strength, maximum elongation, toughness, and modulus in common geotextile units and (on the basis of a nominal thickness of 0.33 mm) in standard engineering units.

Solution: By observation, the strength is

$$T_{max} = 23 \text{ kN/m}$$

and for 0.33 mm thickness,

$$\sigma_{max} = 69{,}700 \text{ kN/m}^2 = 69{,}700 \text{ kPa}$$

The maximum strain is also determined by observation:

$$\varepsilon_f = 69\%$$

The toughness U is then calculated as 1/2 ($T_{max} \times \varepsilon_f$) (actually this is an approximation since it should be the area under the curve):

$$U = \frac{1}{2}(23 \times 0.69)$$

$$= 7.9 \text{ kN/m}$$

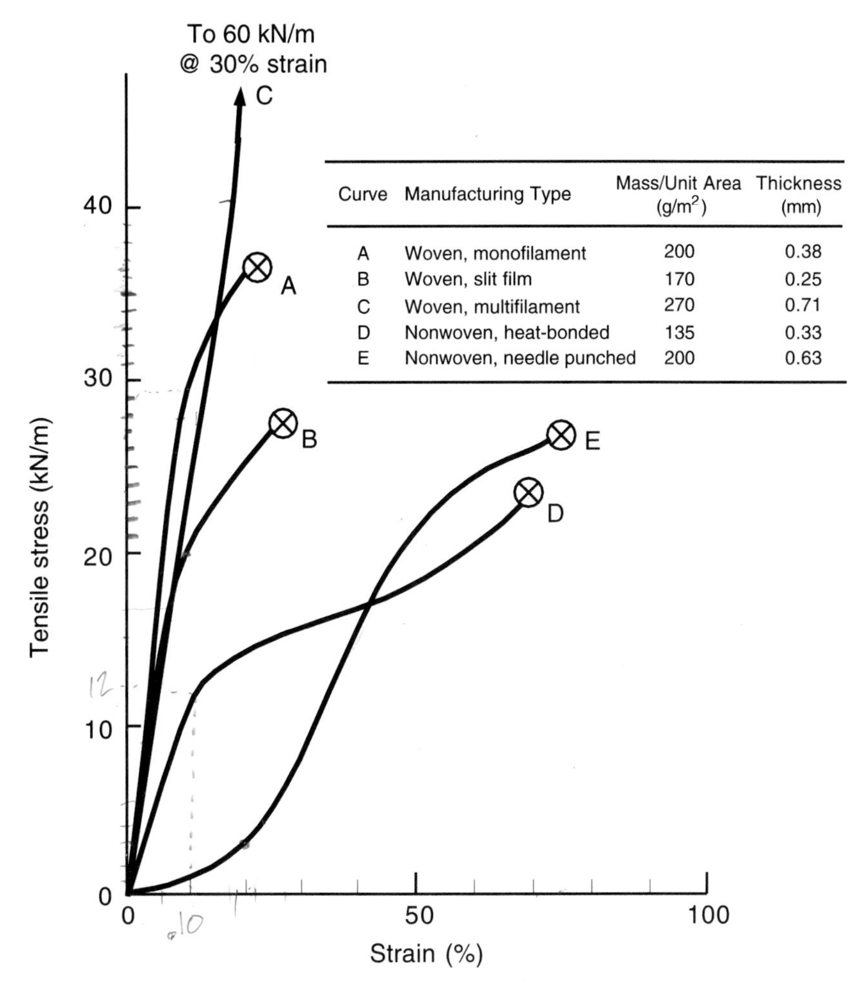

Curve	Manufacturing Type	Mass/Unit Area (g/m^2)	Thickness (mm)
A	Woven, monofilament	200	0.38
B	Woven, slit film	170	0.25
C	Woven, multifilament	270	0.71
D	Nonwoven, heat-bonded	135	0.33
E	Nonwoven, needle punched	200	0.63

Figure 2.7 Tensile test response of various geotextiles manufactured by different processes. All are polypropylene fabrics; test specimens were initially 200 mm wide × 100 mm high.

and for 0.33 mm thickness,

$$U = 24{,}000 \text{ kN/m}^2 = 24{,}000 \text{ kPa}$$

Finally, the modulus is taken from the initial slope of the curve as

$$E = \frac{12}{0.10} = 120 \text{ kN/m}$$

and for 0.33 mm thickness,

$$\begin{aligned} \text{E} &= 364{,}000 \text{ kN/m}^2 = 364{,}000 \text{ kPa} \\ &= 364 \text{ MPa} \end{aligned}$$

There are several features of the tensile test that require further discussion, since they have implications for subsequent design procedures, the major ones being the modulus and the specimen size. Regarding the modulus, several choices are available for measuring the initial slope of the curve. These are the following:

Initial tangent modulus. This is straightforward for many woven geotextiles in both their warp and weft directions and for nonwoven heat-bonded geotextiles. Here the initial slope is linear and (as in conventional soil testing) a reasonably accurate modulus value can be obtained.

Offset tangent modulus. This concept is sometimes used when the initial slope is very low and is typical of nonwoven needle-punched geotextiles (see Figure 2.7, Curve E). To obtain the value, we avoid the initial portion of the curve and essentially shift the y-axis to the right, where it meets the downward extension of the linear portion of the response curve. The slope is then taken from this adjusted axis location.

Secant modulus. To avoid the some arbitrariness of the above mentioned methods, we could stipulate the procedure of obtaining a modulus value, for example, a secant modulus at 10% strain. In such a case, we draw a line from the axes' origin to the designated curve at 10% strain and measure its slope from the origin, irrespective of the actual curve to this point.

Example 2.6 illustrates these procedures.

Example 2.6

For the nonwoven needle-punched fabric E shown in Figure 2.7, determine the initial tangent modulus, offset tangent modulus, and secant moduli at 10% and 35% strain in kN/m and kN/m^2 (kPa) based on an initial thickness of 0.63 mm.

Solution: Scaling directly from the curve,

$$E_T = \frac{4.3}{0.50} = 8.6 \text{ kN/m or } 13{,}600 \text{ kN/m}^2$$

$$E_{OT} = \frac{20}{0.46 - 0.20} = 77 \text{ kN/m or } 122{,}000 \text{ kN/m}^2$$

$$E_{S10} = \frac{1.1}{0.10} = 11 \text{ kN/m or } 17{,}500 \text{ kN/m}^2$$

$$E_{S35} = \frac{11.6}{0.35} = 33 \text{ kN/m or } 52{,}400 \text{ kN/m}^2$$

Regarding the test specimen size (length, width, and aspect ratio or length-to-width ratio), much has been written. ASTM standards D1682, D751, D4632 , and D4595 allow for a number of variations. Figure 2.8 illustrates the current most popular specimen sizes. The grab tensile test D4632 is a very widely used and reported test. The geotextile specimen dimensions are 100 mm wide and 150 mm long, but the jaws of the clamps grip only the central 25 mm of the test specimen. Almost all geotextile manufacturers and geotextile specifications use this value. Narrow strip tests (usually 25 or 50 mm wide) are used in many research and development studies since they use a min-

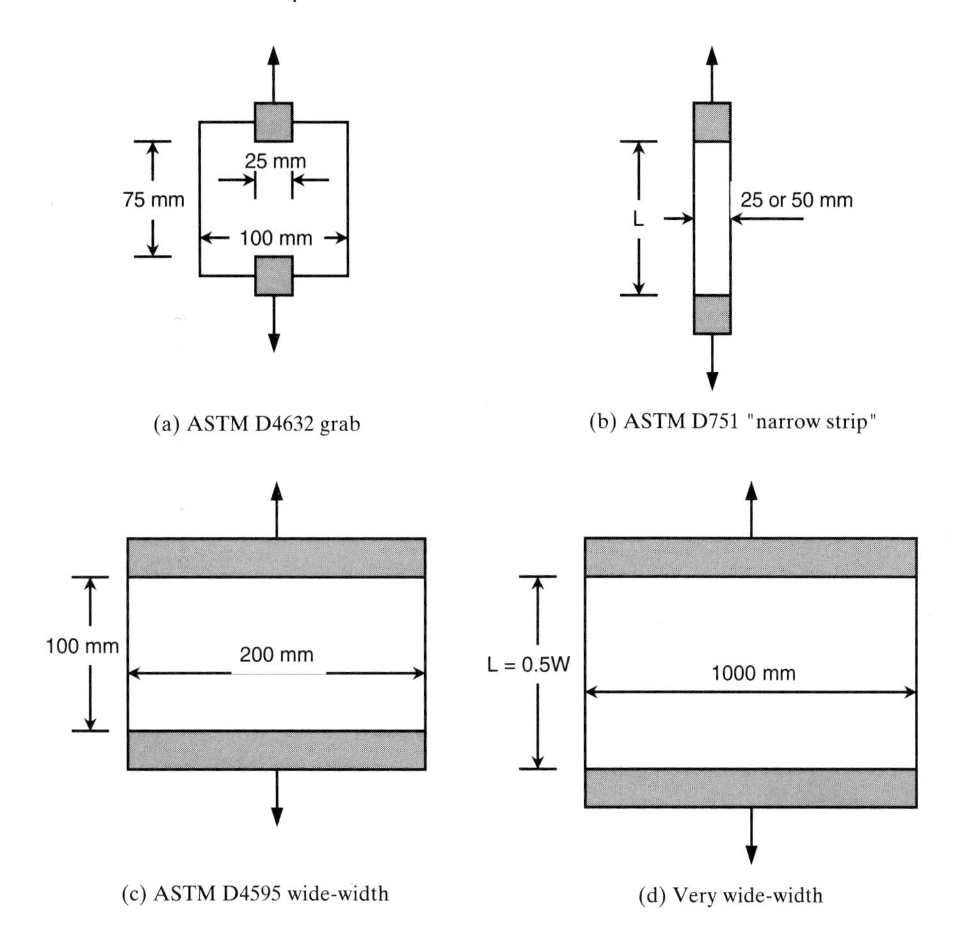

Figure 2.8 Various tensile test specimen sizes used to obtain fabric strength properties.

imum amount of geotextile. The reason wide-width specimens are necessary is that geotextiles (particularly nonwovens) when tensioned tend to have a severe Poisson's ratio effect under increasing stress and they rope-up, giving artificially high values. Thus the tendency for design-related tests is to use wide-width specimens. The most common wide-width test is ASTM D4595 and ISO 10319, both of which use a 200 mm wide specimen 100 mm long between the faces of the opposing grips. Such a test is not intended to be a routine or index test. The grab specimen should continue to be used in this regard (e.g., as a manufacturers' quality-control or conformance test). There are no universal relationships between the different test specimen sizes or shapes, and therefore the choice of specimen size depends on the intended use of the data. The proper identification of the specimen size on the test data is always necessary. Regarding other features of tensile testing of geotextiles (effect of conditioning, load rate, load method, etc.), the applicable standard(s) should be consulted.

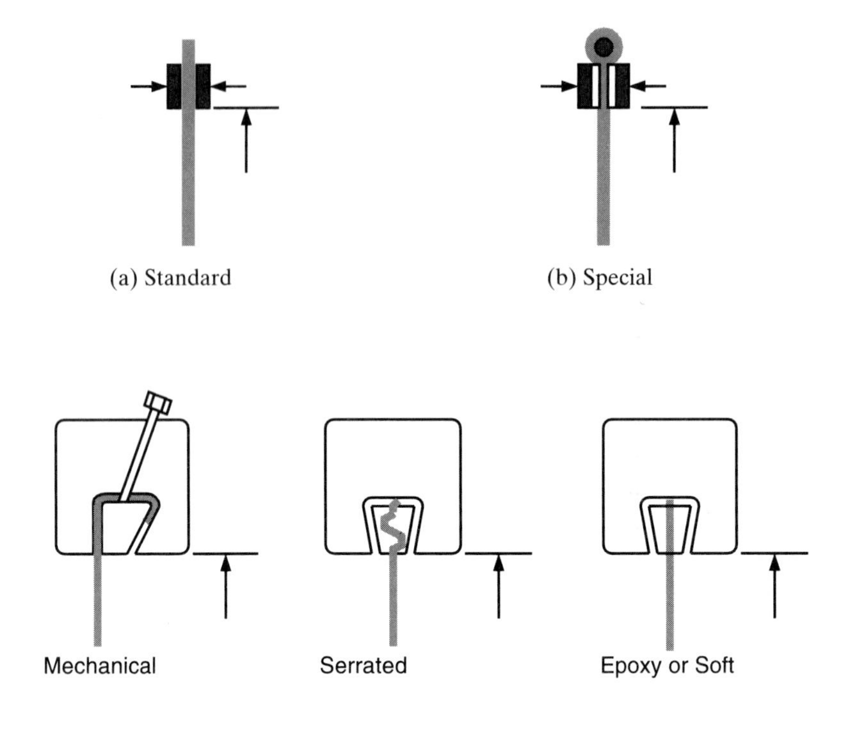

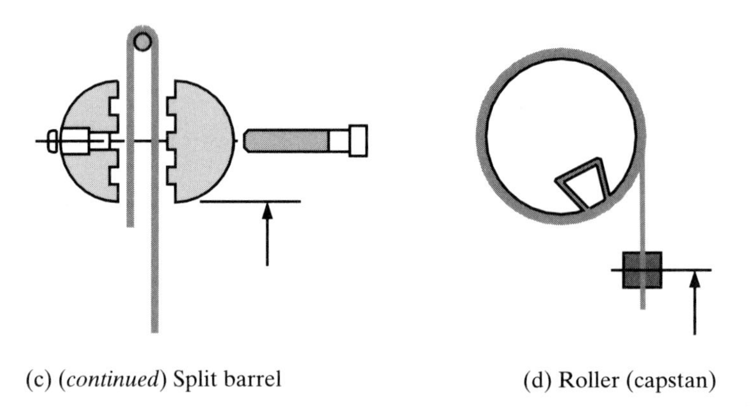

Figure 2.9 Various grip types for testing geotextiles and geogrids. (Adapted from Myles and Carswell [21])

As the strength of the geotextile being tested increases, a number of operational problems arise. The most obvious need is for a higher-capacity tensile testing machine than is used for conventional geotextiles. This is straightforward. The types of devices used to grip the geotextile however, is another matter. Figure 2.9 illustrates the various grip types recommended for use depending on the ultimate strength of the geotextile. With geotextile strengths greater than approximately 50 kN/m, standard clamping jaws

(Figure 2.9a) are not satisfactory. This is due to slippage within the conventional grips or stress concentrations at the face of the grips leading to erroneous values of stress and strain. Standard grips can be made adaptable up to approximately 90 kN/m (see Figure 2.9b). Some type of wedge or split-barrel grips are necessary (see Figure 2.9c). Wedge grips can be made in a number of styles, using mechanical wedges, serrated wedges, or cast metal wedges. However, even these grips become troublesome with geotextile strengths greater than approximately 180 kN/m, due mainly to stress concentration failures at the edge of the upper or lower grips. Here stresses are very high and can only be avoided using a roller or capstan type of grip (see Figure 2.9d). In this case, the geotextile tightens on itself around the rollers and failure is within the test specimen between the opposing set of rollers. Elongation, however, can no longer be read from the testing machine's crosshead separation, since geotextile take-up around the rollers is occurring. This necessitates the use of an external measuring device such as a linear variable differential transformer (LVDT), laser sensor, or infrared sensor. Most use a 100 mm gage distance located in the center of the test specimen between the roller grips. This output is fed into an xy recorder for the x-axis elongation (or strain) reading. The y-axis or load is taken directly from the tensile test machine. Thus one obtains the stress-versus-strain diagram similar to Figure 2.7, albeit with considerable effort and many operational problems.

Confined Tensile Strength. Before finishing the topic of tensile strength, it should be cautioned that all of the tests just described are performed without lateral confining pressure (i.e., they are in-isolation tests). With lateral confinement, which obviously is how geotextiles are eventually used, results can be different. McGown et al. [22, 23], have pioneered in this area using a boxlike chamber separated into two halves, as shown in Figure 2.10, where the geotextile test specimen is sandwiched between lubricated membranes and thin soil layers that have been pressurized by rubber bellows. It is important to allow the test specimen to elongate freely without friction being mobilized by the confining soil adjacent to the lubricated membranes. A friction mobilized pull-out test will be described later. The confinement pressure within the bellows simulates the in situ pressure. The test specimen is 200 mm wide and 100 mm long in the test zone. While tedious and relatively complex, it is the best attempt at obtaining a true tensile strength/elongation response known to the author. It is particularly important for obtaining the modulus value of nonwoven needle-punched geotextiles if finite-element methods of design are used.

In order to assess a wide range of geosynthetic materials tested in this manner, Wilson-Fahmy et al. [24], have evaluated four different styles of geotextiles (see Figure 2.11). It is important to note from this study that *only* the nonwoven needle-punched geotextiles show significantly improved stress-versus-strain behavior under confinement (see Figures 2.11c, d). This apparently comes from confining pressure holding the random fibers in their original positions. Thus the low initial modulus response seen in Figure 2.7 (curve E) is eliminated. For the other geotextiles tested under confining pressure, woven in Figure 2.11a and nonwoven heat-bonded in Figure 2.11b, there is essentially no difference, except as failure is approached. Although not shown in Figure 2.11, there was no measurable improved strength noted with geonets, geomembranes, or GCLs with woven slit-film geotextiles when they were placed under confinement; see [24] for details.

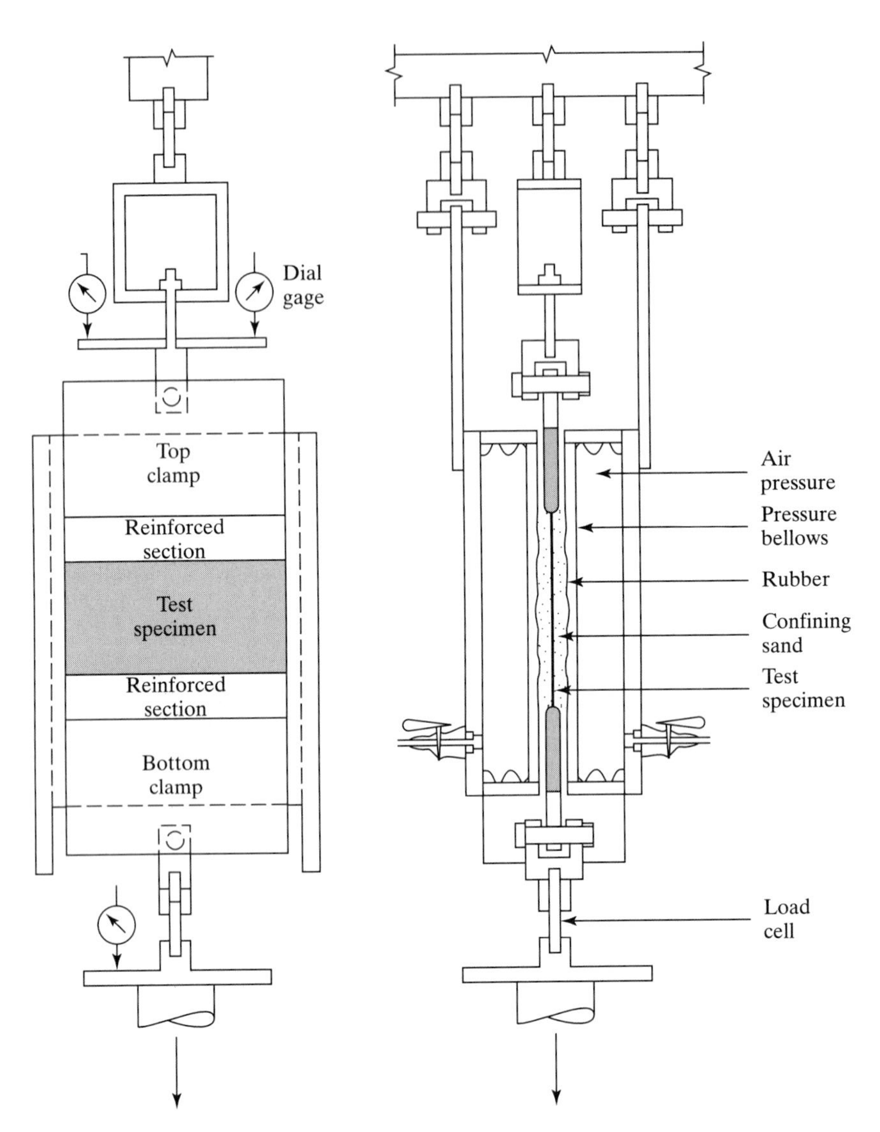

Figure 2.10 Geotextile placed under lateral confinement during tensile testing. (After McGown et al. [22, 23])

Seam Strength. Often the ends or sides of rolls of geotextiles have to be joined together for the purpose of transferring tensile stress. By far the most common method is by sewing. Various styles of sewn seams will be described later, but whatever the type, they must be laboratory-evaluated for their load-transfer capability from one geotextile roll to another. The ASTM D4884 test method calls for the following requirements.

- The shape of the seamed test specimen is 200 mm wide, except at the seam, itself. Here an additional 25 mm of seamed material is allowed to protrude from both sides, that is, at the seam the test specimen is 250 mm wide. This accounts for a cer-

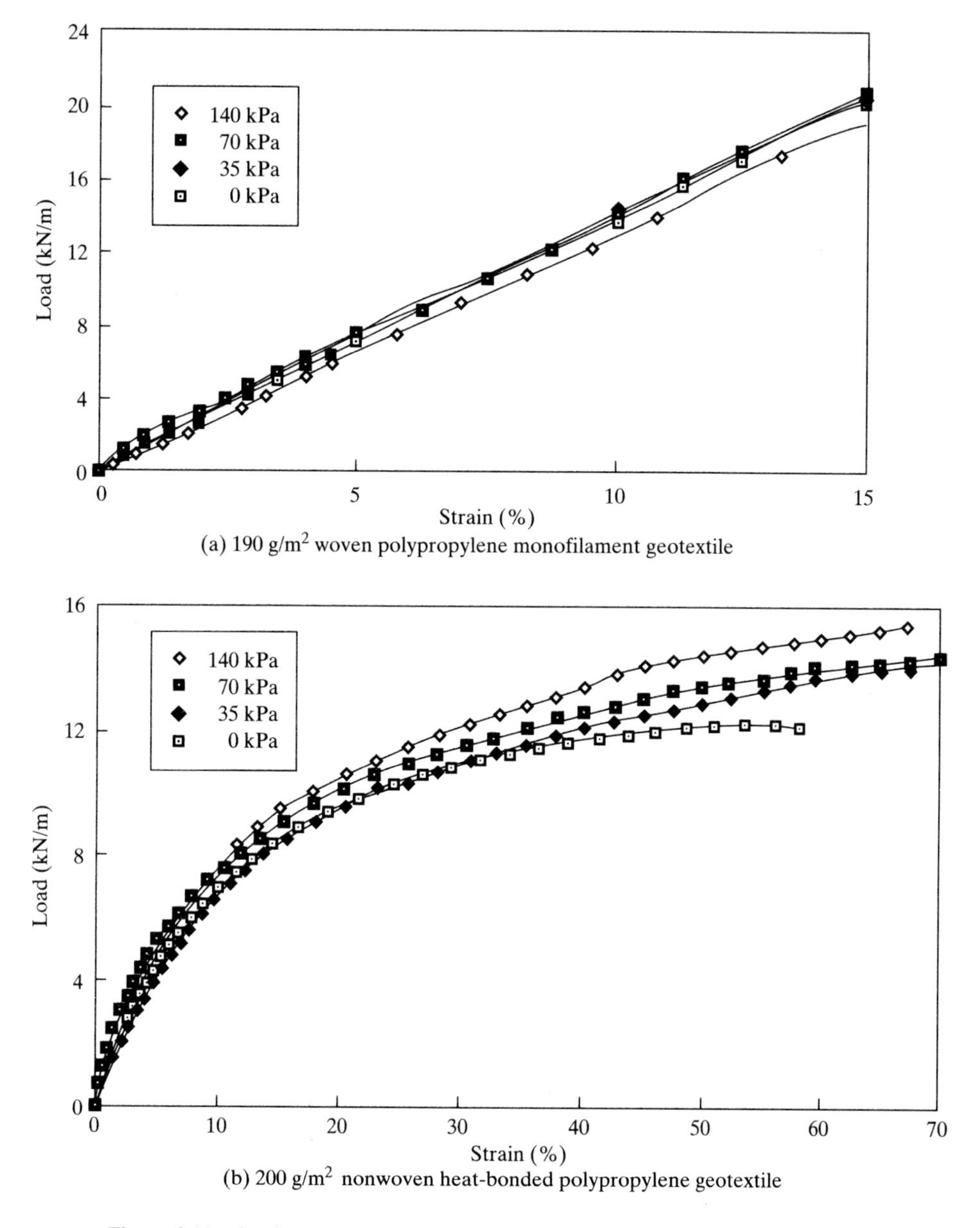

(a) 190 g/m^2 woven polypropylene monofilament geotextile

(b) 200 g/m^2 nonwoven heat-bonded polypropylene geotextile

Figure 2.11 Confined wide-width tensile strength response of different types of geotextiles. (After Wilson-Fahmy et al. [24])

tain amount of loss of seam strength when cutting of the seaming yarns during specimen preparation, but to what degree is uncertain.

- The resulting ultimate load is divided by a 200 mm width and reported in kN/m. The appropriateness of this computational step is questionable.

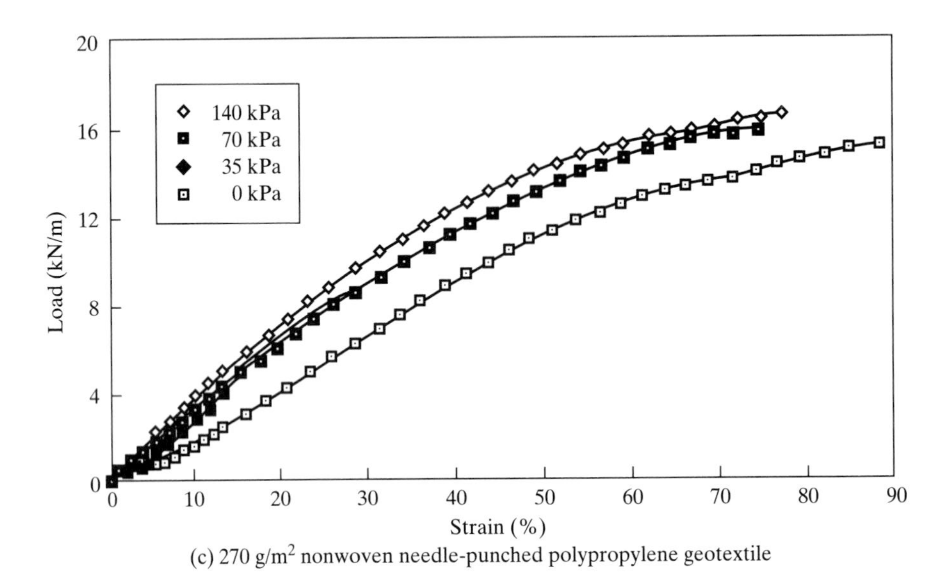

(c) 270 g/m² nonwoven needle-punched polypropylene geotextile

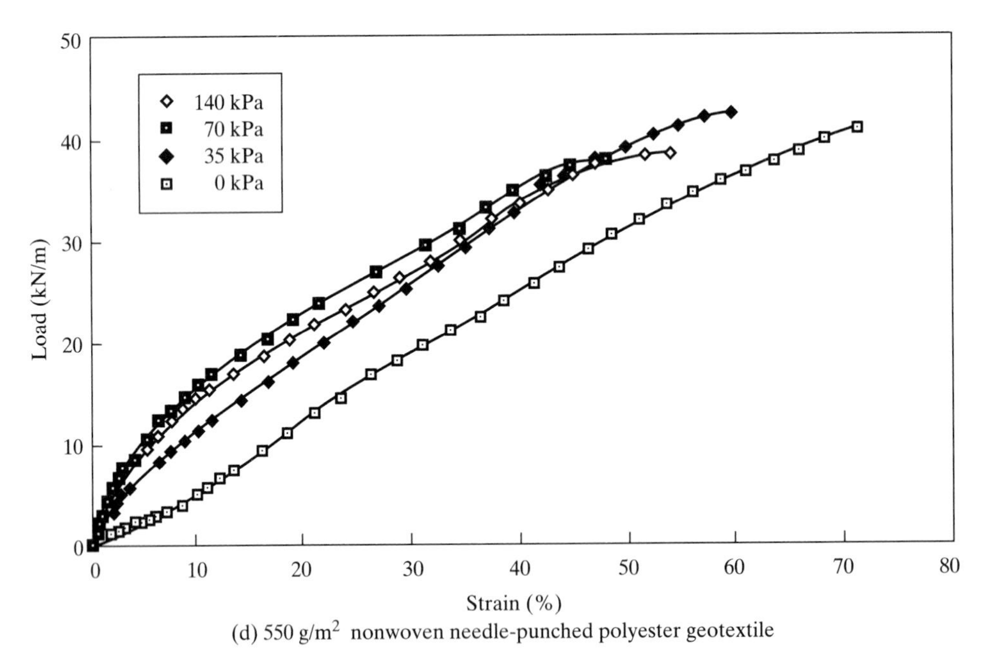

(d) 550 g/m² nonwoven needle-punched polyester geotextile

Figure 2.11 *(continued)*

- The rate of extension is 10%/min.
- Measurement of elongation across the seam is not required, that is, the test only measures tensile strength.

The comparable ISO test is ISO 10321. Test results from the evaluation of well-made sewn seams of geotextiles having wide-width strengths up to approximately 20 kN/m usually results in sewn-strength efficiencies near 100%,

$$E(\%) = \frac{T_{\text{seam}}}{T_{\text{geotextile}}} \times 100 \tag{2.11}$$

where

E = seam efficiency (percent),
T_{seam} = wide-width seam strength via ASTM D4884, and
$T_{\text{geotextile}}$ = wide-width geotextile strength via ASTM D4595 (i.e., unseamed).

As the geotextile strength becomes higher, however, seam strengths become progressively less efficient (see Figure 2.12). Above 50 kN/m, even the best of seams fall below the 100% efficiency line, and beyond 200 to 250 kN/m, the best we can do is approximately 50% seam efficiency. Note that by this point poorly made seams become extremely low in their load-transfer capabilities. The seaming of high-strength geotextiles simply begs for better joining or bonding methods than sewing. Many possibilities are available, such as the use of epoxy resins [25] or mechanical joining.

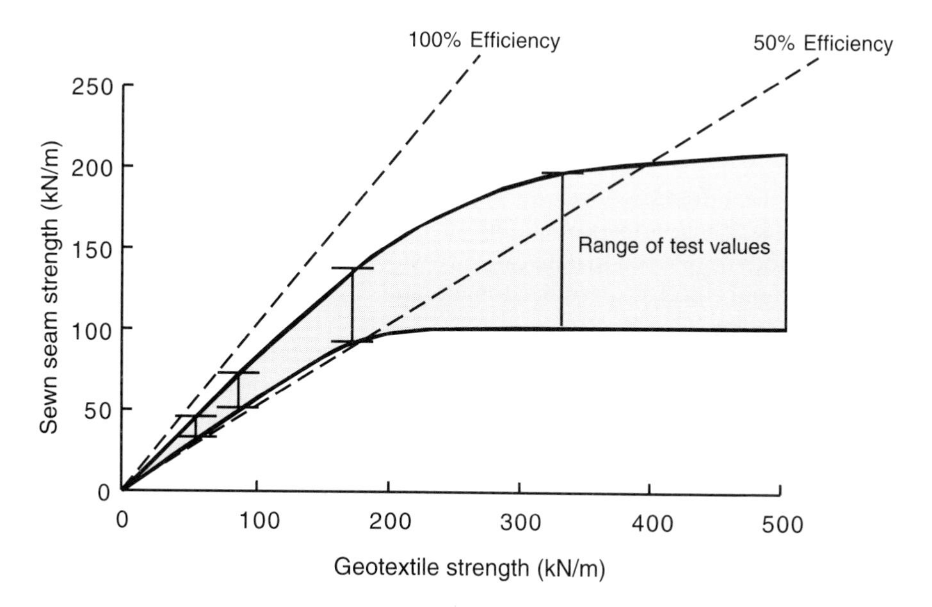

Figure 2.12 Field performance behavior of sewn geotextile seams.

Fatigue Strength. Fatigue strength or fatigue resistance is the ability of a geotextile to withstand repetitive loading before undergoing failure. A tensile test specimen, usually of a wide-width variety, is stressed longitudinally at a constant rate of extension to a predetermined load (less than failure) and then back to a lower or zero load. This cycling is repeated until failure occurs. The resulting cyclic stress-versus-strain response (i.e., the hysteresis loops) can be used to calculate a cyclic modulus that becomes evident after a number of load cycles are applied. Also important is the number of cycles required to bring the geotextile to failure and the respective loads that were applied. The load resulting in failure is converted to stress and this value is usually expressed as a fraction of the quasi-statically applied failure stress (strength) described under Tensile Strength. As expected, the lower the stress level, the larger the number of cycles required before failure.

Although many variables remain to be defined (primarily the decision as to what loads to apply during testing), the test reasonably simulates in situ conditions for applications such as seismic and railroad loadings, and wave or tidal action. Research in this area seems justified; see Ashmawy and Bourdeau [26].

Burst Strength. There are two test methods which stress geotextiles out of plane, thereby mobilizing tension until failure occurs. The most common is the Mullen burst test, which is covered in ASTM D3786. In this test, an inflatable rubber membrane is used to distort the geotextile into the shape of a hemisphere of 30 mm diameter. Bursting of the geotextile occurs when no further deformation is possible. The test is widely used for quality control. It is reported in most manufacturers literature and cited in many specifications.

The alternative test uses a large rectangular test specimen and deforms it by an underlying rubber membrane. Called a diaphragm test by Raumann [27], the central portion of the geotextile (along the minor axis) is very close to plane strain conditions. As such, the pressure-versus-strain response yields a very accurate modulus. It is a difficult test to setup and perform and the current tendency is to utilize wide width tensile tests of the type described previously.

Tear Tests. Often during their installation, geotextiles are subjected to tearing stresses. While a test simulating such situations is important, it will be seen that the methods developed to date can vary widely in their response. There are three tear tests commonly used: trapezoidal, tongue, and Elmendorf.

> *Trapezoidal tear test.* The trapezoidal tearing load is the force required to break individual yarns in a fabric. One such test was originally developed to test the failure of automotive fabrics due to screwdrivers in the back pockets of people sitting down. Since then, it has been discontinued for that use, but has been revised and modified for geotextiles. The new test is ASTM D4533, the trapezoidal tear test. In this test, the geotextile is inserted into a tensile testing machine on the bias, which causes the yarns to tear progressively. An initial 15 mm cut is made to start the process. The load actually stresses the individual yarns gripped in the clamps rather than stressing the fabric structure. The value is reported by all manufacturers and is commonly referred to in specifications as *trap tear*.

Tongue tear test. ASTM D751 uses a 75 mm × 200 mm geotextile specimen with a 75 mm long initiation cut. The geotextile is placed in a testing machine with the cut ends in the jaws of the machine. An increasing tensile force is applied to make the geotextile tear along the initiation cut. The test configuration permits the yarns to rope-up and work together to resist tear propagation. Thus the values resulting from tongue tear tests are usually much higher than those from trapezoidal tear tests.

Elmendorf tear test. The Elmendorf tear test is covered in ASTM D1424 and involves a procedure for the determination of the average force required to propagate a single-rip tongue-type tear starting from a premade cut in a woven geotextile. The cut is then continued by means of a falling pendulum apparatus. The tearing force is the force required to continue the tear previously started in the test specimen. The strength is calculated as the work done in tearing the specimen divided by twice the length of the tear. The test is often used in Europe to measure tear strength. It is generally not used for nonwoven geotextiles.

Impact Tests. Since falling objects can readily create holes in geotextiles (e.g., rocks, tools, and other construction items), a number of tests have been developed to assess the impact resistance of geotextiles. Often such tests use a weighted cone or dart falling freely from a known height above the geotextile. The geotextile is clamped firmly in an empty container such as a CBR mold. The amount of cone penetration into the geotextile is indicative of its resistance to impact stresses [28].

A different test, and one that measures impact resistance directly in energy units (joules), has been developed for an Elmendorf tear apparatus. The impacting cone is attached to the pendulum arm of the Elmendorf tear tester and penetrates the geotextile specimen, which is fixed on the end of the device. The fixture holding the geotextile test specimen is called a Spencer impact attachment. Impact resistance units are read directly from the system. Unfortunately, the limit of most commercially available systems is about 25 J, which is too low for many geotextiles. Thus it is necessary to use impact pendulum systems developed for other materials, such as metals, which have energies of up to 300 J. Such devices are covered under ASTM A370 and ASTM D256. The test specimen holder, however, must be converted to hold geotextiles rather than notched metal bars. See [29] for details and results from this type of test procedure.

Puncture Tests. In addition to a dynamic test, as just described for impact resistance, there is a need for an assessment of geotextile resistance to objects such as stones and stumps under quasi-static conditions. Such a test is described under ASTM D4833. This test uses a penetrating steel rod of 8.0 mm in diameter. The geotextile test specimen is firmly clamped in an empty cylinder with a 45 mm inside diameter and the rod pushed through it via a compression testing machine. Resistance to puncture is measured in force units.

This test is a popular one due to its simplicity and its ability to be automated. It is reported by all manufacturers and listed in most specifications. A considerably large database exists using this test method (e.g., see [29]). It is important to note the exact shape of the end of the metal rod. Three types are in current use: hemispherical, flat, and beveled flat. The interrelationships and differences between these types have not been identified. The last type, with a 0.8 mm 45° bevel around its circumference is specified in D4833.

The small size of the device described above is also of concern. For example, a lightweight nonwoven geotextile can selectively be chosen in a low-density fiber region or in a high-density fiber region. The differences in puncture resistance will be very large.

With such a concern in mind, a larger-size puncture test has been developed [29]. It uses a conventional soil-testing CBR plunger and mold. The penetrating steel rod is 50 mm in diameter and the geotextile is firmly clamped in an empty mold with a 150 mm inside diameter. The circumference of the plunger should be beveled 0.80 mm on a 45° angle so as not to cut the yarns at the edge of the rod. Shown in Table 2.4 are data from this type of test on both woven and nonwoven geotextiles. This test is formalized as ISO/DIS 12236 and in Germany as DIN 54307.

There is a direct relationship between the CBR puncture-resistance value and the wide-width tensile strength of geotextiles. This is because the geotextile between the inner edge of the specimen holder and the outer edge of the puncturing rod is in a state of pure axi-symmetric tension. Cazzuffi and Venesia [31] propose the following empirical equation as a correlation between the breaking force of the CBR test and the wide-width tensile strength for isotropic, nonwoven geotextiles

$$T_f = F_p/2\pi r \tag{2.12}$$

where

T_f = tensile force per unit width of fabric (kN/m),
F_p = puncture breaking force (kN), and
r = radius of the puncturing rod (m).

Both the German and Italian standards have correlations between the CBR test results and the wide-width tensile elongation of the geotextile. According to the German (DIN) standard, the tensile elongation at failure (ε_f) is calculated as:

$$\varepsilon_f = \frac{(x - a)}{a} \times 100 \tag{2.13}$$

where

x = diagonal elongation of the geosynthetic at failure (m), and
a = horizontal distance between the outer edge of the plunger and the inner edge of the mold (m).

The Italian (ENEL) standard uses the following equation to calculate the tensile elongation at failure:

$$\varepsilon_f = \frac{[\pi(R + r)x + \pi r^2 - \pi R^2]}{\pi R^2} \times 100 \tag{2.14}$$

TABLE 2.4 RELATIONSHIP BETWEEN CBR (PUNCTURE) STRENGTH AND WIDE-WIDTH TENSILE STRENGTH OF VARIOUS GEOTEXTILES*

Geotextile Designation	CBR Puncture Strength (kN)	Wide-Width Strength Using Eq. (2.12) (kN/m)	Measured Wide-Width Tensile Strength (kN/m)	Percent Variation (%)	CBR Calculated Elongation (%) Using Eqs. 2.13 and 2.14		Measured Wide-Width Elongation (%)	Percent Variation	
					DIN	ENEL		DIN	ENEL
WGT-1									
fill			36.0	38.1			23.5	15.3	23.4
warp			23.6	5.5			28.8	30.9	37.5
average	3.55	22.3	29.8	25.2	19.9	18.0	26.2	24.0	31.3
WGT-2									
fill			33.1	3.9			24.2	37.2	43.0
warp			50.0	36.4			13.0	16.9	6.2
average	5.06	31.8	41.5	23.4	15.2	13.8	18.6	18.3	25.8
WGT-3									
fill			17.5	41.7			23.0	—	—
warp			9.2	10.9			23.9	—	—
average	1.62	10.2	13.4	23.9	—	—	23.4	—	—
WGT-4									
fill			15.1	13.9			11.9	—	—
warp			10.8	20.4			11.4	—	—
average	2.07	13.0	13.0	0.0	—	—	11.6	—	—
WGT-5									
fill			24.2	17.4			11.6	—	—
warp			20.1	0.5			17.3	—	—
average	3.18	20.0	22.2	9.9	—	—	14.4	—	—
WGT-6									
fill			32.0	16.2			18.7	—	—
warp			30.1	11.0			13.5	—	—
average	4.27	26.8	31.0	13.5	—	—	16.1	—	—
NWGT-1									
MD			5.9	30.5			134.0	—	—
XMD			6.9	11.6			135.2	—	—
average	1.23	7.7	6.4	20.3	—	—	134.6	—	—
NWGT-2									
MD			14.4	22.9			152.8	—	—
XMD			20.7	14.5			162.2	—	—
average	2.72	17.7	17.5	1.1	—	—	157.5	—	—

TABLE 2.4 (continued)

Geotextile Designation	CBR Puncture Strength (kN)	Wide-Width Strength Using Eq. (2.12) (kN/m)	Measured Wide-Width Tensile Strength (kN/m)	Percent Variation (%)	CBR Calculated Elongation (%) Using Eqs. 2.13 and 2.14		Measured Wide-Width Elongation (%)	Percent Variation	
					DIN	ENEL		DIN	ENEL
NWGT-3									
MD			16.7	35.3			167.1	62.1	65.7
XMD			23.9	5.4			152.2	58.4	62.4
average	3.60	22.6	20.3	11.3	63.3	57.2	159.6	60.3	64.2
NWGT-4									
MD			6.5	12.3			44.7	17.9	25.7
XMD			8.1	9.9			60.4	39.2	45.0
average	1.16	7.3	7.3	0.0	36.7	33.2	52.5	30.1	36.8
NWGT-5									
MD			13.0	11.5			51.1	17.6	25.4
XMD			9.8	17.3			45.5	7.5	16.3
average	1.83	11.5	11.4	0.9	42.1	38.1	48.3	12.8	21.1

*Descriptions of the geotextiles tested in this study

Designation	*Type (mass per unit area)*
WGT-1	Woven monofilament polypropylene ($200/gm^2$)
WGT-2	Woven slit-film polypropylene ($240/gm^2$)
WGT-3	Woven monofilament polyvinyl chloride ($390/gm^2$)
WGT-4	Woven slit-film polypropylene ($100/gm^2$)
WGT-5	Woven slit-film polypropylene ($150/gm^2$)
WGT-6	Woven slit-film polypropylene ($200/gm^2$)
NWGT-1	Nonwoven needle-punched polypropylene ($200/gm^2$)
NWGT-2	Nonwoven needle-punched polypropylene ($400/gm^2$)
NWGT-3	Nonwoven needle-punched polypropylene ($600/gm^2$)
NWGT-4	Nonwoven heat-bonded polypropylene ($140/gm^2$)
NWGT-5	Nonwoven heat-bonded polypropylene ($200/gm^2$)

Source: After Murphy and Koerner [30].

where

R = radius of the mold (m), and
r = radius of the puncturing rod (m).

Included in Table 2.4 are the wide-width tensile strengths and elongation at failure for the same geotextiles that are tested in CBR (puncture) strength. Also included according to Eqs. (2.12), (2.13), and (2.14) are the calculated wide-width strengths and elongation at failure, the latter according to both DIN and ENEL equations. From this information a percent variation can be calculated.

It appears that the strength predictions are certainly reasonable, with the nonwovens being more accurate than the wovens. The variation in predicted elongation at failure has more scatter, but is still reasonable. Clearly, the CBR test for puncture strength or as a form of axi-symmetric tensile strength has considerable merit.

Friction Behavior. In many design problems it is necessary to know the soil-to-geotextile friction behavior. The most likely test setup is an adaptation of the direct shear test used in geotechnical engineering [32]. As shown in Figure 2.13a, the geotextile is firmly fixed to one-half of the test device with soil in the other half. After normal stress is applied, a shear force is mobilized until sliding occurs between the geotextile and the soil with no further increase in required shear force. When the test is repeated at different normal stresses, the data is plotted, a trend is established, as shown in Figure 2.13b, and the parameters of the Mohr-Coulomb failure criterion are obtained; recall Eq. (2.5a). From a comparison of the geotextile-to-soil response versus the soil-to-soil response the shear strength efficiencies on the soil's cohesion and friction angle can be obtained; recall Eqs. (2.5b), (2.6), and (2.7). Note that the soil's shear-strength parameters are the upper limit, that is, an efficiency of 1.0. Conversely, some interfaces may result in a drop from peak strength to a lower residual strength. This requires the plotting of a second curve (see Figure 2.13c) that will define the residual shear-strength parameters.

Results from such a test setup by Martin et al. [33] are presented in Table 2.5 for four geotextile types against three different cohesionless soils. Peak soil-to-geotextile friction angles are given, as well as the geotextile efficiency versus the peak soil-friction angle by itself, as per Eq. (2.7). Here it is seen that most geotextiles can mobilize a high percentage of the soil's friction and can be used to advantage in situations requiring this feature. A review and compilation of a number of direct shear tests on various geotextiles against different granular soils is given by Richards and Scott [34]; Williams and Houlihan [35] covers a wider range of soils, including some sands, silts, and mixed soils.

The ASTM D5321 direct shear test (as with many other geosynthetic standardization groups) calls for a shear box of 300 mm × 300 mm in size. While such a large test box is appropriate for geonets, geogrids, and many geocomposites, this author considers it to be excessive for geotextiles (and most certainly for geomembranes). Standard geotechnical engineering laboratory shear boxes (e.g., 100 mm × 100 mm), are satisfactory for geotextile testing and focus should be on more relevant shear-strength testing parameters, such as the following,

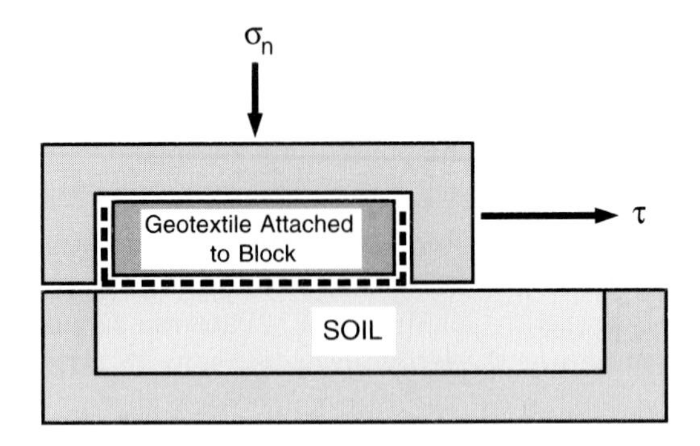

(a) Direct shear test device

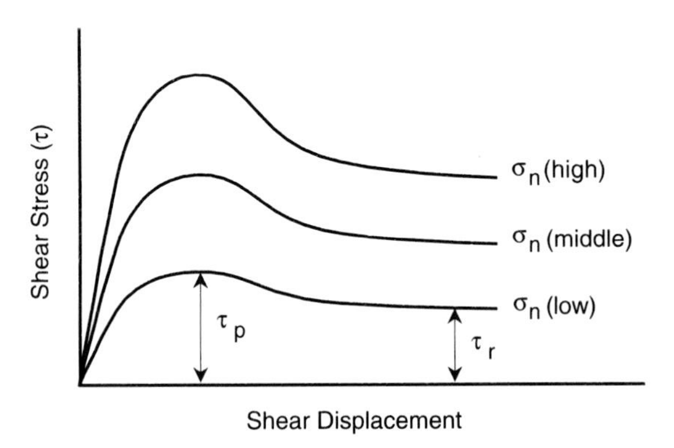

(b) Direct shear test data

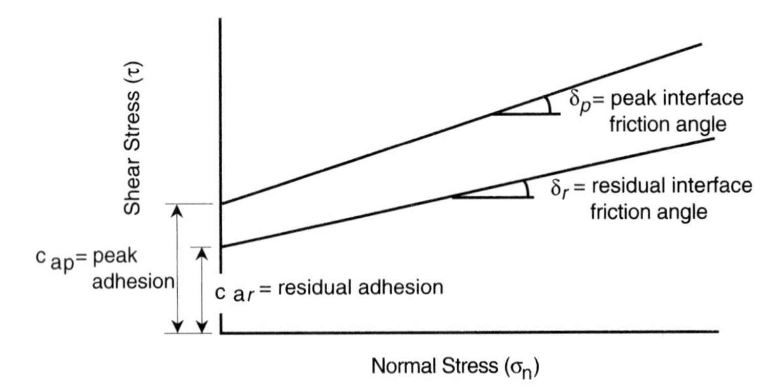

(c) Mohr-Coulomb stress space

Figure 2.13 Test setup and procedure to assess interface shear strengths involving geotextiles.

TABLE 2.5 PEAK SOIL-TO-GEOTEXTILE FRICTION ANGLES AND EFFICIENCIES IN SELECTED COHESIONLESS SOILS*

Geotextile type	Concrete sand (ϕ = 30°)	Rounded sand (ϕ = 28°)	Silty sand (ϕ = 26°)
Woven, monofilament	26° (84%)	—	—
Woven, slit-film	24° (77%)	24° (84%)	23° (87%)
Nonwoven, heat-bonded	26° (84%)	—	—
Nonwoven, needle-punched	30° (100%)	26° (92%)	25° (96%)

*Numbers in parentheses are the efficiencies. Values such as these should *not* be used in final design. Site specific geotextiles and soils must be individually tested and evaluated in accordance with the particular project conditions: saturation, type of liquid, normal stress, consolidation time, shear rate, displacement amount, and so on.

Source: After Martin et al. [33].

- Use of site-specific soil types
- Control of density and moisture content of the as-placed soil
- Geotextile fixity conditions
- Saturation conditions during consolidation and shear testing
- Actual type of saturation fluid (e.g., leachate)
- Use of field anticipated strain rates
- Adequate shear box deformation to achieve residual shear strength

than on the box size.

Pullout (Anchorage) Tests. Geotextiles are often called upon to provide anchorage for many applications within the reinforcement function. Such anchorage usually has the geotextile sandwiched between soil on either side. The resistance can be modeled in the laboratory using a pullout test, which will be discussed in Section 3.1.2 on anchorage strength in geogrids. The pullout resistance of the geotextile is obviously dependent on the normal force applied to the soil, which mobilizes shear forces on both surfaces of the geotextile.

Since the test greatly resembles a direct shear test, albeit with stationary soil on both sides of the tensioned geotextile, a possible design strategy is to take direct shear test results (for both sides of the geotextile) and use these values for pullout design purposes. However, this may not be a conservative practice.

Test results by Collios et al. [36] show a relationship of pullout test results to shear test results with some notable exceptions. For pullout testing, if the soil particles are smaller than the geotextile openings, efficiencies are high; if not, they can be low. In all cases, however, pullout test resistances are less than the sum of the direct shear test resistances. This is due to the fact that the geotextile is taut in the pullout test and exhibits large deformations. This, in turn, causes the soil particles to reorient themselves into a reduced shear strength mode at the soil-to-geotextile interfaces, resulting in lower pullout resistance. The stress state mobilized in this test is a very complex one requiring additional research.

2.3.4 Hydraulic Properties

Unlike the physical and mechanical properties just discussed, traditional tests on textile materials rarely have hydraulic applications; that is, the garment industry usually does not test for liquid flowing through clothing fabrics. As a result, hydraulic testing of geotextiles has required completely new and original test concepts, methods, devices, interpretation, and databases. Certainly, this is a lot to ask of a young technology, but it is nevertheless necessary and tremendously important. Both geotextile tests in-isolation and with soil will be described in this section.

Porosity. As conventionally defined with soils in geotechnical engineering, the porosity of a geotextile is the ratio of void volume to total volume. It is related to the ability of liquid to flow through or within the geotextile but is rarely measured directly. Instead, it is calculated from other properties of the geotextile.

$$n = 1 - \frac{m}{\rho t} \tag{2.15}$$

where

n = porosity (dimensionless),
m = mass per unit area (weight) (g/m^2),
ρ = density (g/m^3), and
t = thickness (m).

Intuitive in Eq. (2.15) is that for a given geotextile's weight and density, the porosity is directly related to thickness. Thickness, in turn, is related to the applied normal stress as shown in Figure 2.6.

Pore size can be measured by careful sieving with controlled-size glass beads (see the AOS test later in this section) using image analyzers [37] or by mercury intrusion [38]. Bhatia et al. [39] have compared these measurement techniques using a variety of geotextiles, illustrating behavioral trends and comparisons. The image analyzer results presented by ICI Fibers [40] for their various weights of geotextiles are instructive in showing that the pore size shifts gradually lower as the geotextile weight increases (see Figure 2.14a). McGown [41] has provided information of the same type comparing different geotextile manufacturing styles (see Figure 2.14b). These results are for the as-manufactured geotextile. However, geotextiles have pore sizes that are sensitive to changes in geotextile thickness due to applied normal stresses, as is typical of in situ conditions.

Percent Open Area. Percent open area (POA) is a geotextile property that has applicability only for woven monofilament geotextiles. POA is a comparison of the total open area (the void areas between adjacent yarns) to the total specimen area. A convenient way to measure the open area is to project a light through the geotextile onto a large poster-sized piece of cardboard. The magnified open spaces (resembling a window screen) can be mapped by a planimeter. Alternatively, a cardboard back-

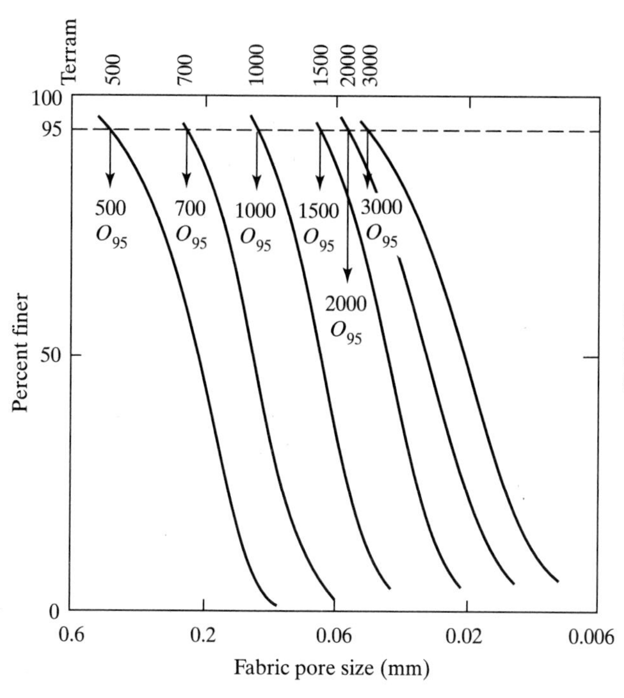

(a) Complete fabric pore size distribution for BBA Nonwovens Inc,geotextiles. Note that geotextile mass per unit area increases progressively from Terram 500 up to Terram 3000

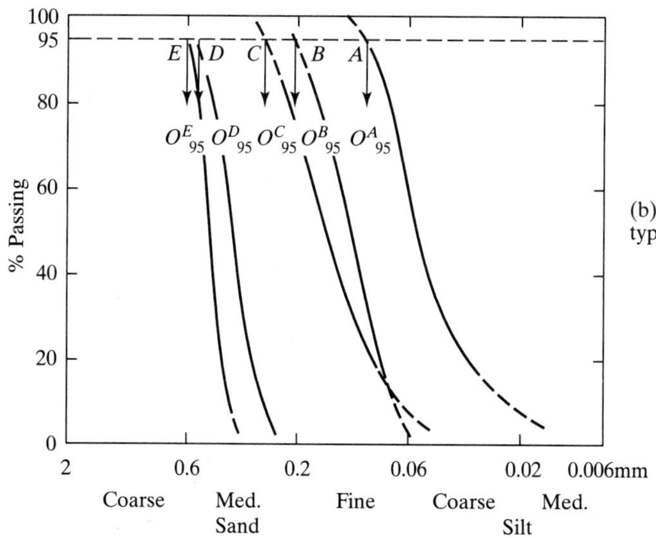

(b) Typical pore size distributions of different types of geotextiles

Legend

A. 250 gm/m^2 resin-bonded
B. 300 gm/m^2 needle-punched
C. 140 gm/m^2 heat-bonded
D. 380 gm/m^2 hessian woven
E. 185 gm/m^2 woven terylene

Figure 2.14 Complete fabric pore size distribution curves. (After McGown [41])

ground that is crosshatched like graph paper can be used. Here the squares are counted and summed up for the open area. The total area (yarns plus voids) must be measured at the same magnification as the voids measurement. A library-type microfiche reader can also be used. Woven monofilament geotextiles vary from essentially a closed structure (POA $\cong$ 0%) to one that is extremely open (POA = 36%), with many commercial woven monofilament geotextiles being in the range of 6 to 12%.

The test is not applicable to nonwovens, since the overlapping yarns block any light from passing directly through the geotextile. Thus a different test method is required to measure void sizes in nonwovens.

Apparent Opening Size (or Equivalent Opening Size). A test for measuring the apparent opening size (AOS) was developed by the U.S. Army Corps of Engineers to evaluate woven geotextiles. The test has since been extended to cover all geotextiles, including the nonwoven types. The AOS or equivalent opening size (EOS) (AOS and EOS are essentially equivalent terms) are defined in CW-02215 as the U.S. standard sieve number that has openings closest in size to the openings in the geotextile. The equivalent ASTM test is designated D4751. The test uses known-diameter glass beads and determines the 0_{95} size by standard dry sieving. Sieving is done using beads of successively larger diameters until the weight of beads passing through the test specimen is 5%. This defines the 0_{95}-size of the geotextile's openings in millimeters. Values of 0_{95} are indicated on the curves of Figures 2.14a and 2.14b. Note, however, that the 0_{95} value only defines one particular void size of the geotextile and not the total pore-size distribution. A conversion from the 0_{95} size in millimeters can then be made using Table 2.6

TABLE 2.6 CONVERSION OF U.S. STANDARD SIEVE SIZES TO EQUIVALENT SQUARE OPENING SIZES

Sieve Size (No.)	Opening (mm)
4	4.750
6	3.350
8	2.360
10	2.000
16	1.180
20	0.850
30	0.600
40	0.425
50	0.300
60	0.250
70	0.210
80	0.180
100	0.150
140	0.106
170	0.088
200	0.075
270	0.053
400	0.037

to obtain the closest U.S. sieve size and its number defines the AOS (or EOS) value. Thus AOS, EOS, and 0_{95} all refer to the same specific pore size, the difference being that AOS and EOS are sieve numbers, while 0_{95} is the corresponding sieve-opening size in millimeters. It should also be noted in the conversion on Table 2.6 that as the AOS sieve number increases, the 0_{95} particle size value decreases; that is, the numbers are inversely related to one another. In this book we will generally use the 0_{95} value since it is the target value for design purposes.

The AOS test is a poor test, having many problems, but the simplicity of the test and inertia seem to sustain its use in the United States. Some of the problems associated with the test are as follows:

- The glass beads can easily get trapped in the geotextile (particularly for thick nonwovens) and not pass through at all.
- Yarns in some geotextiles easily move with respect to one another (as they do in woven slit-film geotextiles), thereby allowing the beads to pass through an enlarged void not representative of the total geotextile test specimen.
- Reproducibility of the test is not good, with temperature, humidity, bead-size variation, and test duration all influencing the test results.
- The test is directed only at the 5% size (equivalent to the 95% passing size), which allows for determination of the 0_{95} size. The remainder of the pore-size curve is not defined.

As alternatives to the dry-sieving test just described, there are a number of wet-sieving methods. In Canada and France, a frame containing the geotextile specimen has a well-graded standard soil placed on it and the frame is repeatedly submerged in water. The soil fraction that escapes is analyzed and a d_{98}-equivalent particle size is obtained. In Germany, the setup is similar but a water spray is used. The soil fraction that escapes is analyzed and an effective opening diameter is calculated. The ISO/DIS 12956 test is also a wet-sieving test and will undoubtedly be seeing greater use than dry-sieving in the future. In general, these wet-sieving tests avoid many of the problems of dry sieving and are more representative of site conditions.

Permittivity (Cross-Plane Permeability). One of the major functions that geotextiles perform is filtration. (Note that most transportation agencies' specifications and some manufacturers' literature incorrectly call this "drainage.") In filtration, the liquid flows perpendicularly through the geotextile into crushed stone, a perforated pipe, or some other drainage system. It is important that the geotextile allow this flow to occur without being impeded. Hence the geotextile's cross-plane permeability must be quantified. As we discussed in the compressibility section, however, fabrics deform under load (recall Figure 2.6). Thus a new term, permittivity (Ψ), was previously defined in Eq. (2.8) (repeated here).

$$\Psi = \frac{k_n}{t} \tag{2.8}$$

where

Ψ = permittivity (s^{-1}),
k_n = permeability (properly called *hydraulic conductivity*) normal to the geotextile (m/s), and
t = thickness of the geotextile (m).

Eq. (2.8) is used in Darcy's formula as follows:

$$q = k_n iA = k_n \frac{\Delta h}{t} A$$

$$\frac{k_n}{t} = \Psi = \frac{q}{(\Delta h)(A)} \tag{2.16}$$

where

q = flow rate (m^3/s),
i = hydraulic gradient (dimensionless),
Δh = total head lost (m), and
A = total area of geotextile test specimen (m^2).

The formulation above is used for constant head tests in a manner identical to soil permeability testing. Typically, the flow rate q is measured at one value of Δh and then the test is repeated at different values of Δh. These different values of Δh produce correspondingly different values of q. When plotted (e.g., ΔhA versus q) the slope of the resulting straight line yields the desired value of Ψ.

The test can also be conducted using a falling (variable) head procedure as is also performed on soils. Here Darcy's formula is integrated over the head drop in an interval of time and used in the following equation:

$$\frac{k_n}{t} = \Psi = 2.3 \frac{a}{A \Delta t} \log_{10} \frac{h_o}{h_f} \tag{2.17}$$

where

Ψ = permittivity (s^{-1}),
a = area of water supply standpipe (m^2),
A = total area of geotextile test specimen (m^2),
Δt = time change between h_o and h_f (s),
h_o = head at beginning of test (m), and
h_f = head at end of test (m).

In either case, the permittivity can be multiplied by the geotextile thickness to obtain the traditional permeability value, if so desired.

If the permeating fluid is not water (e.g., is leachate or waste oil), compensation for differences in density and viscosity must be made (Hausmann [42]). This is done by using a ratio conversion as follows:

$$\Psi_f = \Psi_w \frac{\rho_f \mu_w}{\rho_w \mu_f} \tag{2.18}$$

where

Ψ_f = permittivity of the fluid under consideration,
Ψ_w = permittivity using water,
ρ_f = density of the fluid,
ρ_w = density of water,
μ_w = viscosity of water, and
μ_f = viscosity of the fluid.

The ASTM Method D4491 uses a device as shown in Figure 2.15 to measure the permittivity of geotextile test specimens. It is similar to ISO/DIS 11058. Either constant head or falling head can be used, although the standard is written around the constant

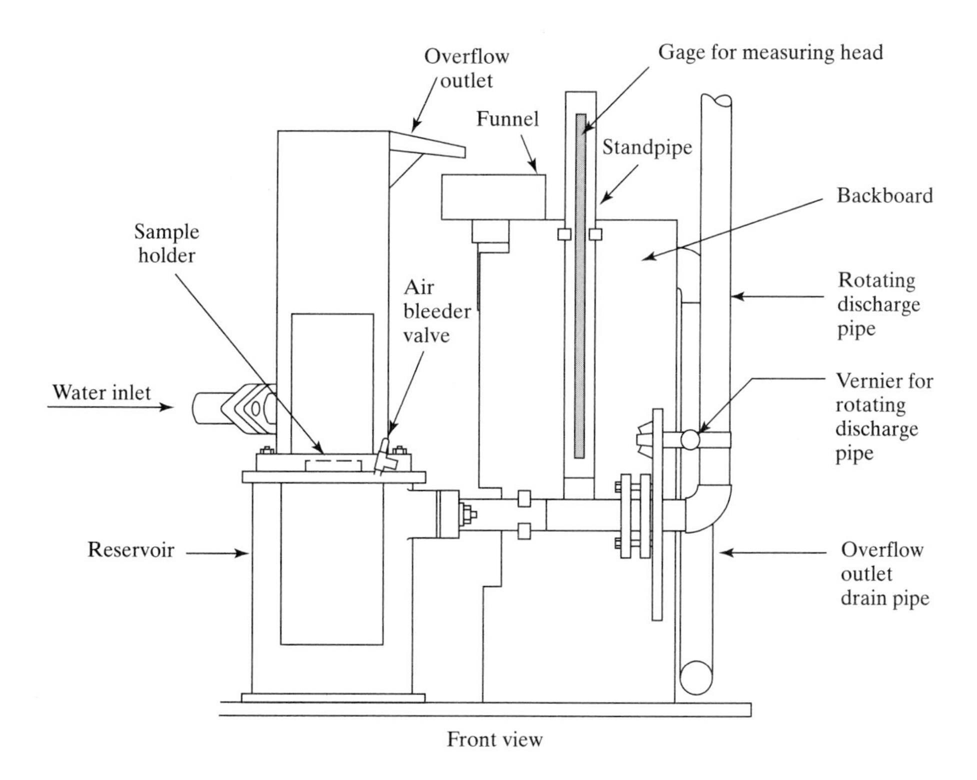

Figure 2.15 Permeability device for measuring geotextile permittivity (cross-plane flow).

head test at a head of 50 mm. As with the permeability of soils, geotextile values of permittivity (and permeability) range over several orders of magnitude,

Permittivity, Ψ: from 0.02 to 2.2 s^{-1}
Permeability, k_n: from 8×10^{-6} to 2×10^{-3} m/s

Some important test considerations are preconditioning of the fabric, temperature, and the use of de-aired water. ASTM D4491 requires a dissolved oxygen content of less than 6.0 mg/l. Tap water is allowed unless disputes arise, in which case de-ionized water should be used. Note that conventional soil-testing permeameters cannot be used to test geotextiles, since the size of their water outlets is rarely large enough to handle the flow coming through most geotextiles. The testing of soil and geotextile systems for long-term flow compatibility will be treated later under endurance properties in Section 2.3.5.

Permittivity Under Load. The previously described permittivity test has the geotextile test specimen under zero normal stress, a situation rarely encountered in the field. To make the test more performance-oriented, numerous attempts to construct a permittivity-under-load device have been made. Generally a number of layers of geotextile (from 2 to 5 layers) are placed upon one another with an open-mesh stainless steel grid on top and bottom. This assembly is placed inside a permeameter and loaded normally via ceramic balls of approximately 12 mm diameter. Thus normal stress is imposed on the geotextile, but flow is only nominally restricted. Loading by soil itself (which would definitely affect flow) is completed avoided. The test has been standardized by ASTM as D5493, and results seem to indicate the following trends in the relationship between standard permittivity and permittivity-under-load:

woven monofilament geotextiles:	no change to a slight increase when under load.
woven slit-film geotextiles:	data scatter is too large to establish trends.
nonwoven heat-bonded geotextiles:	no change to slight decrease when under load.
nonwoven needle-punched geotextiles:	slight decrease to moderate decrease, depending on magnitude of load and weight of geotextile.

Transmissivity (In-Plane Permeability). For the flow of water within the plane of the geotextile (e.g., in the utilization of the drainage function), the variation in geotextile thickness (its compressibility under load) is again a major issue. Thus transmissivity (θ) was introduced in Eq. (2.10); it is used in Darcy's formula as follows:

$$q = k_p iA = k_p i(W \times t) \tag{2.19}$$

$$k_p t = \theta = \frac{q}{iW} \tag{2.20}$$

where

θ = transmissivity of the geotextile (m^2/s or $m^3/s\text{-}m$),

k_p = permeability (hydraulic conductivity) in the plane of the geotextile (m/s),
t = thickness of the geotextile (m),
q = flow rate (m^3/s),
W = width of the geotextile (m),
i = hydraulic gradient (dimensionless) = $\Delta h/L$,
Δh = total head lost (m), and
L = length of the geotextile (m).

If the permeating fluid is not water, (e.g., is turbid water or leachate), the density and viscosity can be accommodated by using Eq. (2.18). Also, note in Eq. (2.20) that θ and q/W carry the same units but are numerically equal *only* at a hydraulic gradient i of unity.

A number of test devices are configured to model the above formulation, where liquid (usually water) flows in the plane of the geotextile (of dimensions $L \times W \times t$) in a parallel flow trajectory; ASTM D4716 and ISO/DIS 12958 use such a device. Koerner and Bove [43] provide a review of such devices, and Table 2.7 gives typical values. These test devices are necessary for high-flow geonets and drainage geocomposites (discussed in Chapters 4 and 8, respectively), but are somewhat awkward for geotextiles. Such devices are large, time-consuming to setup, and difficult to seal against sidewall leaks. This last item is particularly important for geosynthetics with relatively low transmissivity values such as geotextiles. A low flow-rate measurement, with a high (and unknown) potential for leakage results in a relatively uncertain value for transmissivity. Instead, a variation is recommended for geotextile testing whereby radial drainage is achieved. Schematically such a device is shown in Figure 2.16, where liquid enters into the inside of the upper load bonnet, then flows radially through the geotextile and is collected around the outer perimeter of the stationary reaction section of the device. The theory is adapted as follows:

$$q = k_p iA = k_p \frac{dh}{dr}(2\pi rt) \tag{2.21a}$$

$$2\pi(k_p t)\int dh = q\int \frac{dr}{r}$$

$$(k_p t) = \theta = \frac{q \ln (r_2/r_1)}{2\pi\Delta h} \tag{2.21b}$$

TABLE 2.7 TYPICAL VALUES OF TRANSMISSIVITY (IN-PLANE DRAINAGE CAPABILITY) OF GEOTEXTILES*

Type of Geotextile	Transmissivity m^2/s	Permeability Coefficient m/s
Nonwoven, heat-bonded	3.0×10^{-9}	6×10^{-6}
Woven, slit-film	1.2×10^{-8}	2×10^{-5}
Woven, monofilament	3.0×10^{-8}	4×10^{-5}
Nonwoven, needle-punched	2.0×10^{-6}	4×10^{-4}

*Values taken at applied normal stress of 40 kPa.

Source: Data after Gerry and Raymond [14].

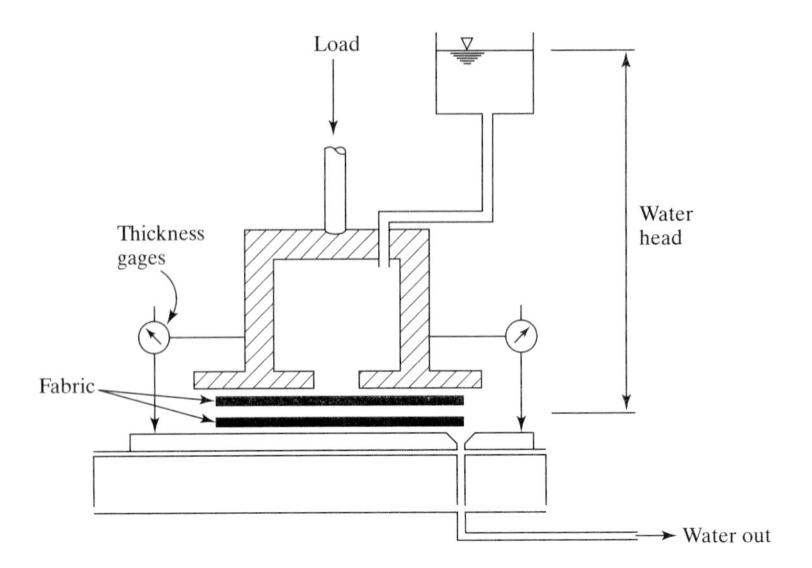

Figure 2.16 Permeability device for measuring geotextile transmissivity (radial in-plane flow).

where

r_2 = outer radius of the geotextile test specimen, and
r_1 = inner radius of the geotextile test specimen.

The thicker nonwoven geotextiles are best suited to convey water in the drainage function, but these are the same geotextiles that are subject to relatively high compression under load. Thus the exponential decrease in transmissivity of the geotextiles shown in Figure 2.17 should come as no surprise. Fortunately, most geotextiles reach constant values after approximately 85 kPa; beyond this the yarn structure is sufficiently tight and dense to hold the load and still convey liquid to the extent shown. Note also the increase in transmissivity with increasing mass per unit area (or number of layers).

Lastly, it should be noted that the radial device shown in Figure 2.16 can readily be adapted to measure the flow of gases (e.g., air, methane, and radon) by placing a shroud around the outside of the load bonnet. The gas is introduced under controlled pressure and measured at the outlet for its flow rate. Typical air transmissivity data is given in Figure 2.18a. The same device can be used under combined air flow through partially saturated geotextiles to assess *permselectivity*, as shown in Figure 2.18b.

Soil Retention: Underwater Turbidity Curtains. One variation of the soil-retention test is directed primarily toward the use of geotextiles as *underwater turbidity (or silt) curtains.* The test device consists of two rectangular tanks that are placed end to end with slide gates facing one another. Between these two slide gates is the geotextile test specimen. The upstream tank (with the slide gate closed) is filled with water that has a known amount of uniformly mixed silt in it. The gate valve at the exit end of the downstream tank is opened. The test begins when the slide gates on each side of the

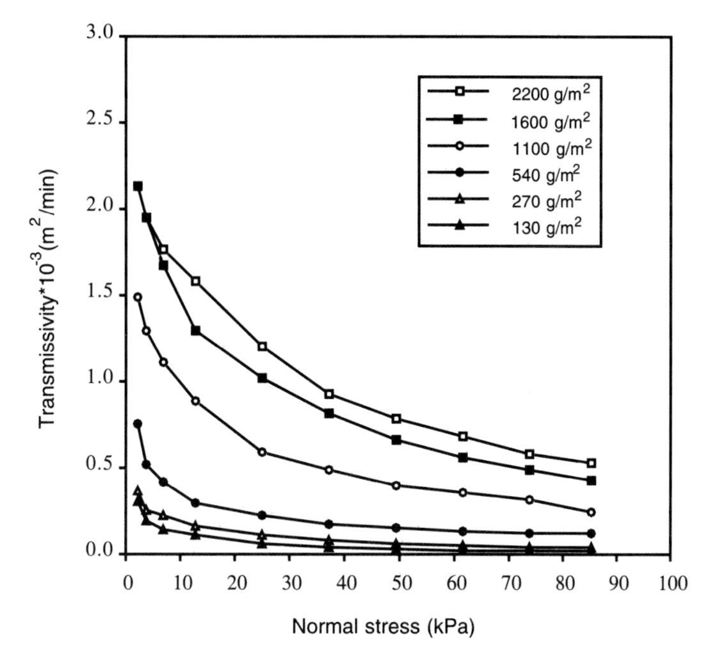

Figure 2.17 Transmissivity test results for different mass per unit area of nonwoven needle-punched geotextiles.

geotextile are lifted up, allowing the (turbid) water to pass through the geotextile, which is acting as a submerged soil filter. Clear water is continually added to the upstream tank to maintain a constant head. Two values are generated:

1. The flow rate and velocity through the geotextile, which is indicative of its void space and the amount of clogging that occurs.
2. The percent of solids passing through the geotextile during the test process, which is indicative of the geotextile's retention capability.

This test, developed by the New York Department of Transportation, is aimed at the reduction in turbidity of rivers and streams during adjacent construction activities.

Soil Retention: Above-Ground Silt Fences. A second variation of the soil-retention test is directed toward the use of geotextiles as *above-ground silt fences.* The test protocol and setup is covered in ASTM D5141. Here the soil (usually a silty sand) is slurried in water and poured into a flume box 1200 mm long × 800 mm wide × 300 mm high. The candidate geotextile, 800 mm × 300 mm, forms the downstream end of the box, which is at an 8% slope. The flow rate of the soil-water mixture is monitored with time and the amount of fines passing through the geotextile is measured to determine the soil retention capability. The process is repeated at least three times to determine the degree of clogging that occurs. Two values are generally reported: the slurry flow rate (l/min-m) and the retention efficiency (%). The recommended procedure, developed by the Virginia Department of Transportation under designation VTM-51, also includes a field method with the same objective of determining the filtering efficiency of the geotextile.

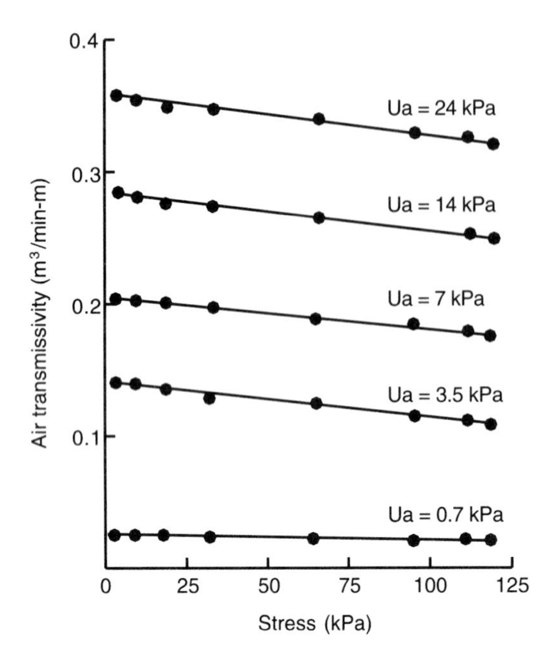

(a) Air transmissivity data

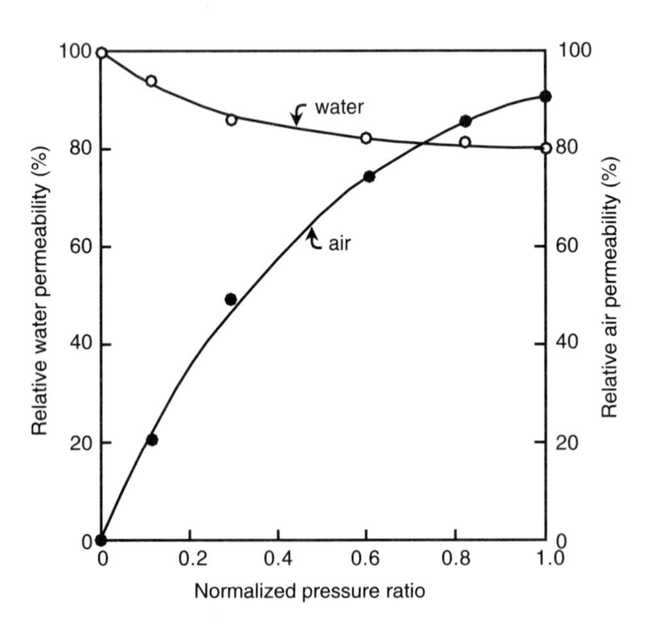

(b) Air versus water transmissivity interaction under 120 kPa normal pressure

Figure 2.18 Radial transmissivity data for air and air/water mixtures on a 550 g/m^2 needle-punched nonwoven geotextile. (After Koerner et al. [44])

2.3.5 Endurance Properties

Thus far the testing of geotextiles has concentrated on the short-term material behavior of the as-manufactured fabric. Yet, the question remains regarding their behavior during service conditions over the design lifetime of the system; in other words, *endurance* is of concern. This section addresses some of the tests that focus on this question. The reader should also see ASTM D5819, which is a guide for selecting durability test methods.

Installation Damage. It should be recognized that harsh installation stresses can cause geotextile damage. In some cases installation stresses might be more severe than the actual design stresses for which the geotextile is intended. There are a number of studies available, but most involve the removal of the geotextile after a considerable time, usually years. In this section, focus is on the immediate damage caused by the contractors' operations during installation.

In order to assess this, 100 field sites were evaluated by removing approximately 1.0 m^2 of the geotextile within hours after placement. The procedure followed closely the ISO/DIS 13437 protocol. Most of the geotextiles were used for highway base separation, but some were for embankments, walls, underdrains, erosion control, staging areas, access ways, and so on. The entire exhumed sample was brought into the laboratory along with an equal size of the unused and uninstalled geotextile for comparison purposes. Test specimens were taken from both the exhumed and unused geotextiles, and were tested. A percent strength reduction was calculated. The following mechanical tests were performed:

- Grab tensile (3 to 6 tests per site)
- Puncture resistance (3 to 6 tests per site)
- Trapezoidal-tear resistance (3 to 6 tests per site)
- Burst resistance (3 to 6 tests per site)
- Wide-width strength in machine direction (1 to 2 tests per site)
- Wide-width strength in cross-machine direction (1 to 2 tests per site)

A hole assessment was also made of the exhumed geotextiles. In Figure 2.19, the relationship between the retained strength (the weighted average of the above tests) and number of holes per square meter is presented. The entire data set was arbitrarily divided into three groups: acceptable, questionable and nonacceptable (Regions A, B, and C, respectively, in Figure 2.19b). While loss of strength can be accommodated via a suitable installation-damage reduction factor (the inverse of the numeric strength-retained value shown in Figure 2.19), we could well wonder how the separation and filtration functions can properly work with the occurrence of so many holes. It should be noted that the recommendations resulting from this study (Koerner and Koerner [45]), suggests that no geotextile less than 270 g/m^2 should be used unless special precautions are taken, such as a sand cushioning layer along with lightweight construction equipment, to avoid installation damage.

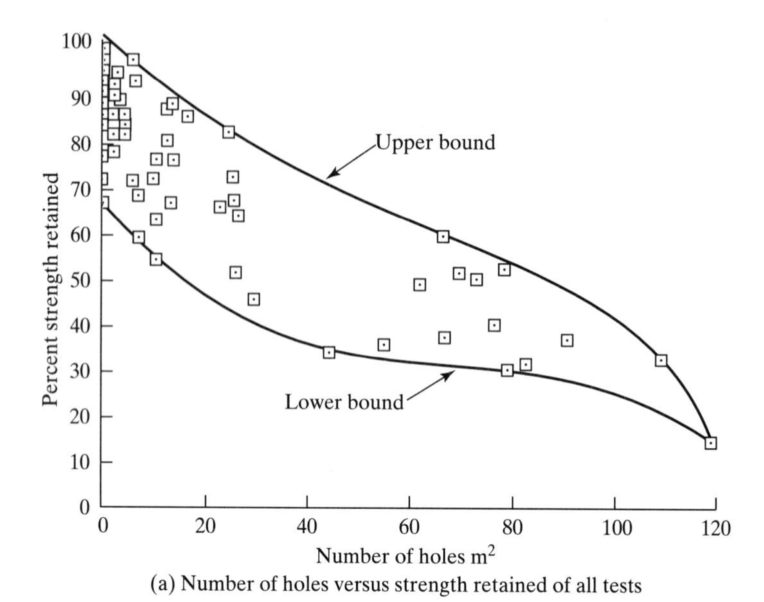

(a) Number of holes versus strength retained of all tests

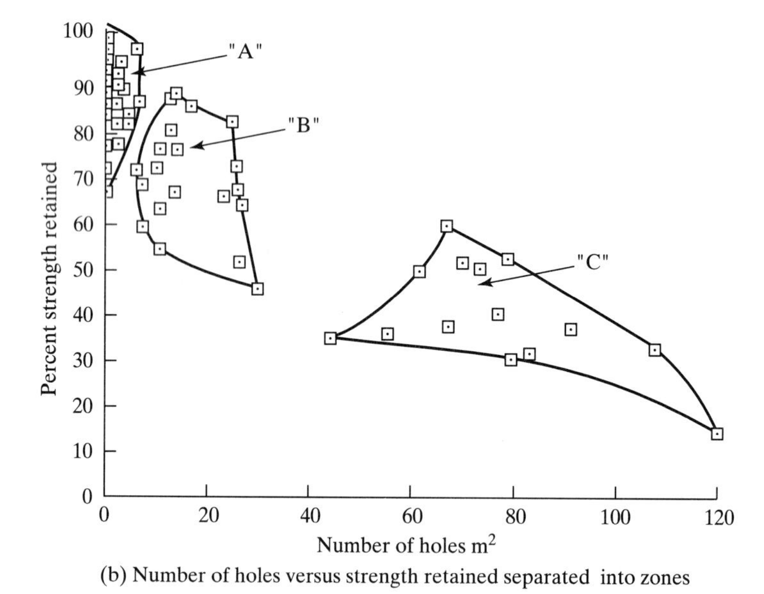

(b) Number of holes versus strength retained separated into zones

Figure 2.19 Results of field exhuming of geotextiles immediately after installation to assess installation damage. (After Koerner and Koerner [45])

Creep (Constant Stress) Tests. *Creep* is the name commonly applied to the elongation of a geotextile under constant load. Since polymers are generally considered creep-sensitive materials, it is an important property to evaluate. The test specimens should be of the wide-width variety (recall Figure 2.8c) and are usually stressed by means of stationary hanging weights. Since the test duration should be long, a number of tests are often conducted simultaneously by cascading the test specimens and their respective loads. The setup can also be horizontal with a number of specimens connected to one another.

Selection of the load is important and it is usually based on a percentage of the geotextile's strength as determined from a conventional test as described in Section 2.3.3. If such a value is considered to be 100%, creep test stresses of 20%, 40%, and 60% are sometimes evaluated. Stresses are commonly applied for 1000 to 10,000 (depending on the particular application) and readings are taken at progressively longer time increments from the beginning of the test (e.g., 1, 2, 5, 10, and 30 min; then 1, 2, 5, 10, 30, 100, 200, 500, 750, and 1000 hr). For creep tests longer than 1000 hr, readings every 500 hr are usually adequate. The elongation or percent strain versus time should be plotted for each stress increment. Both ASTM D5262 and ISO/DIS 13431 describe details of the testing procedure. Creep rates are then calculated from these response curves. Although completely arbitrary, the slope of the terminal portion of the curve (in mm/hr) is then reported. The resulting values can then be empirically compared to maximum allowable values, or used directly in a predictive procedure as described by Shrestha and Bell [46].

Numerous studies are available on the creep behavior of geotextile-forming yarns and fabrics. Perhaps the greatest sensitivity is due to stress level and polymer type (see Figure 2.20). Such information is very important in design, since the inverse of the percentage of quasi-static strength at which no creep occurs is used as a value for the reduction factor necessary to avoid objectionable creep deformation. Values from den Hoedt [47] shown in Figure 2.20 and other collected values from the literature are given in Table 2.8. Care in using these values is suggested, however, since the tests should be geotextile-specific (e.g., polymer type, polymer processing, geotextile type), have similar environmental conditions (e.g., temperature, moisture), model the in situ stresses (e.g., confinement), and be adjusted for the anticipated length of service lifetime. Typically the lower values in Table 2.8 are for short lifetimes while the higher numbers are for long lifetimes.

An emerging procedure that shortens the time needed to obtain creep data of the type just described is to perform *creep-rupture* tests. In this test, loadings are high enough to enter the tertiary creep range of the material whereby actual creep failure occurs (see Ingold et al. [50]). Research currently centers on the adequacy of simulating long-term behavior in this manner and the development of other accelerated techniques.

Confined Creep Tests. As with the confined wide-width tensile test device shown in Figure 2.10, McGown et al. [22, 23] have also performed confined creep tests in such a device. Some of the results obtained are shown in Figure 2.21. The general tendency of these tests is to show that the creep behavior is improved with soil confinement. As with the short-term confined tensile tests of Section 2.3.3, the major improvement is with nonwoven needle-punched geotextiles, followed by other nonwovens, and then by woven geotextiles, the last appearing to show little if any improvement. While expensive and time-consuming to perform, such tests are important

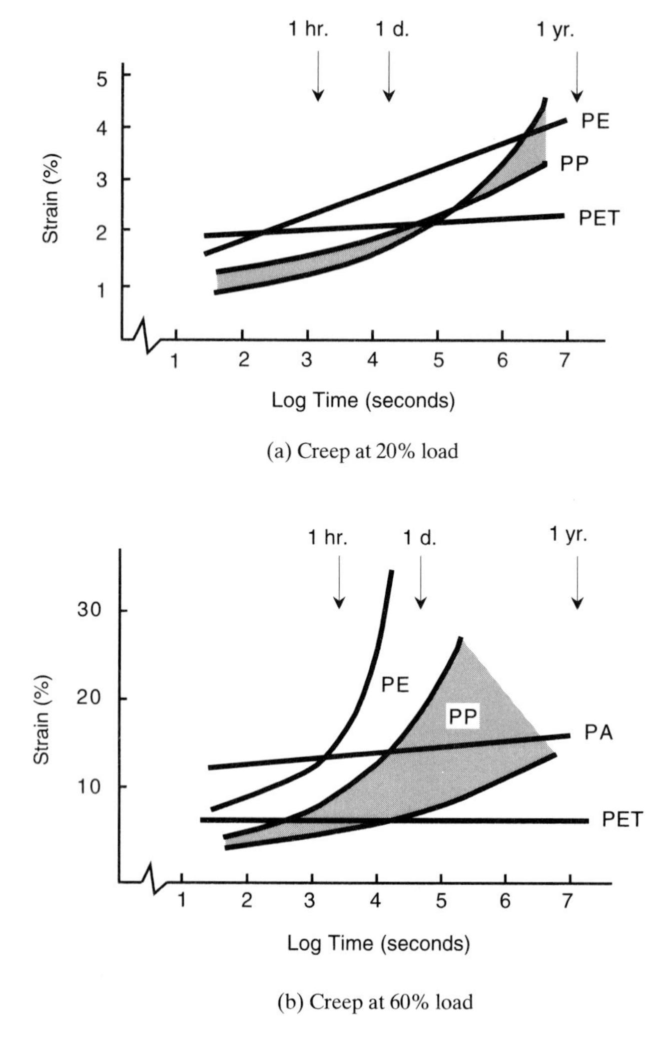

(a) Creep at 20% load

(b) Creep at 60% load

Figure 2.20 Results of creep tests on various yarns of different polymers. (After den Hoedt [47])

to set realistic creep-reduction factors. This implies that the values given in Table 2.8 may be upper-bound values.

Stress Relaxation (Constant Strain) Tests. *Stress relaxation* is the name commonly given to the reduction in stress in a geotextile while it is under a constant deformation. As with creep (to which it is mathematically related), stress relaxation is an important property. Unfortunately, the test setup is difficult, requiring load cells and electronic strain measurements and, as with creep, taking considerable time. The litera-

TABLE 2.8 SELECTED VALUES OF GEOTEXTILE REDUCTION FACTORS AGAINST CREEP DEFORMATION*

Geotextile Fiber/Yarn Type	den Hoedt [47]	Lawson [48]	Task Force #27 [49]	Koerner
Polypropylene†	4.0	2.5 to 5.0	5.0	3.0 to 4.0
Polyethylene†	4.0	2.5 to 5.0	5.0	3.0 to 4.0
Polyamide	2.5	1.5 to 2.5	2.9	2.0 to 2.5
Polyester	2.0	1.5 to 2.5	2.5	2.0 to 2.5

*These values are for use in avoiding creep deformation completely (i.e., the zero-creep condition).

†Refers to polyolefin geotextile yarns, *not* to oriented homogeneous geogrids, which are less sensitive to creep and are discussed in Chapter 3.

ture is notably silent on this property in regard to geotextiles; however, there are geogrid and geomembrane data available [51], which will be reviewed in Chapters 3 and 5.

Abrasion Tests. The abrasion of geotextiles when in service can be the cause of the failure of soil-geotextile systems. The ASTM test methods for abrasion resistance of textile fabrics are designated D1175 and cover six procedures: inflated diaphragm; flexing and abrasion; oscillatory cylinder; rotary platform, double head; uniform abrasion; and impeller tumble. In all cases, abrasion is defined as "the wearing away of any part of a material by rubbing against another surface." There are, however, a large number of variables to be considered in such a test. Results are reported as the percent weight lost or strength/elongation retained under the specified test and its particular conditions.

One of the tests for evaluating the abrasion resistance of geotextiles is the rotary platform, double head (Taber Test, Model 503) method. In the test, both heads are fitted with 1000 g weights and vitrified (CS-17) abrasion wheels. The test specimen is disk-shaped with a 90 mm outer diameter and a 60 mm inner diameter. The specimen is placed on a rubber base on the platform, which is rotated and abraded by the stationary abrasion wheels for up to 1000 cycles. Two strip tensile specimens are cut from the abraded geotextile and tested for their tensile strengths. The average value is then compared to the tensile strength of the nonabraded geotextile and the results are reported as a percentage of strength retained by the geotextile after abrasion. The percentage of elongation retained after abrasion can also be reported. Figure 2.22 gives the results for different types of geotextiles, all of which are approximately 200 g/m^2.

While this particular test is straightforward and many devices are commercially available to perform it, the ASTM and ISO preference is for a uniform (sand paper) abrasion test . The descriptions for this test are D4886 and 13427, respectively.

All of these tests, however, are questionable simulators of field abrasion conditions. In many cases it would be better to use some sort of tumble test, such as the German Test Standard DIN 5385, which is a large test using basalt-stone aggregate abrading a geotextile test specimen within a 1.0 m diameter rotating drum.

Long-Term Flow (Clogging) Test. One of the greatest endurance concerns is that of the long-term flow capability of a geotextile with respect to the hydraulic load

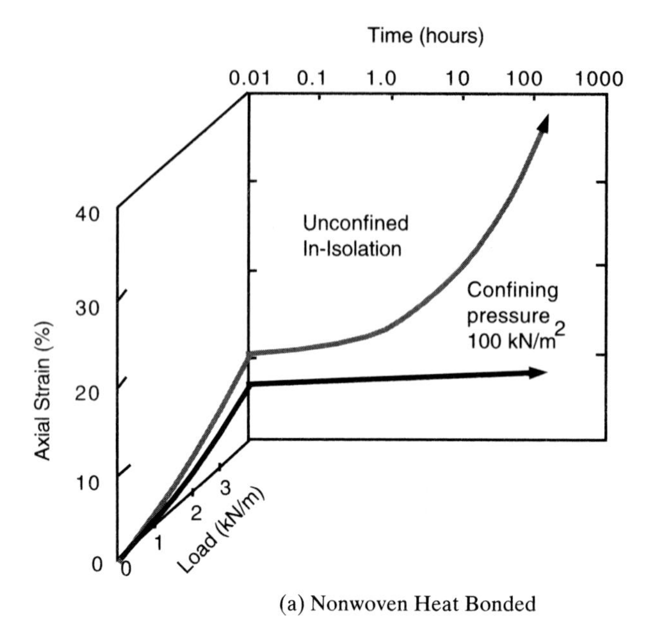

(a) Nonwoven Heat Bonded

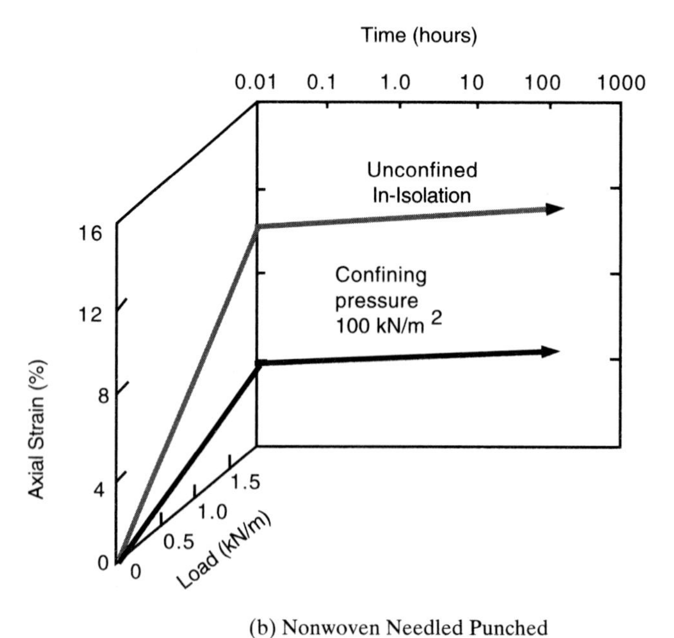

(b) Nonwoven Needled Punched

Figure 2.21 Results of confined creep tests on various geotextiles. (After McGown et al. [22])

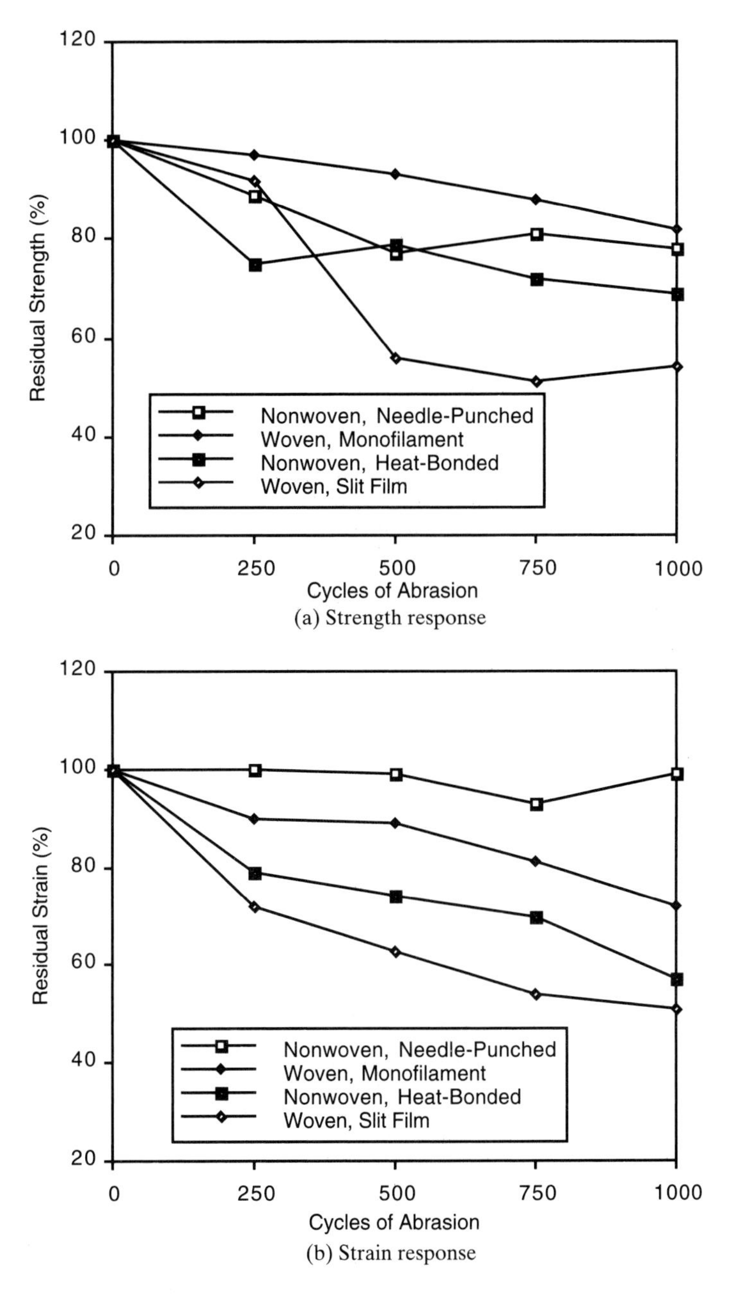

Figure 2.22 Abrasion test results for various geotextiles using the Taber test device model 503.

coming from the upstream soil. Tests are needed to assess the potential of *excessive* geotextile clogging.

The most direct testing approach is to take a sample of the soil at the site and place it above the candidate geotextile, which is fixed in position in a test cylinder. It is

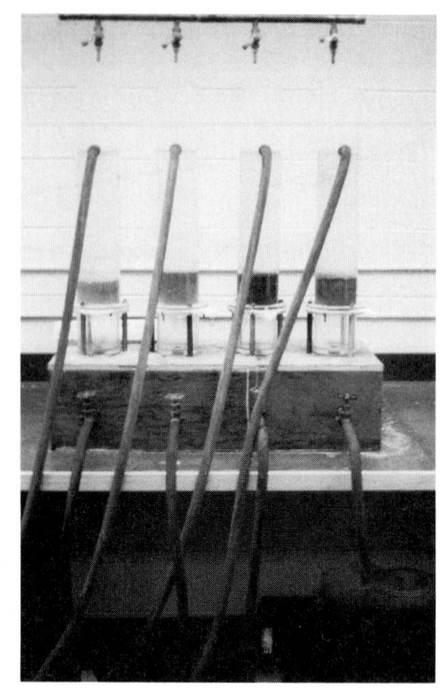

(a) Photograph of setup

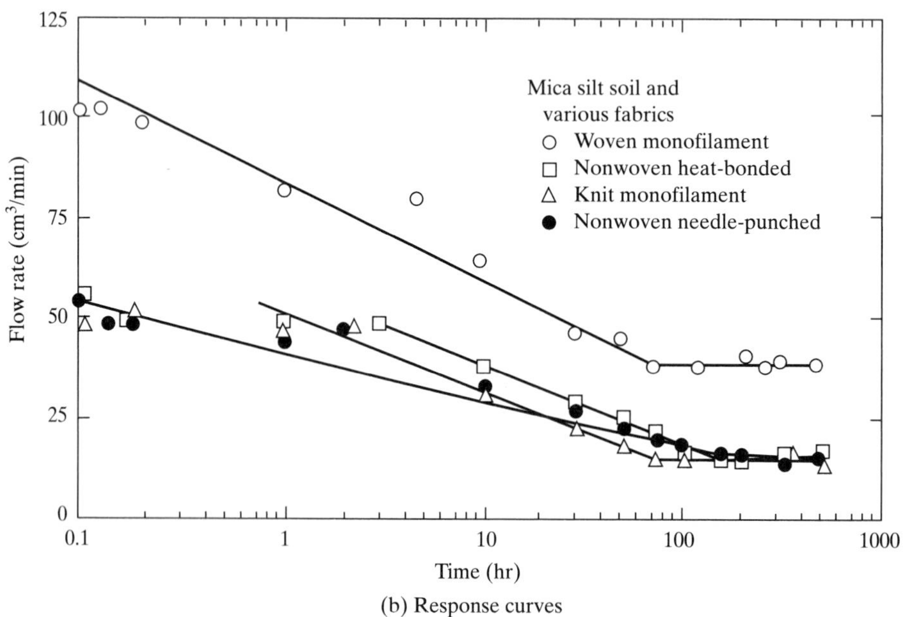

(b) Response curves

Figure 2.23 Long-term flow tests on soil-geotextile systems and typical response curves. (After Koerner and Ko [52])

then evaluated under constant head flow over a long period of time. A set of such test units can be built and a series of candidate geotextiles tested simultaneously (see Figure 2.23a). The general result is piecewise linear, with the initial portion due largely to compaction of the soil and therefore not of direct interest. At a transition time of approximately 10 hr (for granular soils) to 200 hr (for fine-grained soils), the soil-geotextile system will enter its field-simulated behavior. If the slope of the response curve becomes zero after this transition time, the geotextile is compatible with the soil, at least under the imposed test conditions (hydraulic gradient, temperature, water, etc.; see Figure 2.23b). Assuming that the flow rate value at equilibrium is adequate for the situation, the candidate geotextile(s) should be appropriate. If the slope continues to be negative, however, increased clogging is indicated and eventually excessive clogging could occur. In such a case, the geotextile may not be suited for this type of soil and these test conditions. The database for this particular test has been extended for both clear and turbid water using a number of soil-to-geotextile conditions [53].

Although seemingly straightforward in its approach, there are drawbacks to this test, the major one being time. The test should normally run for 1000 hr ($\cong$ 40 days) to establish the slope of the curve beyond the transition time. Note that the time axis in Figure 2.23b is logarithmic. This is unfortunately too long for many real-time situations, where an answer regarding potential clogging is usually needed within a few days. Also, the test chamber can develop bacteria growth in a warm laboratory environment during the required test time and periodic flushing with a detergent is necessary; this could cause changes in the soil-geotextile system. Finally, the question of de-aired and/or deionized water must always be addressed in hydraulic tests of this type.

Gradient Ratio (Clogging) Test. A test that may be performed in a considerably shorter time than the long-term flow test and that is aimed at determining the hydraulic compatibility of a soil-geotextile system is the U.S. Army Corps of Engineers Gradient Ratio Test CW-02215. It has been adopted (with slight variations) by ASTM as the D5101 test method. In this test, the flow configuration setup is similar to the long-term flow test just described. Now, however, instead of measuring flow rates, the hydraulic head at various locations in the soil-geotextile column is measured. Head differences are then converted to hydraulic gradients and finally the GR value, as defined below, is calculated.

$$GR = \frac{\Delta h_{GT+25S}/t_{GT+25}}{\Delta h_{50S}/t_{50}} \tag{2.22}$$

where

Δh_{GT+25S} = head change (mm) from the bottom of the geotextile to 25 mm of soil above the geotextile,

t_{GT+25} = geotextile thickness (mm) plus 25 mm of soil,

Δh_{50S} = head change (mm) between 50 mm of soil above the geotextile, and

t_{50} = 50 mm of soil.

The Army Corps of Engineers suggests that gradient ratio values greater than 3.0 indicate nonacceptable geotextiles for the type of soil under test. Figure 2.24 gives data illustrating various combinations of soil types and geotextiles. The soil types were systematically varied from an "ideal" rounded sand (Ottawa test sand) to controlled mixtures of sand and cohesionless silt, by varying the percentage of silt added; that is, a gap-graded soil of increasing silt content was created. When different geotextile types

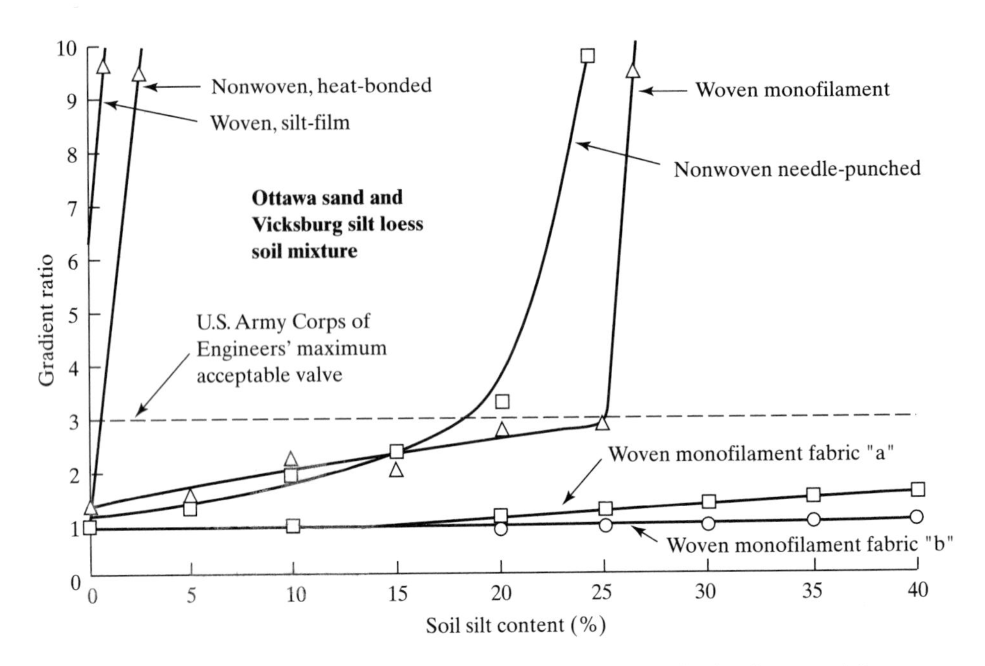

Figure 2.24 Gradient ratio test data used to illustrate geotextile clogging potential. (After Haliburton and Wood [7])

were evaluated with each soil type, the GR response was measured. It is easily seen that the nonwovens and woven slit-film geotextiles failed (GR > 3.0) as higher percentages of silt were added. In contrast, the woven monofilament geotextiles behaved nicely (GR values < 3.0). This type of response is powerful in supporting the use of woven monofilament geotextiles for critical hydraulic applications. However, these are very severe test conditions in which high hydraulic gradients, cohesionless soils, and gap-graded particle size distributions are present. These three conditions appear to lead to excessive soil-clogging when using certain types of geotextiles. Also in these test results, it is important to note that Haliburton and Wood [7] did not report on the amount of silt that passed through the high open-area woven geotextiles that have the low GR values. How the downstream drainage system accommodates the fines that are carried through the geotextile and the possible lack of stability of the upstream soil are both important site-specific design issues.

The test is not without its share of problems and complications, including long-term stability of the gradient ratio value [8], piping along the test cylinder walls, use of de-aired or de-ionized water, and air pockets in the soil, geotextile, and head monitoring system.

Hydraulic Conductivity Ratio (Clogging) Test. Williams and Abouzakhm [54] propose the use of a flexible wall permeameter test to assess not only excessive clogging conditions, but also excessive soil loss and equilibrium conditions. In the hydraulic conductivity ratio (HCR) test the soil column is prepared as per ASTM D5084, the customary flexible wall permeameter test for soils, albeit with the candidate geotextile on the top of the soil column. Complete geotechnical engineering procedures should be deployed, that is, back-pressure saturation, stress state loading conditions, site-specific soil, and site-specific permeating liquid. A hydraulic conductivity (permeability) test on the sample is now performed in two separate modes according to ASTM D5567.

1. The permeant flows down through the clean geotextile and through the soil column, resulting in the soil permeability, k_s.
2. Flow is then reversed and the permeant flows up through the soil column and the covering (now downgradient) geotextile, thereby challenging its behavior and ultimately resulting in the soil/geotextile permeability, k_{sg}.

The upper curves of Figure 2.25 show the anticipated behavior of such a set of tests using four hypothetical soil types. Using data such as this, a HCR is now calculated using the equilibrium values of k_s and k_{sg} as follows and is grouped accordingly; see the lower set of curves in Figure 2.25.

$$\text{HCR} = k_{sg}/k_s \tag{2.23}$$

An interpretation of these curves is suggested by Luettich and Williams [55]: high values of HCR suggest soil loss, low values of HCR suggest excessive clogging, and intermediate values of HCR (e.g., between 0.4 and 0.8) suggest soil-to-geotextile equilibrium. The author feels that this test is the premier laboratory test to assess the potential of excessive geotextile clogging and/or soil retention.

2.3.6 Degradation Properties

"How long will the geotextile last?" This important question for permanent and/or critical applications is asked more frequently than any other in geosynthetics. It will be addressed via different mechanisms and testing procedures in this subsection. At its root are the various potential degradation mechanisms, along with general polymeric aging. It should be noted that all of the mechanisms to be described result in some form of molecular chain scission, bond breaking, cross-linking, or extraction of components. Thus there is a fundamental change (albeit extremely slow in a buried environment) in the polymer at the molecular level from its as-manufactured state. While many chemical fingerprinting methods can be used to detect these changes (recall Table 1.4) they are time-consuming, expensive, and tedious to perform. At the macroscopic level, the mechanical properties will eventually change from the as-manufactured state. For example, a stress-versus-strain curve will show a gradual transition from a plastic to brittle behavior, in that the modulus will increase; the elongation at failure will decrease; and the strength will often temporarily increase, but will eventually decrease.

Sunlight (Ultraviolet) Degradation. Sunlight is an important cause of degradation in all organic materials, including the polymers from which geosynthetics are made. For geosynthetic purposes, energy from the sun is divided into three parts:

Infrared:	wavelengths longer than 760 nm
Visible:	wavelengths between 760 and 400 nm
Ultraviolet (UV):	wavelengths shorter than 400 nm

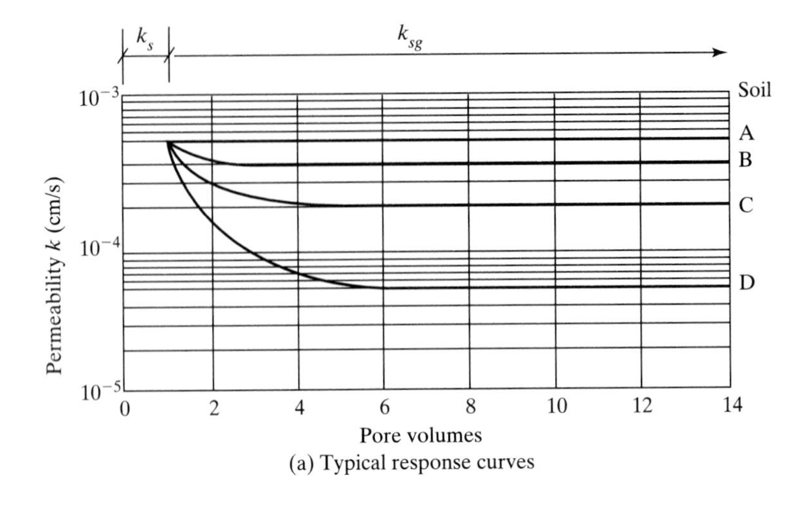

(a) Typical response curves

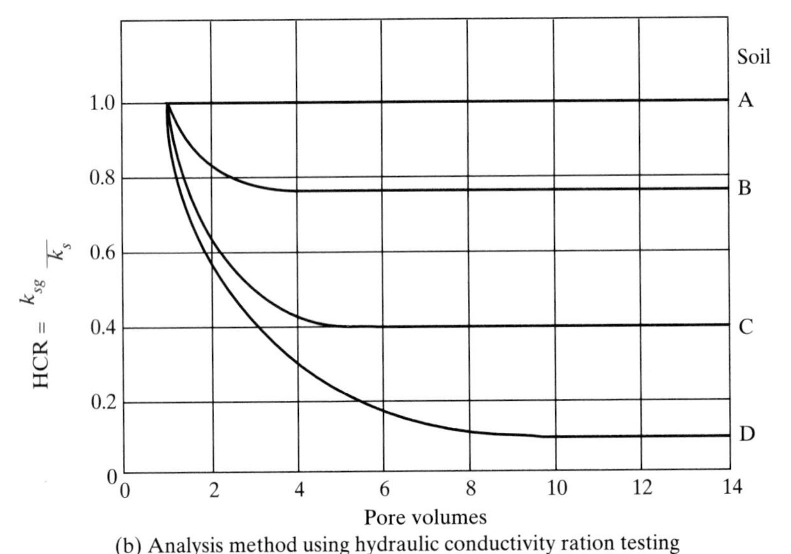

(b) Analysis method using hydraulic conductivity ration testing

Figure 2.25 (After Leuttich and Williams [55])

The UV region is further subdivided into UV-A (400 to 315 nm), which causes some polymer damage; UV-B (315 to 280 nm), which causes severe polymer damage; and UV-C (280 to 100 nm), which is only found in outer space.

From summer to winter there are changes in both the intensity and the spectrum of sunlight (see Figure 2.26); most significant is the loss of the shorter-wavelength UV radiation during the winter months. Other factors in the UV degradation process of polymers are geographic location, temperature, cloud cover, wind, moisture, and atmospheric pollution, and these must be considered in any test method. Laboratory simulations are at best approximate but nonetheless very important.

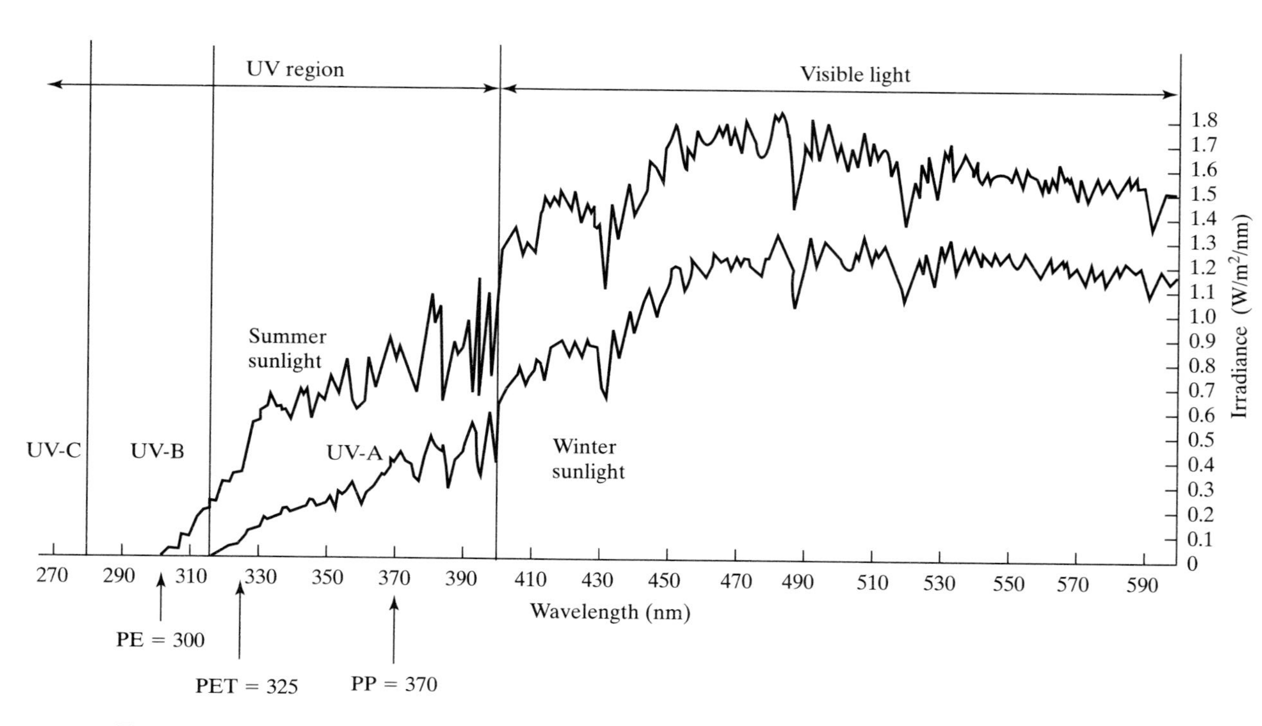

Figure 2.26 The wavelength spectrum of visible and UV solar radiation. (After Q-Panel Co., Cleveland, OH)

For laboratory simulation of sunlight, artificial light sources (lamps) are generally compared with the worst-case condition, the "solar maximum condition." The actual degradation is caused by photons of light breaking the polymer's chemical bonds. For each type of bond there is a threshold wavelength for bond scission; above the threshold the bonds will not break, below it they will. Thus the short wavelengths are critical. The literature [56] shows that polyethylene is most sensitive to UV degradation around 300 nm, polyester around 325 nm, and polypropylene around 370 nm.

Of all the laboratory-exposure devices, xenon-arc exposure is widely used and is often recommended for use on geotextiles. Two features are very important: the filters (used to reduce unwanted radiation) and the irradiance settings (used to compensate for lamp aging). The recommended ASTM test for geotextiles is D4355, which uses either Type BH or C as described in ASTM G26. Specimens are exposed for 0, 150, 300, and 500 hr. The exposure consists of 120-min cycles, light only (102 min) and then water spray and light (18 min). The test procedure allows the user to plot a curve to assess the amount of degradation, but it should be recognized that the device requires significant maintenance, is very expensive, and can be challenged on technical grounds as to its similitude to actual conditions.

Another laboratory exposure device (the type favored by the author) is the Ultraviolet Fluorescent Light Test Method covered under ASTM test methods G53 and D5208. While this method can also be challenged as to its simulation relevancy and use in comparing different polymer types, the device is easy to maintain and quite reasonable in its initial cost. The exposure is similar to that described above, that is, cycles of light only, followed by water spray and light. The test coupons are removed at designated times, cut into strip-tension test specimens and evaluated for their retained strength and elongation. Results are then compared to the unexposed geotextile for percent retained values; see Figures 2.27a,b.

Ultraviolet degradation is also covered by ASTM under the title "Outdoor Weathering of Plastics," designated D1435. This procedure is intended to define site-specific conditions for the exposure of plastic materials to light and weather. It is a comparative test, depending on a defined climate, time of year, atmospheric conditions, and so on. Racks are constructed with the geotextile coupons to be evaluated fixed to them. Samples can be placed at 0, 45, or 90° to the horizontal and in different solar orientations. Exposure test samples should simulate the service conditions of the end-use application as much as is practical. A specific version of this test is available as ASTM D5970, which was used to generate the data of Figures 2.27c,d. Clearly, if a test of site-specific UV degradation is desired at a critical field site where the geotextile is to be exposed for months or years, outdoor weathering tests of this type should be considered.

Whatever the test method used to produce the UV-degradation results, it is clear that geotextiles must be shielded from prolonged UV light exposure. Geotextile rolls are always shipped with a protective plastic covering and only when the material is ready for use should it be unrolled and exposed. The manufacturer's recommendations for "timely cover" (backfilling) must be rigidly met. Timely cover for geotextiles is generally within 14 days (per AASHTO M288-96), with polyesters being allowed longer exposure times. For long-term exposure applications, testing of the type reported in Figures 2.27c,d is necessary; that is, actual outdoor weathering tests performed on a site-specific basis.

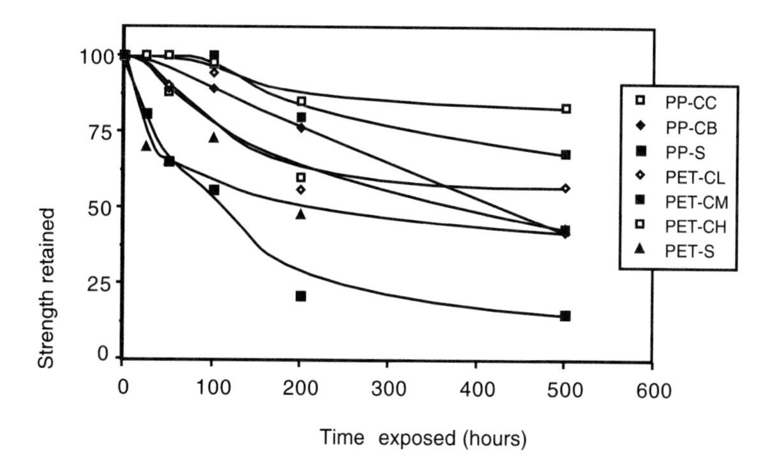

(a) Strength retained after exposure in laboratory ultraviolet fluorescent device, per ASTM D5208.

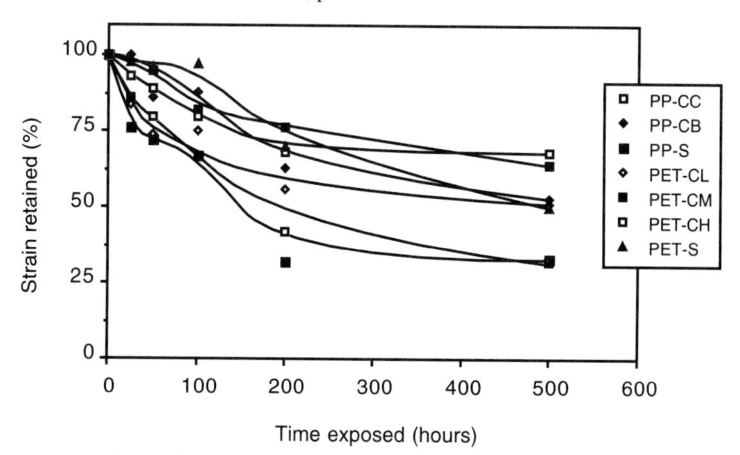

(b) Strain retained after exposure in laboratory ultraviolet fluorescent device, per ASTM D5208.

Figure 2.27 Degradation of various geotextiles due to ultraviolet exposure. (After Koerner et al. [57])

Temperature Degradation. Clearly high temperature causes all polymer degradation mechanisms to occur at an accelerated rate. In fact, at the heart of *time-temperature superposition* lifetime prediction techniques (used in Arrhenius modeling, the rate process method, etc.) is to test laboratory specimens at high temperatures and extrapolate the accelerated degradation down to field-anticipated (lower) temperatures. Thus high temperature is presented as an acceleration phenomenon acting with other degradation mechanisms like sunlight, oxidation, hydrolysis, chemical, radiation,

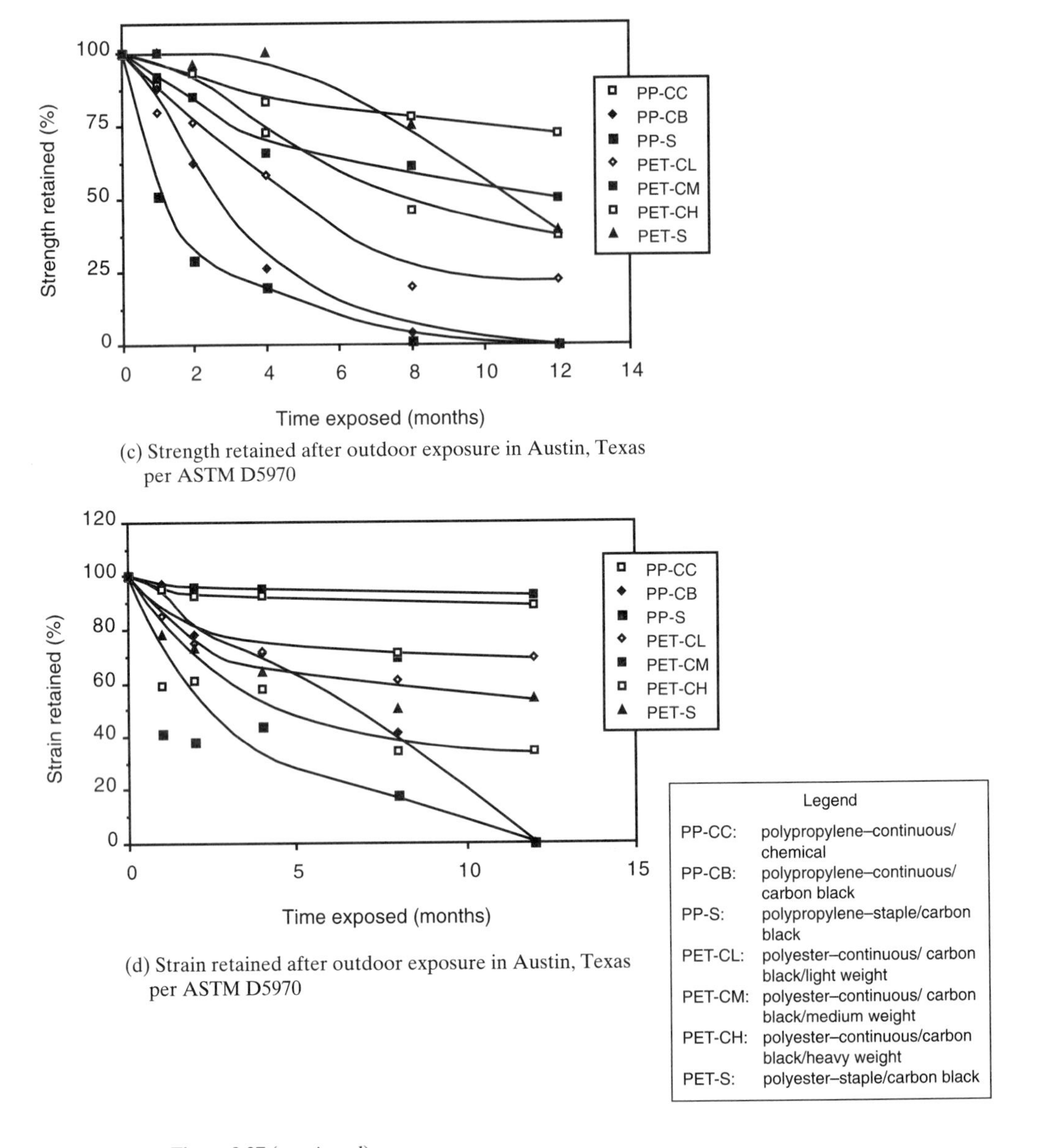

(c) Strength retained after outdoor exposure in Austin, Texas per ASTM D5970

(d) Strain retained after outdoor exposure in Austin, Texas per ASTM D5970

Figure 2.27 *(continued)*

biological, and so on. As such, temperature (per se) as a degradation mechanism will not be discussed separately.

Regarding the mechanical behavior of the plastics (insofar as engineering properties are concerned), hot and cold temperatures cause a softening and stiffening, as would be expected. For geotextiles, high temperatures increase flexibility, and ASTM D1388 can be used to quantify the behavior; recall Section 2.3.2.

ASTM Test Method D746 addresses the effect of cold temperatures on plastics and, in particular, on their brittleness and impact strength. At severely cold temperatures, specimens are tested by a specified impact device in a cantilever-beam test mode. The temperature causing brittleness is defined as "that temperature, estimated statistically, at which 50% of the specimens fail in the specified test." Numerous samples are required for testing, since a statistical value is required.

Oxidation Degradation. While all types of polymers react with oxygen causing degradation, the polyolefins (polypropylene and polyethylene) are generally considered to be the most susceptible to this phenomenon. Hsuan et al. [58] describe the chemical mechanism. It will be presented in detail in Chapter 5 on geomembranes, since it is this geosynthetic that is seeing the most research activity and data is now becoming available.

ASTM recommended practice D794 describes high-temperature oxidation testing for plastics; only the incubation procedure for heat exposure is specified, the test method(s) for assessment being governed by the potential end use. Heat is applied using a forced-air oven with controlled airflow and with substantial fresh-air intake. Two types of incubation are described: continuous heat and cyclic heat. In the former, heat is gradually increased until failure occurs. Failure is defined as a change in appearance, weight, dimension, or other properties that alter the material to a degree that it is no longer serviceable for the purpose. The test may be very short or require months, depending on the rate of temperature increase. The cyclic heat test repeatedly applies heat up to a constant value until failure.

A number of research efforts are ongoing to assess geotextile oxidative behavior using forced-air ovens at constant elevated temperatures; the higher the temperature, the greater the rate of oxidative degradation. Changes in tensile strength, elongation, and modulus are tracked over time. Properly plotted, these trends are back-extrapolated to site-specific (i.e., lower) temperatures to arrive at the predicted lifetimes. The procedure is the essence of time-temperature superposition, followed by Arrhenius modeling.

Caution should be exercised in the incubation of geotextiles at extremely high temperature. Polypropylene melts at 165°C and polyethylene melts at 125°C. These high temperatures should, obviously, be avoided; incubation should be at significantly lower temperatures.

Hydrolysis Degradation. Hydrolysis can cause degradation via either internal or external yarn reactions; see Halse et al. [59]. Polyester resin is particularly affected and especially when the immersion liquid has a high alkalinity.

Table 2.9 gives an indication of trends in degradation behavior insofar as loss of strength is concerned [59, 60]. As seen, extremely high pH values affect some polyesters, while extremely low pH values are harsh on some polyamides. It is important that the polyester resin used for permanent geotextile applications have a high molecular weight (e.g., $> 25{,}000$) and a low carboxyl end group (CEG) concentration (e.g., < 30). These effects are further described by Hsuan et al. [58].

TABLE 2.9 EFFECT OF HYDROLYSIS ON STRENGTH LOSS AT DIFFERENT pH LEVELS ON VARIOUS GEOTEXTILES AT 20°C AFTER 120 DAYS

Geotextile type	Solution	Weight (g/m^2)	pH = 2	pH = 4	pH = 7	pH = 10	pH = 12
PP monofilament woven (Manufacturer 1)	$Ca(OH)_2$	220	—	—	nc	nc	nc
PP needle-punched nonwoven (Manufacturer 2)	$Ca(OH)_2$	770	—	—	nc	nc	nc
PP heat-bonded nonwoven (Manufacturer 3)	$Ca(OH)_2$	100	—	—	nc	nc	incl
PVC monofilament woven (Manufacturer 4)	$Ca(OH)_2$	95	—	—	nc	nc	nc
PET staple needle-punched nonwoven (all white fibers, Manufacturer 5)	$Ca(OH)_2$	550	—	—	nc	nc	nc
PET heat-bonded nonwoven (Manufacturer 3)	$Ca(OH)_2$	100	—	—	nc	nc	nc
PET staple needle-punched nonwoven (mixture of white and black fibers, Manufacturer 5)	Na(OH)	450	—	—	nc	−33%	−53%
PET heat-bonded nonwoven (Manufacturer 3)	Na(OH)	100	—	—	nc	nc	nc
PET staple needle-punched nonwoven A (mixture of white and black fibers, Manufacturer 6)	Na(OH)	150	−18%+	nc	nc	−27%	−32%
PET staple needle-punched nonwoven B (mixture of white and black fibers, Manufacturer 6)	Na(OH)	150	nc	nc	nc	−13%	−16%
PET staple needle-punched nonwoven (all white fibers, Manufacturer 6)	Na(OH)	150	nc	nc	nc	nc	nc
PET needle punched nonwoven A (carbon black blended fibers, Manufacturer 6)	Na(OH)	134	−12%	−15%	nc	nc	nc
PET needle-punched nonwoven B (carbon black blended fibers, Manufacturer 6)	Na(OH)	134	nc	nc	nc	nc	nc
PET staple needle-punched nonwoven (bottle-grade resin, Manufacturer 7)	Na(OH)	264	nc	nc	nc	nc	nc

All fibers are continuous unless otherwise noted.

Abbreviations: "nc" means *no change* within estimated experimental accuracy of ±10%: "incl" means *inconclusive* due to large scatter of the data; "+" means that longer testing time is required to draw a meaningful conclusion.

Source: After Halse et al. [59, 60].

In cases of concern, the candidate geotextile should be incubated in water having the prevailing pH level at 20°C and 50°C and tested for changes in strength. For a baseline comparison, it is important to have a complete parallel set of samples incubated in distilled water (of pH 7) at the same temperatures.

Chemical Degradation. ASTM method D543 covers chemical degradation under the title "Resistance of Plastics to Chemical Reagents." The test method includes provisions for reporting changes in weight, dimensions, appearance, and strength. Provisions are also made for various exposure times and exposure to reagents at elevated temperatures. A list of 50 standard reagents is supplied in order to attempt some sort of standardization.

For example, the Du Pont Company has evaluated most of its yarns (including acetate, dacron, nylon, orlon, rayon, cotton, wool, and silk) under a wide range of chemicals (sulfuric acid, hydrochloric acid, nitric acid, hydrofluoric acid, phosphoric acid, organic acids, sodium hydroxide, bleaching agents, scouring and laundering agents, salt solutions, and organic and miscellaneous chemicals), many at different concentrations and at different temperatures. After the specified exposure, the coupons were rinsed, air dried, and then conditioned at 21°C and 65% relative humidity for 16 hr. Data on the test specimen breaking strength, breaking elongation, and toughness of the exposed yarns were compared to control specimens of the yarns that were not exposed to the chemical. Similar information is available from most raw material suppliers and geotextile manufacturers for their particular yarns in many chemical environments.

Notable exceptions to the use of manufacturers' data are geotextiles used at landfill sites in contact with aggressive types of leachate. Here site-specific tests must be performed and are currently being developed. In this regard, ASTM committee D-35 on geosynthetics has a task group working on chemical degradation assessments of geotextiles (and all other geosynthetics) after either laboratory or field-immersion procedures. The laboratory-incubation procedure is formalized under ASTM D5322 and a field-incubation procedure under D5496. The recommended suite of test methods is still under investigation.

Radioactive Degradation. While there are no references in the open literature on the radioactive degradation of geotextiles, the subject is generally discussed when dealing with radioactive-waste disposal. It is assumed, though clearly not proven, that low-level radioactive exposure (by hospital garments, nuclear power plant tools and clothing, etc.) is orders of magnitude too low to cause chain scission of geosynthetic-related polymers. Conversely, high-level radioactive waste (e.g., spent nuclear fuel rods) is suspect and, if it is in the proximity of geotextiles, could cause radiation degradation. Clearly, this situation must be experimentally evaluated if radioactive conditions are anticipated.

Biological Degradation. For microorganisms such as bacteria or fungi to degrade polymers the organisms must attach themselves to the yarn surface and use the polymer as a feedstock. For the commonly used polymeric resins, this is very unlikely. All the resins used for geosynthetics are very high in molecular weight with relatively few chain endings for the biodegradation process to be initiated. In fact, there is a concerted effort ongoing to develop polymers that *will* biodegrade, with little apparent success.

The additives to the polymer, however, might be somewhat more vulnerable than the resin itself. Plasticizers or processing aids might be vulnerable, although there is no authoritative research on the subject to my knowledge.

Although no formalized procedure or test method for biological degradation exists Ionescu et al. [61] evaluated six geotextiles (four of polypropylene, one of polyester, and a composite) in eight media for 5 to 17 months. The media were distilled water (the test control), seawater, compost, a particular soil, iron bacteria, lavensynthesizing bacteria, desulfavibrios bacteria, and a liquid mineral. Geotextile evaluation was by permeability, wide-width tensile strength, and infrared spectroscopy. The results showed no measurable effect on permeability (although some roots were growing through the geotextile), only small variations in tensile strength, and no structural changes visible using infrared spectroscopy evaluations.

Other Degradation Processes. Other processes that may degrade geotextiles are ozone attack (only in certain climates and when exposed), rodent or termite attack (which is site-specific and certainly possible), and perhaps a combination of these and the previously mentioned degradation mechanisms. Obviously this type of synergistic effect is a complex issue and worthy of further inquiry. In spite of these degradation mechanisms, the performance record for geotextile durability has been quite good, as will be described in the following section.

Geotextile Aging. While specific test standards are not available to measure the aging of geotextiles (due to the complexity of the many mechanisms involved), field-exhuming work is continually being reported. Perhaps the longest functioning geotextile that is periodically exhumed is at the Volcros Dam in France (Gourc et al. [62]). Both mechanical and hydraulic properties are examined and compared to original properties. Losses were generally nominal with maximum values (perhaps installation-related) being 30%.

Numerous studies on the exhuming of geotextiles have been reported in the literature. Tests on these recovered samples show that geotextiles remain in good to excellent condition. This indicates to the author that the proper polymers are at hand and are being utilized in the manufacture of geotextiles. What is needed at this point is to properly design them in an engineering sense, and then to properly install and backfill them.

2.3.7 Summary

This section on geotextile properties is very important and could have been dealt with in greater detail than we have. The subject deserves this attention, for any quantifiable design method will result in numbers to be compared to the candidate geotextile's actual properties. This section dealt with the relevant properties and subsequent test values, how they are obtained, their authenticity, and their reliability.

The section included a mixture of in-isolation (or index) properties and soil-to-geotextile related (or performance) properties. Eventually, the tests for these different properties will sort out into their respective categories and uses, but most organizations are looking at the complete collection of tests as they were presented here.

Table 2.10 is a summary table of geotextile properties. The rapidly changing market and its demands make it difficult to give accurate values, but for typical commercially available geotextiles, Table 2.10 gives the range of current values. For the specific values of specific types of geotextiles, the respective manufacturers should be consulted.

2.4 ALLOWABLE VERSUS ULTIMATE GEOTEXTILE PROPERTIES

It is important to recognize that many of the preceding geotextile test properties represent idealized conditions and therefore result in the maximum possible numeric values when used directly in design; that is, they result in upper-bound values. In the design-by-function concept described in Section 2.1.3, the factor of safety was formulated around an allowable test value (Eqs. 2.2a and 2.2b). Thus, most laboratory test values cannot generally be used directly; they must be suitably modified for the in situ conditions. This could be done directly in the test procedure, for example, by conducting a completely simulated performance test; but in most cases this simply is not possible. Simulating installation damage, performing long-term creep testing, using site-specific liquids, reproducing in situ pore-water stresses, providing complete stress state modeling, and so on, are generally not feasible. To account for such differences between the laboratory measured test value and the desired performance value, two approaches can be taken:

1. Use an extremely high factor of safety at the end of a problem.
2. Use reduction factors on the laboratory-generated test value to make it into a site-specific allowable value.

The latter alternative of *reduction factors*‡ will be used in this book. By doing this, the usual value of the factor of safety can be used in the final analysis. Our approach will be to refer to the general laboratory-obtained value as an *ultimate* value and to modify it by reduction factors to an *allowable* value.

2.4.1 Strength-Related Problems

For problems dealing with geotextile strength, such as in separation and reinforcement applications, the formulation of the allowable values takes the following form. Typical values for reduction factors are given in Table 2.11. Note that these values, however, must be tempered by the site-specific considerations. If the laboratory test includes the mechanism listed, it appears in the equation as a value of 1.0.

$$T_{\text{allow}} = T_{\text{ult}}\left(\frac{1}{\text{RF}_{ID} \times \text{RF}_{CR} \times \text{RF}_{CD} \times \text{RF}_{BD}}\right) \tag{2.24a}$$

$$T_{\text{allow}} = T_{\text{ult}}\left(\frac{1}{\Pi\text{RF}}\right) \tag{2.24b}$$

‡In previous editions of this book, reduction factors were called partial factors of safety. This edition is changed to reflect the current trend in agency specifications and the more appropriate terminology.

TABLE 2.10 TYPICAL RANGE OF PROPERTIES FOR CURRENTLY AVAILABLE GEOTEXTILES

Physical Properties	
Specific gravity	0.9–1.7
Mass per unit area	135–1000 g/m^2
Thickness	0.25–7.5 mm
Stiffness	nil to 25,000 mg-cm
Mechanical Properties	
Compressibility	nil to high
Tensile strength (grab)	0.45–4.5 kN
Tensile strength (wide width)	9–180 kN/m
Confined tensile strength	18–180 kN/m
Seam strength	50–100% of tensile
Cyclic fatigue strength	50–100% of tensile
Burst strength	350–5200 kPa
Tear strength	90–1300 N
Impact strength	14–200 J
Puncture strength	45–450 N
Friction behavior	60–100% of soil friction
Pullout behavior	50–100% of geotextile strength
Hydraulic Properties	
Porosity (nonwovens)	50–95%
Percent open area (wovens)	nil to 36%
Apparent opening size (sieve size)	2.0 to 0.075 mm (#10 to #200)
Permittivity	0.02–2.2 s^{-1}
Permittivity under load	0.01–3.0 s^{-1}
Transmissivity	0.01 to 2.0 $\times$ 10^{-3} m^2/min
Soil retention: turbidity curtains	m.b.e.
Soil retention: silt fences	m.b.e.
Endurance Properties	
Installation damage	0–70% of fabric strength
Creep response	g.n.p if $<$ 40% strength is being used
Confined creep response	g.n.p. if $<$ 50% strength is being used
Stress relaxation	g.n.p. if $<$ 40% strength is being used
Abrasion	50–100% of geotextile strength
Long-term clogging	m.b.e. for critical conditions
Gradient ratio clogging	m.b.e. for critical conditions
Hydraulic conductivity ratio	0.4–0.8 appear to be acceptable
Degradation Properties	
Temperature degradation	high temperature accelerates degradation
Oxidative degradation	m.b.e. for long service lifetimes
Hydrolysis degradation	m.b.e. for long service lifetimes
Chemical degradation	g.n.p. unless aggressive chemicals
Ratioactive degradation	g.n.p.
Biological degradation	g.n.p.
Sunlight (UV) degradation	major problem unless protected
Synergistic effects	m.b.e.
General aging	actual record to date is excellent

Abbreviations: m.b.e.—must be evaluated; g.n.p.—generally no problem.

TABLE 2.11 RECOMMENDED REDUCTION FACTOR VALUES FOR USE IN EQ. (2.24a)

Application Area	Range of Reduction Factors			
	Installation Damage	Creep*	Chemical Degradation	Biological Degradation
Separation	1.1 to 2.5	1.5 to 2.5	1.0 to 1.5	1.0 to 1.2
Cushioning	1.1 to 2.0	1.2 to 1.5	1.0 to 2.0	1.0 to 1.2
Unpaved roads	1.1 to 2.0	1.5 to 2.5	1.0 to 1.5	1.0 to 1.2
Walls	1.1 to 2.0	2.0 to 4.0	1.0 to 1.5	1.0 to 1.3
Embankments	1.1 to 2.0	2.0 to 3.5	1.0 to 1.5	1.0 to 1.3
Bearing capacity	1.1 to 2.0	2.0 to 4.0	1.0 to 1.5	1.0 to 1.3
Slope stabilization	1.1 to 1.5	2.0 to 3.0	1.0 to 1.5	1.0 to 1.3
Pavement overlays	1.1 to 1.5	1.0 to 2.0	1.0 to 1.5	1.0 to 1.1
Railroads (filter/sep.)	1.5 to 3.0	1.0 to 1.5	1.5 to 2.0	1.0 to 1.2
Flexible forms	1.1 to 1.5	1.5 to 3.0	1.0 to 1.5	1.0 to 1.1
Silt fences	1.1 to 1.5	1.5 to 2.5	1.0 to 1.5	1.0 to 1.1

*The low end of the range refers to applications which have relatively short service lifetimes and/or situations where creep deformations are not critical to the overall system performance.

where

T_{allow} = allowable tensile strength,
T_{ult} = ultimate tensile strength,
RF_{ID} = reduction factor for installation damage,
RF_{CR} = reduction factor for creep,
RF_{CD} = reduction factor for chemical degradation,
RF_{BD} = reduction factor for biological degradation, and
ΠRF = value of cumulative reduction factors.

Note that Eq. (2.24a) could have included additional site-specific terms, such as reduction factors for seams and intentionally made holes. It also could have been formulated with fractional multipliers (values ≤ 1.0) placed in the numerator of the equation or on the opposite side of the equation, as with the *load-factor design method*. It has been put in this form following other studies (e.g., Voskamp and Risseeuw [63]). While the equation indicates tensile strength, it can be applied to burst strength, tear strength, puncture strength, impact strength, and so on.

2.4.2 Flow-Related Problems

For problems dealing with flow through or within a geotextile, such as filtration and drainage applications, the formulation of the allowable values takes the following form. Typical values for reduction factors are given in Table 2.12. Note that these values must be tempered by the site-specific conditions, as in Section 2.4.1. If the laboratory test includes the mechanism listed, it appears in the equation as a value of 1.0.

$$q_{\text{allow}} = q_{\text{ult}}\left(\frac{1}{\text{RF}_{SCB} \times \text{RF}_{CR} \times \text{RF}_{IN} \times \text{RF}_{CC} \times \text{RF}_{BC}}\right) \tag{2.25a}$$

TABLE 2.12 RECOMMENDED REDUCTION FACTOR VALUES FOR USE IN EQ. (2.25a)

	Range of Reduction Factors				
Application	Soil Clogging and Blinding*	Creep Reduction of Voids	Intrusion into Voids	Chemical Clogging†	Biological Clogging
Retaining wall filters	2.0 to 4.0	1.5 to 2.0	1.0 to 1.2	1.0 to 1.2	1.0 to 1.3
Underdrain filters	5.0 to 10	1.0 to 1.5	1.0 to 1.2	1.2 to 1.5	2.0 to 4.0
Erosion-control filters	2.0 to 10	1.0 to 1.5	1.0 to 1.2	1.0 to 1.2	2.0 to 4.0
Landfill filters	5.0 to 10	1.5 to 2.0	1.0 to 1.2	1.2 to 1.5	5 to 10‡
Gravity drainage	2.0 to 4.0	2.0 to 3.0	1.0 to 1.2	1.2 to 1.5	1.2 to 1.5
Pressure drainage	2.0 to 3.0	2.0 to 3.0	1.0 to 1.2	1.1 to 1.3	1.1 to 1.3

*If stone riprap or concrete blocks cover the surface of the geotextile, use either the upper values or include an additional reduction factor.

†Values can be higher particularly for high alkalinity groundwater.

‡Values can be higher for turbidity and/or for microorganism contents greater than 5000 mg/l.

$$q_{\text{allow}} = q_{\text{ult}}\left(\frac{1}{\Pi \text{RF}}\right) \tag{2.25b}$$

where

q_{allow} = allowable flow rate,
q_{ult} = ultimate flow rate,
RF_{SCB} = reduction factor for soil clogging and blinding,
RF_{CR} = reduction factor for creep reduction of void space,
RF_{IN} = reduction factor for adjacent materials intruding into geotextile's void space,
RF_{CC} = reduction factor for chemical clogging,
RF_{BC} = reduction factor for biological clogging, and
ΠRF = value of cumulative reduction factors.

As with Eqs. (2.24) for strength reduction, this flow-reduction equation could also have included additional site-specific terms, such as blocking of a portion of the geotextile's surface by riprap or concrete blocks.

2.5 DESIGNING FOR SEPARATION

Application areas for geotextiles used for the separation function were given in Section 1.3.3. There are many specific applications, and it could be said, in a general sense, that geotextiles always serve a separation function. If they do not also serve this function, any other function, including the primary one, will not be served properly. This should not give the impression that the geotextile function of separation always plays a secondary role. Many situations call for separation only, and in such cases the geotextiles serve a significant and worthwhile function.

2.5.1 Overview of Applications

Perhaps the target application that best illustrates the use of geotextiles as separators is their placement between a reasonably firm soil subgrade (beneath) and a stone base course, aggregate, or ballast (above). We say "reasonably firm" because it is assumed that the subgrade deformation is not sufficiently large to mobilize uniformly high tensile stress in the geotextile. (The application of geotextiles in unpaved roads on soft soils with membrane-type reinforcement is treated later in Section 2.6.1.) Thus for a separation function to occur the geotextile has only to be placed on the soil subgrade and then have stone placed, spread, and compacted on top of it. The subsequent deformations are very localized and occur around each individual stone particle. A number of scenarios can be developed showing which geotextile properties are required for a given situation.

2.5.2 Burst Resistance

Consider a geotextile on a soil subgrade with stone of average particle diameter (d_a) placed above it. If the stone is uniformly sized, there will be voids within it that will be available for the geotextile to enter. This entry is caused by the simultaneous action of the traffic loads being transmitted to the stone, through the geotextile, and into the underlying soil. The stressed soil then tries to push the geotextile up into the voids within the stone. The situation is shown schematically in Figure 2.28. Giroud [64] provides a formulation for the required geotextile strength that can be adopted for this application.

$$T_{\text{reqd}} = \frac{1}{2} p' d_v [f(\varepsilon)] \tag{2.26}$$

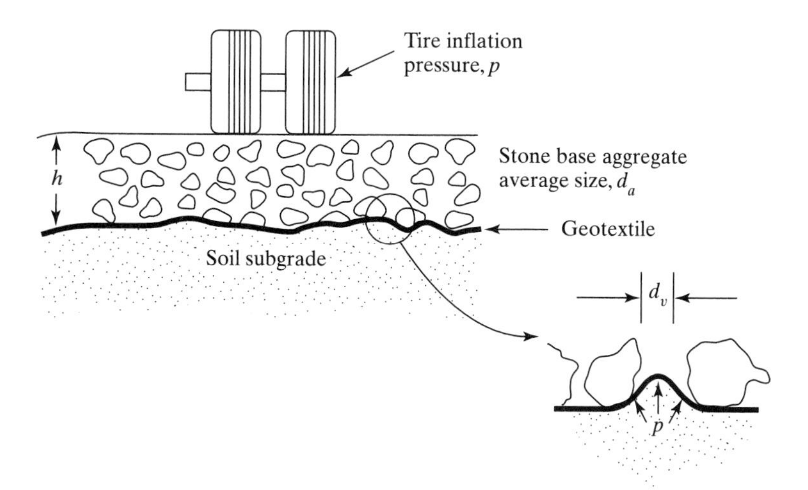

Figure 2.28 Geotextile being forced up into voids of stone base by traffic tire loads.

where

T_{reqd} = required geotextile burst strength;
p' = stress at the geotextile's surface, which is less than or equal to p, the tire inflation pressure at the ground surface;
d_v = maximum void diameter of the stone $\cong 0.33d_a$;
d_a = the average stone diameter,
$f(\varepsilon)$ = strain function of the deformed geotextile
$= \frac{1}{4}\left(\frac{2y}{b} + \frac{b}{2y}\right)$, in which
b = width of opening (or void), and
y = deformation into the opening (or void).

The field situation is analogous to the ASTM D3786 (Mullen) burst test, which has the geotextile being stressed into a gradually increasing hemispherical shape until it fails in radial tension (recall Section 2.3.3). Thus, the adapted form of Eq. (2.26) is:

$$T_{ult} = \frac{1}{2}p_{test}d_{test}[f(\varepsilon)] \tag{2.27}$$

where

T_{ult} = ultimate geotextile strength,
p_{test} = burst test pressure, and
d_{test} = diameter of the burst test device (= 30 mm).

Knowing that $T_{allow} = T_{ult}/(\Pi RF)$, where ΠRF = cumulative reduction factors, we can formulate an expression for the FS as follows:

$$FS = \frac{T_{allow}}{T_{reqd}} = \frac{(p_{test}d_{test})}{(\Pi RF)p'd_v}$$

For example, if $d_{test} = 30$ mm, $d_v = 0.33\,d_a$, and $\Pi FS = 1.5$ (which is not particularly low since creep is not an issue with this application), then the FS is the following, with d_a in mm.

$$FS = \frac{p_{test}(30)}{(1.5)p'(0.33d_a)}$$

$$FS = \frac{60.6p_{test}}{p'd_a} \tag{2.28}$$

Example 2.7

Given a 700 kPa truck tire inflation pressure on a poorly graded stone-base course consisting of 50 mm maximum-size stone, what is the factor of safety using a geotextile with an ultimate burst strength of 2000 kPa and cumulative reduction factors of 1.5?

Solution: Assuming that the tire inflation pressure is not significantly reduced through the thickness of the stone base, we can solve Eq. (2.28) as follows.

$$\text{FS} = \frac{60.6(2000)}{700(50)}$$

$$= 3.5$$

Note that with the cumulative reduction factors of 1.5 already included, the resulting factor of safety value is acceptable.

For a range of stone-base particle diameters (d_a), values of tire inflation pressure (p'), and cumulative reduction factors of 1.5, along with a factor of safety of 2.0, we get the design guide in Figure 2.29. Here it can be seen that stone size is quite significant insofar as the required burst-pressure values are concerned. Note also that these are poorly graded aggregates and that the presence of fines will lessen the severity of the design; hence this approach should be considered to be a worst-case design.

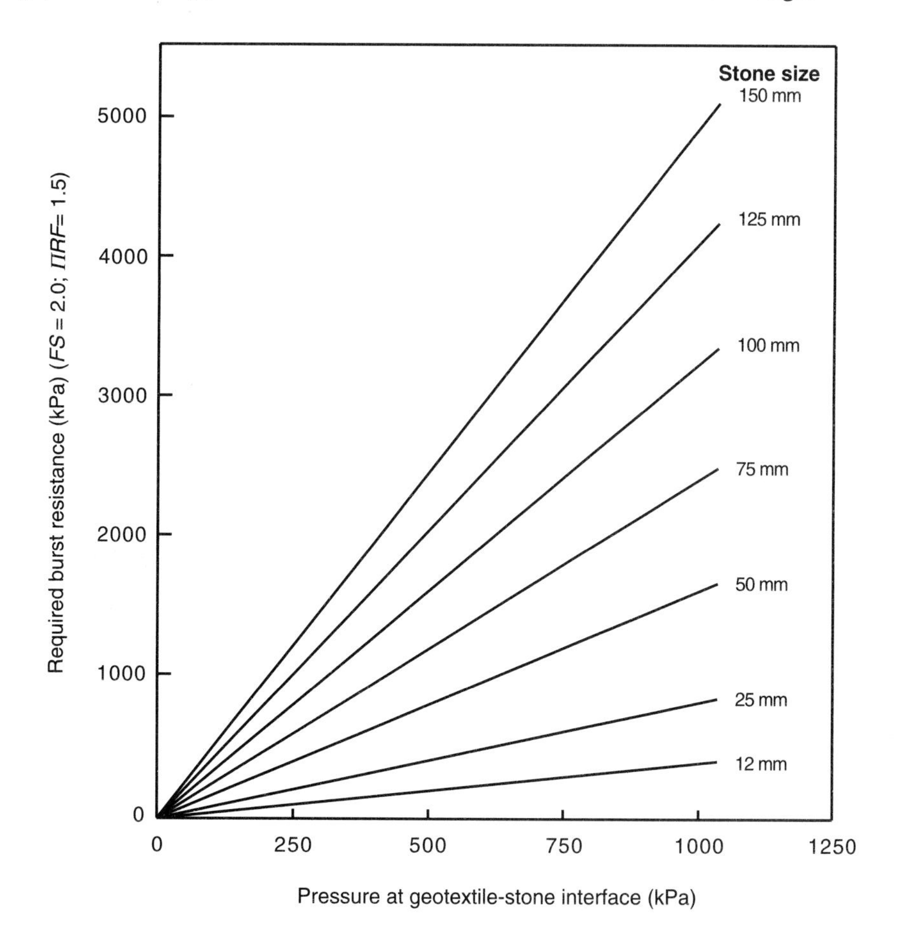

Figure 2.29 Design guide for burst analysis of geotextile used in a separation function based on cumulative reduction factors of 1.5 and a factor of safety of 2.0.

2.5.3 Tensile Strength Requirement

Continuing the discussion of the general problem, there is a process acting on the geotextile simultaneously as its tendency to burst in an out-of-plane mode: tensile stress mobilized by in-plane deformation. This occurs as the geotextile is locked into position by the stone-base aggregate above it and soil subgrade below it. A lateral or in-plane tensile stress in the geotextile is mobilized when an upper piece of aggregate is forced between two lower pieces that lie against the geotextile. The analogy to the grab tensile test can be readily visualized, as illustrated in Figure 2.30. Here we can estimate the maximum strain that the geotextile will undergo as the upper stone wedges itself down to the level of the geotextile. Using the dimensions shown (where $S \sim d/2$ and l_f = deformed geotextile length), the maximum strain with no slippage or stone breakage can be calculated.

$$\begin{aligned}\varepsilon &= \frac{l_f - l_o}{l_o}(100)\\ &= \frac{[d + 2(d/2)] - 3(d/2)}{3(d/2)}(100)\\ &= \frac{4(d/2) - 3(d/2)}{3(d/2)}(100)\\ &= 33\%\end{aligned}$$

Note that the preceding assumptions result in a strain that is independent of particle size. Thus the strain in the geotextile could be as high as 33% given the idealized

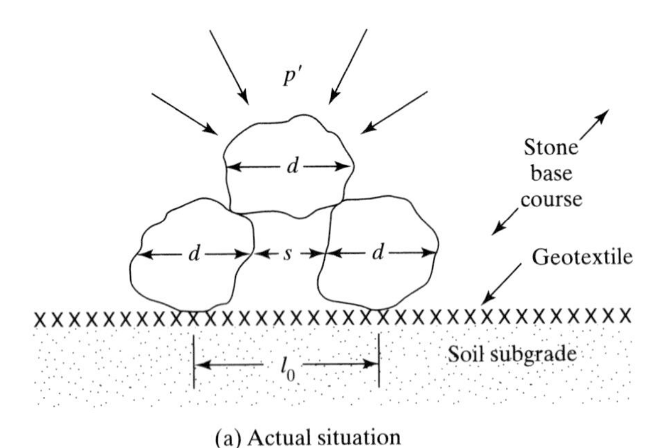

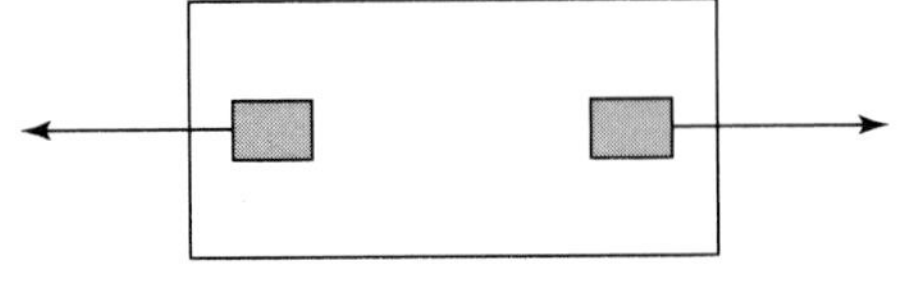

Figure 2.30 Geotextile being subjected to tensile stress as surface pressure is applied and stone base attempts to spread laterally.

(upper-bound) assumptions stated above. The tensile force being mobilized is related to the pressure exerted on the stone as follows [64].

$$T_{\text{reqd}} = p'(d_v)^2[f(\varepsilon)] \tag{2.29}$$

where

T_{reqd} = required grab tensile force;
p' = applied pressure;
d_v = maximum void diameter $\simeq 0.33\, d_a$, where
d_a = average stone diameter; and
$f(\varepsilon)$ = strain function of the deformed geotextile;
$= \frac{1}{4}\left(\frac{2y}{b} + \frac{b}{2y}\right)$, where
b = width of stone void, and
y = deformation into stone void.

Example 2.8 illustrates the design procedure above.

Example 2.8

Given a 700 kPa truck-tire inflation pressure on a stone-base course consisting of 50 mm maximum-size stone with a geotextile beneath it, calculate **(a)** the required grab tensile stress on the geotextile, and **(b)** the factor of safety for a geotextile whose grab strength at 33% is 500 N with cumulative reduction factors of 2.5 and $f(\varepsilon) = 0.52$.

Solution: **(a)** Using an empirical relationship that $d_v = 0.33\, d_a$ and $f(\varepsilon) = 0.52$, the required grab tensile strength from Eq. (2.29) is as follows.

$$\begin{aligned} T_{\text{reqd}} &= p'(d_v)^2(0.52) \\ &= p'(0.33d_a)^2(0.52) \\ &= 0.057\, p'd_a^2 \\ &= 0.057(700)(1000)(0.050)^2 \\ &= 100 \text{ N} \end{aligned}$$

(b) The factor of safety for a 500 N grab tensile geotextile at 33% strain with cumulative reduction factors of 2.5 is as follows.

$$\begin{aligned} \text{FS} &= \frac{T_{\text{allow}}}{T_{\text{reqd}}} \\ &= \frac{500/2.5}{100} \\ &= 2.0 \qquad \text{which is acceptable.} \end{aligned}$$

2.5.4 Puncture Resistance

The geotextile must survive the installation process. This is not just related to the function of separation; indeed, fabric survivability is critical in all types of applications—without it the best of designs are futile (recall Figure 2.19). In this regard, sharp stones,

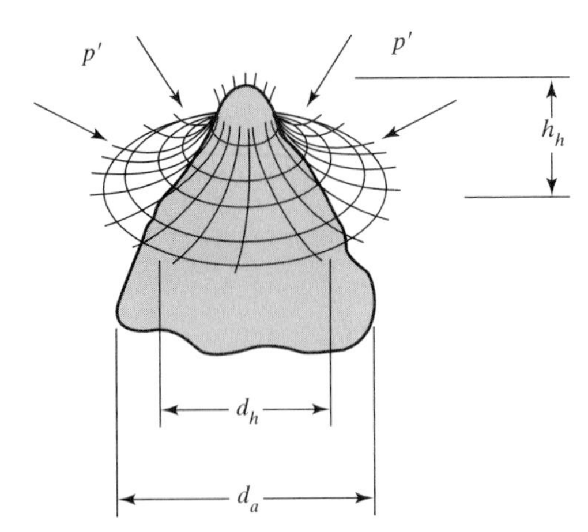

Figure 2.31 Visualization of a stone puncturing a geotextile as pressure is applied from above.

tree stumps, roots, miscellaneous debris, and other items, either on the ground surface beneath the geotextile or placed above it, could puncture through the geotextile after backfilling and traffic loads are imposed. The design method suggested for this situation is shown schematically in Figure 2.31. For these conditions, the vertical force exerted on the geotextile (which is gradually tightening around the protruding object) is as follows:

$$F_{\text{reqd}} = p' d_a^2 S_1 S_2 S_3 \tag{2.30}$$

where

F_{reqd} = required vertical force to be resisted;
d_a = average diameter of the puncturing aggregate or sharp object;
p' = pressure exerted on the geotextile (approximately 100% of tire inflation pressure at the ground surface for thin covering thicknesses);
S_1 = protrusion factor = h_h/d_a;
h_h = protrusion height $\leq d_a$;
S_2 = scale factor to adjust the ASTM D4833 puncture test value (which uses an 8.0 mm diameter puncture probe) to the diameter of the actual puncturing object = d_{probe}/d_a;
S_3 = shape factor to adjust the ASTM D4833 flat puncture probe to the actual shape of puncturing object = $1 - A_p/A_c$, (values for A_p/A_c range from 0.8 for rounded sand, to 0.7 for run-of-bank gravel, to 0.4 for crushed rock, to 0.3 for shot rock);
A_p = projected area of puncturing particle;
A_c = area of smallest circumscribed circle around puncturing particle.

Example 2.9

What is the factor of safety against puncture of a geotextile from a 50 mm stone on the ground surface mobilized by a loaded truck with a tire inflation pressure of 550 kPa traveling on the surface of the base course? The geotextile has an ultimate puncture strength of 200 N, according to ASTM D4833.

Solution: Using the full stress on the geotextile of 550 kPa and the values 0.33, 0.15, and 0.6 for the factors S_1, S_2, and S_3, respectively,

$$
\begin{aligned}
F_{\text{reqd}} &= p'd_a^2 S_1 S_2 S_3 \\
&= (550)(1000)(50 \times 0.001)^2(0.33)(0.15)(0.6) \\
&= 40.8 \text{ N}
\end{aligned}
$$

Assuming that the cumulative reduction factors are 2.0, the factor of safety is as follows:

$$
\begin{aligned}
FS &= \frac{F_{\text{allow}}}{F_{\text{reqd}}} \\
&= \frac{200/2.0}{40.8} \\
&= 2.4 \qquad \text{which is acceptable}
\end{aligned}
$$

Using the following assumptions (which can be modified as desired), a design guide can be developed as shown in Figure 2.32: the geotextile has an angular subgrade

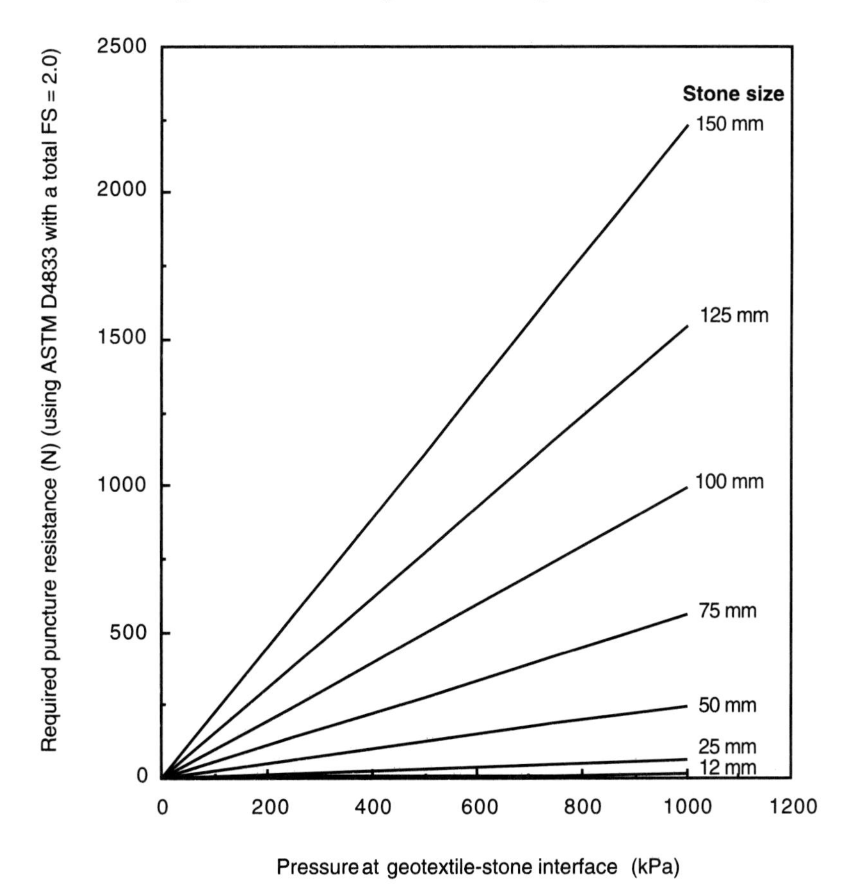

Figure 2.32 Puncture resistance design guide based on cumulative reduction factors of 2.0, a factor of safety of 2.0, and conditions stated in text.

above it such that $S_1 = 0.33$, $S_2 = 0.15$, and $S_3 = 0.5$; the cumulative reduction factors are 2.0; and the factor of safety is also 2.0.

$$F_{\text{reqd}} = p'd_a^2(0.33)(0.15)(0.5)$$

$$= 0.0248p'd_a^2$$

$$\text{FS} = \frac{F_{\text{ult}}/\Pi\text{RF}}{F_{\text{reqd}}}$$

$$2.0 = \frac{F_{\text{ult}}/2.0}{0.0248p'd_a^2}$$

$$F_{\text{ult}} = 0.099p'd_a^2 \qquad \text{which is graphed accordingly.}$$

2.5.5 Impact (Tear) Resistance

As with the puncture requirement just described, the resistance of a geotextile to impact is as much a survivability criterion as it is a separation function. Yet in many applications of separation, the geotextile must resist the impact of various objects. The most obvious one is a rock falling on it, but there are also situations in which construction equipment and materials can cause or contribute to impact damage on geotextiles.

The problem concerns the energy mobilized by a free-falling object of known weight and the height of the drop. Rarely will an object be intentionally impelled onto an exposed geotextile with additional force, so only gravitational energy will be assumed.

To develop a design guide, we assume free-falling stones of specific gravity of 2.60, varying in diameter from 25 to 600 mm and falling from heights of 0.5 to 5 m. Using this data the design curves of Figure 2.33 are developed. The relationship is as follows.

$$E = mgh$$

$$= (V \times \rho)gh$$

$$= [V \times (\rho_w G_s)]gh$$

$$= \left(\frac{\pi(d_a/1000)^3}{6}\right)\left(\frac{1000kg}{m^3}\right)(2.6)(9.81)h$$

$$E = 13.35 \times 10^{-6} d_a^3 h \tag{2.31}$$

where

E = energy developed (joules),
m = mass of the object (kg),
g = acceleration due to gravity (m/s²),
h = height of fall (m),

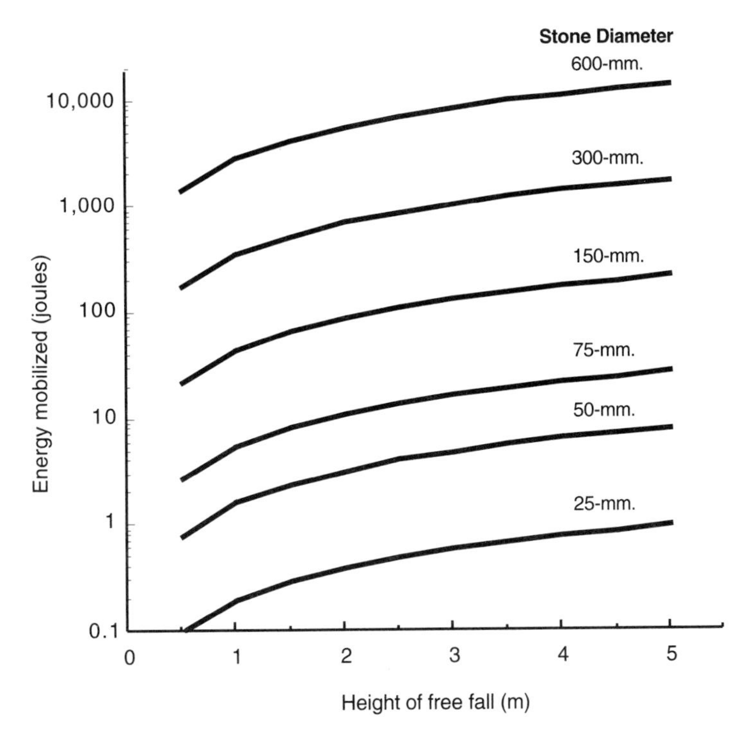

Figure 2.33 Energy mobilization by a free-falling rock on a geotextile with an unyielding support.

V = volume of the object (m^3),
ρ = density of the object (kg/m^3),
ρ_w = density of water (kg/m^3),
G_s = specific gravity of the object (dimensionless), and
d_a = diameter of the object (mm).

Note that these calculated energies are based on the geotextile resting on an unyielding surface, that is, the worst possible condition. As the soil beneath the geotextile deforms, the geotextile can absorb greater amounts of impacting energy. Since this is always the case, the reduction factors of Figure 2.34 are to be used in conjunction with the curves of Figure 2.33. Once the required energy is calculated, it should be compared to the allowable impact strength of the geotextile (e.g., the Elmerdorf tear or other impact test as discussed in Section 2.3.3). Example 2.10 illustrates the procedure.

Example 2.10

What energy is mobilized by a free-falling stone of 300 mm size falling 1.5 m onto a geotextile? The geotextile is supported by a poor subsoil having an unsoaked CBR strength of 4. If the geotextile has an allowable impact strength of 36 J, what is the factor of safety?

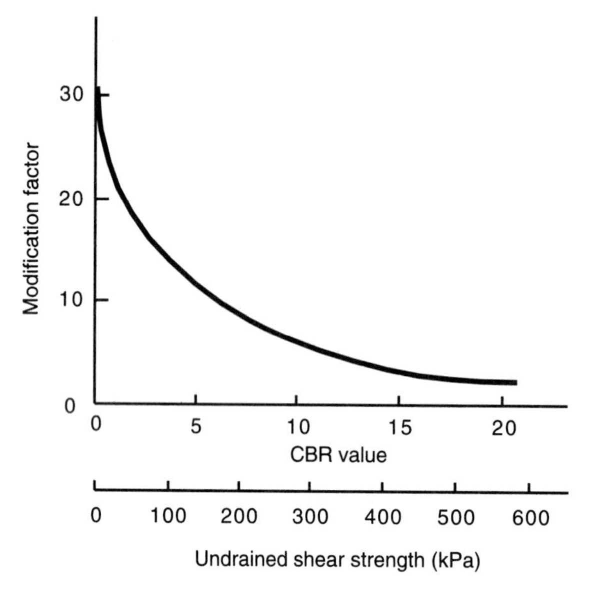

Figure 2.34 Modification factors to be used with energy mobilized by objects falling on geotextiles of varying support resistances characterized by their unsoaked CBR values or undrained shear strength.

Solution: Using Eq. (2.31) one calculates the required impact energy

$$E_{max} = 13.35 \times 10^{-6}(d_a^3)(h)$$
$$= 13.35 \times 10^{-6}(300)^3(1.5)$$
$$E_{max} = 540 \text{ J}$$

Note that this value is substantiated by the design chart in Figure 2.33. Of course, other design charts can be made for different assumptions.

This value is reduced according to the subgrade conditions of Figure 2.34.

$$E_{reqd} = 540/13$$
$$= 41.5 \text{ J}$$

This results in a global factor of safety calculation as follows.

$$\text{FS} = \frac{E_{allow}}{E_{reqd}}$$
$$= \frac{36}{41.5}$$
$$\text{FS} = 0.87 \qquad \text{which is not acceptable.}$$

Thus holes are likely to be formed when free-falling rocks of this size fall directly on the exposed geotextile. Not included in this analysis is the effect of the contact area of the falling object on the geotextile; for a very rounded rock, the effect is much less severe than for a sharp, angular one, which could easily cut through the fabric. A more sophisticated design of this nature for puncture of geomembranes, along with geotextile protection, will be developed in Section 5.6.7.

It should be emphasized that the last two methods of puncture and impact design refer not only to separation per se, but to the construction survivability of geotextiles in general (recall Table 2.2a). In all cases these considerations should be examined, for they are critical in many situations.

2.5.6 Summary

Separation, the *most underrated of all geotextile functions,* was addressed in this section. I say underrated because every use of geotextiles carries with it the separation function, yet rarely is separation designed on its own merit. Hopefully, the designs in this section will allow the engineer to determine quantitatively which geotextile is suitable for a specific situation.

Last, and in a sense most important, is the economic justification for the use of geotextiles in the separation function. It lies in the greater use and service lifetime of the system with geotextiles than without. When a geotextile separator is used in roadway cross sections, geotextiles could well double or triple the lifetime; however, field data for such quantification is sparse and greater efforts should be taken to provide test sites for this. Figure 2.35 is the photograph of a 40 m long driveway test plot, which was subdivided into four elongated quadrants, two with geotextiles and two without. Further, the two geotextiles were different and placed diagonally across from one another. After nine years of service, no cracks have surfaced on the paving and the test is continuing with the objective of providing lifetime data with and without geotextiles, and on which is the preferred type of geotextile. A database of like projects is being developed; see [65].

Figure 2.35 Different separation geotextiles being used to determine pavement (driveway) lifetime contrasted to sections with no geotextile.

2.6 DESIGNING FOR ROADWAY REINFORCEMENT

The combined use of soil (good in compression and poor in tension) and a geotextile (good in tension and poor in compression) suggests a number of situations in which geotextiles have made existing designs work better or provided for the development of entirely new applications. These applications were previewed in Section 1.3, together with a brief history of the original applications in this particular application area. This section focuses on the design methods using geotextiles for various roadway reinforcement applications.

2.6.1 Unpaved Roads

Overview. The application in this section is the use of geotextiles in unpaved roads, in which soft soil subgrades have stone aggregate placed directly above. No permanent surfacing (i.e., concrete or asphalt pavement) is placed immediately on the stone. At most, the road is surfaced with quarry crusher run or chip seal for reasonable ridability. There are many thousands of kilometers of unpaved secondary roads, haul roads, access roads, and the like, with no permanent surfacing on them. At a later time, perhaps years after settlement takes place and ruts are backfilled, an asphalt pavement may be placed.

This particular application triggered the high-volume of use and acceptance of geotextiles in the 1970s, since calculations can be made for the thickness of stone required both without a geotextile and with a geotextile, and finally for the thickness of stone that is saved. By determining the cost of the stone saved versus the cost of the geotextile, the value of using a geotextile is known immediately. The particular design process used in arriving at the respective thicknesses is the focus of this section.

Before beginning, however, it is important to realize that the geotextile must have its tensile modulus or strength mobilized via deformation of the soil subgrade. Although this can be done intentionally by prestressing the fabric, this is usually not the case because of the construction difficulties involved. Instead, the yielding of the soil subgrade is the triggering phenomenon, allowing for geotextile deformation and the mobilization of its tensile properties. How much deformation is necessary with regard to the vehicular loading, the particular geotextile, the time it takes for adequate strength mobilization, and so on, are all pressing questions, but the deformation characteristics of the soil subgrade takes precedence. A soft, yielding soil subgrade is needed to mobilize the geotextile's strength—but how soft? In light of the tremendous variety of situations, we must use a broad generality; in this case it will be based on the California Bearing Ratio (CBR) of the soil subgrade. The CBR test is used throughout the world and standardized accordingly (e.g., see ASTM D1883). The CBR value is a comparison of the soil's resistance to the force of a 50 mm diameter plunger at a given deformation with that of a crushed stone-base material. It is actually a percentage value, although it is rarely expressed as such. The test on the soil subgrade can be performed either at the in situ moisture content or the soil can be saturated for 24 hr and then tested. These two conditions give rise to unsoaked and soaked CBR values, respectively. Typical unsoaked CBR values are given in Table 2.13, where a considerable

TABLE 2.13 CORRELATION CHART FOR ESTIMATING UNSOAKED CBR VALUES FROM SOIL STRENGTH OR PROPERTY VALUES

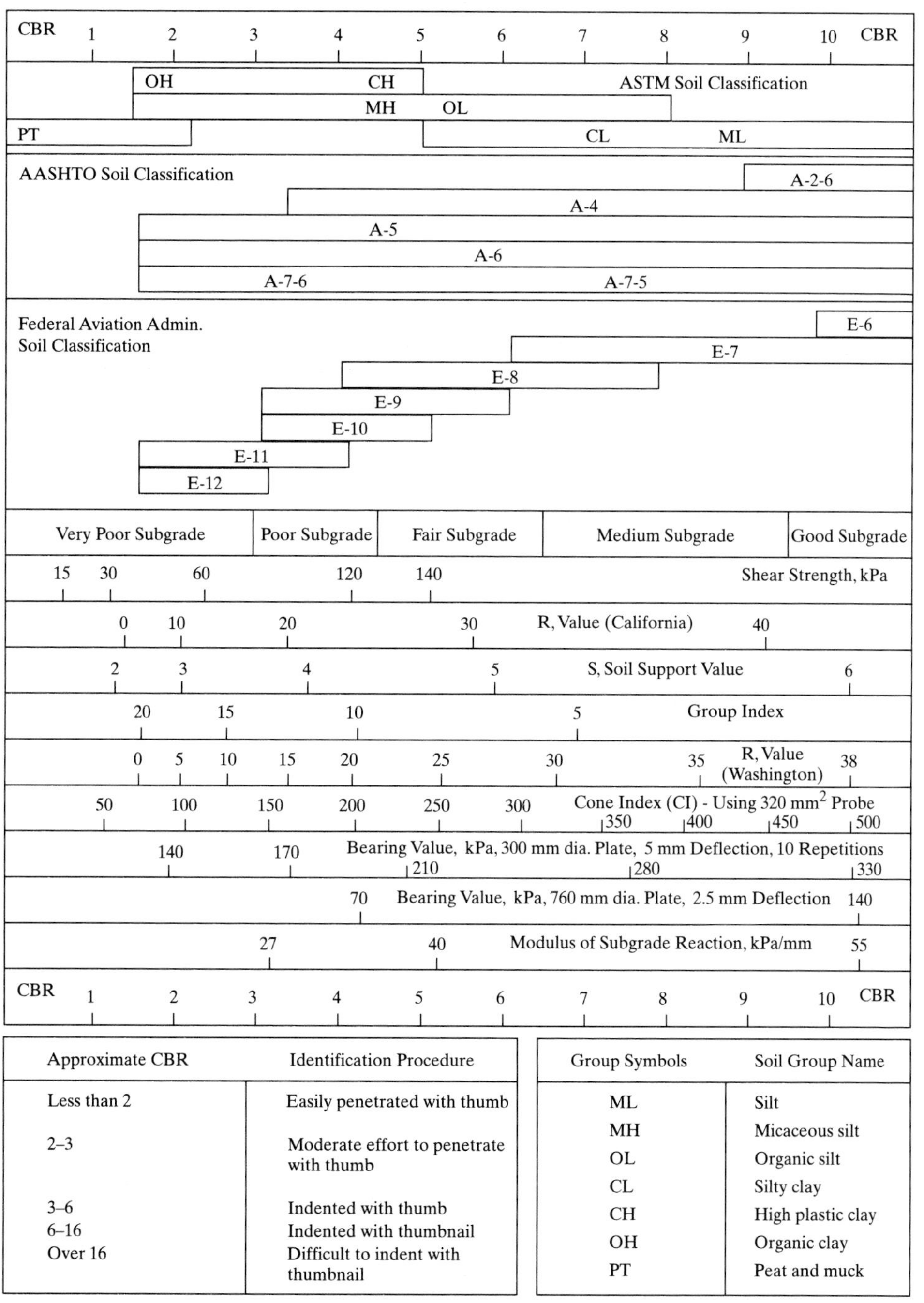

Approximate CBR	Identification Procedure
Less than 2	Easily penetrated with thumb
2–3	Moderate effort to penetrate with thumb
3–6	Indented with thumb
6–16	Indented with thumbnail
Over 16	Difficult to indent with thumbnail

Group Symbols	Soil Group Name
ML	Silt
MH	Micaceous silt
OL	Organic silt
CL	Silty clay
CH	High plastic clay
OH	Organic clay
PT	Peat and muck

Source: After Portland Cement Association and E. I. DuPont literature.

TABLE 2.14 RECOMMENDED SOIL SUBGRADE CBR VALUES TO DISTINGUISH DIFFERENT GEOTEXTILE FUNCTIONS IN ROADWAY CONSTRUCTION

Geotextile Function	CBR Value	
	Unsoaked	Soaked
Separation	$\geqslant 8$	$\geqslant 3$
Stabilization*	8–3	3–1
Reinforcement and separation	$\leqslant 3$	$\leqslant 1$

*A frequently used but poorly defined transition term that always includes separation, some unknown amount of reinforcement, and usually filtration.

body of empirical correlations is presented. Soaked CBR values are generally lower than unsoaked values, but the difference depends on the soil type. Table 2.14 is given as a guide.

For the purpose of using geotextiles in roadway applications on soil subgrades of different strength characteristics, we will subdivide the functions per Table 2.14. Here we can see that with medium to good soil subgrades, CBR (unsoaked) ≥ 8 and CBR (soaked) ≥ 3, the function is uniquely separation. The design for this is described in Section 2.5. For poor to very poor soil subgrades, CBR (unsoaked) $\leqslant 3$ and CBR (soaked) $\leqslant 1$, the function is both reinforcement and separation. This is the topic of this section. The intermediate category is only loosely defined; it is generally called stabilization. This term represents an interrelated group of functions (separation, reinforcement, and filtration) and is essentially a transition category between the two extremes.

Manufacturers' Methods. All of the major geotextile manufacturers have an unpaved-road design method for use with their particular geotextiles. They usually show CBR (or other soil strength values) on the x-axis and the required stone thickness (with and without a geotextile) on the y-axis. All result in logical behavior, with the geotextile providing greater savings in stone aggregate as the soil subgrade becomes weaker. Since most manufacturers have a range of geotextiles available for the reinforcement of unpaved roads, it is also seen that the heavier and stronger geotextiles result in greater stone savings than the lighter and weaker ones. Because each manufacturer's set of curves has its own background (based on theory, laboratory work, field observation, or empirical observation), it is nearly impossible to compare one technique with another. Yet the design methods have served the industry well and generally with excellent success. Their use is certainly acceptable and if only one geotextile is available, its manufacturer's method should continue to be used. If, however, a number of geotextiles are available, a method that views them on the basis of a specific, well-defined property is needed. Such a property could well be the geotextile modulus, which is the basis of design in the procedure to follow. It should be noted, however, that a number of generic techniques are available, and that Hausmann [66] has assessed and compared them to one another.

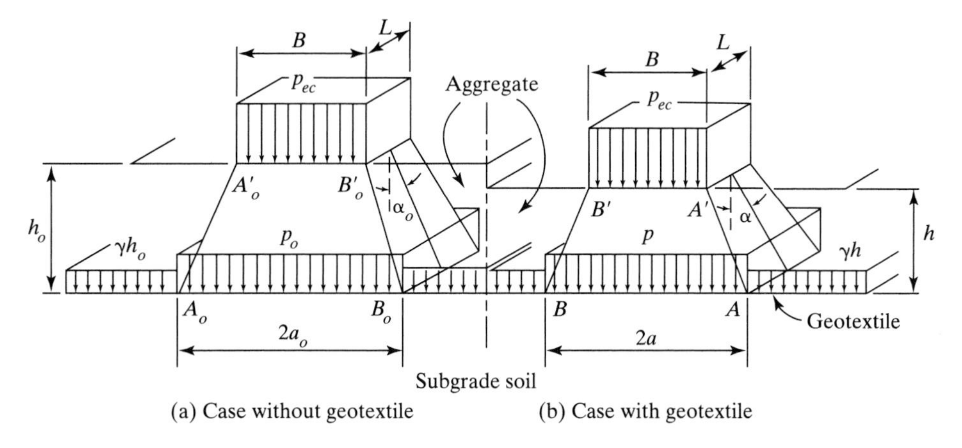

Figure 2.36 Load distribution by aggregate layer. (After Giroud and Noiray [67])

Analytic Method. Giroud and Noiray [67] use the geometric model shown in Figure 2.36 for a tire wheel load of pressure p_{ec} on a $B \times L$ area, which dissipates through h_o thickness of stone base without a geotextile and h thickness of stone base with a geotextile. The geometry indicated results in stress on the soil subgrade of p_o (without geotextile) and p (with geotextile) as follows.

$$p_o = \frac{P}{2(B + 2h_o \tan \alpha_o)(L + 2h_o \tan \alpha_o)} + \gamma h_o \tag{2.32}$$

$$p = \frac{P}{2(B + 2h \tan \alpha)(L + 2h \tan \alpha)} + \gamma h \tag{2.33}$$

where

P = axle load, and
γ = unit weight of the stone aggregate.

Knowing the pressure exerted by the axle load through the aggregate and into the soil subgrade, the shallow-foundation theory of geotechnical engineering can now be used. Assumed throughout the analysis is that the soil is functioning in its undrained condition and thus that its shear strength is represented completely by the cohesion (i.e., $\tau = c$). Thus the tacit assumption is that the soil subgrade consists of saturated fine-grained silt and clay soils. Critical in this design method are the assumptions that without the geotextile the maximum pressure that can be maintained corresponds to the elastic limit of the soil, that is,

$$p_o = \pi c + \gamma h_o \tag{2.34}$$

and that with the geotextile the limiting pressure can be increased to the ultimate bearing capacity of the soil, that is,

$$p^* = (\pi + 2)c + \gamma h \tag{2.35}$$

These assumptions reasonably agree with the earlier findings of Barenberg and Bender [68] using small-scale laboratory tests, where on a deformation basis they found that large-scale ruts began at a $3.3c$ with no fabric reinforcement versus $6.0c$ with fabric (where c is the undrained soil shear strength).

Thus for the case of no geotextile reinforcement, Eqs. (2.32) and (2.34) can be solved to give Eq. (2.36), which results in the desired aggregate thickness response curve without the use of a geotextile.

$$c = \frac{P}{2\pi(\sqrt{P/p_c} + 2h_o \tan \alpha_o)(\sqrt{P/2p_c} + 2h_o \tan \alpha_o)} \tag{2.36}$$

where

c = soil cohesion,
P = axle load,
p_c = tire inflation pressure,
h_o = aggregate thickness, and
α_o = angle of load distribution ($\cong 26°$).

For the case where geotextile reinforcement is used, p^* in Eq. (2.35) is replaced by $(p - p_g)$, where p_g is a function of the tension in the geotextile; hence its elongation is significant. On the basis of the probable deflected shape of the geotextile-soil system,

$$p_g = \frac{E\varepsilon}{a\sqrt{1 + (a/2S)^2}} \tag{2.37}$$

where

E = modulus of geotextile,
ε = elongation (strain),
a = geometric property (see Figure 2.36), and
S = settlement under the wheel (rut depth).

Combining Eqs. (2.33), (2.35), and (2.37) and using $p^* = p - p_g$, gives Eq. (2.38), where h is the unknown aggregate thickness. It can be graphed for various rut-depth thicknesses and various moduli of fabrics.

$$(\pi + 2)c = \frac{P}{2(B + 2h \tan \alpha)(L + 2h \tan \alpha)} - \frac{E\varepsilon}{\alpha\sqrt{1 + (a/2S)^2}} \tag{2.38}$$

With these two sets of equations, the design method is essentially complete, since both h_o (thickness without a geotextile) and h (thickness with a geotextile) can be calculated. From these two values $\Delta h = h_o - h$ can be obtained, which represents the savings in aggregate due to the presence of the geotextile. For convenience, however, the result can be read directly from Figure 2.37. This figure also considers the effects of traffic. In this case, the required thickness h' becomes $h' = h_o' - \Delta h$, which is obtainable from the curves by subtracting the two ordinate values of h_o' and Δh. Note that the effect of service lifetime is in the form of number of vehicle passages.

Two examples follow: Example 2.11 illustrates the general design procedure [67] and Example 2.12 shows a specific example with an economic analysis included. The influence of rut depth has been further evaluated by Holtz and Sivakugan [69].

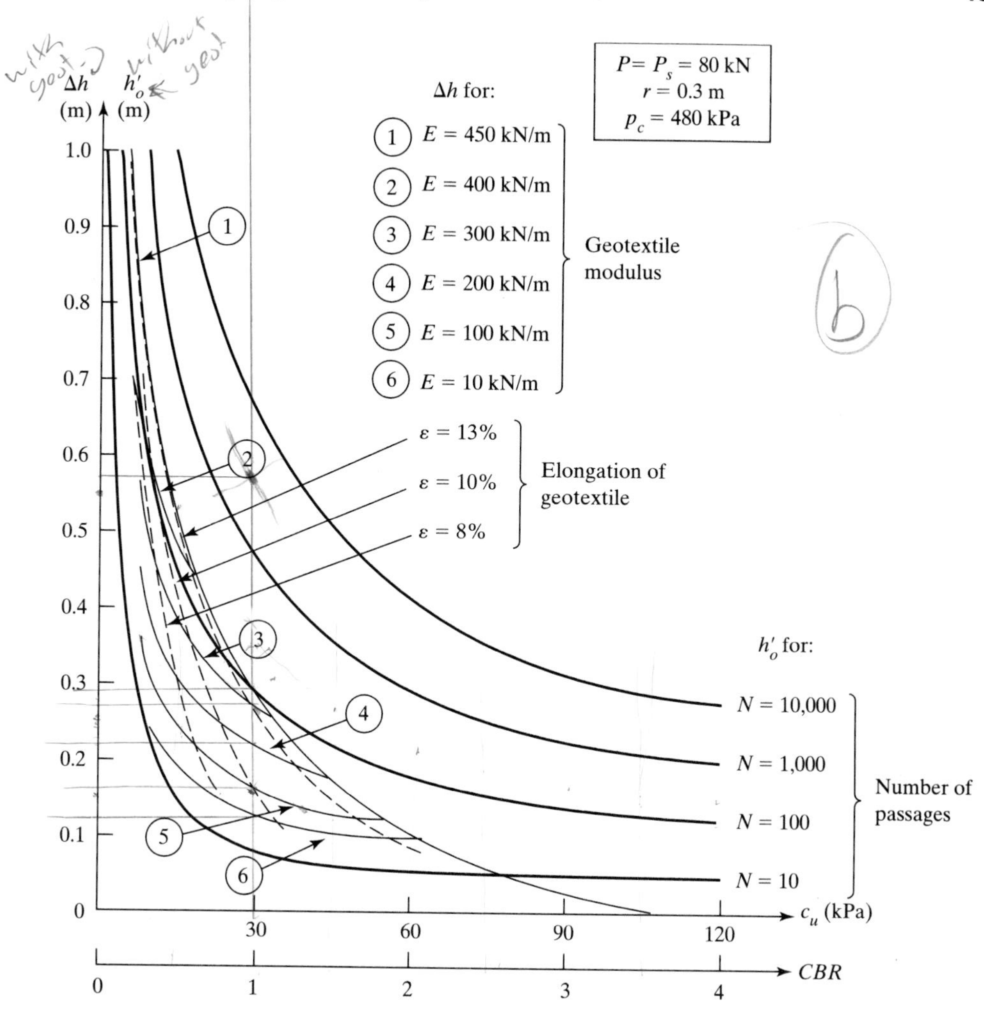

Figure 2.37 Reducing aggregate thickness with a geotextile. Aggregate thickness h'_0 without geotextile when traffic is taken into account; possible change in aggregate thickness (Δh) resulting from the use of geotextile rather than relying on subgrade soil cohesion. Chart related to an on-highway truck with standard-axle load. (After Giroud and Noiray [67])

Example 2.11

Given 340 passages of a 80 kN single axle-load vehicle with tire inflation pressure = 480 kPa, soaked soil CBR = 1, geotextile modulus E = 90 kN/m, and an allowable rut depth = 0.3 m, what is the required aggregate thickness of an unpaved road?

Solution: Figure 2.37 gives h'_o = 0.35 m for N = 340 and CBR = 1 (i.e., thickness when no geotextile is used). It also gives Δh = 0.15 m for E = 90 kN/m and CBR = 1 (i.e., the reduction in aggregate thickness when a geotextile-reinforcement layer is used). The difference between the two values, 0.20 m, is the required aggregate depth when the geotextile is used.

Example 2.12

Given 1000 passages of a 80 kN single-axle-load vehicle with a tire inflation pressure of 480 kPa on a soaked soil CBR = 2, a candidate geotextile modulus E = 170 kN/m, and an allowable rut depth = 0.3 m, **(a)** plot the response curve from Figure 2.37, **(b)** determine the aggregate savings, and **(c)** do an economic analysis based on the distance the project is from the stone quarry and geotextile supplier, respectively, using the following data (the stone unit weight is 20 kN/m^3):

Distance (km)	Aggregate cost (dollars/kN)	Geotextile cost (dollars/m^2)
≤5	0.90	0.72
5–20	1.20	0.76
20–50	1.70	0.78
50–100	2.50	0.84
100–200	3.80	0.90

Solution: (a) The required complementary curve to Figure 2.37 follows.

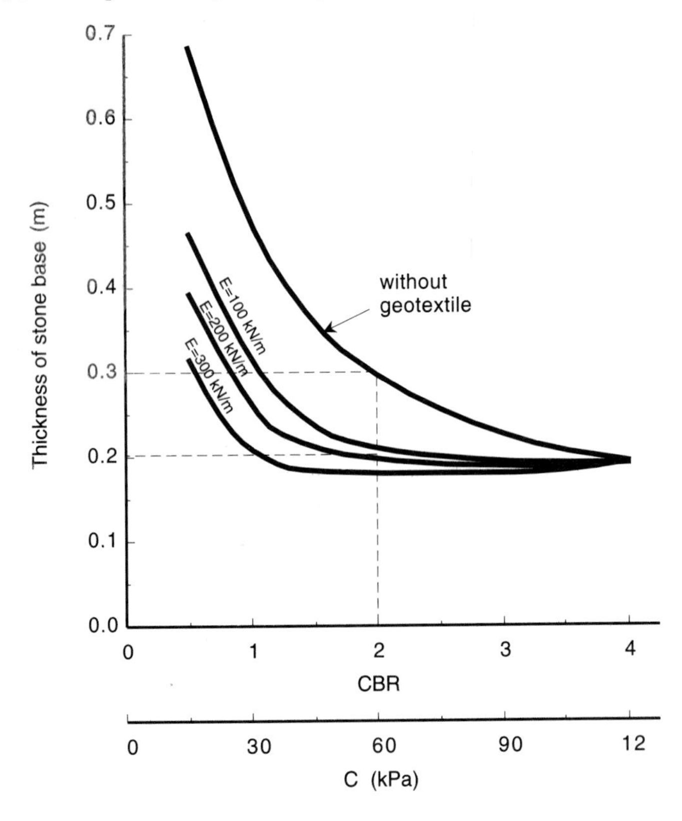

(b) At CBR = 2.0,

$$h'_o \text{ (without geotextile)} = 300 \text{ mm}$$
$$h' \text{ (with geotextile)} = 205 \text{ mm}$$
$$\Delta h \text{ (savings in stone)} = 95 \text{ mm}$$

(c) Based on 20 kN/m^3, this is a 0.020 kN/m^2-mm stone thickness, which results in the following table. It is easily seen is that the use of the geotextile is very economical and becomes more so as the distance from the stone quarry to the project site becomes greater.

Distance (km)	Aggregate cost (dollars/kN)	Aggregate cost (dollars/m^2-mm)	Aggregate savings (dollars/m^2)	Geotextile cost (dollars/m^2)	Geotextile savings (dollars/m^2)
<5	0.90	0.018	1.71	0.72	0.99
5–20	1.20	0.024	2.31	0.76	1.55
20–50	1.70	0.035	3.31	0.78	2.53
50–100	2.50	0.050	4.79	0.84	3.95
100–200	3.80	0.075	7.13	0.90	6.23

Laboratory Method. If laboratory facilities are available, it is possible to model the situation so as to arrive at the reinforcement ratio provided by the geotextile. The procedure is as follows:

1. Take the lower portion of a standard laboratory CBR mold and fill it with the soil in question at its in situ density and water content.
2. Place crushed stone in the upper portion of the mold.
3. With the load piston on top of the stone, perform a load-versus-deflection test at discrete intervals of piston deflection and record the data.
4. Using a CBR mold that has been modified to hold a geotextile at the interface between the soil subgrade and the crushed stone, repeat the test with the candidate geotextile in position and record the data. (The modification can be made by welding flanges to the upper and lower sections of the CBR mold and clamping or bolting the geotextile between the flanges).
5. Calculate the ratio of the loads at each deflection increment. The data in Table 2.15 show this reinforcement ratio for four separate test sets of a geotextile placed on a kaolinite clay at different water contents. Here we see that the reinforcement ratio increases as both the deflection and the water content increase.

TABLE 2.15 LABORATORY-OBTAINED REINFORCEMENT RATIOS FROM MODIFIED CBR TESTS

	Kaolinite Clay* Soil at Water Content			
Deflection (mm)	32%	35%	38%	41%
3.3	1.0	1.0	1.2	1.4
6.7	1.0	1.1	1.3	1.7
10	1.0	1.2	1.5	2.0
13	1.1	1.3	1.7	2.2
25	1.3	1.5	2.0	2.4
37	1.5	1.8	2.4	3.0
50	1.8	2.2	3.0	3.4

*with shrinkage limit, w_s = 18%; plastic limit, w_p = 32%; liquid limit, w_l = 41%

6. Assuming that this reinforcement ratio can be used as a multiplier to the in situ CBR of the soil, a number of accepted design procedures can be used to arrive at an aggregate thickness with and without geotextiles. Example 2.13 illustrates such an approach.

The laboratory test method, although straightforward, is not without its faults. These faults are the influence of geotextile tension before loading begins and the uncertainty involved with scale effects occurring during laboratory tests of this type.

Example 2.13

Using the U.S. Army Corps of Engineers Modified CBR Design Method (WES TR 3-692), calculate the required stone-base thickness for an unpaved road carrying 5000 coverages of 45 kN equivalent single-wheel loads using a tire contact area of 300 × 450 mm for **(a)** stone on a kaolinite clay soil at 41% water content with a CBR = 1.0 with no geotextile reinforcement, and **(b)** the same conditions with a geotextile whose reinforcement ratios are typical of Table 2.15 at a 25 mm deflection. Then **(c)** compare the resulting thicknesses.

Solution: The essential formula is

$$h = (3.24 \log C + 2.21)\left(\frac{P}{36.0 \times \text{CBR}} - \frac{A}{2030}\right)^{1/2} \tag{2.39}$$

where

h = aggregate thickness (mm),
C = traffic in terms of coverages,
P = equivalent single wheel load (N), and
A = tire contact area (mm^2).

This leads to the general equation

$$h = (3.24 \log 5000 + 2.21)\left(\frac{45{,}000}{36.0 \times \text{CBR}} - \frac{(300)(450)}{2030}\right)^{1/2}$$

$$= (14.19)\left(\frac{1250}{\text{CBR}} - 66.5\right)^{1/2}$$

(a) For no geotextile reinforcement and CBR = 1.0, the required thickness is

$$h'_o = (14.19)(34.4)$$
$$= 488 \text{ mm}$$

(b) When using a geotextile that results in an equivalent CBR = 2.4 (from Table 2.15, 2.4 × 1.0 = 2.4), the thickness is

$$h' = (14.19)(21.3)$$
$$= 302 \text{ mm}$$

(c) Thus the savings in stone base (Δh) afforded by using a geotextile is

$$\Delta h = h'_o - h'$$
$$= 488 - 302$$
$$= 186 \text{ mm } (\cong 38\% \text{ savings})$$

Sewn Seams. With the soft compressible soils under consideration in this section on unpaved roads, the matter of geotextile overlap for transferring stress across

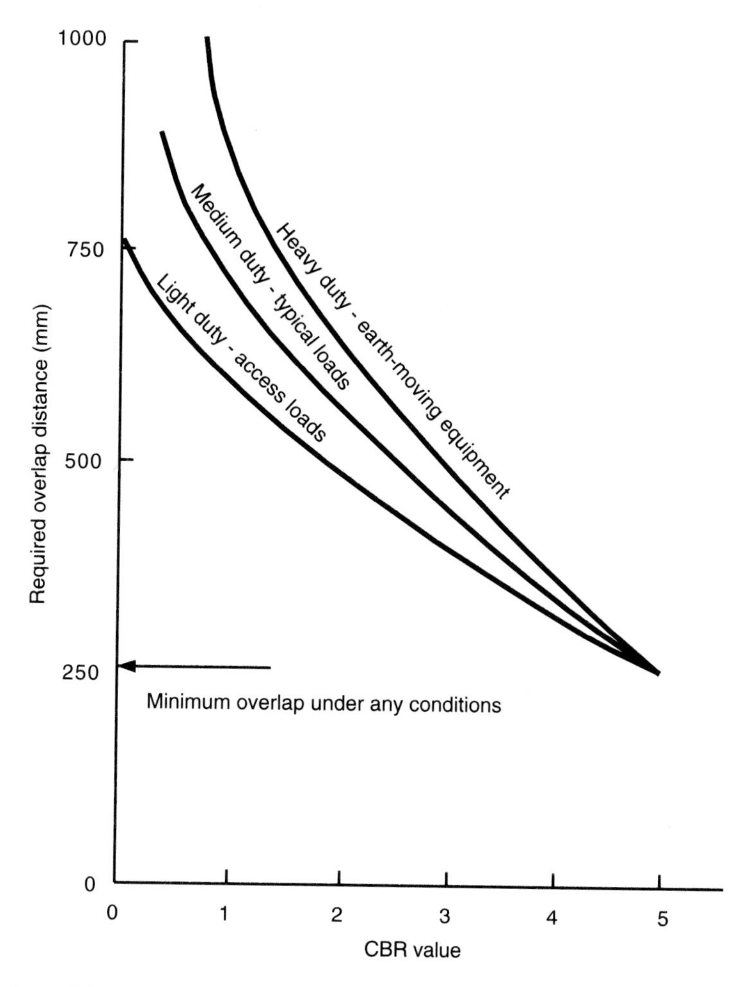

Figure 2.38 Recommended overlap for geotextiles used in unpaved roads as a function of unsoaked soil subgrade CBR value.

rolls becomes an issue. This overlap affects both the longitudinal sides and the transverse ends of the geotextile rolls. As expected, the softer the soil, the greater the necessary amount of overlap. Figure 2.38 gives a guide for different types of use on the basis of the amount of overlap required. It is easily seen that large overlap distances are required for low-strength soils. Not only is this wasted geotextile but it necessitates the calculation of geotextile-to-geotextile friction (recall the shear tests of Section 2.3.3). As a result, field-sewing of the geotextiles is generally preferred; see Figure 2.39. Example 2.14 illustrates how field-sewing of the geotextile can be very economical.

Example 2.14

Using the overlap guide in Figure 2.38, calculate when geotextile sewing becomes more economical than geotextile overlap for a single seam down the center of a 2000 m long access road. The costs are $1.75/m^2$ for a heavy geotextile, $1.37/m^2$ for a medium geotextile,

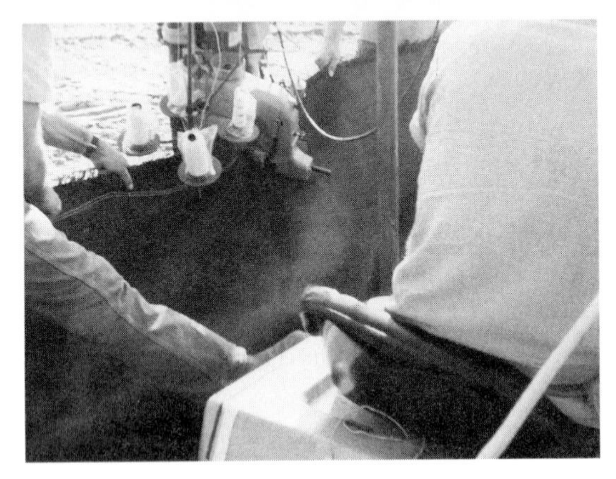

Figure 2.39 Field sewing of geotextiles. (Compliments of Union Special Corp.)

and $1.00/m² for a light geotextile. The sewing costs are $400 per day for sewing machine rental and thread and $175 per day each for three laborers, who can easily sew a 2000 m seam in one day.

Solution: The following chart is constructed from data in Figure 2.38 and the costs given.

Soil Subgrade CBR	Type of Loads	Overlap Required (m)	Geotextile Required (m²)	Overlap Geotextile Cost ($/2000 m)		
				Heavy ($1.75/m²)	Medium ($1.37/m²)	Light ($1.00/m²)
1.0	Heavy	0.86	1720	3010	2360	1720
	Medium	0.71	1420	2480	1940	1420
	Light	0.58	1160	2030	1590	1160
2.5	Heavy	0.57	1140	1990	1560	1140
	Medium	0.51	1020	1780	1400	1020
	Light	0.44	880	1540	1200	880
4.0	Heavy	0.37	740	1300	1010	740
	Medium	0.35	700	1220	960	700
	Light	0.33	660	1150	900	660

Since the total sewing costs are $925 per day, it is seen that only on the lower-right portion of the chart (on relatively strong subgrades with light geotextiles for light- and medium-duty vehicles) will sewing not pay for itself.

When field-sewing geotextiles, a number of details must be addressed. They are:

Thread type. The choices are polyester, polypropylene, and polyamides. Consideration should be given to using the same thread type as geotextile fiber type.

Thread tension. This is usually adjusted in the field so as to be sufficiently tight without cutting the geotextile.

Stitch density. Two, three, or four stitches per inch are customary.

Stitch type. The choices are prayer, J-type, or butterfly (see Figure 2.40), the strongest being the butterfly type;

Number of rows. One, two, or three are customary; generally two are recommended (see Figure 2.40).

Type of chainstitch. The 401 two-thread is recommended.

The sewing of geotextiles has rapidly advanced to the point where all fabric construction on soft ground should consider its use. A geotextile sewing guide is provided in [71]. Tensile seam strengths of 170 kN/m have been attained (recall Figure 2.12) and productivity has reached a point where sewing is no longer an obstacle impeding the rapid progress of the work.

2.6.2 Membrane-Encapsulated Soils

Concept and Overview. It is beyond question that well-graded angular sands and gravels make the best base course materials for both paved and unpaved roads. Such coarse-grained soils provide good structural stability and adequate drainage in both vertical and horizontal directions. Yet in many areas of the world, such soil types are simply not available. In abundance, however, are fine-grained silts, clays, and related mixed soils. Although such materials can be placed in a stable condition (by careful control of their moisture content), their performance over time is generally poor. During wet seasons they take up moisture and become excessively wet and soft, while during dry seasons they lose moisture and become friable and weak. Their behavior is particularly poor in cold regions, where such fine-grained soils exhibit marked frost susceptibility, which involves excessive heave while frozen and rapid collapse during thaw periods.

The idea of encapsulating such fine-grained soils in a state just lower than their optimum moisture content and preventing moisture migration from occurring has created a good deal of interest in places that lack good-quality stone base. Beginning in 1930 with the Bavarian Highway Department's use of prefabricated asphalt panels [72], the concept was extended by Casagrande in 1937 to the use of bitumen-coated jute fabric in the same area of Germany. It should also be noted that the Asphalt Institute [73] had recommended an asphalt barrier, with or without membranes, as capillary cutoffs even before 1930. The use of geomembranes (i.e., plastic sheets) for encapsulation

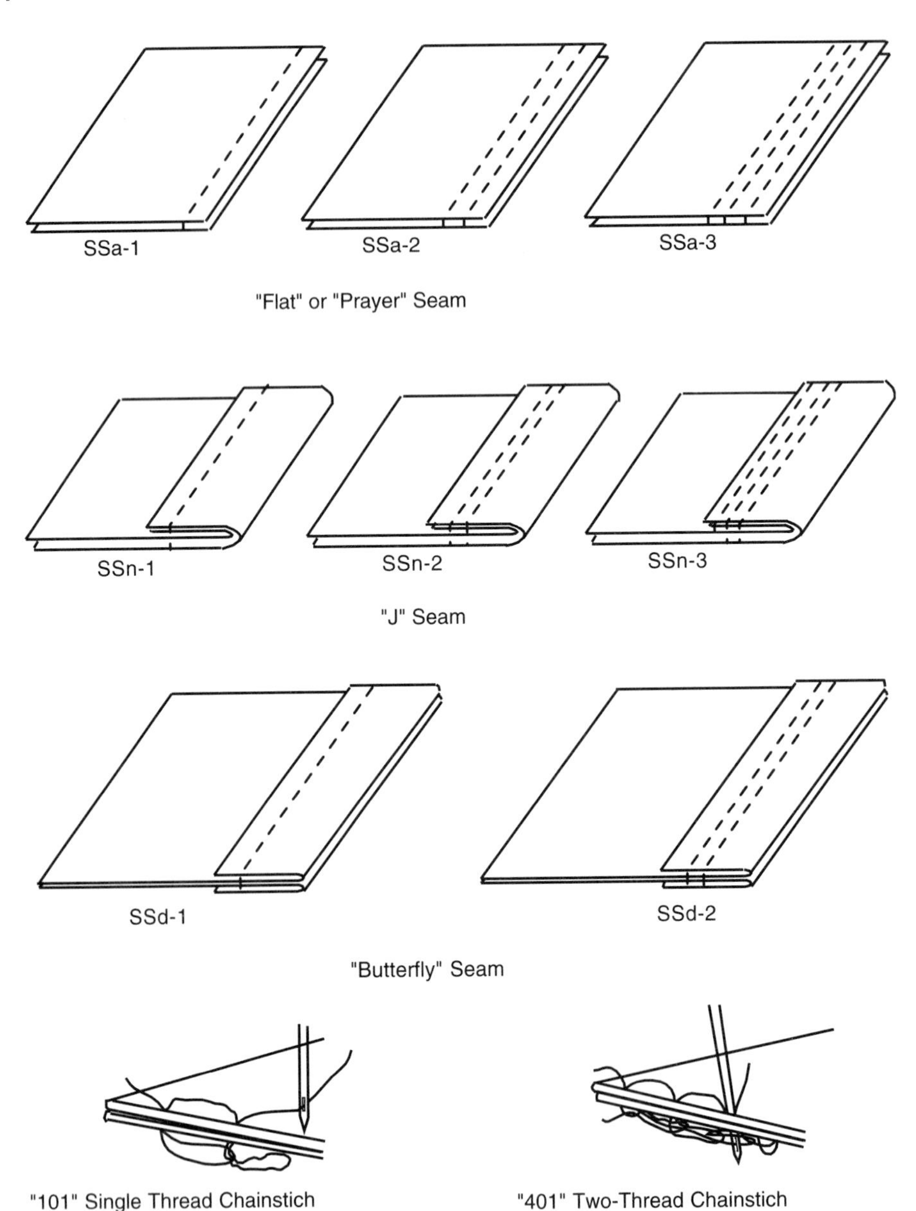

Figure 2.40 Various types of sewn seams for joining geotextiles. (After Diaz [70])

began about 1953 by Bell and Yoder [74] at Purdue University. They used both polyethylene and polyvinyl chloride sheets that were sealed with adhesive at their overlapping seams. Later systems developed by the U.S. Army Cold Regions Research and Engineering Laboratory [75] were hybrids, in that a geomembrane was used beneath and on the sides of the soil to be encapsulated, and an asphalt emulsion-impregnated geotextile was used as a cover.

All of these early attempts at encapsulating poor-quality soils were aimed at providing a moisture-tight barrier beneath the roadway's wearing surface. Thus, in the context of this book, the primary function of the geosynthetic materials so used is clearly as a *moisture barrier*, although there is a secondary function of containment reinforcement as well. Today the major applications are the following:

- To encapsulate frost-sensitive silt soils from problems associated with freeze-thaw cycles.
- To encapsulate expansive and heavy clays from volume changes associated with wet-dry cycles.
- To protect subgrade soils from exposure to water from rainfall or snowmelt.
- To minimize the amount of granular soil used to bridge over weaker or moisture-sensitive subsoils.
- To make it possible to use a substandard type of granular soil when its saturated strength is too low for the imposed traffic loads or its gradation is improper.

As currently practiced, membrane-encapsulated soil layers (MESLs) use geotextiles as the base material, which are then impregnated with a sprayed-on bituminous or elastomeric product (see Figure 2.41), thereby forming bottom, sides, and top around the enclosed soil. There are good reasons for using geotextiles over other materials to solve this problem:

- Geotextiles, as they are made from plastic materials, do not degrade with time as do jute or cotton fabrics.
- Geotextiles mobilize good friction resistance and have high survivability characteristics (puncture, tear, and impact resistance) compared with geomembranes.
- Bitumen- or polymeric-impregnated geotextiles, although not watertight under hydrostatic pressure, are adequate for eliminating water migration across them under most circumstances (to provide an essentially watertight barrier is usually not necessary).

Figure 2.41 Spraying of a geotextile to form the bottom of a membrane encapsulating soil layer. (Photo compliments of Chevron Research Co.)

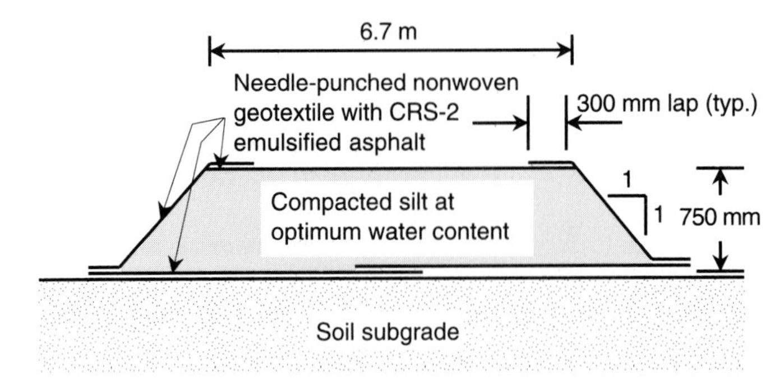

(a) Above-ground construction in cold regions (after Smith and Pazsint [76])

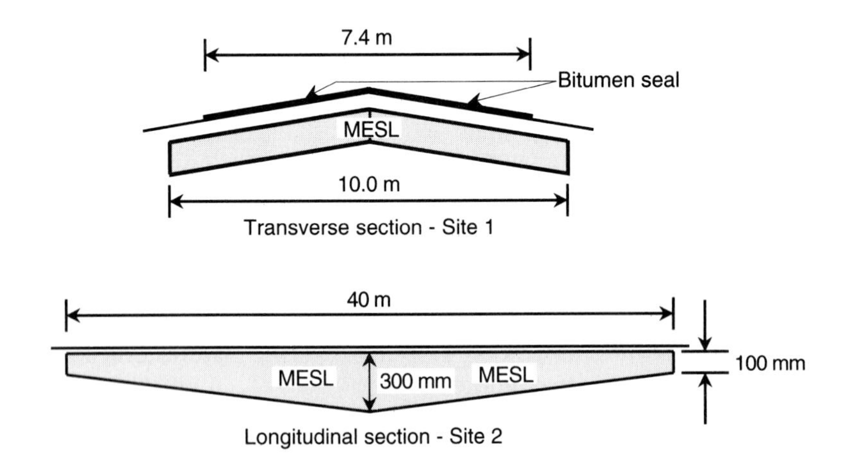

(b) Below-ground construction in semiarid regions (after Lawson and Ingles [77])

Figure 2.42 Various cross sections of membrane-encapsulated soil layers.

- Overlapped or sewn seams can easily be made using geotextiles.
- Geotextiles seem to offer the most cost-effective solution.

Design Methods. MESL designs can be characterized as being either above grade or below grade. An example [76] of an above-grade scheme is shown in Figure 2.42a, in which a frost-sensitive silt soil near Fairbanks, Alaska, has been encapsulated in a nonwoven needle-punched geotextile sprayed with CRS-2-emulsified asphalt. A MESL approach was made necessary by a general lack of granular soils in the arctic and subarctic areas together with the problem of permafrost. Conversely, moisture-sensitive silt is very available. Upon thawing, these silt soils weaken considerably, causing considerable surface distortion (deep ruts) and local bearing-capacity failures.

The subgrade was prepared, rolled, and covered with the geotextile. CRS-2-emulsified asphalt was used to seal the geotextile and included the overlapped centerline joint. A road grader was used to mix the silt so as to bring it to its optimum water content. The silt was then placed in three lifts of 250 mm each to a total height of 750 mm. The upper surface and sides were covered with the same type of geotextile, and sealing was done using the same type of emulsified asphalt. A 40 mm layer of gravely sand was spread over the top surface so that traffic would not travel directly on the geotextile.

Using the same MESL concept in a semiarid region of Australia, Lawson and Ingles [77] report on two below-grade case histories (see Figure 2.42b). Both sites were subjected to water infiltration into the subgrade and had low unsoaked CBR values, typically 3 to 4 in the wet season and about 7 in the dry season. The soil encapsulated in the nonwoven heat-bonded geotextile was more granular than the subgrade, and the goal was a reinforcing function, in addition to the waterproofing. A sprayed-on bituminous product was used as the sealant. Note that the longitudinal section of Site 2 had a varying thickness of 300 mm at the center, tapering to 100 mm at each end. The transverse section was uniform, as was the case for the entire Site 1.

Regarding a design methodology for providing a water barrier using MESLs, it is essential that the geotextiles be impregnated with a proper sealant. This is usually bitumen, although coal tar [77] and various elastomers [78] have also been used. Sometimes polyester geotextiles are used rather than polypropylene, due to their smaller tendency to swell when subjected to hydrocarbons. The saturation process is similar to the use of geotextiles beneath bituminous overlays to prevent reflection cracking, which will be treated in Section 2.10.2. To saturate geotextiles, Murray [79] has found that 0.25 to 1.5 l/m^2 is necessary for geotextiles in the weight range of 100 to 200 g/m^2. Unlike the reflective cracking application, more sealant can be used than is required for saturation, and the values above should be considered as minimum values. Recommended sealants [80] are asphaltic cement (AC-10 or AC-20), cationic asphalt emulsion (CRS-2 or CRS-1h), and anionic asphalt emulsion (RS-2 or RS-1). Cutback asphalts should not be used with polypropylene fabrics, since the solvent in them reacts with the polymer under high temperatures and/or during long durations. The cure period for most of these sealants is 30 min to 4 hr; however, extremely cold temperatures will greatly extend the time required.

Properly done, this saturation of the geotextile will exhibit a wetting pressure within it greater than the water pressure in the surrounding soil. Thus water movement into and through the barrier cannot occur. Water *vapor* transport, however, will occur. To quantify this value of water-vapor transport, the governing equation that can be used is Fick's first law, which assumes steady-state conditions:

$$\frac{\partial W}{\partial t} = -KA\frac{\partial C}{\partial x} \tag{2.40}$$

where

$\frac{\partial W}{\partial t}$ = rate of water diffused by weight,

$\frac{\partial C}{\partial x}$ = vapor gradient across the barrier,
A = surface area of the barrier, and
K = diffusion constant of the barrier.

The critical parameter is K, which is also known as the water-vapor transmission constant (not to be confused with k, the permeability coefficient, which is related to water flow) and can be found experimentally using the water-vapor transmission test, ASTM E96. The procedure is given in Lord and Koerner [81] together with other diffusion and transmission measurement techniques, and values are given by Lawson and Ingles [82].

The first-order differential equation [Eq. (2.40)] is integrated over boundary conditions of

At $x = 0$: $C = C_1$
At $x = L$: $C = C_2$

which yields

$$W = \frac{KA(C_2 - C_1)\Delta t}{L}$$

and if $K' = K/L$

$$W = K'A\, \Delta C\, \Delta t \tag{2.41}$$

where

K' = permeance,
K = water-vapor transmission,
L = thickness,
A = surface area,
ΔC = change in vapor pressure,
Δt = elapsed time, and
W = amount of water, via vapor transmission, gained or lost over time Δt.

The design procedure becomes one of calculating the water weight gained or lost within the MESL to see if an acceptable condition exists. Example 2.15 illustrates this procedure.

Example 2.15

Given a below-grade MESL constructed from a bitumen impregnated geotextile with a permeance $K' = 5 \times 10^{-5}$ day^{-1}, the following cross section, and a vapor-pressure gradient over the four month wet season of 6.0 kPa, what will be the increase in the moisture content of the soil within the MESL if it is placed at a unit weight of 18 kN/m^3 and at 12% moisture content?

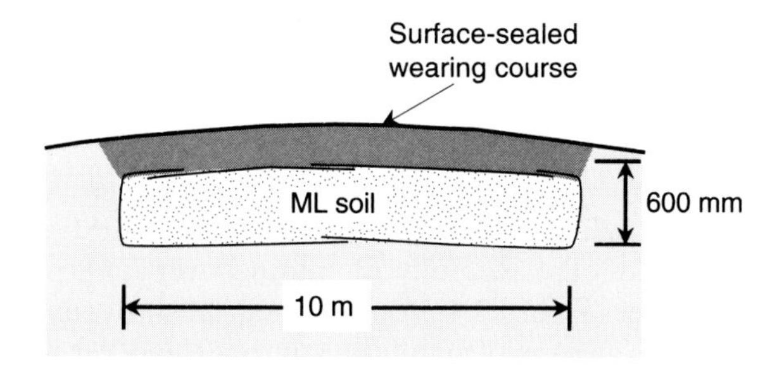

Solution: The weight of water gained, W, is:

$$W = K'A\,\Delta C\,\Delta t$$
$$= (5 \times 10^{-5})(21.2)(6.0)(1000)(121)$$
$$W = 770 \text{ N/m}$$

Now using standard soil mechanics principles, calculate the change in water content.

As constructed:

$$\text{Total weight wet soil} = 10 \times 0.6 \times 18{,}000 = 108{,}000 \text{ N/m} = W_W + W_S$$
$$\text{but } W_W/W_S = 0.12$$
$$\text{Therefore, total weight water} = 11{,}600 \text{ N/m}$$

After the wet season:

$$\text{Total weight water} = 11{,}600 + 770$$
$$= 12{,}400 \text{ N/m}$$
$$\text{Total weight dry soil} = 108{,}000 - 11{,}600 = 96{,}400 \text{ N/m}$$
$$\text{Water content} = \frac{12{,}400}{96{,}400}$$
$$= 12.9\%$$

Thus the water content of the soil has increased from 12.0% to 12.9% after four months of exposure under these stated conditions. Clearly, the bitumen-impregnated geotextile is performing its intended function as a barrier.

Some comments on the design procedure and its assumptions using this approach are in order. First, there is a problem concerning the units of permeance. Water-vapor transmission is given in units of g/m^2-day, which when divided by the vapor-pressure gradient gives metric-perms. Its units are

$$\left(\frac{\text{g}}{\text{m}_2\text{-day}}\,\frac{1}{\text{mm Hg}}\right)$$

This must be converted to day^{-1} to use the design procedure described above. Second, the driving force in the problem, the anticipated vapor-pressure gradient in the field,

ΔC, is very difficult to estimate in the absence of tensiometers and field measurements. In addition, the time frame for the calculations is an assumption. A factor of safety can be effectively used in the design in light of the statements above; a value of 3.0 is recommended.

Field Behavior. The cross sections of the MESLs shown in Figure 2.42 are all test sections and were carefully monitored over long periods. The above-grade silt MESL in Alaska (Figure 2.42a) performed satisfactorily, although two problems occurred. One was an area of significant rutting; the other was a section where the seam separated. Both occurred when the geotextile was exposed to surface water; thus both problems were related to the cover material—the 40 mm of gravely sand—rather than to the performance of the MESL itself. Clearly, a more durable surface course is warranted in this situation. An additional measure would be to sew all the seams, particularly those on the top and sides of the MESL, where overburden pressures from cover soils are low.

A large amount of information was gained from the below-grade MESLs in Australia shown in Figure 2.42b. Observations were made over a period of seven years, corresponding to 300,000 standard axle loads, which is approximately one-half of its design life. Moisture content and surface deflections were taken periodically on both MESL sections and on the adjacent control sections that did not have MESLs. Several general observations can be made.

First, no degradation of the bitumen-sealed geotextile was observed. even though it is a relatively thin, nonwoven heat-bonded geotextile. Second, there was good bond between the adjacent soil and the impregnated geotextile. At Site 2 surface cracking was successfully arrested by the MESL. Third, the moisture content of the MESL soil was better than in the control sections, which in turn led to lower surface deflections of the MESL sections. The overall performance of the MESL sections was unquestionably superior to the control sections. Severe crack crazing and edge failures occurred in control sections on both sides of the MESL of Site 2, but none within the MESL section itself. Finally, no estimate of service-life equivalence could be obtained, since even the thinnest MESL of 100 mm has continued to perform impeccably for seven years under traffic.

2.6.3 Paved Roads

Whenever geotextile use in unpaved roads is discussed, the question of the material's use in paved roads usually follows. To address this properly, we must focus on the general characteristics of the situation. It is most important to recognize that if the road is to be paved with concrete or asphalt immediately (i.e., during construction), it cannot be placed on an excessively yielding soil subgrade. If the subgrade yields, the road section will deform and the surfacing will simply crack up after a few load repetitions. Many agencies put the *lower limit* of acceptable unsoaked CBR values in the range 10 to 15. As just discussed, however, the geotextile must deform in order to mobilize its strength and the *upper limit* of soil subgrade strength for such mobilization as suggested in Table 2.14 is an unsoaked CBR value of 3 to 8. This contradiction begs the question of how the geotextile is to reinforce if it is not significantly deforming. Advocates of a reinforcement function in paved roads on firm soil subgrades will suggest that the geotextile deformation around the coarse-aggregate base course (when heavily rolled) is sufficient to mobilize the geotextile's strength. Thus the design can proceed in a manner similar to that for unpaved roads.

Those who feel this is not the case still desire a geotextile under the stone base, but for reasons other than soil subgrade reinforcement. Here the primary function becomes separation, discussed in Sections 2.2.1 and 2.5, or filtration, handled in Sections 2.2.3 and 2.8. The economic justification is the longer service lifetime with a geotextile separator than without. Recall the discussion in Section 2.5.6 and Figure 2.35.

When separation is the primary function in paved road applications, it is important to recognize where the geotextile should be located with respect to the pavement cross sections and applied loads. In a trial test site with 40 mm of asphalt paving, 150 mm of base course, and 100 mm of large crushed stone, a lightweight geotextile (150 g/m^2) failed under 165,000 repetitions of a standard 80 kN axle load [83]. This premature geotextile failure was evidenced by abrasion of the yarns followed by fines pumping up from the subgrade into the stone base. Although no specific design is available to explain the situation, it does illustrate that a minimum set of geotextile properties is required in most situations. In other words, a survivability criterion is required to ensure adequate performance in general situations.

To specifically add reinforcement for paved roads on firm subsoils, a geotextile pre-tensioning system is required. By pretensioning the geotextile, the stone base will be placed in compression (i.e., thereby providing a lateral confinement) and will effectively increase its modulus over the nonreinforced case. Some of these concepts are discussed in Section 2.7.4, however, they are extremely difficult to implement.

2.7 DESIGNING FOR SOIL REINFORCEMENT

This section continues the discussion of the use of geotextiles in the primary function of reinforcement. Since this was the topic of the preceding section involving road systems, it could easily have been incorporated into that section. However, this type of soil reinforcement raises a unique set of applications, whereby the geotextile in layers and the interspersed soil form a system rather than acting as discrete material elements. The problems here involve wall reinforcement, embankment (slope) reinforcement, foundation reinforcement, and in situ slope reinforcement.

2.7.1 Geotextile Reinforced Walls

Background. Conventional gravity and cantilever-wall systems made from masonry and concrete resist lateral earth pressure by virtue of their large mass. They act as rigid units and have served the industry well for centuries. However, a new era of retaining walls was introduced in the 1960s by H. Vidal with Reinforced Earth®. Here metal strips extending from exposed facing panels back into the soil serve the dual role of anchoring the facing units and being restrained through the frictional stresses mobilized between the strips and the backfill soil. The backfill soil both creates the lateral pressure and interacts with the strips to resist it. The walls are relatively flexible compared to massive gravity structures. They offer many advantages, including significantly lower cost per square meter of exposed surface. A steady series of variations followed Vidal's, all of which can be put into the flexible retaining-wall category:

- facing panels with metal strip reinforcement
- facing panels with metal wire mesh reinforcement
- solid panels with tieback anchors
- anchored gabion walls
- anchored crib walls
- geotextile-reinforced walls (described here)
- geogrid-reinforced walls (described in Chapter 3).

In all cases, the soil behind the wall facing is said to be *mechanically stabilized earth (MSE)* and the wall system is generically called an MSE wall.

Construction Details. Critical in the proper functioning of a geotextile-reinforced MSE wall is its proper construction, which is done on a planned sequential basis. After preparing an adequate soil foundation, which consists of removing unsuitable material and compacting in situ or replacement foundation soils, the wall itself is begun. Note that there is no footing of any kind for these walls, and the lowest geotextile layer is placed directly on the foundation soil. An iterative construction sequence, developed by the U.S. Forest Service and illustrated in Figure 2.43, is followed.

1. A wooden form of a height slightly greater than the individual soil-layer thickness, called *lift height*, is placed on the ground surface or on the previously placed lift after the first layer is completed. This form is nothing more than a series of metal L brackets with a continuous wooden brace board running along the face of the wall.
2. The geotextile is then unrolled and positioned so that approximately 1.0 m extends over the top of the form and hangs free. If it is sufficiently wide, the geotextile can be unrolled parallel to the wall. In this way the geotextile's cross machine direction is oriented in the maximum stress direction. This will depend on the required design length and geotextile strength, which will be discussed later. If a single roll is not wide enough, two of them can be sewn together. The sewn strength is then a governing factor. Alternatively, the geotextile can be deployed perpendicular to the wall and the adjacent roll edges can be overlapped or sewn. In this way the geotextile's machine direction is oriented in the maximum stress direction.
3. Backfill, which should be free-draining granular sand and should not be angular gravel (which causes high installation damage), is now placed on the geotextile for 1/2 to 3/4 of its lift height and compacted. This is typically 200 to 400 mm and is done with light-weight construction equipment.
4. A windrow is made 300 to 600 mm from the face of the wall with a road grader or is dug by hand. Care must be exercised not to damage the underlying geotextile.
5. The free end of the geotextile, that is, its *tail*, is folded back over the wooden form into the windrow.

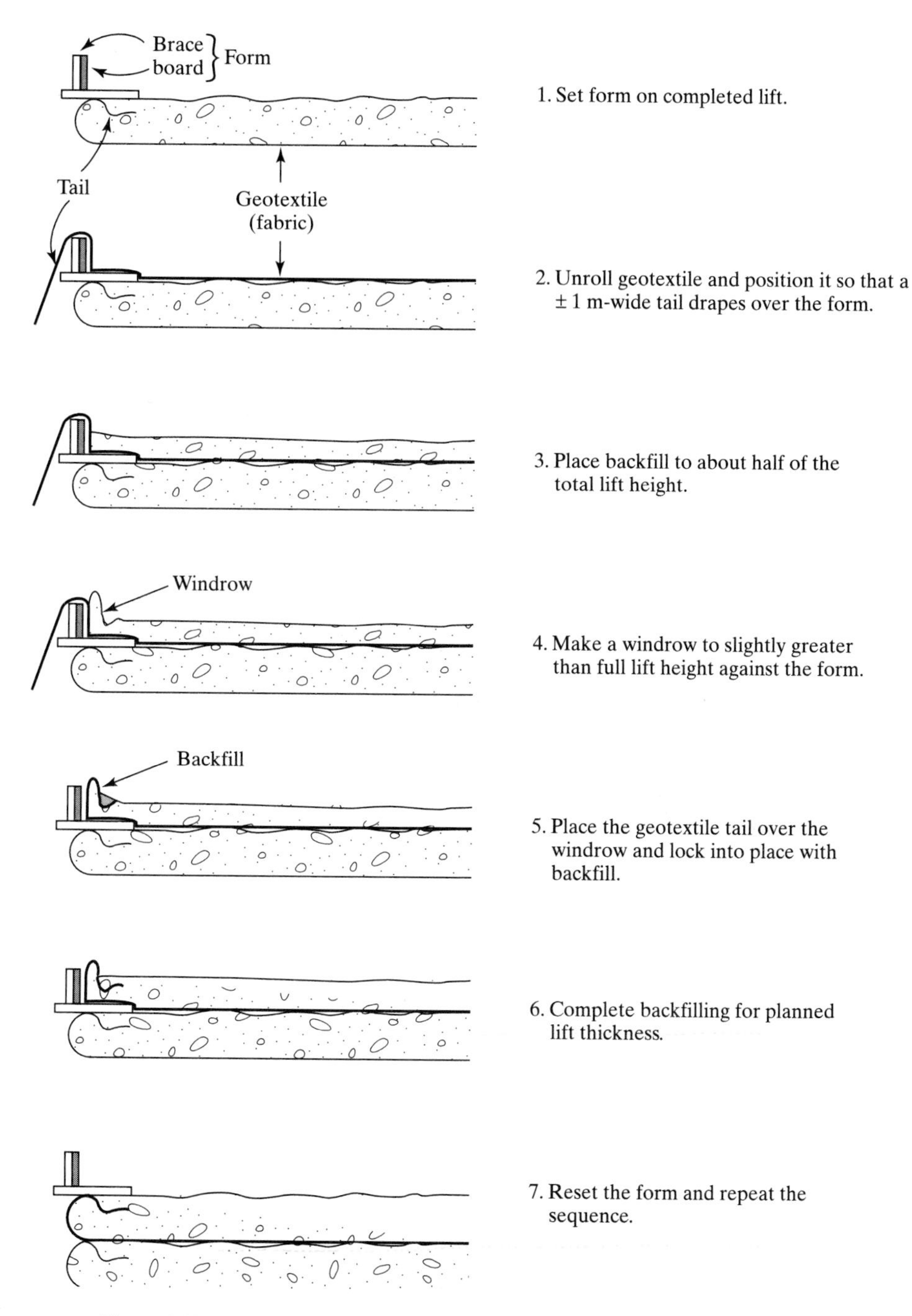

Figure 2.43 Construction sequence for geotextile walls followed by U.S. Forest Service.

6. The remaining lift thickness is then completed to the planned lift height and suitably compacted.
7. The wooden form is then removed from in front of the wall, and the metal brackets from beneath the lift, and the assembly is reset on top in preparation for the next higher lift. Note that it is usually necessary to have scaffolding in front of the wall when the wall is higher than 1.5 or 2.0 m.

When completed, this sequence provides walls such as in Figure 2.44. The exposed face of the wall must now be covered to prevent the geotextile's weakening due to UV exposure (recall Section 2.3.6) and possible vandalism. Bituminous emulsions or other asphalt products have been used for covering the wall face and have the advantage of being flexible, as is the wall itself. Unfortunately, oxidation of the bitumen causes deterioration after a few years and it must be periodically reapplied. Alternatively, the surface of wrap-around geotextile walls have been covered with shotcrete (wet-mixed cement/sand/water paste with air supplied at the nozzle) or gunite (dry cement/sand mix with water and air supplied at the nozzle). A wire mesh anchored between the geotextile layers may be necessary to keep the coating adhered to the vertical face of the wall.

Design Methods. There are two somewhat different approaches to the design of geotextile walls: that used by Broms [84] and that used by the U.S. Forest Service, Steward et al. [85], and Whitcomb and Bell [86]. The latter method will be followed in this book. This method follows the work that Lee et al. [87] did on reinforced earth with

Figure 2.44 Geotextile walls. (Compliments of Crown Zellerbach Corp.)

metallic strips and was originally adapted to geotextile walls by Bell et al. [88]. The design progresses in parts, as follows:

1. Internal stability is first addressed to determine geotextile spacing, geotextile length, and overlap distance.
2. External stability against overturning, sliding, and foundation failure is investigated and the internal design is verified or modified accordingly.
3. Miscellaneous considerations, including wall-facing details, are addressed.

To determine the geotextile layer separation distances, earth pressures are assumed to be linearly distributed using Rankine active *earth pressure* conditions for the soil backfill and *at rest* conditions for the surcharge. A prediction conference at the Canadian Royal Military College, however, has shown that the entire design to be presented here is quite conservative (see Jarrett, et al. [89]). Therefore, active earth pressure (K_a) conditions will be used throughout. An even less conservative approach would be to use a Coulomb analysis for the earth pressure values. This will be discussed later. Boussinesq elastic theory for live loads on the soil backfill is used. As shown in Figure 2.45, the following earth pressures result:

$$\sigma_{hs} = K_a \gamma z \tag{2.42}$$

$$\sigma_{hq} = K_a q \tag{2.43}$$

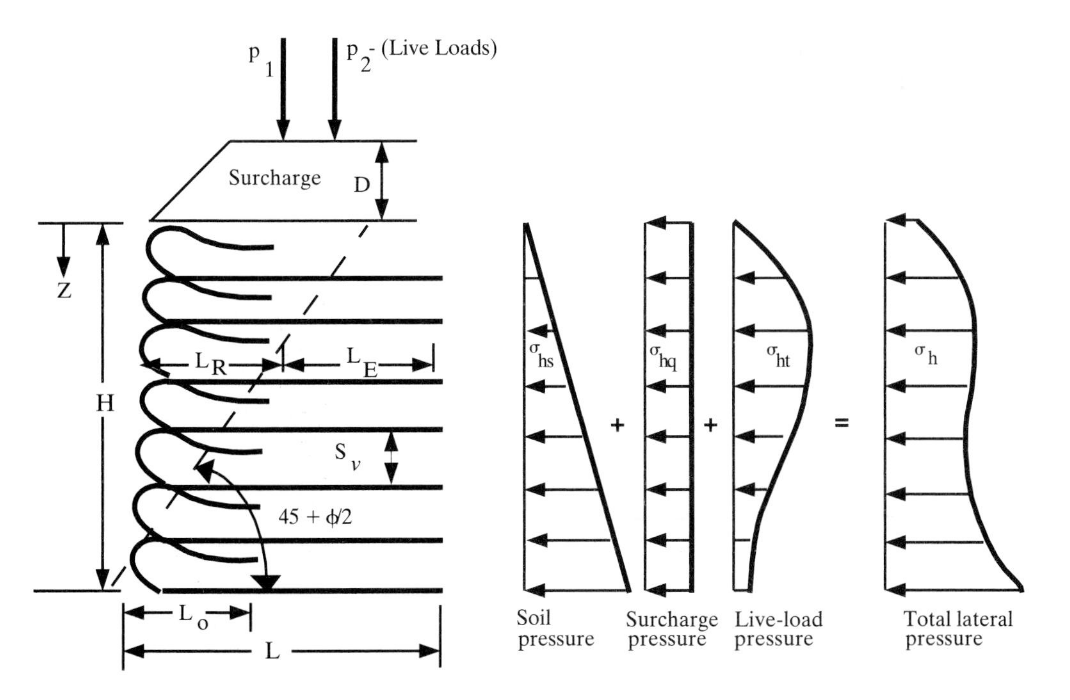

Figure 2.45 Earth pressure concepts and theory for geotextile walls.

$$\sigma_{hl} = P\frac{x^2 z}{R^5} \tag{2.44}$$

$$\sigma_h = \sigma_{hs} + \sigma_{hq} + \sigma_{hl} \tag{2.45}$$

where

σ_{hs} = lateral pressure due to soil;
$K_a = \tan^2(45 - \phi/2)$ = coefficient of active earth pressure;
ϕ = angle of shearing resistance of backfill soil;
γ = unit weight of backfill soil;
z = depth from ground surface to layer in question;
σ_{hq} = lateral pressure due to surcharge load;
$q = \gamma_q D$ = surcharge load on ground surface, where
γ_q = unit weight of surcharge soil, and
D = depth of surcharge soil;
σ_{hl} = lateral pressure due to live load;
P = concentrated live load on backfill surface;
x = horizontal distance load is away from wall; and
R = radial distance from load point on wall where pressure is being calculated.

The calculations of σ_{hs} and σ_{hq} are quite straightforward, but σ_{hl} presents problems, particularly for multiwheeled truck loads where superposition of each wheel must be performed. Figure 2.46 greatly aids in such calculations.

By taking a free body at any depth in the total lateral pressure diagram and then summing the forces in the horizontal direction, Eq. (2.46) for the lift thickness is obtained.

$$\sigma_h S_v = \frac{T_{\text{allow}}}{\text{FS}}$$

$$S_v = \frac{T_{\text{allow}}}{\sigma_h \text{FS}} \tag{2.46}$$

where

S_v = vertical spacing (lift thickness),
T_{allow} = allowable stress in the geotextile (recall Eq. (2.24) and Table 2.11),
σ_h = total lateral earth pressure at depth considered, and
FS = factor of safety (use 1.3 to 1.5 when using T_{allow} as determined above).

The same free-body approach can be taken to obtain the length of embedment of the geotextile layers in the anchorage zone, L_e. Note that when these values are obtained they must be added to the nonacting lengths (L_R) of the geotextile within the active zone for the total geotextile lengths (L); that is,

$$L = L_e + L_R \tag{2.47}$$

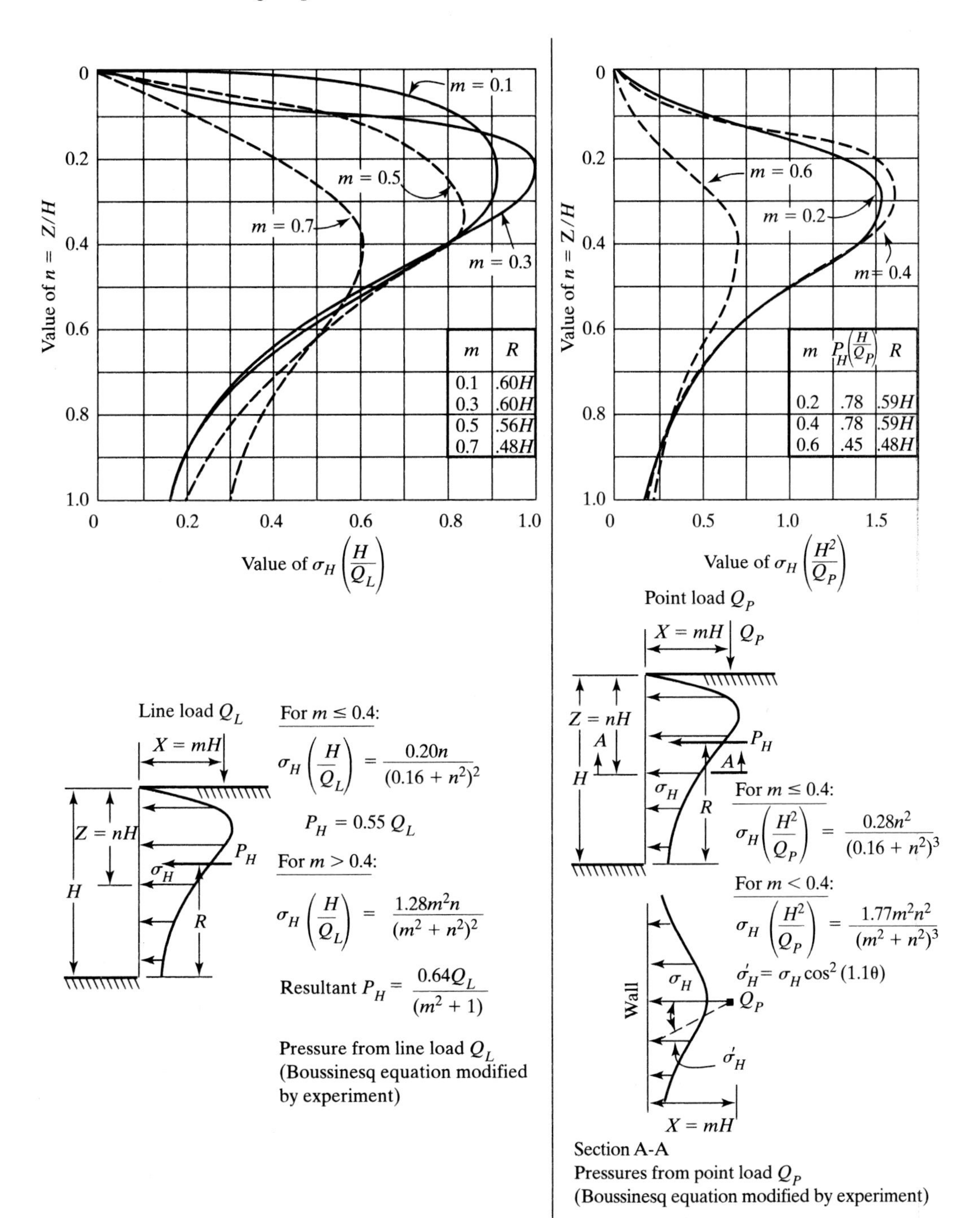

Figure 2.46 Lateral earth pressure due to a surface load. Left side is for line load; right side is for point load. (After NAVFAC [90])

where

$$L_R = (h - z) \tan \left(45 - \frac{\phi}{2}\right) \tag{2.48}$$

and

$$S_v \sigma_h \text{FS} = 2\tau L_e$$

$$= 2(c_a + \sigma_v \tan \delta) L_e$$

$$= 2(c_a + \gamma z \tan \delta) L_e$$

$$L_e = \frac{S_v \sigma_h \text{FS}}{2(c_a + \gamma z \tan \delta)} \tag{2.49}$$

where

τ = shear strength of the soil to the geotextile,
L_e = required embedment length (minimum is 1 m),
S_v = vertical spacing (lift thickness),
σ_h = total lateral pressure at depth considered,
FS = factor of safety,
c_a = soil adhesion between soil and geotextile (zero if granular soil is used),
γ = unit weight of backfill soil,
z = depth from ground surface, and
δ = angle of shearing friction between soil and geotextile.

Finally, the overlap distance L_o is obtained in a manner similar to that above with a few exceptions, namely that the distance Z should be measured to the middle of the layer and σ_h is not as large as illustrated in Figure 2.45. It is reasonably well-established that the stress in reinforcement elements is maximum near the failure plane and falls off sharply to either side [91]. As an approximation, $0.5\sigma_h$ will be used, which results in Eq. (2.50).

$$L_o = \frac{S_v \sigma_h \text{FS}}{4(c_a + \gamma z \tan \delta)} \tag{2.50}$$

where L_o is the required overlap length (the minimum is 1 m).

Next, we must consider external stability of the geotextile wall, which includes overturning, sliding, and foundation failures. These are illustrated in Figure 2.47. These features are common to all wall systems and can be treated in exactly the same way as gravity or crib walls. They are generally site-specific insofar as the calculations are concerned. In the U.S., the Federal Highway Administration has recommended the following: for overturning and foundation-bearing capacity the FS value $\geqslant 2.0$ and for sliding the FS value $\geqslant 1.5$.

The miscellaneous considerations that must be addressed are generally the facing details; facing connections (if applicable); seaming methods (if necessary); drainage be-

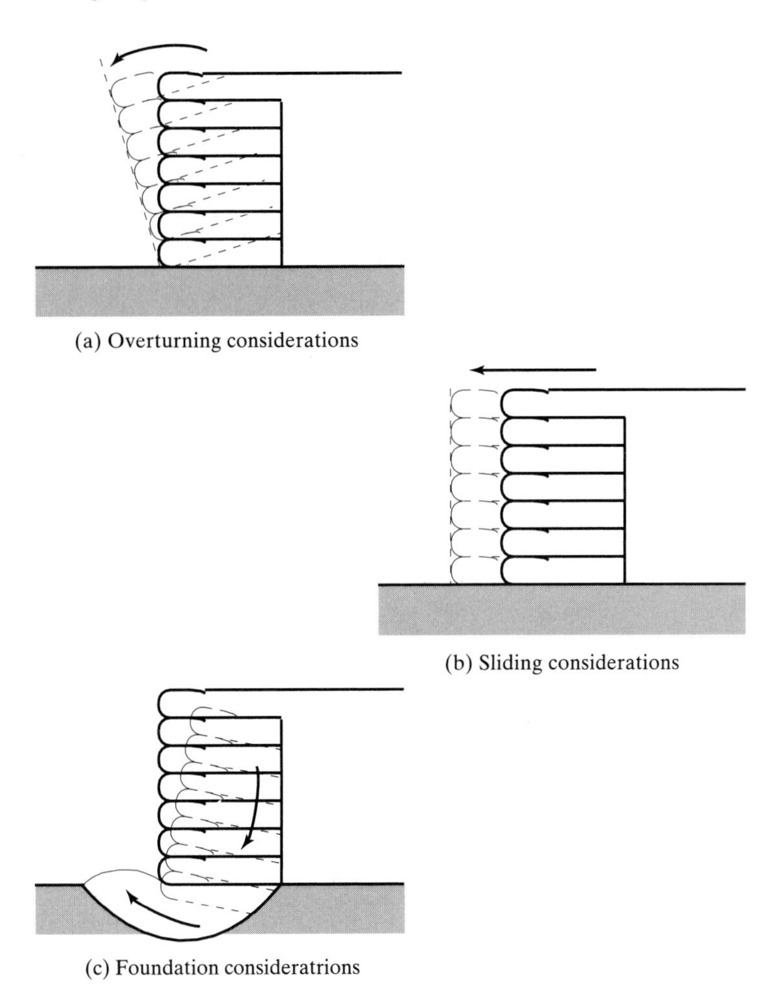

Figure 2.47 External stability considerations for geotextile walls.

hind, beneath, and in front of the wall; erosion above and in front of the wall, guard posts, light posts, fencing, and other appurtenances.

Example 2.16

Design a 6-m-high wrap-around type of geotextile wall that is to carry a storage area of equivalent dead load of 10 kPa. The wall is to backfilled with a granular soil (SP) having the properties of $\gamma = 18$ kN/m^3, $\phi = 36°$, and $c_a = 0$. A woven slit-film geotextile with warp (machine) direction ultimate wide-width tensile strength of 50 kN/m and friction angle with granular soil of $\delta = 24°$ (see Table 2.5) is intended to be used in its construction. The orientation of the geotextile is perpendicular to the wall face and the edges are to be overlapped or sewn to handle the weft (cross machine) direction. A factor of safety of 1.4 is to be used along with site-specific reduction factors.

Solution: (a) Determine the horizontal pressure as a function of the depth z in order to calculate the spacing of the individual layers, i.e., the S_v value.

$$K_a = \tan^2 (45 - \phi/2)$$
$$= \tan^2 (45 - 36/2)$$
$$= 0.26$$

$$\sigma_h = \sigma_{hs} + \sigma_{hq}$$
$$= K_a \gamma z + K_a q$$
$$= (0.26)(18)(z) + (0.26)(10)$$
$$= 4.68z + 2.60$$

and the allowable geotextile strength with the following reduction factors:

$$T_{\text{allow}} = T_{\text{ult}}\left[\frac{1}{\text{RF}_{ID} \times \text{RF}_{CR} \times \text{RF}_{CD} \times \text{RF}_{BD}}\right]$$
$$= 50\left[\frac{1}{1.2 \times 2.5 \times 1.15 \times 1.1}\right]$$
$$= \frac{50}{3.79}$$
$$= 13.2 \text{ kN/m}$$

Now using Eq. (2.46) for varying depths, calculate the geotextile layer spacings.
At $z = 6$ m:

$$S_v = \frac{T_{\text{allow}}}{\sigma_h(\text{FS})}$$
$$= \frac{T_{\text{allow}}}{[4.68(z) + 2.60]1.4}$$
$$= \frac{13.2}{[4.68(6.0) + 2.60]1.4}$$
$$S_v = 0.307 \text{ m} \qquad \text{use } 0.30 \text{ m}$$

By trial and error, see if the spacing can be opened up to 0.50 m at $z = 3.3$ m.

$$S_v = \frac{13.2}{[(4.68)(3.3) + 2.60]1.4}$$
$$S_v = 0.52 \text{ m} \qquad \text{use } 0.50 \text{ m.}$$

By trial and error, see if the spacing can be further opened up to 0.65 m at $z = 1.3$ m.

$$S_v = \frac{13.2}{[(4.68)(1.3) + 2.60]1.4}$$

$$= 1.08 \text{ m} \qquad \text{use } 0.65 \text{ m}$$

Thus, the layers and their spacings are as shown below. At this point there is an MSE mass, actually a geosynthetically stabilized earth mass, that is self-standing within itself.

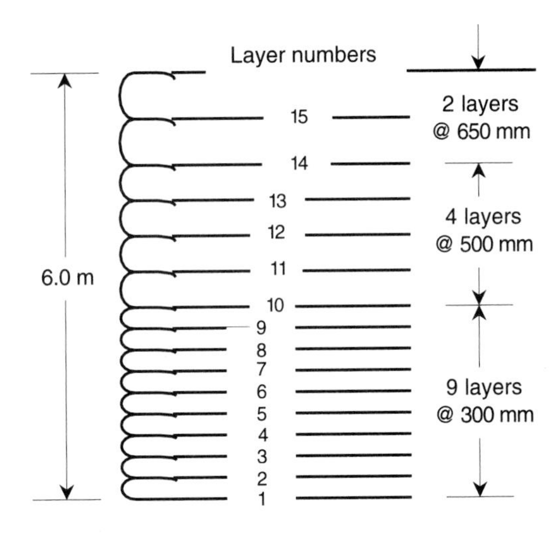

Due to the critical nature of some MSE-walls a *serviceability* criterion can be imposed, limiting the possible outward deflection of the wall to some acceptable value, for example, 20 mm. This is done utilizing the wide-width stress-versus-strain response of the reinforcement at the design (allowable) stress in the reinforcement. This latter value includes both reduction factors and the factor of safety.

(b) Determine the length of the fabric layers (L) using Eq. (2.49) for L_e with $\delta = 24°$ and $c_a = 0$. Note that L_R uses a Rankine failure plane and is calculated from Eq. (2.48).

$$L_e = \frac{S_v \sigma_h (\text{FS})}{2(c + \gamma z \tan \delta)}$$

$$= \frac{S_v(4.68z + 2.60)1.4}{2(0 + 18z \tan 24°)}$$

$$L_e = \frac{S_v(6.55z + 3.64)}{16.0z} \quad \text{and}$$

$$L_R = (H - z) \tan \left(45 - \frac{36}{2}\right)$$

$$L_R = (6.0 - z)(0.509)$$

Layer no.	Depth, z (m)	Spacing, S_v (m)	L_e (m)	L_e min. (m)	L_R (m)	L_{calc} (m)	L_{spec} (m)
15	0.65	0.65	0.49	1.0	2.72	3.72	use 4.0
14	1.30	0.65	0.38	1.0	2.39	3.39	
13	1.80	0.50	0.27	1.0	2.14	3.14	
12	2.30	0.50	0.26	1.0	1.88	2.88	use 3.0
11	2.80	0.50	0.25	1.0	1.63	2.63	
10	3.30	0.50	0.24	1.0	1.37	2.37	
9	3.60	0.30	0.14	1.0	1.22	2.22	
8	3.90	0.30	0.14	1.0	1.07	2.07	
7	4.20	0.30	0.14	1.0	0.92	1.92	use 2.0
6	4.50	0.30	0.14	1.0	0.76	1.76	
5	4.80	0.30	0.14	1.0	0.61	1.61	
4	5.10	0.30	0.14	1.0	0.46	1.46	
3	5.40	0.30	0.14	1.0	0.31	1.31	
2	5.70	0.30	0.14	1.0	0.15	1.15	
1	6.00	0.30	0.13	1.0	0.00	1.00	

Note that the calculated L_e values are very small (this is typically the case with geotextile walls) and the minimum value of 1.0 m should be used. When this is added to L_R for the total length, you should round up to a even number of meters. Also, the important consideration of total geotextile width must be addressed. Three situations can be envisioned.

Case 1. If the geotextile rolls are wide enough they can be deployed parallel to the wall and the weft or cross machine direction is the important property insofar as its wide-width strength is concerned. While this is possible for the lower fabric layers, it is not for the uppermost, since, for example, 4.0 + 0.65 + 1.0 = 5.65 m, which is wider than current commercially available geotextiles.

Case 2. Alternatively, two adjacent rolls of fabric can be used parallel to the wall, but a sewn seam, or large overlap, must be used for the uppermost fabric layers. If sewn seams are used, an appropriate partial factor of safety must be used.

Case 3. The fabric layers can be deployed perpendicular to the wall, thereby utilizing their warp or machine direction wide-width strength in the major principal stress directions. This is the case posed in this example. This requires sewn seams or overlaps in the opposite direction. However, in this (the minor principal stress) direction the required forces are significantly lower, for example, 35 to 50% of the major principal stress direction.

(c) Check the overlap length L_o, to see if it is less than the 1.0 m recommended value using Eq. (2.50).

$$L_o = \frac{S_v \sigma_h(\text{FS})}{4(c_a + \gamma z \tan \delta)}$$

$$= \frac{S_v[4.68(z) + 2.60]1.4}{4[0 + (18)z \tan 24°]}$$

maximum at the upper layer at $z = 0.65$ m,

$$L_o = \frac{0.65[4.68(0.65) + 2.60]1.4}{4[0 + (18)(0.65)\tan 24°]}$$

$$= 0.25 \text{ m} \qquad \text{OK use 1.0 m}$$

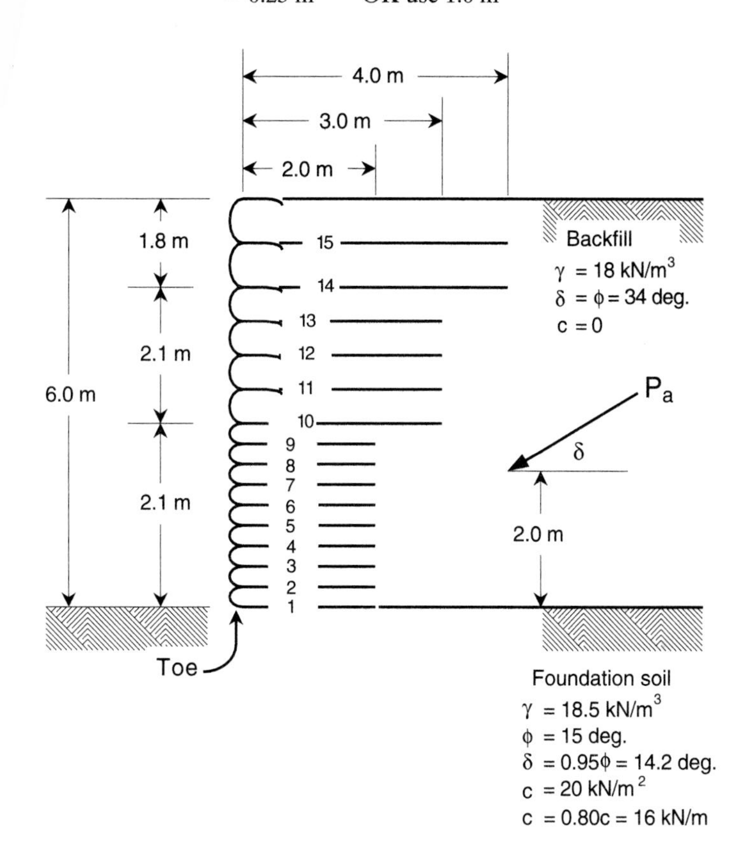

(d) Since the internal stability of the wall has been provided for, focus now shifts to external stability. Standard geotechnical engineering concepts are used to analyze overturning, sliding and bearing capacity. From the figure above,

$$K_a = \tan^2(45 - \phi/2) = \tan^2(45 - 34/2)$$

$$= 0.28$$

$$P_a = 0.5\gamma H^2 K_a$$

$$= 0.5(18)(6)^2(0.28)$$

$$= 90.7 \text{ kN/m}$$

$$P_a \cos 34 = 75.2 \text{ kN/m}$$

$$P_a \sin 34 = 50.7 \text{ kN/m}$$

Now, for overturning, moments are taken about the toe of the wall to generate a factor of safety value.

$$FS_{OT} = \sum \frac{\text{resisting moments}}{\text{driving moments}}$$

$$= \frac{w_1x_1 + w_2x_2 + w_3x_3 + P_a \sin \delta(4)}{P_a \cos \delta(2)}$$

$$= \frac{(6)(2)(18)(1) + (3.9)(1)(18)(2.5) + (1.8)(1.0)(18)(3.5) + (50.7)(4)}{(75.2)(2.0)}$$

$$= 4.7 > 2.0 \qquad \text{which is acceptable}$$

This high value of calculated factor of safety is very typical of walls of this type. Even further, overturning is not a likely failure mechanism since this is a very flexible MSE system that cannot support bending stresses. Thus many engineers do not even include an overturning calculation in the design process.

For sliding, horizontal forces at the bottom of the wall are summed to calculate another factor of safety:

$$FS_s = \sum \frac{\text{resisting forces}}{\text{driving forces}}$$

$$= \frac{\left[c_a + \left(\frac{w_1 + w_2 + w_3 + P_a \sin \delta}{2}\right) \tan \delta\right] 2}{P_\alpha \cos \delta}$$

$$= \frac{\left[16 + \left(\frac{216 + 70.2 + 32.4 + 50.7}{2}\right) \tan 14.2\right] 2}{75.2}$$

$$FS_s = 1.7 > 1.5 \qquad \text{acceptable (but barely)}$$

Finally, check for a foundation failure using shallow-foundation bearing capacity theory (e.g., [92]).

$$P_{\text{ult}} = cN_c + qN_q + 0.5\gamma BN_\gamma$$

$$= (20)(10.98) + 0 + 0.5(18.5)(2)(2.65)$$

$$= 219.6 + 49.0$$

$$= 269 \text{ kN/m}^2$$

$$P_{\text{act}} = (18)(6) + (10)$$

$$= 118 \text{ kN/m}^2$$

$$FS_{BC} = \frac{p_{\text{ult}}}{p_{\text{act}}}$$

$$= \frac{269}{118}$$

$$= 2.3 > 2.0 \qquad \text{which is acceptable}$$

The internal design is now complete. It uses fifteen layers of fabric (the lowest nine at 0.30 m spacing; the middle four at 0.50 m spacing; the upper two at 0.65 m spacing). The fabric lengths are 3.3 m (2 + 0.3 + 1) at the lowest level, 4.5 m (3 + 0.5 + 1) at the intermediate level, and 5.65 m (4 + 0.65 + 1) at the upper level.

Example 2.16 illustrates a geotextile retaining wall design completely except for the incorporation of live loads as produced by traffic. To illustrate how this is done, Example 2.17 is given, but just to the point of calculating the additional horizontal stress distribution against the wall. Beyond this, the design proceeds as in Example 2.16.

Example 2.17

For the 200 kN dual-tandem-axle truck whose eight wheel dimensions are shown below, calculate the horizontal wall stresses for a 6 m high wall at 1 m increments. Use Figure 2.46 for your calculations.

Solution: Using Figure 2.46, with $n = z/H$, $m = x/H$, $H = 6$ m, $Q_p = 25$ kN, and $\sigma_h' = \sigma_h \cos^2(1.1\theta)$, each wheel gives the following tabulated horizontal stresses (in kN/m²).

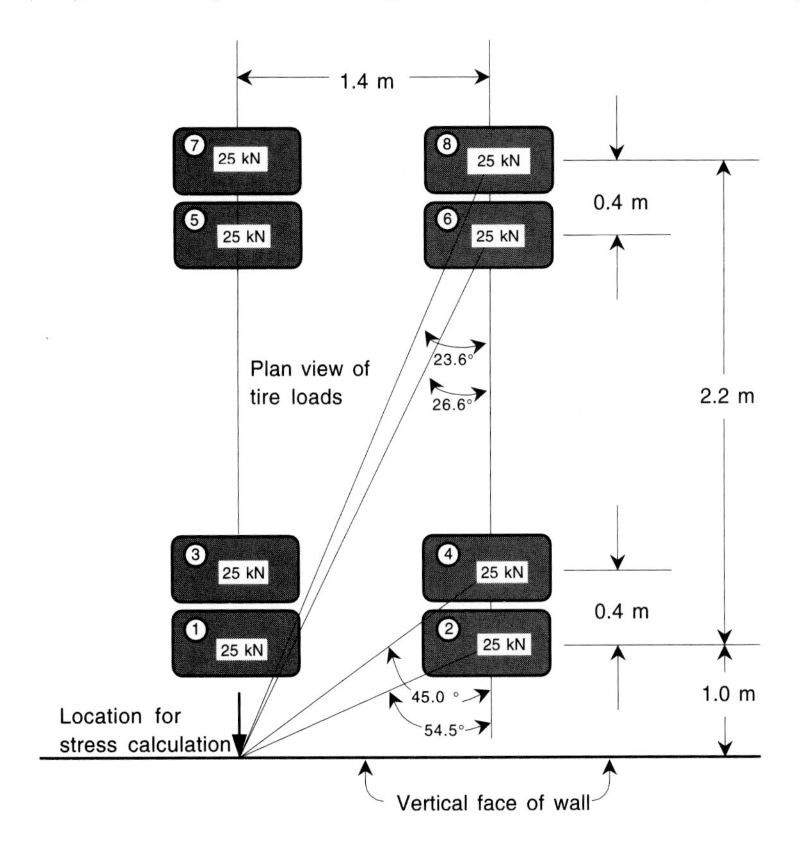

The stress distribution diagram on page 197 for the live load is now added to those from the soil and surcharge (if any) to obtain the resultant horizontal pressure distribution as shown in Figure 2.45.

	Wheel 1					Wheel 2
z	$n = z/H$	x	$m = x/H$	$\sigma_h H^2/Q_p$	σ_h	$\sigma'_h = 0.22\ \sigma_h$
0	0.00	1.0	0.17	0.0	0.00	0.00
1	0.17	1.0	0.17	1.2	0.82	0.18
2	0.33	1.0	0.17	1.6	1.08	0.24
3	0.50	1.0	0.17	1.0	0.71	0.16
4	0.67	1.0	0.17	0.6	0.39	0.09
5	0.83	1.0	0.17	0.3	0.22	0.05
6	1.00	1.0	0.17	0.2	0.12	0.03

	Wheel 3					Wheel 4
z	$n = z/H$	x	$m = x/H$	$\sigma_h H^2/Q_p$	σ_h	$\sigma'_h = 0.22\ \sigma_h$
0	0.00	1.4	0.23	0.0	0.00	0.00
1	0.17	1.4	0.23	1.2	0.82	0.34
2	0.33	1.4	0.23	1.6	1.08	0.46
3	0.50	1.4	0.23	1.0	0.71	0.30
4	0.67	1.4	0.23	0.6	0.39	0.16
5	0.83	1.4	0.23	0.3	0.22	0.09
6	1.00	1.4	0.23	0.2	0.12	0.05

	Wheel 5					Wheel 6
z	$n = z/H$	x	$m = x/H$	$\sigma_h H^2/Q_p$	σ_h	$\sigma'_h = 0.22\ \sigma_h$
0	0.00	2.8	0.47	0.0	0.00	0.00
1	0.17	2.8	0.47	0.7	0.50	0.38
2	0.33	2.8	0.47	1.2	0.84	0.64
3	0.50	2.8	0.47	0.9	0.65	0.50
4	0.67	2.8	0.47	0.6	0.41	0.31
5	0.83	2.8	0.47	0.4	0.24	0.19
6	1.00	2.8	0.47	0.2	0.15	0.11

	Wheel 7					Wheel 8
z	$n = z/H$	x	$m = x/H$	$\sigma_h H^2/Q_p$	σ_h	$\sigma'_h = 0.22\ \sigma_h$
0	0.00	3.2	0.53	0.0	0.00	0.00
1	0.17	3.2	0.53	0.5	0.32	0.26
2	0.33	3.2	0.53	0.9	0.63	0.51
3	0.50	3.2	0.53	0.8	0.57	0.46
4	0.67	3.2	0.53	0.6	0.40	0.33
5	0.83	3.2	0.53	0.4	0.26	0.21
6	1.00	3.2	0.53	0.2	0.16	0.13

	$\Sigma\ (\sigma_h + \sigma'_h)$
z	stress kPa
0	0.00
1	3.62
2	5.47
3	4.05
4	2.48
5	1.47
6	0.89

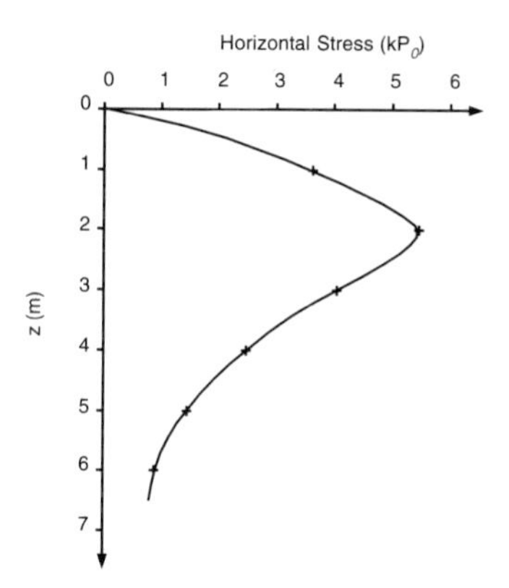

Summary. It is easily seen that the geotextile wall design just completed (with or without live loads) is not trivial, and to do such designs continuously is a very time-consuming task. In the case of a manufacturer of a particular geotextile style it would be preferred to develop design guides by systematically varying certain parameters in the analysis (e.g., the height of the wall and the slope angle of the wall face). Several innovative design graphs can be generated; an example using Polyfelt's TS style of geotextiles is shown in Figure 2.48. Graphs for different geotextiles can be similarly developed, or the type of loading could be included as a separate variable. The variations are essentially limitless. There are numerous computer codes available from manufacturers for their particular products as well as generic programs that are commercially available.

Regarding the performance of geotextile walls, one of the most carefully developed, constructed, and monitored walls was built in 1982 near Glenwood Springs, Colorado. It is 5 m high and 90 m long, with the length consisting of ten 9.0 m segments. These segments consist of different nonwoven geotextiles supplied by separate manufacturers. They were sewn together in order to form the 90 m length. Two of the segments were purposely underdesigned to provide for controlled failure. When such a failure did not occur, 5.2 m of surcharge was added, but still there was no failure. Each segment of wall continues to remain serviceable, with no noticeable creep, even though part of the wall is founded on soft soil that has settled more than 600 mm [93]. Clearly, geotextile walls are intrinsically sound in concept, but this case history suggests that the design method may be quite conservative. Several other noteworthy geotextile-reinforced walls appear in the literature. For example, a 12.6 m high wall was constructed to form support for an additional 5 m of surcharge fill in Seattle, Washington (Allen et al. [94]) and a 12.2 m high vertical wall was used to support a high bridge approach while its adjacent section was being constructed (Stevens and Souiedan [95]).

When we compare geotextile walls to gravity walls (and, to a lesser extent, to other types of flexible walls), there are the following advantages and disadvantages to using geotextiles:

- Advantages: a flexible wall system is created; a minimum excavation is needed behind face of wall; there are no corrosion problems; the backfill can contain fines; drainage can occur using certain geotextiles; unskilled labor can be used; no

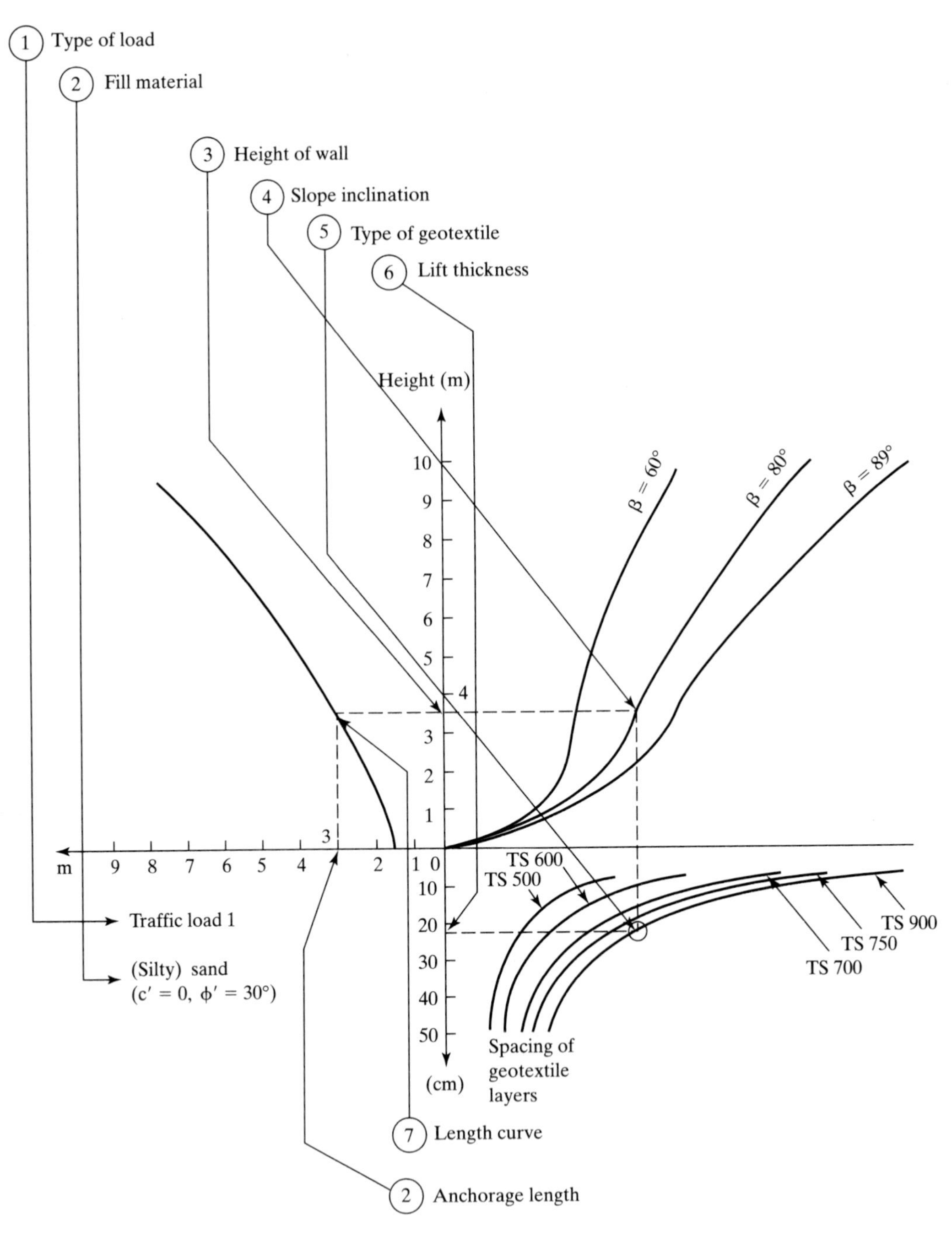

Figure 2.48 Design guide for geotextile walls using Polyfelt geotextiles. (After G. Werner and J. A. Studer, available from Polyfelt, Inc.)

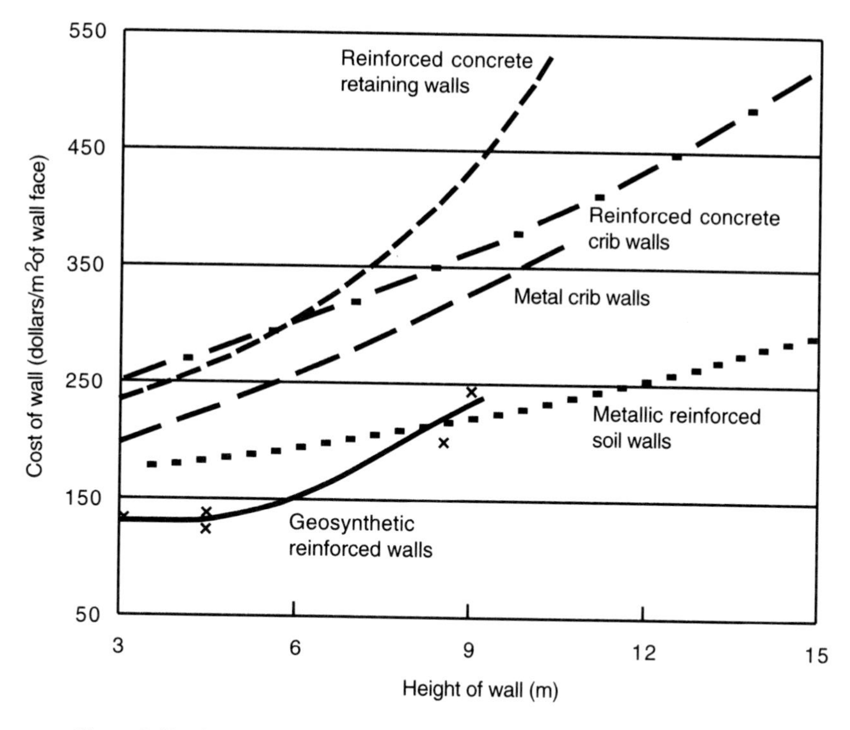

Figure 2.49 Cost comparison of reinforced wall systems. (After FHWA [96])

heavy equipment is required; and the cost per square meter of exposed wall is very low (see Figure 2.49).

- Disadvantages: The design method appears to be quite conservative; the geotextile interaction in the analysis (perhaps via arching theory) is not currently considered; creep is potentially a problem, thereby requiring a relatively high reduction factor; the wall face must be coated to prevent UV degradation and vandalism; and the coatings (shotcrete, gunite or asphalt) are not particularly attractive.

The disadvantages regarding the wall facing of wrap-around walls are being dealt with ever more effectively. For instance, it is becoming popular to use wall facing panels made from large-sized timbers, gabions, or concrete panels. The geotextiles are fixed to the rear of the facing panels and extend into the backfill soil exactly as described and designed in this section. For walls less than 5 m in height, timber can be used for the facing, as illustrated in Figure 2.50. The attachment detail is important, as described by Richardson and Behr [97].

However, the type of wall (both reinforced and nonreinforced) growing fastest in popularity consists of concrete-block facing elements. See Figure 2.51a. These precast blocks, weighing up to 400 N each, are laid in a stacked configuration as is a masonry wall; however, they are laid dry, that is, no mortar is used. Small walls of this type (less than 1.0 m in height) can be nonreinforced. Larger walls have the reinforcement placed between the blocks as shown in Figure 2.51b. When geotextiles (or flexible geogrids)

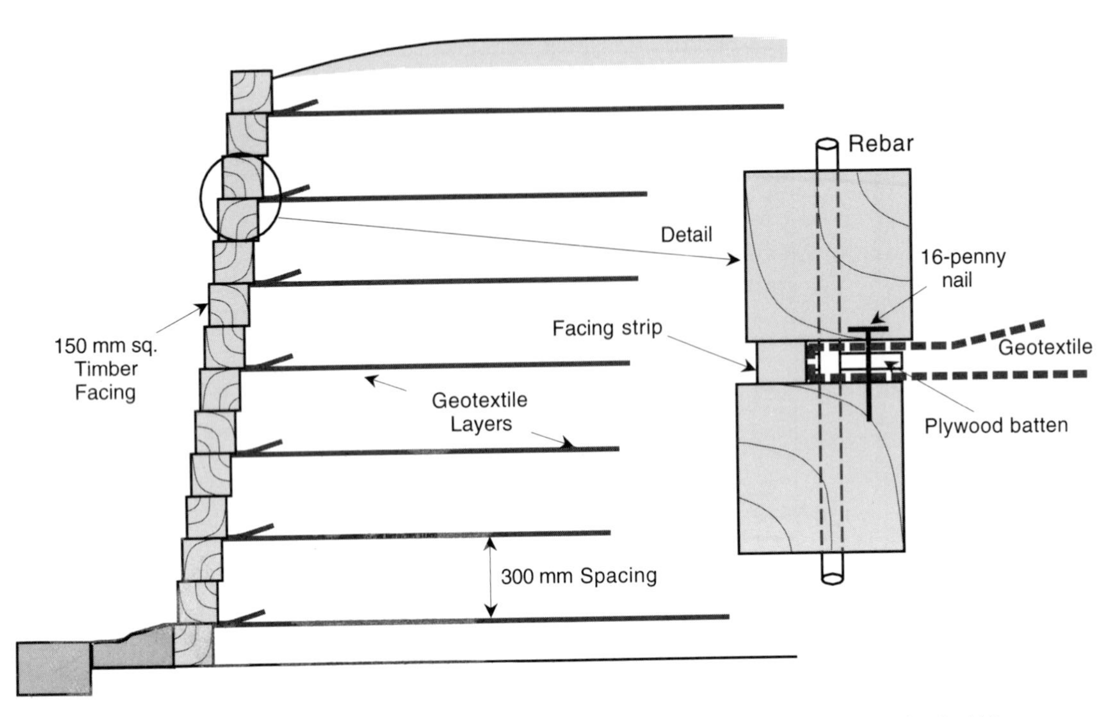

Figure 2.50 Geotextile-reinforced, timber-faced wall, showing wall connection details. (After Richardson and Behr [97])

are used for the reinforcement, the mated surfaces of the upper and lower blocks provide an interlocking mechanism. This greatly aids the frictional resistance of the reinforcement on the surfaces of the blocks. Stiff geogrids (described in Chapter 3) can rely on mechanical anchorage using pins or connections in the lower and/or upper blocks. The various block/reinforcement systems are sometimes called modular block walls (MBW) and are becoming more common due to the aesthetics of the very attractive surface finishes on the blocks, their adaptability to curving around trees, ponds, poles and other obstructions, the ease of construction by nonskilled labor, the lack for need of heavy construction equipment (e.g., cranes), the publication of a design manual specifically focused on MBWs (Simac et al. [99]), and all the advantages of flexible walls mentioned earlier.

Perhaps the greatest economy can be realized in the nature of the backfill soil. Large stone (e.g., AASHTO #8 or #57) should not be used, since the installation damage is likely to be excessive; rather a sand backfill, with sufficient permeability for drainage, is recommended. Table 2.16 gives a suggested gradation. In many cases, sand is locally available and is generally significantly less expensive than quarried stone or river gravel. Even backfill soils that are finer-grained than sand is possible, but then a geosynthetic drainage system within the backfill must be considered. This can be handled by geotextiles with adequate transmissivity (recall Figure 2.17), with various geonets (described in Chapter 4) or with various geocomposite drains (described in Chapter 8).

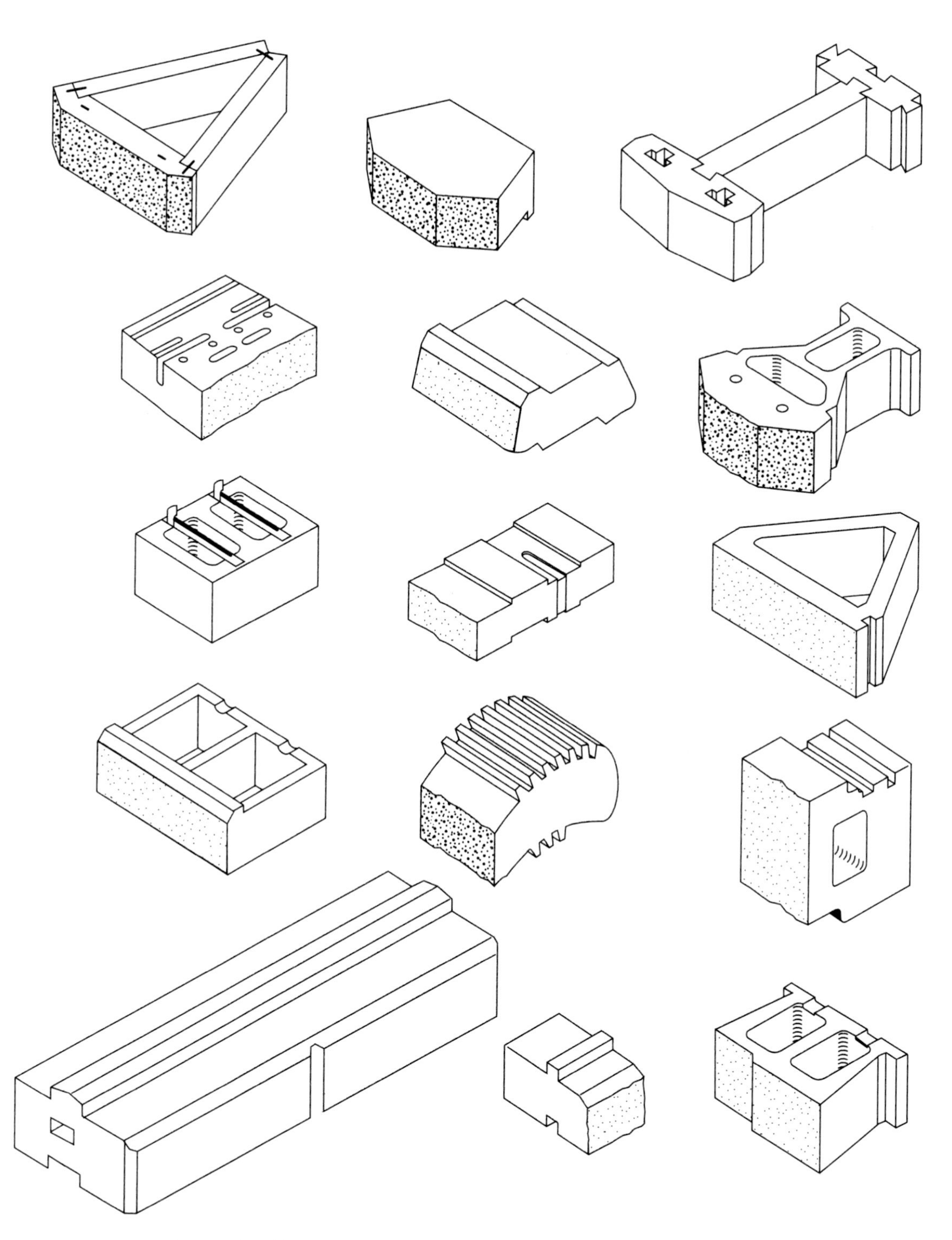

Figure 2.51 (a) Examples of commercially available modular blocks for retaining walls. Note: many of the blocks are proprietary or patented. (After Bathurst and Simac [98]).

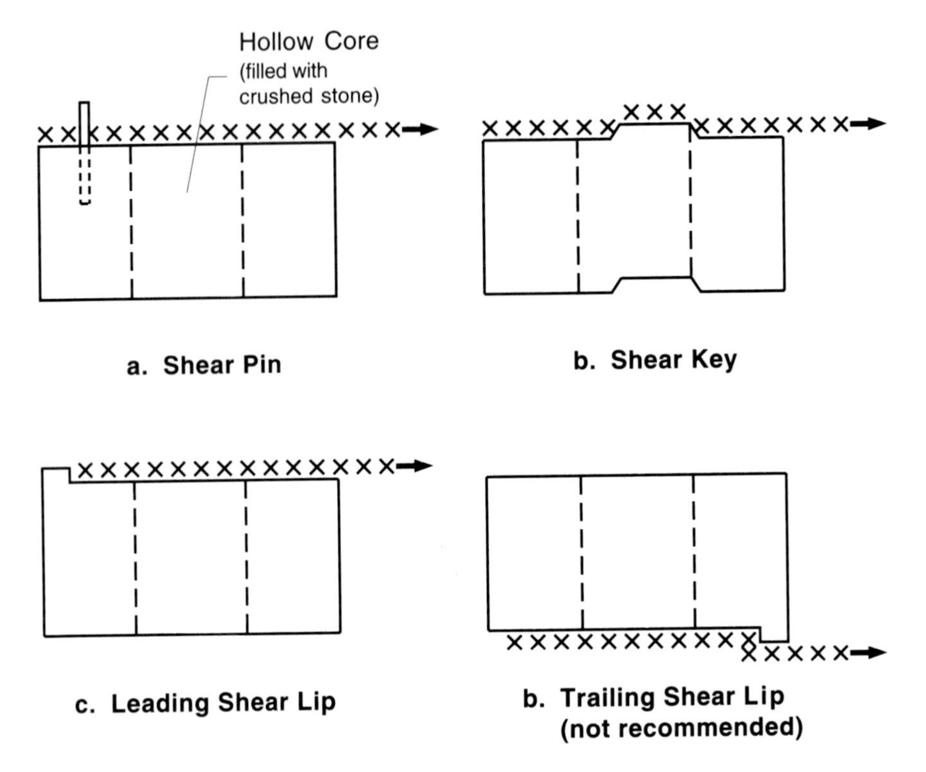

Figure 2.51 (b) Details of various modular blocks illustrating how geosynthetic reinforcement is connected mechanically or by friction. (Modified from Simac et al. [99])

2.7.2 Geotextile Reinforced Embankments

Background. It should come as no surprise that if vertical walls can be built using geotextiles, soil embankments can be stabilized by them also. In fact, as the slope angle with the horizontal (β) decreases, a wall essentially transitions to an embankment, albeit one in which the exposed face is not covered with anything except vegetation aided by some type of erosion-control material. In this section, geotextiles will be

TABLE 2.16 RECOMMENDED SOIL BACKFILL GRADATION FOR GEOTEXTILE AND GEOGRID REINFORCEMENT APPLICATIONS (WALLS AND SLOPES) TO AVOID EXCESSIVE INSTALLATION DAMAGE

Sieve Size (No.)	Particle Size (mm)	Percent Passing
4	4.76	100
10	2.0	90–100
40	0.42	0–60
100	0.15	0–5
200	0.075	0

Source: After Koerner et al. [100].

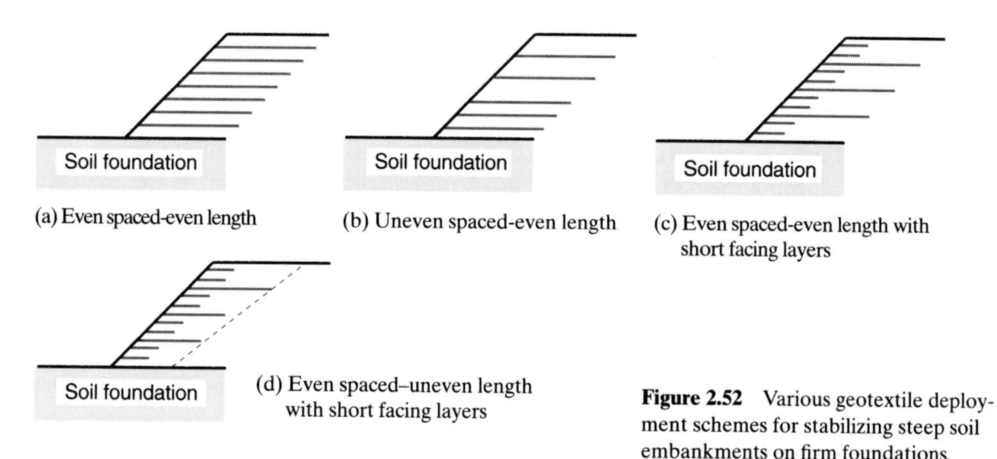

Figure 2.52 Various geotextile deployment schemes for stabilizing steep soil embankments on firm foundations.

deployed in horizontal layers with no upturned facing treatment or hard-wall faced surface. When this is the case, the design methodology transitions from lateral earth pressure theory to slope stability analyses. At which β-value the transition occurs is an interesting issue, albeit largely an academic one. Various geotextile deployment schemes for embankments are shown in Figure 2.52. The (a) pattern is typical. The uneven spacing pattern of (b) reflects those cases where stresses are higher in the lower regions of the slope than in the top. The short edge strips shown in (c) and (d), sometimes called *secondary* reinforcement, represent compaction aids, necessary since high compaction at the edge of the slope is difficult to achieve. These short geotextile layers also eliminate shallow sloughing failures between widely-spaced reinforcement layers. Note that all of these schemes require the embankment to be built at the same time as placement of the geotextile proceeds; that is, they are not in situ stabilization schemes (these are discussed later in this section).

Construction Details. Geotextile placement in embankment stabilization situations is relatively simple in that the sheets are usually horizontally placed as directed by the design. When using woven geotextiles (more so than with nonwovens) it is important to recognize the direction of maximum stress. For two-dimensional plane strain cases this is typically in the direction of the embankment face. If the geotextile is sufficiently wide, it can be used parallel to the face of the slope and the weft, or cross-machine direction, must carry the major principal stresses; it must be designed and specified accordingly. If the geotextile is not sufficiently wide and is still to be used parallel to the face of the slope, sewn seams will be required, which will probably be the limiting feature in the design.

If the geotextile is oriented perpendicular to the face of the slope, the warp or machine direction will carry the major principal stress. Now the weft or cross-machine direction of these rolls will have to carry the minor principal direction, which is typically 35 to 50% of the major principal stress and can be handled by sewing or by an overlap sufficient to mobilize the required strength by frictional resistance.

Although either method can effectively transmit the mobilized tensile stresses in the minor principal direction from one geotextile sheet to the next, the labor cost of sewn seams usually becomes small when many seams are required. Thus the cost comparison in Section 2.6.1 swings even further in the favor of sewn seams.

Limit Equilibrium Design. The usual geotechnical engineering approach to slope-stability problems is to use limit equilibrium concepts on an assumed circular arc failure plane, thereby arriving at an equation for the global factor of safety. Alternatively, a two-part wedge analysis can be used and will be illustrated in Chapter 3 on geogrids. The resulting equations for a circular arc failure for total stresses and effective stresses, respectively, are given below, corresponding to Figure 2.53. It is illustrated for the case of several layers of geotextile reinforcement.

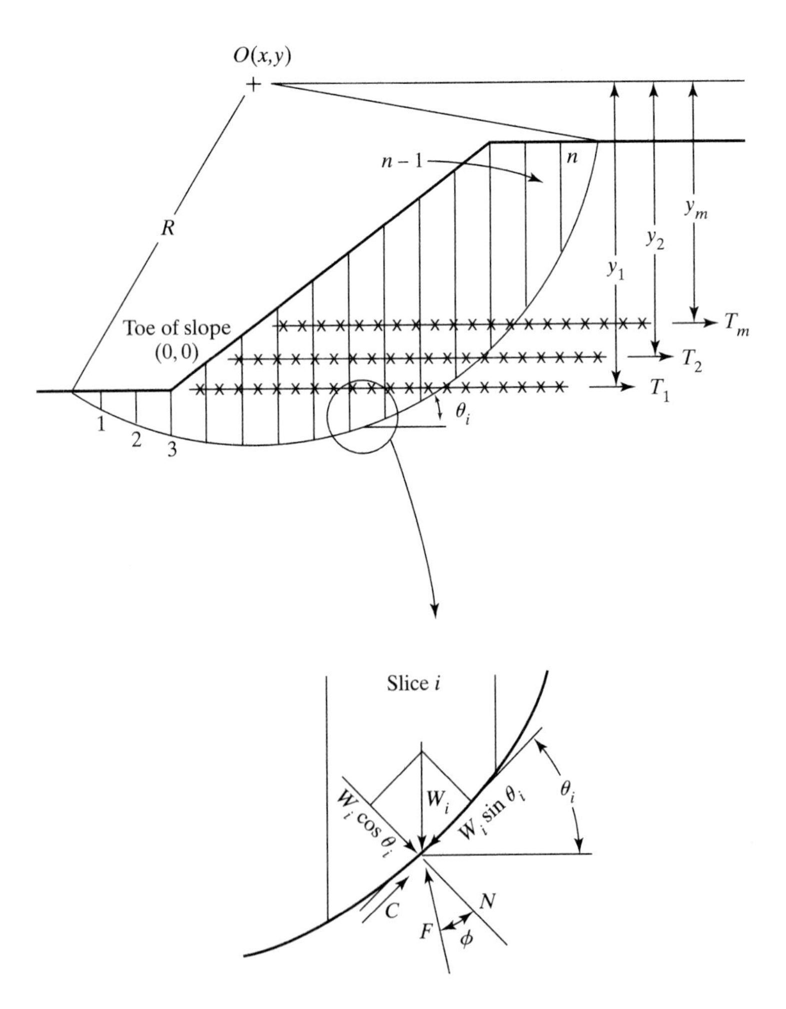

Figure 2.53 Details of circular arc slope stability analysis for (c, ϕ) shear strength soils.

$$FS = \frac{\sum_{i=1}^{n}(N_i \tan \phi + c\Delta l_i)R + \sum_{i=1}^{m} T_i y_i}{\sum_{i=1}^{n}(W_i \sin \theta_i)R} \tag{2.51}$$

$$FS = \frac{\sum_{i=1}^{n}(\overline{N}_i \tan \overline{\phi} + \overline{c}\Delta l_i)R + \sum_{i=1}^{m} T_i y_i}{\sum_{i=1}^{n}(W_i \sin \theta_i)R} \tag{2.52}$$

where

FS = factor of safety;
$N_i = W_i \cos \theta_i$;
W_i = weight of slice;
θ_i = angle of intersection of horizontal to tangent at center of slice;
Δl_i = arc length of slice;
R = radius of failure circle;
$\phi, \overline{\phi}$ = total and effective angles of shearing resistance, respectively;
$c, \overline{c}$ = total and effective cohesion, respectively;
T_i = allowable geotextile tensile strength;
y_i = moment arm for geotextile (note that in large-deformation situations this moment arm could become equal to R, which is generally a larger value);
n = number of slices;
m = number of geotextile layers; and
$\overline{N} = N_i - u_i \, \Delta x_i$, in which
$u_i = h_i \gamma_w$ = pore-water pressure,
h_i = height of water above base of circle,
Δx_i = width of slice, and
γ_w = unit of weight of water.

Use of the total stress analysis, Eq. (2.51), is recommended for embankments where water is not involved or when the soil is at less than saturation conditions. The effective stress analysis, Eq. (2.52), is for conditions where water and saturated soil are involved-conditions typical of earth dams and delta areas involving fine-grained cohesive soils.

These equations are tedious to solve, and when additional consideration is given to finding the minimum value of factor of safety by varying the radius and coordinates of the origin of the circle, the process becomes unbearable to do by hand. Many computer algorithms exist that can readily be modified to include the contribution of the geotextile reinforcement, the $T_i y_i$ term. When the search for critical radius and coordinates is not included as part of the algorithm, the analysis portion easily fits on most personal computers.

The moment arm y has been the topic of discussion in that its initial placement as shown is likely to be distorted as rotational deformation occurs. In the limit, this distortion could orient the geotextile along the potential failure arc, changing (and increasing) the moment arm from "y" to "R." Kaniraj [101] has found that this transition can add as much as 45% to the resisting moment of the geotextile. While this is certainly possible, it is quite site-specific and the more conservative value of y is preferred.

For fine-grained cohesive soils whose shear strength can be estimated by undrained conditions, the entire analysis becomes much simpler. (Recall that this is the same assumption we used in Section 2.6.1 on unpaved roads.) Here slices need not be taken, since the soil strength does not depend on the normal force on the shear plane. Figure 2.54 gives the details of this situation, which results in Eq. (2.53). Examples 2.18 and 2.19 illustrate its use.

$$\text{FS} = \frac{cL_{\text{arc}}R + \sum_{i=1}^{m} T_i y_i}{WX} \tag{2.53}$$

where

FS = global factor of safety,
c = cohesion = 0.5 q_u,
q_u = unconfined compression strength of soil,
L_{arc} = length of the failure arc,
R = radius of the failure circle,
T_i = allowable tensile strength of various geotextile layers,
y_i = moment arm for geotextile(s),
W = weight of failure zone, and
X = moment arm to center of gravity of failure zone.

Example 2.18

For the 10 m high, 50° angle slope shown opposite, which consists of a silty clay embankment (γ = 19 kN/m^3, ϕ = 0°, c = 15 kPa, area = 60 m^2, center of gravity as indicated) on a

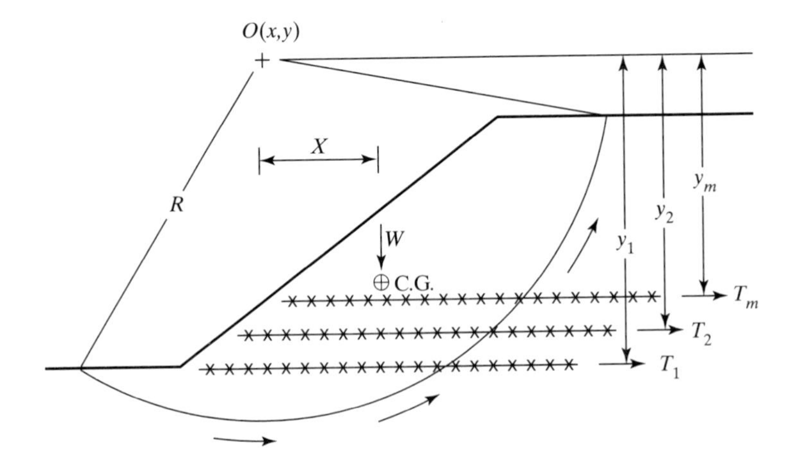

Figure 2.54 Details of circular arc slope stability analysis for soil strength represented by undrained conditions.

silty clay foundation ($\gamma = 20$ kN/m^3, $\phi = 0°$, $c = 18$ kPa, area = 55 m^2, center of gravity as indicated) determine **(a)** the factor of safety with no geotextile reinforcement, **(b)** the factor of safety with a geotextile of *allowable* tensile strength 40 kN/m (note that with a cumulative reduction factor of 3.0, this is an ultimate strength geotextile of 120 kN/m) placed along the surface between the foundation soil and the embankment soil, and **(c)** the factor of safety with ten layers of the same geotextile placed at 1 m intervals from the foundation interface to the top of the embankment. Assume that sufficient anchorage behind the slip circle shown is available to mobilize full geotextile strength.

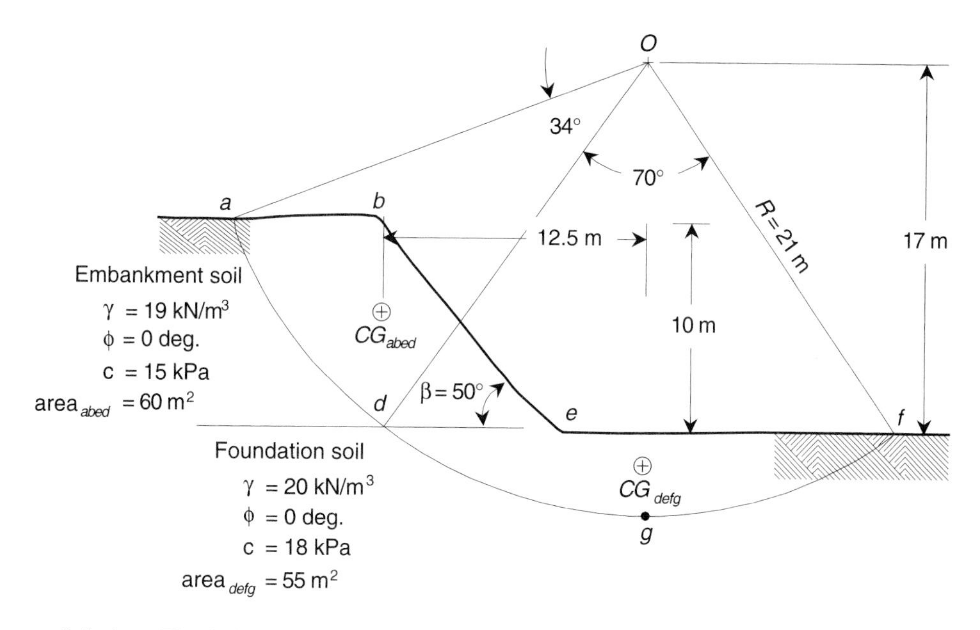

Solution: The following computational data are needed in all parts of the problem:

$$W_{abed} = (60)(19) = 1140 \text{ kN/m}$$

$$W_{defg} = (55)(20) = 1100 \text{ kN/m}$$

$$L_{ad} = 2(21)\pi\left(\frac{34}{360}\right) = 12.5 \text{ m}$$

$$L_{df} = 2(21)\pi\left(\frac{70}{360}\right) = 25.7 \text{ m}$$

(a) Slope as shown (with no geotextile reinforcement):

$$\text{FS} = \sum\frac{\text{resisting moments}}{\text{driving moments}}$$

$$\text{FS} = \frac{(c_e L_{ad} + c_f L_{df})R}{W_{abed}(12.5) + W_{defg}(0)}$$

$$= \frac{[(15)(12.5) + (18)(25.7)]21}{1140(12.5) + 0}$$

$$= \frac{13650}{14250} = 0.96 \qquad \text{not acceptable; failure is indicated}$$

(b) Slope with a geotextile along surface *ed* with sufficient anchorage beyond point *d*:

$$\text{FS} = \frac{13650 + 40(17)}{14250}$$

$$= 1.01 \qquad \text{still not acceptable; failure is incipient}$$

(c) Slope with ten layers at 1 m intervals from surface ed upward, all of which have sufficient anchorage behind the slip surface:

$$\text{FS} = \frac{13650 + 40(17 + 16 + 15 + \cdots + 9 + 8)}{14250}$$

$$= 1.31 \qquad \text{acceptable, but barely!}$$

Example 2.19

For the slope in Example 2.19, **(a)** determine how much embedment (or anchorage length) is required *behind* the potential slip circle in order to mobilize the allowable tensile strength of the geotextile. Assume that the transfer efficiency of the geotextile to the shear strength of the soil is 0.80 and base the calculation on a FS = 1.5, **(b)** determine the total length of the geotextile, using 8.0 m as the maximum distance from the slope face to the failure plane, and **(c)** comment on the effect of the possible orientations of the geotextile rolls.

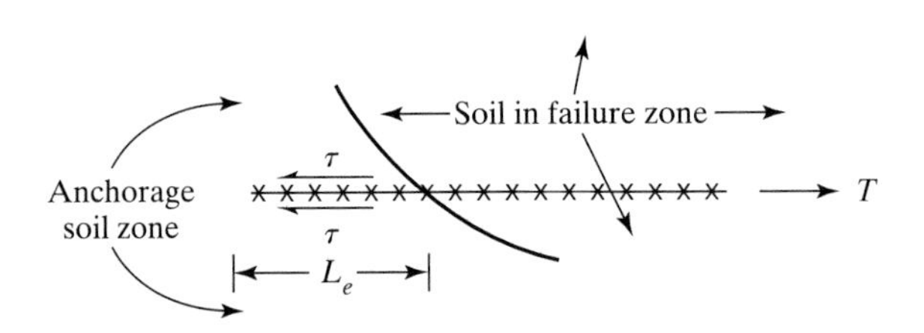

Solution: (a) When the anchorage test was explained in Section 2.3.3, it was assumed that the resistance was uniformly distributed over the geotextile's embedment and that the geotextile was entirely mobilized. This is almost certainly not the case. It appears that the concentration decreases rapidly as the embedment length increases and that separate mobilized and fixed portions of the geotextile exist. For this example however, a linear distribution will be assumed over a continuous displaced length, since it results in a conservative length. Taking force summation in the x-direction results in the following equation:

$$2\tau E L_e = T_{\text{allow}}(\text{FS})$$

$$2(15)(0.8)L_e = 40(1.5)$$

$$L_e = 2.5 \text{ m},$$

which is the required embedment length beyond the potential slip circle for sufficient anchorage of each geotextile layer.

(b) The total length of each geotextile layer will be 2.5 + 8.0 = 10.5 m.

(c) Since the typical widths of commercially available geotextiles range from 3 to 4.5 m, seams will be necessary in at least one direction. If the geotextile rolls are oriented parallel to the slope face at least two seams will be required and they will be oriented perpendicular to the major stress direction. If this is the case, a reduction factor for seam strength will have to be included, giving the value of T_{allow}. Figure 2.12 will provides guidance in this regard. However, if the geotextile is oriented perpendicular to the slope face, sewn seams will be required along the edges of each roll of geotextile and will therefore be in the minor principal stress direction. Overlaps could also be considered, but they would probably not be an economic solution for the higher-strength geotextiles used in this problem; for these, sewn seams are the logical choice.

Finite-Element Analysis. Finite-element methods (FEM) have been used to study the performance of geotextile-reinforced embankments in both analysis and design situations [102, 103]. Although these sophisticated computer-based methods might not be routinely used for noncritical situations, they do give great insight into the behavior of geosynthetically reinforced systems.

To illustrate the results of the technique, Rowe and Soderman [104] evaluated two instrumented test embankments on very soft soils in Holland. The embankments, one with geotextile reinforcement and one without, were purposely brought to failure. The one without geotextile reinforcement failed as the height was brought to 1.75 m (see Figure 2.55a). The other, reinforced with a geotextile, reached a height of 2.75 m before failure (see Figure 2.56a). Using a plane strain nonlinear elastoplastic FEM program with over 1000 triangular elements, the velocity field and the plastic region are as shown in Figures 2.55b,c for a fill height of 1.8 m for the case without reinforcement. Both figures clearly show that continuous failure is mobilized at approximately the height when it did indeed fail (i.e., 1.75 m versus the predicted 1.8 m). To adapt the FEM to the soil-geotextile interface for the reinforced section, the displacement of the soil and geotextile are assumed to be compatible until the shear stress reaches the limiting shear stress defined by the Mohr-Coulomb failure criterion. Once this shear stress is attained, slip occurs. Note from Figure 2.56b, the FEM-generated plastic zone at a height of 1.8 m, that a continuous plastic region does not yet exist (i.e., the embankment has not yet failed). The plastic zone begins to be continuous at a height of 2.05 m (Figure 2.56c), and is fully mobilized at a height of 2.66 m. This FEM-predicted height of 2.66 m corresponds nicely with the actual failure height of 2.75 m.

Techniques such as this certainly are the way of the future for geotextile-related designs of permanent and/or critical situations; see Rowe and Soderman [105].

Summary. Geotextile reinforcement of embankments has been shown to be a practical expedient in many situations. When reinforced, slope angles can be significantly increased over the nonreinforced situation. Designwise, the process involves modifications to limit equilibrium procedures that are within the realm of geotechnical engineering practice and seem to be a rational approach. There are uncertainties in the analysis, however, and additional research is warranted in the following areas:

- What reduction factors should be used to adapt an ultimate strength value to an allowable value?
- Should the reduction factors be used on the other side of the equation, as in the load factor design (LFD) methods?

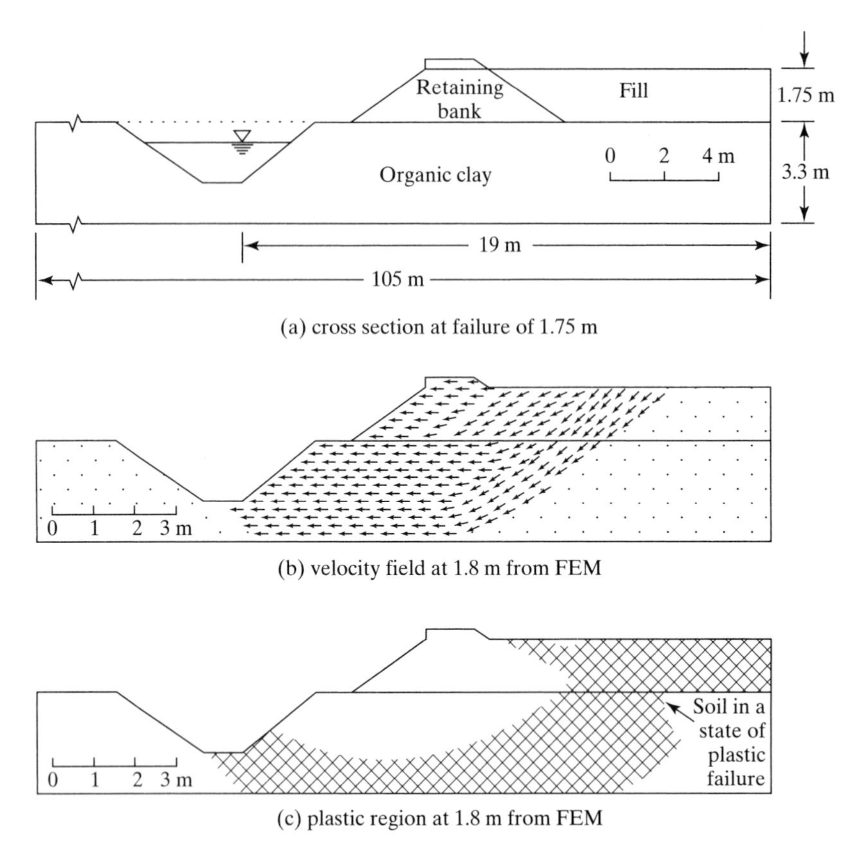

Figure 2.55 Test embankment on a soft clay soil with corresponding FEM velocity field and plastic region. (After Rowe and Soderman [104])

- Should the global factor of safety be included on the soil and the geotextile and, if so, should they be the same values?
- What moment arm should be used?
- How are shear stresses transmitted from the soil to the geotextile?
- Is there interaction between closely spaced geotextile layers?
- What anchorage is needed, and how is it mobilized?
- How is the strain compatibility of the soil and geotextile(s) considered?
- What type of surface treatment is necessary to provide erosion protection?

For the geotextile reinforcement of *granular* embankment soils, that is, those with a frictional component, the design method involves taking slices and making the modifications described. This is certainly possible but it requires a computer algorithm with search capabilities to find the critical arc radius and coordinates. Note that even in the undrained example presented, this same type of search is required, although the calculations are much simpler because slices are not necessary.

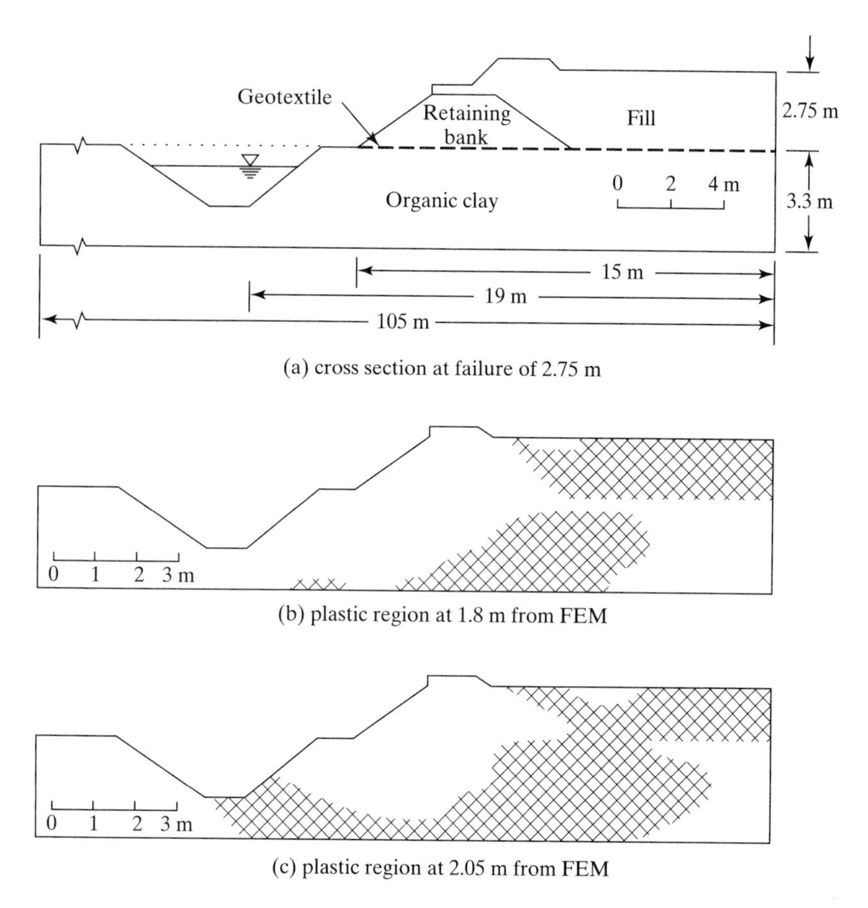

Figure 2.56 Geotextile-reinforced test embankment shown in Figure 2.55 with corresponding FEM plastic regions. (After Rowe and Soderman [104])

In this section the focus was on the stabilization of the embankment soil and not of its foundation soil. While it is very similar in its concept to the reinforcement of embankments, geotextile reinforcement of foundations consisting of very weak soils is a rapidly growing area in its own right and will be treated separately in the next section.

2.7.3 Geotextile Reinforced Foundation Soils

The purpose of this section is to focus on stable embankment soils or other structures placed above unstable and/or weak foundation soils. This is in contrast to Section 2.7.2, in which the embankment itself consisted of unsuitable or weak soils or the situation required very steep and/or high slopes.

Background. Fine-grained saturated soils exist near most river estuaries and harbor areas around the world. Unfortunately, these are also often the areas where industrialization is the heaviest. Buildings, factories, freight yards, stockpiles, storage tanks, railroads, equipment, and other appurtenances of industry are all incompatible with such weak foundation soil conditions. The traditional foundation options given to the owners of such facilities have been to (a) drive deep foundations through the unsuitable soils, thereby avoiding them altogether, (b) excavate and replace the soil with suitable soil, (c) stabilize the soil with injected additives, or (d) surcharge and wait until natural consolidation occurs. All of these methods have a degree of applicability, but all suffer from being either expensive or time-consuming.

The newly emerging alternate method, which is the focus of this section, is to deploy a high-strength geotextile (or geogrid) over the site, place a sand drainage layer/working blanket above it, install vertical wick drains (described in Chapter 8) to the bottom of the soft foundation layer, and then complete the surcharge fill up to the design grade. The sand blanket and surcharge layers can be placed by conventional earthmoving equipment or by dredging. As will be illustrated later, the entire process can be accomplished under water if the need arises.

Clearly, the geotextile acts as a reinforcement material, since the shear strength of the foundation soils are often less than 10 kPa, which will hardly support the weight of an individual. The vertical wick drains (also called prefabricated vertical drains or strip drains) are geosynthetic composites and are used to drain the excess pore water from the foundation soil as it is mobilized by the surcharge fill. The surcharge fill usually consists of locally available soils.

There are two somewhat different variations for the configurations of these projects. One is a large *areal fill* in which the length and width of the site are approximately equal. In such cases there is no clearly defined principal stress direction, and the strength of the geotextile must be equally balanced in all directions. This, of course, is required of the seams in both directions, as well as for the material itself. Actual situations in this category are often industrial- or building-site development projects. The second variation is one in which the length of the fill is much larger than the width, called a *linear fill*. In these cases, the major principal stress direction can be identified and the geosynthetic reinforcement can be aligned accordingly. Seams can often be avoided or placed in the minor principal stress directions. Situations in this category are roadway embankments and containment dikes. Both situations will be illustrated by case histories.

Construction Details. The U.S. Army Corps of Engineers has been the leading force behind the use of high-strength woven geotextiles to reinforce very soft foundation soils [106, 107]. Quite often the task has been to construct permanent linear dikes for dredged soil containment. As such, the necessity for high-strength seams can be somewhat avoided by placing the warp (strong) direction transverse to the dike's alignment, in the direction of the major principal stress. This allows the weft (weaker) direction to be seamed and placed in alignment with the minor principal stress. See Sprague and Koutsourais [108] for a compilation of twenty-eight projects of this type. The foundation soil strengths for these projects were generally very low, from 1 to 8 kPa. The geotextiles used had wide-width tensile strengths 80 to 700 kN/m. They were all relatively heavy woven fabrics made from polypropylene or polyester. The embankments placed above the geotextiles varied in height from 2 to 7 m. Post-construction consolidation settlements varied from 1 to 5 m. To my knowledge all have been successful with only one known problem: at one site a propagating tensile failure occurred in the geotextile between instrumentation holes as surcharge was near full height. This illustrates the importance of reducing geotextile strength to allow for planned holes, such as those made in the fabric for the installation of instrumentation devices or wick drains [109].

The Wilmington Harbor South Disposal Area project [110] is an example of a *linear reinforced foundation fill*. The project consisted of the construction of a U-shaped dike from land out into the Delaware River to provide storage capacity for dredged material from maintenance dredging.

The foundation soils, which are under as much as 5 m of water, consist of the weak, highly compressible silts and clays that form the tidal flats and shallows. Unconfined compression shear strengths range from nil to 10 kPa for depths averaging 27 m, where firm sands and gravels are eventually encountered.

The poor foundation conditions just described, the limited quantity of granular borrow soil available, and environmental considerations led to the adoption of a wide-bermed embankment to enclose this disposal area. The concept behind the design involves floating the dike on the soft foundation soil with the use of a high-strength geotextile for tensile reinforcement. The chosen geotextile employed on this project was woven of high-tenacity polyester yarn and was specifically designed for this application. The geotextile specifications are given in Table 2.17. Following the placement of the high-strength geotextile, the dike was constructed of dredged granular soil placed in two stages. The first stage averaged approximately 3 m deep × 180 m wide and formed the wide berm of the dike section. This first stage construction consisted of five separate hydraulic fills (see Figure 2.57). When constructing the embankment, the outer two fills, designated Fills 1 and 5, were placed concurrently, so as to contain the foundation soil and prevent its lateral extrusion. This also placed the central section of

TABLE 2.17 COMPARISON OF GEOTEXTILE SPECIFICATIONS FOR TWO HIGH-STRENGTH STABILIZATION PROJECTS

Geotextile Specification Values	Linear Fill (Containment Dike; Wilmington Harbor; U.S. Army Corps of Engineers)	Areal Fill (Industrial Development; Seagrit, Maryland; Maryland Port Admin.)
Polymer Type	PET	PP; PET
Tensile Strength (kN/m)	260	180
Modulus (kN/m)	3300	500
Elongation (%)	10–35	15–35
Stiffness (mg-cm)	—	30,000
Friction Angle	30°	30°
Seam Strength		
warp (kN/m)	none	105
weft (kN/m)	140	105
Seam Type	J	J
Seam Thread Type	PET	PA; PET

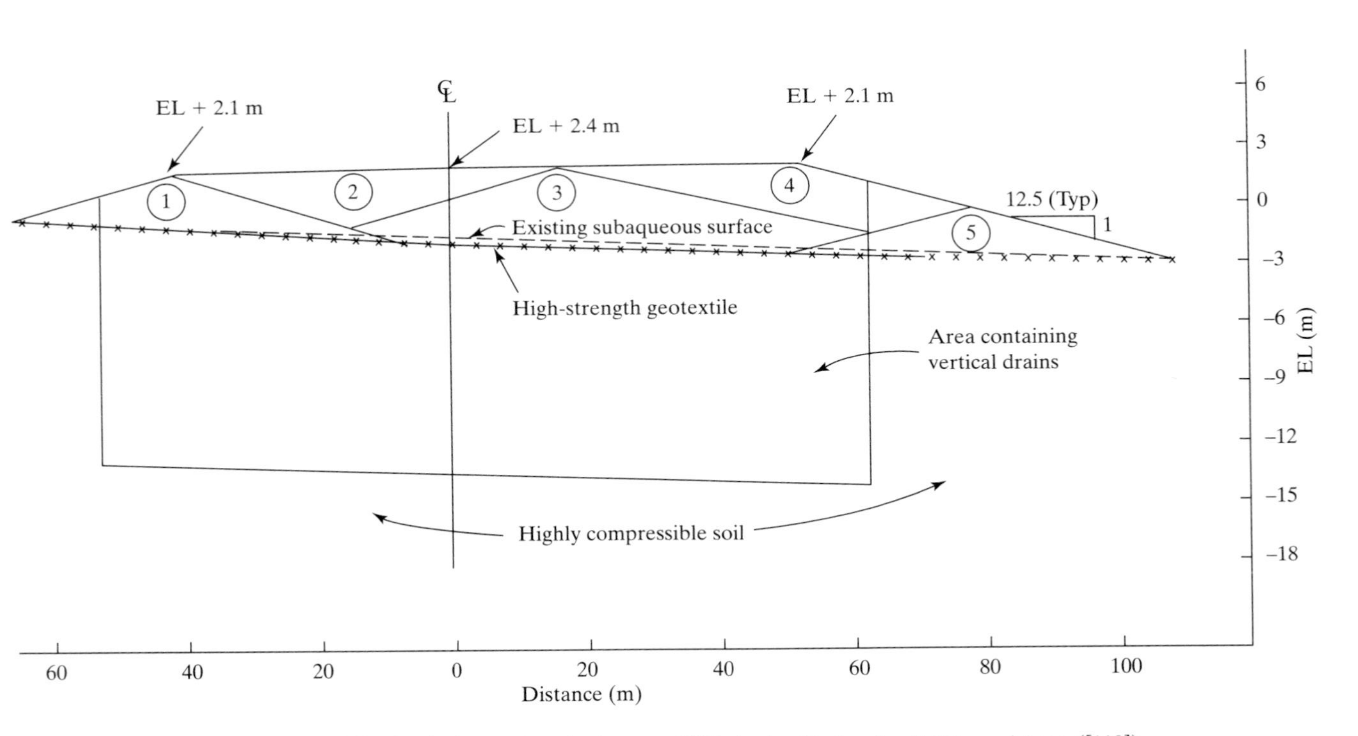

Figure 2.57 Typical stage 1 embankment of Wilmington Harbor South Disposal Area. ([110])

the geotextile in tension and provided added support for Fills 3, 2, and 4. Fills 1 and 5 proceeded approximately 30 to 46 m in advance of Fill 3 while Fills 2 and 4 trailed Fill 3 by about 15 m. The loads on the geotextile were balanced by keeping the total weights of Fills 1 and 5 approximately equal and their leading edges even with each other as filling progressed. The same thing was done with Fills 2 and 4, although this latter requirement was not considered as essential as for the first two fills. Fill 3 was critical in its applying load whenever the geotextile rose above the level of the water due to high pressures from the underlying foundation soils. The first stage reached an average top-of-fill elevation of 2.1 m at the completion of its initial placement. This was approximately 1.0 m above mean high water.

Prefabricated vertical wick drains were installed through the granular fill and geotextile to a depth of 12 m. The drains were in a triangular pattern of 3 m. Upon the completion of the primary consolidation settlement (measured via piezometers, settlement anchors, and inclinometers), the second-stage embankment was placed on top of stage one up to an elevation of +4.6 m. The outboard slope has a riprap erosion control system protecting it, and the inboard slope has a silty soil liner.

The contractor chose to fabricate the high-strength geotextile tensile reinforcement mat from rolls measuring 3.7 and 5.2 m in width. These rolls were continuous for their full length in the warp direction, since the contract specifications allowed no seams in the warp direction (recall Table 2.17). The geotextile supplier seamed four rolls together to form a 17.4-m-wide panel from two 5.2-m-wide and two 3.7-m-wide rolls in an off-site location. The seamed panels were then rolled onto axles for shipment to the actual construction site. Upon arrival at the site, the axles were attached to wheels, loaded onto barges, unrolled, and then field-seamed to one another. This resulted in an accordion-folded mat of the proper (and unseamed) length in the warp direction and continuous (albeit seamed) length in the weft direction.

When the field-seaming of the panels on the barges was completed, the now-continuous geotextile mat was deployed on the soft foundation subsoil beneath the river's surface. The normal means of accomplishing this was to move the interconnected laying barges forward along the centerline of the embankment, letting the geotextile feed off the barge over rollers and down into place on the river bottom. When starting off a new leg of the embankment, this process was reversed with the geotextile being pulled from the barge, which was temporarily anchored. The hydraulic placement of dredged granular fill was begun once the high-strength geotextile was in place. The contractor chose to branch five lines off of a primary supply line to allow the selective placement of the granular material on the five fill segments (recall Figure 2.57).

Prefabricated vertical wick drains were installed following the completion of the stage-one fill, as explained earlier. The subcontractor's wide-tracked installation crane was rigged to install the drains to depths of 18 m. The mandrel point and flat-anchor system employed by the drain installer was very satisfactory. (Details of the wick drain aspect of the project are described in Chapter 8.)

An example of an *areal fill* on dredged soil using a high-strength geotextile was at Seagirt, Maryland, for the Maryland Port Administration [111]. This 45-ha site contains 6 to 11 m of dredged soil at water contents of 50% to 150% above the liquid limit. The goal was to prepare the site for ground surface loads of approximately 30 kPa within the extraordinarily short time of six months. This goal was achieved by deploying a

1000 g/m² woven geotextile having a wide-width strength of approximately 210 kN/m. The required seam strength was 105 kN/m, which was the limiting design constraint. See Table 2.17 for a comparison of the geotextile specification of this project with the previously described linear fill project. In the areal fill project, sewn seams were particularly critical since the direction of the maximum principal stress was not known. Thus seam strength dominated the design (recall Figure 2.12 for seam efficiencies of high-strength geotextiles that must be included in Eq. (2.24) as an additional reduction factor). A single 0.75 m thick lift of granular soil, serving as the drainage blanket/working platform, was placed and wick drains were installed from this at 1.5 m centers to the bottom of the in situ soil. Note that the allowable geotextile strength must make accommodation for these holes and must be included as still another reduction factor in Eq. (2.24). The final operation was to place a 2.5 m surcharge fill over the entire site. Settlement plates and piezometers were the main control instruments from which the surcharge fill-placement rate and dwell time were controlled. The project was very successful in that consolidation occurred to the anticipated degree within the desired six-month time frame. The area is currently paved and used for heavy truck storage, loading, and unloading.

Design Methods. In considering an embankment placed upon very soft foundation soil and supported by a geotextile (or geogrid), a number of design elements, all of which are potential failure scenarios, are present. Figure 2.58 illustrates these possi-

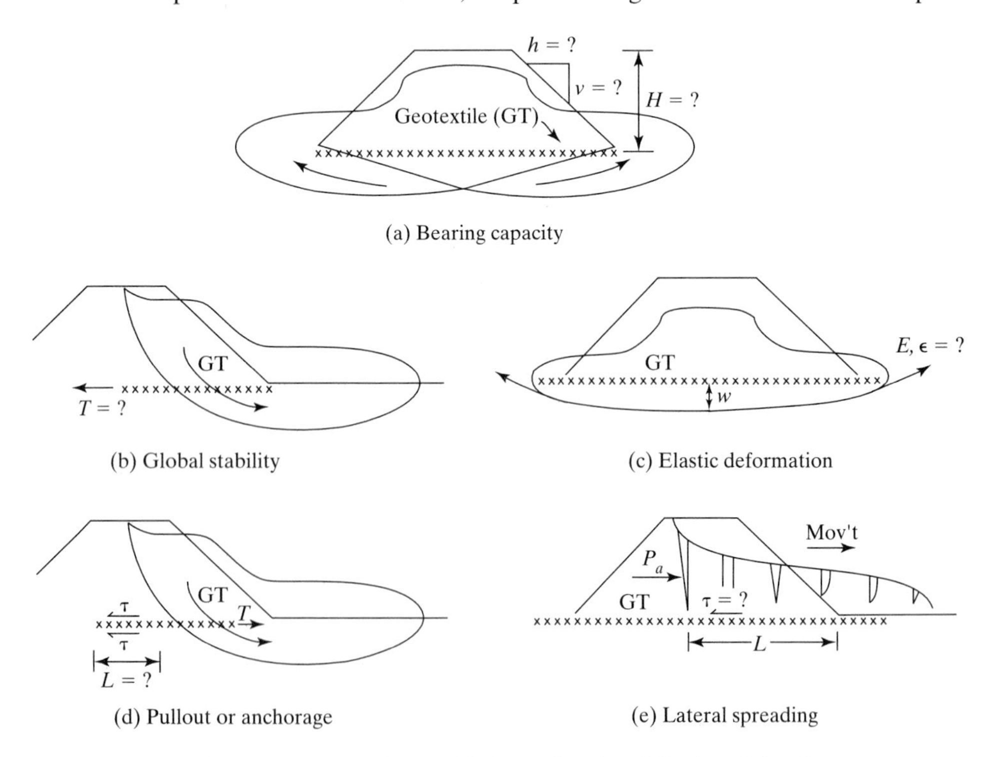

Figure 2.58 Geotextile design models for use in soft stabilization. (After Koerner et al. [109])

bilities. In sequentially going from one element to the next, the overall geotextile/embankment design gradually becomes more defined.

Bearing capacity	for overall embankment geometry
Global stability	for strength design in the major principal stress direction
	for strength design in the minor principal stress direction
Elastic deformation	for modulus and maximum strain in the major principal stress direction
	for modulus and maximum strain in the minor principal stress direction
Pullout or anchorage	for anchorage length behind slip plane(s)
Lateral spreading	for frictional properties of backfill soil against the geotextile

These quantifiable characteristics are then viewed in light of more qualitative considerations, such as the geotextile's stiffness, length, width, and weight of rolls, to arrive at a final design. As will be discussed later, this final design must be modified in light of manufacturing the geotextile so that a balanced design results that will be both safe and cost-effective. The discussion will be framed around the use of geotextiles, although it should be recognized that in many cases geogrids can be used in an equivalent manner.

Bearing Capacity. Regarding the bearing capacity as shown in Figure 2.58a, the limiting embankment height that can be placed on a given foundation soil is essentially independent of the geotextile. If a mass failure occurs beyond the limits of the reinforced zone, the geotextile will be carried along en masse. Thus conventional geotechnical engineering theory can be used directly.

$$q_{\text{allow}} = \frac{cN_c}{\text{FS}} \tag{2.54}$$

where

$q_{\text{allow}} = \gamma H_{\text{allow}}$ = allowable bearing capacity, in which
γ = unit weight of embankment soil, and
H_{allow} = allowable height of embankment;
c = cohesion of the foundation soil;
N_c = bearing capacity factor (= 3.5 to 5.7); and
FS = factor of safety.

Calculations based on Eq. (2.54) are surely worst-case situations, for soil strength invariably increases with depth; see Humphrey [112].

Global Stability. Figure 2.58b shows the type of global stability model that results in the required strength of the geotextile. It is precisely the same as formulated in Eq. (2.53), since the soil foundation strength can generally be estimated by its undrained shear strength and a single reinforcement layer is usually being placed. The strength of the embankment soil above the geotextile is another matter. Depending upon the soil's thickness, this shear strength is often taken as being zero—the assumption being that if tension cracks occur in the soil due to lateral deformation, the shear strength can easily be lost. Shown in Figures 2.59 a,b,c are the required geosynthetic strength values for cases of surcharge weight only, surcharge weight plus placement bulldozer, and surcharge weight plus wick drain installation crane, respectively. For weak foundation soils and typical slope angles in the 20 to 40° range, the required tensile strength of the geotextile is seen to be quite high. Two key points to remember in this regard are that the strength of the reinforcement system is that of its weakest element (e.g., seams or fabric with holes from wick drains) and that it must be compared to an allowable strength as per Eq. (2.24).

Example 2.20

What are the required, allowable, and ultimate wide-width geotextile strengths for a 4.0-m-high reinforced embankment whose face is on a slope of 3(H) to 1(V) using a factor of safety of 1.3 and reduction factors from Table 2.11. The undrained shear strength of the foundation soil is 2 kPa.

Solution: Using a slope angle of 18.4° and Figure 2.59a results in

$$T_{\text{reqd}} = 66 \text{ kN/m}$$

from which

$$\text{FS} = \frac{T_{\text{allow}}}{T_{\text{reqd}}}$$

$$1.3 = \frac{T_{\text{allow}}}{66}$$

$$T_{\text{allow}} = 86 \text{ kN/m}$$

Using Eq. (2.24) and Table 2.11 gives the necessary ultimate wide-width laboratory test strength.

$$T_{\text{allow}} = T_{\text{ult}}\left(\frac{1}{\text{RF}_{ID} \times \text{RF}_{CR} \times \text{RF}_{CD} \times \text{RF}_{BD}}\right)$$

$$86 = T_{\text{ult}}\left(\frac{1}{1.2 \times 2.0 \times 1.2 \times 1.0}\right)$$

$$T_{\text{ult}} = 86(2.88)$$

$$T_{\text{ult}} = 250 \text{ kN/m}$$

Note that the above value is for the geotextile itself. If seams are involved in the direction of the principal stress, see Figure 2.12 for typical efficiencies. Here it is seen that the maximum efficiency for a 250 kN/m geotextile seam is approximately 0.75. This is comparable to a reduction factor of 1.33. If holes are involved, for the insertion of wick drains, the loss of strength is approximately linear with the hole dimension [109]. Thus

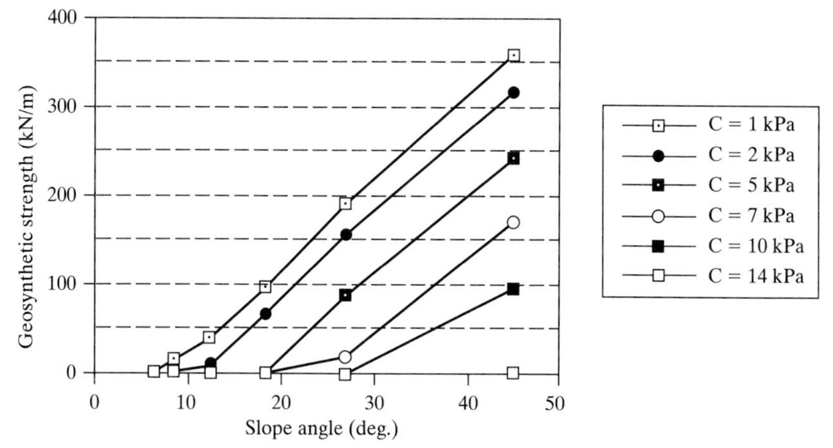

(a) Required geosynthetic strength based on *FS* = 1.3. Chart reflects soil surcharge height of 4m.

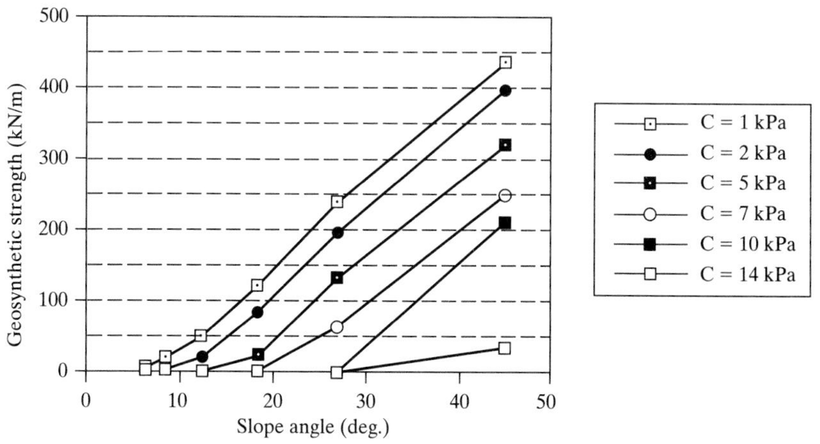

(b) Required geosynthetic strength based on *FS* = 13. Chart reflects soil surcharge height of 4 m plus 13 kPa dozer on embankment.

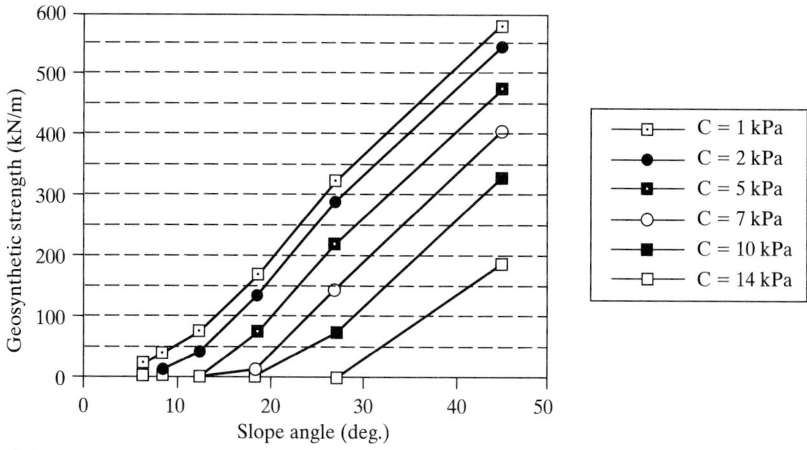

(c) Required geosynthetic strength based on *FS* = 1.3. Chart reflects soil surcharge height of 4 m plus 42 kPa wick drain installation equipment on embankment.

Figure 2.59 Parametric studies for global stability.

wick drains at 3 m centers making geotextile holes of 300 mm size will require another reduction factor in the formulation of 10/9 = 1.11. Using these two additional factors makes the formulation as follows:

$$T_{\text{allow}} = T_{\text{ult}}\left[\frac{1}{\text{RF}_{ID} \times \text{RF}_{CR} \times \text{RF}_{CD} \times \text{RF}_{BD} \times \text{RF}_{\text{Seams}} \times \text{RF}_{\text{Holes}}}\right]$$

$$86 = T_{\text{ult}}\left[\frac{1}{1.2 \times 2.0 \times 1.2 \times 1.0 \times 1.33 \times 1.11}\right]$$

$$T_{\text{ult}} = 86(4.25)$$

$$= 365 \text{ kN/m}$$

Elastic Deformation. The amount of elastic deformation allowed by the geotextile will govern the deformation of the embankment, as shown in Figure 2.58c. Obviously, too great an amount will cause unwanted embankment deformation and loss of underlying foundation soil. Thus relatively high modulus values of the geosynthetic are required. Unfortunately, "relatively high" is a poorly defined term. The U.S. Army Corps of Engineers' desired value of maximum strain at the required stress is approximately 10%; thus

$$E = \frac{T_{\text{reqd}}}{\varepsilon_f} \tag{2.55}$$

$$E = \frac{T_{\text{reqd}}}{0.10}$$

$$E_{\text{reqd}} = 10T_{\text{reqd}} \tag{2.56}$$

However, to obtain this E_{reqd} value requires a significantly stronger geosynthetic than T_{reqd} without this condition. The modulus requirement will dominate over the strength requirement if this condition is imposed. Note that these comments are based on the geosynthetic itself and do not consider seamed areas. The latter situation is difficult to consider and only recently has the deformation monitoring of seams been attempted (Guglielmetti et al. [113]).

Pullout or Anchorage. With the mobilization of all or part of the geotextile reinforcement's strength comes an equal and opposite requirement that the soil behind the slip zone resist pullout. As shown in Figure 2.58d, the situation is one in which an anchorage problem can be envisioned. Extending the work of Section 2.7.2, Eq. (2.57) can be formulated as follows:

$$T_{\text{act}} = 2\tau L$$

$$= 2(c_a + \sigma_v \tan \delta)L$$

and

$$L_{\text{reqd}} = \frac{T_{\text{act}}}{2(c_a + \sigma_v \tan \delta)} \tag{2.57}$$

or

$$L_{\text{reqd}} = \frac{T_{\text{act}}}{2E(c_a + \sigma_v \tan \delta)} \tag{2.58}$$

where

L_{reqd} = required anchorage length behind the slip plane;
T_{act} = actual stress in the geosynthetic;
c = cohesion of the soil;
c_a = adhesion of the soil to the geosynthetic;
ϕ = friction angle of the soil;
δ = friction angle of the soil to the geosynthetic;
σ_v = average vertical stress = γH, in which
γ = unit weight of embankment soil, and
H = average height of embankment above geosynthetic; and
E = anchorage, or pullout, efficiency of geosynthetic-to-soil.

For geotextiles, $E = 0.8$ to 1.2; for geogrids, $E = 1.3$ to 1.5. If insufficient anchorage distance is available to mobilize the required strength of the geosynthetic, physical methods of attachment—rolling of the material around stone or attachment to timber cribbing—is necessary.

Lateral Spreading. On occasion, tension cracks have been observed on the surface of embankments, as shown schematically in Figure 2.58e. The situation can be analyzed using the following equation for active earth pressure and considering granular soil fills to be above the geosynthetic.

$$P_a = \tau L$$

$$P_a = (\sigma_{\nu(\text{avg.})} \tan \delta)L$$

$$0.5\gamma H^2 K_a = (0.5\gamma H \tan \delta)L$$

$$\tan \delta_{\text{reqd}} = \frac{HK_a}{L} \tag{2.59}$$

where

δ_{reqd} = required friction angle of geosynthetic-to-soil;
H = embankment height;
K_a = coefficient of active earth pressure = $\tan^2 (45 - \phi/2)$;
L = length of zone involved in spreading; and
$\tan \delta_{\text{reqd}}$ = $E(\tan \phi)$, in which
ϕ = friction angle of embankment soil, and
E = shear, or frictional, efficiency of geosynthetic-to-soil, for geotextiles, $E = 0.6$ to 0.8; for geogrids, $E = 1.0$ to 1.5.

Geotextile Implications. The previous designs focus entirely on quantitative analyses and procedures that lead to the ultimate selection of the reinforcement geosynthetic. However, there are other considerations, many of which are qualitative, that must be brought into focus. These include the specific gravity and rigidity (or stiffness) of the geosynthetic, and the size and weight of the rolls.

If the site under consideration is at or under the surface of water, buoyancy is usually not a desirable feature. The geosynthetic should not float and its specific gravity should be greater than 1. Rigidity, or stiffness, of the geosynthetic is desirable for providing some type of working platform for deployment. The ASTM stiffness test described in Section 2.3.2 can be used for specifications. The minimum value is related to the CBR of the foundation soil (recall Table 2.3). This is an important feature and an area where additional investigation is warranted. The size and weight of the geosynthetic rolls must be considered by everyone involved in the process. It is obviously a site-specific situation, but one that is paramount in the success of the project. Designs that can not be reasonably constructed *should not* be designed to begin with.

Summary. It is the author's perception that soft soil stabilization using geosynthetics has been implemented from two different perspectives. The first is by the geotechnical engineer, who has modified traditional design methods to accommodate the inclusion of a reinforcement; the other is by the manufacturer, who has provided a means for accomplishing an end. In this case, the means is a high-strength geosynthetic, either a geotextile or geogrid.

For designs and installations that are both economical and safe, the perspectives of engineer and manufacturer must be brought together. The flow chart of Figure 2.60 attempts to achieve this goal. Beginning with the design elements illustrated in this section, the flow in Figure 2.60 is to the right as the stress-strain characteristics of the geosynthetic are progressively defined. The initial technical design is modified by other site-specific considerations, the applicable yarns, and fabrication of the geotextile or geogrid. The next, and most significant, question is whether the result is a balanced design. For example, where is the critical aspect of the design? If it is the field-sewn seams, then everything (warp, weft, modulus, elongation, etc.) should be formulated from this point. Last, there is the question of whether the resulting geosynthetic is constructible in light of the actual site situation. Here considerations of workability and, concomitantly, survivability are very important.

In the final analysis, we should be able to arrive at a geosynthetic design that is both optimally safe and economical. It is indeed a very worthwhile pursuit, for finally the profession has a technique whereby we can almost walk on water.

2.7.4 Geotextiles for Improved Bearing Capacity

With the recognition that multiple layers of geotextiles and/or high-strength geotextiles can reinforce walls, slopes, and foundations, it follows that soils beneath rigid walls, footings, piers, and so on, that have poor bearing capacity should also be a target for improved performance using geotextiles. What are the benefits in bearing capacity of placing horizontal layers of closely spaced geotextiles in compressible and/or soft-soil foundations? The logic behind the question follows the work of Binquet and Lee [114,

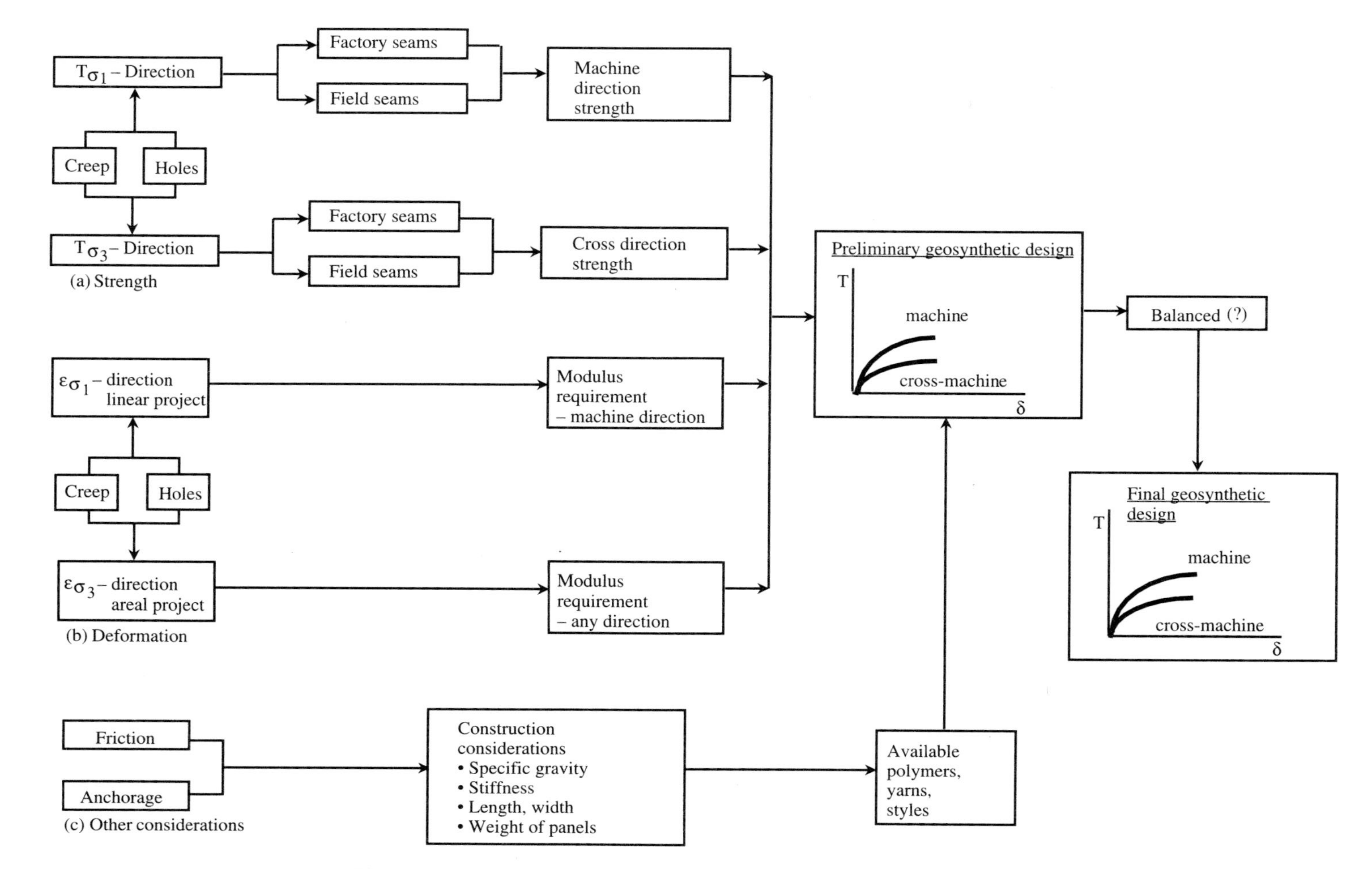

Figure 2.60 Required geosynthetic considerations in order to optimize the soft foundation soil design process.

115] on the improved bearing capacity of compressible sands using metal strips (thereby creating reinforced earth). They found definite improvement, which was further evidenced by an economic analysis showing cost savings. However, when corrosion was considered, the economic benefits were essentially lost. With noncorroding geotextiles as the reinforcement, the problem of corrosion is obviated. What remains is the research needed to quantify the possible improvements.

Laboratory studies by Guido et al. [116] using layers of geotextiles on loose sand produce the results in Figure 2.61a. Here it is seen that multiple (up to three) layers produced beneficial results, but only after a measurable settlement had occurred. This is to be expected, since the geotextile-soil system must deform before its reinforcing benefit is realized. Their tests used a nonwoven heat-bonded geotextile and varied a number of parameters, including the distance to the upper geotextile, the spacing between layers, and the distance the geotextile extends beyond the edge of the footing.

Laboratory studies at the Geosynthetic Research Institute on soft compressible fine-grained soils at saturations above their plastic limit produced the curves shown in Figure 2.61b. Here, using a woven slit-film geotextile, similar behavioral trends are noticed; some improvement in bearing pressure is noted throughout, but only at large deformations is the improvement noteworthy. A convenient measure of the improvement is

$$\text{BCR} = \frac{q}{q_o} \tag{2.60}$$

where

BCR = bearing capacity ratio,
q_o = bearing pressure of the nonreinforced soil at a given settlement, and
q = bearing pressure of the reinforced soil at a corresponding settlement.

It can be seen that for both of the studies portrayed in Figure 2.61, a method of prestressing the geotextile would be an advantage as it would eliminate the required deformation before a BCR improvement is noted. How to do this in a cost-effective manner, however, is not known at this time. In lieu of prestressing the geotextile, design must consider improved bearing capacity only after relatively large settlement. Also, the laboratory-generated curves shown must be used with considerable caution, since the scale effects are essentially unknown. In proceeding with a design, four modes of failure must be considered. They are shown schematically in Figure 2.62 and are explained below.

Bearing capacity failure of the soil above the upper geotextile. This is probably avoidable if the upper geotextile is within 300 mm of the ground surface.

Insufficient embedment length. This causes anchorage pullout,which is avoidable if the geotextile extends far enough beyond the potential failure zone to mobilize the required resisting anchorage force. It is described in Sections 2.7.2 and 2.7.3.

Tensile failure of the geotextile(s) due to overstressing. This is the main design element and uses information such as that presented in Figure 2.61.

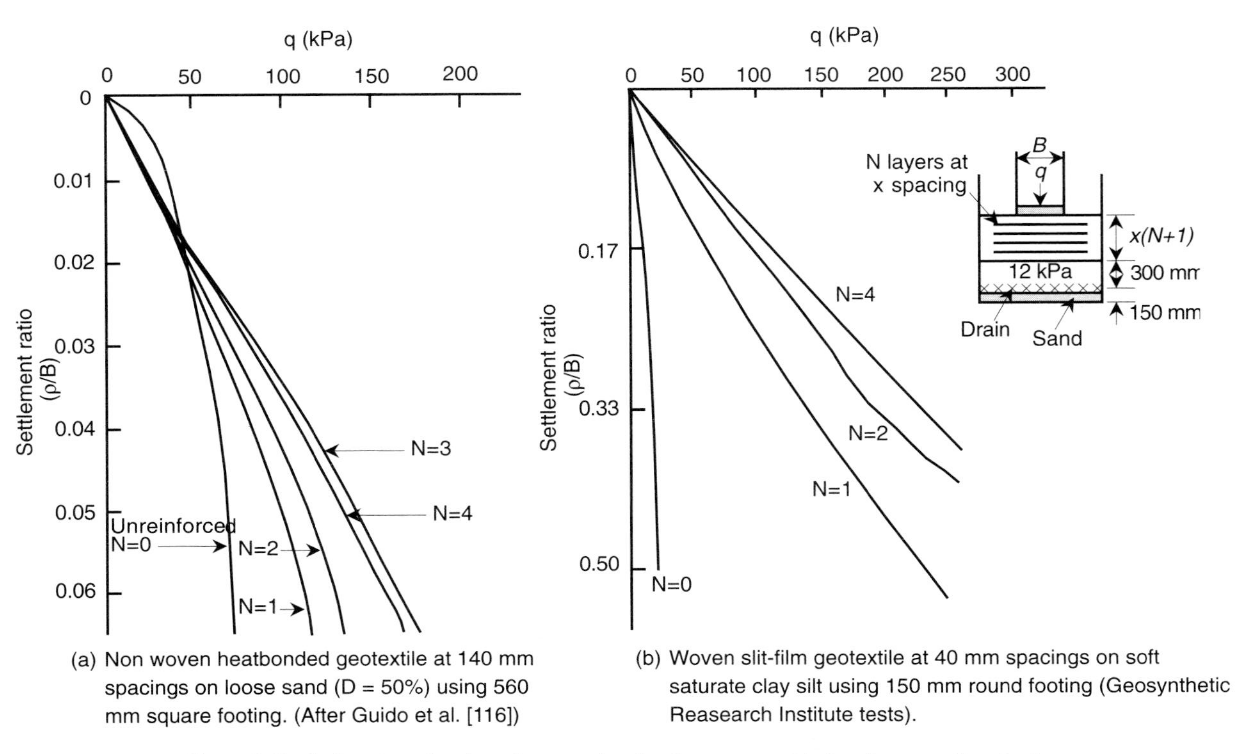

(a) Non woven heatbonded geotextile at 140 mm spacings on loose sand (D = 50%) using 560 mm square footing. (After Guido et al. [116])

(b) Woven slit-film geotextile at 40 mm spacings on soft saturate clay silt using 150 mm round footing (Geosynthetic Reasearch Institute tests).

Figure 2.61 Laboratory-developed curves showing improvement in bearing capacity of soils using geotextiles. ρ is the footing settlement and B is the footing width.

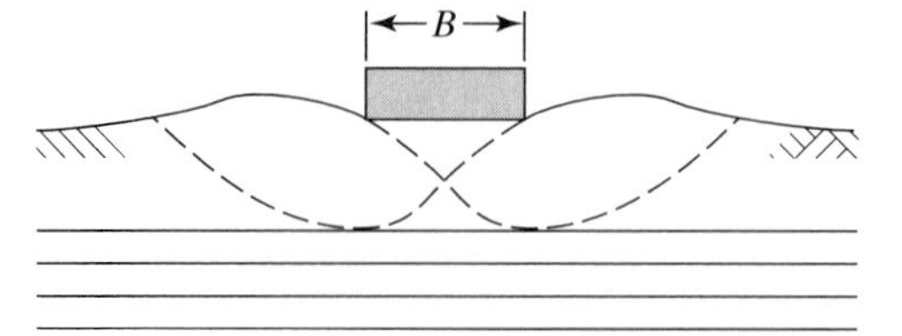

(a) bearing capacity failure above upper geotextile layer

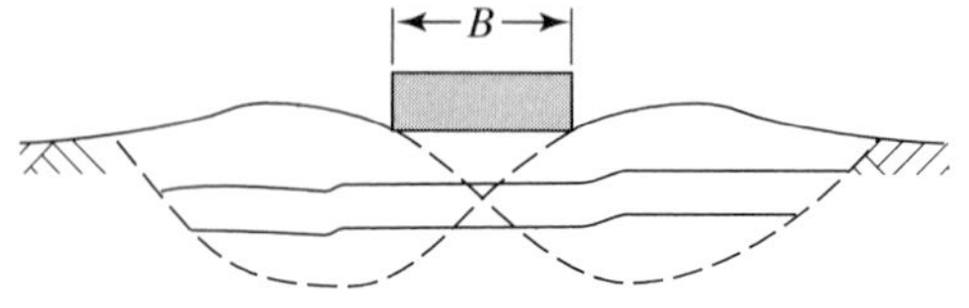

(b) anchorage pullout of geotextiles due to insufficient embedment length

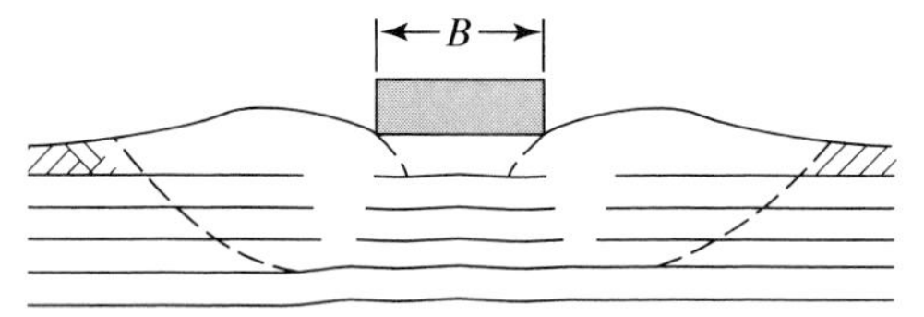

(c) tensile failure (breaking) of geotextiles

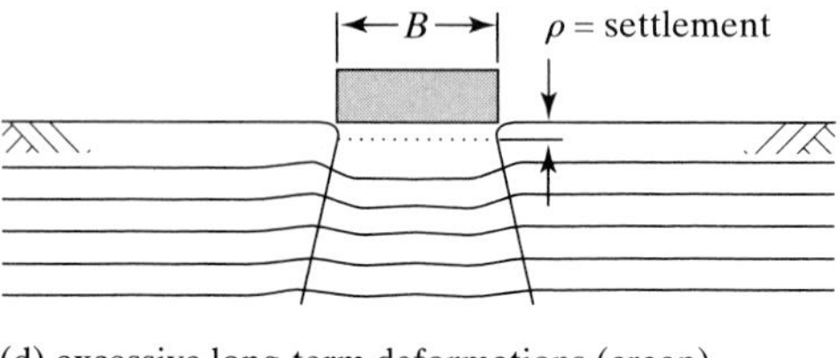

(d) excessive long-term deformations (creep)

Figure 2.62 Possible modes of failure of geotextile-reinforced shallow foundations.

Excessive long-term (creep) settlement. This is due to the sustained surface loads and subsequent geotextile stress relaxation, which can be avoided if low-enough allowable stresses in the geotextile are used. Allowable tensile strength values should be chosen very conservatively (see Eq. 2.24 and Table 2.11).

2.7.5 Geotextiles for In Situ Slope Stabilization

Background. The methods described in this section thus far are all oriented toward new construction, where the geotextiles can be placed along with the earthwork involved. However, there are many situations in which existing soil slopes and embankments are at or near to their failure state. Oftentimes homes or other structures are precariously close to the edge of the slope. Landslides cause large annual losses of lives and losses of billions of dollars; see Table 2.18 for data in the U.S. The usual indicators of slope instability are bulges of soil at the toe of the slope, tension cracks in the soil slope or at the crest, springs near the toe of the slope, and vegetative growth that is

TABLE 2.18 ESTIMATED ANNUAL LOSSES FROM, AND ANNUAL RESEARCH FUNDS FOR, SELECTED GROUND FAILURE HAZARDS IN THE U.S.*

Hazard	Annual Loss of Life	Annual Economic Loss (millions of dollars)	Annual Research Funds (millions of dollars)
Landslides	25–50	1000–2000	3–5
Permafrost	0	20	2
Subsidence	n/a	500	10
Swelling soils	0	6000	2
Frost action	0	n/a	n/a
Rock deformation	1	n/a	3
Earthquakes	15	100	53
Volcanoes	1	10	10

*A major earthquake or volcanic eruption affecting a large urban area could greatly affect the estimated annual life and economic losses from these hazards. The other hazards in the table are less episodic, and the estimates given are probably more reliable as estimates of future annual losses.

Source: [117].

leaning toward the downstream toe. The basic problem in such cases is insufficient shear strength of the soil with respect to the slope angle and height of the embankment. This often suggests low relative density in granular soils and high water content in cohesive soils. Both of these situations (low density and high water content) could be positively influenced by some type of in situ stabilization system. The addition of a surface-deployed geotextile that is "nailed" into the slope could provide such a system. In *soil nailing* long steel rods are driven into the ground and then shotcreted or gunited to form vertical temporary retaining walls (see Shen et al. [118]). Instead of using a rigid impermeable facing such as shotcrete or gunite, the designer could consider using a geotextile reinforced locally at the points where it is nailed to the soil slope.

Such a system, developed by the author [119, 120], is shown in Figure 2.63. Here the anchored geotextiles (also called *anchored spider netting*, since the surface appears quilted and tucked into the soil in radial patterns) are used in the compaction and/or consolidation of the in situ soils. With both the geotextile netting and the steel-rod nails in tension, the set of free-body diagrams of the system is as shown in Figure 2.64. Shown in Figure 2.65 is a sketch of the anchor assembly at the surface of the geotextile. The system must be reanchored periodically, since loss of pore volume (either air or water) will result in a relaxed tensile stress of the geotextile net. During this time, of course, the soil itself is gaining in shear strength either by increased densification and/or by consolidation. This improvement in shear strength is precisely what is necessary to reinstate the slope's stability. Once the soil properties are sufficiently improved, the geotextile no longer serves a useful strengthening purpose. Thus UV degradation is not a severe detriment. If, however, erosion control is to be an added feature of the system, UV inhibitors must be included in the geotextile during its manufacture.

Construction Details. The current guidelines for anchored geotextiles are as follows. Geotextiles of 35 kN/m wide-width tensile strength and opening size tight

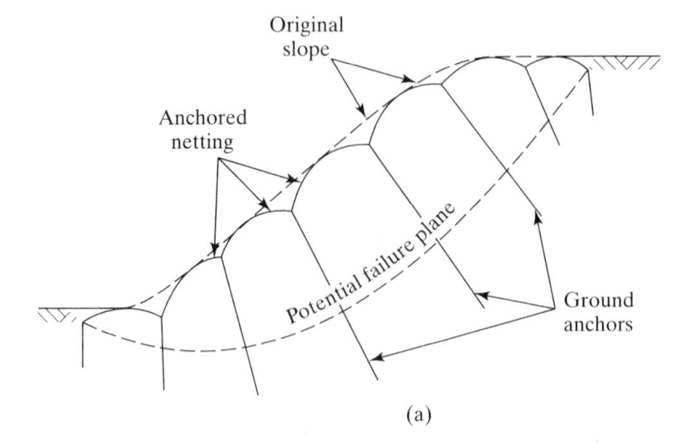

(a)

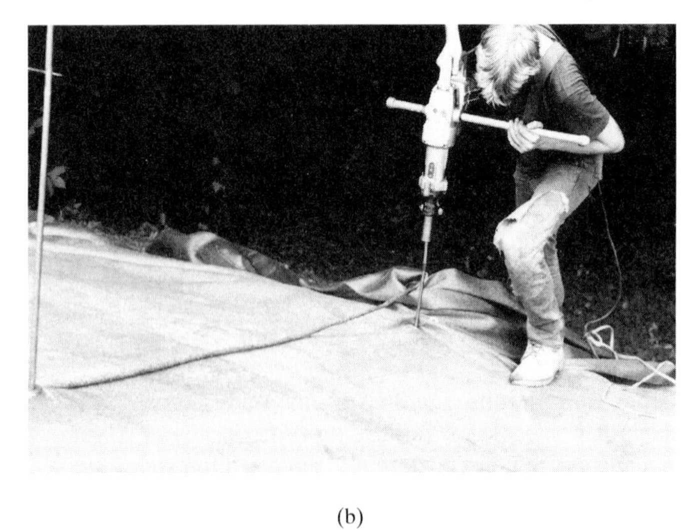

(b)

(c)

Figure 2.63 Cross section and photographs of in situ slope stabilization using anchored geotextiles. (After Koerner [119, 120])

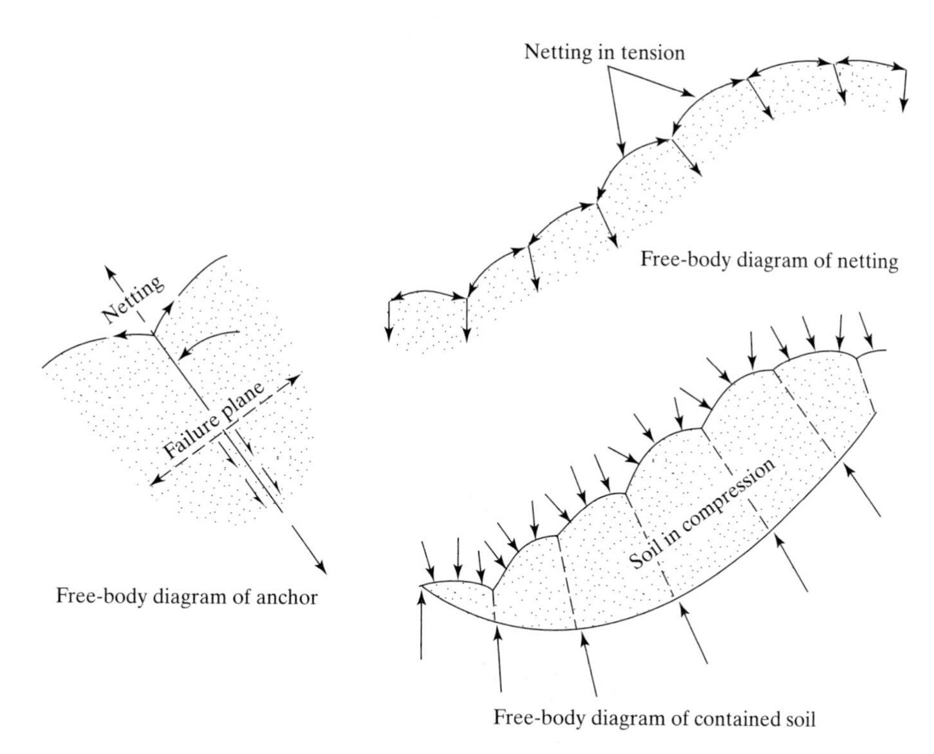

Figure 2.64 Free-body diagrams of various components of anchored geotextiles. (After Koerner [119, 120])

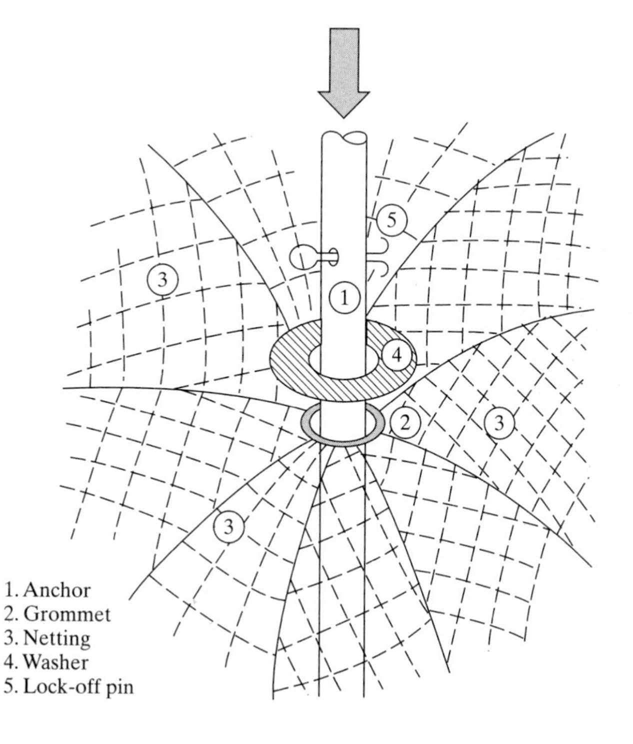

Figure 2.65 Sketch of anchor lock-off assembly at points where anchor passes through grommet in netting. (After Koerner [120])

enough so as not to lose soil are used. Local reinforcement of the geotextiles at anchor points equivalent to a 90 kN/m wide-width tensile strength for 150 mm around the anchor are necessary. Grommets should be prefabricated into the geotextile unless an alternative is provided; such an alternative could be localized fabric reinforcement at the anchor locations. Anchor points should be at 1.5 m centers on a triangular or square pattern. Steel-rod anchors should be 13 mm in diameter or larger. Anchor lengths should exceed the potential failure plane by at least 1.5 m, which for base failures can mean rods of great length. The option of a prefabricated grid or net with localized reinforced zones has also been used.

The following steps illustrate the usual manner of deploying anchored geotextile netting on a soil slope in need of stabilization. Note that it is done without the aid of heavy construction machinery or equipment, which is important for quasi-stable slopes.

1. The slope should be rough-graded to eliminate abrupt high spots and to fill in holes and sharp depressions. This can usually be done by hand, depending on existing site conditions.
2. The netting, with its prefabricated grommets or localized reinforced zones in place, is unrolled and spread on the slope from the upper levels downward and horizontally across it.
3. If wind is a problem, or if part of the slope is beneath water, it may be necessary to staple the netting to the soil. L-shaped nails or U-shaped pins are recommended. They should be approximately 300 mm long.
4. Beginning at the upper part of the slope, steel anchor rods are placed through the anchor holes and are hand-driven as far as possible. If threaded rod or pipe is being used, it will be necessary to have an adapter put over the threads so as not to damage them during driving. These rods should be driven to within 75 to 90% of their intended depth. In most cases it will be necessary to use an impact hammer, paving breaker, or other type of mechanized driving tool or hammer.
5. When this depth is reached, fix the washer and lock-off assembly to the anchor; see Figure 2.65.
6. Continue driving the anchor into the soil. At this time, the netting is being pulled along with the anchor and it is being stressed (tensioned) in a radial manner.
7. Temporarily stop driving a short distance (approximately 5%) from the final design depth or refusal.
8. Repeat Steps 4 through 7 on each of the immediately adjacent anchors.
9. Return to the original anchor and drive it to design depth or refusal.
10. Continue in this manner down the slope and across it, progressing in a uniform manner.
11. As an alternative to Steps 4 through 10, it should be possible to explosively drive the anchors into the soil, carrying the net along with it. Myles and Briddle [121] have developed such a method for installing nails and anchors into soil.
12. At the end of each roll of netting, positive fixity (versus overlapping) is required; this means that sewing or clamping of the geotextile netting is required. At least 90% of the basic strength of the netting should be available at all seams.
13. The top, bottom, and extreme ends of the total stabilized slope should end with anchors of the same type as the interior of the system.

14. Depending on the site conditions, the slope can be seeded either before or after the placement of the geotextile netting, although seeding before is generally preferred. Growth of vegetation through the netting is considered to be an advantage in the long-term stabilization of the slope.
15. For long-term slope stabilization, particularly with high water-content cohesive soils (silts, clays, and their respective mixtures) it is necessary to return to the slope periodically to redrive the anchors. This is required because of the long-term consolidation characteristics of high water-content fine-grained soils, as described previously. This aspect of the system must be carefully tuned to the local conditions but will result in a greatly stabilized site.

Design Methods. While the role of the surface-tensioned geosynthetic is clear (along with its compressive influence on the enclosed soil mass), the role of the nails protruding into and beyond the potential failure plane is not. Their contribution toward stabilizing the slope is difficult to assess. Using concepts taken from soil nailing, which allows temporary vertical slopes to be retained during construction, there are a number of potential benefits of the rods. Some of these are: friction along the surface of the rods, adhesion along the surface of the rods, suction at the rod ends if saturated conditions exist, bearing at any protrusions along the surface of the rods, the bending resistance of the rods due to the downward movement of the soil mass, and the torsional resistance of the rods due to any out-of-plane bending forces. These are very difficult phenomena to include in a theoretical analysis and will be lumped together into one single soil-modification parameter in the following analysis.

For the effect of the tensioned net on the soil's surface, the analysis is much more understandable and analytically tractable. The purpose of the tensioned net is clearly to place the encapsulated soil mass in compression. This will add normal forces to the slope, which will increase its short-term stability. Over the longer term it will also cause the soil to densify. For granular soils this densification takes the form of compaction where pore air or pore water is expelled. For saturated cohesive soils, this densification takes the form of consolidation where the pore water is expelled, but within a considerably longer time frame than when dealing with granular soil masses. In both cases, the need for a porous netting at the soil's surface is obvious. Use of an impervious geomembrane is not applicable for this method.

The time-dependent densification process just described will ultimately cause an increase in soil shear strength by means of an increase in friction and/or cohesion, thereby increasing the slope's stability. This densification (and hence volume reduction) process will require the redriving of the anchors to greater depths on a periodic basis. Eventually the increased shear strength parameters will allow for the slope to support itself without the need the anchored spider netting at all.

In the analysis to follow, the modified Bishop method based on effective stresses is used [122]. Effective stress analyses are necessary since the slope soils invariably have a moisture component and require use of the method of slices. Both moment and vertical force equilibrium are satisfied in the following analysis (after Koerner and Robbins [123]), resulting in the following equations.

$$\bar{S}_i = \frac{(\bar{c}l_i + \bar{\sigma}l_i \tan \bar{\phi})}{\text{FS}} \tag{2.61}$$

$$\bar{\sigma} l_i = (W_i - \mu_i l_i \cos \theta_i - \bar{S}_i \sin \theta_i) \sec \theta_i \tag{2.62}$$

Solved simultaneously, Eqs. (2.61) and (2.62) result in the following equation for the global factor of safety.

$$\text{FS} = \sum_{i=1}^{n} \frac{\bar{c} l_i + (W_i - \mu_i l_i \cos \theta) \tan \bar{\phi} \sec \theta_i}{(W_i \sin \theta_i)\left(1 + \dfrac{\tan \bar{\phi} \tan \theta_i}{\text{FS}}\right)} \tag{2.63}$$

Note in Eq. (2.63) that the factor of safety is not an explicit function and an iterative solution is necessary. Added to the complexity of the equation is the necessity of summing each of the individual slices and finding the minimum factor of safety. Thus a computer solution is required.

With the addition of anchored spider netting on the surface of the slope, as shown in Figure 2.63, a number of features are added. Moment and force equilibrium now yield the following equations:

$$\bar{S}_i = \frac{[(1 + f)(\bar{c}_m l_i + \bar{\sigma} l_i \tan \bar{\phi}_m)]}{\text{FS}} \tag{2.64}$$

$$\bar{\sigma} l_i = (W_i + F_i \cos \beta_i - \mu l_i \cos \theta_i - \bar{S}_i \sin \theta_i) \sec \theta_i \tag{2.65}$$

which when solved simultaneously result in Eq. (2.66) for the desired factor of safety.

$$\text{FS} = (1 + f) \sum_{i=1}^{n} \frac{\bar{c}_m l_i + (W_i + F_i \cos \beta_i - \mu_i l_i \cos \theta_i) \tan \bar{\phi}_m \sec \theta_i}{[W_i \sin \theta_i - (F_i d_i / R)]\left[1 + \dfrac{(1 + f) \tan \bar{\phi}_m \tan \theta_i}{\text{FS}}\right]} \tag{2.66}$$

where

$\bar{c}$ = effective cohesion,
$\bar{\phi}$ = effective angle of shearing resistance,
W_i = slice weight,
l_i = arc length of slice,
μ_i = pore water pressure in the slice,
θ_i = angle that the midpoint of the slice makes with the horizontal, and
n = number of slices, which is arbitrary.

By now comparing Eq. (2.63) to Eq. (2.66), the influence of the anchored spider netting can be seen. These features (all of which positively influence the factor of safety) are as follows:

$\bar{c}_m$ = modified effective cohesion (where $\bar{c}_m \geqslant \bar{c}$),
$\bar{\phi}_m$ = modified angle of shear resistance (where $\bar{\phi}_m \geqslant \bar{\phi}$),
$(1 + f)$ = contribution of the anchors (nails) penetrating the failure plane to stability,

$(F_i d_i/R)$ = moment due to the pressure of the stressed net at the ground surface, and

$(F_i \cos \beta_i)$ = contribution of the stressed net at the bottom of the slice (where equilibrium equations are taken) to stability.

In order to investigate the numeric influence of these added terms to the slope's factor of safety, a computer-based sensitivity analysis has been performed [123]. The analysis uses a uniform slope of height 7.6 m, a slope angle of 55°, a soil unit weight of 16.8 kN/m^3, a cohesion of 9.5 kPa, and a friction angle of 20°. The factor of safety of the slope using Eq. (2.63) without anchored spider netting is 0.967.

Now using anchored spider netting on the slope and only the influence of the rods in Eq. (2.66), with $(1 + f)$ varying from 0 to 25%, the factor of safety increases as shown in Figure 2.66a. Using only the influence of the surface loading term $\Sigma(F_i/d_i)/R$ in Eq. (2.66), with the netting pressure σ varying from 0 to 2700 Pa, the factor of safety increases as shown in Figure 2.66b. Using only the influence of an increased normal force at the base of the slice, $F_i \cos \beta_i$, in Eq. (2.66), with σ varying from 0 to 2700 Pa, the factor of safety increases as shown in Figure 2.66c. Finally, taking all of the short-term gains collectively in Eq. (2.66) gives Figure 2.66d. Here the increase in the global factor of safety of a slope due to a properly designed and constructed anchored spider net is

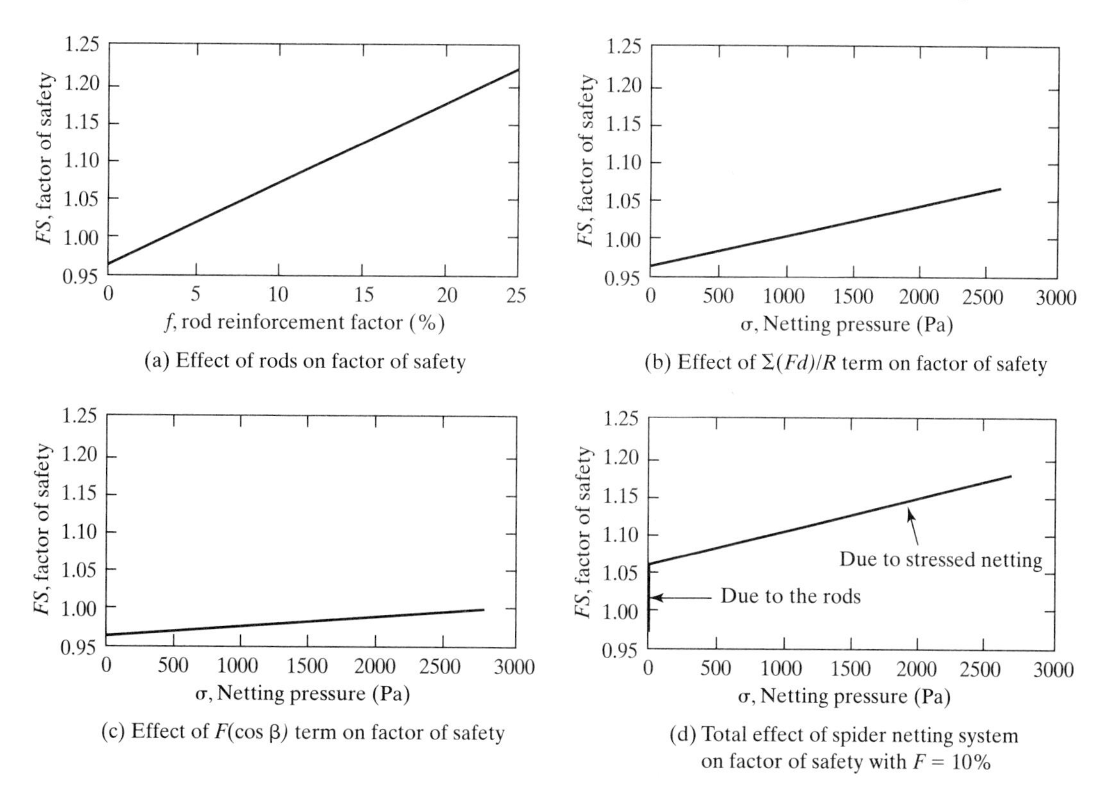

Figure 2.66 Parametric study of factors influencing soil slope stability when using anchored spider netting. (After Koerner and Robins [123])

quite obvious and, indeed, very beneficial. The theoretical aspects of the anchored netting concept has been extended and modeled in large-scale laboratory tests; see Ghiassian et al. [124]. In a companion paper, Ghiassian et al. [125] have investigated the seepage considerations in the soil slope beneath the geosynthetic using a number of surface deformation patterns.

Note that the analysis could also be extended to account for long-term gains in the shear strength parameters ($\bar{c}_m$) and ($\bar{\phi}_m$). However, their influence is typical of strength gains in any slope-stability analysis. They are significant and well understood by geotechnical engineers.

Summary. Anchored spider netting is a soil slope stabilization technique aimed at the enormously large landslide problem area (recall Table 2.18). It is an in situ technique in which a geosynthetic material (generally a geotextile) is placed on the unstable or questionable slope and anchored to it with long steel rod nails. These nails must be long enough to penetrate the actual or potential failure surface. Upon being suitably deployed, anchored spider netting offers a number of advantages in arresting slope failures: the steel rods, in penetrating the failure surface, increase stability; the stress caused by the netting at the ground surface increases stability; the surface netting stress-mobilizes normal stress at the base of the failure surface, which increases stability; and the entire system causes soil densification, which increases the soil's shear strength parameters, increasing stability.

Each of these features was illustrated analytically by modifying a limit equilibrium slope stability procedure. Furthermore, a computer-based sensitivity analysis was performed showing the relative and collective contribution of each of the factors to slope stabilization. Large-scale laboratory strength and seepage modeling data are also available.

A number of case histories are the subject of ongoing studies, the most recent types using a bitumen-coated, knitted nylon net. Knitting allows for localized reinforcement zones for the anchor rods and avoids considerable field-oriented detail.

2.8 DESIGNING FOR FILTRATION

When liquid flows *across the plane* of the geotextile, the geotextile acts and is designed as a filter. Unfortunately, most literature, such as many highway specifications and numerous manufacturers brochures, identifies this topic incorrectly as drainage. Drainage, a distinct and separate topic, will be treated later in Section 2.9.

2.8.1 Overview of Applications

In Section 1.3.3 a number of applications were presented in which geotextiles are used adjacent to soil for the purpose of allowing liquid to pass through them while retaining the soil on the upstream side. Generally, the situations represent liquid moving in one direction only, but in some cases reversing the flow was mentioned, for example, in tidal areas. Furthermore, the situations that will be discussed here all involve flow conditions designed on a worst-case scenario basis. This should come as no surprise, since it is the same type of conservatism that is used in all civil engineering design. Time-dependent random or dynamic flow situations will only be considered peripherally, because too little information is currently available for handling such situations. The factor of safety can always be increased to account for such considerations. Hopefully, the designs to follow cover most of the commonly encountered situations. The specific designs to be treated are the following:

- Geotextile filters behind retaining walls
- Geotextile filters wrapped around underdrains
- Geotextile filters used beneath erosion control structures
- Geotextile filters used as silt fences

2.8.2 General Behavior

The general behavior of a filter simultaneously demands adequate permeability and proper soil retention. Permeability is required to allow the liquid to pass the filter so as not to build up excess hydrostatic pore pressure. At the same time, it is necessary to retain the majority of the upstream soil to avoid soil piping. By soil piping, we mean the gradual loss of upstream soil fines, which results in higher flow rates, which in turn causes more soil loss, and so on. Thus we are asking for a design of an open geotextile structure that allows liquid to pass and, at the same time, a closed geotextile structure that retains the soil on the upstream side of the geotextile. While these are indeed contradictory demands, such a geotextile filter design is possible because the amount of liquid flow through the soil is related primarily to particle size. For example, a commonly used empirical relationship between the permeability coefficient and soil particle size is

$$k = Cd_{10}^2 \tag{2.67}$$

where

k = permeability coefficient (hydraulic conductivity) of the soil,
C = site-specific constant,
d_{10} = effective soil particle size (i.e., size at which 10% of the soil is finer).

Thus large particle-size soils can generate high-flow conditions (requiring, geotextiles with relatively large voids), while small particle-size soils are associated with low-flow conditions (requiring geotextiles with relatively small voids). This results in a completely possible design situation, which is addressed in this section.

Before beginning with the actual designs it must be repeated that long-term soil-to-geotextile compatibility is also a necessary requirement. This third aspect of the design is treated in Section 2.2.3, and the particular details will not be repeated here.

2.8.3 Geotextiles behind Retaining Walls

Behind conventional reinforced-concrete retaining walls there must be a vertical drainage layer, typically consisting of granular soil that serves as a flow path allowing water from the backfill soil to escape into an underdrain system (Figure 2.67a) or

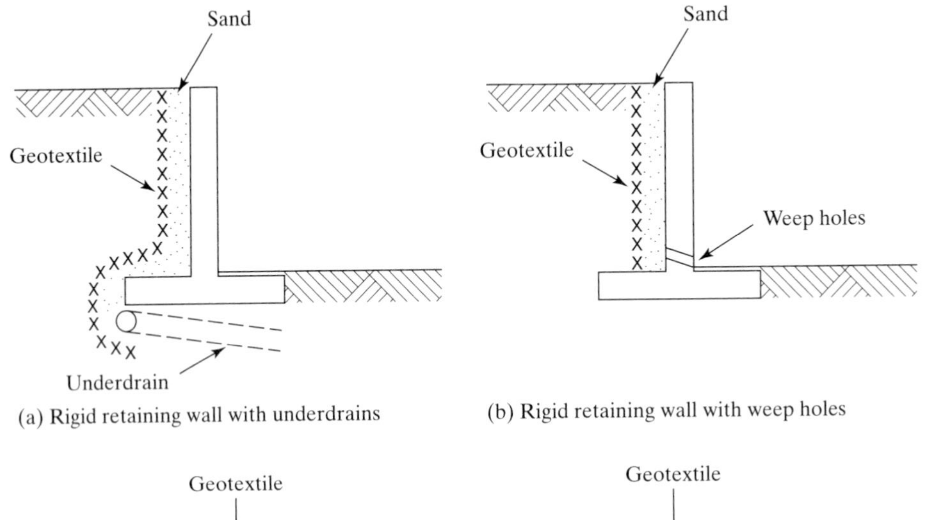

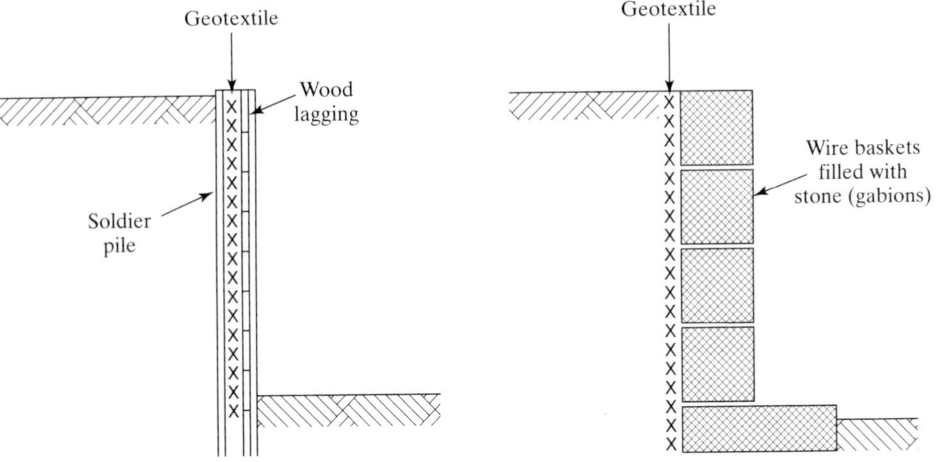

Figure 2.67 Various types of retaining walls in which geotextiles can be used as filters.

through weep holes (Figure 2.67b). Without this drainage layer, hydrostatic pressures will build up and, together with the horizontal soil pressure, could easily cause failure. Such hydrostatic pressures, if not dissipated by adequate drainage, can double the pressure against the wall. It must also be considered that the drainage sand must stay free-draining for the lifetime of the wall. If it excessively clogs with backfill soil within this time span, the sand becomes as useless as if it was not there at all. Thus it must be protected by a soil filter (which is expensive and difficult to place) or by a geotextile filter.

An identical situation, as far as the geotextile is concerned, is in the construction of flexible wall systems that are free-draining in themselves but would, without a soil filter or geotextile filter, allow the backfill soil to move into and through the open spaces.

Such walls are illustrated in Figures 2.67c,d in which the gabion style consists of wire baskets filled with 100 mm and larger stones. To backfill against such walls with no filter medium (soil or geotextile) would be sheer idiocy. Since the soil filter is difficult to place in a vertical or near-vertical manner and may even require a series of graded filter soils, the geotextile filter becomes very attractive. Geotextile filter design is illustrated in Example 2.21.

Example 2.21

Given a 3.5 m high gabion wall consisting of three $1 \times 1 \times 3$ m long baskets sitting on a $0.5 \times 2 \times 3$ m long mattress as shown below, the backfill soil is a medium-dense silty sand of $d_{10} = 0.03$ mm, $C_u = 2.5$, $k = 0.0075$ m/s, and $D_R = 70\%$. Check the adequacy of three candidate geotextiles whose laboratory test properties are given below. Use the cumulative reduction factors in Eq. (2.25) of 15.0, in order to adjust the ultimate laboratory-obtained permittivity value to an allowable field-oriented value.

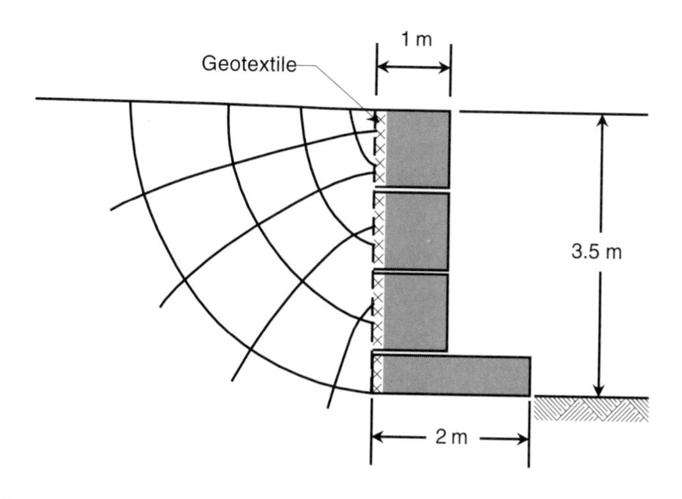

Geotextile		Permittivity	AOS*
No.	Type	(s^{-1})	(mm)
1	nonwoven needle-punched	2.0	0.30
2	woven monofilament	1.2	0.42
3	nonwoven heat-bonded	0.4	0.21

*Note that if the AOS is given in sieve size number, it must be converted to O_{95} in mm using Table 2.6.

Solution: The design is in two stages, with the first being a determination of the flow factor of safety of the geotextile; the second being a consideration of the AOS

(a) The first is done by calculating the required permittivity, Ψ, which is $\Psi = k/t$.

(1) Calculate the actual flow rate using a flow net.

$$q = kh\left(\frac{F}{N}\right)$$

$$= (0.0075)(3.5)\left(\frac{4}{5}\right)$$

$$= 0.021 \text{ m}^2/\text{s}$$

(2) Calculate the required permittivity.

$$q = kiA$$

$$q = k\frac{\Delta h}{t}A$$

$$\frac{k}{t} = \frac{q}{(\Delta h)(A)}$$

$$\Psi_{\text{reqd}} = \frac{0.021}{(3.5)(3.5 \times 1)}$$

$$= 1.71 \times 10^{-3}\, s^{-1}$$

(3) Check against the allowable permittivity of the candidate geotextiles.

Geotextile 1 (nonwoven needle punched):

$$\Psi_{\text{ult}} = 2.0 \text{ s}^{-1}$$

$$\Psi_{\text{allow}} = \Psi_{\text{ult}}\left(\frac{1}{\text{RF}_{SCB} \times \text{RF}_{CR} \times \text{RF}_{IN} \times \text{RF}_{CC} \times \text{RF}_{BC}}\right)$$

$$= \frac{2.0}{15.0}$$

$$= 0.13 \text{ s}^{-1}$$

$$\text{FS} = \Psi_{\text{allow}}/\Psi_{\text{reqd}}$$

$$= \frac{0.13}{0.00171}$$

$= 76$ acceptable; the geotextile has a high factor of safety

Geotextile 2 (woven monofilament):

$$\Psi_{\text{ult}} = 1.2 \text{ s}^{-1}$$

$$\Psi_{\text{allow}} = \frac{1.2}{15.0}$$

$$= 0.080 \text{ s}^{-1}/\text{m wall}$$

$$= \Psi_{\text{allow}}/\Psi_{\text{reqd}}$$

$$= \frac{0.080}{0.00171}$$

$= 47$ acceptable; this geotextile is also permeable enough

Geotextile 3 (nonwoven heat bonded):

$$\Psi_{\text{ult}} = 0.40\ \text{s}^{-1}$$

$$\Psi_{\text{allow}} = \frac{0.40}{15.0}$$

$$= 0.027\ \text{s}^{-1}$$

$$\text{FS} = \Psi_{\text{allow}}/\Psi_{\text{reqd}}$$

$$= \frac{0.027}{0.00171}$$

$$= 16 \qquad \text{acceptable; this geotextile is also adequate}$$

The above shows that many commercially available geotextiles can easily handle the required flow.

(b) The second part of the design relates to the geotextile's opening size, so as to prevent excessive soil loss. The three candidate geotextiles have AOS values of 0.30, 0.42, and 0.21 mm, respectively.

(1) The appropriate criterion for opening size must first be selected. Since this is a noncritical situation, Carroll's criterion (recall Section 2.2.3) will be used. This calls for the following:

$$O_{95} < 2.5\, d_{85}$$

since $d_{10} = 0.03$ mm and $C_u = 2.5$, and an approximate value of $d_{85} = 0.15$ mm. (Note, that the d_{85} value should be obtained directly by sieving the upstream soil.)

$$O_{95} < 2.5(0.15)$$

$$< 0.375\ \text{mm}$$

(2) Check against the candidate geotextiles' AOS values:

Geotextile 1: $AOS = 0.30\ \text{mm} < 0.375\ \text{mm}$ acceptable with FS = 1.25
Geotextile 2: $AOS = 0.42\ \text{mm} \geqslant 0.375\ \text{mm}$ not acceptable with FS = 0.89
Geotextile 3: $AOS = 0.21\ \text{mm} < 0.375\ \text{mm}$ acceptable with FS = 1.79

Thus Geotextile 2 is too open and will experience excessive soil loss based upon this soil-retention criterion. (If such a geotextile style were used, it would have to have a tighter AOS value, which is possible since the permittivity factor of safety was so high, i.e., 47.) Candidate Geotextiles 1 and 3 are acceptable. The technical decision as to which of these geotextiles to use will be based on the site-specific concern as to which mechanism (permittivity or soil retention) is more important. The nontechnical, but important, final decision is based on cost and availability.

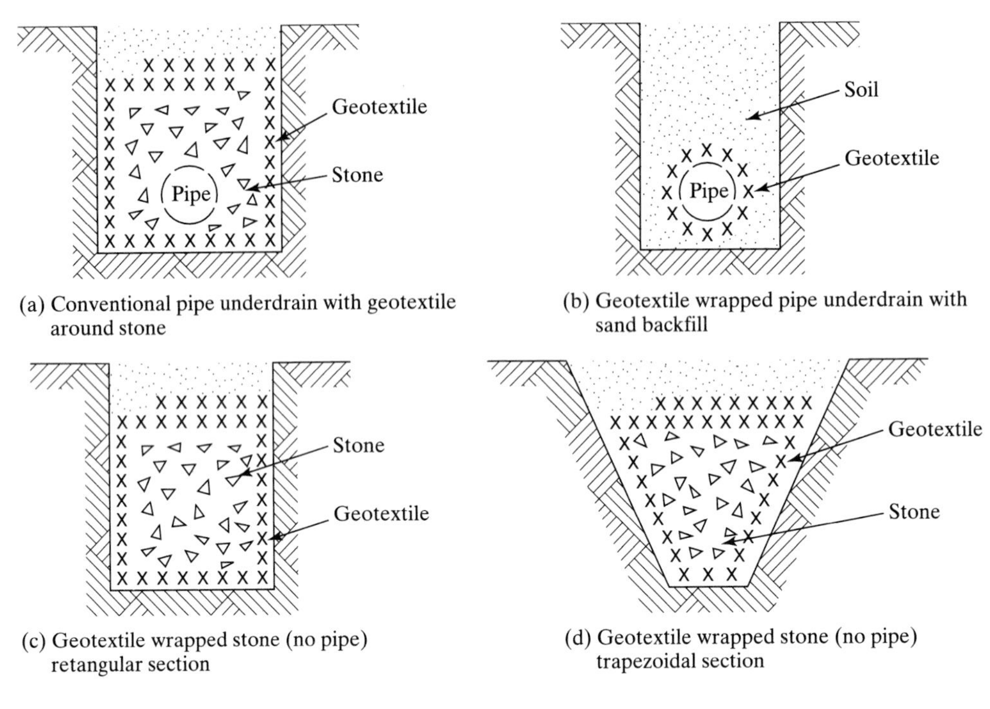

(a) Conventional pipe underdrain with geotextile around stone

(b) Geotextile wrapped pipe underdrain with sand backfill

(c) Geotextile wrapped stone (no pipe) retangular section

(d) Geotextile wrapped stone (no pipe) trapezoidal section

Figure 2.68 Typical cross sections of underdrain systems with and without perforated pipes.

2.8.4 Geotextiles around Underdrains

Geotextiles have been found to make excellent replacements for uniform or graded soil filters around perforated pipe underdrains. Highways, airfields, and railroads are the major application areas. There are numerous design schemes possible, as shown in Figure 2.68. Figures 2.68a,b show the geotextile acting as a filter to protect the stone surrounding the perforated pipe or to protect the pipe directly. In the latter case, the pipe is often wrapped with a geotextile stocking in the manufacturing plant and is shipped as a complete unit to the job site. These sketches raise the question, "Why have the pipe at all?" Indeed, the transmissivity of an open-graded stone is adequate to handle many flow situations, as long as its long-term protection against fine soil contamination is ensured. This point is precisely why the geotextile is involved. By wrapping the stone as shown in Figures 2.68c,d, no pipe at all is required. This is the design for a French drain system, but unlike those in the past, this one should remain free from fine soil contamination if the geotextile is properly designed.

Example 2.22

Design the geotextile filter surrounding an open-graded stone aggregate that in turn surrounds a perforated-pipe underdrain, as shown in the sketch below. Flow will enter through the stone base from the upper part of the underdrain, while soil infiltration will come from the surrounding native soil. This soil is a dense sandy silt (ML) with the relevant properties of 15% nonplastic fines with $C_c = 2.0$, $C_U = 5.0$, $I_D = 80\%$, $d_{50} = 0.035$ mm and $k = 1 \times 10^{-5}$ m/s. The geotextile being considered is nonwoven needle-punched with laboratory-tested values of $\Psi = 1.5\ \text{sec}^{-1}$ and AOS = 0.212 mm.

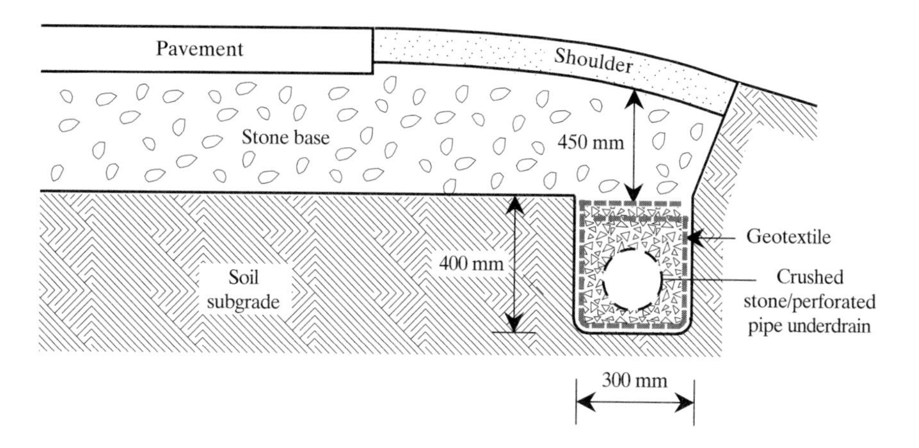

Solution: The solution is in two parts, one for adequate flow capability, the other for adequate soil retention capability.

(a) First, the flow aspects.

(1) Estimate the maximum flow coming to the geotextile. This will be through the 450 mm stone base above the underdrain. Cedegren [126] has numerous design charts, from which we have selected a relatively high value of 15 m³/day-m, due to the current tendency to use open-graded base courses.

(2) Calculate the required permittivity.

$$q = kiA = k\frac{\Delta h}{t}A$$

$$\frac{k}{t} = \frac{q}{\Delta h A}$$

$$\Psi = \frac{15}{(0.45)(0.30 \times 1)}$$

$$\Psi = 111\ \text{day}^{-1} = 1.3 \times 10^{-3}\ \text{s}^{-1}$$

(3) Check this required permittivity against the allowable permittivity of the geotextile. Using data from Table 2.12 in the adapted form of Eq. (2.25) gives the allowable permittivity (the values of partial factors of safety were estimated):

$$\Psi_{\text{allow}} = \Psi_{\text{ult}}\left(\frac{1}{\text{RF}_{SCB} \times \text{RF}_{CR} \times \text{RF}_{IN} \times \text{RF}_{CC} \times \text{RF}_{BC}}\right)$$

$$= 1.5\left[\frac{1}{7.0 \times 1.2 \times 1.1 \times 1.4 \times 2.0}\right]$$

$$= 1.5 \frac{1}{25.9}$$

$$= 0.058\ s^{-1}$$

$$FS_{flow} = \Psi_{allow}/\Psi_{reqd}$$

$$= \frac{0.058}{0.0013}$$

$$= 45 \quad \text{acceptable.}$$

(b) The second part of the design has to do with soil retention of all of the soils surrounding the geotextile-enclosed drainage stone. Since the stone base course above the geotextile is of no real concern, the finer soil subgrade adjacent to the geotextile and beneath it becomes the focus of attention. Also to be considered is that siltation of the perforated pipe within the enclosure is a possibility, therefore the drainage stone must not become contaminated. This situation is considered to be critical.

The criterion we will use for soil retention will be taken from Figure 2.4a (recall Section 2.2.3). From this diagram for the site-specific soil we select the following equation:

$$0_{95_{reqd}} < \frac{18}{C_U} d_{50}$$

$$0_{95_{reqd}} < \frac{18}{5.0}(0.035)$$

$$< 0.126 \text{ mm}$$

Since the candidate geotextile has a given opening size $0_{95_{act}} = 0.212$ mm, the candidate geotextile is not acceptable (i.e., FS = 0.59). The openings are too large and soil will not be retained properly. Another candidate geotextile with a tighter pore structure must be selected.

(c) The design procedure for the alternative geotextile follows these same lines. Note that it is entirely possible to find a suitable geotextile since the permittivity factor of safety is quite high (i.e., 45) in the analysis. By using a tighter geotextile the value of 45 will be reduced, but will come into conformance with the soil retention criterion. This type of tuning the geotextile's properties should be done on the side where the greatest safety is needed, that is, either permittivity or soil retention.

2.8.5 Geotextiles Beneath Erosion Control Structures

Geotextiles have been used beneath erosion-control structures since the late 1950s. Section 1.3.1 mentions some of the original applications. In these applications, both rock riprap and precast concrete blocks were placed on the geotextile, and the geotextile was referred to as a *filter fabric*. In other designs, depending primarily on the care exercised by the contractor in placing the riprap, a sand cushion may be needed to protect the geotextile from impact damage during installation or abrasion damage during its lifetime (e.g., due to wave action agitating the rock riprap). If precast concrete blocks are being used, a sand layer is often used, not so much as a cushion (since these blocks are placed by hand or carefully lowered into position), but as a pore water dissipater, since a major part of the geotextile can be directly covered by the blocks. This feature will be in the design in Example 2.23. Figure 2.69 shows photographs of the geotextile filter in place with a riprap cover. It also shows a failed articulated precast-block system that had inadequate permittivity due to the direct covering of the geotextile filter by the blocks.

Example 2.23

Evaluate the filtration adequacy of a candidate geotextile for placement beneath a rock riprap erosion-control system in a coastal inlet area with 1 m tides (i.e., reversing flow conditions with mild water currents) as shown in the following sketch. The candidate geotextile laboratory properties are $\Psi = 0.5\ \text{s}^{-1}$ and AOS $= 0.21$ mm. The in situ soil is a beach sand (SP) with $C_u = 3.5$, $d_{50} = 0.10$ mm, $d_{90} = 0.40$ mm and porosity $= 0.40$.

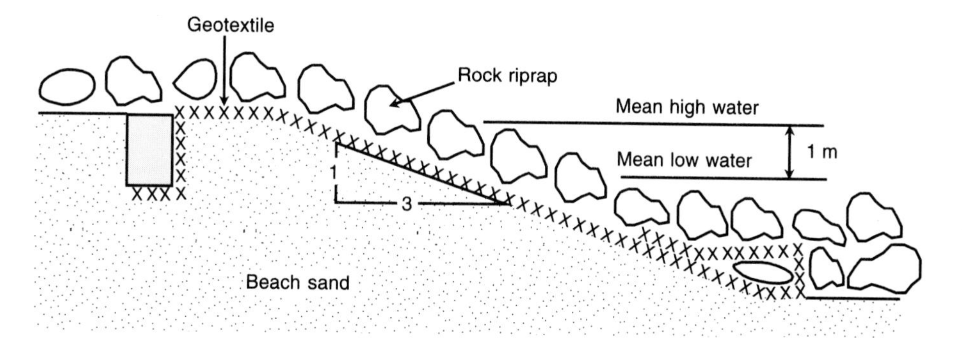

Solution: As with all filtration designs this is a two-part problem, the first for adequate flow and the other for soil retention.

(a) For adequate flow, the procedure is as follows.

(1) Estimate the maximum flow rate due to the 1 m tidal lag. If we assume a water profile as follows:

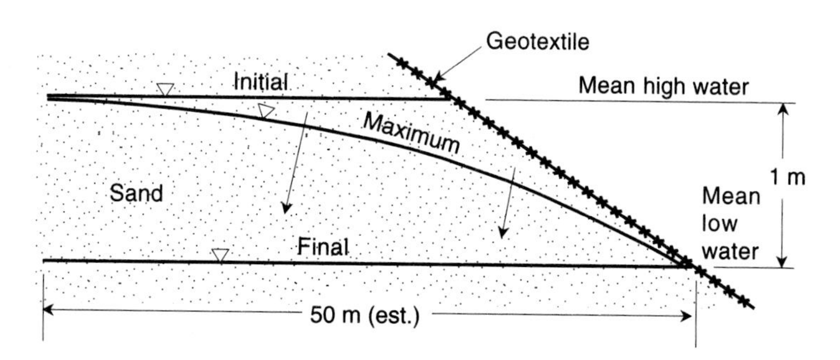

With the tide receding at a maximum rate during an initial 2 hr period as shown, then

$$d_{\text{max}} = \frac{50 \times 1 \times 1}{2} \times \frac{0.4}{2}$$

(a) Partial coverage by rip-rap

(b) Complete coverage by paving blocks

(c) Problem with complete coverage

Figure 2.69 Photographs of geotextiles being used as filters beneath erosion-control structures.

$$= 5.0\ \text{m}^2/\text{hr}$$
$$= 1.39 \times 10^{-3}\ \text{m}^2/\text{s}$$

(2) Calculate the required permittivity.

$$q = kiA = k\frac{\Delta h}{t}A$$

$$\frac{k}{t} = \frac{q}{\Delta h A}$$

$$= \frac{0.00139}{(1.0)(3.16)}$$

$$\Psi_{\text{reqd}} = 0.00044\ \text{s}^{-1}$$

(3) Since the candidate geotextile has a laboratory-obtained ultimate permittivity of $0.5\ \text{s}^{-1}$, it must be modified with reduction factors for site-specific conditions.

(4) The allowable permittivity is found from Eq. (2.25) and values in Table 2.12, where the reduction factor for blinding is used as its maximum value of 10.0 since the rock will cover a large portion of the geotextile's surface area.

$$\Psi_{\text{allow}} = \Psi_{\text{ult}}\left(\frac{1}{\text{RF}_{SCB} \times \text{RF}_{CR} \times \text{RF}_{IN} \times \text{RF}_{CC} \times \text{RF}_{BC}}\right)$$

$$= 0.50\left[\frac{1}{10.0 \times 1.2 \times 1.2 \times 2.5 \times 3.0}\right]$$

$$= 0.50\left[\frac{1}{108}\right]$$

$$= 0.0046\ \text{s}^{-1}$$

(5) The factor of safety is then

$$\text{FS} = \frac{\Psi_{\text{allow}}}{\Psi_{\text{reqd}}} = \frac{0.0046}{0.00044}$$

$$= 10 \qquad \text{acceptable}$$

(b) The geotextile is now evaluated as to its adequacy to retain the soil beneath it.

(1) Since these ocean-control structures are destroyed when the contained soil passes through the geotextile voids (resulting in subsidence and loss of stability of the riprap) and the flow regime is pulsating and cyclic, we will use Figure 2.4b, which results in the following criterion.

$$d_{50} < 0_{95} < d_{90}$$

or

$$d_{50} < \text{AOS} < d_{90}$$

(2) Check this against the AOS of the candidate geotextile, which is 0.21 mm.

$$0.10\ \text{mm} < 0.21\ \text{mm} < 0.40\ \text{mm} \qquad \text{OK}$$

Since 0.21 mm is within the proper bounds, excessive soil loss will not occur and the geotextile is proper as far as soil retention is concerned. The filtration design is essentially complete with the candidate geotextile's flow and retention properties being adequate.

(c) Another part of the design, however, is to see that the geotextile has adequate strength properties to withstand the impact of falling rip-rap and/or puncture from equipment moving on the surface of the rip-rap. These strength-related design issues are described in Section 2.5.

2.8.6 Geotextile Silt Fences

Silt fences consist of geotextiles placed vertically on posts to prevent sediment-carrying sheet runoff from entering downstream creeks, rivers, or sewer systems. Since all construction activities must have an associated sedimentation and erosion-control plan, this concept is used regularly and has replaced the bales of straw, hay, and other makeshift methods. The bottom of the silt fence is embedded in a small anchor trench; the posts, to which the geotextile is attached, are usually at 1.5 to 3.0 m spacings. Sometimes a geogrid backup is required on the geotextile to provide additional support; in this case, the geogrid is attached to the posts first, followed by the geotextile. Since the geotextile is exposed to sunlight, it must be UV-stabilized. A number of manufacturers market geotextiles for this specific purpose. See Figure 2.70 for photographs of typical installations showing the silt fence (a) as placed, (b) working as designed and intended, and (c) improperly constructed without an adequate anchor trench and being undermined.

The physical model utilized in the original development was developed by Bell and Hicks [127] and has sediment-carrying runoff water depositing soil particles within and on the surface of the geotextile as it first acts as a true filter, and then eventually clogs with soil particles in its lower part so that water can no longer pass freely through it. The reservoir created by this action causes coarse sediment to settle behind the clogged portion of the silt fence and the water carrying finer particles to try to reach higher levels on the silt fence above the clogged region. With time, an equilibrium saturation is established. This situation is shown in Figure 2.71.

Richardson and Middlebrooks [128] provide a design method based upon the required site-specific storage volume to be contained by the silt fence. The method assumes that the ground surface is smooth and bare, with sheet erosion (versus rill or gully erosion) being the predominant erosion mechanism. The recommended procedure, somewhat modified here, follows a sequential series of calculations.

1. The maximum slope length that can be contained by a single silt fence is obtained from the following equation for a given slope angle under the above assumed conditions.

$$L_{max} = 36.2e^{-11.1\alpha} \tag{2.68}$$

where

L_{max} = slope length (m),

(a) cascading silt fences after installation

(b) properly functioning silt fence

(c) silt fence that has been undercut by erosion

Figure 2.70 Geotextile silt fence and examples of field performance.

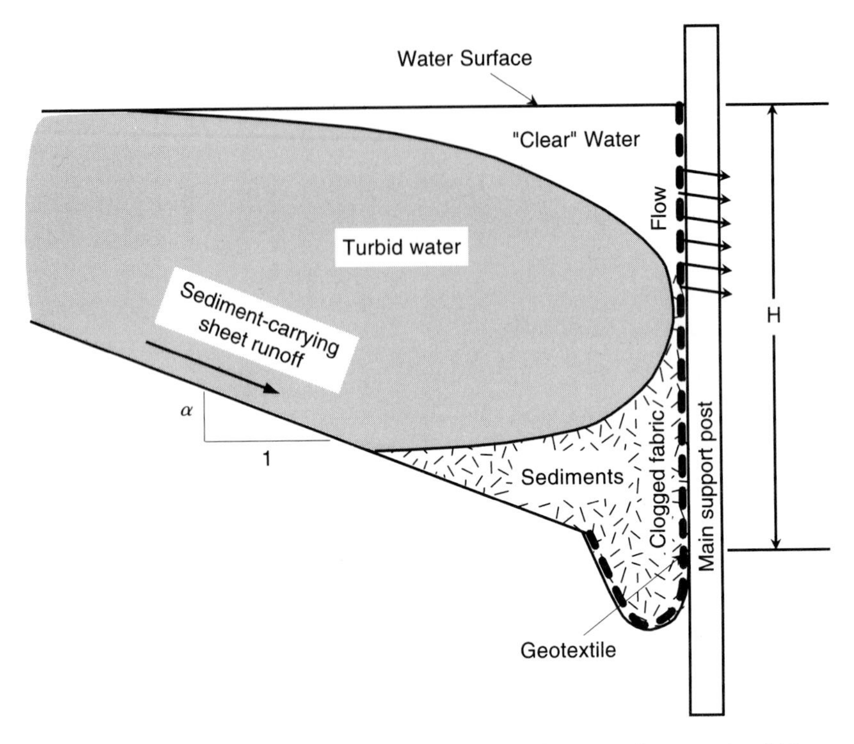

Figure 2.71 Cross section of geotextile silt fence and suggested manner in which system functions.

and

α = slope inclination as measured by its steepness, i.e., vertical rise to horizontal length (dimensionless).

For greater slope lengths, a set of cascading silt fences must be used, each of which is individually designed.

2. The runoff flow rate (water plus sediment) is calculated using the site-specific value of rainfall intensity. This is somewhat subjective and may be controlled by local regulations. It is recommended that the hourly rainfall based on a 10-year recurrence interval be used.

$$Q = C\,I\,A \tag{2.69a}$$

where

Q = runoff flow rate (m^3/hr),
C = surface runoff coefficient (dimensionless),
I = rainfall intensity (m/hr), and
A = area (m^2).

The recommended surface runoff coefficient for smooth and bare soil surfaces is 0.5. Thus the above equation can be rewritten as follows:

$$Q = 5 \times 10^{-4} (I)(A) \tag{2.69b}$$

where

Q = runoff flow rate (m^3/hr),
I = rainfall intensity (mm/hr), and
A = area (m^2).

3. The sediment flow rate can be estimated to ensure that it does not exceed the total runoff flow rate, which can only occur after repeated storms. Stated differently, the sediment accumulates after every storm and remains behind the silt fence, whereas the clear water flows through the silt fence. This calculation should follow the Uniform Soil Loss Equation (described in Chapter 8). Typically, silt fences are designed to contain sediments for at least three major storms. After that, the sediment must be removed and/or the silt fence replaced.
4. The height of the silt fence is then determined using the value of Q obtained previously on the basis of single-storm intensity. A factor is then applied that represents the number of recurring storm events (i.e., a FS = 3 represents three similar storms). Thus

$$V = QT = H\left(\frac{H}{\alpha}\right) \tag{2.70}$$

$$H = [(Q)(t)(\alpha)]^{1/2}$$

where

V = total runoff volume (m^3),
Q = runoff flow rate (m^3/hour),
t = storm duration (hr) [assumed to be 1.0 hr based on the value of I selected in the calculation of Q],
H = silt fence height to contain a single storm (m), and
α = slope inclination (vertical-to-horizontal ratio).

5. The spacing of the silt fence posts is then arbitrarily selected and is integrated into the design per the next two steps.
6. The geotextile is selected on the basis of its ultimate wide-width tensile strength in the weakest direction. Figure 2.72a can be used for this determination. A reduction factor can be included. Note that the opening size of the geotextile is not a governing criterion since excessive clogging via turbidity is the issue and with individual soil particle sediment, this will occur. Thus woven slit-film geotextiles predominate in this particular application; they are rapidly clogged, yet possess high tensile strength.
7. Lastly, the type of fence post is selected. This can be done using the guide given in Figure 2.72b.

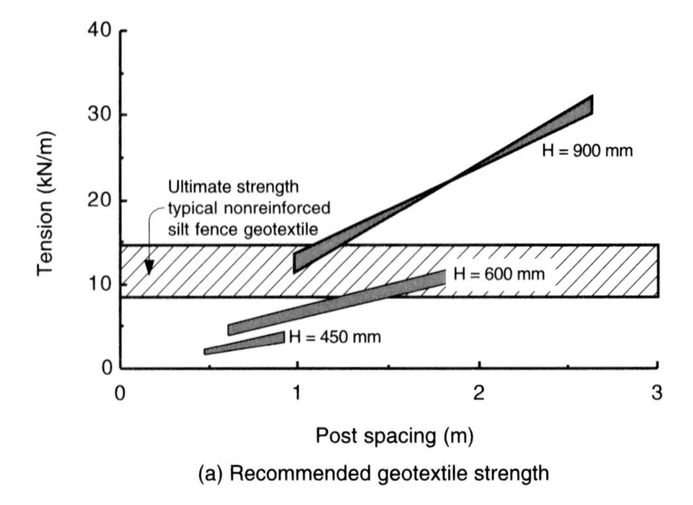

(a) Recommended geotextile strength

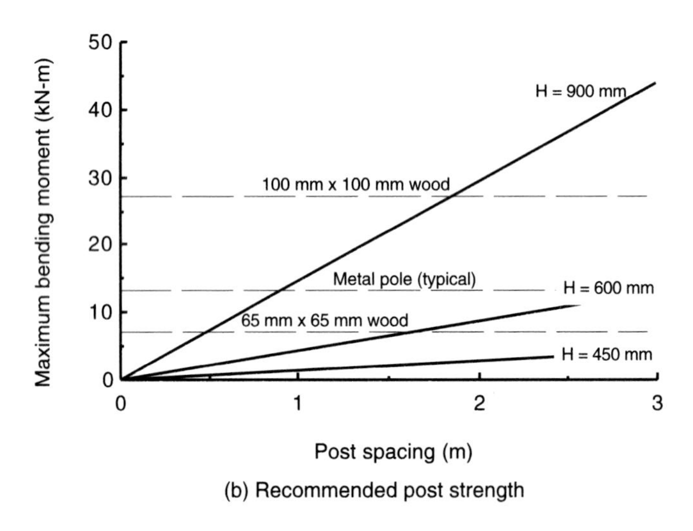

(b) Recommended post strength

Figure 2.72 Design recommendations for silt fence geotextile and post strengths. (After Richardson and Middlebrooks [128])

Example 2.24

Design a silt fence for a 60 m long relatively smooth surface construction site where topsoil has been stripped and the average slope inclination is 5%. The 10 year recurring single-storm intensity is 100 mm/hr.

Solution: Using the procedure just described, the problem is solved in successive steps.

(a) Calculate the maximum slope length per silt fence.

$$L_{\max} = 36.2e^{-11.1\alpha}$$

$$= 36.2e^{-11.1(0.05)}$$

$$L_{max} = 21 \text{ m} \qquad \text{use 20 m as the maximum slope length}$$

Therefore, three cascading silt fences will be needed for this construction site.

(b) Calculate the runoff flow rate.

$$Q = 5 \times 10^{-4}(I)(A)$$
$$= 5 \times 10^{-4}(100)(20 \times 1)$$
$$= 1.0 \text{ m}^3/\text{hr}$$

(c) Either calculate the sediment flow rate, or assume the number of storms that the silt fence must contain. Here, we select:

$$\text{number} = 3 \text{ events}$$

This establishes the factor that will be used on the calculated height of the silt fence.

(d) Calculate the height of the silt fence.

$$H = [(Q)(t)(\alpha)]^{1/2}$$
$$= [(1.0)(1.0)(0.05)]^{1/2}$$
$$= 0.22 \text{ m}$$

For three storms,

$$3 \times H = 3 \times 0.22$$
$$= 0.66 \text{ m}$$
$$= 660 \text{ mm}$$

(e) The spacing of the silt fence posts is assumed to be

$$S = 1.5 \text{ m}$$

(f) The required strength of the geotextile is taken from Figure 2.72a.

$$T_{reqd} = 10 \text{ kN/m}$$

which is adequate without geogrid backup support. This can be increased for damage due to stapling/attachment to the fence post,

$$T_{ult} = \text{FS } T_{reqd}$$
$$= 1.2(10)$$
$$T_{ult} = 12 \text{ kN/m}$$

(g) Finally, the type of post is selected from Figure 2.72b.

Use 65 mm $\times$ 65 mm wooden posts.

2.8.7 Summary

Presented in this section were a series of designs in which the geotextile is serving in the filtration function. In such cases the liquid (usually water) is moving across the plane of the geotextile. Thus, the designs allow for a required flow capacity that is expressed in terms of permittivity Ψ, a term that includes both the permeability coefficient and the geotextile's thickness. However, more than just required flow capability is necessary—soil retention is also necessary. Since these two demands are contradictory (flow requiring large geotextile voids and sufficient soil retention requiring small geotextile voids), problems illustrating the criticality of each demand were presented.

The fine soils used behind flexible walls (Section 2.8.3) and adjacent to underdrains (Section 2.8.4) showed that adequate permittivity is easily achieved, yet small opening-size values are required. This means that the fibers or yarns have to be close to one another, resulting in a relatively tight or dense geotextile structure.

Conversely, the geotextiles beneath erosion-control structures (Section 2.8.5) usually occur with free-draining sands. Thus the permittivity of the geotextile is challenged by a lower factor of safety, whereas the opening-size requirement is easily met. Note that if the erosion-control system lies directly on the geotextile and covers a high percentage of it (as with precast-concrete blocks fastened to the geotextile), the permittivity design always becomes critical. In this case, the percentage of covered geotextile (sometimes as high as 70%) was included via a high reduction factor, i.e., $RF_{SCB} = 10$. A simple proportional factor is appropriate, since flow rate should be a linear function of the area remaining open to flow.

In all of these problems the long-term compatibility of the soil to the geotextile should be considered. In most cases the designs illustrated should suffice; however, when the situation is critical, laboratory flow tests described in Section 2.3.5 should be considered.

Finally, the geotextile silt fence of Section 2.8.6 was considered. This design simply begs for complete clogging of the geotextile, which must happen so that the sedimentation reservoir can form behind it. The design then becomes a structural design of the strength of the geotextile, and its supporting post spacings and type. It was discussed in this section because the geotextile above the clogged area is indeed still serving as a true filter.

2.9 DESIGNING FOR DRAINAGE

In the context of this book, *drainage* refers to planar flow within the structure of the geosynthetic. Geotextiles acting as drains will be considered in this section; geonets and drainage geocomposites will be discussed in Chapters 4 and 8.

2.9.1 Overview of Applications

Although closely related to the filtration function just described, geotextile drainage occurs in the in-plane direction rather than cross-plane. The transitive verb *to drain,* according to *Webster's New Collegiate Dictionary,* means "to draw off (liquid) gradually...," and it is to precisely this action, as it is performed by geotextiles, that the present section is devoted (see Figure 2.73). Note that all geotextiles are capable of draining liquid in their in-plane directions, but to widely varying degrees. In order of *in-*

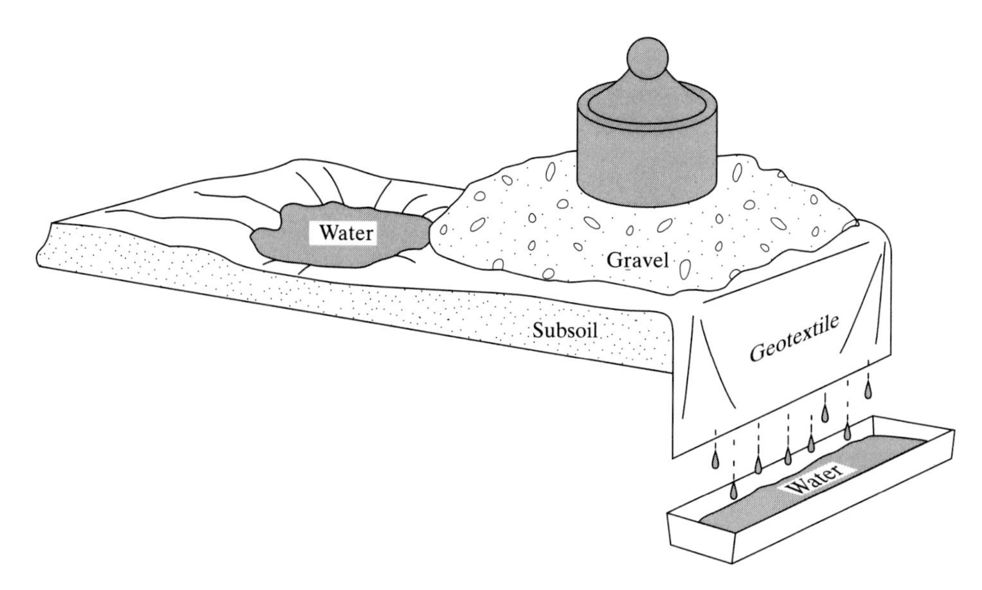

Figure 2.73 Sketch of drainage of water within geotextiles. (Compliments of Monsanto Co.)

creasing in-plane drainage capability, the various manufactured styles are ranked as follows:

- Woven, slit-film
- Woven, monofilament
- Nonwoven, heat-bonded
- Nonwoven, resin-bonded—increasing with increasing weight and decreasing resin
- Nonwoven, needle-punched—increasing with increasing weight (recall Figure 2.17)
- Hybrid drainage systems—particularly geonets and drainage geocomposites (see Chapters 4 and 8)

Although all geotextiles possess some in-plane drainage capability, the first three types of geotextiles listed above have too little to take advantage of. The nonwoven resin-bonded and needle-punched groups, however, can be made sufficiently thick to have meaningful quantities of liquid flow within them. These geotextiles, particularly the needle-punched variety, are often 5 mm thick and can be made much thicker in a cost-effective manner. The nonwoven needle-punched geotextiles will be the focus of the designs included in this section.

For the flow capability of geotextiles, we will be considering two general categories: gravity flow (Section 2.9.3) and pressure flow (Section 2.9.4). The distinction will become obvious in the sections to follow. Some selected applications in each category are listed below:

- Gravity drainage: chimney drains and drainage galleries in earth and earth/rock dams, pore water dissipaters behind retaining walls, flow interceptors (as in fin drains) beneath a geomembrane-lined reservoir for water drainage or gas conveyance
- Pressure drainage: as vertical drains for rapid soil consolidation, within the soil backfill of reinforced earth walls, within earth embankments and dams, beneath surcharge fills

2.9.2 General Behavior

Except for the direction of flow (in-plane rather than cross-plane), the design similarities of this drainage section with the preceding filtration sections will be obvious. There are three aspects of design: adequate flow capacity, proper soil retention, and long-term soil-to-geotextile flow equilibrium. Since soil retention and long-term soil-to-geotextile flow equilibrium are discussed in other sections, these aspects will not be fully treated in this section; only some brief comments will be included. The reader is referred to Sections 2.3.5 and 2.8 for full details.

For the hydraulic design parameter of major concern, we will focus on the transmissivity of the geotextile θ in Eq. (2.10) (repeated here),

$$\theta = k_p t \tag{2.10}$$

where

θ = transmissivity,
k_p = in-plane geotextile permeability, and
t = geotextile thickness.

The relationship will appear as $\theta = kt$, the k being understood here to be in-plane permeability (rather than the filtration-related cross-plane permeability). Transmissivity will be used in conjunction with Darcy's formula under the assumption that laminar flow exists within the geotextile. This is generally valid, but when very thick geotextiles are used together with high hydraulic gradients, the assumption of a laminar flow regime becomes questionable. (Indeed, with the geonets and drainage geocomposites discussed in Chapters 4 and 8, flow is generally turbulent and Darcy's formula should not be used. For geonets and drainage geocomposites a related design based on flow rates will be used.)

2.9.3 Gravity Drainage Design

For gravity drainage problems involving liquid flow in geotextiles, the driving force is merely the slope at which the geotextile is placed. Using the geometry of the particular situation under consideration, a required transmissivity can be calculated using Darcy's formula. This value is then compared to the allowable transmissivity of the candidate geotextile for calculation of a factor of safety. Depending on the severity of the situation, these values of factor of safety should be quite high, depending on how the allowable transmissivity is obtained [recall Eq. (2.25) and Table 2.12].

Note that the allowable geotextile transmissivity is the value at the particular normal stress that is acting upon it. This is usually calculated on the basis of the effective stress of the soil placed above it, if the geotextile is horizontal. If the geotextile is vertical, the normal stress is the vertical effective stress times the appropriate coefficient of earth pressure. This can usually be taken as the at-rest value, and the relationship $K_o = 1 - \sin \phi$ is often used. If the friction angle of the soil (ϕ) is not known, K_o can be taken approximately equal to 0.5. It should be recalled from the testing section involving hydraulic properties (Section 2.3.4) that the transmissivity decreases substantially with applied normal stress on the geotextile. Figure 2.17 illustrated this behavior for various mass per unit area nonwoven needle-punched geotextiles. However, a near-constant value of transmissivity is reached above approximately 85 kPa. Example 2.25 illustrates the use of these concepts.

Example 2.25

Given a 10 m high zoned earth dam for use as an irrigation reservoir, the dam has a cross section as shown below. A geotextile is being considered as a chimney drain and drainage gallery. The geotextile under consideration is a 2000 g/m^2 nonwoven needle-punched geotextile with $\theta_{ult} = 15 \times 10^{-4}$ m^2/min. Use cumulative reduction factors of 3.0 to convert this to θ_{allow}. What factor of safety does this geotextile have for the amount of flow seeping through the core wall, which is a clayey silt of permeability 1×10^{-7} m/s?

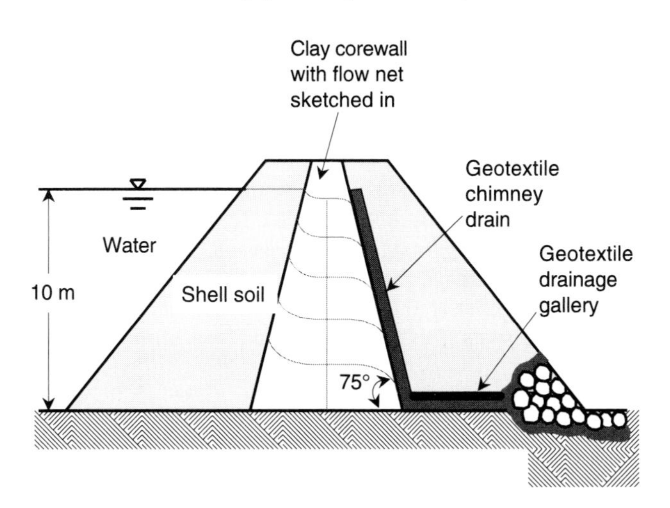

Solution: In stages, the solution is as follows:

(a) Calculate the maximum seepage coming through the clay core wall that the geotextile must carry. The use of a flow net (as shown in the sketch) gives

$$q = kh\left(\frac{F}{N}\right)$$

$$= (1 \times 10^{-7})(10)\left(\frac{5}{2}\right)$$

$$= 2.50 \times 10^{-6}\ \text{m}^2/\text{s}$$
$$= 1.50 \times 10^{-4}\ \text{m}^2/\text{min}$$

(b) Calculate the gradient of flow in the geotextile.

$$i = \sin 75°$$
$$= 0.97$$

(c) Calculate the required transmissivity θ_{reqd} using Darcy's formula.

$$q = kiA$$
$$= ki(t \times W)$$
$$= (kt)(i \times W)$$
$$kt = \frac{q}{i \times W}$$
$$\theta_{\text{reqd}} = \frac{1.50 \times 10^{-4}}{(0.97)(1.00)}$$
$$= 1.55 \times 10^{-4}\ \text{m}^2/\text{min}$$

(d) Determine the global factor of safety.

$$\text{FS} = \frac{\theta_{\text{allow}}}{\theta_{\text{reqd}}} = \frac{\theta_{\text{ult}}/\Pi RF_p}{\theta_{\text{reqd}}}$$
$$= \frac{(15 \times 10^{-4})/3.0}{1.55 \times 10^{-4}}$$
$$= 3.2$$

Due to the critical nature of this application, this FS value is too low and a minimum value of 5.0 is recommended. Two options present themselves: one is to use multiple layers of geotextile (to increase θ_{allow}) in the lower part of the chimney drain and in the drainage gallery (the upper part of the chimney drain could still use one layer); the other is to use the FS = 5.0 and back-calculate the necessary geotextile's transmissivity. This latter suggestion is illustrated as follows.

$$\theta_{\text{allow}} = \theta_{\text{reqd}} \times \text{FS}$$
$$= (1.55 \times 10^{-4}) \times 5.0$$
$$\theta_{\text{allow}} = 7.75 \times 10^{-4}\ \text{m}^2/\text{min}$$

This, in turn, requires a geotextile to have an ultimate (or as-manufactured) transmissivity considerably in excess of θ_{allow}. If the cumulative reduction factor is 3.0

$$\theta_{\text{allow}} = (7.75 \times 10^{-4}) \times 3.0$$
$$= 23.2 \times 10^{-4}\ \text{m}^2/\text{min}$$

As seen in Figure 2.17, this is possible only by selecting an extremely thick nonwoven needle-punched geotextile. Alternatively, geonets or geocomposites can be considered (Chapters 4 and 8, respectively).

(e) We must now do a soil retention analysis to see that soil particles do not embed in the geotextile and decrease its transmissivity. The analysis is the same as in Section 2.8.

(f) Finally, long-term soil-to-geotextile compatibility must be addressed. Here within an earth dam is where long-term flow tests, gradient ratio tests, or hydraulic conductivity ratio tests have applicability. See Section 2.3.5 for details.

Example 2.26

Calculate the factor of safety of a 500 g/m² geotextile required to drain water from behind an 8-m-high concrete cantilever retaining wall if it has an allowable transmissivity of $\theta_{\text{allow}} = 0.15 \times 10^{-3}$ m²/min of geotextile. The soil backfill is a silty sand (ML-SW) with $k = 5 \times 10^{-5}$ m/s.

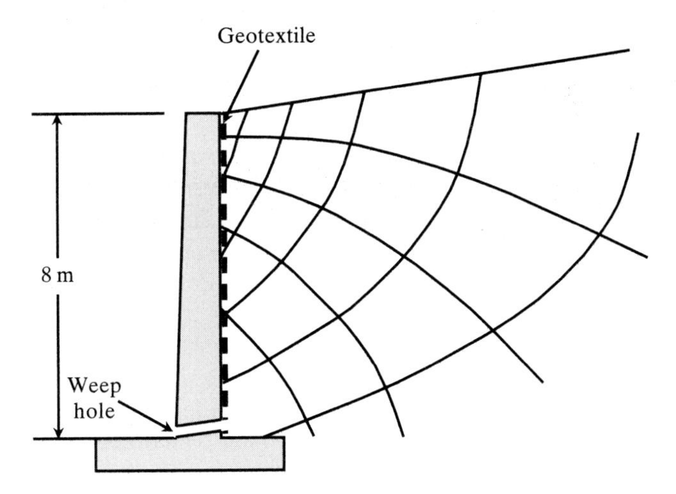

Solution: As before, we proceed in parts:

(a) Calculate the maximum flow rate coming to the geotextile. From the flow net above, we have

$$q = kh\left(\frac{F}{N}\right)$$

$$= (5.0 \times 10^{-5})(60)(8)\left(\frac{5}{5}\right)$$

$$= 0.024 \text{ m}^2/\text{min}$$

(b) Determine the flow gradient within the geotextile.

$$i = \sin 90^\circ$$

$$= 1.0$$

(c) Calculate the required transmissivity.

$$q = kiA$$

$$= ki(t \times W)$$

$$= (kt)(iW)$$

$$kt = \frac{q}{i \times W}$$

$$= \frac{0.024}{1.0 \times 1.0}$$

$$\theta_{\text{reqd}} = 0.024 \text{ m}^2/\text{min}$$

(d) Compare this value to the actual geotextile's transmissivity to obtain a factor of safety.

$$\text{FS} = \frac{\theta_{\text{allow}}}{\theta_{\text{reqd}}}$$

$$= \frac{0.00015}{0.024}$$

$$= 0.0062 \qquad \text{not acceptable!}$$

It is easily seen that this application is not suited for geotextiles. It is, however, a perfect situation for geonets or drainage geocomposites which have much greater in-plane flow capacity. Geonets are the subject of Chapter 4, and geocomposites are the subject of Chapter 8; there we will repeat this exact problem as Examples 4.5 and 8.5 and note the substantial increases in the factor of safety.

2.9.4 Pressure Drainage Design

Geotextile transmissivity is the key parameter in both gravity and pressure drainage; in this sense the two topics are quite similar. The difference is that for pressure drainage water will flow from locations of higher pressure to locations of lower pressure regardless of the geotextile's orientation. Thus, flow direction depends on each specific situation; some of these situations have been identified. The following equation is formulated for a geotextile placed beneath a surface fill on a fine-grained compressible foundation soil, after Giroud [64].

$$\theta_{\text{reqd}} = k_p t = \frac{B^2 k_s}{(c_v T)^{1/2}} \tag{2.71}$$

where

θ_{reqd} = geotextile transmissivity,
k_p = in-plane permeability coefficient of the geotextile,
t = thickness of the geotextile,
B = width of the surcharge layer,
k_s = permeability coefficient of the foundation soil,
c_v = vertical coefficient of consolidation of the foundation soil, and
T = time for the surcharge fill to be placed.

Example 2.27 illustrates use of the formula.

Example 2.27

Given a variable-width surcharge fill placed in 10 days (14,400 min) on a foundation soil of 1×10^{-9} m/s permeability and 4.6×10^{-6} m²/min coefficient of consolidation, as shown below: **(a)** determine the required geotextile transmissivity as a function of base width of the surcharge fill and graph the result; **(b)** using a ultimate geotextile transmissivity of 0.75×10^{-3} m²/min and cumulative reduction factors of 5.0, find the maximum width of surcharge that can be used under these conditions.

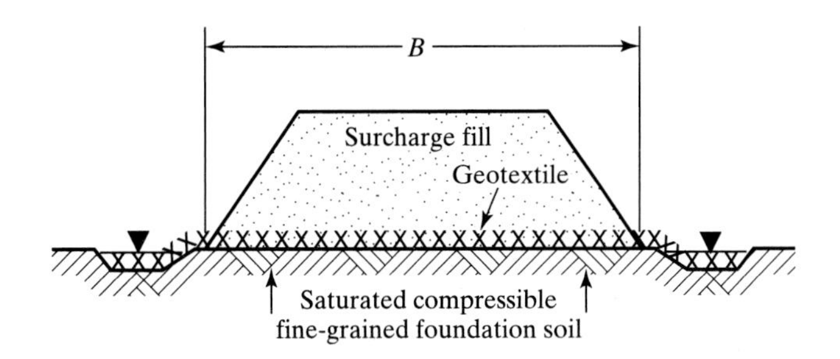

Solution:

(a) To determine the graph of B versus θ

$$\theta_{\text{reqd}} = \frac{B^2 k_s}{(c_v T)^{1/2}}$$

$$= \frac{(1 \times 10^{-9})(60)B^2}{[(4.6 \times 10^{-6})(0.0144 \times 10^6)]^{1/2}}$$

$$= 2.33 \times 10^{-7}\, B^2 \qquad \text{where } B \text{ is in m}$$

When different values of B are plotted, the following curve results:

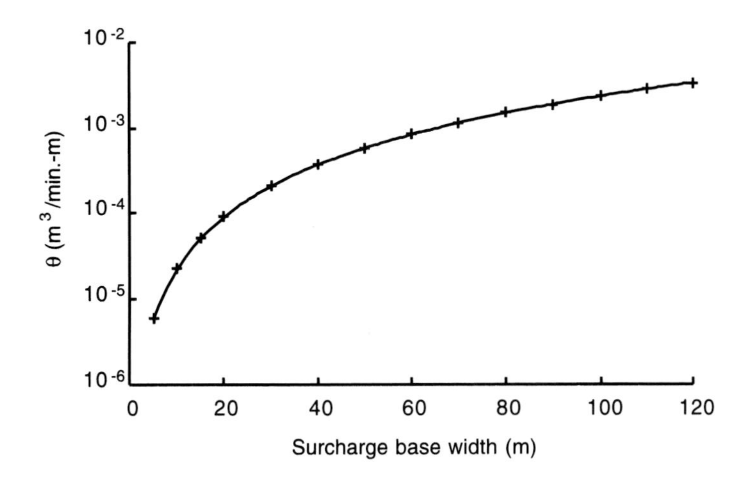

(b) Using the above graph,

$$\theta_{\text{ult}} = 0.00075 \text{ m}^2/\text{min}$$

$$\theta_{\text{allow}} = \theta_{\text{ult}}/\Pi\text{RF}_p$$

$$= 0.00075/5.0$$

$$= 0.00015 \text{ m}^2/\text{min}$$

and from the chart the value is as follows:

$$B_{\text{max}} = 30 \text{ m}$$

2.9.5 Capillary Migration Breaks

The upward movement of water in fine-grained soils has been known to present problems in two distinctly different areas of the world. One is in cold regions, where temperature gradients below freezing cause moisture to rise above the stationary water table. This rise occurs in the capillary zone and can eventually result in frozen layers (ice lenses), which will expand, lifting any structure placed above it. The phenomenon is called *frost heave* and is well-documented in the geotechnical literature.

Remedies for frost heave usually involve a capillary break or cut-off placed horizontally at a depth beneath the lowest elevation of the freezing isotherm. Sands and gravels have been used, but geotextiles offer an attractive and cost-effective alternative. A thick nonwoven needle-punched geotextile is easily placed and can be readily graded to drain the rising water away from the area of immediate concern. Several laboratory studies and studies of actual use of geotextiles in this manner are available [129, 130].

It should be mentioned that geotextiles are generally hydrophobic, that is, they repel water, so there is no wicking action across the plane of the geotextile. This is important to note because there is a popular misconception that geotextiles can wick water, as does the wick of a candle with wax. This is not so, since both polypropylene and polyester repel water. Until the geotextile voids are saturated, there is only minor intimate contact of the water with the fibers. This is just the opposite of water in soil, where soil particles are hydrophilic, and the water nests itself near points of contact of adjacent soil particles. Once geotextiles are completely saturated, however, they can be used to siphon water in a continuous manner, thereby maintaining flow within them [131].

In a completely different part of the world, namely in deserts and arid regions, there is a similar problem involving capillary rise. Here, as groundwater rises it brings dissolved salts with it [132]. As this salt-ladened water comes near the ground surface, it kills all vegetation, which takes up the salt water through the root system. Equally severe is the salt attacking building foundations of both stone and reinforced concrete, making these usually-adequate structural materials very friable.

As with the frost cutoff, this application area of salt-migration cutoff can effectively use a geotextile exhibiting proper in-plane drainage characteristics. The design procedure uses the transmissivity parameter, as in the case of gravity drainage. Once again laminar flow is assumed, so that Darcy's formula is employed in the problem solution. Example 2.28, which follows, actually suggests gravity and pressure situations but is solved completely by a gravity approach.

Example 2.28

Given a storage building for frozen foods that is to be founded on the site illustrated below, with a capillary break beneath the building's foundation, a geotextile is being considered as a solution. Will a 700 g/m^2 nonwoven needle-punched geotextile having an allowable transmissivity $\theta = 0.0007$ m^2/min at 25 kPa normal stress be adequate using a global FS = 3.0?

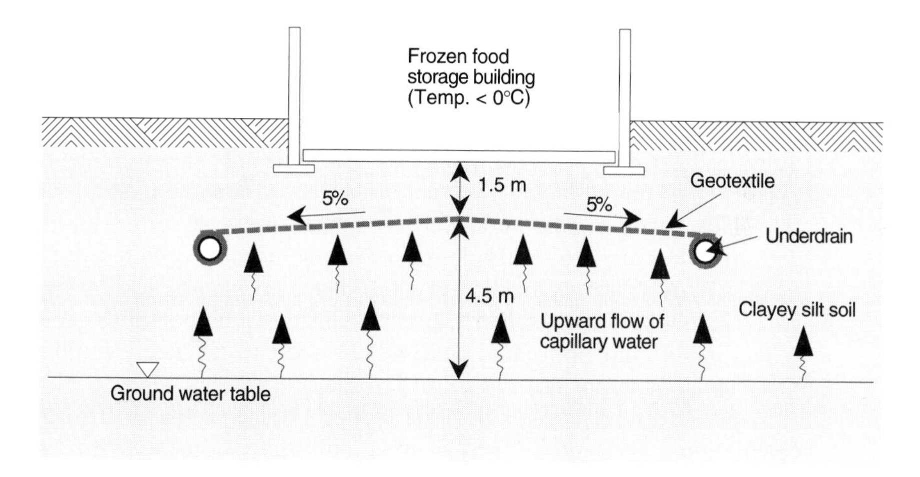

Solution: The problem is similar to the chimney drain example of Section 2.9.3 with some obvious exceptions.

(a) Determine the flow rate of upward water migration to the geotextile, which is a function of the soil's permeability and the thermal gradient drawing the water upward. Guides are given in [92], where a conservative value is selected for this example.

$$q = 2.7 \times 10^{-5} \text{ m}^2/\text{mm}$$

(b) Calculate the gradient of flow in the geotextile.

$$5 \text{ percent slope} = 0.05 \text{ gradient}$$

(c) Calculate the required transmissivity, θ_{reqd}.

$$q = kiA$$
$$= ki(t \times W)$$
$$= (kt)(i \times W)$$

$$kt = \frac{q}{i \times W}$$

$$\theta_{\text{reqd}} = \frac{0.000027}{(0.05)(1.0)}$$
$$= 0.00054 \text{ m}^2/\text{min}$$

(d) Determine the factor of safety

$$FS = \frac{\theta_{allow}}{\theta_{reqd}}$$

$$= \frac{0.00070}{0.00054}$$

$$= 1.3 < 3.0 \qquad \text{not acceptable}$$

Therefore, at least a triple layer of geotextile is needed (or a geonet or drainage geocomposite, as discussed in Chapters 4 and 8).

2.9.6 Summary

Geotextiles used in the drainage function are quite attractive for use as long as the required planar flow rates are relatively low. Certainly the drainage of fine-grained soil masses (like silts and clays) is possible, along with some examples shown in this section. Nonwoven needle-punched geotextiles are best suited for this function. High mass per unit areas or multiple layers can be used to obtain sequentially higher flow rates. Although not shown herein, their flow rates can be further augmented by being constructed with high-denier fibers. Significantly higher flow capacity, however, requires a different approach. That will be seen to be nicely fulfilled by geonets and drainage geocomposites.

Some of the problems of this section used an allowable transmissivity for their calculations. Thus no reduction of an ultimate transmissivity value by means of Eq. (2.25) and Table 2.12 was illustrated. Other problems used relatively low reduction factors to obtain an allowable transmissivity from the laboratory-test value. This is because the current transmissivity test procedure, ASTM D-4716 and ISO/DIS 12958, is often conducted as a performance test. Site-specific conditions and loads can be readily replicated, and the resulting value will then be an allowable value (or near-allowable) rather than an ultimate one. If not, the procedures set forth in reducing an ultimate to an allowable value by using reduction factors must be used.

2.10 DESIGNING FOR MULTIPLE FUNCTIONS

2.10.1 Introduction

In Sections 2.5 to 2.9, the primary function of the geotextile was readily apparent. This was to orient the reader's attention to the design-by-function concept, and also because most applications lend themselves to a readily definable primary function design. Nevertheless, there are other applications in which the geotextile must be designed for multiple functions. In these cases a single dominant (primary) function cannot always be identified. Thus there are primary, secondary, tertiary, and perhaps even quaternary functions that may vary in a particular application. Furthermore, these functions might vary from site to site. Such multiple-function applications are the focus in this section. They should not be taken lightly or be considered of lesser importance than those discussed previously. Some of the major uses of geotextiles are included in this section on multiple-function applications.

2.10.2 Reflective Crack Prevention in Bituminous Pavement Overlays

Overview. The resurfacing of existing pavements that have excessive cracks in them represents an ongoing and expensive task for all federal, state, local, and private organizations that own and maintain roads. Such resurfacing is usually done with bituminous (asphaltic cement) overlays ranging in thickness from 25 to 100 mm. Particularly exasperating to the road owners (and to the users and their automobiles, as well) is when the cracks in the original pavement reflect up through the new overlay earlier than anticipated. To combat this, thicker overlays than desirable are used but at the cost of added expense, lower curb heights, and excessive weight and thickness on the subgrade system. Due in part to the magnitude of this problem and the potential market that it represents, the use of geotextiles to remedy this has been attempted in a number of ways. In some instances strips of geotextile have been placed over the crack, spanning it by 150 to 600 mm on each side, and the overlay placed above. Polyester, polypropylene, and fiberglass geotextiles, as well as geogrids (see Chapter 3), have all been used in this regard. By far the major use, however, has been to place full-width geotextile sheets over the entire pavement surface, which has been waterproofed with asphalt cement or asphalt emulsion, and then overlaid with the final bituminous surfacing. The goal of such a process is to either (a) decrease the thickness of the overlay while keeping a lifetime equivalent to not using a geotextile, or (b) increase the lifetime of the overlay while using the same thickness as without the use of the geotextile.

A tremendous market has developed for this use, amounting to approximately 70 million m^2 in the U.S. in 1995. It is interesting to note that users in other countries are not nearly as involved in this application as those in the U.S. This is probably due to better ongoing road maintenance using conventional techniques (i.e., cleaning and filling of cracks when they are small) than that generally practiced in the U.S., but this is not known to be a fact. Finally, it should be noted that this technique is used only for existing bituminous pavements, not for portland-cement concrete pavements. The significantly sharper edges of concrete would generally puncture and tear the lightweight geotextiles customarily used for this application.

As with other topics in this section, a clear-cut primary function is difficult to identify. It might involve the reinforcement from one side of the existing crack to the other (via the geotextile's tensile strength), but this is greatly aided by the waterproofing provided against water potentially moving through the pavement and into the subgrade, via the impregnation of the geotextile with asphalt cement or asphalt emulsion. Since either function is possible; both design concepts will be explored after the construction and related details are addressed.

Construction. The use of full-pavement-width geotextiles in reflective-crack prevention for bituminous pavement overlays is seemingly quite simple and straightforward. The general process follows; however, each step has its own peculiarities and subtleties.

1. The existing failures in the foundation soils must be repaired before any resurfacing is done. The waterproofed geotextile of 160 to 320 g/m^2 cannot be expected to hold up highway vehicles traveling only a few millimeters above it when the foundation soil beneath the stone base is unacceptable. In some instances these primary repairs must be very extensive.
2. Cracks in the existing bituminous pavement must be filled (see Figure 2.74a). Cracks up to approximately 6 mm thick are filled with hot liquid crack filler, while larger cracks are filled with asphalt, hot mix, or cold patch.

(a) Filling cracks in existing bituminous pavement

(b) Spraying asphalt sealant over existing pavement

(c) Geotextile being placed by mechanical equipment

(d) Hot mix bituminous overlay being placed

(e) Asphalt pavement core showing the crack-arresting feature offered by the geotextile with the new overlay placed above

Figure 2.74 Construction procedures and equipment for using geotextiles in reflective crack prevention in bituminous overlays. (Compliments of Amoco Fabrics and Fibers Co.)

3. An asphaltic sealant is then uniformly sprayed over the existing pavement (see Figure 2.74b). The amount ranges from 0.2 to 2.3 l/m^2, depending on the porosity of the existing pavement and the absorbency of the geotextile. Recommended sealants are asphaltic cement (AC-5 or AC-20), cationic asphalt emulsions (CRS-2 or CRS-1h), and anionic asphalt emulsions (RS-2 or RS-1). When using the asphalt emulsions, care is required to ensure that they cure adequately before the geotextile is placed. Curing takes from 30 min to 4 hr, depending on the temperature and humidity. Cutback asphalts cannot be used with polypropylene geotextiles, since the solvent in them reacts with the polymer at high temperatures.
4. The geotextile is then placed on the sealant by hand or with mechanical equipment (see Figure 2.74c). Excessive wrinkles or folds in the geotextile must be cut open and laid flat. Stiff brooms are used to obtain a good bond with the sealant and to smooth out the surface as well. Joint overlaps of 25 to 75 mm are generally used. Additional sealant should be applied at these joints. If sealant comes through the geotextile, sand can be spread over it to absorb the excess.
5. The hot mix overlay is placed directly on the geotextile as soon as possible (see Figure 2.74d). The temperature of the mix should be about 150°C, with a maximum of 165°C. Care must be taken to avoid movement or damage to the geotextile from the turning of the paver or truck movement, since these vehicles are riding directly on the waterproofed geotextile.
6. The resulting system should appear as the core shown in Figure 2.74e.

A few points in the procedure need additional comment. The amount of asphalt sealant used is very important. Too little sealant leaves the geotextile unsaturated (thus permeable to infiltrating water), and too much sealant leaves an excess above or below the geotextile (thus forming a potential slip layer when the overlay is placed). Phenomenologically, the hot overlay draws the sealant up into the geotextile, just saturating it. (Note that for this reason, cold patch cannot be used as the overlay material.) Just how much sealant should be used depends on both the quality of the existing pavement and the type of geotextile being used. Button et al. [133] present the following equation for the quantity of sealant (also called a tack coat) to be used.

$$Q_d = 0.36 + Q_s \pm Q_c \tag{2.72}$$

where

Q_d = design sealant quantity (l/m^2),
Q_s = saturation content of the geotextile being used (l/m^2), and
Q_c = correction based on sealant demand of the existing pavement surface (l/m^2).

Concerning the value of Q_s, most manufacturers have specific geotextiles for this application and are familiar with the required amount. In the absence of this information we can take a flat pan with the candidate geotextile placed in it and experimentally determine the required amount. The geotextile is first saturated in asphaltic cement at 120°C for 1 min. It is allowed to cool and then is pressed with a hot iron between two absorbent papers to remove the excess asphalt. The value of Q_s is measured accordingly. A related procedure can be done in the field with a completely flat piece of sheet

TABLE 2.19 SEALANT DEMAND OF EXISTING BITUMINOUS PAVEMENT SURFACES

Surface Condition	Q_c (l/m^2)
Flushed	−0.09 to 0.09
Smooth, nonporous	0.09 to 0.23
Slightly porous, slightly oxidized	0.23 to 0.36
Slightly porous, oxidized	0.36 to 0.50
Badly pocked, porous, oxidized	0.50 to 0.59

Source: After Button et al. [133].

metal beneath the geotextile. The quantity Q_s depends mainly on the geotextile's thickness and to a lesser extent on other manufacturing details. For the value of Q_c, see Table 2.19, where it is shown that the older more oxidized pavement surfaces require greater amounts of sealant.

Much has been written concerning geotextile selection. On the basis of the majority of field projects completed to date, lightweight nonwoven needle-punched geotextiles prevail. Such low initial modulus geotextiles make us wonder about the reinforcement possibilities made for this application. They do indeed saturate well with sealant, so that waterproofing may be the dominant function. Laboratory modeling, however, suggests something very different. Using small cross-section test specimens consisting of a cracked asphalt base layer, waterproofed geotextile above it, and finally the overlay, Murray [79] has found a strong relationship between geotextile effectiveness with 5% secant modulus. The data in Table 2.20 show the results based on the number of dynamic cycles until the lower crack reflected through the overlay. The data plot

TABLE 2.20 LABORATORY-TEST RESULTS OF DYNAMIC CYCLE LIFE SHOWING EFFECT OF GEOTEXTILE AND MODULUS

Geotextile	Type	Mass/unit area (g/m2)	Secant modulus* (N)	Cycles to failure	Standard deviation	Fabric effectiveness factor†
—	Control, no geotextile	—	—	480	50	1.0
B	Nonwoven needle-punched polypropylene	150	590	1000	55	2.1
C	Nonwoven needle-punched polyester	200	540	2300	880	4.8
D	Nonwoven needle-punched polypropylene	200	930	3260	610	6.8
E	Woven slit-film polypropylene/polyester	170	1600	2760	570	5.8
A	Nonwoven heat-bonded polyester	108	2000	7650	575	15.9

*Values are for force at 5% strain using a grab tensile test.

†FEF is the ratio of the cycles to the failure of the geotextile (fabric)-reinforced specimens to that of the control specimen without fabric.

Source: After Murray [79].

is close to linear, indicating that higher-modulus geotextiles, via a primary reinforcement function, outperform the lower-modulus geotextiles.

Since the decision as to which is the primary function is by no means clear cut, two separate design methods will be presented. The first is based on reinforcement as the primary function, and the second is based on waterproofing as the primary function.

Reinforcement-Based Design. The key to the reinforcement-based design of geotextiles in reflective-crack prevention in bituminous pavement overlays is the fabric's effectiveness as determined by laboratory testing or by experience. Quantitatively, it is defined as follows:

$$\text{FEF} = \frac{N_r}{N_n} \tag{2.73}$$

where

FEF = fabric effectiveness factor,
N_r = number of load cycles to cause failure in the geotextile reinforced case, and
N_n = number of load cycles to cause failure in the nonreinforced case.

Values of FEF vary widely when based on laboratory tests (as they usually are), the range being from 2.1 to 15.9, as shown in Table 2.20. Upon having this value, however, design can be approached by a number of procedures. Majidzadeh et al. [134] use a mechanistic design procedure influenced by both rutting (distortion) and fatigue (cracking). Another approach, however, is merely to modify existing asphalt overlay design methods. In this regard the *design traffic number* (DTN), on which overlay designs are based, will be modified as follows.

$$\text{DTN}_r = \frac{\text{DTN}_n}{\text{FEF}} \tag{2.74}$$

where

DTN_r = design traffic number in the fabric-reinforced case,
DTN_n = design traffic number in the nonreinforced case, and
FEF = fabric effectiveness factor.

Using the Asphalt Institute's overlay design procedure [135], Example 2.29 illustrates the procedure. It is based on an arbitrarily selected FEF value of 3.0 and uses Figure 2.75 as the basic design guide. The procedure is as follows.

1. Determine the soil subgrade strength value as represented by its CBR value.
2. Determine the initial traffic number (ITN), as discussed in [135]. This is a combination of each vehicle's weight and the respective number of load repetitions based on traffic counting.
3. Determine the adjustment factor for the desired design period and estimate traffic annual growth rate, as described in [135].

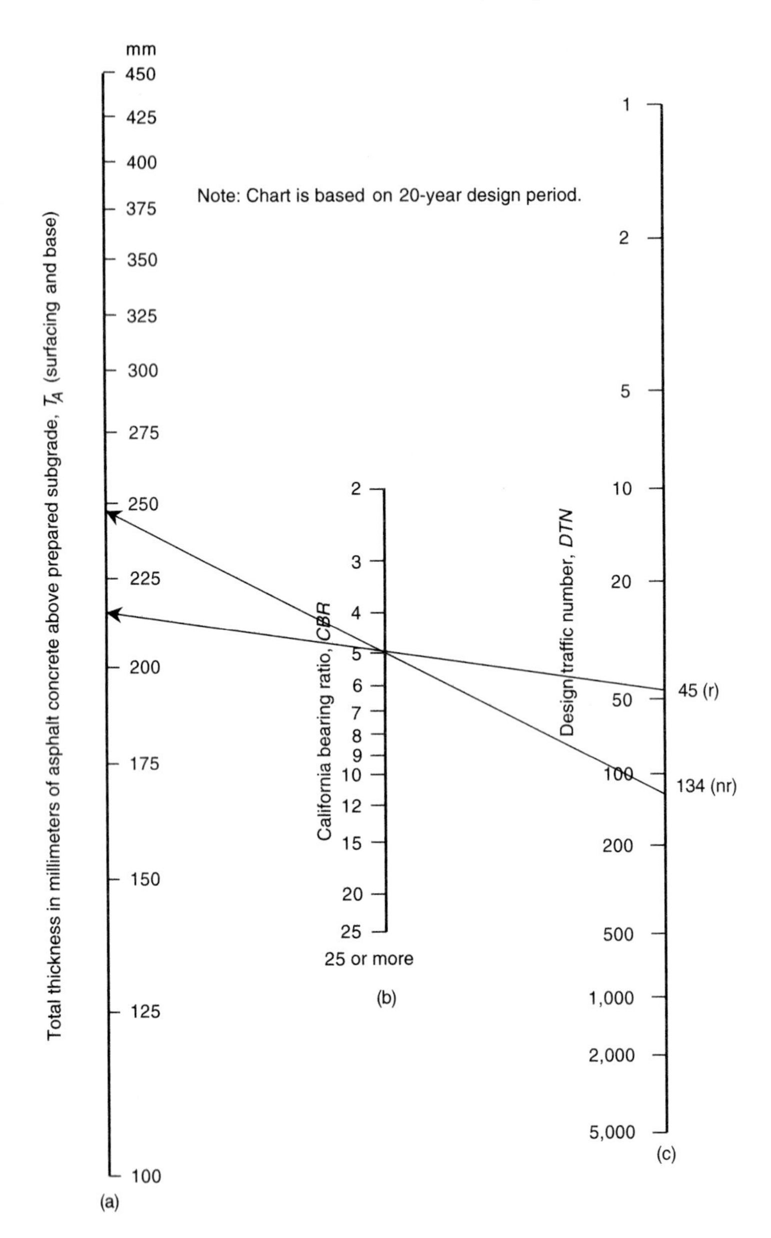

Figure 2.75 Thickness requirements for asphalt pavement structures using unsoaked subgrade soil CBR. (After Asphalt Institute [135])

4. Multiply the ITN by the adjustment factor to obtain the DTN for use in the thickness design chart.
5. Use Figure 2.75 (or its equivalent) to determine the full-depth asphalt-to-pavement thickness, t_{An}, needed for the design subgrade strength value, the DTN, and the selected design period.

6. Determine the effective thickness, t_e, of the existing pavement as discussed in [135].
7. The thickness of asphalt concrete overlay required, then, is equal to $t_{An} - t_e$.
8. This process is repeated for the geotextile-reinforced case using Eq. (2.74), which results in a thickness t_{Ar}.
9. The resulting two thicknesses (nonreinforced and geotextile-reinforced) are then compared ($t_{An} - t_{Ar}$) to note the savings in asphalt overlay thickness Δt.

Example 2.29

Given an interurban two-lane highway carrying an average of 4000 vehicles per day, 400 (10%) of which are heavy trucks of 135 kN average gross mass. The single-axle load limit is 80 kN. Traffic growth rate is 4% annually. The existing pavement consists of 75 mm of asphalt concrete surface and 200 mm of crushed stone base on a soil whose CBR = 5.0. The pavement is in generally good condition, but visual evaluation indicates the need for an overlay. Find the overlay thickness for a 20-year design period **(a)** without using geotextile, and then **(b)** using a geotextile with FEF = 3.0. **(c)** Compare the two overlay thicknesses. This problem is from [135].

Solution: **(a)** Determine that the initial traffic number = 90, and the adjustment factor = 1.49, resulting in a DTN for the nonreinforced fabric case of

$$\begin{aligned} \text{DTN}_n &= 90 \times 1.49 \\ &= 134 \end{aligned}$$

Using this and a CBR = 5, Figure 2.75 gives in a full-depth nonreinforced asphalt pavement thickness (t_{An}) of

$$t_{An} = 245 \text{ mm}$$

The existing pavement effective thickness (t_e) calculation uses a weighing factor of 0.8 on the existing asphalt and 0.4 on the existing stone base.

$$\begin{aligned} t_e &= 75(0.8) + 200(0.4) \\ &= 140 \text{ mm} \end{aligned}$$

Therefore, the required overlay thickness (t_{on}) without using a geotextile (i.e., nonreinforced) is

$$\begin{aligned} t_{on} &= t_{An} + t_e \\ &= 245 - 140 \\ &= 105 \text{ mm} \end{aligned}$$

(b) The solution varies for the case of using a geotextile-reinforcement layer as follows.

$$\begin{aligned} \text{DTN}_\mathbf{r} &= \frac{\text{DTN}_\mathbf{n}}{\text{FEF}} \\ &= \frac{134}{3.0} \\ &= 45 \end{aligned}$$

which, from Figure 2.75, results in

$$t_{Ar} = 215 \text{ mm}$$

and a required overlay thickness with reinforcement of

$$t_{or} = 215 - 140$$
$$= 75 \text{ mm}$$

(c) Thus the savings in asphalt overlay thickness using a geotextile layer (and based on a reinforcement hypothesis) is

$$\Delta t_o = t_{on} - t_{or}$$
$$= 105 - 75$$
$$\Delta t = 30 \text{ mm}$$

Note that this same result can be reached by using the values of t_{An} and t_{Ar} directly,

$$\Delta t = t_{An} - t_{Ar}$$
$$= 245 - 215$$
$$\Delta t = 30 \text{ mm}$$

Waterproofing-Based Design. Again using the Asphalt Institute's techniques [135], we can develop an alternate design procedure for geotextiles used in asphalt overlay situations; this time one that is based on a waterproofing hypothesis. This concept should not come as a surprise since adequate subgrade drainage of pavements has long been suspected as being the key factor for extending conventional pavement lifetimes. Figure 2.76 clearly illustrates this type of improved pavement lifetime. The particular procedure we will use, adopted from Bell [137], utilizes field-measured rebound

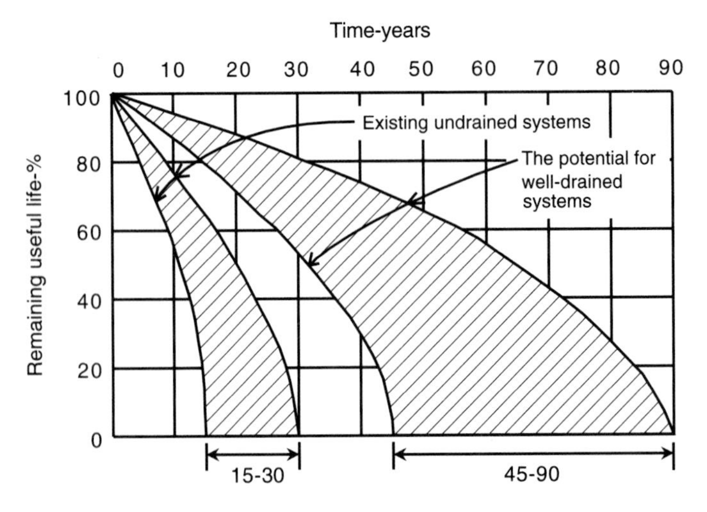

Figure 2.76 Comparison of the potential life expectancies of well-drained pavements with the conventional undrained pavements. (After Cedegren [136])

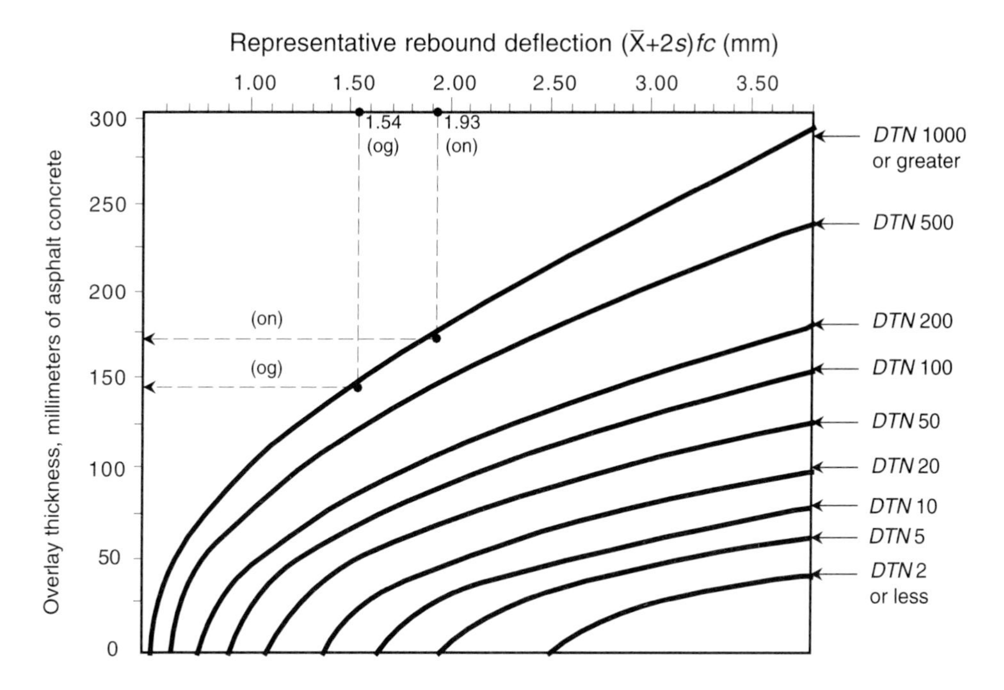

Figure 2.77 Asphalt concrete overlay thickness required to reduce pavement deflection from a measured to a design deflection value (a rebound test). (After Asphalt Institute [135])

deflections of the existing pavement system along with the design guide of Figure 2.77. The individual steps in the design are as follows.

1. Determine the representative rebound deflection as discussed in [135], which is based on Benkelman beam field deflection tests.
2. Determine the ITN as discussed in [135]. It is a combination of each vehicle's weight and respective number of load repetitions based on traffic counting.
3. Determine the initial traffic number adjustment factor for the desired design period as described in [135].
4. Multiply the ITN by the ITN adjustment factor to obtain the DTN, for use in the overlay thickness chart.
5. Enter the overlay thickness chart (Figure 2.77) at the representative rebound deflection and move down vertically to the curve representing the DTN (interpolate if necessary). Move horizontally to the overlay thickness scale and read the thickness of overlay required.
6. For the case of a geotextile included in the pavement cross-section and of it being suitably waterproofed, we can appropriately modify the representative rebound deflection (RRD).

$$\text{RRD} = (\bar{X} + 2s)fc \tag{2.75}$$

where

$\overline{X}$ = arithmetic mean of measured Benkelman beam values,
s = standard deviation,
f = temperature adjustment (see [135]), and
c = critical period adjustment factor, which is largely influenced by moisture in the subgrade system (this is the term that will be empirically adjusted in design Example 2.30).

7. The design process is then repeated as with the nongeotextile case, and the resulting two thicknesses are compared to note the savings in asphalt overlay using the geotextile.

Example 2.30

Given a four-lane interurban highway with an average of 16,000 vehicles per day, 2400 (15%) of which are heavy trucks with an average gross mass of 145 kN; the design lane estimated to carry 45% of the heavy trucks; the traffic growth rate 5% annually, and the legal single-axle load limit is 80 kN, some cracking of the pavement surface is evident. High deflections indicate the need for an overlay. Find the overlay thickness required for a 20-year design period **(a)** without using a geotextile, and then **(b)** using a geotextile that changes the value of c in Eq. (2.75) from 1.25 to 1.00. **(c)** Compare the two overlay thicknesses. In the analysis use ITN = 590 and an adjustment factor of 1.67. This problem is also taken from [135].

Solution: (a) First calculate the representative rebound reflection from Eq. (2.75) using field-gathered data of

$$\overline{X} = 1.55 \text{ mm},$$

$$s = 0.10 \text{ mm},$$

$$f = 0.88,$$

$$c = 1.25, \text{ and}$$

$$\begin{aligned} \text{RRD} &= (1.55 + 0.20)0.88 \times 1.25 \\ &= 1.93 \text{ mm} \end{aligned}$$

The design traffic number is also needed:

$$\begin{aligned} \text{DTN} &= \text{ITN} \times \text{adjustment factor} \\ &= 590 \text{ x } 1.67 \\ &= 985 \end{aligned}$$

Using Figure 2.77, the required overlay thickness without a geotextile is

$$t_{on} = 170 \text{ mm}$$

(b) For the case with a geotextile as a waterproofing layer, the constant c is changed from 1.25 to 1.00 and the process is repeated:

$$\begin{aligned} \text{RRD} &= (1.55 + 0.20)0.88 \times 1.00 \\ &= 1.54 \text{ mm} \end{aligned}$$

Using this value and a DTN = 985, the required overlay waterproofed geotextile thickness from Figure 2.77 is

$$t_{og} = 140 \text{ mm}$$

(c) Based on these asphalt overlay thickness values, the saving using a geotextile is as follows:

$$\Delta t = t_{on} - t_{og}$$

$$= 170 - 140$$

$$\Delta t = 30 \text{ mm}$$

Note that the equivalent thickness of the existing pavement system is the same in both cases, and the resulting saving if a constant value were added to both t_{on} and t_{og} is still 30 mm.

Summary. As mentioned in Section 2.10.1, multiple-function geotextile applications are difficult to come to grips with, since a clear-cut primary function cannot easily be identified. This topic of crack reflection prevention in bituminous pavement overlays illustrates the dilemma perfectly. Using two completely different hypotheses (one based on reinforcement and the other based on waterproofing), two completely different designs can be developed. It simply begs the question to ask where the truth actually lies. Considering these two extremes, it might be that a combination of the two phenomena are working together! Further, a separation interlayer may be another function that is yet to be developed.

Clearly, well-instrumented, well-monitored, well-analyzed, and well-reported case histories are needed in this application area. Perhaps, then, a decision as to which is the primary function will evidence itself, pointing the way to the correct design methodology. The case histories that are currently available on the topic not only do not give this identification but cast doubt on where the technique can be used.

Beginning in 1968, thirty-seven projects aimed at evaluating geotextiles used for the control of reflective cracking in bituminous overlays have been initiated by the U.S. Department of Transportation, Federal Highway Administration, under its National Experimental and Evaluation Program (NEEP). The most comprehensive programs were conducted by Arizona, California, Florida, and North Dakota. One finding was that there is no strong evidence that the geotextile used provides a generalized mechanism for extending the crack-free life of an overlay. Most reports have identified the problem as being very complex and being influenced by the type and degree of pavement distress, working versus nonworking cracks, type of pavement, loads, climate, geotextile-asphalt bonding (tack coat), geotextile type and properties, and asphalt type and mixing properties. The general conclusion of the Federal Highway Administration is that geotextiles may be used in the control of reflection cracking, but that the best results can be achieved where the existing pavement has experienced fatigue-associated alligator cracking with cracks 3 mm wide or less and located in relatively mild climates.

To be sure, the summary statements above challenge the universal use of geotextiles in the control of reflection cracking in bituminous overlays. In general, however, it is felt that reports are definitely on the positive side. Based on experiences to date, the following recommendations are offered:

- For those states that are using proprietary or competition-limiting specifications for geotextiles, the adaptation of Texas specifications for state use is encouraged.
- Prior to placement of a geotextile overlay system, the condition of the existing roadway should be documented. When an unstable roadway is suspected, deflection tests are recommended. While limiting deflection values have yet to be established for geotextile systems, it is important that the data be obtained that could assist in their eventual evaluation.
- Since the geotextiles presently being marketed are not equivalent in physical properties, agencies should conduct the tests identified in the Texas Specification, including the asphalt-retention test, so as to develop documentation that may be useful later in assessing relative geotextile performance.
- Rather than placing the geotextile on the cracked existing pavement, construct an asphalt leveling course first so as to provide a relatively unblemished surface for applying the tack coat and the geotextile. This will assure more complete and uniform impregnation of the geotextile by the tack coat and will also assist in determining the type of tack coat to be used and the proper application rate.
- For pavement rehabilitation projects that include pavement widening with new asphaltic concrete overlays, geotextiles placed longitudinally over the shoulder-pavement and/or widening joint should be considered. The state of Maine has had success in this regard. Table 2.21 shows that both longitudinal and transverse joints were greatly retarded when using a high strength geotextile directly spanning the cracks in question.

TABLE 2.21 RESULTS OF REFLECTIVE CRACKING PREVENTION STUDY IN PARIS, MAINE, USING HIGH-STRENGTH GEOTEXTILE STRIP REINFORCEMENT

Control	Strip Reinforcement	Improvement
Transverse Joints (% Cracks Reflected)		
77.8	10.7	7.3
85.4	19.0	4.5
96.7	32.3	3.0
100	43.3	2.3
100	46.6	2.1
		Avg. = 3.8 (380%)
Longitudinal Joints (% Cracks Reflected)		
1.2	0.3	4.0
1.2	0.3	4.0
7.0	1.5	4.7
8.1	6.6	1.2
17.1	7.4	2.3
		Avg. = 3.2 (320%)

Source: [138].

- Over jointed portland-cement concrete pavements, no evidence has been provided to support placing a geotextile system across the full roadway width in a continuous mat. Instead, the use of heavy-duty geotextile systems in strips over transverse and pavement edge joints and cracks is presently recommended.

A summary report by the U.S. Army Corps of Engineers on this particular application of geotextiles has arrived at similar conclusions [139]. Of importance insofar as current research and development is concerned is a series of RILEM conferences focused specifically on reflective cracking in pavements [140].

2.10.3 Railroad Applications

Geotextiles are often used in railroads beneath the stone ballast upon which the wooden or concrete tie system is placed. As will be discussed, a critical aspect of the design is the depth at which the geotextile is placed beneath the bottom of the tie (i.e., the thickness of overlying ballast). First, however, it is necessary to gain a perspective of the function(s) of the geotextile under various possible conditions.

Overview. It is virtually impossible to identify a unique primary function for geotextile use in railroad applications. Site-specific conditions will vary the primary function among a number of possibilities. In listing these possibilities it is important to keep in mind whether the railroad is being newly constructed or being rehabilitated. In new construction, the material beneath the geotextile will probably be the in situ soil; in rehabilitation, the material beneath the geotextile will be previously placed ballast (now contaminated with soil), which has been driven into the soil over the past working history of the railroad. Considering both of these situations the possible geotextile functions are the following.

- Separation, in new railroads, between in situ soil and new ballast
- Separation, in rehabilitated railroads, between old contaminated ballast and new clean ballast
- Lateral confinement-type reinforcement in order to contain the overlying ballast stone
- Lateral drainage from water entering from above or below the geotextile within the geotextile
- Filtration of soil pore water rising up from the soil beneath the geotextile, due to rising water conditions or the dynamic pumping action of the individual wheel loads, across the plane of the geotextile

Irrespective of the difficulty of identifying a single function of the geotextile (it is obviously a multifunction application), the acceptance of geotextiles by railroad companies is large and increasing. Newby [141] reports that as far back as 1982 the Southern Pacific Railroad used geotextiles in over 1600 km of trackage.

Specific Design. A review of the geotextile literature on railroad applications shows they are somewhat inconsistent. Railroad specifications seem to favor relatively heavy nonwoven needle-punched geotextiles because of their high flexibility and in-plane drainage (transmissivity) characteristics. The logic behind high flexibility is apparent, since the geotextiles must deform around the relatively large ballast and not fail or form a potential slip plane. In-plane drainage, however, is not itself a dominant function, because any geotextile acting as a proper separator and filter will preserve the integrity of the drainage of the ballast stone itself, where ample void volume is always present. Nevertheless, geotextile drainage is invariably emphasized in railroad specifications [141].

Laboratory work, even large-scale model testing, seems to be directed at a reinforcing and stiffening function of the geotextile. Work by Eisenmann and Leykauf [142] (which also includes the filtration function), Saxena and Wang [143], and Bosserman [144] clearly illustrates the reinforcement benefits of using a geotextile beneath the ballast stone. In the author's opinion, however, such membrane-type reinforcement can be gained only after quite high deformations of the subgrade soils. For the majority of existing railroads, as in rehabilitation work, such deformations do not seem possible or desirable, since densification of the subsoil has usually occurred many years before rehabilitation is necessary.

For these reasons, the design of geotextiles beneath ballast in railroad applications can be addressed in the following steps.

1. Design the geotextile as a separator—this function is always required. The procedures in Section 2.5 have direct applicability here. Burst strength, grab tensile strength, puncture resistance, and impact resistance all have significance in this particular application.
2. Design the geotextile as a filter—this function is also usually required. The general procedures illustrated in Section 2.8, in particular those illustrated for walls and underdrains, are relevant in this application. The general requirements of adequate permeability, soil retention, and long-term soil-to-geotextile flow equilibrium are needed as in all filtration designs. A note of caution, however: railroad loads are dynamic; thus pore pressures must be rapidly dissipated. For this reason high factors of safety are required in the permittivity design of the process.
3. Consider geotextile flexibility if the cross section is raised above the adjacent subgrade. Here a very flexible geotextile is an advantage in laterally containing the ballast stone in its proper location. Quantification of this type of lateral confinement reinforcement is, however, very subjective (recall Section 2.3.2).
4. Consider the depth of the geotextile beneath the bottom of the tie. The very high dynamic loads of a railroad acting on ballast imparts accelerations to the stone that are gradually diminished with depth. If the geotextile is not deep enough, it will suffer from abrasion at the points of contact with the ballast. Raymond [145] has evaluated a number of geotextiles beneath Canadian and U.S. railroads and found many that are pockmarked with abrasion holes (see Figures 2.78a,b). In fact, there were so many cases that he has quantified the situation (see Figure 2.79). Here it is seen that major damage occurs within 250 mm of the tie, and only deeper than 350 mm does the situation become acceptable. From this curve it can be concluded that the minimum depth for geotextile placement is 350 mm plus 50 mm for track settlement for a total 400 mm. If this depth is considered ex-

(a)

(b)

Figure 2.78 Abrasion failures of geotextiles placed too close to the track structure. (Compliments of G. Raymond)

cessive in view of the large amount of ballast stone required, a highly abrasion-resistant geotextile must be used. Raymond [145] recommends a resin-dipped nonwoven needle-punched geotextile (which is a very stiff fabric) that has been forced-air dried to reestablish its porosity.

5. The last step in the design is to consider the geotextile's survivability during installation. To compact ballast under ties the railroad industry customarily uses a series of vibrating steel prongs forced into the ballast. Considering both the forces exerted and the vibratory action, high geotextile-puncture resistance is required. Hence, and in keeping with Step 4, it is necessary to keep the geotextile deep or to use a special high-puncture resistant geotextile (recall Section 2.3.5).

Summary. Geotextile use in railroads beneath the ballast offers a number of possible benefits. These are separation, filtration, reinforcement, and drainage. Unfortunately, it cannot be categorically stated that one function dominates over the others

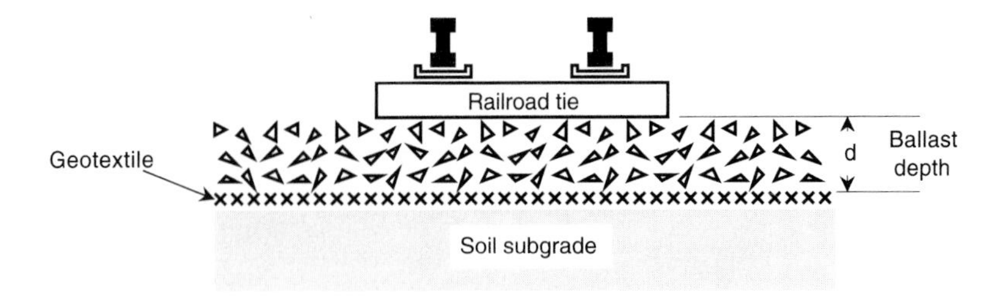

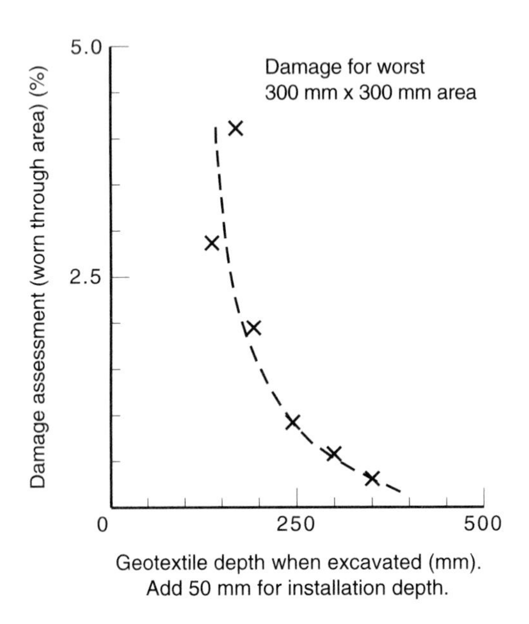

Figure 2.79 Observed geotextile abrasion damage as a function of depth beneath bottom of railroad tie. (After Raymond [145])

in all cases. This application is indeed a multifunctioned one that must be handled on a site-specific basis. Yet some functions (e.g., separation and filtration) are always present. A recommended design procedure has been outlined.

It must be emphasized that a geotextile beneath a railroad is a very demanding application. Avoiding both puncture during installation and abrasion during service lifetime requires deep placement ($\geq$ 400 mm) beneath the bottom of the tie or specially treated geotextiles designed for high-abrasion resistance and puncture resistance.

In closing, more well-planned, well-documented, and well-reported case histories of the type reported by Chrismer and Richardson [146] are recommended. Only with quantified results under specific conditions can a definitive function be associated with a geotextile. At that point the design process can advance in a position of strength.

2.10.4 Flexible Forming Systems

The traditional method of formwork for concrete or grout is to use wood or metal. These rigid forms are properly positioned and fixed in location until the material placed in them adequately cures and has sufficient strength of its own. While the constraint of a rigid form is an obvious advantage in building a concrete wall or footing to exact line and grade, it is a decided disadvantage in a number of other applications. These situations, which can benefit from the use of flexible forms made from geotextiles, are explored in this section.

Overview. It is easy to visualize that a geotextile, in the form of a tube or bag, could be used as a flexible form into which concrete, grout, or soil could be placed or pumped. Such a system would work as well under water (where water is displaced from within the fabric) as it would above ground (where air is displaced from within the fabric). Upon curing, the shape of the solidified mass takes the shape of the deformed geotextile; the number of possible shapes is enormous. Additionally, the geotextiles, being flexible, can be inserted in difficult-to-reach locations and filled after proper positioning. This concept of flexible-forming systems with geotextiles can be used in many practical situations.

As a historical note, Terzaghi was the first to use geotextiles as flexible forms at the Mission Dam (now Terzaghi Dam) in British Columbia, Canada, in 1955 [147]. A seepage cutoff wall was being placed within an existing dam and a separation was required between the temporary bentonite slurry on one side and the permanent cement-bentonite grout on the other. Since the walls on each side of cutoff wall were not parallel (one side was a multicurved concrete surface and the other was steel piling), a pipe or beam separator was not possible. Instead, Terzaghi used a nylon-reinforced canvas bag (today, a geotextile) of 21 m length and 450 mm diameter and wrapped it around a 25-mm diameter grout pipe. The bag-surrounded grout pipe was placed in its proper position and then inflated with cement-bentonite grout. To be sure the system would work properly, a 6 m prototype was first constructed to see whether the gap would be completely filled and whether the fabric was sufficiently strong. The success was immediately obvious, and the system was used throughout this phase of the project. Since that first use, the use of geotextiles as flexible forms has grown rapidly and in many different directions.

Columns for Mine and Cavern Stability. Abandoned mines and caverns are obvious problems when any type of structure is to be founded on or near them. The obvious preconstruction remedy is to fill them before subsidence can occur; however, they often either have no access or the access is blocked and unavailable. Under these conditions, alternative methods are hydraulic backfilling, grouting, or the construction of conventional caissons to the adequate depth, all of which are quite expensive. Geotextiles, however, can be used as flexible forms (in much the same way as in the Terzaghi Dam) without the necessity of entering the mine or cavity. The technique consists of drilling 100 to 150 mm diameter holes to intercept the roof of the mine and then carrying these holes to the floor of the mine, penetrating the floor approximately 300 to

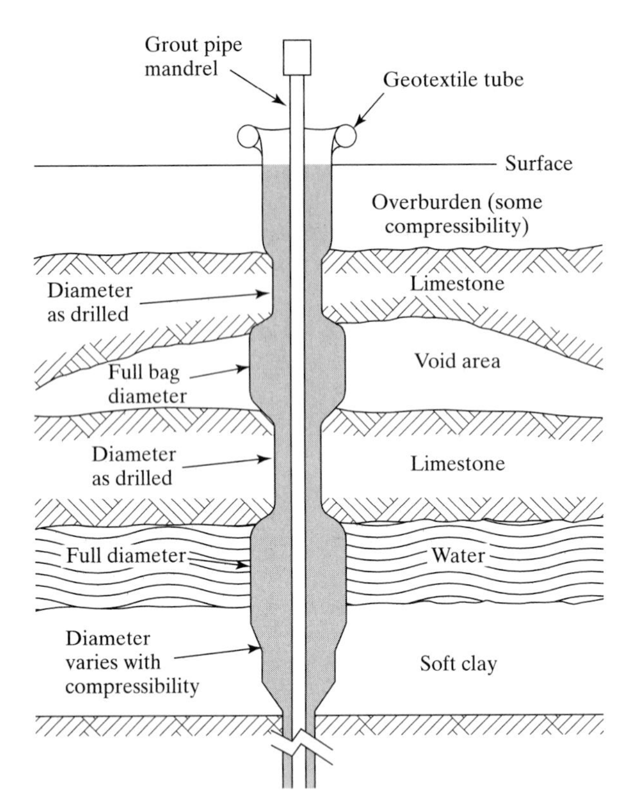

Figure 2.80 Idealized sketch of complete cross section of Fabriform mortar piles showing grout pipe within a geotextile sleeve placed in a previously drilled 150 mm diameter hole. (After B. A. Lamberton, Intrusion Prepakt)

450 mm (see Figure 2.80). A prefabricated and sewn tube of geotextile of approximately 500 mm diameter is wrapped around a grout pipe and is snaked down the drill hole into the keyhole at the floor of the opening. Then fine-aggregate concrete is injected under controlled pressure as the grout pipe is withdrawn. Expansion occurs to the full tube diameter where no resistance is met, that is, where voids exist (see Figure 2.80). Thus support shoulders are created beneath the competent rock strata. The tube of geotextile has to be supported at the surface or through rings at the top of the geotextile with a cable system to the ground surface. Each application requires a determination of how much pressure the geotextile can withstand. It may be necessary to pump the tube in multiple lifts. Reinforcing steel can be placed either in the void area only or for the full length of the column. The critical point in this application is to get the maximum support of the column of concrete under the roof of the individual rock strata. Where the cavity or mine is dry, it is feasible to observe the expanded column with a television camera from an adjacent hole. In areas where the opening is fully or partially filled with soil or other compressible or objectionable materials, it is possible to jet out an opening in this material by having the grout pipe extend through the bottom of the tube and, while jetting through this pipe, maintain an adequate head of bentonite or grout in the geotextile form. This technique has the advantage over other methods of forming grout columns in that a positive form can be installed with a rela-

tive economy, as large quantities of concrete are not lost in a wasteful base that would be needed to build up the angle of repose of the concrete.

Using this concept it is also possible to create a bulkhead in an underground mine. By drilling holes on a predetermined line and pumping alternate columns initially, the geotextile can then be placed in the secondary or intermediate locations and expanded with concrete to interlock between the originally placed columns. Two parallel walls can be created to form a bulkhead or cutoff. This technique has been used in relatively shallow mine and limestone applications. See Koerner and Welsh [80] for several case histories. The method is very similar to the construction of secant-pile or tangent-pile cutoff walls in conventional foundation engineering.

As with other topics in this section, the geotextile to be used serves multiple functions (filtration, separation, and reinforcement through containment). The design centers on the following steps.

1. The opening size of the geotextile must be designed as a filter with emphasis on grout retention. A recommended criterion in this regard is the following:

$$0_{95} \leqslant 2d_{50} \tag{2.76}$$

   where

   0_{95} = 95% opening size of the geotextile (mm), and
   d_{50} = average (50%) particle size of the cement used in the grout (mm).

   Note that some bleeding of the grout will occur, but this should not significantly decrease its strength. It is necessary to have an open geotextile when construction is underwater so as to expel the water within the fabric.
2. The required amount of shoulder area underneath the roof of the mine or cavern must be estimated on the basis of the surface loads to be imposed. This, together with the diameter of the geotextile column, will allow for a rough estimate of the support strength. The tensile strength must not be exceeded by the grout pressure, yet sufficient pressure must be generated to obtain the required shoulder area.
3. Since the geotextile must be longitudinally seamed, the sewing procedure is critical. In this application the seam strength must be equal to the required strength.
4. A detailed example of the use of geotextiles as flexible forms will be given at the end of Section 2.10.4, after other flexible-form applications are presented.

Horizontal Bags and Tubes (aka Geotubes). The use of fabrics to make sandbags has been widely practiced for hundreds of years. Rather than making the bags from degradable natural fibers, however, the use of UV-stabilized polymeric fibers has added new life into such installations. With controlled strength and deformation, the bags can be filled with sand, grout, or even concrete and can be specially tuned to the particular application at hand. When used for erosion-control work, as they often are, a high-strength woven nylon geotextile sewn together with monochord polyester or Kevlar thread is often used. The bags are placed in a staggered pattern and weigh many tons when filled (see [80] for details). Note that both vandalism and UV degradation must be considered when the bags are filled with sand. This is not the case when they are mortar-filled, which is of course a much more expensive alternative.

Rather than using discrete bags of finite length in forming erosion-control barriers, tubes of unlimited length can also be used. Flexible sand-filled tubes were made as early as 1957, but they were not very successful. Eventually in 1967, a patent was granted to a Danish firm, Aldek A.S., in conjunction with the Danish Institute of Applied Hydraulics. Aldek's system, as explained below, was developed further in 1970 [148].

An inner tube of 0.25 mm impermeable polyethylene (a geomembrane) is manufactured in 0.7, 1.0, and 1.7 m diameters. The outer material is a woven flexible polyethylene geotextile with UV-stabilizing additives. The tubes are hydraulically filled with sand at the site by attaching a steel inlet drum to one end and a regulating outlet drum at the other end. A diaphragm pump mixes sand and water; this mix is then pumped to fill the tubes. A typical 1.0 m tube, 100 m long, can be filled in 3 to 4 hr. Fine sands to coarse gravels (up to 5 to 10% by volume) have been successfully pumped. General configurations for the tubes are shown in Figure 2.81.

Geocontainers. An environmentally friendly method of removing river/harbor bottom sediments from shipping channel areas to proper disposal areas using

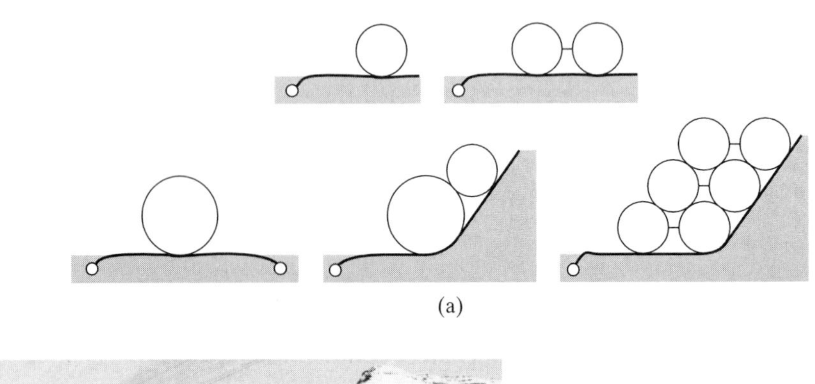

(a)

(b)

Figure 2.81 Typical cross sections of grout-filled tubes. Note geotextile use along with small 250 mm-diameter anchor tubes.

(a) Geotextile within closed barge

(b) Geotextile being filled

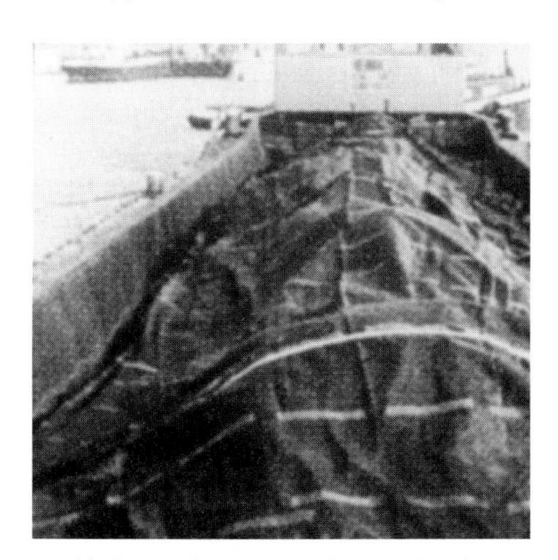

(c) Geotextile sewn over backfill forming geocontainer

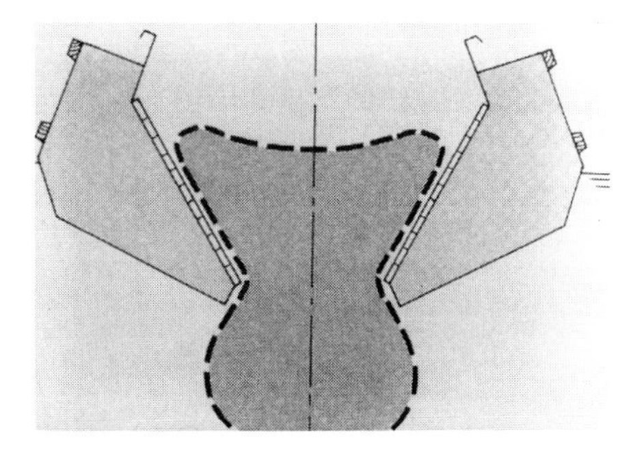

(d) Geocontainer being dropped out of bottom of barge at disposal site

Figure 2.82 Geocontainers for removal of river and harbor sediment (sometimes contaminated) to construct underwater containment areas using bottom-dump barges. (Compliments TC Nicolon-TC Mirafi Corp.)

geotextiles has been rapidly growing [149, 150, 151, 152]. It was developed by the Nicolon-Mirafi-Ten Cate Corporation and utilizes high-strength geotextiles (greater than 50 kN/m tensile strength) and bottom-dump barges (see Figure 2.82). The geotextile is placed in the empty barge and then filled with the bottom sediments. When full, the geotextile ends are folded over the top and sewn together, thereby completing the container. The barge is towed to the disposal area and, when properly positioned, the split hull of the barge is opened and the sediment-filled container drops to the bottom. Subsurface embankments are being formed by this technique, which has the significant advantage that the sediments (particularly when they are contaminated) never leave the estuary or harbor. Furthermore, additional storage is possible behind the embankments, now reinforced by the geocontainers themselves. The technique has worldwide implications and is in a rapid stage of development. Particularly intriguing is the design of the geotextile vis-a-vis its deformed shape, which includes the dynamic placement of the dredged sediment and the impact stresses of its final positioning (see Leschinsky and Leshchinsky [153]).

Restoration of Piles (Pile Jacketing). All piles in a marine environment suffer deterioration at varying rates. The deterioration is caused by normal marine exposure, wet-dry cycles, freeze-thaw cycles, and chemical, industrial, and sanitary wastes. Moreover, each type of pile has its own particular problems.

> *Wood piles* in a marine environment can be subject to attack by bores. There are three basic types: teredo and bankia (both mollusks), and the arthropod limnoria. Although the limnoria is a surface eroder and the mollusks are internal borers, it is extremely difficult to detect damage to wood piles by any of them by visual inspection until serious deterioration has occurred.
>
> *Concrete piles* are subject to deterioration in the water and splash fluctuation zone, caused by the wet-dry, freeze-thaw cycles. Some concrete piles (both precast and cast-in-place) deteriorate below the mudline because of poor original placement techniques. This comes about from permeable concrete, which allows corrosion of the reinforcing steel and the subsequent expansion and spalling of the concrete. It is also sometimes caused by a sulfite reaction of the concrete.
>
> *All types of steel pipes* are subject to corrosion, with average corrosion rates being 0.13 mm per year under normal conditions. However, this value can increase drastically under certain conditions. Values of 0.36 mm per year have been measured [154]. Still higher values could conceivably result under certain stress corrosion conditions.

The methods used for the rehabilitation of piles are varied and constantly increasing in number. The oldest technique is to use metal forms, such as corrugated steel in half sections joined together by angles attached to each of the half sections. Dockworkers and divers place, join, and seal the sections, which are used to contain cement grout, which bonds to the deteriorated pile section. This highly labor-intensive operation has led to the development of other, more economical techniques. Rigid plastic forms have been proposed as not only an economical form but as high-strength envelope to prevent further deterioration to the piling system. The annular space between the form and pile can be filled with either concrete grout or a specially formulated epoxy. Another pile-jacketing technique utilizes bituminized-fiber forms. However, all of these systems have a problem with bottom closure and sealing when they are acting as a form and the pile is not to be jacketed down to the firm subsoils. Also, complex configurations are all but impossible to form when using rigid enclosures. It should be noted that built-up layers of epoxy or bituminous coatings have also been utilized to protect pilings from deterioration; however, most of these techniques involve hand-placement by divers, and a thorough coating of the piles is difficult and expensive to obtain.

In the 1960s, a technique was developed that utilizes geotextiles as a concrete-forming system [80]. Basically, this concept uses as a concrete form a jacket of geotextile, the edges of which are connected by a heavy industrial zipper prefabricated into the geotextile. The ends of the geotextile above and below the deteriorated pile zone are banded to the pile. These flexible geotextile forms possess economic advantages over other concrete-forming systems because of their light weight, ease of installation, adaptability to any configuration, relatively low cost, and ease of connection onto the piles at any location above the mudline. The geotextile is so designed that when concrete is injected into it, the excess water bleeds through the voids of the geotextile without allowing the cementatious portion to escape. This lowering of the water/cement ratio produces a dense surface of concrete that resists further deterioration of the pile. Typical installation procedures are illustrated in Figure 2.83. Case histories are presented in Koerner and Welsh [80].

The design procedure for this application is similar to that of other cases using geotextiles as flexible-forming systems and will be described later. One departure worth noting, however, is that elongation of the geotextile under load should be kept to a minimum. Thus high-strength woven geotextiles are often used. As with other situations where grout is being pumped, high-strength seams comparable to the required design strength are necessary.

Bridge Pier Underpinning. The analysis and design to estimate erosion of shallow foundation bridge piers in rivers and streams is extremely difficult [155]. Even for bridge piers founded on rock, the rock often deteriorates and is scoured away during floods that are accompanied by high-velocity water. The problem is so severe that divers sometimes find that they can swim beneath the bridge pier itself [80]. (We have to wonder about the structural factor of safety under such circumstances!)

In a related problem, the Ambursen Hydraulic Co. constructed a number of hollow-core, reinforced-concrete slab-and-buttress dams throughout the eastern and midwest U.S. from 1910 to 1940. The dams consisted of flat concrete slabs at 45°, which were supported by vertical buttresses at 3.0 to 4.5 m. Both the slabs and buttresses were relatively thin (e.g., 300 to 600 mm of reinforced concrete). Today, many of these water reservoir dams are in need of repair, particularly at their buttress footing regions where compressive stresses are the highest.

Using fabrics as flexible forms, some very clever solutions to these difficult problems have been developed. Shown in Figure 2.84 is a solution used by Welsh [156] for a number of scoured bridge piers. A geotextile tube is prefabricated to fit around the perimeter of the pier between the top of the stable foundation material and the bottom of the pier foundation. As grout inflation of the geotextile proceeds, pipes are placed to communicate from the outside of the pier to within the enclosure. After the curing of the perimeter tube, an injection of high-strength grout into the inside of the perimeter

(a) Cleaning pile and placing reinforcement

(b) Bonding geotextile to lower, sound pile

(c) Geotextile ready for filling

(d) Completed pile restoration

Figure 2.83 Procedure for using geotextiles as flexible forms in pile jacketing rehabilitation work.

tube reestablishes the bearing capacity of the pier foundation. The previously installed pipes serve the dual function of allowing grout to enter the enclosure and allowing entrapped water to be displaced. The concrete cures as does typical tremie concrete placed under water.

Uniform Pressure Distribution Beneath Caissons. When placing prefabricated caissons on uneven bearing strata, stress concentrations can be excessively high. Typically, this occurs in areas where uneven rock is encountered. Use of geotextiles as

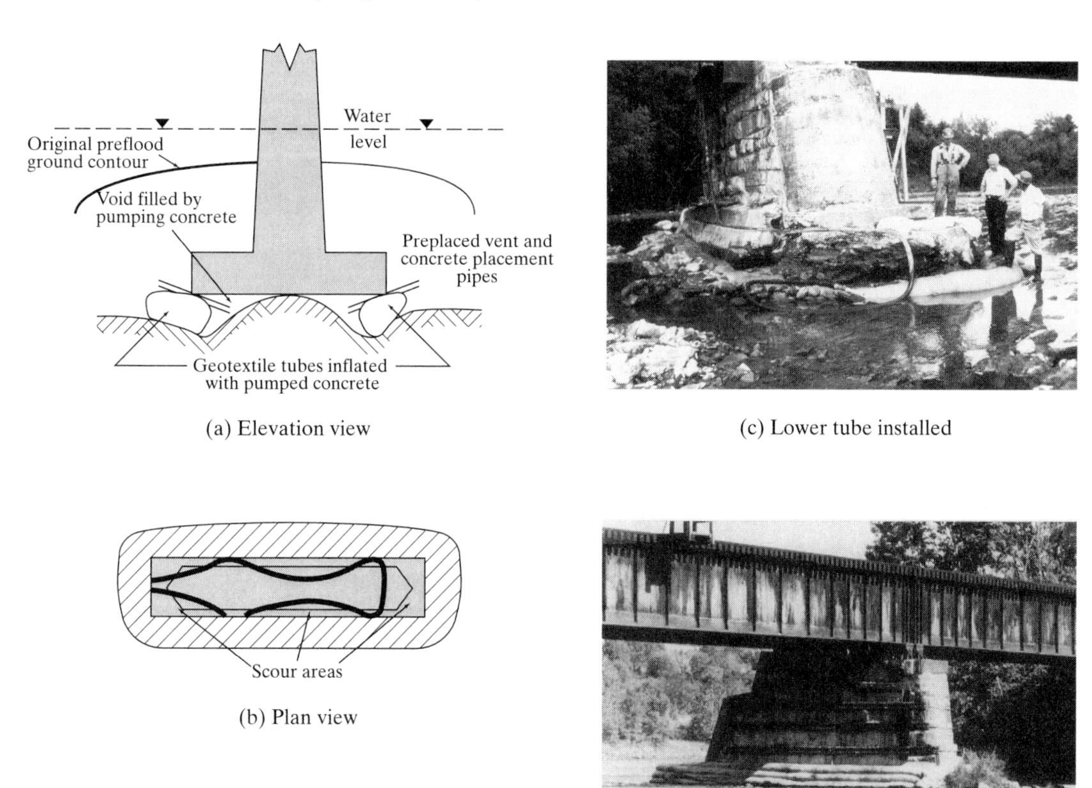

(a) Elevation view

(b) Plan view

(c) Lower tube installed

(d) Completed system

Figure 2.84 Underpinning of scoured bridge pier using grout-filled geotextile forms. (After Welsh [156])

a form into which high-strength cement grout is placed can alleviate the problem. A test case by den Hoedt and Mouw [157] nicely illustrates the technique; see Figure 2.85. Here a prefabricated high-strength geotextile was placed beneath a 1/10th scale prototype model concrete caisson and was pumped with grout. After curing, the caisson was lifted off the rock foundation and was seen to have the exact mirror image of the rock surface. The full-scale project was constructed in a similar manner.

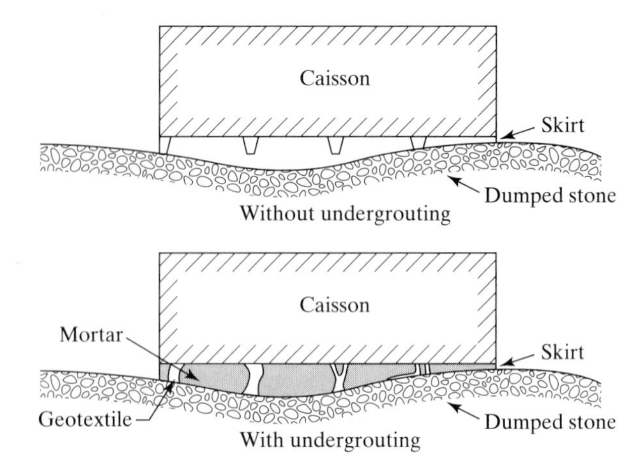

Figure 2.85 Prototype for evaluating use of geotextiles as forms for establishing uniform pressure distribution beneath concrete foundations. (After den Hoedt and Mouw [157])

Erosion-Control Mattresses. By taking two sheets of geotextile and joining them at discrete points, a form will result that can be pumped with grout to form a mattress that will conform to essentially any subsoil condition. The thickness and geometry are controlled by internal spacer threads woven into the upper and lower sheets of geotextile. Thicknesses of up to 500 mm have been made with various configurations. Shown in Figure 2.86 are two common styles: a uniform cross-section surface and an undulating surface where filter points are constructed at uniform spacings. These filter points serve to dissipate pore-water pressure trying to escape from the subsoil. They are available in varying diameters and spacings but are limited in their ability to control large amounts of subsoil seepage (recall Section 2.8.5).

Design Procedures for Flexible-Forming Systems. As shown in the preceding seven subsections, there are many applications for grout-filled geotextile forms. There are subtle differences between the designs, yet the basic design philosophy is quite similar.

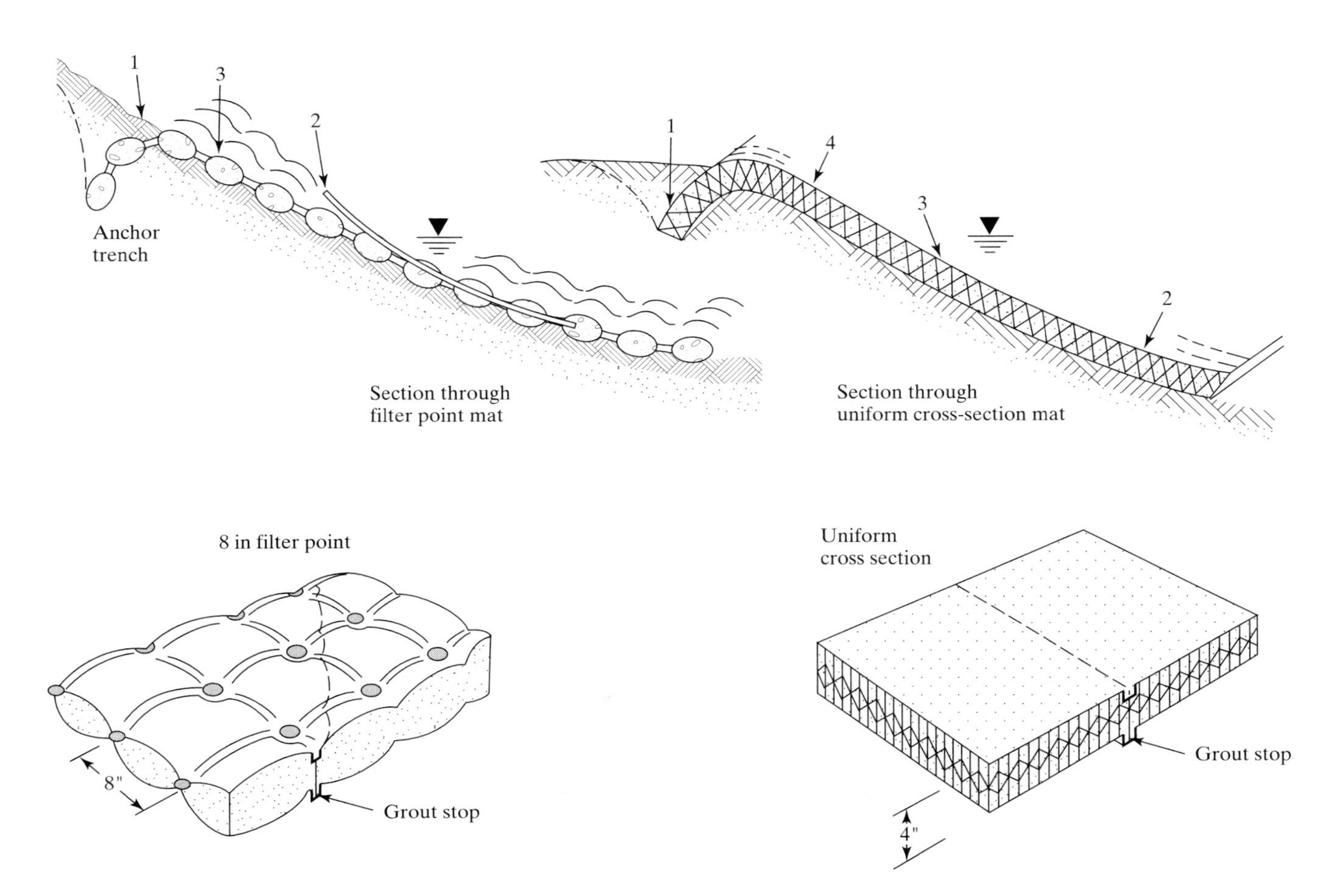

Figure 2.86 Typical cross sections and schematics of erosion-control mattresses formed by grout-filled geotextiles. (After Construction Techniques, Inc.)

The type of fill material for a particular application is obviously critical. Many applications use a cement/fly ash grout of the following composition:

Cement: 5.2 to 5.8 kN/m^3
Sand: 11.6 to 12.8 kN/m^3
Fly Ash: 0.58 kN/m^3
Water: 3.4 to 3.5 kN/m^3

which results in a water/cement ratio of 0.63 to 0.61. Under many conditions, water loss will be from 0.89 to 1.0 kN, resulting in a final water/cement ratio of about 0.39. Obviously, other fill materials from sand to ordinary concrete can also be used.

For the geotextile design, the following points must be considered: sufficient permeability or permittivity must be available to allow the removal of water (standing in forms or from within grout during curing); the geotextile must have a proper opening size to prevent loss of cement and sand; the geotextile and its seams must have adequate strength to prevent rupture; and the geotextile must allow adequate elongation to fulfill the specific need (e.g., columns for mine or cavern stability) without failure. Example 2.31 illustrates a typical application.

Example 2.31

Design a geotextile for pile jacketing in 3 m of water as shown below.

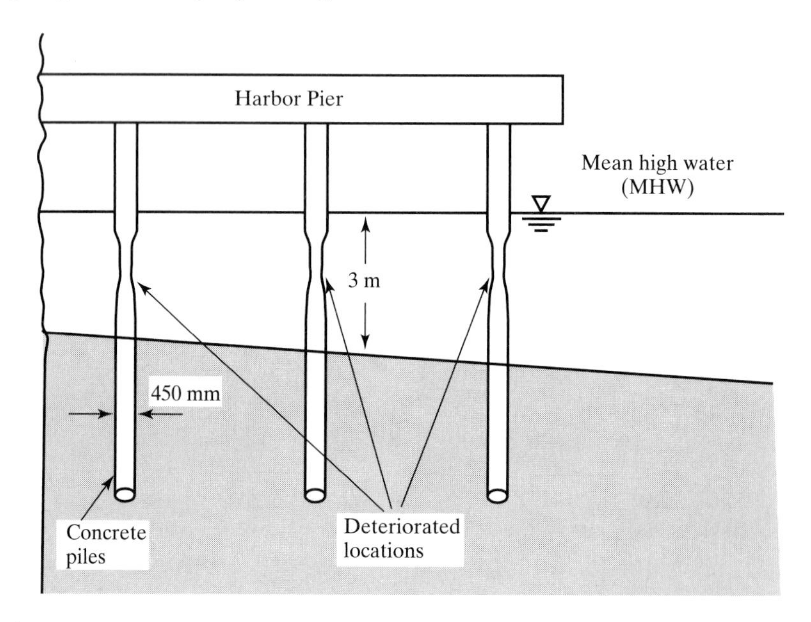

Solution: **(a)** For adequate permeability/permittivity, an estimate of the flow of water out of the geotextile during grouting is necessary. If we use an estimated value of 0.2 m^3/min

$$q = kiA = \frac{k\Delta hA}{t}$$

$$\Psi = \frac{k}{t} = \frac{q}{\Delta h A}$$

$$= \frac{0.2}{(3.0)\pi(0.45)(3.0)(60)}$$

$$= 2.6 \times 10^{-4}\ \text{s}^{-1}$$

Since most geotextiles are in the range of 10^{-1} to 10^{-3} for their actual permittivity, there are many commercial geotextiles available.

(b) For the retention of the grout mix, assume that the d_{50} of sand is 0.4 mm and that it has a $C_U = 4.0$; use

$$0_{95} < \frac{9(d_{50})}{C_U} \quad \text{(it is actually a fluid)}$$

$$< \frac{(9)(0.40)}{4.0}$$

$$< 0.9\ \text{mm}$$

Most geotextiles are in the AOS range of 0.1 to 0.5 mm; thus a wide range is available. If the cement paste is to be retained, its proportions must be estimated. The design is straightforward but will result in a geotextile of much smaller openings.

(c) For strength and elongation, the stress-versus-strain curve of the candidate geotextile(s) must be known to match up with the grouting pressure and required elongation. This is very much a site-specific situation. Note that woven nylon geotextiles with a grab strength of 1000 N and an elongation at failure of 15% are often used for geotextile flexible forms of this type.

Summary. To be sure, the use of geotextiles as flexible-forming systems is an exciting concept simply waiting for new and innovative applications. Contrasted to other geotextile uses, the geotextile is sacrificial in most cases (e.g., when the grout or concrete is placed within the geotextile form). Thus UV degradation is no problem in these cases. For sand-filled bags and tubes, however, it is a definite problem that must be considered.

The design of the geotextile follows nicely along lines of other geotextile systems. Strength and elongation considerations are invariably necessary along with proper filtration; it has multiple functions. Thus the topic is being considered in this section. A good deal of future activity will undoubtedly be seen in this particular application of geotextiles.

2.11 CONSTRUCTION METHODS AND TECHNIQUES USING GEOTEXTILES

2.11.1 Introduction

Although this book is devoted primarily to design issues, it is critically important to consider the constructability of the final design. All too often adequate designs have been negated by the inability to construct them or by the use of improper construction methods. Of course, either situation can be disastrous as far as the final system is concerned.

Construction with geotextiles is not particularly difficult as long as it is remembered that the textile product being dealt with typically has a mass per unit area from 150 to 600 g/m^2—that is, it is not a steel-wire blasting mat! Most building contractors, heavy-construction contractors, land developers, and federal, state, and local construction forces that deal with other types of construction materials are well-equipped for handling geotextiles. In fact, it is most interesting to note the adaptability of these groups in devising new and clever geotextile deployment and installation procedures. Freely available literature by manufacturers is also extremely helpful in this regard. There are, however, certain areas where new and unique techniques are required, and these have been discussed in their specific sections: the reinforcement area (walls and embankments), the reflective-crack prevention area, and flexible-forming systems.

One area that requires constant vigilance is that of UV-light susceptibility. Contractors often fail to recognize that geotextiles can be literally destroyed by exposing them to prolonged sunlight, especially in southern climates. The work of Koerner et al. [57], originally commented on in Section 2.3.6, is reemphasized here to bring attention to this susceptibility. Figures 2.27c,d, the reader will recall, shows the drastic strength and elongation reduction of both polypropylene and polyester fabrics when exposed to sunlight. In some cases more than 50% of the strength is lost within a few months. Clearly, there is the utmost need for the contractor to keep the geotextile in its protective cover for as long as possible and even perhaps to keep it in an enclosure. Once the roll is opened and the geotextile is placed in its final position, it must be backfilled *in a timely manner*. Unused portions of rolls or sampled rolls must be rerolled and suitably protected. The specification must be clear and the inspection rigid in this regard. There are numerous studies available (e.g., Hsuan et al. [158]), as well as agency guidelines. The latter guideline is usually 2 to 4 weeks as the maximum exposure times, unless the geotextile is of a unique type and/or stabilized. When the geotextile is UV-stabilized (generally using carbon black and/or a chemical inhibitor in the polymer mixture) UV susceptibility is greatly reduced, but certainly not eliminated.

2.11.2 Geotextile Survivability

Geotextile survivability refers to the ability of the geotextile to withstand the handling and installation stresses it will receive prior to its being placed in its final position. It is related to construction equipment, construction technique, substrate material, substrate condition, backfill material, backfill size and shape, and so on. Recall the levels of installation damage shown in Figure 2.19 and discussed in Section 2.3.5. Table 2.22 considers these features and rates the general geotextile requirements of survivability as (in increasing order): low, moderate, high, or very high. Depending on, and keyed into, these categories are a set of survivability requirements that are considered to be the minimum geotextile properties for necessary placement in the intended and final position. The numeric values of these survivability properties are given in Table 2.2a. The classes correspond to the requirements in Table 2.22 as follows.

Class 1 (highest values) $\simeq$ very high to high survivability
Class 2 (intermediate values) $\simeq$ high to moderate survivability
Class 3 (lowest) values $\simeq$ moderate to low survivability

TABLE 2.22 REQUIRED DEGREE OF SURVIVABILITY AS A FUNCTION OF SUBGRADE CONDITIONS AND CONSTRUCTION EQUIPMENT*

Subgrade Conditions	Low ground-pressure equipment ($\leq$ 25 kPa)	Medium ground-pressure equipment ($>$ 25 kPa, $\leq$ 50 kPa)	High ground-pressure equipment ($>$ 50 kPa)
Subgrade has been cleared of all obstacles except grass, weeds, leaves, and fine wood debris. Surface is smooth and level so that any shallow depressions and humps do not exceed 450 mm in depth or height. All larger depressions are filled. Alternatively, a smooth working table may be placed.	Low	Moderate	High
Subgrade has been cleared of obstacles larger than small to moderate-sized tree limbs and rocks. Tree trunks and stumps should be removed or covered with a partial working table. Depressions and humps should not exceed 450 mm in depth or height. Larger depressions should be filled.	Moderate	High	Very high
Minimal site preparation is required. Trees may be felled, delimbed, and left in place. Stumps should be cut to project not more than $\pm$150 mm above subgrade. Fabric may be draped directly over the tree trunks, stumps, large depressions and humps, holes, stream channels, and large boulders. Items should be removed only if placing the fabric and cover material over them will distort the finished road surface.	High	Very high	Not recommended

*Recommendations are for 150 to 300 mm initial lift thickness. For other initial lift thicknesses:

300 to 450 mm: reduce survivability requirement one level;
450 to 600 mm: reduce survivability requirement two levels;
$>$ 600 mm: reduce survivability requirement three levels

For special construction techniques such as prerutting, increase the fabric survivability requirement one level. Placement of excessive initial cover material thickness may cause bearing failure of the soft subgrade.

Source: After Christopher, Holtz, and DiMaggio [159].

It should be emphasized that if the values in Table 2.22 exceed those calculated on the basis of functional design (as they sometimes will), the values in the table must be used. Thus calculated design properties do not always prevail.

2.11.3 Cost and Availability Considerations

Of prime importance to all involved with the candidate geotextile is its installed cost. Although this has been noticeably absent in the book because of changing price indices, site and climate variations, type and quantity of geotextile, and so on, a few comments are in order.

The cost of the geotextile itself is reasonably related to its mass per unit area. Heavier geotextiles cost proportionately more than lighter ones. Note, however, that the installation cost may not be significantly higher for the heavier geotextiles. The type of manufacture is also a factor, with woven slit-film types generally being the least expensive, followed by nonwoven heat-bonded and nonwoven needle-punched, and woven monofilament types, which are the most expensive on the basis of an equivalent mass per unit area. These comments, however, should in no way sway a design toward the preference of one geotextile over another. They are offered only to give a feeling for the costs involved. As of this writing these costs ranged from $0.60 to $2.00 per square meter for geotextiles in the range 150 to 500 g/m^2, with installation costs being an additional $0.15 to $0.50 per square meter, depending on the site conditions, quantity involved, and particular application.

It should also be recognized that geotextile availability is sometimes very important. In aggressively marketed areas, many geotextiles are available and the free-market system will sort things out. In more remote areas, however, where only one or two geotextiles are available, design must necessarily reflect this situation. It is totally unrealistic to think that manufacturers will tailor-make a geotextile to a design specification if it involves only a small quantity in a remote area. In a similar vein, labor unions have been known to affect costs, as has patent infringement in certain select areas.

2.11.4 Summary

At the heart of any well-designed facility is its proper and careful attention to details. In my personal investigations, geotextile-related failures fall into the following groups:

Construction related. This is by far the largest group, with excessive ultraviolet degradation, installation damage, poorly constructed seams, and lack of intimate contact being the major problems. This last issue is particularly important for filtration and erosion-control applications.

Design/specification related. Design failures, per se, have been relatively few. Some walls and embankments on soft soils have had excessive deformations, but they have generally been repaired on site, without insurance or litigation problems. Specifications, however, have often been lax and in many instances contributed to some of the construction-related failures mentioned.

Testing related. There have been many testing-related problems, but no serious failures to my knowledge. The major concern in this regard is the use of literature values for interface shear strengths, instead of properly simulated direct-shear tests. This practice must cease before major problems arise. In spite of the concern over the lack of true performance tests, conservative reduction factors probably compensate for the shortcoming. In this regard, a designer should never apologize for using high reduction factors or global factors of safety.

Product related. Other than supplying the wrong product, and then accepting and installing it, product failures are sparse. The use of the MARV concept is certainly warranted and provides a needed safeguard against product variability. It also challenges the manufacturer to decrease product variability to the minimum possible.

Overall, the situation vis-a-vis geotextile failures is good. We certainly have an engineering material worthy of being considered with other materials conventionally used by civil engineers and related professions. Field performance to date, after more than 20 years of service life in a multitude of applications, has been excellent.

REFERENCES

1. Broms, B. B., "Triaxial Tests with Fabric-Reinforced Soil," *C. R. Coll. Inst. Soils Text.* (Paris), Vol 3, 1977, pp. 129–133.
2. Taylor, D. W., *Fundamentals of Soil Mechanics,* New York: John Wiley and Sons, 1948.
3. Christopher, B. R. and Fischer, G. R., "Geotextile Filtration Principles, Practices and Problems," *J. Geotextiles and Geomembranes,* Vol. 11, Nos. 4–6, 1992, pp. 337–354.
4. *Report on Task Force 25,* Joint Committee Report of AASHTO-AGC-ARTBA, American Association of State, Highway, and Transportation Officals, Washington, DC, Jan. 1991.
5. Carroll, R. G., Jr., "Geotextile Filter Criteria," *Engineering Fabrics in Transportation Construction,* TRR 916, TBR, Washington, DC, 1983, pp. 46–53.
6. Luettich, S. M., Giroud, J. P., and Bachus, R. C., "Geotextile Filter Design Guide," *J. Geotextiles and Geomembranes,* Vol. 11, No. 4–6, 1992, pp. 19–34.
7. Haliburton, T. A., and Wood, P. D., "Evaluation of U.S. Army Corps of Engineers Gradient Ratio Test for Geotextile Performance," *Proc. 2d Intl. Conf. on Geotextiles,* St. Paul, MN: IFAI, 1982, pp. 97–101.
8. Halse, Y., Koerner, R. M., and Lord, A. E., Jr., "Filtration Properties of Geotextiles under Long Term Testing," *Proc. ASCE/PennDOT Conf. Advances in Geotechnical Eng.,* Harrisburg, PA: PennDOT, 1987, pp. 1–13.
9. Williams, N. D., and Abouzakhm, M. A., "Evaluation of Geotextile/Soil Filtration Characteristics Using the Hydraulic Conductivity Ratio Analysis," *J. Geotextiles and Geomembranes,* Vol. 8, No. 1, 1989, pp. 1–26.
10. Koerner, G. R., Koerner, R. M., and Martin, J. P., "Design of Landfill Leachate Collection Systems," *J. Geotechnical Eng.,* Vol. 120, No. 10, 1994, pp. 1792–1803.
11. McGown, A., "The Properties of Nonwoven Fabrics Presently Identified as Being Important in Public Works Applications," *Index 78 Programme,* University of Strathclyde, Glasgow, Scotland, 1978.
12. Heerten, G., "A Contribution to the Improvement of Dimensioning Analogies for Grain Filters and Geotextile Filters," *Proc. Intl. Conf. Filters, Filtration and Related Phenomena,* Karlsruhe Technical University, Karlsruhe, Germany, 1992, pp. 110–122.
13. Giroud, J. P., "Granular Filters and Geotextile Filters," *Proc. Geofilters '96,* ed. J. LaFleur and A. Rollin, Ecole Polytechnique, Montreal, Canada, 1996, pp. 565–680.
14. Gerry, G. S., and Raymond, G. P., "The In-Plane Permeability of Geotextiles," *Geotech. Test. J.,* Vol. 6, No. 4, 1963, pp. 181–189.
15. "Textiles: Yarns, Fabrics and General Test Methods," *Annual Book of Standards,* Parts 31 and 32, Philadelphia, PA: ASTM.
16. Kaswell, E. R., *Handbook of Industrial Textiles,* New York: West Point Peperell, 1963.
17. Booth, J. E., *Principles of Textile Testing,* London: Newnes-Butterworths, 1968.
18. Morton, W. E., and Heart, J. W., *Physical Properties of Textile Fibers,* New York: John Wiley and Sons, 1975.

19. van den Berg, C. and Myles, B., eds., "Geotextile Testing Inventory," Brussels: IGS Secretariat, 1986.
20. Haliburton, T. A., Fowler, J., and Langan, J. P., "Design and Construction of a Fabric Reinforced Test Section at Pinto Pass, Mobile, Alabama," *Trans. Res. Record #79,* Washington, D.C., 1980.
21. Myles, B., and Carswell, I., "Tensile Testing of Geotextiles," *Proc. 3d Intl. Conf. Geotextiles,* Vienna: Austrian Engineering Society, 1986, pp. 713–718.
22. McGown, A., Andrawes, K. Z., and Kabir, M. H., "Load-Extension Testing of Geotextiles Confined in Soil," *Proc. 2d Intl. Conf. Geotextiles,* IFAI, 1982, pp. 793–796.
23. McGown, A., Andrawes, K. Z., and Murray, R. T., "The Load-Strain-Time-Temperature Behavior of Geotextiles and Geogrids," *Proc. 3d Intl. Conf. Geotextiles,* Vienna: Austrian Engineering Society, 1986, pp. 707–712.
24. Wilson-Fahmy, R. F., Koerner, R. M., and Fleck, J. A., "Unconfined and Confined Wide Width Testing of Geosynthetics Used in Reinforcement Applications," in *Geosynthetic Soil Reinforcement Testing Procedures,* ed. S. C. J. Cheng, ASTM STP 1190, Philadelphia, PA: ASTM, 1993, pp. 49–63.
25. Wayne, M. H., Carey, J. E., and Koerner, R. M., "Epoxy Bonding of Geotextiles," *J. Geotextiles and Geomemranes.,* Vol. 9, No. 4–6, 1990, pp. 559–564.
26. Ashmawy, A. K., and Bourdeau, P. L., "Geosynthetic Reinforced Soils Under Repeated Loadings: A Review and Comparative Study," *Geosynthetics Intl.,* Vol. 2, No. 4, 1995, pp. 643–678.
27. Raumann, G., "A Hydraulic Tensile Test with Zero Transverse Strain for Geotechnical Fabrics," *Geotechnical Testing J.,* Vol. 2, No. 2, 1979, pp. 69–76.
28. Alfheim, S. L., and Sorlie, A., "Testing and Classification of Fabrics for Application in Road Construction," *C. R. Coll. Intl. Soils Textiles,* Vol. 2 (Paris), 1977, pp. 333–338.
29. Koerner, R. M., Monteleone, M. J., Schmidt, R. K., and Roethe, A. T., "Puncture and Impact Resistances of Geosynthetics," *Proc. 3d Intl. Conf. Geotextiles,* Vienna: Austrian Engineering Society, 1986, pp. 677–682.
30. Murphy, V. P., and Koerner, R. M., "CBR Strength (Puncture) of Geosynthetics," *J. Geotechnical Testing J.,* Vol. 11, No. 3, 1988, pp. 167–172.
31. Cazzuffi, D., and Venesia, S., "The Mechanical Properties of Geotextiles: Italian Standard and Interlaboratory Test Comparison," *Proc. 3d Intl. Conf. Geotextiles,* Vienna: Austrian Engineering Society, 1986, pp. 695–700.
32. Ingold, T. S., "Some Observations on the Laboratory Measurement of Soil-Geotextile Bond," *Geotechnical Testing J.,* Vol. 5, No. 3–4, 1982, pp. 57–67.
33. Martin, J. P., Koerner, R. M., and Whitty, J. E., "Experimental Friction Evaluation of Slippage between Geomembranes, Geotextiles and Soils," *Proc. Intl. Conf. Geomembranes,* St. Paul, MN: IFAI, 1984, pp. 191–196.
34. Richards, E. A. and Scott, J. D., "Soil Geotextile Frictional Properties," *2d Canadian Symp. Geotextiles and Geomembranes,* Ottawa, Canada: Canadian Geotech. Soc., 1985, pp. 13–24.
35. Williams, N. D., and Houlihan, M. F., "Evaluation of Interface Friction Properties Between Geosynthetics and Soil," *Proc. Geosynthetics '87,* St. Paul, MN: IFAI, 1987, pp. 616–627.
36. Collios, A., Delmas, P., Gourc, J. P., and Giroud, J. P., "Experiments of Soil Reinforcement with Geotextiles," *Proc. Symp. Use of Geotextiles for Soil Improvement,* New York: ASCE, 1980, pp. 53–73.
37. Gourc, J. P., Faure, Y., Rollin, A., and LeFleur, J., "Standard Tests of Permittivity and Application of Darcy's Formula," *Proc. 2d Intl. Conf. Geotextiles,* Vol. 1, St. Paul, MN: IFAI, 1982, pp. 149–154.
38. Holtz, R. D., *Mercury Intrusion Characterization of Geotextile Pore Size Distribution,* Report to Hoechst-Celanese Corp., Spartanburg, SC, 1988.

39. Bhatia, S. K., Smith, J. L., and Christopher, B. R., "Geotextile Characterization and Pore Size Distribution: Part III. Comparison of Methods and Applications to Design," *Geosynthetics Intl.*, 1996, Vol. 3, No. 3, pp. 301–328.
40. ICI Fibres, Ltd., "Designing with Terram," design brochure, Gwent, U.K.
41. McGown, A. W., "The Properties and Uses of Permeable Fabric Membranes," *Residential Workshop on Materials and Methods for Low Cost Roads and Reclamation Works*, Leura, Australia, 1976, pp. 663–710.
42. Hausmann, M., *Engineering Principles of Ground Engineering*, New York: McGraw-Hill, 1991.
43. Koerner, R. M., and Bove, J. A., "In-Plane Hydraulic Properties of Geotextiles," *Geotechnical Testing J.*, Vol. 6, No. 4, 1983, pp. 190–195.
44. Koerner, R. M., Bove, J. A., and Martin, J. P., "Water and Air Transmissivity of Geotextiles," *J. Geotextiles and Geomembranes*, Vol. 1, No. 1, 1984, pp. 57–73.
45. Koerner, G. R., and Koerner, R. M., "The Installation Survivability of Geotextiles and Geogrids," *Proc. 4th IGS Conference on Geotextiles, Geomembranes and Related Products*, Rotterdam: A. A. Balkema, 1990, pp. 597–602.
46. Shrestha, S. C., and Bell, J. R., "Creep Behavior of Geotextiles Under Sustained Loads," *Proc. 2d Intl. Conf. Geotextiles*, St. Paul, MN: IFAI, 1982, pp. 769–774.
47. den Hoedt, G., "Creep and Relaxation of Geotextile Fabrics," *J. Geotextiles and Geomembranes*, Vol. 4, No. 2, 1986, pp. 83–92.
48. Lawson, C. R., "Geosynthetics in Soil Reinforcement," *Proc. Symp. on Geotextiles in Civil Engineering*, Institution of Engineers Australia, Newcastle, 1986, pp. 1–35.
49. Task Force #27, "Guidelines for the Design of Mechanically Stabilized Earth Walls," AASHTO-AGC-ARTBA Joint Committee, Washington, DC, 1991.
50. Ingold, T. S., Montanelli, F., and Rimoldi, P., "Extrapolation Techniques for Long Term Strengths of Polymeric Geogrids," *Proc. 5th IGS Conf.*, Singapore: Southeast Asia Chapter, IGS, 1994, pp. 1117–1120.
51. Koerner, R.M., Hsuan, Y., and Lord, A. E., Jr., "Remaining Technical Barriers to Obtain General Acceptance of Geosynthetics," 1992 Mercer Lecture, *J. Geotextiles and Geomembranes*, Vol. 12, No. 1, 1993, pp. 1–52.
52. Koerner, R. M., and Ko, F. K., "Laboratory Studies on Long-Term Drainage Capability of Geotextiles," *Proc. 2d Intl. Conf. Geotextiles*, St. Paul, MN: IFAI, 1982, pp. 91–95.
53. Wayne, M. H., and Koerner, R. M., "Correlation Between Long Term Flow Testing and Current Geotextile Filtration Design Practice," *Proc. Geosynthetics '93*, St. Paul, MN: IFAI, 1993, pp. 501–517.
54. Williams, N. D., and Abouzakhm, M. A., "Evaluation of Geotextile/Soil Filtration Characteristics Using the Hydraulic Conductivity Ratio Analysis," *J. Geotextiles and Geomembranes*, Vol. 8, No. 1, 1989, pp. 1–26.
55. Luettich, S. M., and Williams, N. D., "Design of Vertical Drains Using the Hydraulic Conductivity Ratio Analysis," *Proc. Geosynthetics '89*, St. Paul, MN: IFAI, 1989, pp. 95–103.
56. Van Zanten, R. V. ed., *Geotextiles and Geomembranes in Civil Engineering*, Rotterdam: A. A. Balkema, 1986.
57. Koerner, G. R., Hsuan, Y., and Koerner, R. M., "Outdoor vs. Accelerated Weathering of Various Geotextiles," *J. Geotechnical Eng. Div., ASCE*, under review.
58. Hsuan, Y. G., Koerner, R. M., and Lord, A. E., Jr., "A Review of the Degradation of Geosynthetic Reinforcement of Materials and Various Polymer Stabilization Methods," ASTM STP 1190, ed. S.C. J. Cheng, Philadelphia, PA: ASTM, 1993, pp. 228–244.
59. Halse, Y., Koerner, R. M., and Lord, A. E., Jr., "Effect of High Alkalinity Levels on Geotextiles. Part I, $Ca(OH)_2$ Solutions," *J. Geotextiles and Geomembranes*, Vol. 5 No. 4, 1987, pp. 261–282.

60. Halse, Y., Koerner, R. M., and Lord, A. E., Jr., "Effect of High Alkalinity Levels on Geotextiles. Part II, NaOH Solution," *J. Geotextiles and Geomembranes,* Vol. 6, No. 4, 1987, pp. 295–305.
61. Ionescu, A. et al., "Methods Used for Testing the Bio-Colmatation and Degradation of Geotextiles Manufactured in Romania," *Proc. 2d Intl. Conf. Geotextiles,* St. Paul, MN: IFAI, 1982, pp. 547–552.
62. Gourc, J.-P., and Faure, Y.-H., "Soil Particles, Water and Fibers—A Fruitful Interaction Now Controlled," *Proc. 4th IGS Conf.,* Rotterdam: A. A. Balkema, 1990, pp. 949–972.
63. Voskamp, W., and Risseeuw, P., "Method to Establish the Maximum Allowable Load under Working Conditions of Polyester Reinforcing Fabrics," *J. Geotextiles and Geomembranes,* Vol. 6, Nos. 1–3, 1987, pp. 173–184.
64. Giroud, J. P., "Designing with Geotextiles," *Mater. Const.* (Paris), Vol. 14, No. 82, 1981, pp. 257–272; *Geotextiles and Geomembranes, Definitions, Properties and Designs,* St. Paul, MN: IFAI, 1984.
65. Koerner, G. R., "Long-Term Benefit Cost Performance and Analysis of Geotextile Separators in Pavement Systems," *Proc. Geosynthetics '97,* IFAI, pp. 701–714.
66. Hausmann, M. R., "Fabric Reinforced Unpaved Road Design Methods—Parametric Studies," *Proc. 3d Intl. Conf. Geotext.,* IFAI, 1986, pp. 19–24.
67. Giroud, J. P., and Noiray, L., "Design of Geotextile Reinforced Unpaved Roads," *J. Geotechnical Eng. Div., ASCE,* Vol. 107, No. GT9, 1981, pp. 1233–1254.
68. Barenberg, E. J., and Bender, D. A., "Design and Behavior of Soil-Fabric-Aggregate Systems," paper presented at 57th Annual Meeting, TRB, Washington, D.C., 1978.
69. Holtz, R. D., and Sivakugan, K. "Design Charts for Roads with Geotextiles," *J. Geotextiles and Geomembranes,* Vol. 5, No. 3, 1987, pp. 191–200.
70. Diaz, V., "Thread Selector for Geotextiles," *Geotechnical Fabrics Rpt.*, Vol. 3, No. 1, 1985, pp. 15–19.
71. *Field Seaming of Geotextiles,* St. Paul, MN: IFAI, 1989.
72. Proksch, H., "German Experiences with the Replacement of Granular Frost Blankets by Other Types of Construction," *Proc. 3d Intl. Conf. Structural Design of Asphalt Pavements,* Vol. 1, London: Inst. Civil Engineers, 1972, pp. 115–125.
73. Sayward, J. M., *Evaluation of MESL Membrane—Puncture, Stiffness, Temperature, Solvents,* CRREL Rep. 76-22, Hanover, NH: U.S. Army Corps of Engineers, 1976.
74. Bell, J. R., and Yoder, E. J., "Plastic Moisture Barrier for Highway Subgrade Protection," *Proc. Highway Research Board,* Vol. 36, Washington, D.C., 1957, pp. 713–735.
75. Smith, N., "Techniques for Using MESL in Roads and Airfields in Cold Regions," *ASCE Conf. Cold Regions Tech.,* New York: ASCE, 1978.
76. Smith, N., and Pazsint, D. A., *Field Test of a MESL Road Section in Central Alaska,* Tech. Rpt. 260, CRREL, Hanover, NH: U. S. Army Corps of Engineers, 1975.
77. Lawson, C. R., and Ingles, O. G., "Long-Term Performance of MESL Road Sections in Australia," *Proc. 2d Intl. Conf. Geotextiles,* Vol. 2, St. Paul, MN: IFAI, 1982, pp. 535–539.
78. Meader, A. L., Jr., "Construction of Geomembranes in Place by Spraying an Elastomer over a Geotextile," *Proc. 1st Intl. Conf. Geomembranes,* Vol. 2, St. Paul, MN: IFAI, 1984, pp. 389–393.
79. Murray, C. D., "Simulation Testing of Geotextile Membranes for Reflection Cracking," *Proc. 2d Intl. Conf. Geotextiles,* Vol. 2, St. Paul, MN: IFAI, 1982, pp. 511–516.
80. Koerner, R. M., and Welsh, J. P., *Construction and Geotechnical Engineering Using Synthetic Fabrics,* New York: John Wiley and Sons, 1980.
81. Lord, A. E., Jr., and Koerner, R. M., "Fundamental Aspects of Chemical Degradation of Geomembranes," *Proc. 1st Intl. Conf. Geomembranes,* Vol. 2, St. Paul, MN: IFAI, 1984, pp. 293–298.

82. Lawson, C. R., and Ingles, O. G., "Water Repellency Requirements for Geomembranes," *Proc. 1st Intl. Conf. Geomembranes,* Vol. 2, St. Paul, MN: IFAI, 1984, pp. 169–173.
83. Hoffman, G. L., and Shamon, M. E., "Premature Failure of Permeable Subbase Pavement Sections Incorporating Geotextiles," Paper 6, PennDOT, Harrisburg, PA, 1984.
84. Broms, B. B., "Design of Fabric Reinforced Retaining Structures," *Proc. Symp. Earth Reinforcement,* Pittsburgh, PA: ASCE, 1978, pp. 282–303.
85. Steward, J. E., Williamson, R., and Mohney, J., "Earth Reinforcement," in *Guidelines for Use of Fabrics in Construction and Maintenance of Low Volume Roads,* U.S. Forest Service, Portland, OR, 1977.
86. Whitcomb, W., and Bell, J. R., "Analysis Techniques for Low Reinforced Soil Retaining Walls," *Proc. 17th Eng. Geol. Soils Eng. Symp.,* Moscow, ID: Idaho DOT, 1979, pp. 35–62.
87. Lee, K. L., Adams, B. D., and Vagneron, J. M. J., "Reinforced Earth Retaining Walls," *J. Soil Mech. Fdtn. Eng. Div.,* Vol. 99, No. SM10, 1973, pp. 745–764.
88. Bell, J. R., Stilley, A. N., and Vandre, B., "Fabric Retained Earth Walls," *Proc. 13th Eng. Geol. Soils Eng. Symp.,* Moscow, ID: Idaho DOT, 1975, pp. 46–57.
89. Jarrett, P. W., and McGown, A., eds., *Proc. Appl. of Polymeric Reinforcement in Soil Retaining Structures,* Royal Military College, Kingston, Ontario, 1988.
90. *NAVFAC, DM-7.2,* Bureau of Yards and Docks, U. S. Navy, Apr. 1982.
91. Broms, B. B., "Polyester Fabric as Reinforcement in Soil," *C. R. Coll. Intl. Soil Textiles,* Vol. 1 (Paris), 1977, pp. 129–135.
92. Koerner, R. M., *Construction and Geotechnical Methods in Foundation Engineering.* New York: McGraw-Hill, 1984.
93. Barrett, R. K., "Geotextiles in Earth Reinforcement," *Geotechnical Fabrics Rpt.,* Vol. 3, No. 2, 1985, pp. 15–19.
94. Allen, T. M., Christopher, B. R., and Holtz, R. D., "Performance of a 12.6 m High Geotextile Wall in Seattle, Washington," *Proc. Geosynthetic Reinforced Retaining Walls,* ed. J. T. H. Wu, Rotterdam: A. A. Balkema, 1992, pp. 81–100.
95. Stevens, J. B., and Souiedan, B., "Geotextile Wall Aids Bridge Construction," *Geotechnical Fabrics Rpt.,* Vol. 8, No. 3, 1990, pp. 10–15.
96. *Geotextile Design and Construction Guidelines,* Federal Highway Administration, FHWA-HI-90-001, Washington, D.C., 1989.
97. Richardson, G. N., and Behr, L. H., Jr., "Geotextile-Reinforced Wall: Failure and Remedy," *Geotechnical Fabric Rpt.,* Vol. 6, No. 4, 1988, pp. 14–18.
98. Bathurst, R. J., and Simac, M. R., "Geosynthetic Reinforced Segmental Retaining Wall Structures in North American, *Proc. 5th IGS Conf.,* Keynote Lecture Vol., Singapore: Southeast Asia Chapter, IGS, 1994, pp. 29–54.
99. Simac, M. R., Bathurst, R. J., Berg, R. R., and Lothsperch, S. E., *Design Manual for Segmental Retaining Walls (Modular Concrete Block Retaining Wall Systems),* Herndon, VA: National Concrete Masonry Association, 1993.
100. Koerner, G. R., Koerner, R. M., and Elias, V., "Geosynthetic Installation Damage under Two Different Backfill Conditions," *Geosynthetic Soil Reinforcement Testing Procedures,* ASTM STP 1190, ed. S. C. J. Cheng, Philadelphia, PA: ASTM, 1993, pp. 163–184.
101. Kaniraj, S. R., "Direction and Magnitude of Reinforcement Force in Embankments on Soft Soil," *Earth Reinforcement,* ed. H. Ochiai, N. Yasufuku, and K. Omine, Rotterdam: A. A. Balkema, 1996, pp. 221–225.
102. Andrawes, K. Z., McGown, A., Wilson-Fahmy, R. F., and Mashhour, M. M., "The Finite Element Method of Analysis Applied to Soil-Geotextile Systems," *Proc. 2d Intl. Conf. Geotextiles,* Vol. 2, St. Paul, MN: IFAI, 1982, pp. 690–700.
103. Rowe, R. K., "Reinforced Embankments: Analysis and Design," *J. Geotechnical Eng. Div., ASCE,* Vol. 110, No. 2, 1984, pp. 231–246.

104. Rowe, R. K., and Soderman, K. L., "Comparison of Predicted and Observed Behavior of Two Test Embankments," *Intl. J. Geotextiles and Geomembranes,* Vol. 1, 1984, pp. 143–160.
105. Rowe, R. K., and Soderman, K. L., "The Role of Finite Element Analyses in Soft Soil Stabilization Using Geosynthetics," *J. Geotextiles and Geomembranes,* Vol. 6, No. 1–3, 1987, pp. 53–80.
106. Haliburton, T. A., Fowler, J., and Langan, J. P., *Design and Construction of a Fabric Reinforced Test Section at Pinto Pass, Mobile, Alabama,* TRR 79, Washington, D.C., 1980.
107. Fowler, J., "Theoretical Design Considerations for Fabric Reinforced Embankments," *Proc. 2d Intl. Conf. Geotextiles,* St. Paul, MN: IFAI, 1982, pp. 665–676.
108. Sprague, C. J., and Koutsourais, M., "The Evolution of Geotextile Reinforced Embankments," Geotechnical Special Pub. 30, ed. R. H. Borden, R. D. Holtz, and I. Juran, New York: ASCE, 1992, pp. 1129–1141.
109. Koerner, R. M., Hwu, B-L., Wayne, M. H., "Soft Soil Stabilization Designs Using Geosynthetics," *J. Geotextiles and Geomembranes,* Vol. 6, No. 1–3, 1987, pp. 33–52.
110. Koerner, R. M., and Uibel, B. L., "Hydraulic Fill Embankments Utilizing Geosynthetics," *Proc. ASCE Conf. on Hydraulic Fill Structures '88,* New York: ASCE, 1988.
111. Koerner, R. M., Fowler, J., and Lawrence, C. A., *Soft Soil Stabilization Study for Wilmington Harbor South Dredge Material Disposal Area,* U.S. Army Engineer Waterways Experiment Station, Misc. Paper GL-86-38, 1986.
112. Humphrey, D. N., "Discussion 'Current Design Methods,'" *J. Geotextiles and Geomembranes,* Vol. 6, No. 1–3, 1987, pp. 89–92.
113. Guglielmetti, J. L., Koerner, G. R., and Battino, F. S., "Geotextile Reinforcement of Soft Landfill Process Sludge to Facilitate Final Closure," *J. Geotextiles and Geomembranes,* Vol. 14, Nos. 7–8, 1996, pp. 377–392.
114. Binquet, J., and Lee, K. L., "Bearing Capacity Tests in Reinforced Earth Slabs," *J. Geotechnical Eng. Div., ASCE,* Vol. 101, No. GT12, 1975, pp. 1241–1255.
115. Binquet, J., and Lee, K. L., "Bearing Capacity Analysis of Reinforced Earth Slabs," *J. Geotechnical Eng. Div., ASCE,* Vol. 101, No. GT12, 1975, pp. 1257–1276.
116. Guido, V. A., Biesiadecki, G. L., and Sullivan, M. J., "Bearing Capacity of a Geotextile Reinforced Foundation," *Proc. 11th ISSMFE,* Vol. 3, New York: ASCE, 1985, pp. 1777–1780.
117. National Resource Council, *Reducing Losses from Landsliding in the United States,* Committee on Ground Failure Hazards, Washington, D.C., National Academic Press, 1985.
118. Shen, C. K., Bang, S., and Herrman, L. R., "Ground Movement Analysis of Earth Support System," *J. Geotechnical Eng. Div., ASCE,* Vol. 107, No. GT12, 1981, pp. 1609–1624.
119. Koerner, R. M., "Slope Stabilization Using Anchored Geotextiles: Anchored Spider Netting," *Proc. Spec. Geotech. Eng. for Roads and Bridges Conf.,* Harrisburg, PA: PennDOT, 1984, pp. 1–11.
120. Koerner, R. M., "In-Situ Soil Slope Stabilization Using Anchored Nets," *Proc. Conf. Low Cost and Energy Saving Construction Materials,* Bethlehem, PA: Envo Pub., 1984, pp. 465–478.
121. Myles, B., and Bridle, R. J., "Fired Soil Nails," *Proc. Earth Reinforcement Practice,* ed. Ochiai, Hayashi and Otani, Rotterdam: A. A. Balkema, 1992, pp. 509–514.
122. Lambe, T. W., and Whitman, R. V., *Soil Mechanics,* New York: John Wiley and Sons, 1969.
123. Koerner, R. M., and Robins, J. C., "In-Situ Stabilization of Soil Slopes Using Nailed Geosynthetics," *Proc. 3d Conf. Geotextiles,* Vienna: Austrian Engineering Society, 1986, pp. 395–399.
124. Ghiassian, H., Hryciw, R. D., and Gray, D.H., "Laboratory Testing Apparatus for Slopes Stabilized by Anchored Geosynthetics," *Geotechnical Testing J.,* Vol. 19, No. 1, 1996, pp. 65–73.
125. Ghiassian, H., Gray, D. H., and Hryciw, R. D., "Seepage Considerations and Stability of Sandy Slopes Reinforced by Anchored Geosynthetics," *Proc. Geosynthetics '97,* St. Paul, MN: IFAI, 1997, pp. 581–593.
126. Cedegren, H. R., *Seepage, Drainage and Flow Nets,* New York: John Wiley and Sons, 1967.

127. Bell, J. R., and Hicks, R. G., *Evaluation of Test Methods and Use Criteria for Geotechnical Fabrics in Highway Applications,* Final Report, FHwA, Contract No. DOT-FH-119353, Oregon State Univ., 1984.
128. Richardson, G. R., and Middlebrooks, P., "A Simplified Design Method for Silt Fences," *Geosynthetics '91 Conf.,* St. Paul, MN: IFAI, 1991, pp. 879–885.
129. Roth, H., "Filter Fabric for Improving Frost Susceptible Soils," *Proc. Int. Conf. Use of Fabrics in Geotechnics,* Vol. 1, Paris: Association Amicale des Ingénieurs, 1977, pp. 23–28.
130. Anderson, O., "The Use of Plastic Fabric for Pavement Protection During Frost Break," *Proc. Intl. Conf. Use of Fabrics in Geotechnics,* Vol. 1, Paris: Association Amicale des Ingénieurs, 1977, pp. 143–149.
131. Gamski, K., and Rigo, J. M., "Geotextile Soil Drainage in Siphon or Siphon-Capillarity Conditions," *Proc. 2d Intl. Conf. Geotextiles,* Vol. 1, St. Paul, MN: IFAI, 1982, pp. 145–452.
132. Clough, I. R., and French, W. J., "Laboratory and Field Work Relating to the Use of Geotextiles in Arid Regions," *Proc. 2d Intl. Conf. Geotextiles,* St. Paul, MN: IFAI, 1982, pp. 447–452.
133. Button, J. W., Epps, J. A., Lytton, R. L., and Harmon, W. S., "Fabric Interlayer for Pavement Overlays," *Proc. 2d Intl. Conf. Geotextiles,* St. Paul, MN: IFAI, 1982, pp. 523–528.
134. Majidzadeh, K., Luther, M. S., and Skylut, H., "A Mechanistic Design Procedure for Fabric-Reinforced Pavement Systems," *Proc. 2d Intl. Conf. Geotextiles,* St. Paul, MN: IFAI, 1982, pp. 529–534.
135. The Asphalt Institute, *Asphalt Overlays and Pavement Rehabilitation,* Manual Series No. 17 (MS-17), College Park, MD, 1977.
136. Cedergren, H. R., "Seepage, Drainage and Flow Nets," New York: John Wiley & Sons, 1989.
137. Bell, J. R., Jr., "Designing with Geosynthetics," unpublished course notes of R. M. Koerner and J. R. Bell, 1983–1985.
138. Personal communication, Maine Department of Transportation, 1994.
139. Ahlrich, R. C., "Evaluation of Asphalt Rubber and Engineering Fabrics as Pavement Interlayers," GL-86-34, Vicksburg, MS: U.S. Army Corps of Engineers, 1986.
140. *Reflective Cracking in Pavements,* Conference proceedings, Paris: RILEM Intl. Union of Testing and Research Laboratories for Materials and Structures, 1989, 1993, and 1996.
141. Newby, J. E., "Southern Pacific Transportation Co. Utilization of Geotextiles in Railroad Subgrade," *Proc. 2d Intl. Conf. Geotextiles,* St. Paul, MN: IFAI, 1982, pp. 467–472.
142. Einsenmann, J., and Leykauf, G., "Investigation of a Nonwoven Fabric Membrane in Railway Track Construction," *Proc. Intl. Conf. Use of Fabrics in Geotechnics,* Paris: Association Amicale des Ingénieurs, 1977, pp. 41–45.
143. Saxena, S. K., and Wang, S., "Model Test of a Rail-Ballast-Fabric-Soil System," *Proc. 2d Intl. Conf. Geotextiles,* St. Paul, MN: IFAI, 1982, pp. 495–500.
144. Bosserman, B., "Reviewing Geotextiles at FAST," *Railroad Track and Structures,* June 1981, pp. 42–58.
145. Raymond, G., "Geotextiles for Railroad Bed Rehabilitation," *Proc. 2d Intl. Conf. Geotextiles,* St. Paul, MN: IFAI, 1982, pp. 479–484.
146. Chrismer, S. M., and Richardson, G. R., "In-Track Performance of Geotextiles at Caldwell, Texas," Transportation Research Board 1071, TRB, Washington, D.C., 1986, pp. 72–80.
147. Terzaghi, K., and Lacroix, Y., "Mission Dam: An Earth and Rockfill Dam on a Highly Compressible Foundation," *Geotechnique,* Vol. 14, No. 1, 1964, pp. 13–50.
148. Zirbel, R., "Sand Filled Tubes Used in Beach Protection Plan," *World Dredging Marine Constr.,* Dec. 1975, pp. 24–29.
149. Ockels, R., "Innovative Hydraulic Engineering with Geosynthetics," *Geosynthetics World,* Vol. 2, No. 4, 1991, pp. 26–27.
150. Fowler, J., "Geotubes and Geocontainers for Hydraulic Applications," *Proc. Cleveland Section ASCE,* Cleveland, OH: ASCE, 1995.

151. den Adel, H., Henrickse, C. S. H., and Pilarczyk, K. W., "Design and Application of Geotubes and Geocontainers," *Proc. Geosynthetics: Application, Design and Construction,* ed. M. B. deGroot, G. den Hoedt, and R. J. Termaat, Rotterdam: A. A. Balkema, 1996, pp. 925–932.
152. Pilarczyk, K. W., "Application and Design Aspedts of Geocontainers," *Proc. Geosynthetics '97,* St. Paul, MN: IFAI, pp. 147–160.
153. Leshchinsky, D., and Leshchinsky, O., "Geosynthetic Confined Pressurized Slurry (GeoCoPS)," Tech. Rpt., CPAR-GL-96-1, Washington, D.C.: U.S. Army Corps of Engineers, 1996.
154. Escalante, E. M., and Iverson, R. J., "Corrosion of Steel Piles," *Mater. Perform.,* Vol. 17, No. 10, 1978, pp. 9–15.
155. National Academy of Engineering, *Scour at Bridge Waterways,* NCHRP Publ. 5, Highway Research Board, Washington, D. C., 1970.
156. Welsh, J. P., "PennDOT Uses a New Method for Solving Scour Problems beneath Bridge Piers," *Highway Focus,* Vol. 9, No. 1, 1977, pp. 72–81.
157. den Hoedt, G., and Mouw, K. A. G., "The Application of High Strength Woven Fabrics in Hydraulic Engineering Construction," *Proc. 7th Intl. Congr. Koninklyke Vlaamse Ingenieursvereniging,* Vol. 2, Rotterdam: V. Z. W., May 1978, pp. 11–19.
158. Hsuan, Y. G., Koerner, R. M., and Soong, Y.-T., "Behavior of Partially Ultraviolet Degraded Geotextiles," *Proc. 5th IGS Conf.,* Singapore: Southeast Asia Chapter, IGS, 1994, pp. 1209–1212.
159. Christopher, B. R., Holtz, R. D., and DiMaggio, J. A., *Geotextile Engineering Manual,* U.S. DOT, FHWA Contract No. DTFH 61-80-C-00094, Washington, D.C., 1984.

PROBLEMS

2.1. Assume that a shopping center site developer has \$15,000 available to purchase a geotextile to be used as a separator between subgrade soil and a stone base, and the total area to be covered is 2.5 ha, which of the following would the developer probably use?
- **(a)** 165 g/m^2 fabric at \$0.75/m^2
- **(b)** 200 g/m^2 fabric at \$0.85/m^2
- **(c)** 250 g/m^2 fabric at \$0.95/m^2

2.2. Given available funds of \$7500, how many linear meters of pipe underdrain trench could be covered with geotextile if it is 2.5 m wide and costs \$0.95/m^2 in place?

2.3. Using the Pennsylvania Department of Transportation specification given in Table 2.1,
- **(a)** for erosion-control geotextiles, when do you use Type A versus Type B?
- **(b)** for sediment-control geotextiles when do you use Type A versus Type B?

2.4. Assuming that the values listed in the specifications of Table 2.1 are MARV and that one standard deviation of a particular geotextile is 8%, convert them to:
- **(a)** average (mean) roll values
- **(b)** minimum values at $\bar{x} - 3s$

2.5. Regarding the statistics involved in the discussion of Figure 2.1:
- **(a)** Define mean and standard deviation.
- **(b)** For a normally distributed behavior, what percentage of total occurrences falls between $\bar{x} \pm s$; $\bar{x} \pm 2s$; and $\bar{x} \pm 3s$?

2.6. Each class of geotextile in Table 2.2a was subdivided according to the elongation at break of 50%. Where do the following types of fabrics fall in this regard?
- **(a)** woven slit-film
- **(b)** woven monofilament

(c) nonwoven heat-bonded
(d) nonwoven needle-punched

2.7. Compare the Pennsylvania Department of Transportation Specification values given in Table 2.1 to the AASHTO M288-96 Class 2 values listed in Table 2.2a for the following:
(a) subsurface filtration versus Table 2.2b
(b) separation versus Table 2.2c
(c) erosion control, Type A versus Table 2.2e
(d) sediment control, Type A versus Table 2.2f

2.8. Using the M288-96 specifications in Table 2.2, determine what geotextile properties are needed for the following applications under severe installation conditions:
(a) woven monofilament erosion-control geotextile under stone rip-rap for soil subgrade of 65% passing the 0.075 mm sieve size
(b) unsupported woven geotextile silt fence
(c) nonwoven needle-punched geotextile for prevention of reflective cracking as a paving fabric

2.9. In the description of designing-by-function (Section 2.1.3), what would be the estimated range of required factors of safety for the following situations:
(a) a highway underdrain in the shoulder area of a secondary road
(b) a filter fabric behind a retaining wall, above which is a paved parking lot
(c) a reinforcing fabric for an access road on which cranes will be setting structural steel
(d) a filter to protect a chimney drain within an earth dam where a small town is located downstream

2.10. In designing-by-function (Section 2.1.3) the focus is on the primary function. How does one know what is primary? (*Hint:* Would the resulting factors of safety from various aspects of the problem be indicative in any way?)

2.11. In Section 2.2, separation was distinguished from reinforcement. In the case of geotextiles placed on soil subgrades beneath stone base courses for highways, when does the geotextile act as a separator vis-a-vis reinforcement?

2.12. **(a)** In Section 2.2, filtration was distinguished from drainage. Describe these two functions and how they are different.
(b) In handling flow for these two different functions, Ψ and θ were defined. Why is this necessary and why is the thickness of the geotextile included in each term?

2.13. For steady-state flow conditions, using Figure 2.4a, what is the appropriate opening size formula for the following conditions?
(a) Gravel with $C'_U = 5$ in a dense condition
(b) Stable sand favoring retention with $C'_U = 2.5$ in a loose condition
(c) Sandy silt that is nonplastic favoring permeability, widely graded $C'_U > 15$ and in a medium-density state.

2.14. For dynamic-flow conditions, using Figure 2.4b, what is the appropriate opening size formula for the following conditions?
(a) Clayey (plastic) soil that is nondispersive
(b) Silty sand that is widely graded ($C_U = 13$) under a mild water current
(c) Gravel under severe wave attack

2.15. In geotextile testing:
(a) What is an index test?
(b) What is a performance test?
(c) How can typical laboratory-test values be made into allowable values for a design-by-function procedure?

2.16. Calculate the compressibility modulus and the compressibility coefficient of the four geotextiles shown in Figure 2.6. (Note that the thickness change must be converted to strain using the original fabric thickness measured at 2.0 kPa.)

2.17. Determine the strength, maximum elongation, toughness, and initial tangent modulus of fabrics A, B, C, and E shown in Figure 2.7 in both geotextile units and standard engineering units.

2.18. Concerning the different geotextile tensile test specimen shapes shown in Figure 2.8:
(a) Rank the tensile strength in units of kN/m for a woven geotextile for the four shapes shown.
(b) Rank the tensile strength in units of kN/m for a nonwoven needle-punched geotextile for the four shapes shown.

2.19. Comparing results of wide-width tests in isolation versus under confinement, calculate the modulus of elasticity, the maximum stress (strength), and the elongation at failure for each of the geotextiles shown in Figures 2.11a,b,c,d. Prepare your answers in table form, calculate the percentages of improvement from the in-isolation case, and describe why the major improvement is with the nonwovens.

2.20. Concerning the type of puncture probe used in evaluating the puncture resistance of geotextiles, as described in Section 2.3.3, why should (a) a large-diameter probe be used for nonwovens and (b) a beveled probe be used for wovens?

2.21. Given the following set of data from a soil-geotextile friction test:
(a) Plot the Mohr failure envelope.
(b) Obtain the friction angle.
(c) Calculate the fabric efficiency based on a soil friction angle of 38°.

Normal Stress (kPa)	Shear Strength (kPa)
17	8.6
35	20
70	36
140	75

2.22. Calculate the porosity of the following geotextiles.

Geotextile	Mass/unit area (g/m^2)	Density (kN/m^3)	Thickness (mm)
A	135	4.7	0.33
B	200	6.0	0.38
C	350	5.5	0.63
D	600	6.0	0.32

2.23. What is an image analyzer, and how could it be used to determine fabric porosity?

2.24. Given the following projection of a woven monofilament geotextile, determine its percent open area (POA). (*Hint:* You will have to enlarge the figure considerably. A photocopier used a number of times will get you a size convenient to use with cross-section paper or a planimeter.)

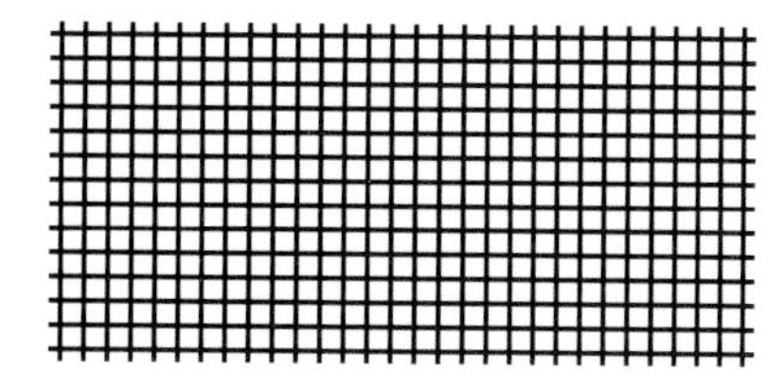

2.25. The apparent opening size test described in Section 2.3.4 is very controversial. There are three versions described in the literature: dry sieving (ASTM D4751), wet sieving (ISO/DIS 12956) and hydrodynamic sieving (used in Germany). Make a table describing each test with the advantages and disadvantages.

2.26. If a series of glass bead-sieve tests (either dry or wet) gave the following 0_{95} values in mm, what would be the AOS sieve-size value?

Geotextile	0_{95} (mm)
A	0.87
B	0.415
C	0.079

2.27. Given the following data for a constant-head cross-plane flow of water through a 50 mm diameter, 0.30 mm thick geotextile. Calculate the permittivity (s^{-1}) and coefficient of permeability (cm/s).

Δh (mm)	q (cm³/min)
31	300
62	680
125	1010
250	1400

2.28. Given the following constant-head data for planar flow of water in a 1.50 mm thick geotextile that is 300 mm wide × 600 mm long. Calculate the transmissivity in m³/min-m of fabric and then the planar coefficient of permeability in cm/s.

Δh (mm)	q (cm³/min)
75	21
150	41
225	60
300	79

2.29. Given the following constant-head data set for a radial flow of water in a 1.02 mm thick geotextile that has a 57 mm outer radius and a 28.7 mm inner radius. Calculate the transmissivity in m^3/min-m of fabric and the planar coefficient of permeability in cm/s.

Δh (mm)	q (cm^3/min)
150	1400
300	2900
450	4500
600	6000

2.30. Assume the data from Problem 2.29 were taken at normal pressure on the fabric of 7 kPa and the test was repeated at 14, 28, and 56 kPa, giving the additional data given below. Plot the transmissivity (m^3/min-m) versus the applied normal pressure (kN/m^2) response curve.

Pressure (kPa)	Δh (mm)	q (cm^3/min)
14	150	820
	300	1730
	450	2500
	600	3500
28	150	570
	300	1220
	450	1730
	600	2250
56	150	540
	300	1080
	450	1650
	600	2200

2.31. What is the range of factors of safety for installation damage for the data shown in Figure 2.19b in Regions A, B, and C?

2.32. Calculate the strain of the polyester fabric shown in Figure 2.20 at the end of 30 years, assuming a linear extension of the data for both 20% and 60% of failure loads.

2.33. Concerning long-term flow tests as in Section 2.3.5:

(a) Describe the physical phenomena that accompany the various portions of the long-term flow behavior shown in Figure 2.23(b).

(b) Which of these phenomena is most important in the possibility of the response shutting off flow completely?

2.34. Concerning the gradient ratio tests as in Section 2.3.5:

(a) In the gradient ratio test, head is measured via plastic tubing attached to a manometer board. Since it is this value that is most important in calculating the gradient ratio, what experimental problems do you envision?

(b) In Figure 2.24, why is the value 3.0 used to separate acceptable and nonacceptable fabrics?

2.35. Concerning the clogging of geotextiles in the field:
(a) What are the main conditions known where soils are likely to cause excessive clogging of geotextiles?
(b) In such instances, what is the logical recommendation?

2.36. The type of clogging described in Section 2.3.5 has to do with various soils upstream of the fabric and water as the permeant. If the permeant is a liquid with a large amount of suspended solids and microorganisms in it (e.g., farm runoff or leachate), what possible mechanisms of clogging could occur?

2.37. How would you evaluate the chemical resistance of a geotextile to a hazardous-waste leachate for which there is no standard test data (recall Section 2.3.6)?

2.38. What specific portions of the ultraviolet-light spectrum are damaging to polymeric materials (recall Section 2.3.6)? To what wavelength are PE, PP, and PET most sensitive?

2.39. How do manufacturers avoid or minimize the harmful effects of ultraviolet degradation to exposed geotextiles?

2.40. How would you predict the exposed lifetime of a geotextile on the basis of accelerated weathering (ultraviolet, temperature, and moisture) laboratory-test data?

2.41. Regarding the lifetime prediction of geosynthetics that are covered in a timely manner:
(a) Devise an experiment for evaluating the long-term lifetime of a PP geotextile to be used for a reinforced wall application.
(b) Repeat part (a) for a PET geotextile.

2.42. If the ultimate strength of a geotextile from an index-type test is 45 kN/m, what would be the allowable strength for the design purposes, according to Table 2.11, for
(a) separation?
(b) reinforced embankment?
(c) increased bearing capacity?
(d) flexible-forming systems?

2.43. If the ultimate flow rate of a geotextile from an index-type test is 18 l/min-m, what would be the allowable flow rate for design purposes according to Table 2.12 for
(a) gravity drainage problems?
(b) pressure drainage problems?

2.44. If the ultimate permittivity of a geotextile from an index-type test is 1.2 s^{-1}, what would be the allowable value for design purposes according to Table 2.12 for a
(a) retaining-wall filter?
(b) highway-underdrain filter?
(c) filter beneath rock riprap?
(d) filter above leachate collection system?

2.45. What are the mechanical properties of a geotextile that are of most importance when using it as a separator in an unpaved road with only a stone base above it and relatively soft soil below it?

2.46. Regarding geotextiles used in separation:
(a) What is the required burst pressure of a geotextile supporting 75 mm maximum-size stone and heavy trucks with a tire inflation pressure of 1000 kPa? (Use $p \approx 0.75\, p_a$, a cumulative reduction factor of 2.0, and a factor of safety of 2.0.)
(b) What is the required burst pressure of a geotextile under the conditions in part (a), except that now the road will haul only light vehicles with tire inflation pressure of 500 kPa.

2.47. What would be the influence on a burst analysis if well-graded stone base were used instead of poorly-graded aggregate materials?

2.48. Redo Example 2.8 (in Section 2.5.3) assuming that slippage does occur between the stone and geotextile. In your analysis assume that the strains mobilized are 75%, 50%, and 10%. Plot these values together with the 33% strain versus required grab tensile strength.

2.49. Using the design guide of Figure 2.32, determine what styles of commercially available geotextiles meet the puncture criterion of 700 kPa pressure and 50 mm stone.

2.50. Develop a design chart like that in Figure 2.32 for grab tensile strength, per Section 2.5.3. Use the same reduction factors and factor of safety assumptions as in Example 2.8.

2.51. Regarding geotextile puncture resistance:

(a) What is the required puncture resistance using ΠRF = 2.0 and FS = 1.5 of a geotextile whose average opening size is a No. 70 sieve (d_i = 0.21 mm) from a rock of 75 mm size with a vehicle tire pressure of 700 kPa on the aggregate surface?

(b) Redo part (a) for a puncture from a 75 mm diameter tree stump.

2.52. Regarding impact resistance, what energy is mobilized by rock of size 150 mm falling out of a dump truck 1.5 m to the geotextile, if the geotextile rests on

(a) a soft soil of unsoaked CBR = 3?

(b) a firm soil of unsoaked CBR = 9?

(c) a hard soil of unsoaked CBR = 16?

2.53. What energy is mobilized by a 250 N jackhammer falling 2.0 m out of a dump truck onto a geotextile on a hard soil of unsoaked CBR = 20? Would this situation cause a problem?

2.54. Regarding geotextiles for use in unpaved roads:

(a) When using geotextiles as reinforcement for unpaved roads on soft subsoils, do they also act as separators?

(b) If the answer to part (a) is yes, of what benefit is it?

(c) From product literature, or simply your gut feeling, what does this separation function amount to as far as thickness of stone base saved (if any) is concerned?

2.55. Regarding soil subgrade characteristics for roadway construction:

(a) Describe the details of the CBR test (both the unsoaked and soaked versions).

(b) Using Table 2.13 for a soil whose CBR = 1.0, what is the equivalent in: **(i)** shear strength? **(ii)** R value (California)? **(iii)** S, soil support value? **(iv)** group index? **(v)** R value (Washington)? **(vi)** cone index (320 mm^2 probe)? **(vii)** bearing value (300 mm plate, 5 mm deflection)? **(viii)** bearing value (760 mm plate, 2.5 mm deflection)? **(ix)** Modulus of subgrade reaction?

(c) In Table 2.13 why do the ASTM, AASHTO, and FAA classifications not go down as far as CBR = 1.0?

2.56. Given a 80-kN axle-load vehicle with tire inflation pressure of 480 kPa and an allowable rut depth of 0.3 m for an unpaved road to be designed (i) without a geotextile, (ii) then with a geotextile using Figure 2.37, and (iii) ultimately resulting in a thickness of stone aggregate to be saved:

(a) For 340 passages without and with a 90-kN/m modulus geotextile, draw the response curve to varying soil CBR values from 0.5 to 4.0.

(b) For 340 passages and a CBR = 1.0, draw the response curve for varying geotextile moduli from 450 kN/m to 10 kN/m.

(c) For a soil of CBR = 1.0 and a geotextile modulus of 90 kN/m, draw the response curve to a varying number of vehicle passages from 10 to 10,000.

2.57. Evaluate the sensitivity of varying rut depths [the value S in Eq. (2.38)] on the required thickness of stone base using typical values for unpaved road problems, and plot your results. (*Hint:* It would help to go to the original study by Giroud and Noiray [67] to gain insight into this aspect of the development.)

2.58. Plot the required thickness of stone base of an unpaved road as a function of CBR varying from 0.50 to 30 using Eq. (2.39) in Example 2.13 (Section 2.6.1). Use 25 kN equivalent single-wheel loads for 10,000 coverages on a 300 $\times$ 450-mm tire contact area in your solution.

2.59. Regarding the sewing of geotextile seams:
(a) What type of test would you use to evaluate the strength of a sewn seam?
(b) In writing a specification for the sewing of seams, what items would you include?
(c) If you recommended a minimum tensile strength for the seam, what percentage of the minimum tensile strength of the geotextile would you recommend?
(d) How would you conduct field tests for strength of sewn seams?

2.60. MESLs are usually constructed by in situ impregnation of geotextiles.
(a) Why is in situ spraying preferred over prefabricated systems?
(b) Why are geomembranes (Chapter 5) generally not used?
(c) Why are needle-punched nonwovens generally used over other manufactured styles of fabrics?

2.61. Regarding the waterproofing of geotextiles for MESL construction:
(a) What is an asphalt emulsion and how does it waterproof after being sprayed onto a geotextile?
(b) What is an asphaltic cement and how does it work vis-a-vis an emulsion?
(c) What are cutbacks and why are they not recommended for impregnating MESLs?
(d) How do you determine the proper quantity needed to saturate the geotextile?

2.62. For Example 2.15 (Section 2.6.2) concerning MESLs, the water content of the encapsulated soil varies only 0.8% over the conditions stated. If this soil were known to become unstable at 18% moisture content, what variations of input parameters could bring it about?

2.63. Look up and briefly describe the water-vapor transmission test (ASTM E96) and include comments on how the test would be different when evaluating a relatively thick impregnated geotextile to be used in construction of a MESL as opposed to when evaluating thin geomembranes.

2.64. For geotextiles used to reinforce paved roads on firm soil subgrades, the geotextile must somehow be prestressed (recall Section 2.6.3).
(a) Why is this the case?
(b) Sketch some methods for prestressing geotextiles for such an application.
(c) In light of your answer above, consider creep and possible stress relaxation, and postulate the road system's long-term performance.

2.65. The concept of flexible wall systems, as opposed to rigid concrete and masonry walls, is very much in style. List advantages and disadvantages of each type.

2.66. Regarding geotextile-reinforced walls:
(a) Design a 5.5 m high geotextile wall carrying a road consisting of a 300 mm stone base ($\gamma = 22$ kN/m^3) and 150 mm asphalt ($\gamma = 24$ kN/m^3) for 180 kN dual-tandem-axle loads whose wheel pattern is shown in Example 2.17 (Section 2.7.1). The wall is to be backfilled with SW-ML soil of $\gamma = 18$ kN/m^3, $\phi = 35°$, and $c = 0$. The geotextile to be used is a nonwoven heat-bonded fabric of 200 g/m^2 with a grab tensile strength of 1400 N.

(b) Check and/or modify your answer to part (a) for external stability, considering that the foundation soil is ML-CL with $\gamma = 19\ \text{kN/m}^3$, $\phi = 15°$, $\delta = 0.95\phi$, $c = 24$ kPa, and $c_a = 0.90c$.

2.67. Develop a design chart similar to that in Figure 2.48 for the wheel load and backfill of Problem 2.66, for vertical-faced walls ($\beta = 90°$) and for wall heights varying from 1.5 to 8.0 m and geotextile tensile strengths of 20, 40, 60, and 80 kN/m. [*Note:* This is an extremely long problem and requires a computer algorithm to be generated].

2.68. For Examples 2.18 and 2.19 (Section 2.7.2) using geotextiles to stabilize embankments, the fabric's allowable tensile strength is 40 kN/m. Repeat this problem using fabrics whose strength varies from 20, 40 (the example), 80, 150, and 300 kN/m and plot the resulting FS value against allowable strength.

2.69. Concerning the allowable geotextile strength to be used in stabilization problems (for both walls and embankments):

(a) What test method should be used?

(b) What considerations enter into your choice of allowable strength?

(c) How does creep enter into the situation?

2.70. Given a 15 m high embankment at a slope angle of $\beta = 40°$, the soil strength parameters are $\phi = 22°$, and $c = 15$ kPa in both the embankment and foundation sections. The unit weight is 16 kN/m^3. For a failure circle located at coordinates (+3, +18) with respect to the toe at (0, 0) (see Figure 2.53) and a radius of 21 m, what is the factor of safety? How many layers of geotextiles spaced 300 mm apart and having an allowable tensile strength of 55 kN/m placed at the interface of the foundation and the embankment are required to raise this factor of safety to 1.40?

2.71. For Problem 2.70, find the minimum factor of safety for both the nonreinforced and reinforced conditions. [*Note:* This is a very long problem requiring a search of both variation in radius and center of circle. A computer program is necessary.]

2.72. In using high-strength geotextiles for areal stabilization projects as described in Section 2.7.3, how do you sew the transverse seams after the larger panels are already deployed? Furthermore, how do you sew the intersection where the four panels come together?

2.73. In placing fill on a high-strength geotextile used to stabilize very soft soils by the linear-fill method, Figure 2.57 shows the edges being placed first. Why is this necessary? What would happen if the entire fill were advanced together? What would happen if the center were advanced first?

2.74. Describe by means of a sketch how you would instrument a high-strength geotextile of the type shown in Figure 2.57 to verify the design models shown in Figure 2.58.

2.75. There is considerable appeal in the use of geotextiles to improve the bearing capacity of shallow foundations (Section 2.7.4). Most efforts are aimed at showing improvement in bearing capacity as indicated in Figure 2.61.

(a) Why are the improvements low at low deformations and considerably better at high deformations?

(b) How could the low-deformation behavior be improved?

(c) By observing these data, would you consider using only one high-strength geotextile instead of a number of lower-strength layers?

(d) If the answer to part (c) is yes, how would you join the geotextile ends and sides?

2.76. Regarding improved bearing capacity:

(a) Would a geotextile placed under a large mat foundation measuring 30 × 30 m on a compressible soil prevent (total) settlement?

(b) If the answer to part (a) is no, would it be of any help insofar as reinforcement is concerned?

2.77. Regarding the spider netting in Section 2.7.5:

(a) Illustrate the various mechanisms that soil nails can provide in soil slope stabilization.

(b) What are the high-stress regions of the net as currently configured?

(c) Why must the nails be continually driven into the ground after the initial installation?

2.78. The geotextile is exposed on the surface of the slope in spider netting. What advantages and disadvantages occur?

2.79. What are the three essential features that must be addressed in filtration design?

2.80. What is the difference in total active earth pressure between the two retaining wall cases shown below? One has good drainage; the other has no drainage and full hydrostatic head.

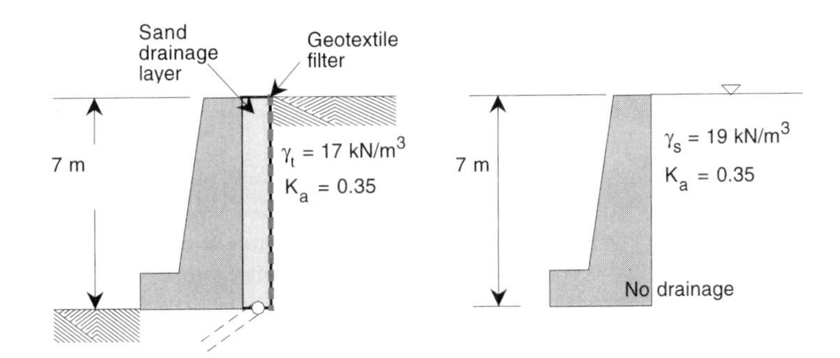

2.81. **(a)** In considering hydraulic designs involving geotextiles, why is permittivity used in filtration and transmissivity used in drainage, rather than just the respective permeability (hydraulic conductivity) coefficients?

(b) In the laboratory determination of a specific geotextile's permittivity or transmissivity, at what pressure should the geotextile's thickness be measured?

2.82. In wrapping the stone aggregate around a highway underdrain with a geotextile:

(a) What is the major function of the geotextile?

(b) If properly designed, what should be the long-term condition of the stone base?

(c) If the stone base has sufficient open space to transmit the entering water, of what necessity is the perforated pipe?

(d) What is a French drain?

2.83. A geotextile filter is being considered to protect the stone aggregate drain behind a cantilever retaining wall, as shown in Figure 2.67b. The wall stem is 7.5 m high, retaining an ML soil with $k = 2.5 \times 10^{-4}$ m/s, $d_{50} = 0.05$ mm, $C_U = 4.8$, and $D_R = 85\%$. The candidate geotextile is a heat-bonded nonwoven with a permittivity $\Psi = 0.01$ s^{-1} with an AOS = No. 70 sieve. What is the factor of safety against flow and the adequacy as far as soil retention is concerned?

2.84. If the required permittivity of a sandy soil beneath a rock riprap-protected slope is 0.052 s^{-1} and riprap covers 75% of the geotextile's surface, would a geotextile of $k = 3.5 \times 10^{-4}$ m/s and 0.65 mm thick be adequate? What is the factor of safety in this case?

2.85. Concerning geotextile silt fences:

(a) While geotextile silt fences function as filters, Section 2.8.6 focused mainly on strength considerations. Why is strength so important?

(b) How do geotextile silt fences filter the turbid water and retain the suspended soil particulates?
(c) Discuss UV-degradation of silt fences in light of their use as erosion- and sedimentation-control systems.

2.86. Recalculate Example 2.24 (Section 2.8.6) using the same values for the silt fence, except varying the storm intensity from 10, 50, 100 (the example problem), and 200 mm/hour to determine:
(a) the height of the silt fence
(b) the strength of the geotextile
(c) the type of posts to be used for support

2.87. Regarding the drainage capability of geotextiles:
(a) Which manufactured style is best suited to convey water in its plane?
(b) What conditions are required to satisfy Darcy's formula?
(c) What is the driving force for water flow in gravity drainage situations?
(d) What are some driving force mechanisms for water flow in pressure drainage situations?

2.88. Geotechnical engineers are generally reluctant to use drainage geotextiles as chimney drains and drainage galleries, as shown in the earth dam example of Section 2.9.3. What are some reasons for this reluctance?

2.89. For the 8 m high concrete cantilever retaining wall in Example 2.26 (Section 2.9.3), recalculate the soil's permeability to determine what value is required to have $\theta_{\text{allow}} = 0.00015\ \text{m}^2/\text{min}$ be adequate with FS = 4.0 (i.e., work the problem backwards).

2.90. Repeat Example 2.27, illustrating pressure drainage of the consolidating soil beneath a surcharge fill (Section 2.9.4), with a $B = 30$ m and vary the time for the surcharge fill to be placed from 1 day to 1 year. Plot the results and show on the graph the acceptable zone based on FS = 5.0.

2.91. When using a geotextile as a capillary-migration break as in Section 2.9.5:
(a) What is to prevent the water from continuing across the barrier as though it were not there?
(b) Does the fact that geotextiles are hydrophobic play a role in this situation?
(c) If soil particles become clogged in the fabric structure and accumulate, does the situation change?
(d) How is the situation in part (c) prevented from occurring?

2.92. There are some areas where full-pavement-width geotextiles should not be used in reflective crack prevention, as per Section 2.10.2. Describe why this is so for each case:
(a) where subsoil foundation problems exist beneath the stone base
(b) for reinforced-concrete pavements
(c) in areas of rapid and harsh cyclic freeze-thaw temperature conditions

2.93. Regarding the proper quantity of sealant for geotextile in reflective cracking applications:
(a) Why is sealant in excess of saturation a problem?
(b) Why is sealant less than saturation a problem?

2.94. In reflective crack prevention, assuming reinforcement to be the primary function, the FEF is of paramount interest.
(a) What is FEF?
(b) How is it determined?
(c) What geotextile property is it mainly dependent on?
(d) What are the dangers of taking laboratory-generated data and using it to project field performance?

2.95. Regarding geotextiles in reflective cracking, assuming reinforcement as the primary function:

(a) Using $DTN_N = 200$ and a CBR value varying from 2 to 20, what is T_A using Figure 2.75?

(b) Using $DTN_N = 200$ and a geotextile resulting in FEF $= 4.0$, what is T_A if the CBR varies from 2 to 20?

(c) Plot the two resulting curves on a graph of T_A versus CBR and comment on the results and the differences between the curves.

2.96. Postulate on the mechanism(s) that might occur in assuming waterproofing to be the major function in using geotextiles to prevent reflective cracks in asphalt pavements.

2.97. Regarding geotextiles in reflective cracking, assuming moisture barriers as the primary function:

(a) Using $DTN_N = 500$ and $x = 1.57$ mm, $s = 0.15$ mm, $f = 0.80$, and $c = 1.25$, determine the asphalt overlay thickness according to Figure 2.77.

(b) Using a fabric assumed to function as a waterproofing barrier, redo part (a) for c varying from 2.5 to 0.5 in Eq. (2.75).

(c) Plot the results on a graph of T_o versus c and comment on the curves and the differences.

2.98. Design and sketch a field experiment for determining whether reinforcement or waterproofing is the major function in using fabrics as preventing reflective cracks in asphalt overlays.

2.99. For reflective-crack prevention using narrow strips of reinforcement:

(a) Fiberglass geotextiles have some distinct advantages and potential disadvantages over polymeric materials. What are they?

(b) How do you anchor the sides of the geotextile strip on each side of the crack?

(c) Which are the more troublesome cracks in old pavements, transverse or longitudinal, and why?

2.100. List the functions (in order of priority) that you feel are acting when geotextiles are placed beneath railroad ballast in the following situations:

(a) new railroad track construction

(b) remediation of existing railroad trackage

2.101. Comment on why most railroad specifications call for thick needle-punched nonwovens and most laboratory-generated research papers use relatively thin wovens or heat-bonded nonwoven fabrics.

2.102. Regarding geotextiles used as railroad ballast separators:

(a) Of the geotextile photographs shown in Figure 2.78, a few of the holes were probably not caused by abrasion. What was their most likely cause, and which ones were they?

(b) What is the minimum depth at which a geotextile should be placed beneath the bottom of a railroad tie to prevent abrasion problems?

2.103. In the use of geotextiles as flexible forms for columns in mine stabilization or pile jacketing (Section 2.10.4), will the resulting shape be circular? If not, why?

2.104. For the fabric used as a flexible form in mine stabilization as shown in Figure 2.80, what type, dimensions, and properties of geotextile should be considered?

2.105. What is the major vulnerability of erosion-control mattresses of the type shown in Figure 2.86?

2.106. Regarding the use of geotextiles as flexible forms:

(a) Is there any limit to the depth at which tremie concrete or grout can be placed under water?

(b) When placed in geotextile forms, how is the water displaced and the concrete or grout kept in?

(c) After the concrete or grout is set, what is the function of the geotextile?

2.107. Concerning the geotextile survivability concepts discussed in Section 2.11.2 and Table 2.22, what minimum mechanical properties would be required in the following cases using the specification of Table 2.2a?

(a) 35 kPa construction equipment using average site preparation

(b) same as in part (a), except no site preparation has been prepared

(c) 20 kPa construction equipment on a dredged site with no vegetation or growth

2.108. What general and specific methods does the design engineer have with respect to the contractor regarding the proper care and handling of geotextiles during the installation process?

3 Designing with Geogrids

3.0 INTRODUCTION

The geotextiles discussed in Chapter 2 and the geogrids discussed in this chapter compete for use in various reinforcement applications. They are also designed by similar methods but they differ in their manufacture, appearance, and placement. A geogrid can be defined as follows:

> **geogrid**, *n.* a geosynthetic material consisting of connected parallel sets of tensile ribs with apertures of sufficient size to allow strike-through of surrounding soil, stone, or other geotechnical material.

Thus, geogrids are matrix-like materials with large open spaces called *apertures*, which are typically 10 to 100 mm between the *ribs*, called *longitudinal* and *transverse*, respectively. The ribs themselves can be manufactured from a number of different materials, and the rib cross-over joining or junction-bonding methods can vary. The primary function of geogrids is clearly reinforcement; thus the subdivisions within this chapter are not by function but by type of reinforcement application.

Shown in Figure 3.1 are photographs of geogrids currently on the North American and European markets. Each will be explained in some detail, since the testing results to follow relate somewhat to the method of manufacture.

Figure 3.1a shows products of Netlon Ltd. and the Tensar Corporation. Both companies market uniaxially and biaxially oriented geogrids. Each style begins as a polyolefin polymer sheet that has a uniform and controlled pattern of prepunched holes. The prepunched sheet is then sent over and under a number of rollers, each going faster than the one before it, thus inducing longitudinal stress in the remaining sheet, now called *ribs*. This stress causes the ribs to deform and elongate in the direction of movement. In the uniaxially deformed products, circular holes punched in the high-density

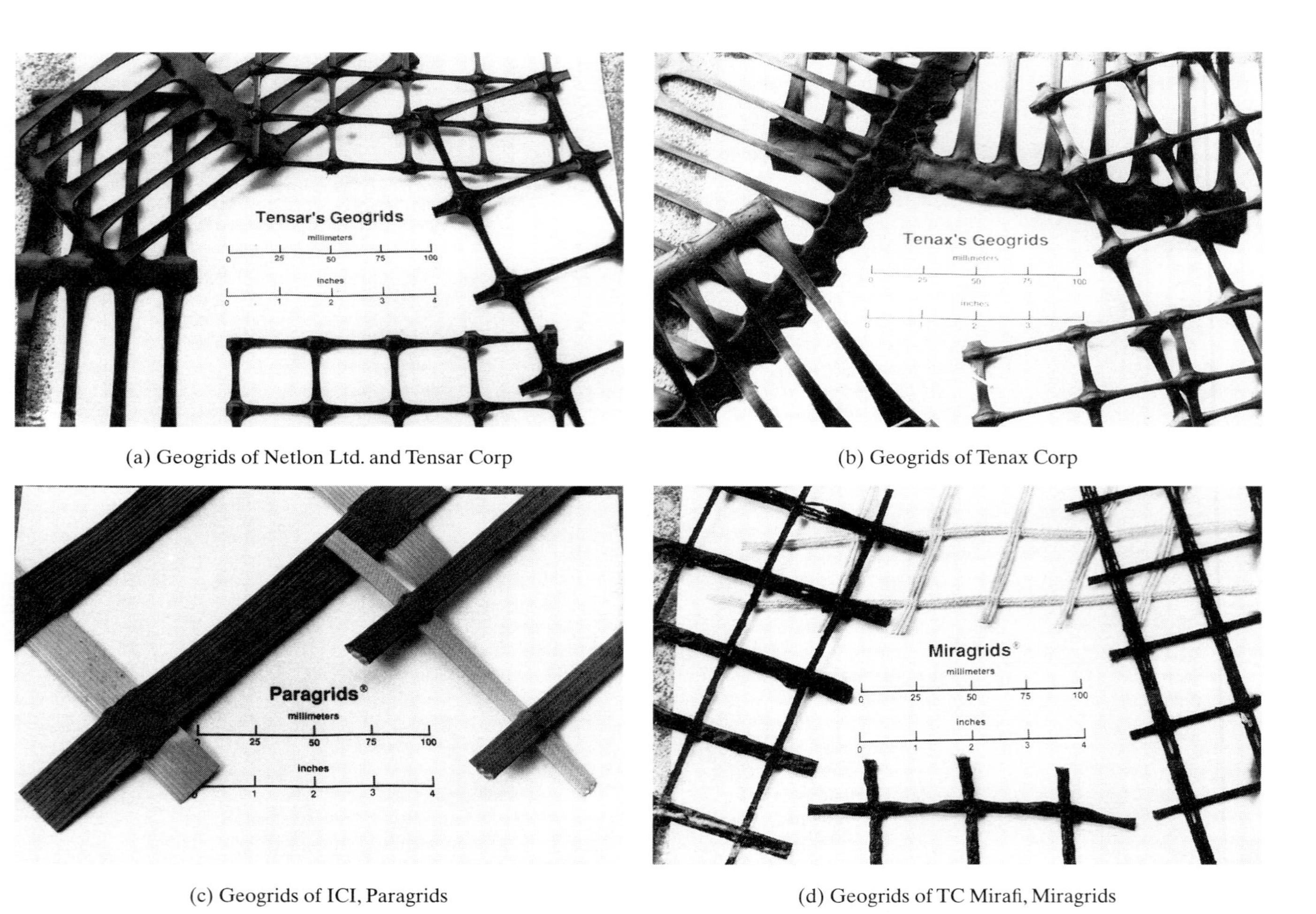

(a) Geogrids of Netlon Ltd. and Tensar Corp

(b) Geogrids of Tenax Corp

(c) Geogrids of ICI, Paragrids

(d) Geogrids of TC Mirafi, Miragrids

Figure 3.1 Types of geogrids.

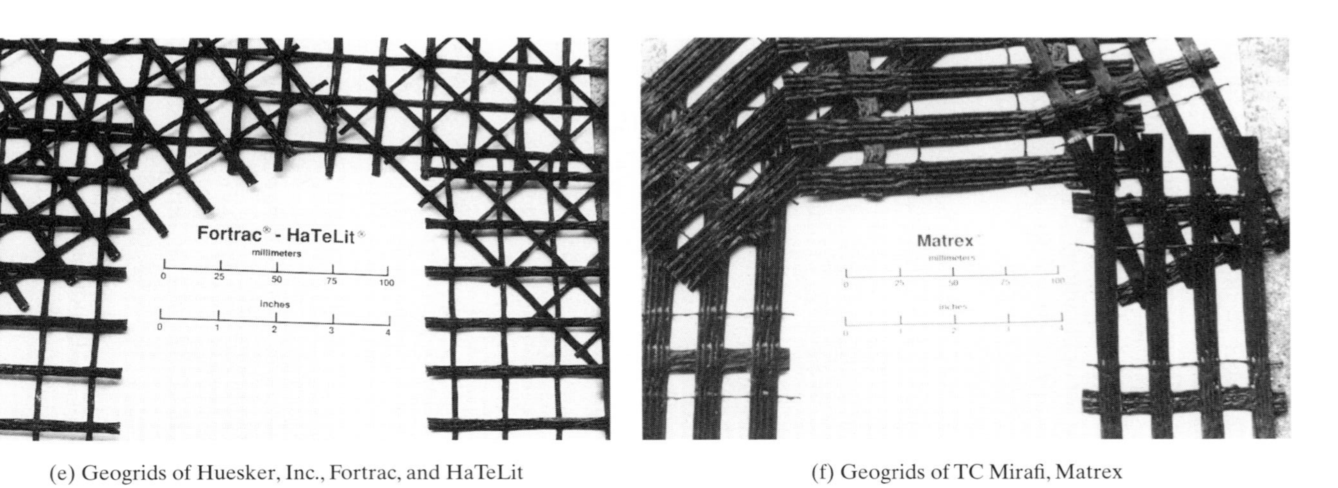

(e) Geogrids of Huesker, Inc., Fortrac, and HaTeLit

(f) Geogrids of TC Mirafi, Matrex

(g) Geogrids of Strata Systems, Inc., Stratagrid

(h) Other geogrids

Figure 3.1 *(continued)*

are available. In the biaxially deformed products, squares are punched in polypropylene sheet, which is drawn longitudinally (using rollers) and then transversely (using a stretcher), forming near-square or rectangular apertures. Doing so increases strength in both directions in the biaxial product. At least seven different styles are available. The uniaxial products are for applications in which the major principal stress direction is known and the biaxial products are for applications in which mobilized stresses are essentially random.

The Tenax Corporation manufacturers geogrids that appear to be similar to those of Netlon/Tensar and are also available in both uniaxial and biaxial styles (see Figure 3.1b). Polymerwise they are also the same (polyethylene for uniaxial and polypropylene for biaxial). A basic difference is the method of manufacturing: Tenax draws their punched sheets mechanically rather than using production by continuous rolling.

Paragrids are products of ICI Fibres Ltd. and consist of many high-tenacity polyester fibers gathered together by an encompassing polypropylene sheath (see Figure 3.1c). At the junctions where the longitudinal and transverse ribs cross, the contacting polypropylene sheaths are melt-bonded to one another, forming the connection. There are a number of styles and strengths available with different rib sizes and hence different junction contact areas.

The TC-Mirafi organization manufactures and markets Miragrid geogrids, shown in Figure 3.1d. The basic structure is formed by an open-weaving process using high-tenacity polyester fibers to form ribs that are subsequently knit-stitched together at the junctions. The product is then latex-coated in its final processing step. A number of strengths are available.

Huesker Synthetic Company makes geogrids under the trademarks Fortrac and HaTeLit. Both are made from woven high-tenacity polyester yarns, entangled at the junctions, and then coated with either PVC or a bituminous finish. Several of the available styles are shown in Figure 3.1e. They are marketed by the Huesker and also the Akzo, Wellman, and ACF organizations. There are a number of different strengths of Fortrac and HaTeLit geogrids available.

The TC-Mirafi organization also manufactures Matrex geogrids, which are marketed in the U.S. by the Reinforced Earth Company. These are woven rib structures using high-tenacity polyester yarns as the base material. They are coated with a bituminous material and are shown in Figure 3.1f. A number of different strengths are available.

Strata Systems, Inc. manufactures and markets a geogrid called Stratagrid in a number of different styles. It is a woven high-tenacity polyester product entangled by knitting at its junctions and then coated with bitumen. These geogrids are shown in Figure 3.1g.

As shown in Figure 3.1h there are an assortment of other geogrids that are available or are in the development stage. In the polyester group, Raugrid (produced by Lückenhaus GmbH and Lückenhaus N.A., Inc.) has been recently introduced as has a geogrid by Synteen, Inc. In Japan there are a number of geogrids made from polyamide yarns. Also within this group are geogrids made with fiberglass yarns, for e... Bay Mills Company has a product called Glasgrid and PPG Industries has a nu... prototype products; fiberglass geogrids have also been used in Japan. The yar... joined at their intersections and then coated with a polymer, latex, or bitumen.

3.1 GEOGRID PROPERTIES AND TEST METHODS

In contrast to the entire range of test methods described in the geotextile chapter, only those involved in reinforcement applications will be addressed here. (Thus properties relating to filtration, drainage, and containment applications are not included.) There are, however, a number of unique aspects to geogrid tests in comparison to geotextiles.

3.1.1 Physical Properties

Many of the physical properties of geogrids can be measured directly and are relatively straightforward. These include the type of structure, junction type, aperture size, and thickness. Other properties that are of interest are mass per unit area, which varies over a tremendous range from 200 to 1000 g/m^2, and percent open area, which varies from 40% to 95%.

An additional physical property of geogrids (important in constructability) is its stiffness. This can be measured using ASTM D1388, a test for flexural rigidity. This test method slides a geogrid test specimen over an inclined plane at an angle of 41.5° with the horizontal. When the geogrid bends and eventually touches the surface of the inclined plane, its distance is measured and then related to the mass per unit area. The test is described in Section 2.3.2. While the test leaves much to be desired, it is a quantitative method to assess geogrid stiffness and can be used to qualitatively separate the geogrids shown in Figure 3.1 into two groups. *Stiff* geogrids, generally made from polyethylene or polypropylene, are characterized by having flexural rigidity values greater than 1000 g-cm in this test. They are also sometimes called *homogeneous* geogrids. *Flexible* geogrids, generally made by a textile weaving process and involving polyester, nylon, and fiberglass yarns, are characterized by having flexural rigidity values less than 1000 g-cm in this test.

3.1.2 Mechanical Properties

The mechanical properties of the geogrids covered in this section all relate directly to their use in reinforcement applications. Some are index tests, while others are clearly performance-oriented.

Single Rib and Junction (Node) Strength. The initial tendency toward assessing a geogrid's tensile strength is to pull a single rib in tension until failure and note its behavior. A secondary tendency is to evaluate the in-isolation junction strength by pulling a longitudinal rib away from its transverse rib's junction. It is important to state *in-isolation* since there is no normal stress on the junction; thus the test will not represent performance conditions. The latter test must be done with the entire geogrid structure contained within soil embedment. This is a much more complicated test and will be covered in this section under anchorage strength from soil pullout.

A *single rib tension strength* test merely uses a constant rate-of-extension testing machine to pull a single rib to failure. For uniaxial geogrids, this is most likely a longitudinal rib. For biaxial geogrids, both longitudinal and transverse ribs require evaluation. By knowing the repeat pattern of the ribs, an equivalent wide-width strength could be calculated. However, usually it is not done in this manner and a number of ribs are tested simultaneously to obtain a more statistically accurate value for the wide-width strength (see below).

An *in-isolation junction or node strength* test can also be performed. The test method uses a clamping fixture that grips the transverse ribs of the geogrid immediately adjacent to and on each side of the longitudinal rib (see Figure 3.2). The lower portion of the longitudinal rib is gripped in a separate clamp, and each clamp is mounted in a tensile testing machine, where they are pulled apart. The strength of the junction, in force units, is obtained. Table 3.1 gives the junction strength for a number of commercially available geogrids. Note that the rib strength can also be evaluated as described previously. Having both sets of data, a junction strength efficiency can be calculated. As Table 3.1 shows, geogrid junction strength efficiencies vary from essentially 100% to 7%. It is important to note that these tests have the junction in an unconfined status. Thus it is an index type of test. The effect of simulated normal stress on the individual junction by means of soil confinement cannot be determined in this test.

Wide-Width Tensile Strength. Clearly the large-scale tensile strength of a geogrid, in its machine direction for uniaxial geogrids and in both machine and cross-

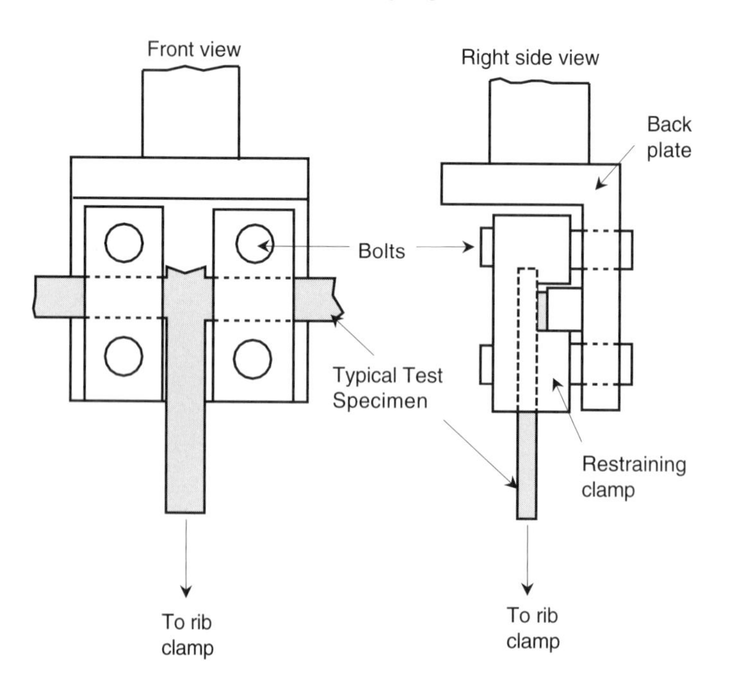

Figure 3.2 Test fixture for measuring geogrid junction strength in isolation.

TABLE 3.1 IN-ISOLATION GEOGRID JUNCTION AND RIB STRENGTH TEST RESULTS

Geogrid Designation	Avg. Junction Strength* (kN)	Avg. Rib Strength (kN)	Junction Efficiency (%)
TU −1	1.12	1.16	97
−2	1.86	2.04	91
−3	2.60	2.49	104
−4	1.31	1.33	99
−5	2.14	2.17	99
−6	2.68	2.77	97
TB −1	0.57	0.61	93
−2	0.72	0.75	96
−3	1.44	1.51	95
TA	1.02	1.03	99
TB	1.38	1.49	93
SA	1.96	4.16	47
PA	0.29	4.40	7
MA	0.18	1.42	13

*The junction strength tests were not laterally confined and were evaluated in a device of the type shown in Figure 3.2.

machine directions for biaxial geogrids, is of prime importance. This implies that longitudinal ribs are the focus for uniaxial products, while both longitudinal and transverse ribs are important for biaxial products. When testing the products, the test clamps grip a larger test specimen than described above, including a number of repeating rib units in the width direction and perhaps repeating length sections as well. The resulting data gives strength values in units of force per unit width, which is calculated by using the repeat distance of the actual geogrid structure. Obviously there is an extremely wide range in product behavior depending on type of polymer, thickness or number of yarns per rib, spacing of ribs, and so on. The strength of geogrids with respect to geotextiles is also of interest. Geogrid strengths fall between conventional geotextiles and those geotextiles specifically made for high-strength applications.

There are two procedural test methods used to evaluate the wide-width tensile strength of geogrids. One choice is to use ASTM D4595, the wide-width strength test for geotextiles, and modify it for the testing of geogrids. The major modifications required (since the test method is silent on the issue of geogrid testing) are the width of the test specimen, its length, its clamping mechanism, its strain rate, and the method for measuring deformation. If this test is selected, these items must be agreed upon by the parties involved. Alternatively, we can use ISO 10319 for wide-width strength testing of geogrids. In this procedure the width and length of the test specimen is prescribed and clamping should conform to the guide shown in Figure 2.9. Deformation monitoring in almost all situations must be based on an external measurement system (e.g., LVDT, optical, laser, etc.).

The information gained from a wide-width tension test on a geogrid comprises: the tensile strength at which the test specimen fails (kN/m); the tensile elongation at which the test specimen fails (i.e., its failure strain (%)); the tensile stress at different elongations prior to specimen failure (e.g., stress at 5% or stress at 10% (kN/m)); and the tensile modulus taken from the initial portion of the strength-versus-elongation curve, or possibly other defined modulus values (kN/m).

Manufacturers' literature is available on these data for the particular products and styles that are in current production. Some compilations of comparative data are available in the literature, but the situation changes regularly and direct inquiries to the manufacturer are recommended.

In the tension testing just discussed, the geogrid is evaluated in isolation, that is, with no soil adjacent to or surrounding it. With soil pressure adjacent to the geogrid, the material may show an improvement in its strength characteristics, although the effect is felt to be nominal at best. This comment is based on the work of Wilson-Fahmy et al. [1], in which a variety of geosynthetic materials were tested without and then with lateral confinement; only the nonwoven needle-punched geotextiles showed significant improvement under confining pressure. Unfortunately, geogrids were not evaluated in the study. When such tests are conducted on geogrids under confined conditions, it is important to isolate friction-induced contributions by the pressurizing medium (usually soil) by careful lubrication of both surfaces of the geogrid test specimen.

Shear Strength. One type of performance test that has been used on geogrids is an adapted form of a conventional geotechnical engineering direct shear test. In such a test (its analog for geotextiles is discussed in Section 2.3.3), the geogrid is fixed to a block and is forced to slide over stationary soil in a shear box while being subjected to normal stress (see Figure 3.3a). The maximum shear stress is obtained (see Figure 3.3b). Then a new test with a replicate geogrid specimen and soil (but now at a different normal stress) is conducted. This process is repeated sufficiently often to develop a set of strength-versus-normal stress points, which are plotted as shown in Figure 3.3c. The resulting line defines what is known as the failure envelope, properly called the Mohr-Coulomb failure envelope. From this graph the shear strength parameters of the geogrid to the particular soil are obtained, the values c_a and δ. Note that if a strain-softening response of the stress-versus-displacement curves occurs, two sets of values result, peak and residual. If the shear strength parameters of the soil by itself (c and ϕ) are also determined in their own separate tests with soil in both halves of the shear box, a comparison or *efficiency* can be calculated as follows:

$$E_c = \left(\frac{c_a}{c}\right) \times 100 \tag{3.1}$$

$$E_\phi = \left(\frac{\tan \delta}{\tan \phi}\right) \times 100 \tag{3.2}$$

where

E_c = efficiency on cohesion,
E_ϕ = efficiency on friction,
c_a = adhesion of soil-to-geogrid,
c = cohesion of soil-to-soil,
δ = angle of shearing resistance soil-to-geogrid, and
ϕ = angle of shearing resistance soil-to-soil.

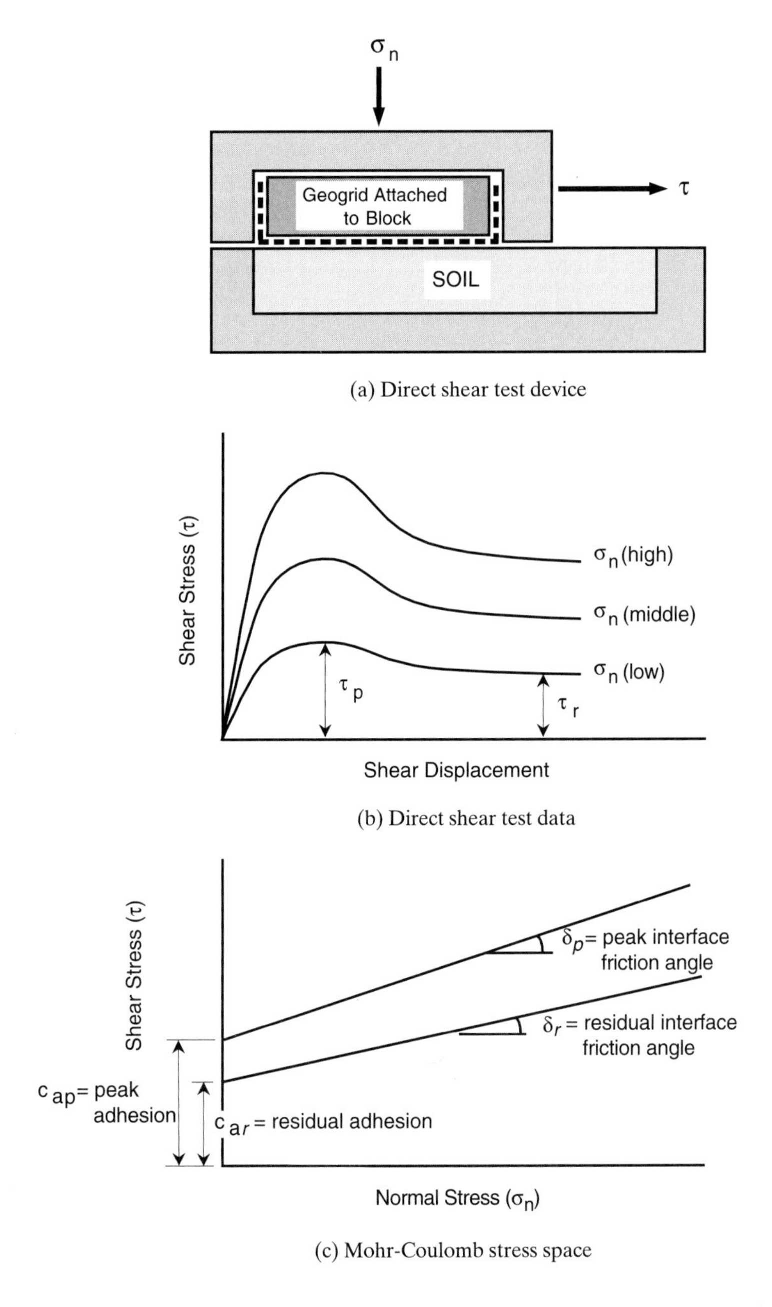

Figure 3.3 Test setup and procedure to assess interface shear strengths involving geogrids.

A large shear box must be used for geogrid testing in order to minimize scale effects. A rule-of-thumb used in soil testing is that the shear testing device must be more than 10 times the size of the largest soil particle. If an analogy is made to the geogrid's apertures, this would generally require a 300 × 300 mm box or larger for geogrid shear

TABLE 3.2 RESULTS OF DIRECT SHEAR TESTS USING VARIOUS GEOGRIDS*

Test Condition	Test 1		Test 2	
	Friction Angle (deg.)	Efficiency (%)	Friction Angle (deg.)	Efficiency (%)
Soil-to-Soil	44	100	44	100
Soil-to-Biaxial Geogrid 1	43	96	44	100
Soil-to-Biaxial Geogrid 2	45	103	45	103
Soil-to-Biaxial Geogrid 3	46	107	46	107
Soil-to-Uniaxial Geogrid 1	35	72	37	78
Soil-to-Uniaxial Geogrid 2	37	78	39	84
Soil-to-Uniaxial Geogrid 3	42	93	43	96

*The geogrids were firmly attached to a wooden platen in the movable portion of the shear box and slid over the stationary soil in the bottom of the shear box.

testing. ASTM D5321 for direct shear testing of geosynthetics requires this size of test device.

Test results using an 450 × 450 mm shear box are shown in Table 3.2. The soil in all cases consisted of a well-graded angular sand (SW) in the dry condition and in a dense compaction state (≅ 90% relative density). The cohesion of the soil was zero. The resulting peak friction angle of the soil by itself was 44°. The efficiencies were calculated on the basis of Eq. (3.2). Note that the efficiencies are all quite high, with many as high as the soil itself. This is understandable since the general configuration of geogrids (with their rather large apertures and relatively thick ribs) forces the failure plane into the soil itself. If any reduction in the soil's strength occurs, it is only along the surface of the geogrid's ribs. Conversely, improvement in the soil's shear strength might be gained by having an element of bearing occur against the surface of the transverse ribs of the geogrid. While it is generally prudent to do product-specific and soil-specific shear testing, the shear strength of most soils small enough to fit into the geogrids apertures will be fully mobilized by most geogrids.

An investigation of the influence of aperture size versus soil particle size on the frictional efficiency of a number of geogrids is available from Sarsby [2]. He finds that the optimum transfer of shear stress, that is, the highest efficiency, occurs when

$$B_{GG} > 3.5\, d_{50} \tag{3.3}$$

where

B_{GG} = the minimum width of geogrid aperture, and
d_{50} = the average particle size of the backfilling soil.

This is an important consideration when selecting the type of backfill to be used around geogrids. Fortunately, the criterion can readily be accommodated by a wide selection of soil types for backfilling purposes.

Anchorage Strength from Soil Pullout. The intrinsic merit of geogrids comes about by their anchorage strength or pullout resistance, which can far exceed the direct shear strength that has just been discussed. Interesting comparison tests between steel grids, steel plate, polymer geogrids, and polymer geonets are reported by Ingold [3]. This behavior comes about by virtue of the large apertures in the geogrid allowing for soil strike-through from one side of the geogrid to the other. Obviously, the soil particles must be sufficiently small to allow for full penetration, thus the d_{50} value in Eq. (3.3) represents the recommended maximum particle size for a particular geogrid's minimum aperture width.

The anchorage strength or pullout resistance is a result of three separate mechanisms, as illustrated in Figure 3.4. One is the shear strength along the top and bottom of the longitudinal ribs of the geogrid. The second is the shear strength contribution along top and bottom of the transverse ribs. The third mechanism is the passive resistance against the front of the transverse ribs. In the last mechanism the soil goes into a passive state and resists pullout by means of bearing capacity. It has been analytically shown that this bearing capacity can be a major contributor to the overall anchorage strength of geogrids [4]. It indeed is a geogrid's forte and can be used admirably in this behavioral mode. Experimental evidence follows the same trends. Example 3.1 shows this effect.

Example 3.1

For an idealized relatively stiff geogrid, as shown in Figure 3.4, subjected to a normal stress of 25 kPa, calculate the anchorage capacity of each of its strength components and the percent contribution due to bearing capacity (BC). The arbitrary dimensions are $L = 900$ mm $\times$ $W = 300$ mm. The longitudinal ribs are at 50 mm spacings and the transverse ribs are at 100 mm spacings. All the ribs are 15 mm wide $\times$ 3.5 mm thick. In the analysis use a shear strength of 25 tan 30° = 14.4 kPa and a bearing capacity 800 kPa based on a soil friction angle of 35°.

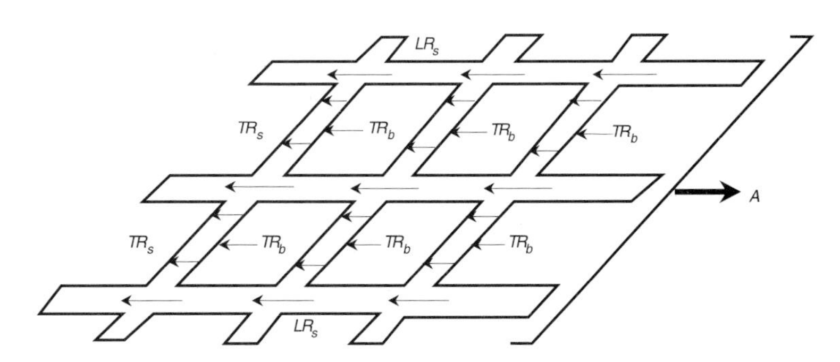

Legend

A_s = Total Anchorage (Pullout) Strength
LR_s = Longitudinal Rib Shear Strength
TR_s = Transverse Rib Shear Strength
TR_b = Transverse Rib Bearing Strength

Figure 3.4 Mechanisms involved in geogrid anchorage strength. ([4, 5])

Solution: The formulation based on Figure 3.4 is as follows. Note that no deduction was taken at rib cross-over points. The anchorage force at failure is:

$$A = 2(\Sigma LR_s + \Sigma TR_s)\tau + (\Sigma TR_b)q_o$$

$$= 2[(0.015 \times 0.900)6 + (0.015 \times 0.300)9]14.4 + [(0.0035 \times 0.300)9]800$$

$$= 2.33 + 1.17 + 7.56$$

$$A = 11.06 \text{ kN}$$

and the percent bearing capacity is:

$$\% \text{ BC} = \left(\frac{7.56}{11.06}\right)100$$

$$= 68\% \text{ of the total anchorage force}$$

Note that the degrees of mobilization of the three components of anchorage resistance during pullout are functions of the load-extension properties of the longitudinal ribs and of the flexibility and load-extension properties of the transverse ribs [5].

The following considerations are important for a soil pullout test setup to determine anchorage strength:

- The test box must be deep enough to permit soil deformation above and below the geogrid as it pulls out of the soil mass. For gravel-size soils, this probably requires 300 mm of soil above and below the geogrid.
- The test box must be long enough to allow for the applied stress on the geogrid to dissipate fully. The number of transverse ribs that are required is dependent on both the geogrid and soil type. A box *at least* 1.0 m long is necessary.
- With such a large size test box, functioning at a high normal stress, the total forces involved can be enormous. This requires a very strongly braced and supported system.
- The geogrid must be gripped from within the encapsulated soil mass. If it is gripped from outside of the test box, passive pressure will be set up against the face of the box, which will impose additional (and quite unknown) normal stresses on the front portion of the test specimen.
- Geogrids, being quite strong, will require a high-strength withdrawal system for the actual geogrid pullout (or tensile failure) to occur.
- To monitor the geogrid's deformation behavior, a number of telltales on different parts of the embedded geogrid are necessary—that is, the incremental movement should be monitored. These telltales are often steel wires attached to the geogrid's nodes and extending out through the rear of the box to which dial indicators are attached.

Figures 3.5a, b show a soil pullout box (schematic diagrams and photographs) for testing the anchorage behavior of geogrids. While the above-described test is clearly not simple (the test is probably the most sophisticated and expensive of all geosynthetic

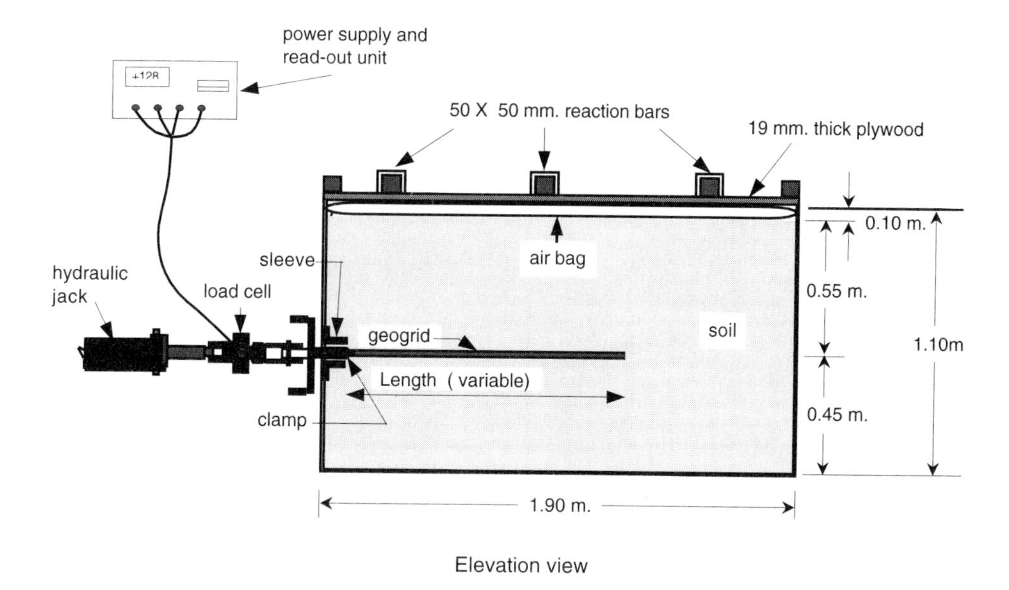

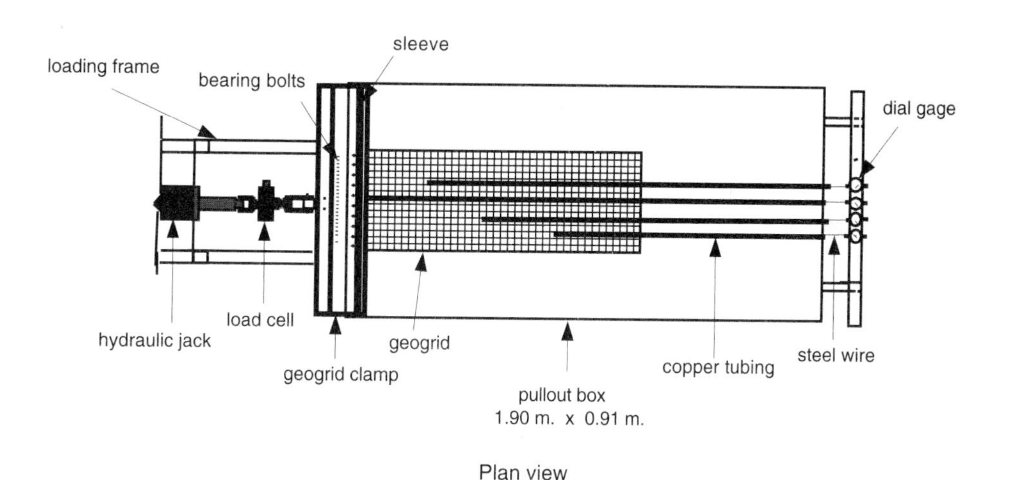

Figure 3.5 Diagrams of soil pullout box for evaluating the anchorage behavior of geogrids.

performance tests at this time), many such tests have been conducted. Some results from the above type of soil pullout testing are shown in Figure 3.6.

Analysis of the type of data shown in Figure 3.6 leads to the determination of an interaction coefficient C_i, which can be used for the design of a specific type of geogrid embedded in the anchorage zone behind a potential failure plane. However, the value of C_i is soil type and test parameter specific. For example, if a geogrid anchorage test is conducted to failure by sheet pullout, the following equation can be formulated.

$$T = 2C_i L_e \sigma_n' \tan \phi' \tag{3.4}$$

Figure 3.5 (*continued*) Photographs of front of pullout box (showing hydraulic jack and instrumentation) and back of pullout box (showing dial gages).

where

T = anchorage capacity per unit width (kN/m),
C_i = interaction coefficient (dimensionless),
L_e = length of geogrid embedment (m),
σ'_n = effective normal stress in the geogrid (kPa), and
ϕ' = effective soil friction angle (deg.).

Note that the value of ϕ' is for the soil alone and not the soil-to-geogrid value. Also, Eq. (3.4) could be modified to handle cohesive soils, but usually granular soils are se-

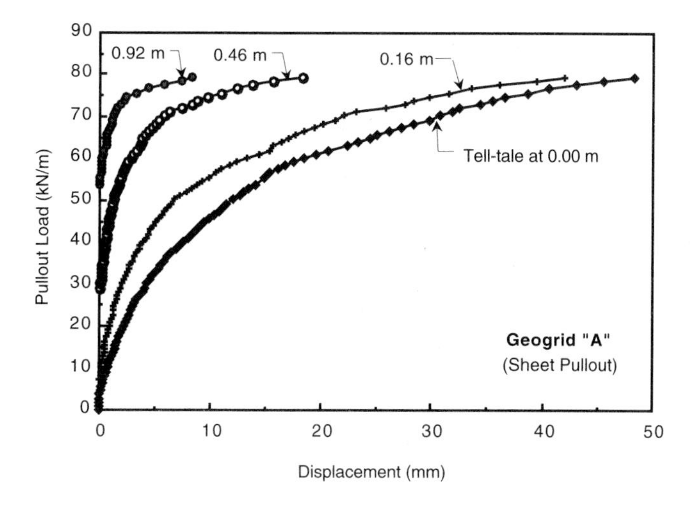

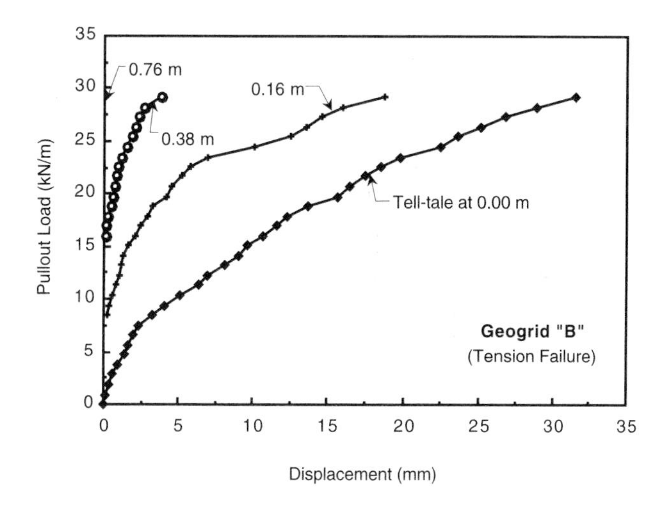

Figure 3.6 Results from selected geogrid pullout tests in a well-graded concrete sand at 69 kPa normal stress at a pullout rate of 1.5 mm/min and geogrid length of 0.92 m.

lected for backfill materials and if not, the omission of a cohesion term leads to a conservative design.

Example 3.2

Given the data of Figure 3.6 (the upper curve), where the length is 0.92 m, the normal stress is 300 mm of 19 kN/m^3 soil plus a 70 kPa surcharge and the effective friction angle is 35°, what is the interaction coefficient for this geogrid in this particular type of well-graded sandy soil?

Solution: First σ'_n is calculated.

$$\sigma'_n = (0.3)(19) + (70)$$
$$= 75.7 \text{ kPa}$$

Then the interaction coefficient is calculated using $T_{\text{ult}} = 80$ kN/m (from Figure 3.6).

$$80 = 2C_i(0.92)(75.7) \tan 35°$$

$$80 = 97.5C_i$$

$$C_i = 0.82$$

Note that the consideration of a soil pullout test, such as described here, completely avoids the issue of in-isolation junction strength (recall the preceding discussion of single rib and junction strength). The geogrid junctions in this test are challenged the same way they are in situ, by having a specific normal stress applied to them through soil embedment. If the junctions are inadequate, the system will fail at a low tensile stress and this will be reflected by a relatively low value of interaction coefficient.

Anchorage Strength from Wall Connections. When geogrids are used to construct soil-reinforced walls, the front edge generally terminates with a wall facing panel (mechanical connection) or modular block facing (friction and/or mechanical connection). The capability of the geogrid's connection to the wall facing should generally be evaluated. There is no standardized test for this, but Kliethermes et al. [6] and Bathurst et al. [7] have evaluated many different situations. Using modular concrete blocks with a geogrid in its proper location and orientation, a predetermined normal stress is imposed. The geogrid is tensioned until failure. Failure can come about in numerous ways: geogrid tension failure, geogrid connection failure, geogrid slippage, or block wall failure. The test nicely exposes the mode of failure and also the ultimate strength of the entire anchorage system. Recognize however that there are at least 50 types of modular-block wall systems, many with their own geogrid type and wall type. Thus this type of experiment must use the specific materials that will be used in the actual wall construction. Furthermore, substitutions cannot be allowed at the time of bidding or during construction via value engineering or the like.

3.1.3 Endurance Properties

As geogrids are used in critical reinforcement applications, some of which require long service lifetimes, it is generally necessary to evaluate selected endurance properties. Installation damage, creep, and stress relaxation will be addressed.

Installation Damage. As with all geosynthetics, the placement of geogrids in the field requires a considerable degree of planning and care. As happens all too often with careless field construction crews and heavy machinery, immediate installation damage of the geogrid can occur. Other uncertainties in this same area are coarse soil impingement, falling objects, and other accidents that occur before the geogrid is covered. A few studies have been conducted in which geogrids have been exhumed after installation and tested, with comparisons being made to the as-received material. Loss in strength has often occurred. In-house investigations (recall Figure 2.19, which had nine geogrids in the total study), show that strength reductions of 0 to 30% are possible (see Koerner et al. [8]). Generally, though, the higher strength-loss values come about where large, poorly graded, quarried aggregate is used and heavy construction equipment performs the placement and compaction. If it is necessary to use such materials and methods, it is prudent to first place a cushioning layer of sand adjacent to the geogrid.

Tension Creep Behavior. A major endurance property involving geogrids is their sustained-load deformation or tension creep. Since all polymers used in the manufacturing of geogrids consist of long-chain molecules arranged in crystalline regions with interspersed amorphous regions, the creep response reflects upon the percent crystallinity. In general, this structure is reflected in the following manner.

For *nonoriented polyolefins* (polypropylene and polyethylene), the molecular chains slip along one another within the crystalline regions. Note that polyolefins are, in general, highly crystalline.

For *oriented polyolefins* the orientation creates a molecularly fibrous (or *affine*) structure, and when creep occurs it does so between fibers in the oriented, and stressed, direction.

For *polymers like polyester and polyamide*, creep slippage hardly occurs in the crystalline region. Here the chains break at the interface of the crystalline and amorphous regions.

Apart from the molecular structure, creep is predominantly a function of stress level, time, temperature, and a number of environmental factors to be discussed later. This property has been extensively evaluated on many geogrids, with results for a particular product given in Figure 3.7. Note that up to approximately 29 kN/m, which is 41% (29/72) of the breaking load of this particular geogrid, the creep deformation is negligible. This corresponds to a creep-strain limit of 10%, which has been extrapolated as being equivalent to a 120 year design life. The allowable, or working, stress will reflect this type of information in that the reduction factor against creep will be the inverse of 41%, which is 2.4. Note in Table 2.8 that PE and PP geotextile yarns require creep reduction factors that are generally higher (i.e., 3.0 to 4.0), due to the lack of geogrid orientation effects, as occurs in the manufacture of homogeneous PE and PP geogrid materials. The above discussion on creep refers to the wide-width creep behavior of the longitudinal ribs of the geogrids. The tension creep test has been adopted by ASTM as test method D5262 and by ISO as ISO/DIS 13431. Creep test data (unconfined or confined) on polyolefin geogrids can be portrayed as isochronous creep data, also shown in Figure 3.7. The creep behavior is readily observed in such a graph.

A variation of the tension creep test just described is the creep rupture procedure presented by Ingold et al. [10]. In this procedure higher stresses are imposed on the test specimens, causing failure to occur in a relatively short time. The data points produce a curve on log-time scale. The desired service lifetime of the system is then used to select the percentage of short term failure load from the creep rupture data. The inverse of this value produces a creep reduction factor for allowable strength (see Miyato [11]).

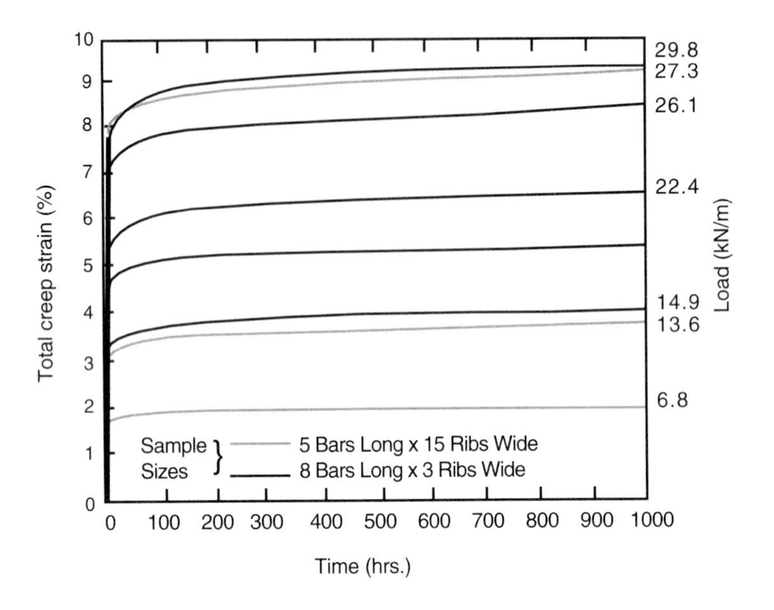

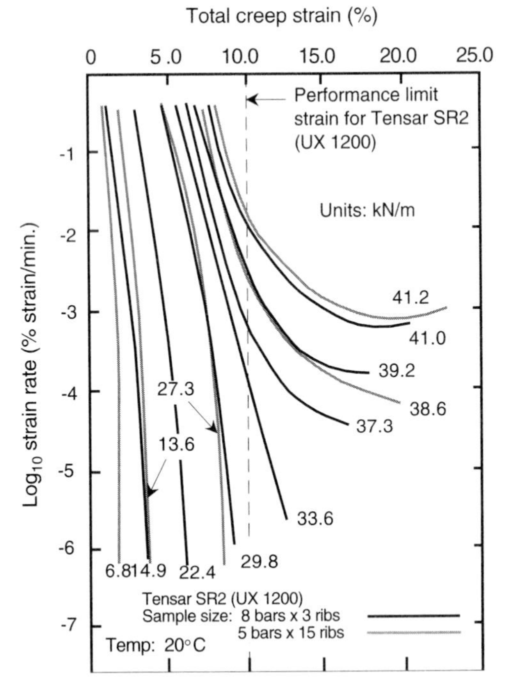

Figure 3.7 Constant stress deformation (creep) test results of UX 1200 geogrid. (After McGown et al. [9])

Stress Relaxation Behavior. Greenwood [12] has evaluated stress relaxation in HDPE geogrids. He plotted experimental data and from this produced isochronous stress-strain curves. These data were then compared to the isochronous stress-strain curves from creep data on the same type of geogrids with reasonable agreement.

Greenwood concludes that in absence of actual data, geogrid stress relaxation response can be reasonably estimated in this manner.

3.1.4 Degradation Properties

For geogrids (both polyolefins and polyester) being used in permanent reinforcement applications, it is generally necessary to evaluate selected degradation properties. Briefly many of these issues are as follows.

Temperature effects. Given the temperature ranges of typical environments, temperature extremes (hot or cold) should have no serious adverse effects on geogrids. The one caution is that high temperatures can exacerbate strains arising from creep and/or stress relaxation. This requires actual testing at the anticipated temperatures, time-temperature superposition assumptions, or the use of higher-than-typical reduction factors on conventional creep information.

Oxidation effects. This long-term mechanism, which is applicable to polyolefin degradation, is discussed in section 2.3.6.

Hydrolysis effects. This long-term mechanism, which is applicable to polyester degradation, is discussed in section 2.3.6.

Chemical effects. Polyolefins and polyester have shown excellent resistance to a wide range of chemicals. If unusual conditions exist, however, the situation may dictate specific testing in the actual chemical environment, that is, such conditions as landfill leachates. Recognize, however, that the incubation involved and the subsequent tests are difficult and expensive, to say the least.

Radioactive effects. Unless high-level radioactive materials are in the immediate vicinity, low-level, and mixed radioactive materials should pose no problem to geogrids.

Biological effects. The discussion of the general lack of biological degradation of geotextiles in Section 2.3.6 is applicable for geogrids, with the possible exception of the coatings on flexible geogrids. Latex, bitumen, or plasticizers in PVC may be sensitive to microorganisms, but no studies on geogrids are available to my knowledge. Even if such an attack on the coatings were possible, the high-crystallinity polyester fibers (to which these geogrids owe their strength) would probably remain unaffected.

Sunlight (UV) effects. As with all polymeric materials, ultraviolet degradation can occur over time and product degradation will follow. The discussion in Section 2.3.6 applies to geogrids as well as to geotextiles. Regarding the problem of timely cover, however, exposure time for geogrids can be considerably longer than for geotextiles, due to the thickness of the ribs of the polyolefin geogrids and the protective coating on the polyester types. A specification should not be left open-ended, however, and a suggested 30-day maximum exposure before covering is recommended.

Stress-crack resistance. Highly crystalline polymers are sometimes sensitive to brittle cracking while under stress. The test used to evaluate this tendency is

ASTM D5397, the notched constant tensile load (NCTL) test and the single-point version (SP-NCTL) described in the Appendix to D5397. Both are explained in Chapter 5 on geomembranes. Only highly crystalline polyethylene is of concern and a review with respect to HDPE geogrids is given by Wrigley [13]. It is not known to be a problem insofar as field problems using geogrids are concerned.

3.1.5 Allowable Strength Considerations

The basis of the design-by-function concept is the establishment of a global factor of safety. For geogrids, where reinforcement is the primary function, this factor of safety takes the following form:

$$FS = \frac{T_{\text{allow}}}{T_{\text{req}}} \tag{3.5}$$

where

FS = factor of safety; (to accommodate unknown loading conditions or uncertainties in the design method),
T_{allow} = tensile strength from laboratory testing, and
T_{req} = required tensile strength as obtained from design for the particular field situation.

The allowable value comes from a tensile test of the type described in Section 3.1.2, and we must compare the test setup with the intended field situation. If the test method is not completely field-simulated, the laboratory value must be suitably adjusted. This will generally be the case. Thus the laboratory-generated tensile strength is usually an ultimate value, which must be reduced before being used in design, that is,

$$T_{\text{allow}} < T_{\text{ult}}$$

One way of accomplishing this is to place reduction factors on each of the items not adequately modeled in the laboratory test. For example, the following equation should be considered [14].

$$T_{\text{allow}} = T_{\text{ult}}\left[\frac{1}{RF_{ID} \times RF_{CR} \times RF_{CD} \times RF_{BD}}\right] \tag{3.6}$$

where

T_{ult} = ultimate tensile strength from a standard in-isolation wide-width tensile test,
T_{allow} = allowable tensile strength to be used in Eq. (3.5) for final design purposes,
RF_{ID} = reduction factor for installation damage,
RF_{CR} = reduction factor for avoiding creep over the duration of the structure's lifetime,

RF_{CD} = reduction factor against chemical degradation, and
RF_{BD} = reduction factor against biological degradation.

Note that some of these values may be 1.0 or slightly above 1.0, and may therefore be inconsequential. Still others, not specifically mentioned in Eq. (3.6), may be included as the situation warrants. For example, reduction factors against ultraviolet degradation, RF_{UV}, field seams, RF_{seam} or penetrations, RF_{pen}, may be included on a site specific basis. Guidelines for various reduction-factor values are given in Table 3.3. Note that some of these values are preliminary and are based on relatively sparse information. It is necessary to consider each item individually and make a conscious decision as to how important it is for the site-specific situation. In Examples 3.3 and 3.4 the values used are assumed on the basis of a hypothetical project and construction method.

Example 3.3

What is the allowable geogrid tensile strength to be used in the construction of a permanent wall adjacent to a major highway if the ultimate strength of the geogrid is 70 kN/m?

Solution: Using estimated values from Table 3.3 in Eq. (3.6) gives

$$T_{allow} = T_{ult}\left[\frac{1}{RF_{ID} \times RF_{CR} \times RF_{CD} \times RF_{BD}}\right]$$

$$= 70\left[\frac{1}{1.2 \times 2.5 \times 1.3 \times 1.1}\right]$$

$$= 70\left[\frac{1}{4.29}\right]$$

$$T_{allow} = 16.3 \text{ kN/m}$$

Example 3.4

What is the allowable geogrid tensile strength to be used in the construction of a temporary embankment separating different zones within a landfill if the ultimate strength of the geogrid is 80 kN/m?

Solution: Using estimated values from Table 3.3 in Eq. (3.6) gives

$$T_{allow} = T_{ult}\left[\frac{1}{RF_{ID} \times RF_{CR} \times RF_{CD} \times RF_{BD}}\right]$$

TABLE 3.3 RECOMMENDED REDUCTION-FACTOR VALUES FOR USE IN EQ. (3.6) FOR DETERMINING ALLOWABLE TENSILE STRENGTH OF GEOGRIDS

Application Area	RF_{ID}	RF_{CR}	RF_{CD}	RF_{BD}
Unpaved roads	1.1 to 1.6	1.5 to 2.5	1.0 to 1.5	1.0 to 1.1
Paved roads	1.2 to 1.5	1.5 to 2.5	1.1 to 1.6	1.0 to 1.1
Embankments	1.1 to 1.4	2.0 to 3.0	1.1 to 1.4	1.0 to 1.2
Slopes	1.1 to 1.4	2.0 to 3.0	1.1 to 1.4	1.0 to 1.2
Walls	1.1 to 1.4	2.0 to 3.0	1.1 to 1.4	1.0 to 1.2
Bearing capacity	1.2 to 1.5	2.0 to 3.0	1.1 to 1.6	1.0 to 1.2

$$= 80\left[\frac{1}{1.2 \times 2.0 \times 1.0 \times 1.0}\right]$$

$$= 80\left[\frac{1}{2.4}\right]$$

$$T_{\text{allow}} = 33.3 \text{ kN/m}$$

Note that these examples could just as well have been framed so as to yield an ultimate strength from a given allowable value. This would be the case if we were working from an analytical method that generated a design value. This design value (as with the allowable value) would have to be increased by reduction factors to arrive at a required (or ultimate) tensile strength.

3.2 DESIGNING FOR GEOGRID REINFORCEMENT

The primary function of geogrids is invariably reinforcement; this section discusses several reinforcement application areas. The order will parallel that of Sections 2.6 and 2.7 on geotextile reinforcement, with the addition of several areas that are unique to geogrids.

3.2.1 Paved Roads—Base Courses

The use of geogrids in paved road base courses is an area in which the large aperture size of geogrids provide an excellent advantage. Here the geogrids are placed within the granular base course, typically crushed stone, with the intention of providing an increased modulus, hence a lateral confinement to the system. This lateral confinement is intended to resist the tendency for base courses to *walk out* from beneath the repetitive traffic loads imposed on the concrete- or bitumen-pavement surface. The situation is applicable for railroad systems as well, perhaps even more so due to nature and intensity of the dynamic loads.

A number of laboratory tests have been conducted to assess the potential benefits and mechanisms involved, most significantly the work of Haas [15] and Abd El Halim [16, 17]. In a large test track measuring 4.0 m long × 2.4 m wide × 2 m deep and using 10 kN loads applied sinusoidally at a frequency of 10 Hz on a 300 mm diameter circular plate, five test series (called loops) were performed (see Table 3.4). Loop 1 compared the response of nonreinforced and reinforced sections using both dry (strong) and saturated (weak) subgrade conditions. Failure appeared in the nonreinforced sections earlier than the reinforced sections under both conditions. Loop 2 provided considerable data, some of which are shown in Figure 3.8. Figure 3.8a shows little difference in elastic deflection among the four trials. More significant is the angle of curvature shown in Figure 3.8b and the elastic strain at the bottom of the asphalt shown in Figure 3.8c. Both indicate a 50% reduction for the reinforced sections, thereby indicating a significant load-spreading phenomenon. Figure 3.8d shows that the permanent surface deformation of the reinforced section is substantially improved over the nonreinforced section. At a 20-mm failure assumption, the nonreinforced section carried

TABLE 3.4 TEST LOOPS AND CONTROLLED VARIABLES

Loop Number	Test Number	Asphalt Thickness (mm)	Subgrade Condition	Description
1	1	150	Dry	Control
	2	150	Dry	Reinforced
	3	150	Saturated	Reinforced
	4	150	Saturated	Control
2	1	165	Dry	Control
	2	165	Dry	Reinforced
	3	165	Dry	Reinforced
	4	165	Dry	Control
3	1	250	Saturated	Control
	2	150	Saturated	Reinforced
	3	150	Saturated	Reinforced
	4	200	Saturated	Control
	5	250	Saturated	Control
	6	150	Saturated	Reinforced
4	1	200	Dry	Reinforced
	2	200	Dry	Reinforced
	3	250	Dry	Control
	4	250	Dry	Control
	5	250	Saturated	Control
	6	250	Saturated	Control
	7	200	Saturated	Reinforced
	8	200	Saturated	Reinforced
	9	200	Saturated	Reinforced
5	1	115	Dry	Control
	2	115	Dry	Control
	3	115	Dry	Reinforced
	4	115	Dry	Reinforced

110,000 load repetitions, compared to 320,000 for the reinforced case. In the context of the discussion on geotextiles used in the control of reflective cracking of paved roadways, this could be called a geogrid effectiveness factor (GEF) equal to 2.9.

Loop 3 investigated the equivalent thickness that can be attributed to the reinforcement. The results, shown in Figure 3.8e, indicate that the 150-mm reinforced section carried about 80,000 load cycles compared to only 34,000 load cycles for the 200-mm nonreinforced and 92,000 loads cycles for the 250-mm nonreinforced. In other words, 150 mm of reinforced asphalt nearly compares to 250 mm of nonreinforced asphalt. Loop 4 confirmed these results, in that reinforced sections result in a savings of 50 to 100 mm of nonreinforced asphalt. Loop 5 involved pressure cells in the soil subgrade and confirmed the load-spreading capability of the reinforcement.

Studies such as this indicate that a reinforcement function is provided to the pavement system by the geogrid, albeit by a rather complex set of mechanisms. Some possible contributors include the following: increasing initial stiffness, decreasing long-term vertical deformation, decreasing long-term horizontal deformation, increasing tensile strength, reducing cracking, improving cyclic fatigue behavior, and holding *the*

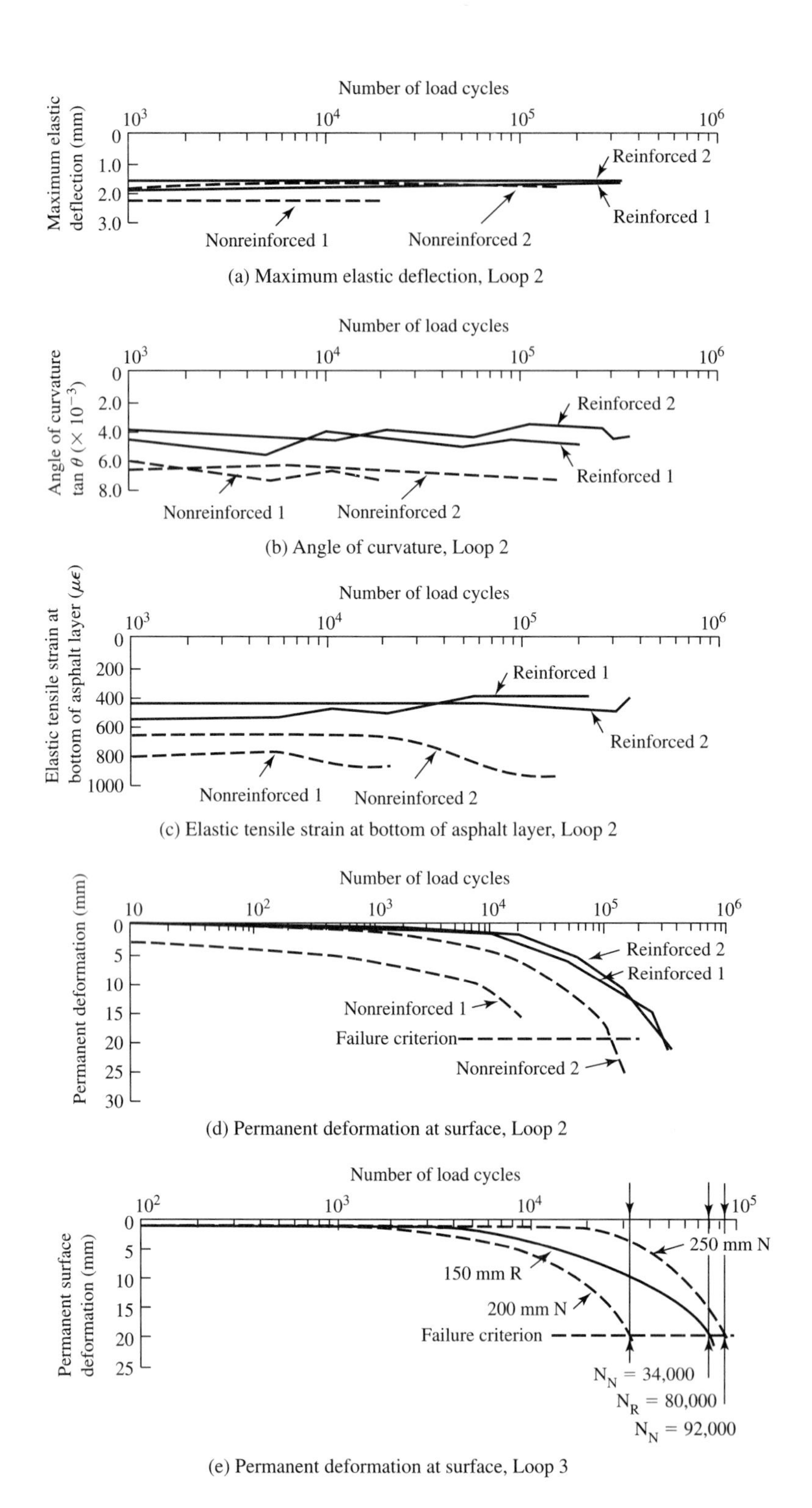

Figure 3.8 Geogrid-reinforced asphalt pavement test results. (After Haas [15]; Adb El Halim [16, 17])

system together. This leads to design difficulties as far as a specific methodology is concerned. On a simplified basis we could use a geogrid effectiveness factor (GEF) and divide it into the design traffic number to determine a modified value and design accordingly, that is,

$$\mathrm{DTN}_R = \frac{\mathrm{DTN}_N}{\mathrm{GEF}} \tag{3.7}$$

where

DTN_R = design traffic number for the geogrid-reinforced case,
DTN_N = design traffic number under standard (nonreinforced) conditions (e.g., using the Asphalt Institute's procedures), and
GEF = geogrid effectiveness factor $\simeq$ 3.0 for the type of geogrid evaluated to provide the data of Figure 3.8.

Carroll et al. [18], have further refined the technique using the same experimental database to calculate a structural number as per AASHTO [19]. Using the concept of structural number, the nonreinforced (control) section is

$$\mathrm{SN} = 25a_1d_1 + 25a_2d_2 \tag{3.8}$$

where

SN = structural number,
a_i = layer coefficients (0.40 for asphalt and 0.14 for granular stone base), and
d_i = thickness (mm) of each layer.

Using a soil subgrade support value S, obtained from a CBR test, the number of 80 kN single-axle equivalents for any cross section can be calculated. A load-correction factor is then calculated for the geogrid-reinforced sections within each experimental test loop. An estimate of the reinforced-pavement SN is derived, and a ratio for reinforced-to-nonreinforced sections is generated. When plotted against the actual reinforced base course thickness, this ratio is seen to be linear. Different values, but the same trend, are seen for geogrids placed in the middle and at the bottom of the base course. A design chart that enables a conventional nonreinforced base course thickness to be converted to a geogrid-reinforced section is given in Figure 3.9. Note that a transition occurs at 375 mm, where the geogrid can be placed either in the middle or at the bottom of the base course. It is important to recognize that this curve is based on experimental data for the specific geogrid used. We cannot use this same information with other geogrids. Furthermore, an equivalency between geogrids is difficult to even suggest. Longitudinal and transverse rib strength, modulus in both directions, and junction strength are all included in the reinforcement mechanisms, but to what degree needs product-specific investigation.

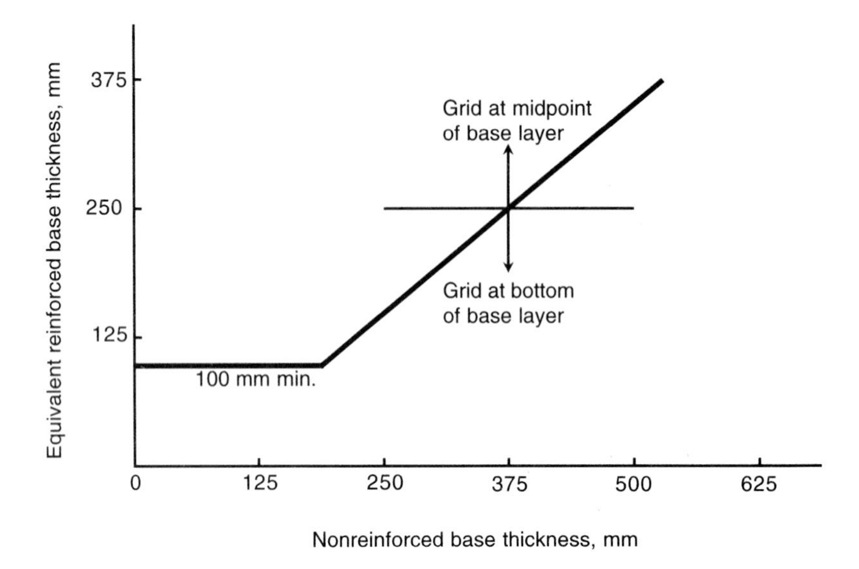

Figure 3.9 Geogrid-reinforced base course for paved highway section using Tensar BX 1100 geogrids. (After Carroll et al. [18])

3.2.2 Paved Roads—Pavements

There is considerable ongoing work on the placement of geogrids directly within the pavement material (bitumen or concrete). This section pertains to both new construction and the rehabilitation of existing pavements (i.e., the retardation of reflective cracks). Brown et al. [20] have reported that at high deformations of the wearing surface, geogrids clearly minimize rutting. However, at low deformations, the improvement is nominal. Keeping the geogrid tight during its placement (and possibly even prestressing it) appears to be logical, but it is clearly difficult to achieve. Equipment and techniques are described by Kennepohl and Kamel [21]. The material and type of geogrid is very important, since asphalt will not easily bond to the surfaces of polyethylene or polypropylene, but might easily do so to flexible geogrids particularly if bitumen-coated. The influence of geogrid shrinkage during the placement of hot asphalt may be a problem for molecular stress relaxation and loss of strength or modulus for highly oriented geogrids. In spite of the above, success has been reported in the prevention of reflective cracking using geogrids as crack arresters [20].

The use of geogrids to retard and minimize the reflective cracks of old pavements from propagating through newly placed asphalt overlays is a topic of great interest. The results of laboratory testing by Molenaar and Nods [22] suggest the use of a power law to calculate the rate of crack propagation through the new overlay thickness.

$$\frac{dc}{dN} = AK^n \tag{3.9}$$

where

$$\frac{dc}{dN} = \text{crack propagation rate per number of load cycles,}$$
K = stress intensity factor, and
A, n = experimentally obtained constants.

Example 3.5 illustrates how Eq. (3.9) can be used in the prediction of overlay lifetime without, and then with, different types of geogrids and a geotextile.

Example 3.5

A 100 mm asphalt overlay is to be placed on top of a severely cracked pavement having a cement-treated base. The DTN for the pavement is 100,000 load repetitions (cycles) per year. The combined overlay, existing asphalt layer and base profile, yields a design stress intensity factor (K) of 10 $N/mm^{1.5}$ and constants A of 1.0×10^{-8} and n of 4.3. **(a)** Calculate the average rate of crack growth of the new asphalt overlay. At a full-propagation failure assumption, what is the lifetime (in terms of number of cycles) of the new asphalt overlay without reinforcement? **(b)** Redo the problem using the inclusion of various geosynthetic reinforcement materials as follows:

nonwoven geotextile:	$A_{GT} = 0.50\ (A_{\text{nonreinf.}})$	author estimate
polypropylene geogrid:	$A_{PP} = 0.35\ (A_{\text{nonreinf.}})$	author estimate
polyester geogrid:	$A_{PET} = 0.50\ (A_{\text{nonreinf.}})$	[22]
fiberglass geogrid:	$A_{FG} = 0.25\ (A_{\text{nonreinf.}})$	author estimate

Solution: (a) Using the power law in Eq. (3.9), the crack-propagation rate is calculated, from which the number of cycles and the lifetime are obtained. The crack-propagation rate is

$$\frac{dc}{dN} = AK^n$$

$$= 1 \times 10^{-8} \times (10)^{4.3}$$

$$= 0.0002 \text{ mm/cycle}$$

from which the number of load cycles (non reinforced) is

$$N = \frac{T}{(dc/dN)}$$

$$= \frac{100}{0.0002}$$

$$= 500{,}000 \text{ cycles or 5 years}$$

(b) Using the various modified A-values for different types of geosynthetic reinforcement produces the following table:

Reinforcement	Crack growth rate (mm/cycle)	Lifetime (cycles/yr)
none	2.0×10^{-4}	500,000/5 yr
geotextile	1.0×10^{-4}	1,000,000/10 yr
PP geogrid	0.7×10^{-4}	1,400,000/14 yr
PET geogrid	6.6×10^{-5}	1,500,000/15 yr
FG geogrid	5.0×10^{-5}	2,000,000/20 yr

The technique is very intriguing and warrants the additional research that is ongoing at several universities and research institutions.

While many field trials are currently ongoing, it is important to note that geogrids are also being used directly in new pavements and in asphalt overlays to resist thermally induced stresses. This is a very important consideration when designing an asphalt pavement or its overlay. Between night and day and from season to season, the change in temperature can be as much as 55°C. The contraction during this temperature shift is considerable; see Example 3.6.

Example 3.6

Calculate the contraction of a 25 m long section of asphalt pavement undergoing a decrease in temperature of 55°C, assuming that the coefficient of expansion/contraction of the asphalt pavement is 12×10^{-6} per 1°C.

Solution: The calculation is as follows.

$$\Delta L = (25)(55)(12 \times 10^{-6})$$

$$= 0.0165 \text{ m}$$

$\Delta L = 16.5$ mm, which could be in the form of a single crack or many smaller cracks. This depends upon the condition of the pavement, primarily on its oxidation since the original placement.

Biaxial geogrids are placed on existing pavements, generally with an adhesive attached or placed separately, and then covered with a bituminous overlay. Clearly, the tensile strength of the geogrid is mobilized by such thermally induced contraction stresses, probably in the very localized region(s) where the cracks initiate. Thus the necessity for high-tensile strength is apparent. All types of polymeric geogrids are used in this application, as are geogrids made from fiberglass. Fiberglass has some excellent tensile strength characteristics in this regard (e.g., high strength, low elongation, high modulus, and low creep). In addition, there are some ongoing attempts at using geogrids to reinforce portland-cement concrete pavements in a manner similar to that described here with asphaltic pavements.

3.2.3 Unpaved Roads

The use of geogrids to reinforce soft and/or compressible foundation soils for unpaved aggregate roads is a major application area. Many successes have been reported, together with several attempts at a design method. The most advanced analytical method, and the one that will be used here, is that of Giroud, Ah-Line, and Bonaparte [23]. The method follows along lines similar to those described in the geotextile section on unpaved roads. First the nonreinforced situation is handled, and then new concepts are developed for the reinforced case. Here the mechanisms of reinforcement are increased soil strength, load spreading, and membrane support via controlled rutting. The difference in the required thickness of stone base is obtained and then compared to the cost of the installed geogrid. If the latter is less expensive (as it usually is for soft-soil subgrades), it is recommended for use.

For the nonreinforced case, Giroud et al. [23] have adapted a U.S. Army Corps of Engineers empirical formula that includes the number of vehicle passages. For the geogrid-reinforced case, new concepts are developed that include the above-mentioned beneficial mechanisms attributed to inclusion of the geogrid. Their effects are as follows.

An increase in soil subgrade strength from the nonreinforced case to the reinforced case as indicated by a comparison of the following equations:

$$p_e = \pi c_{uN} + \gamma h_o$$

$$p_{\text{lim}} = (\pi + 1)c_{uN} + \gamma h$$

where

p_e = bearing capacity pressure based on the elastic limit (nonreinforced case),
p_{lim} = bearing capacity pressure based on the plastic limit (reinforced case),
c_{uN} = undrained soil strength at the Nth vehicle passage,
γ = unit weight of aggregate,
h_o = aggregate thickness without reinforcement, and
h = aggregate thickness with reinforcement.

An improved load distribution to the soil subgrade due to load spreading, which is quantified on the basis of pyramidal geometric shape. Note Figure 3.10, which

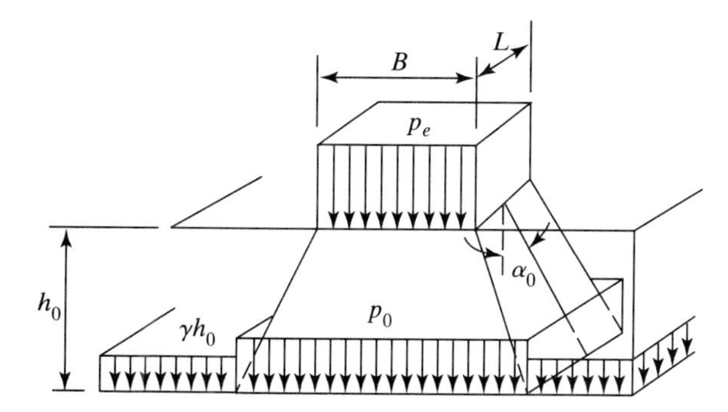

Figure 3.10 Concept of pyramidal load distribution.

shows the angle α_0 for the nonreinforced case versus a similar construction for the reinforced case where the new and larger angle is defined as α. The ratio of reinforced to nonreinforced situations is expressed as a ratio of tan α/tan α_0, which is greater than 1.0.

A tensioned membrane effect, which is a function of the tensile modulus and elongation of the geogrid and the deformed surface of the subgrade soil (i.e., the rut depth).

By taking the combined effect of the first two above-mentioned geogrid-reinforcement mechanisms and comparing it to the nonreinforced case, Giroud et al. [23] have developed the design chart shown in Figure 3.11. The membrane effect has been conservatively neglected. Here (on the right side of the graph) for a standard axle load of 80 kN and any number of vehicle passes from 10 to 10,000, a thickness of nonreinforced stone base (h_0) can be obtained if the soil subgrade strength is known. The rut depth turns out to be relatively insignificant. This value is then extended to the left side of the figure, where it intersects with either

- Curve 1, for BX 1200 geogrids, which assumes a large number of vehicle passes ($N > 1000$) where there is a significant likelihood of aggregate contamination without the geogrid;
- Curve 2, for BX 1200 geogrids, which assumes a low number of vehicle passes and low likelihood of aggregate contamination; or
- Curve 3, for UX 1200 geogrids, which assumes a low number of vehicles passes and low likelihood of aggregate contamination.

This results in an R value that is used in the following equations to determine the aggregate thickness using geogrid reinforcement h. The difference between h_o (nonreinforced) and h (reinforced) is the amount of aggregate saved, Δh.

$$h = Rh_o \qquad \text{for } r < 150 \text{ mm and no channelized traffic pattern}$$

$$h = 0.9\, Rh_o \qquad \text{for } r \geqslant 150 \text{ mm with a channelized traffic pattern}$$

Example 3.7

Given a soil subgrade whose CBR strength is 1.0 and is to carry 1000 standard-axle vehicle passes with a maximum rut depth of 75 mm, what is the required aggregate depth without a geogrid, the aggregate depth with BX 1200 geogrids with a low likelihood of aggregate contamination, and the difference in aggregate thickness between the two cases?

Solution: Using Figure 3.11, the nonreinforced case gives

$$h_o = 0.60 \text{ m}$$

For the geogrid-reinforced case, Curve 2 gives $R = 0.50$ and

$$h = Rh_o$$

$$h = (0.50)(0.60)$$

$$= 0.30 \text{ m}$$

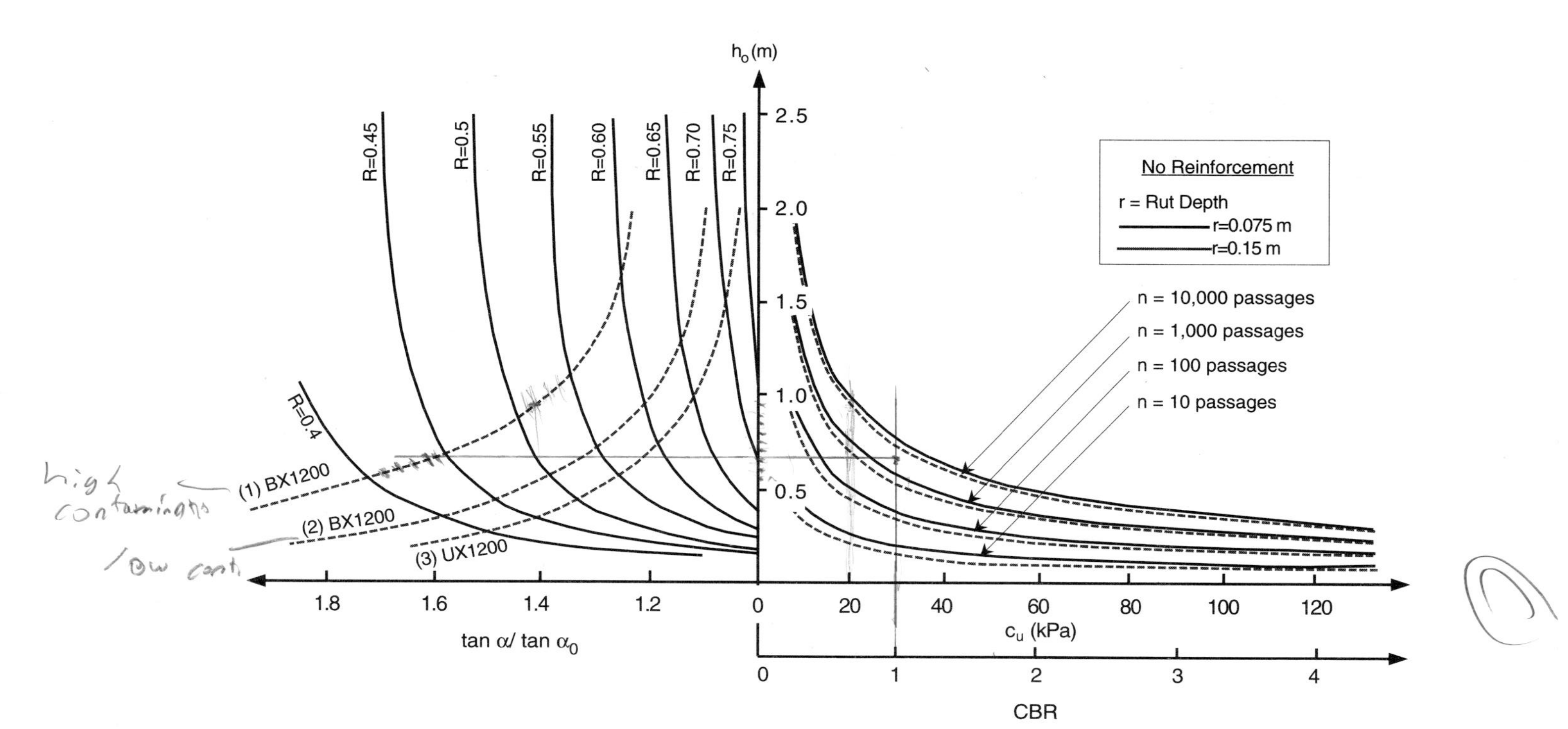

Figure 3.11 Design chart for geogrid-reinforced (left side) and nonreinforced (right side) unpaved roads. (After Giroud et al. [23])

The aggregate saved is

$$\Delta h = 0.60 - 0.30$$
$$= 0.30 \text{ m}$$

3.2.4 Embankments and Slopes

The use of geogrids to reinforce embankments on soft foundations or steep soil slopes directly parallels the techniques and designs that were developed using geotextiles. For embankments on soft foundation soils the methods developed in Section 2.7.2 for geotextiles are exactly the same as those for geogrids, except that the allowable strength is for the specific geogrid product. These examples will not be repeated. The use of geogrids for weak embankment soils or for steepening soil slopes will be the focus of this section.

The use of limit equilibrium methods via a circular arc failure plane, thereby intercepting the various layers of reinforcement, is illustrated in Figures 2.53 and 2.54. This allows for the formulation of a factor of safety expression as follows.

$$FS = \frac{M_R + \sum_{i=1}^{n} T_i y_i}{M_D} \tag{3.10}$$

where

M_R = moments resisting failure due to the soil's shear strength;
M_D = moments causing failure due to gravity, seepage, seismic, dead, and live loads;
T_i = allowable reinforcement strength, providing a force(s) resisting failure;
y_i = appropriate moment arm(s); and
n = number of separate reinforcement layers.

Forsyth and Bieber [24] used this approach to design the reconstruction of a failed slope in California. They selected a desired factor of safety along with a given type of geogrid and calculated the number of layers of reinforcement to realize this value. As the following example illustrates, they used a very low allowable strength for the UX 1200 geogrids (i.e., 6.67 kN/m), which is only 8.4% of the ultimate value. In the context of reduction factors in Section 3.1.5, this is equivalent to 11.8.

Example 3.8

For a failed soil slope of known centroid and radius resulting in a resisting moment of 2010 kN/m and a driving moment of 2570 kN/m, determine **(a)** the factor of safety without reinforcement, and **(b)** the number of layers of UX 1200 geogrid with an ultimate strength of 78.7 kN/m and reduction factors of 11.8. The average centroid of the reinforcement is 14.3 m and the global factor of safety required is 1.4.

Solution: **(a)** The factor of safety for the nonreinforced case is the following.

$$FS = \frac{M_R}{M_D}$$

$$= \frac{2010}{2570}$$

$$= 0.78 \qquad \text{which indicates failure, as indeed did happen}$$

(b) The geogrid-reinforced case is as follows.

$$T_{\text{allow}} = T_{\text{ult}}/(\Pi \text{RF})$$
$$= 78.7/11.8$$
$$= 6.67 \text{ kN/m}$$

$$\text{FS} = \frac{M_R + (n)(T_{\text{allow}})(Y_{\text{avg}})}{M_D}$$

$$1.4 = \frac{2010 + (n)(6.67)(14.3)}{2570}$$

$$n = 16.6 \qquad \text{use 17 layers}$$

When constructing such remediated slopes, or when building steep soil slopes, the main reinforcement layers (as illustrated in Example 3.8) are usually interspersed with secondary reinforcement layers. These layers aid in compaction at the face of the slope and also tend to reduce surface erosion. Figure 3.12 shows a case of widening a major highway to accommodate an additional 3 m wide paved shoulder in existing fill sections. A 45° backcut was made into the existing slope, which allowed for the construction of the reinforced slope.

Example 3.8 did not go into the details of determining the location of the reinforcement layers or their lengths. For these details, slope stability methods are nicely adapted to computer modeling and have resulted in a number of excellent design charts. Schmertmann et al. [26] have developed a summary of the various investigators and some of the assumptions that are made in their studies (see Table 3.5). All of them used limit equilibrium methods to determine the reinforcement (geogrid or geotextile) spacing. However, only Jewell et al. [29], Ruegger [32], and Schmertmann et al. [26] have given charts for the required reinforced lengths. The former two methods use constant-length reinforcement layers placed parallel to the slope face, whereas Schmertmann et al. use gradually decreasing lengths as the layers proceed higher in the slope. For high embankments, where the potential failure surface is curved (e.g., a logarithmic spiral), this is both accurate and more practical. However, for low and medium walls, the more conservative approach of Jewell et al. [29] is favored by the author and will be used here.

For reinforced slopes placed on adequately strong, level foundations, limit equilibrium methods can be used in the form of a two-part wedge surface as shown in Figure 3.13. The design charts include varying soil properties, slope angles, pore-water pressures, and geometric considerations (see Figures 3.14a, b, c, each of which is for a different value of pore-water ratio of the backfill soils). Example 3.9 illustrates the use of one of these charts.

Example 3.9

We plan to construct a soil embankment at a 70° slope angle with the horizontal and 10 m height, to be reinforced with geogrids having an ultimate strength of 180 kN/m and

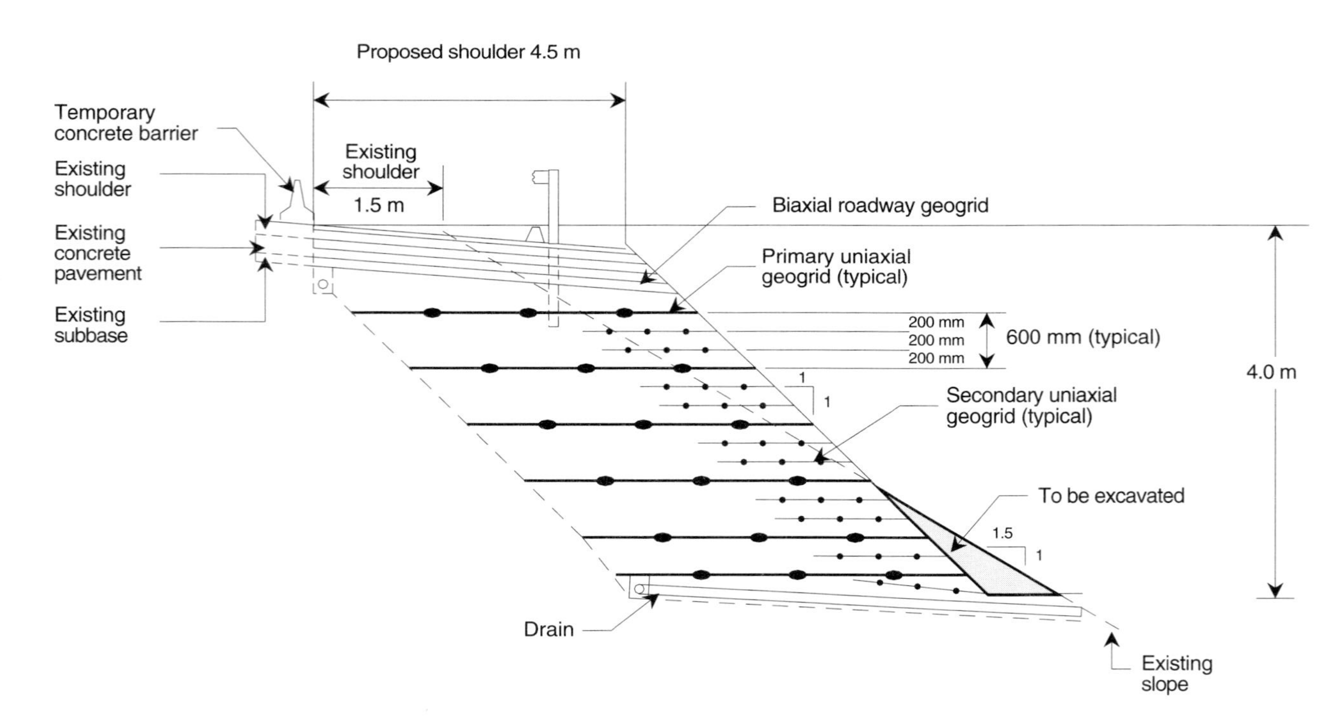

Figure 3.12 Shoulder widening of Pennsylvania Turnpike using geogrid reinforced steep soil slopes. (After Berg et al. [25])

TABLE 3.5 SUMMARY OF SLOPE REINFORCEMENT DESIGN CHARTS

Investigator	Limit Equilibrium Model	Soil Strength	Pore Pressure	Slope Angle (degrees)	Length Reinforcement	Comments
Ingold [27]	Infinite slope	$\phi' = 30°$	N. A.	30–80	N. A.	Surficial stability only
Murray [28]	Two-part wedge	ϕ'	r_u (0, 0.25, 0.5)	10–40	N. A.	Reinforcement load-elongation relationship included in model
Jewell et al. [29]	Two-part wedge	ϕ'	r_u (0, 0.25, 0.5)	30–80	Parallel truncation	Basis for Schmertmann et al. [26] study
Leshchinsky and Reinschmidt [30]	Log spiral	c', ϕ'	N. A.	15–90	N. A.	Rigorous equlibrium model
Schneider and Holtz [31]	Two-part wedge	c', ϕ'	r_u	0–40	N. A.	Extended the work of Murray [28]; charts do not give critical surfaces
Ruegger [32]	Circular	ϕ'	N. A.	30–90	Parallel truncation	For FS = 1.3 only
Schmertmann et al. [26]	Two-part wedge	ϕ'	N. A.	30–80	General truncation	This study extended work of Jewell et al. [29]

N. A.—Not addressed

Source: After Schmertmann et al. [26].

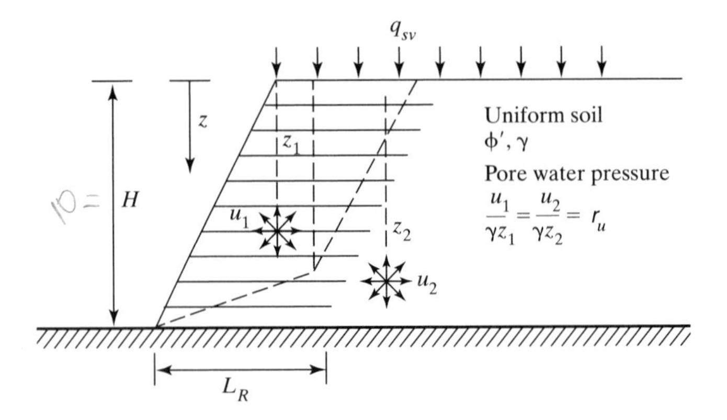

Figure 3.13 Definitions for analysis of steep reinforced soil slopes. (After Jewell [29])

combined reduction factors of 4.12. The global factor of safety is to be 1.4. The soil is granular, with $\gamma = 18 \text{ kN/m}^3$, $\phi = 30°$, $c = 0 \text{ kN/m}^2$, and has no pore-water pressure (i.e., $r_u = 0$). Determine the number, spacing, and length of the individual geogrid layers.

Solution: By observation, this slope at 70° to the horizontal without reinforcement is in a failure state (i.e., FS $<$ 1.0) and is in need of some type of reinforcement. The design procedure is given in steps.

(a) Calculate the allowable strength on the basis of the given reduction factors and then calculate the design strength, which includes the factor of safety

$$T_{\text{ult}} = 180 \text{ kN/m}$$

$$T_{\text{allow}} = \frac{180}{4.12}$$

$$= 43.7 \text{ kN/m}$$

$$T_{\text{des}} = \frac{43.7}{1.4}$$

$$= 31.2 \text{ kN/m}$$

(b) Determine the necessary values from Figure 3.14a for $r_u = 0$, $\beta = 70°$ and $\phi'_d = 30°$. This results in the following.

$$K_{\text{reqd}} = 0.19$$

$$\left(\frac{L_R}{H}\right)_{\text{ovrl}} = 0.51$$

$$\left(\frac{L_R}{H}\right)_{\text{ds}} = 0.38$$

(c) Calculate the spacing S_v at the base of the slope where the stresses are greatest

$$S_v = \frac{T_{\text{des}}}{K_{\text{reqd}} \gamma z_{\max}}$$

$$= \frac{31.2}{(0.19)(18)(10)}$$

$$= 0.91 \text{ m}$$

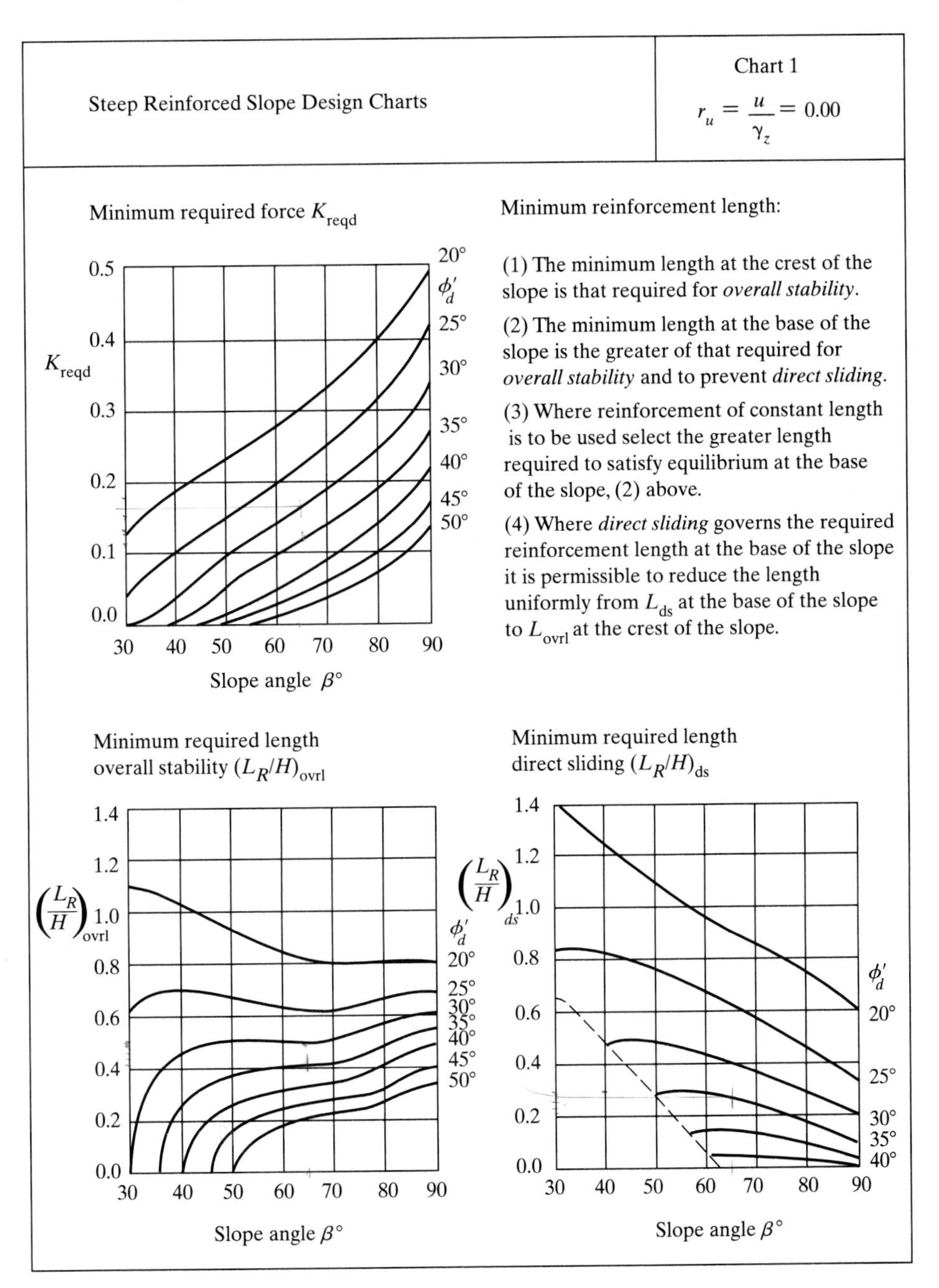

Figure 3.14 (a) Steep reinforced slope design charts for $r_u = u/\gamma z = 0$. (After Jewell [29])

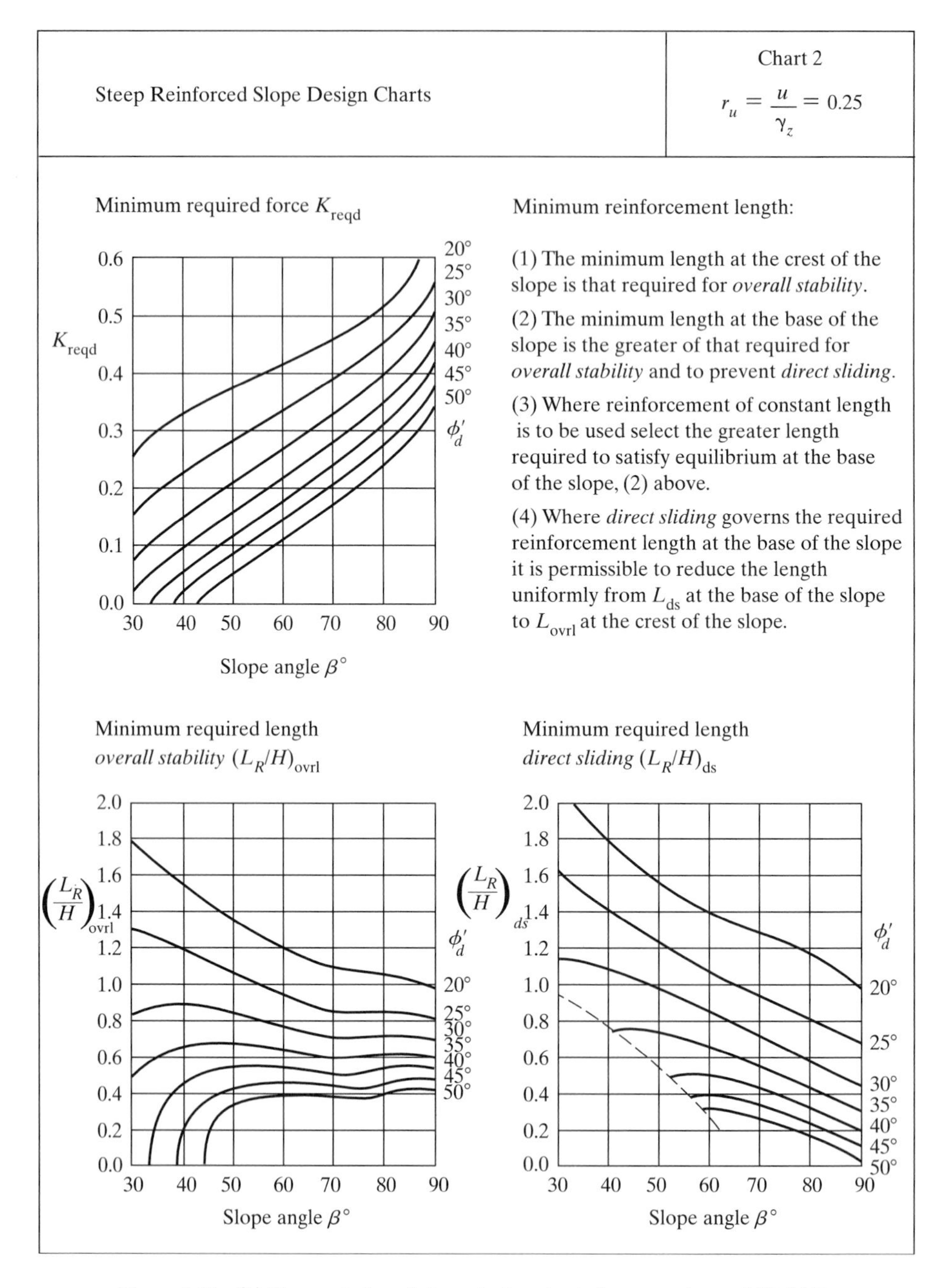

Figure 3.14 (b) Steep reinforced slope design charts for $r_u = u/\gamma z = 0.25$. (After Jewell [29])

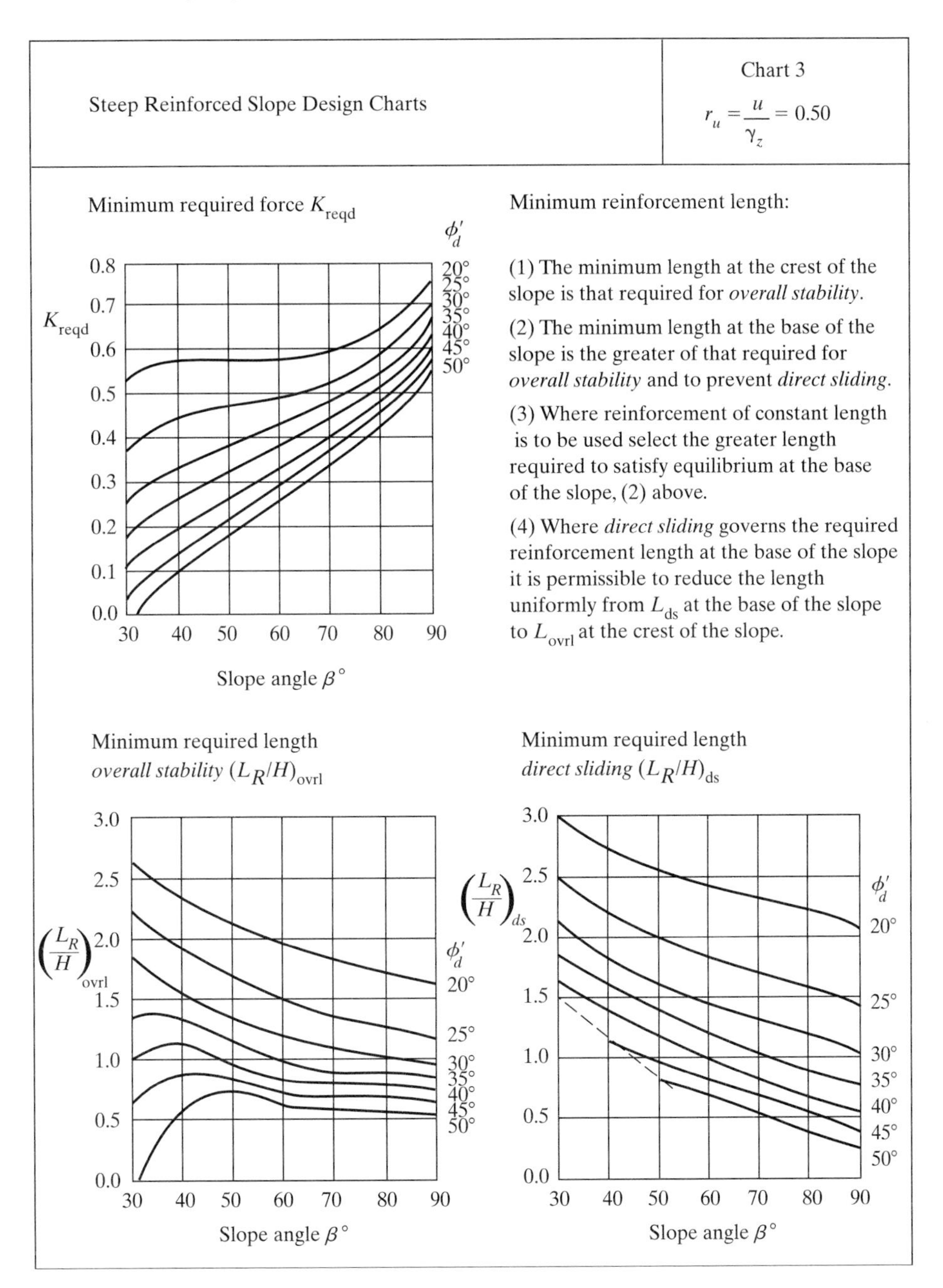

Figure 3.14 (c) Steep reinforced slope design charts for $r_u = u/\gamma z = 0.50$. (After Jewell [29])

If evenly spaced, the required number of geogrid layers will be

$$n = \frac{H}{S_v}$$

$$= \frac{10}{0.91}$$

$$= 11 \text{ layers}$$

(d) Select the reinforcement length.

if $(L_R/H)_{\text{ovrl}} > (L_R/H)_{\text{ds}}$	use constant length $= (L_R/H_{\text{ovrl}})$
if not	use constant length $= (L_R/H)_{\text{ds}}$ or
	taper the lengths from $(L_R/H)_{\text{ds}}$ at the base to $(L_R/H)_{\text{ovrl}}$ at the crest

Since $0.51 > 0.38$, use $L_R/H = 0.51$.

$$\therefore L_R = 5.1 \text{ m throughout}$$

(e) The length at the base can be checked by conventional methods using the entire MSE mass, or according to the equations set forth in [29].

(f) Check among the different geogrid behaviors in the anchorage zone behind the hypothetical shear plane. Such differences must be considered using experimental results, as described in Section 3.1.2. If there is concern about the use of one geogrid product versus another, the designer always has the option of lengthening the geogrids over what is required by Figures 3.14a,b,c.

(g) Sketch the final reinforced slope and provide for miscellaneous details as shown below, that is, use short (secondary) geogrids between the primary reinforcements and adjacent to the slope for compaction aid and surface erosion control (recall Figure 3.12).

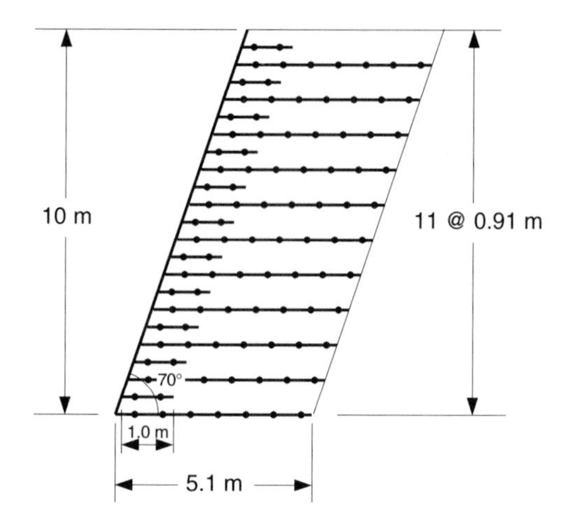

3.2.5 Walls

A tremendous number of geogrid-reinforced walls have been constructed in the past 10 years. Most have been designed for private owners and developers; the entry into the public sector has been through *value engineering*. This is the concept whereby the low-bid general contractor offers the state or federal agency an option for some particular segment of the project, for example a geogrid-reinforced wall in place of a conventional reinforced-concrete wall or steel-reinforced segmental wall. If the option is acceptable to the agency, the financial savings from the difference in the costs of the two different kinds of walls is shared equally between the agency and the contractor. It is a very effective vehicle for the introduction of new products and concepts like geogrid (and geotextile) reinforced walls. As of 1997, however, polymer-reinforced permanent walls, bridge abutments, wing walls, and so on, have been approved for routine design and use by AASHTO in the U.S.

The types of permanent geogrid-reinforced wall facings are as follows.

Articulated precast concrete panels are discrete precast concrete panels with inserts for attaching the geogrid. Many aesthetically pleasing facing designs are possible.

Full height precast panels are concrete panels temporarily supported until backfilling is complete. Note, however, that these types of walls are questionable due to vertical stresses that develop on the geogrid connections to the wall after removal of the panel support.

Cast-in-place concrete panels are often wrap-around walls that are allowed to settle and, after 0.5 to 2.0 years, are covered with a cast-in-place facing panel. These walls are currently favored in Japan [33], where the ends of the geogrid reinforcement are embedded in gabions, which then have a concrete facing panel poured against them.

Masonry block facing walls are an *exploding* segment of the industry with many different types currently available, all of which have the geogrid embedded between the blocks and held by pins, nubs, and/or friction (recall Figure 2.51 a, b).

Timber facings are railroad ties or other large treated timbers with the geogrid attached by batten strips and/or held by friction when placed between the timbers.

Gabion facings are polymer or steel-wire baskets filled with stone, having a geogrid held between the baskets and fixed with rings and/or friction.

Welded wire-mesh facings are similar to gabion facings, but are often used for small temporary walls. The geogrids are attached to the mesh with metal rings.

Wrap-around facings are the same as those illustrated in Section 2.7.1 with geotextile walls. Note that protection is needed against ultraviolet light and vandalism, thus a bitumen or concrete coating is usually applied.

The design of the above-described walls must not be considered trivial, for many of them are critical structures that are meant to be permanent, possessing service lifetimes in excess of 75–100 years. Design centers around external stability of the entire reinforced mass (sliding, overturning, and bearing capacity) and internal stability within the

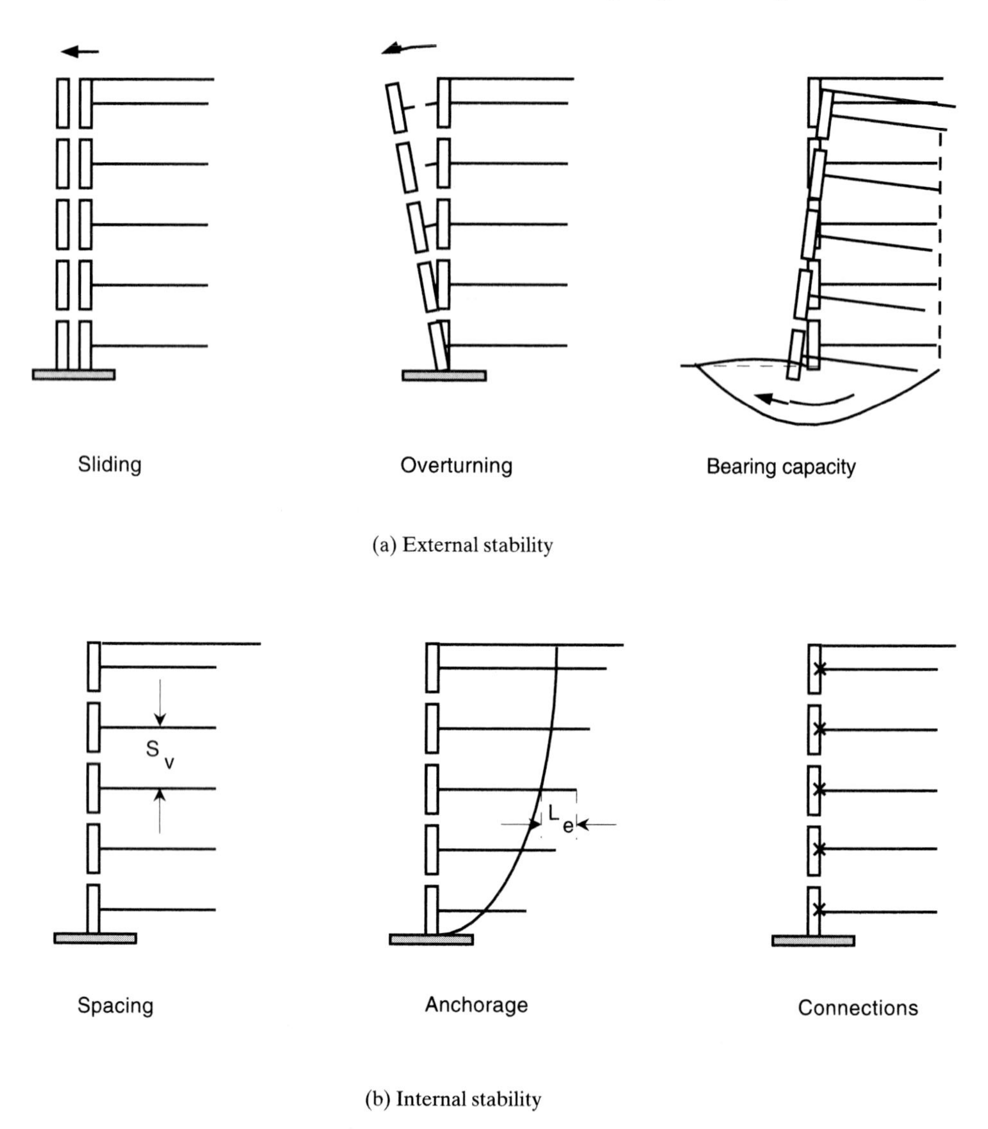

Figure 3.15 Elements of geogrid (or geotextile) reinforced wall design.

reinforced mass (geogrid spacing, anchorage length, and connection strength). Figure 3.15 illustrates these concepts. Each kind of stability must be treated in its own right and used together. The final design must be capable of being done with reasonable construction practices. Example 3.10 illustrates this using a geogrid-reinforced wall with precast-concrete facing panels.

Example 3.10

Design a 7 m high geogrid-reinforced wall where the reinforcement spacing must be at 1.0 m spacings, since the wall facing is of the precast segmented concrete facing panel type.

The coverage ratio is 0.8 (i.e., geogrids do not cover the entire ground surface at each lift, they are slightly separated). The length-to-height ratio of the reinforced soil wall should not be less than 0.7 (i.e., $L \geqslant 4.9$ m). Additional details of the problem, including soil and geogrid data are given in the following figure.

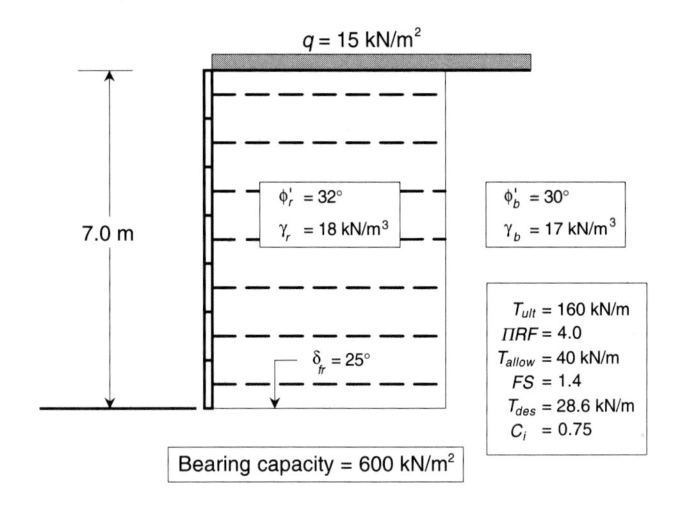

Solution:

(a) Calculate the external stability.

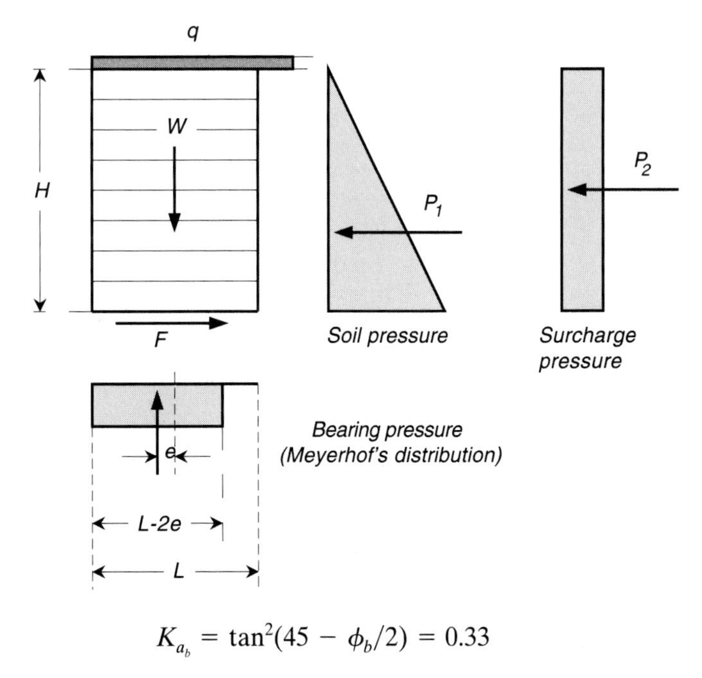

$$K_{a_b} = \tan^2(45 - \phi_b/2) = 0.33$$

the coefficient of active earth pressure of backfill soil behind reinforced zone.

$$P_1 = 0.5 \times \gamma_b \times H^2 \times K_{a_b} = 0.5 \times 17 \times (7)^2 \times 0.33 = 137 \text{ kN/m}$$

$$P_2 = qK_{a_b} \times H = 15 \times 0.33 \times 7 = 34.7 \text{ kN/m}$$

Total force, $P = 137 + 34.7 = 172$ kN/m

(1) For the sliding stability

F = resisting force $= W \times \mu = \gamma_r \times H \times L \times \tan \delta$ (neglecting effect of surcharge):]

$$= 18 \times 7 \times 4.9 \times \tan 25° = 288 \text{ kN/m}$$

FS_s = factor of safety against sliding

$$= \frac{F}{P} = \frac{288}{172} = 1.67 > 1.5 \qquad \text{OK}$$

(2) Overturning stability is really not an issue, since this type of mechanically stabilized wall is not subject to overturning because it cannot mobilize bending due to its inherent flexibility. The following calculation illustrates the conservative aspect of this mechanism.

$$M_s = \text{stabilizing moment} = W \times \frac{L}{2} = (18 \times 7 \times 4.9) \times \left(\frac{4.9}{2}\right)$$

$$= 1513 \text{ kN/m}$$

$$M_{\text{ov}} = \text{overturning moment} = P_1 \times \frac{7}{3} + P_2 \times \frac{7}{2}$$

$$= \left(137 \times \frac{7}{3}\right) + \left(34.7 \times \frac{7}{2}\right) = 441 \text{ kN/m}$$

FS_{ov} = factor of safety against overturning

$$= \frac{1513}{441} = 3.43 > 2.0 \qquad \text{OK}$$

(3) For the stresses on the foundation soil

$$e = \text{eccentricity} = \frac{M_{\text{ov}}}{(W + q \times L)}$$

$$= \frac{441}{(18 \times 7 \times 4.9 + 15 \times 4.9)} = 0.64\text{m}$$

Now, this eccentricity cannot be outside of the central one-third of the footing. That is,

$$e < \frac{L}{6} = \frac{4.9}{6}$$

$0.64 < 0.82$ OK, thus no tension beneath the footing

Acting length (Meyerhof's distribution) $= L - 2 \times e = 4.9 - 2 \times 0.64 = 3.62$ m

$$\text{Bearing pressure} = [(18 \times 7) + 15] \times \left(\frac{4.9}{3.62}\right) = 191\text{kPa}$$

$$\text{FS}_b = \text{factor of safety against bearing capacity failure}$$

$$= \frac{600}{191} = 3.14 > 2.0 \text{ OK}$$

(b) Calculate the internal stability.

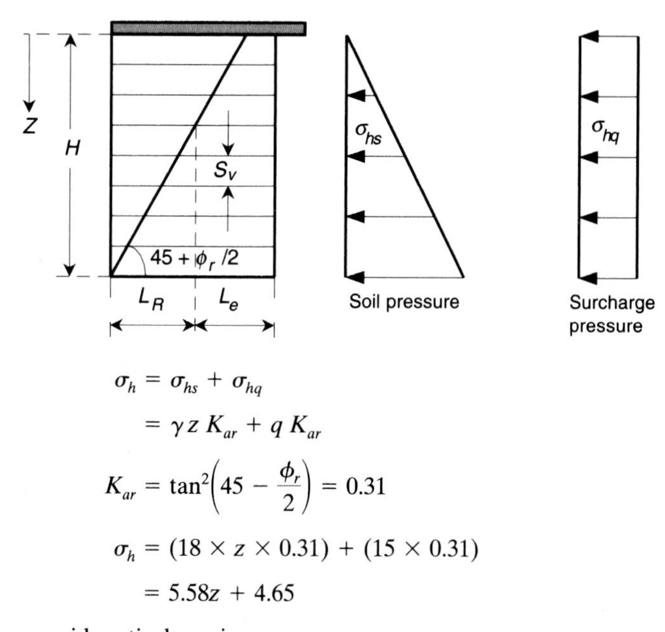

$$\sigma_h = \sigma_{hs} + \sigma_{hq}$$
$$= \gamma z \, K_{ar} + q \, K_{ar}$$

$$K_{ar} = \tan^2\left(45 - \frac{\phi_r}{2}\right) = 0.31$$

$$\sigma_h = (18 \times z \times 0.31) + (15 \times 0.31)$$
$$= 5.58z + 4.65$$

(1) For geogrid vertical spacing

$$T_{\text{des}} = s_v \sigma_h / C_r \quad (\text{where } C_r = \text{ coverage ratio})$$

$$28.6 = s_v \frac{(5.58z + 4.65)}{0.8}$$

$$s_v = \frac{22.9}{5.58z + 4.65}$$

Maximum depth for $s_v = 1$ m.

$$1.0 = \frac{22.9}{5.58z + 4.65} \Rightarrow z = 3.27 \text{ m}$$

Maximum depth for $s_v = 0.5$ m.

$$0.5 = \frac{22.9}{5.58z + 4.65} \Rightarrow z = 7.37 \text{ m}$$

The layout pattern can now be developed based upon the above-calculated maximum spacing values and the type and dimensions of the facing panels. Thus, the reinforcement for the upper 3 facing panels is at 1.0 m spacing and the remaining 4 facing panels are at 0.5 m spacing. The left-hand sketch below gives these details. It is based on the symmetry of geogrid connections, that is, no tension eccentricity is allowed on any facing panel. Hence, small sections of geogrid are required for the top facing panels of the wall, where the spacing interval is 1.0 m. Also note that for cross sections located one facing panel adjacent to the illustrated design cross section, the top and bottom facing panels will be half-height, but the reinforcement spacing must be maintained as calculated above. See the right-hand sketch.

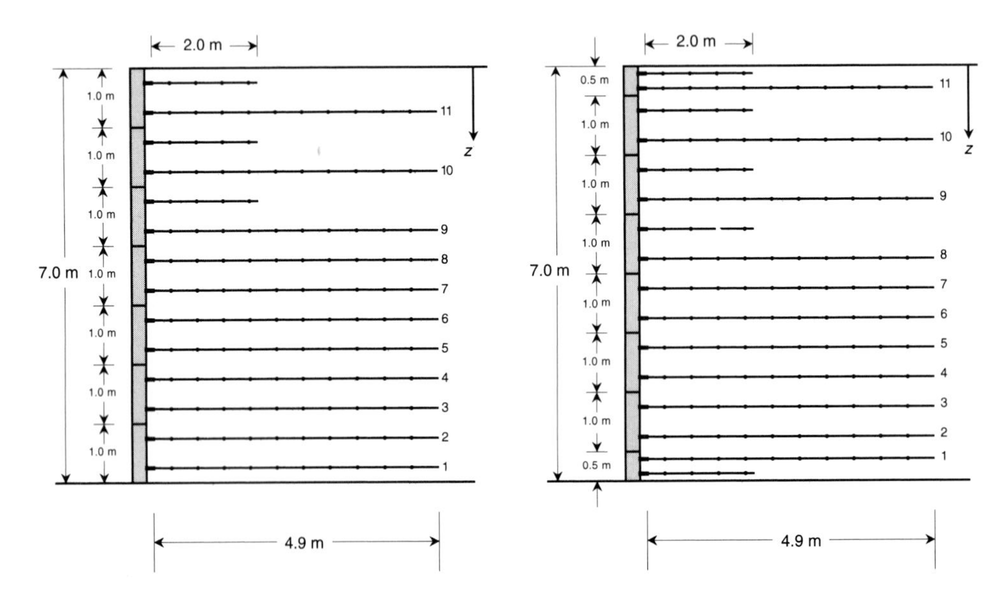

(**2**) For the total length we consider the embedment plus the nonacting Rankine length.

For the embedment length:

$$s_v \times \sigma_h \times \text{FS}_{\text{pullout}} = 2 \times L_e \times C_i \times \sigma_v \tan\phi' \times C_r$$

$$s_v(5.58z + 4.65)1.5 = 2L_e(0.75)(18z)(\tan 32)(0.8)$$

$$L_e = \frac{s_v(5.58z + 4.65)1.5}{(2)(0.75)(18z)(\tan 32)(0.8)}$$

$$L_e = \frac{(0.62z + 0.516)s_v}{z}$$

For the nonacting Rankine length

$$L_R = (H - z)\tan\left(45 - \frac{\phi}{2}\right)$$

$$= (7 - z)\tan\left(45 - \frac{32}{2}\right)$$

$$L_R = 3.88 - 0.554z$$

The above relationships lead to the following table.

Layer Number	Depth z (m)	Spacing s_v (m)	L_e (m)	$L_{e_{min}}$ (m)	L_R (m)	L_{calc} (m)	L_{reqd} (m)
11	0.75	0.75	0.98	1.0	3.46	4.46	5.0
10	1.75	1.00	0.92	1.0	2.91	3.91	5.0
9	2.75	1.00	0.81	1.0	2.36	3.36	5.0
8	3.25	0.50	0.39	1.0	2.08	3.08	5.0
7	3.75	0.50	0.38	1.0	1.80	2.80	5.0
6	4.25	0.50	0.37	1.0	1.52	2.52	5.0
5	4.75	0.50	0.36	1.0	1.25	2.25	5.0
4	5.25	0.50	0.36	1.0	0.97	1.97	5.0
3	5.75	0.50	0.36	1.0	0.69	1.69	5.0
2	6.25	0.50	0.35	1.0	0.42	1.42	5.0
1	6.75	0.50	0.35	1.0	0.14	1.14	5.0

(3) For connection strength, theoretically the tensile stress at the connection of the reinforcement to the facing panels should be very small. This is indeed the case if the panels, reinforcement, and soil backfill stay in horizontal alignment in the same manner as the wall is constructed. Unfortunately, settlement of the backfill (and perhaps of the foundation soil) usually occurs, thereby deforming the reinforcement and imposing stress on the wall panel connection. The amount of this stress is dependent on the backfill soil type, density, moisture content, compactive effort, and foundation conditions. See Soong and Koerner [34].

Due to this uncertainty, a relatively conservative approach is to use the design strength of the reinforcement as the required connection strength. For our example

$$\begin{aligned} T_{conn} &\geqslant 1.0T_{des} \\ &\geqslant 1.0(28.6) \\ &\geqslant 28.6 \text{ kN/m} \end{aligned}$$

See section 3.1.2 for the experiment necessary to obtain the allowable connection strength.

(4) For drainage behind the wall, since MSE walls are generally not designed for hydrostatic stresses, efficient drainage through the backfill and wall itself must be assured. Depending on the site-specific conditions, there are a number of possible strategies: a sand or gravel backfill layer directly behind the wall facing; a chimney drain against the excavated soil face with a connecting blanket drain to the wall footing; an impermeable drainage swale along the backfilled surface behind the wall; or piping systems to drain nearby roofs, parking lots, and roadways away from the wall and its footing.

Note that many geogrid manufacturers have developed design charts, graphs, or computer algorithms for wall designs of the type just illustrated. If a specific geogrid is to be used, these guides can be considered, since their technical background is usually very good. However, it must be remembered that they are product-specific in their assumptions about the reduction factors used to arrive at allowable strength, global factor of safety, and selected other details.

With the growing number of geogrid-reinforced permanent walls, the field behavior of such walls requires examination. Two walls, both with precast facing panels, have been monitored and reported on by Berg et al. [35]. One wall in Tucson, Arizona, was 4.6 m high, used a cumulative reduction factor of 2.6 on ultimate strength for allowable strength, and a value of 1.5 as a global factor of safety. A tie back wedge analysis using three assumptions (Rankine, Meyerhof, and trapezoidal) was used for design. The second wall was in Lithonia, Georgia, and was 6.0 m high. It used the same factors and design method. Figure 3.16 presents the results for both walls shortly after construction was complete. Note the horizontal wall pressures at various wall heights where a large overprediction of measured values is seen for each wall, that is, the wall designs that were used appear to be quite conservative.

This same trend of overprediction of measured stresses was also seen in a controlled large-scale laboratory wall constructed and monitored at the Royal Military College in Kingston, Ontario, Canada. The 3.0 m high wall was carefully constructed and monitored so as to obtain the following information.

- Tensile stress in each of the four geogrid layers at three to six locations on each of the walls along their 3 m lengths,
- Tensile force at the wall facing panels at each of the four geogrid layers,

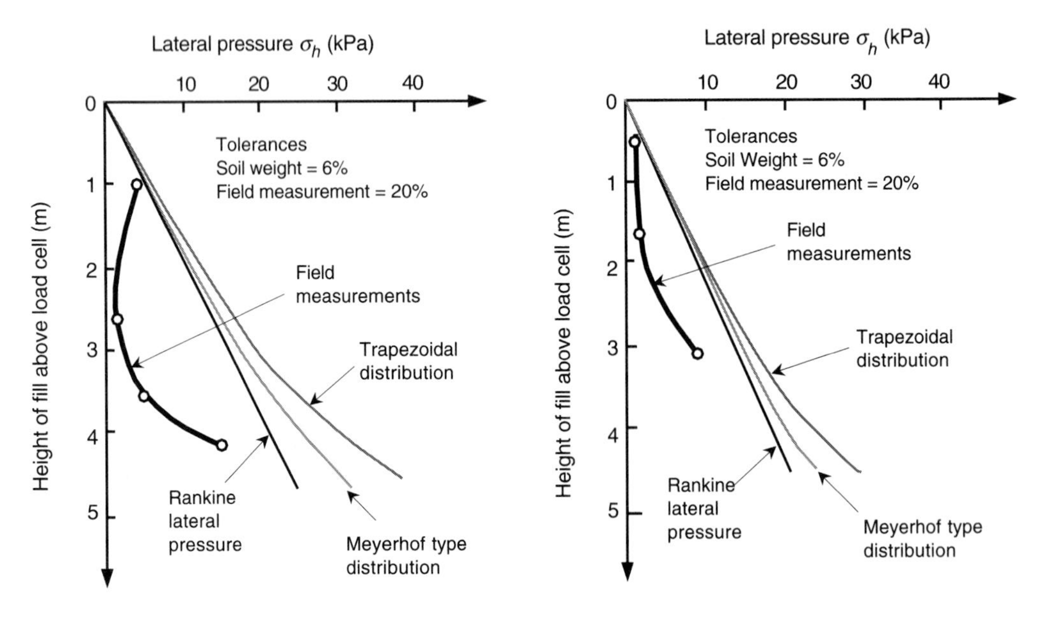

(a) Results of Tucson, Arizona, wall (b) Results of Lithonia, Georgia, wall

Figure 3.16 Comparison of measured stresses to design stresses for two geogrid-reinforced walls. (After Berg et al. [35])

- Vertical pressure at the base of the wall and extending into the reinforced earth fill zone at seven locations,
- Horizontal wall movement of the facing panels at locations of the four geogrid-reinforcement layers, and
- Vertical movement of the soil backfill surface.

All of the above were monitored at various values of surcharge load, from zero to eventual failure. Particularly interesting in this case was that a number of academicians and practitioners were invited to predict each of the above items without knowing the results; that is, they were to give a Class A prediction. As can be reviewed in the very interesting publication, Jarrett and McGown [36], all 10 predictors who used the limit equilibrium method vastly overpredicted the stresses, loads, and movements—*many of them by an order of magnitude.* Only those predictors who used intuition or elastic methods were close to the actual values. Of this latter group, Murray's predictions, which were based on finite-element methods, were very accurate. It should also be noted that the wall never failed, even with the surcharge pressure exceeding the limits of the restraining system! Clearly, the design methods presented in this section are conservative, at least over the relatively short-time frames that the various walls have been in service. The major decision to make is what allowable strength to use in light of creep, installation damage, durability, and aging considerations. The discussion in Sections 3.1.3 and 3.1.4 are important in this regard.

While the use of all types of geogrid-supported wall systems is steadily increasing, the modular block reinforced wall systems are literally exploding in their rise and acceptance. Modular block walls of the types shown in Figure 3.17 are used with geogrid reinforcement even more than with geotextile reinforcement. Figures 2.51a,b showing the facing block details could easily have been placed in this chapter. The design of such walls follows exactly that of the segmental wall just presented. It will be a conservative design if performed in the suggested manner.

Obviously, there are other design approaches. Noteworthy in this regard is the procedure reported by Simac et al. [37] for the National Concrete Masonry Block Institute. Their approach differs from Example 3.10 in a number of significant aspects.

- Instead of Rankine earth pressures, the method of Coulomb is utilized.
- This allows wall facings with a batter (i.e., nonvertical walls) to be used in the calculations, resulting in lower lateral-earth pressures.
- This also allows for sloping backfill surfaces to be included, which generally give lower stresses than an equivalent uniform surcharge load assumption.
- A log-spiral failure surface is assumed, which results in significantly shorter lengths of reinforcement, particularly for the higher walls.
- Connection strengths to the masonry block walls do not consider the possibility of soil-backfill settlement, which would add to the suggested low wall facing connection requirements, based on field measurements to date.

Notwithstanding the above comments, all of the assumptions made are *technically accurate* and the performance of such masonry block walls has been excellent. Additionally, they are the most aesthetically pleasing walls imaginable.

(a)

(b)

Figure 3.17 Geogrid-reinforced modular block walls. (Compliments of TC Mirafi Corp.)

An important closing comment for all types of geogrid-reinforced walls is necessary regarding installation damage. The area of geogrid-reinforced walls follows closely behind mechanically stabilized walls using steel strips or welded wire mesh. The backfill specification for such metallic reinforcement is generally large gravel, typically AASHTO #3, #5, or #57 size stone. This is required to avoid long-term corrosion of the steel reinforcement. However, with polymeric reinforcement (geogrids or geotextiles) this is not only unnecessary, *it is actually dangerous.* Such large stone sizes will surely cause installation damage to the geogrids (or geotextiles) and since corrosion is not an issue, the recommended soil backfill must be reduced in size drastically. In fact, the

TABLE 3.6 RECOMMENDED SOIL BACKFILL GRADATION FOR GEOGRID- OR GEOTEXTILE-REINFORCEMENT APPLICATIONS (WALLS AND SLOPES) TO AVOID EXCESSIVE INSTALLATION DAMAGE

Sieve Size (No.)	Particle Size (mm)	Percent Passing
4	4.76	100
10	2.0	90–100
40	0.42	0–60
100	0.15	0–5
200	0.075	0

backfill particle size should come all the way down to the point where its particle size is dictated by adequate drainage of the pore water in the backfill soil. For example, sandy soils of the SW, SP, and even SM classifications, which have permeabilities 0.1 to 0.001 cm/s, should be used. See Table 3.6 for a recommended soil particle size backfill guide to be used with geogrid (or geotextile) reinforced walls and slopes.

3.2.6 Bearing Capacity via Base Reinforcement

Geogrids have been used to increase bearing capacity of poor foundation soils in different ways: as a continuous layer, as multiple closely-spaced continuous layers with granular soil between layers, and as mattresses consisting of three-dimensional interconnected cells. The technical database for the single-layer continuous sheets has been reported by Jarrett [38] and by Milligan and Love [39]; in both cases large-scale laboratory tests are used. Figure 3.18 presents some of Milligan and Love's work graphed in the conventional nondimensionalized q/c_u versus ρ/B manner and also as $q/\sqrt{c_u}$ versus ρ/B. The latter graph is not conventional but does sort out the data nicely. Clearly shown in both instances is the marked improvement in load-carrying capacity using geogrids at high deformation and only a nominal beneficial effect at low deformation. Beyond these observations, a precise design formulation is not currently available.

Instead of focusing on a global increase in bearing capacity, it is quite likely that a layer or layers of geogrid (or geotextile) will aid in minimizing or eliminating differential settlement. Here localized settlements due to abruptly settling or subsiding weak zones can be spanned by the layer of reinforcement. This is known as base reinforcement. Notable in this regard is the construction of new landfills above existing landfills, a technique called *piggybacking*. The approach is to use arching theory in the calculation of the vertical stress arising from localized subsidence (i.e., differential settlement) and provide suitably strong reinforcement.

It should be recognized that arching in natural soils overlying a locally yielding foundation is well established. In the 1930s, both Terzaghi in Austria (calculating stresses on deep tunnels) and Marston in the U.S. (calculating stresses on buried pipelines) developed the analytic theory. Their work resulted in the following simplified formula for vertical stress on the surface of the particular underground structure (tunnel or pipe, respectively).

$$\sigma_z = 2\gamma_{avg}R[1 - e^{-0.5H/R}] + qe^{-0.5H/R} \qquad (3.11)$$

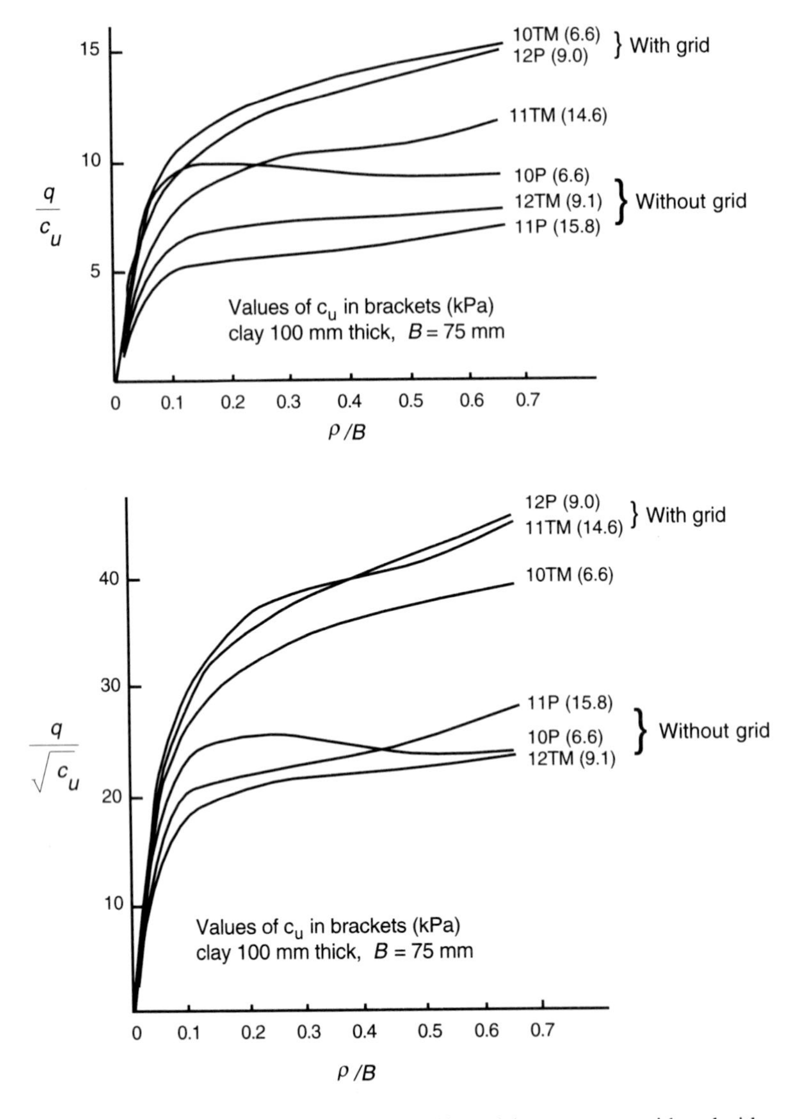

Figure 3.18 Load-versus-settlement curves of large laboratory tests with and without geogrid reinforcement. (After Milligan and Love [38])

where

σ_z = vertical stress on the reinforcement layer,
γ = unit weight of soil (or solid waste) above the settlement area,
R = radius of differential settlement zone,
H = total height above the settlement area, and
q = surcharge pressure placed at the ground surface.

Note that for large values of H (typically $H \geqslant 6R$) the formula reduces to the following value of constant vertical stress.

$$\sigma_z = 2\gamma_{avg}R \tag{3.12}$$

Having a method to calculate the vertical stress, we can now use the value to calculate the stress in the reinforcement layer for a new landfill placed over an existing one. Note that the reinforcement can be either a geogrid or a geotextile. For support over a differential settlement area, the value of T_{reqd} is calculated as follows.

$$T_{reqd} = \sigma_z R\Omega \tag{3.13}$$

where

$\Omega = 0.25[(2y)/B + B/(2y)]$, where
B = width of settlement void, and
y = depth of settlement void.

Giroud et al. [40] have combined the above equations to develop a design chart that can be used to avoid direct calculation; see Figure 3.19. Note that the chart can be used for either circular voids or long extended voids.

Once the value of T_{reqd} is determined, it must be compared to T_{allow} using Eq. (3.6), which includes the site-specific reduction factors. Example 3.11 illustrates the technique.

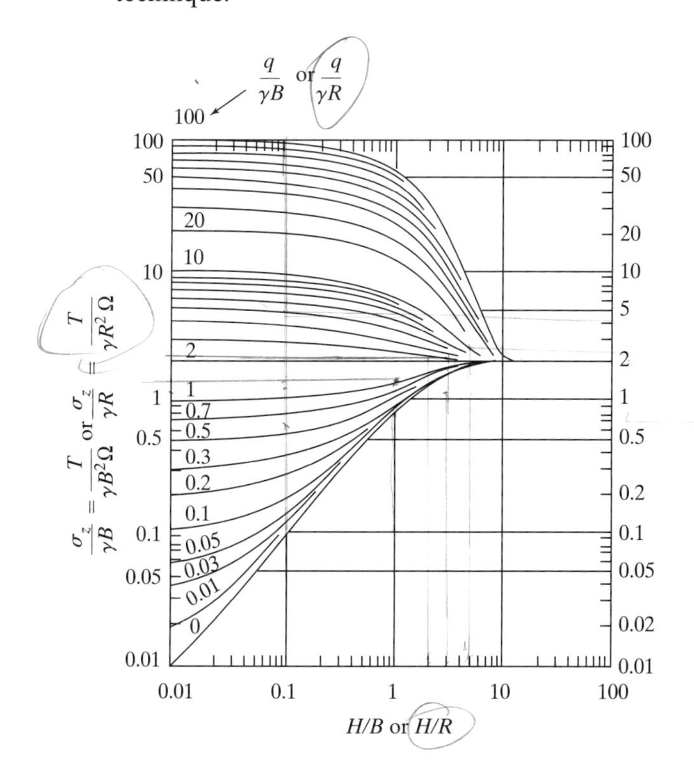

Figure 3.19 Curves of geosynthetic stress and tension that can be used for R (radius of circular void) or B (width of long voids). (After Giroud et al. [40])

Example 3.11

Using the Terzaghi/Marston formulation for calculating vertical stress above localized subsidence, in this case differential settlement in an old landfill of radius 1 m, **(a)** calculate the required wide-width strength of a reinforcement layer if a new 30 m high landfill is to be placed upon the existing one. The compacted unit weight of the waste is 12 kN/m^3. **(b)** Check your calculated value against Figure 3.19. Then, **(c)** calculate the global factor of safety for a geogrid with ultimate wide-width tensile strength of 125 kN/m. In the calculations use cumulative reduction factors of 5.0.

Solution:

(a) The formula for vertical stresses in arching situations under a deep fill (such as in this example) reduces to Eq. (3.12). Therefore, the vertical stress is calculated as

$$\begin{aligned}\sigma_z &= 2\gamma_{avg}R \\ &= 2(12)(1.0) \\ &= 24 \text{ kPa}\end{aligned}$$

To transfer this vertical stress into a horizontal force, we use Eq. (3.13)

$$T_{reqd} = \sigma_z R\Omega$$

where Ω = strain criterion [recall Section 2.5.2]

$$\begin{aligned}\Omega &= 0.97 \text{ at } 5\% \text{ strain} \\ &= 0.73 \text{ at } 10\% \text{ strain}\end{aligned}$$

Assuming $\Omega = 0.73$

$$\begin{aligned}T_{reqd} &= 24 \times 1.0 \times 0.73 \\ &= 17.5 \text{ kN/m}\end{aligned}$$

(b) Check this against Figure 3.19.

$$\frac{H}{R} = \frac{30}{1} = 30$$

$$\therefore \frac{T}{\gamma R^2 \Omega} = 2.0$$

$$\begin{aligned}T_{reqd} &= 2.0(12)(1)^2(0.73) \\ &= 17.5 \text{ kN/m} \qquad \text{which checks}\end{aligned}$$

(c) The factor of safety on a geogrid with 125 kN/m ultimate strength (at 10% strain) is as follows.

$$\begin{aligned}T_{allow} &= T_{ult}\left[\frac{1}{\Pi RF_p}\right] \\ &= \frac{125}{5} \\ &= 25\text{kN/m}\end{aligned}$$

and

$$FS = \frac{T_{allow}}{T_{reqd}}$$
$$= \frac{25}{17.5}$$
$$FS = 1.43 \qquad \text{which is acceptable}$$

In a somewhat different context, but still focused on the improvement of bearing capacity, Edgar [41] reports on a three-dimensional *geogrid mattress* where geogrids are placed vertically and interconnected (see Figure 3.20). Gravel is placed within the geogrid mattress as it is constructed over soft fine-grained soils. Edgar reports on a 15 m high embankment that was successfully constructed above the mattress. It was felt that the nonreinforced slip plane was forced to pass vertically through the mattress and therefore deeper into the stiffer layers of the underlying subsoils. This improved the stability to the point where the mode of failure was probably changed from a circular arc to a less critical plastic failure of the soft clay. The design was considered to be a successful and economic one. Another example of a 1 m high geogrid mattress was constructed to support a 30 m high landfill over extremely soft mine tailings in Hausham, Germany [42]. The mattress was filled with gravel (see Figure 3.21) and the liner system constructed above it. The foundation soil was so soft that a nonwoven geotextile and a biaxial geogrid had to be initially placed to provide a stable working area for the construction of the three-dimensional mattress.

In the design of such three-dimensional geogrid mattresses it is felt that a number of phenomena are occurring, all of which improve the foundation soil stability (see Figure 3.22).

Global slope stability is improved by forcing the potential failure plane through the mattress and deeper into the foundation soil. It is also possible that the foundation soil may improve in strength characteristics at greater depths.

Bearing capacity is improved in a similar manner to the point where it becomes a nonissue for mattresses greater than approximately 30 m in width.

Lateral extrusion (or squeeze-out) is undoubtedly decreased because stress concentrations have been largely eliminated via a uniform pressure distribution applied through the relatively stiff geogrid mattress.

In the absence of global instability, this last item is particularly important. Squeeze-out of the foundation soil is the likely service-limiting mechanism giving rise to excessive deformations. Robertson and Gilchrist [43] and Jenner et al. [44] have used slip line fields to predict the principle stresses in the soft foundation soils. Both studies give actual case histories and the monitoring feedback as to the validity of the design assumptions.

3.2.7 Veneer Cover Soils

Whenever a lined slope (geomembrane, GCL, or compacted clay) is covered with soil, a stability calculation should be made to assess the potential for sliding failure of the soil on the barrier layer. Situations that come to mind are: landfill liners with leachate

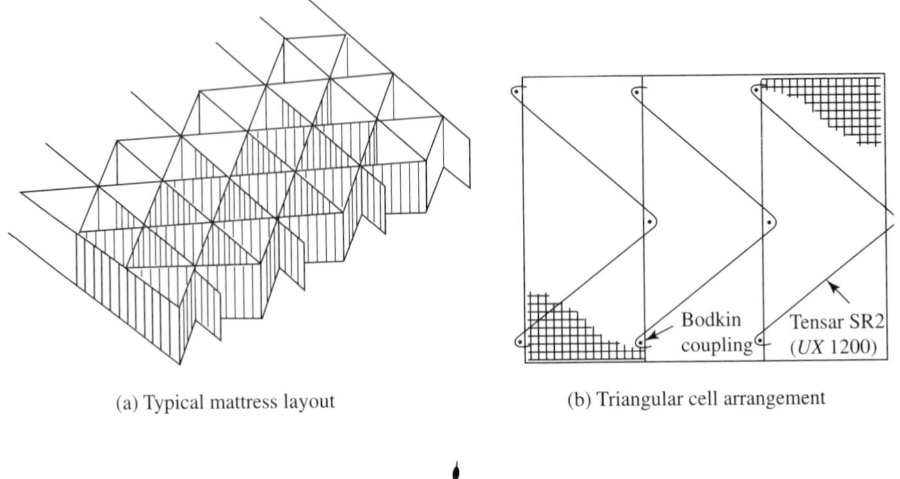

(a) Typical mattress layout (b) Triangular cell arrangement

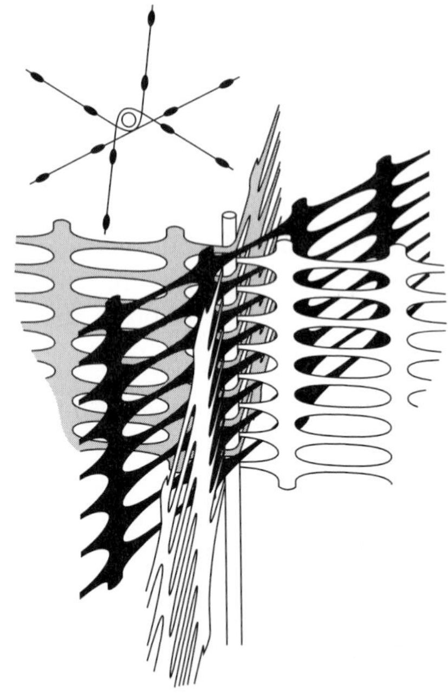

(c) Coupling of diaphrams

Figure 3.20 Construction of embankment stabilization mattress. (After Edgar [41])

collection sand or gravel above them, until such time that the solid waste acts as a passive resistance restraint; surface impoundment liners where the cover soil is placed over the geomembrane to shield it from ultraviolet light, heat degradation, and equipment damage; landfill covers that have topsoil and protection soil placed over the geomembrane. Due to the typically low shear strength of the covering soil to the containment material, stability problems have arisen. Boschuk [45] reports on 11 failure case histories for varied reasons, including cover soils with low shear strength, low-

Figure 3.21 Three-dimensional geogrid mattress being filled with gravel for landfill bearing capacity over soft mine tailings in Hausham, Germany.

friction containment materials, clay-cover soil subsidence, methane-gas leakage, and improper drainage; although not listed seismic activity could certainly trigger cover soil slides.

Koerner and Soong [46] have used limit equilibrium and a finite slope model as shown in Figure 3.23 to analyze the general situation. Consider a cover soil placed directly on a geomembrane (or other barrier layer) at a slope angle β. Two discrete zones can be visualized, as shown in Figure 3.23. There is a small passive wedge resisting a long thin active wedge extending the length of the slope. It is assumed that the cover

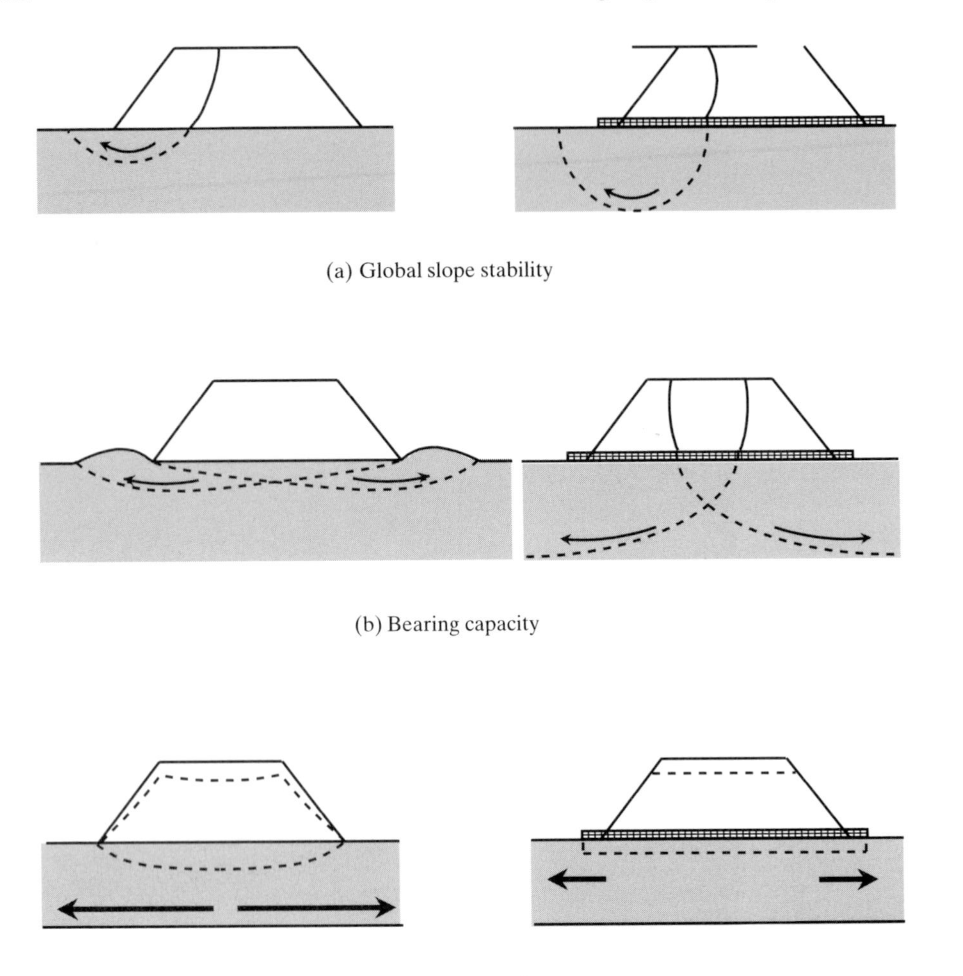

Figure 3.22 Potential improvement of an embankment on soft foundation soils via three-dimensional geogrid mattresses.

soil is of uniform thickness and constant unit weight. At the top of the slope or at an intermediate berm, we anticipate that a tension crack in the cover soil will occur, thereby breaking continuity with the remaining cover soil at the crest.

Resisting the tendency for the cover soil to slide is the interface friction and/or adhesion of the cover soil to the specific type of underlying geomembrane. The values of δ and c_a must be obtained from a laboratory direct shear test as described earlier. Note that the passive wedge is assumed to move on the underlying cover soil so that the shear strength parameters ϕ and c, which come from soil-to-soil friction tests, will also be required.

By taking free bodies of the passive and active wedges with the appropriate forces being applied, the formulation for the global factor of safety results. The

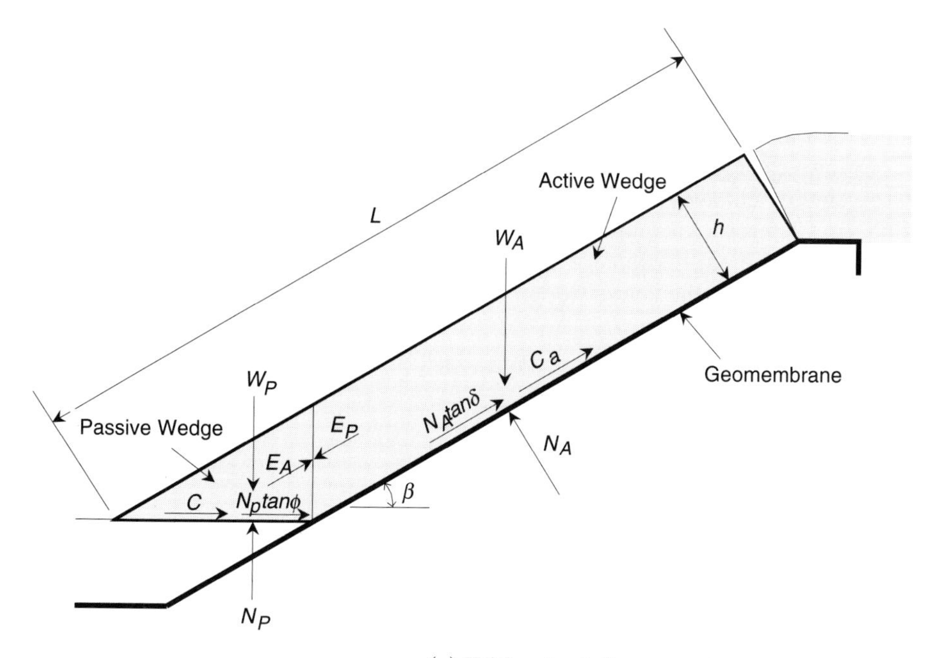

(a) Without reinforcement

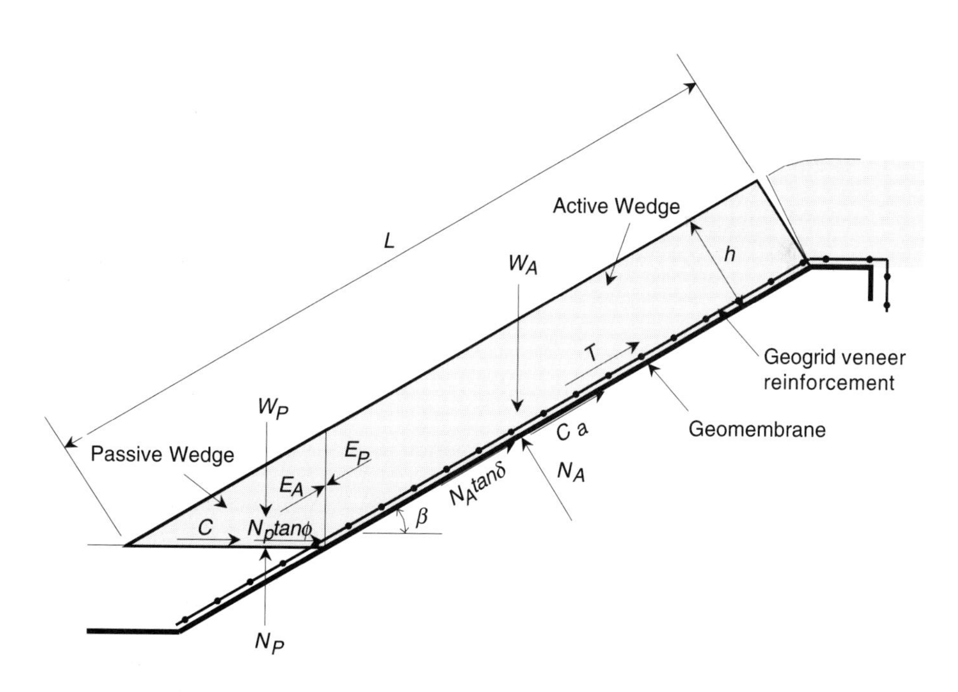

(b) With the use of veneer reinforcement

Figure 3.23 Limit equilibrium forces involved in a finite-length slope analysis for a uniformly thick cover soil.

resulting equation is not an explicit solution for the FS, and must be solved using the quadratic equation. The complete development of the equation is given in [46]. It follows closely the analyses of Giroud and Beech [47] and Koerner and Hwu [48], and is slightly changed from earlier editions of this book because it will be built upon further in Chapter 5 for multiple geosynthetically lined slopes.

The expression for determining the factor of safety can be derived as follows. Considering the active wedge,

$$W_A = \gamma h^2\left(\frac{L}{h} - \frac{1}{\sin\beta} - \frac{\tan\beta}{2}\right) \tag{3.14}$$

$$N_A = W_A \cos\beta \tag{3.15}$$

$$C_a = c_a\left(L - \frac{h}{\sin\beta}\right) \tag{3.16}$$

By balancing the forces in the vertical direction, the following formulation results.

$$E_A \sin\beta = W_A - N_A \cos\beta - \frac{N_A \tan\delta + C_a}{\text{FS}} \sin\beta$$

Hence the interwedge force acting on the active wedge is

$$E_A = \frac{(\text{FS})(W_A - N_A\cos\beta) - (N_A\tan\delta + C_a)\sin\beta}{\sin\beta\,(\text{FS})}$$

The passive wedge can be considered in a similar manner:

$$W_P = \frac{\gamma h^2}{\sin 2\beta} \tag{3.17}$$

$$N_P = W_P + E_P \sin\beta \tag{3.18}$$

$$C = \frac{(c)(h)}{\sin\beta} \tag{3.19}$$

By balancing the forces in the horizontal direction, the following formulation results.

$$E_P \cos\beta = \frac{C + N_P \tan\phi}{\text{FS}}$$

Hence the interwedge force acting on the passive wedge is

$$E_P = \frac{C + W_P \tan\phi}{\cos\beta(\text{FS}) - \sin\beta\tan\phi} \tag{3.20}$$

By setting $E_A = E_P$, the following equation can be arranged in the form of $ax^2 + bx + c = 0$, which in our case, using FS values is

$$a(\text{FS})^2 + b(\text{FS}) + c = 0 \tag{3.21}$$

where

$$a = (W_A - N_A \cos \beta) \cos \beta,$$
$$b = -[(W_A - N_A \cos \beta) \sin \beta \tan \phi + (N_A \tan \delta + Ca) \sin \beta \cos \beta + \sin \beta (C + W_p \tan \phi)], \text{ and}$$
$$c = (N_A \tan \delta + C_a) \sin^2 \beta \tan \phi.$$

The resulting FS value is then obtained from the following equation.

$$\text{FS} = \frac{-b + \sqrt{b^2 - 4ac}}{2a} \quad (3.22)$$

where (in Figure 3.23a and the above analysis)

W_A = total weight of the active wedge,
W_P = total weight of the passive wedge,
N_A = effective force normal to the failure plane of the active wedge,
N_A = effective force normal to the failure plane of the passive wedge,
γ = unit weight of the cover soil,
h = thickness of the cover soil,
L = length of slope measured along the geomembrane,
β = soil slope angle beneath the geomembrane,
ϕ = friction angle of the cover soil,
δ = interface friction angle between cover soil and geomembrane,
C_a = adhesive force between cover soil of the active wedge and the geomembrane,
c_a = adhesion between cover soil of the active wedge and the geomembrane,
C = cohesive force along the failure plane of the passive wedge,
c = cohesion of the cover soil,
E_A = interwedge force acting on the active wedge from the passive wedge,
E_P = interwedge force acting on the passive wedge from the active wedge, and
FS = factor of safety against cover soil sliding on the geomembrane.

When the calculated FS value falls below 1.0, a stability failure of the cover soil sliding on the geomembrane is to be anticipated. Thus a value of greater than 1.0 must be targeted as being the minimum factor of safety. How much greater than 1.0 the FS value should be is a design and/or regulatory issue. Example 3.12 illustrates the procedure.

Example 3.12

Given a cover soil slope of $\beta = 18.4°$ (i.e. 3 to 1), $L = 30$ m, $h = 900$ mm, $\gamma = 18$ kN/m^3, $c = 0$, $\phi = 30°$, $c_a = 0$, $\delta = 18°$, determine the resulting factor of safety.

Solution:

$$W_A = \gamma h^2 \left(\frac{L}{h} - \frac{1}{\sin \beta} - \frac{\tan \beta}{2} \right)$$

$$= (18.0)(0.90)^2\left(\frac{30}{0.90} - \frac{1}{\sin 18.4} - \frac{\tan 18.4}{2}\right)$$

$$= 14.58(33.3 - 3.17 - 0.17)$$

$$= 437 \text{ kN/m}$$

$$N_A = W_A \cos \beta$$

$$= 437 \cos 18.4$$

$$= 415 \text{ kN/m}$$

$$W_P = \frac{\gamma h^2}{\sin 2\beta}$$

$$= \frac{(18.0)(0.90)^2}{\sin 36.8}$$

$$= 24.3 \text{ kN/m}$$

$$a = (W_A - N_A \cos \beta) \cos \beta$$

$$= (437 - 415 \cos 18.4) \cos 18.4$$

$$= 41.0$$

$$b = -[(W_A - N_A \cos \beta) \sin \beta \tan \phi + (N_A \tan\delta + C_a) \sin \beta \cos \beta$$

$$+ \sin \beta(C + W_P \tan \phi)]$$

$$= -[(437 - 415 \cos 18.4) \sin 18.4 \tan 30 + (415 \tan 18 + 0) \sin 18.4 \cos 18.4$$

$$+ \sin 18.4(0 + 24.3 \tan 30)]$$

$$= -[7.84 + 40.4 + 4.43]$$

$$= -52.7$$

$$c = (N_A \tan \delta + C_a) \sin^2 \beta \tan \phi$$

$$= (415 \tan 18 + 0) \sin {}^2 18.4 \tan 30$$

$$= 7.8$$

$$\text{FS} = \frac{-b + \sqrt{b^2 - 4ac}}{2a}$$

$$= \frac{52.7 + \sqrt{(52.7)^2 - 4(41.0)(7.8)}}{2(41.0)}$$

$$= 1.11$$

which is too low for a final cover and an appropriate design option is to consider the use of veneer reinforcement.

Illustrated in Figure 3.23b is a sketch of the growing application of geogrid reinforcement, under the generic classification of *veneer reinforcement.* The geogrid embedded in its own anchor trench at the top of the slope is placed directly on the containment layer shown here as a geomembrane. Soil backfilling (with lightweight construction equipment) proceeds from the toe to the crest of the slope. As backfill is placed, the geogrid reinforcement is tensioned and, depending on the strength of the reinforcement, some or all of the gravitational stress of the soil is resisted. In the analysis that follows, the soil is assumed to be in contact with the geomembrane (through the apertures of the geogrid), the reinforcement is acting at an allowable value (hence, reduction factors must be applied to the ultimate value), and the active wedge has included in it an additional vector, namely the allowable tension T. For the active wedge, we balance the forces in the vertical direction and the following formulation results.

$$E_A \sin\beta = W_A - N_A\cos\beta - \left(\frac{N_A\tan\delta + C_a}{\text{FS}} + T\right)\sin\beta$$

Hence the interwedge force acting on the active wedge is

$$E_A = \frac{(\text{FS})(W_A - N_A\cos\beta - T\sin\beta) - (N_A\tan\delta + C_a)\sin\beta}{\sin\beta(\text{FS})}$$

Again, by setting $E_A = E_P$ [recall Eq. (3.20) for E_P], the resulting formulation can be arranged in the form of Eq. (3.21) where

$$\begin{aligned} a &= (W_A - N_A\cos\beta - T\sin\beta)\cos\beta, \\ b &= -[(W_A - N_A\cos\beta - T\sin\beta)\sin\beta\tan\phi + (N_A\tan\delta + C_a)\sin\beta\cos\beta \\ &\quad + \sin\beta\,(C + W_P\tan\phi)], \text{ and} \\ c &= (N_A\tan\delta + C_a)\sin^2\beta\tan\phi. \end{aligned}$$

Again, the resulting FS value can be obtained using Eq. (3.22). Example 3.13 illustrates the use of the above analysis.

Example 3.13

Let us continue Example 3.12, using a geogrid with $T_{\text{ult}} = 150$ kN/m and cumulative reduction factors amounting to 4.50. What is the resulting factor of safety for this case of veneer reinforcement?

Solution: Using the formulae just presented, the W_A, N_A, and W_P values stay the same as in Example 3.12. The allowable tensile strength of the geogrid reinforcement is

$$\begin{aligned} T &= \frac{T_{\text{ult}}}{\Pi\text{RF}} \\ &= \frac{150}{4.5} \\ &= 33.3 \text{ kN/m} \end{aligned}$$

$$\begin{aligned} a &= (W_A - N_A\cos\beta - T\sin\beta)\cos\beta \\ &= (437 - 415\cos 18.4 - 33.3\sin 18.4)\cos 18.4 \\ &= 31.4 \end{aligned}$$

$$b = -[(W_A - N_A\cos\beta - T\sin\beta)\sin\beta\tan\phi$$

$$+ (N_A \tan \delta + C_a) \sin \beta \cos \beta + \sin \beta (C + W_P \tan \phi)]$$

$$= -[(437 - 415 \cos 18.4 - 33.3 \sin 18.4) \sin 18.4 \tan 30$$

$$+ (415 \tan 18 + 0) \sin 18.4 \cos 18.4 + \sin 18.4(0 + 24.3 \tan 30)]$$

$$= -50.8$$

$$c = (N_A \tan \delta + C_a) \sin^2 \beta \tan \phi$$

$$= (415 \tan 18 + 0) \sin^2 18.4 \tan 30$$

$$= 7.8$$

$$\text{FS} = \frac{-b + \sqrt{b^2 - 4ac}}{2a}$$

$$= \frac{50.8 + \sqrt{(-50.8)^2 - 4(31.4)(7.8)}}{2(31.4)}$$

$$= 1.45 \quad \text{which is acceptable}$$

This solution for veneer reinforcement agrees well with other methods in the literature [47, 48] and with a finite-element solution (Wilson-Fahmy and Koerner [49]).

A significant issue, however, is the input variables for the analysis. This is particularly the case for the δ value and for the reduction factors on the geosynthetic reinforcement. Also, if a high-strength geotextile is being used, the δ value will be for the geotextile to the geomembrane, since geotextiles do not allow for strike-through of the backfill soil. Concerning an acceptable value of the resulting factor of safety, the site-specific situation must be considered. For leachate collection soils in landfills, relatively low values of FS may be acceptable since the solid waste will provide a buttressing effect as it is placed in the landfill. Conversely, for final cover soils in the closure of landfills, quite high values of FS should be considered since the time frames for service life can be extremely long.

3.3 DESIGN CRITIQUE

The design sections just presented use geogrids in their primary function, which is as soil reinforcement. This primary function comes about because of a number of features provided by geogrids.

Economy, as in the reduction of base course thickness in unpaved roads

Practicality, as in geogrid-reinforced embankments and walls

Necessity, as in the veneer reinforcement of cover soils on geomembranes where traditional methods of construction are not adequate

The design methods in each of the above instances were direct adaptations of traditional geotechnical engineering methods—only now the designs include a reinforcement material, namely geogrids. The liberties taken by making this change seem reasonable and justifiable on the basis of field performance and monitoring. In fact, the feedback seems to indicate that the design methods just presented are conservative. However, before we conclude that more liberal designs are in order, the nature of the actual situations must be considered. For example, embankment slopes, reinforced walls, and landfill reinforcement applications are generally permanent structures requiring long service lifetimes, which simply demand a conservative approach.

More uncertain than the design methods are the allowable strength of the geogrid and the proliferation of a wide variety of different geogrid types. Considerations of allowable strength lead directly to the subsequent factor of safety (Section 3.1.5). For this, Bonaparte and Berg [50] subdivide applications into *permanent* versus *temporary* and *critical* versus *noncritical.* It is a very perceptive approach, wherein the permanent and critical systems require the greatest amount of concern and caution. This, of course, is up to the designer on a site-specific basis. Regarding the variety of geogrid types, concern is warranted when a design is made and then an "or equal" specification is written. The introduction to this chapter has shown that little is equal from one geogrid to another. Hence a geogrid specification must be written around a set of performance characteristics. This, too, is a problem, since most of our experience has been in writing specifications for geotextiles and not for geogrids. While the current situation could be easily dismissed as merely growing pains, this does little good for a designer who is unsure of a product or for a manufacturer who is *very* sure of his or her product. At the minimum, a geogrid specification should contain the following:

- Minimum tensile strength in machine direction
- Minimum tensile strength in cross-machine direction (if biaxial stresses are present)
- Minimum modulus at some value of strain in the machine direction
- Minimum modulus at some value of strain in the cross-machine direction (if biaxial stresses are present)
- Minimum aperture size (in both directions)
- Maximum aperture size (in both directions)
- Durability limits and criteria on the type of polymers (in at least a generic sense for temporary/noncritical situations but with specifics for permanent/critical structures)
- Creep data for 10,000 hours on the same product or on one that is shown to be chemically equivalent to the specified product
- An indication of ultraviolet light stability
- An indication of chemical stability
- An indication of biological stability
- An interaction coefficient based on short-term anchorage tests and, if necessary, based on creep anchorage tests (depending on the particular application)
- Minimum seam (or joint) strength (if allowed) in the machine direction
- Minimum seam (or joint) strength (if applicable) in the cross-machine direction (if biaxial stresses are present)
- Connection strength (if applicable) to the mechanical system being considered

- Minimum roll width
- Minimum roll length
- Maximum roll weight

3.4 CONSTRUCTION METHODS

As with geotextiles, geogrids come to the job site in rolls. However, they are often narrower than geotextiles. Typically geogrid roll widths are from 1 to 3 m wide for stiff unitized geogrids and 3 to 4.5 m wide for flexible textile-like geogrids. Their deployment is straightforward unless some sort of tensioning or prestressing is desired. Because of their large aperture size, sewing is not possible to join the sides or ends together and some type of mechanical system is generally employed. Unitized unidirectional geogrids can be bent and the bent end inserted into the opening of an adjacent sheet. By placing a rod or bar down the slot that is formed, excellent load transfer is obtained (see Figure 3.19c). The rod is often an HDPE pipe of 12 mm diameter or a tapered bar called a *bodkin.* A number of joining techniques are under development with the flexible textile-like geogrids. Of course, adequate overlap can also be used to mobilize the tensile strength via shear stresses. Wire cutters will suffice to cut or trim geogrids, but a circular saw is quicker and more efficient. The flexible textile-like geogrids can generally be cut with a sharp knife.

Stiff unitized geogrids have been used to anchor wall facing panels made from concrete in a manner similar to the metal strips of reinforced earth. The attachment of geogrids to facing panels involves the casting of small geogrid sections or metal hooks into the concrete panels during their fabrication. The reinforcement geogrids are mechanically connected, directly or by means of a steel dowel running lengthwise behind the hooks and attached to the ends of the geogrids. Alternatively, geogrids can be used between layers of wall sections, such as gabions, concrete cells, or concrete blocks, to anchor the walls and reduce the earth pressure on the wall itself. Connection is by polymer or fiberglass dowels for stiff unitized geogrids or by friction (sometimes between lock-and-key sections) for textile-like geogrids (recall Figure 2.51b).

During the installation of geogrids for reinforcement purposes, the initial slack in the product must be removed before backfilling. This is generally a difficult task. Usually, laborers will use a crowbar, pick, or steel rod to pull the geogrid taut while it is being backfilled. The amount of tension is essentially a guess. It should be realized that too much tension might not be advisable, especially when the geogrid is attached to wall facing elements that are only temporarily supported. Clearly, tensioning is done on a trial-and-error basis. Its mobilization must be discussed by the parties involved before construction begins.

The above details are almost always geogrid product-specific. Thus the manufacturer or manufacturer's representative must be consulted to be assured that the construction details are appropriate for the proposed material system, that is, the particular geogrid and wall facing type. All geogrid manufacturers, to the author's knowledge, have well-trained geotechnical engineers on their full-time staff to advise consultants and owners on the details and nuances of their products.

REFERENCES

1. Wilson-Fahmy, R., Koerner, R. M., and Fleck, J. A., "Unconfined and Confined Wide Width Testing of Geosynthetics," in *Geosynthetic Soil Reinforcement Testing Procedures,* ASTM STP 1190, ed. S. J. Cheng, Philadelphia, PA: ASTM, pp. 49–63.
2. Sarsby, R. W., "The Influence of Aperture Size/Particle Size on the Efficiency of Grid Reinforcement," *Proc. 2d Canadian Symp. Geotextiles and Geomembranes,* Edmonton, Canada: The Geotechnical Society of Edmonton, 1985, pp. 7–12.
3. Ingold, T. S., "Laboratory Pull-Out Testing of Grid Reinforcement in Sand," *Geotechnical Testing J.,* Vol. 6, No. 3, 1983, pp. 212–217.
4. Koerner, R. M., Wayne, M. H., and Carroll, R. G., Jr., "Analytic Behavior of Geogrid Anchorage," *Proc. Geosynthetics '89,* St. Paul, MN: IFAI, 1989, pp. 525–536.
5. Wilson-Fahmy, R., and Koerner, R. M., "Finite Element Modeling of Soil-Geogrid Interaction in a Pullout Loading Conditions," *J. Geotextiles and Geomembranes,* Vol. 12, No. 5, 1993, pp. 479–501.
6. Kliethermes, J. C., Buttry, K., McCullough, E., and Wetzel, R., "Modular Concrete Retaining Wall and Geogrid Reinforcement Performance and Laboratory Modeling," *Proc. Geosynthetics '91*, St. Paul, MN: IFAI, 1991, pp. 951–964.
7. Bathurst, R. J., and Simac, M. R., "Geosynthetic Reinforced Segmental Retaining Wall Structures in North America," *Proc. 5th IGS Conf.,* Special Volume, Singapore: Southeast Asia Chapter, IGS, 1994, pp. 29–54.
8. Koerner, G. R., Koerner, R. M., and Elias, V., "Geosynthetic Installation Damage under Two Different Backfill Conditions," *Geosynthetic Soil Reinforcement Testing Procedures,* ASTM STP 1190, ed. S. J. Cheng, Philadelphia, PA: ASTM, 1993, pp. 163–183.
9. McGown, A., Andrawes, K. Z., and Kabir, M. H., "Load-Extension Testing of Geotextiles Confined in Soil," *Proc. 2d Intl. Conf. on Geotextiles,* St. Paul, MN: IFAI, 1982, pp. 93–96.
10. Ingold, T. S., Montanelli, F., and Rimoldi, P., "Extrapolation Techniques for Long Term Strengths of Polymeric Geogrids," *Proc. 5th Intl. Conf. on Geosynthetics,* Singapore: Southeast Asia Chapter, IGS, 1994, pp. 1117–1120.
11. Miyata, K., "Walls Reinforced with Fiber Reinforced Plastic Geogrids in Japan," *Geosynthetics International,* Vol. 3, No. 1, 1996, pp. 1–11.
12. Greenwood, J. H., "The Creep of Geotextiles," *Proc. 4th Intl. Conf. Geotextiles, Geomembranes and Related Products,* Rotterdam: A.A. Balkema, 1991, pp. 645–650.
13. Wrigley, N. E., "Durability and Long-Term Performance of Tensar Polymer Grids for Soil Reinforcement," *Materials Science and Technology,* Vol. 3, No. 4, 1987, pp. 161–170.
14. Koerner, R. M., Lord, A. E., Jr., and Halse, Y., "Allowable Geosynthetic Strength and Flow Considerations," *Proc. ASCE/Penn DOT Conf.,* Harrisburg, PA: Central Pennsylvania Section, ASCE, 1988, pp. 1–19.
15. Haas, R., "Structural Behavior of Tensar Reinforced Pavements and Some Field Applications," *Proc. Symp. Polymer Grid Reinforcement in Civil Eng.,* London: ICE, 1984, pp. 166–170.
16. Abd El Halim, A. O. "Geogrid Reinforcement of Asphalt Pavements," Ph.D. thesis, University of Waterloo, Ontario, Canada, 1983.
17. Abd El Halim, A. O., Haas, R., and Chang, W. A., "Geogrid Reinforcement of Asphalt Pavements and Verification of Elastic Layer Theory," Research Board Record No. 949, TRB, 1983, pp. 55–65.
18. Carroll, R. G., Jr., Walls, J. G., and Haas, R., "Granular Base Reinforcement of Flexible Pavements Using Geogrids," *Proc. Geosynthetics '87,* St. Paul, MN: IFAI, 1987, pp. 46–57.
19. "Guide for Design of Pavement Structures," Washington, DC: AASHTO, 1986.

20. Brown, S. F., Brodrick, B. V., and Hughes, D. A. B., "Tensar Reinforcement of Asphalt: Laboratory Studies," *Proc. Symp. Polymer Grid Reinforcement in Civil Eng.,* London: ICE, 1984, pp. 158–165.
21. Kennepohl, G. J. A., and Kamel, N. I., "Construction of Tensar Reinforced Asphalt Pavements," *Proc. Symp. Polymer Grid Reinforcement in Civil Eng.,* London: ICE, 1984, pp. 171–175.
22. Molenaar, A. A. A., and Nods, M., "Design Method for Plain and Geogrid Reinforced Overlays on Cracked Pavements," *Proc. 3rd Intl. RILEM Conferrence,* ed. L. Francken, E. Beuving, and A. A. A. Molenaar, London: E & FN Spon., 1996, pp. 311–320.
23. Giroud, J.-P., Ah-Line, C., and Bonaparte, N., "Design of Unpaved Roads and Trafficked Areas with Geogrids," *Proc. Symp. Polymer Grid Reinforcement in Civil Eng.,* London: ICE, 1984, pp. 116–127.
24. Forsyth, R. A., and Bieber, D. A., "La Honda Repair with Geogrid Reinforcement," *Proc. Symp. Polymer Grid Reinforcement in Civil Eng.,* London: ICE, 1984, pp. 54–57.
25. Berg, R. R., Anderson, R. P., Rose, R. J., and Chouery-Curtis, V. E., "Reinforced Soil Highway Slopes," *Proc. TRB 69th Annual Meeting,* TRB, Washington, DC: TRB, 1990, 46 pgs.
26. Schmertmann, G. R., Chouery-Curtis, V. E., Johnson, R. D., and Bonaparte, R., "Design Charts for Geogrid-Reinforced Soil Slopes," *Proc. Geosynthetics '87,* St. Paul, MN: IFAI, 1987, pp. 108–120.
27. Ingold, T. S., "An Analytical Study of Geotextile Reinforced Embankments," *Proc. 2d Intl. Conf. Geotextiles,* St. Paul, MN: IFAI, 1982, pp. 683–689.
28. Murray, R. T., "Reinforcement Techniques in Repairing Slope Failures," *Proc. Conf. Polymer Grid Reinforcement,* London: ICE, 1984, pp. 47–53.
29. Jewell, R. A., "Application of Revised Design Charts for Steep Reinforced Slopes," *J. Geotextiles and Geomembranes,* Vol. 10, No. 3, 1991, pp. 203–233.
30. Leshchinsky, D., and Reinschmidt, A. J., "Stability of Membrane Reinforced Slopes," *J. Geotechnical Eng.,* ASCE, Vol. 111, No. 11, 1985, pp. 1285–1297.
31. Schneider, H. R., and Holtz, R. D., "Design of Slopes Reinforced with Geotextiles and Geogrids," *J. Geotextiles and Geomembranes,* Vol. 3, No. 1, 1986, pp. 29–52.
32. Reugger, R., "Geotextile Reinforced Soil Structures on Which Vegetation Can Be Established," *Proc. 3d Intl. Conf. Geotextiles,* Vienna: Austrian Society of Engineers, 1986, pp. 453–458.
33. Tatsuoka, F., Murata, O., and Tateyama, M., "Permanent Geosynthetic Reinforced Soil Retaining Walls Used for Railway Embankments in Japan," *Proc. Geosynthetic Reinforced Soil Retaining Walls,* ed. J. T. H. Wu, Rotterdam: A.A. Balkema, 1992, pp. 101–130.
34. Soong, T.-Y., and Koerner, R. M., "On the Required Connection Strength of Geosynthetically Reinforced Walls," *J. Geotextiles and Geomembranes* (to be published).
35. Berg, R. R., Bonaparte, R., Anderson, R. P., and Chouery, V. E., "Design Construction and Performance of Two Tensar Geogrid Reinforced Walls," *Proc. 3d Intl. Conf. Geotextiles,* Vienna: Austrian Society of Engineers, 1986, pp. 401–406.
36. Jarrett, P. M. and McGown, A., eds., *Proc. NATO Workshop on Application of Polymeric Reinforcement in Soil Retaining Structures,* June, 1987, Royal Military College of Canada, Kingston, Ontario, Canada, Kluwer Acad. Publ., 1988.
37. Simac, M. R., Bathurst, R. J., Berg, R. R., and Lothspeich, S. E., *Segmental Retaining Wall Design Manual,* Herndon, VA: National Concrete Masonry Assoc., 1993.
38. Jarrett, P. M., "Evaluation of Geogrids for Construction of Roadways over Muskeg," *Proc. Symp. Polymer Grid Reinforcement in Civil Eng.,* London: ICE, 1984, pp. 149–153.
39. Milligan, G. W. E., and Love, J. P., "Model Testing of Geogrids under an Aggregate Layer in Soft Ground," *Proc. Symp. Polymer Grid Reinforcement in Civil Eng.,* London: ICE, 1984, pp. 128–138.

40. Giroud, J.-P., Bonaparte, R., Beech, J. F., and Gross, B. A., "Design of Soil Layer-Geosynthetic Systems Overlying Voids," *J. Geotextiles and Geomembranes,* Vol. 9, No. 1, 1990, pp. 11–50.
41. Edgar, S., "The Use of High Tensile Polymer Grid Mattress on the Mussleburgh and Portobello Bypass," *Proc. Symp. Polymer Grid Reinforcement in Civil Eng.,* London: ICE, 1984, pp. 103–111.
42. Rueff, H., Stoffers, U., and Leicher, F., "Deponie auf schwierigsten Untergrund," Ernst & Sohn, Bautechnik 69, Heft 5, 1992.
43. Robertson, J., and Gilchrist, A. J. T., "Design and Construction of a Reinforced Embankment across Soft Lakebed Deposits," *Proc. Symp. Polymer Grid Reinforcement in Civil Eng.,* London: ICE, 1984.
44. Jenner, C. G., Bush, D. I., and Bassett, R. H., "The Use of Slip Line Fields to Assess the Improvement in Bearing Capacity of Soft Ground Given by a Cellular Foundation Mattress Installed at the Base of an Embankment," *Proc. Theory and Practice of Earth Reinforcement,* Rotterdam: A.A. Balkema, 1988, pp. 209–214.
45. Boschuk, J., Jr., "Landfill Covers: An Engineering Perspective," *Geotech. Fabrics Rpt.,* Vol. 9, No. 2, 1991, pp. 23–34.
46. Koerner, R. M., and Soong, T.-Y., "Cover Soil Slope Stability under a Wide Variety of Situations," *Proc. 6th IGS Conf.,* St. Paul, MN: IFAI, 1998 (to be published).
47. Giroud, J.-P., and Beech, J. F., "Stability of Soil Layers on Geosynthetic Lining Systems," *Proc. Geosynthetics '89,* St. Paul, MN: IFAI, 1989, pp. 35–46.
48. Koerner, R. M., and Hwu, B.-L., "Stability and Tension Considerations Regarding Cover Soils in Geomembrane Lined Slopes," *J. of Geotextiles and Geomembranes,* Vol. 10, No. 4, 1991, pp. 335–355.
49. Wilson-Fahmy, R. F., and Koerner, R. M., "Finite Element Analysis of Cover Soil on Geomembrane Lined Slopes," *Proc. Geosynthetics '93,* St. Paul, MN: IFAI, pp. 1425–1437.
50. Bonaparte, R., and Berg, R., "Long-Term Allowable Tension for Geosynthetic Reinforcement," *Proc. Geosynthetic '87,* St. Paul, MN: IFAI, 1987, pp. 181–192.

PROBLEMS

3.1. List and describe the basic similarities and differences between the properties of geogrids and geotextiles.

3.2. List and describe the basic similarities and differences between the properties of geogrids and geonets.

3.3. Flexible geogrids are coated with latex, bitumen, or polyvinyl chloride. What is the reason for such coatings?

3.4. Derive an anchorage resistance formula on the basis of Figure 3.4 to include all three components of pullout resistance. (*Hint:* See [4])

3.5. From the anchorage analysis of Problem 3.4:

(a) Using the formula derived in Problem 3.4 for a geogrid in granular soil of $\phi = 35°$ at 19.5 kN/m^3 and 2.0 m depth, what is the ultimate pullout resistance for a length of 600 mm? The physical properties of the geogrid are: 6.0 mm wide ribs at 50 mm centers in both directions, the rib thickness is 3.0 mm, and the friction angle mobilized by the rib's surface to the soil is 30°.

(b) What is the relative proportion of the three components of anchorage strength calculated in part (a)?

(c) If the wide-strip tensile strength of the above geogrid was 45 kN/m, could the anchorage force calculated in part (a) be mobilized? If not, what portion of it could be?

3.6. Repeat the calculations of Problem 3.5(a) using a rib width of 8.0 mm, 10.0 mm, and then 12.0 mm. Plot the anchorage strength versus rib width.

3.7. What junction strength is required for the results of Problems 3.5(a) and 3.6 if all of the available anchorage strength is mobilized? (*Hint:* The transverse rib shear and transverse rib bearing must both be transmitted through the junction.)

3.8. Discuss the usefulness (or uselessness) of performing an in-isolation junction strength test as described in Figure 3.2 and tabulated in Table 3.1. If you feel it has no relevancy, what alternate test do you suggest to evaluate the significance of the transverse ribs of geogrids?

3.9. In light of the geogrid junction strength data of Table 3.1:

(a) What effect would normal stress on the junction have on the test results?

(b) What is the role of long-term stress on the junctions?

(c) How can long-term junction strength be evaluated?

3.10. Regarding stiff unitized polyolefin geogrids:

(a) Phenomenologically describe what happens to the molecular structure of polyethylene and polypropylene at or slightly above room temperature, when they are uniformly stretched in a continuous direction.

(b) What influence does cold working have on the mechanical properties of the final product?

3.11. What is the effect of high temperature on the following mechanical properties of polyolefin and polyester geogrids?

(a) modulus

(b) tensile strength

(c) elongation at failure

(d) creep behavior

3.12. If the ultimate tensile strength of a geogrid evaluated in wide-strip tensile strength test is 60 kN/m, what allowable strength should be used in design for the following situations?

(a) temporary unpaved access road for construction-vehicle use for approximately one year

(b) stone-base reinforcement in a permanent paved road for 30 year lifetime

(c) differential settlement reduction beneath a wall footing (i.e., improved bearing capacity) for a 50 to 100 year lifetime

3.13. In using geogrids for the reinforcement of paved roads, a possible mechanism involving bearing capacity is often mentioned. On a conceptual basis, how does this work?

3.14. How does the geogrid effectiveness factor (GEF) in Section 3.2.1 relate to the fabric effectiveness factor (FEF) in Section 2.10.2?

3.15. For the power law, Eq. (3.9), used to calculate the crack propagation through pavement overlays, what type of laboratory experiment would you devise to arrive at the K, A, and n values required in the analysis?

3.16. Regarding geogrid-reinforced unpaved roads:

(a) What is the required aggregate thickness for an unpaved road carrying ten thousand 80-kN vehicle passes with rut depths of 0.15 m on a soil subgrade whose undrained shear strength is 20 kN/m^2 (use Figure 3.11)?

(b) Using SS2 (BX 1200) geogrid reinforcement, channelized traffic, and high likelihood of aggregate contamination, how much aggregate can be saved?

(c) What increase in load-spreading angle does this represent (assuming that the nonreinforced case is $\alpha_0 = 26.6°$)?

3.17. Repeat Problem 3.16 assuming that the soil subgrade has CBR = 2.0.

3.18. Repeat Problem 3.16 using SR2 (UX 1200) geogrids.

3.19. Determine the factor of safety against sliding and overturning (i.e., the external stability) for the wall below, carrying a surcharge load of 7.2 kPa.

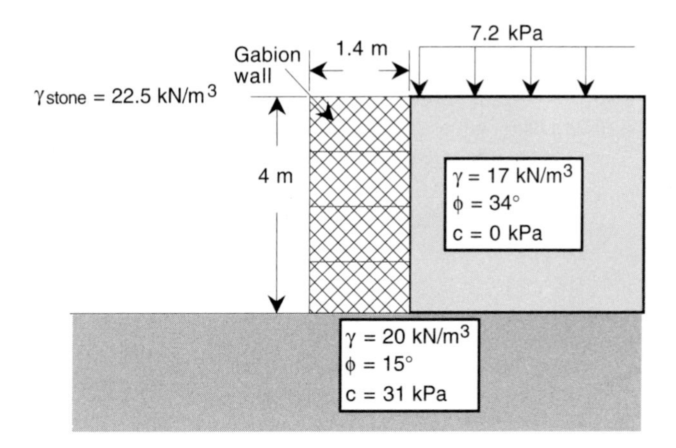

3.20. Using the approach indicated by Figures 3.13 and 3.14, determine the number, spacing, and length of the geogrids needed to stabilize the embankment below using a factor of safety of 1.3. Use combined reduction factors of 4.3 on the ultimate strength of the candidate geogrid of 180 kN/m to arrive at an allowable strength.

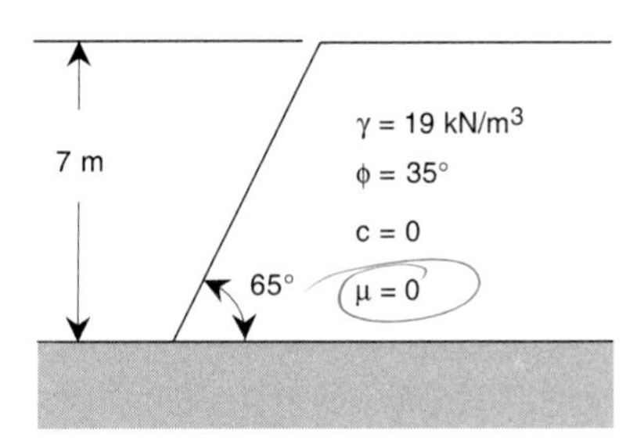

3.21. Figure 3.18 shows that at high deformations the load-carrying capacity of shallow footings can be increased considerably using geogrids. How can this feature be used without having the footing undergo large settlement? (Draw some sketches of how this could be accomplished.)

3.22. Using the design guide of Figure 3.19, solve for T_{reqd} as in Example 3.11 (Section 3.2.6) for the following parametric variations (i.e., hold all values constant except the target parameter).

(a) vary H from 1 to 30 m

(b) vary R from 1 to 10 m

(c) vary q from 7 to 100 kPa at $H/B = 0.1$

3.23. For Example 3.12, concerning the nonreinforced cover soil (Section 3.2.7), recalculate the FS for δ values of 13, 15, 21, and 24° and plot the response along with that for 18° given in the example.

3.24. Recalculate Problem 3.23 varying the cover soil thickness, using 300, 600, and 1200 mm to compare with that given in Example 3.12 for 900 mm (use $\delta = 18°$).

3.25. For Example 3.13, for geogrid-reinforced cover soil (Section 3.2.7), recalculate the factor of safety for T_{ult} values using 100, 200, and 250 kN/m and plot the results along with that for 150 kN/m given in the example.

3.26. What are the various ways of transferring load from one sheet of geogrid to the next for
(a) stiff geogrids?
(b) flexible geogrids?

3.27. When using geogrids for the reinforcement of masonry block facing retaining walls, what are the various ways of attaching the geogrids to such block facings?

4 Designing with Geonets

4.0 INTRODUCTION

A geonet can be defined as follows:

> **geonet**, *n.* a geosynthetic material consisting of integrally connected parallel sets of ribs overlying similar sets at various angles for biaxial or triaxial drainage of liquids or gases. Geonets are often laminated with geotextiles on one or both surfaces and are then referred to as drainage geocomposites.

As originally described in Section 1.5, geonets are formed by a continuous extrusion process into a netlike configuration of parallel sets of homogeneously interconnected ribs. Figure 1.18 shows how the extruded core is opened up into the final net-like configuration. Figures 4.1, 4.2, and 4.3 give a closer view of the details of various types of geonets. Each figure shows an overview, a closeup of the interconnected ribs, and a photomicrograph of the rib intersection. The types illustrated are

- *Biaxial geonets consisting of extruded solid ribs*, by far, the most common type of geonet (Figure 4.1)
- *Biaxial geonets consisting of extruded foamed ribs*, which results in greater overall thickness, and hence greater flow rate (Figure 4.2)
- *Triaxial geonets consisting of extruded solid ribs*, which allows for high preferential flow and the capability of sustaining high normal stresses (Figure 4.3)

Note, however, that there is considerable new product variation in this category of geosynthetics and it is still evolving (see Austin [1]).

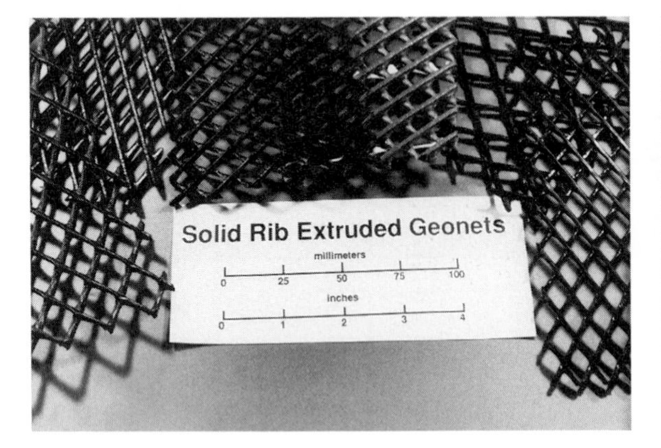

(a) Overview

(b) Closeup of rib intersections

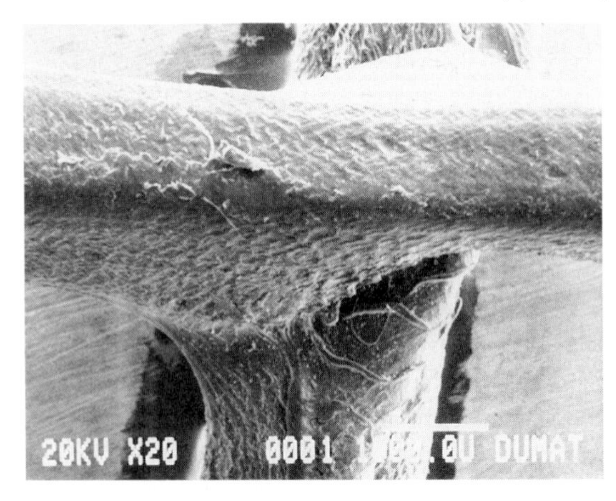

(c) Scanning electron micrograph of a rib intersection (magnification × 20)

Figure 4.1 Biaxial geonets consisting of solid extruded ribs.

All of the geonets currently available are made from polyethylene resin. The density varies from 0.94 to 0.96 mg/l, with the higher values forming the more rigid products. The resin is formulated with 2.0 to 2.5% carbon black (usually in a concentrated form mixed with a polyethylene carrier resin) and 0.25 to 0.75% additives that serve as processing aids and anti-oxidants. With these additives, even the lowest density resins fall into a HDPE category, which is how we will refer to all geonet products in this book.

4.1 GEONET PROPERTIES AND TEST METHODS

Since the primary function of a geonet is to convey liquid within the plane of its structure, the in-plane hydraulic flow rate, or transmissivity, is of major importance (see Williams et al [2]). However, other features, which may influence this value over the

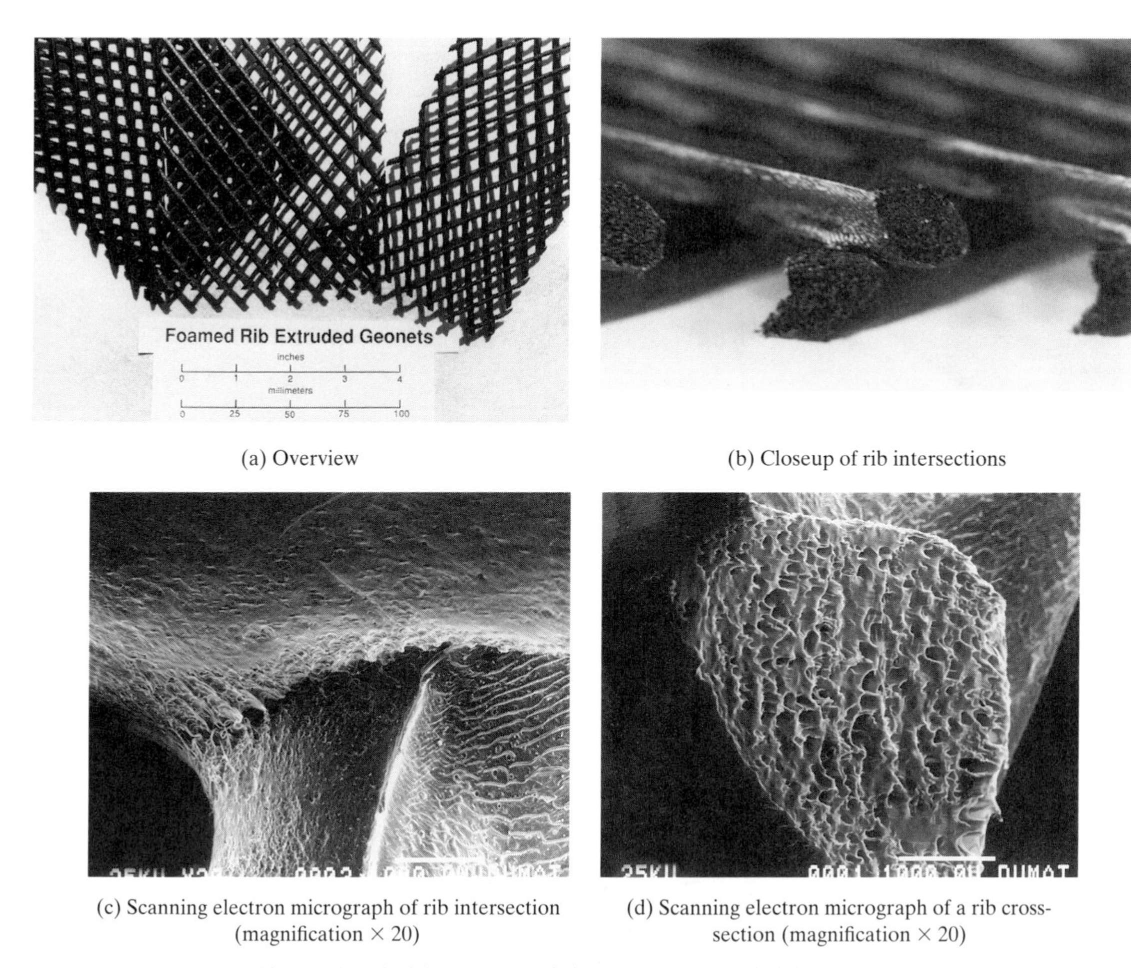

(a) Overview

(b) Closeup of rib intersections

(c) Scanning electron micrograph of rib intersection (magnification × 20)

(d) Scanning electron micrograph of a rib cross-section (magnification × 20)

Figure 4.2 Biaxial geonets consisting of foamed extruded ribs.

service lifetime of the geonet, are also of importance. Thus a number of physical, mechanical, endurance, and environmental properties will also be presented in this section.

4.1.1 Physical Properties

The density or specific gravity of the polymer is an important property and it can be evaluated either by ASTM D1505 or D792. The former is preferred if an accuracy of at least 0.005 mg/l is required. Another physical property needed to characterize a geonet is its thickness, which can be determined using ASTM D5199. While there is no listing for geonets as such, it is recommended that geonet thickness be measured under a normal pressure of 20 kPa. (This is the same pressure that is recommended to measure geomembrane thickness.) Note that geonets are not nearly as sensitive to thickness

(a) Overview

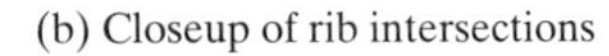

(b) Closeup of rib intersections

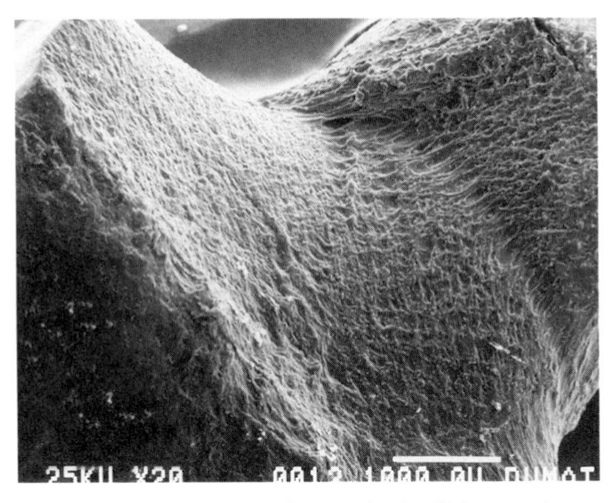

(c) Scanning electron micrograph of a rib intersection (magnification × 20)

Figure 4.3 Triaxial geonets consisting of solid extruded ribs.

variation under normal pressure as are geotextiles; they are similar to geomembranes in this regard.

Mass per unit area can be determined using ASTM D5261. For a 5.0 mm thick, solid-rib extruded geonet, the mass per unit area is usually in the range of 700 to 1000 g/m^2. It is not a design property per se, but it is informative from a manufacturer's point of view.

Other physical properties such as rib dimensions, planar angles made by the intersecting ribs, cross-planar angles made at the junction locations, aperture size and shape, and so on, can be measured directly and are relatively straightforward to obtain.

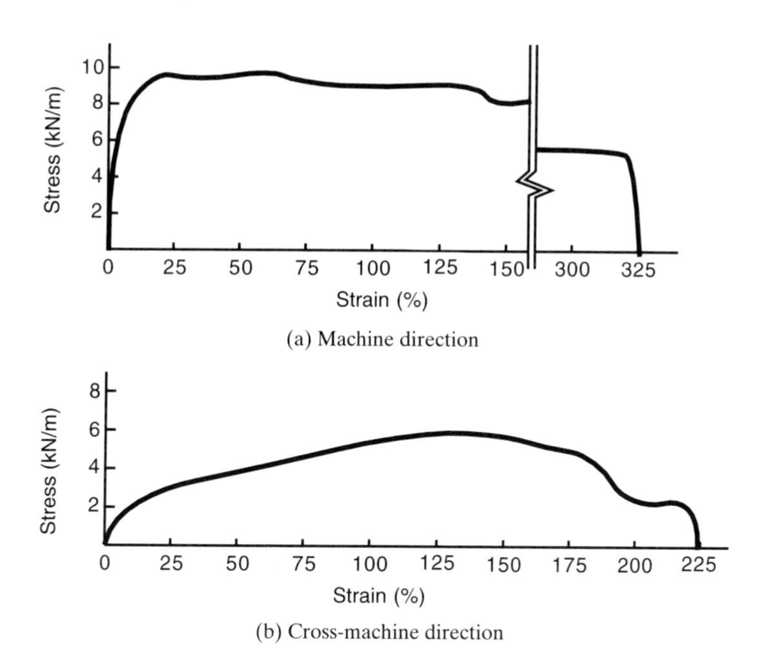

(a) Machine direction

(b) Cross-machine direction

Figure 4.4 Tensile strength behavior of 5.0 mm thick biaxial solid-rib extruded geonet.

4.1.2 Mechanical Properties

A number of strength properties of geonets are important to consider.

Tensile Strength. The wide-width tensile strength curves of Figure 4.4 results from testing of a 5.0 mm thick solid-rib extruded biaxial geonet in the machine and cross-machine directions. A 200 mm wide × 100 mm long test specimen was used in these tests. The strain rate was 10 mm/mm. Note the differences in behavior, suggesting that there is a preferential direction in strength between the machine and cross-machine directions. If site-specific conditions warrant a higher-strength direction, the geonet should be oriented with its machine direction positioned accordingly. Also note that geonets possess a low, but measurable, tensile strength. For the solid-rib geonet illustrated in Figure 4.1 and shown in Figure 4.4, the average of a series of unconfined wide-width tensile tests gave the following information:

- Machine direction: peak strength = 120 kN/m, strain at peak = 23%, strain at failure = 290%
- Cross-machine direction: peak strength = 68 kN/m, strain at peak = 170%, strain at failure = 240%

If a greater machine-direction tensile strength of a geonet is desired, consideration should be given to triaxial geonets oriented in the proper direction.

Compressive Strength. Of greater importance than the above-described in-plane tensile strength tests on geonets is the cross-plane compressive strength. This is because of the influence that compressive deformation or collapse has on the ability of the geonet to conduct liquid within its planar structure. There are a number of approaches to measuring a geonet's compressive strength. Using 150 mm square test specimens normally loaded under a constant strain rate load of 0.05 mm/min, the curves of Figure 4.5 are produced. Here it is shown that both the solid-rib extruded biaxial and the foamed-rib extruded biaxial geonets are initially very stiff but begin to deform at normal stress between 800 and 600 kPa. This occurs because the parallel sets of ribs making up the respective geonets are not exactly perpendicular to one another at the junctions. Thus at high compressive stress, there is a lay-down or roll-over tendency, which gives rise to the characteristic behavior shown. Note that the geonet can still convey liquid beyond this point, but to a somewhat diminished degree. The reason for lay-down stress on the foamed-rib geonet being slightly lower than that on the solid-rib geonet is due to the relatively thicker ribs and hence slightly higher flow rate capacity. The solid-rib extruded triaxial geonet (with a still higher flow rate capacity) shows no inflection point since the central and thicker set of ribs is perpendicular to the normal stress. Thus its response to normal stress increases with compressive strain is in direct proportion to the density of the resin.

Note that the long-term creep strength is not indicated by these short-term tests. High normal stress, indentation, and creep are discussed in Corcoran et al. [3].

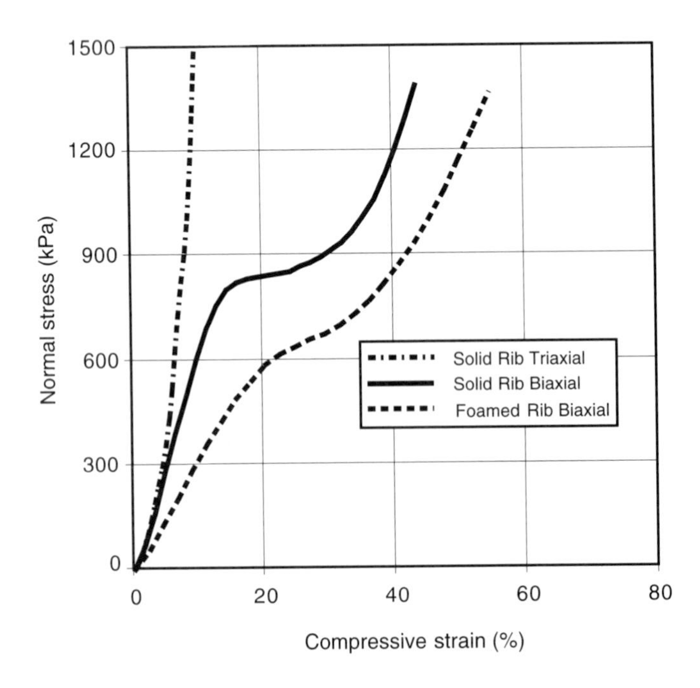

Figure 4.5 Compressive test data of different geonets.

Shear Strength. A geonet's capacity for sustaining planar shear stress within its own thickness without collapsing may be of concern in certain situations. This arises when there are high opposing shear stresses acting on the top and bottom of the geonet, tending to put the material into a state of pure shear. This is not felt to be much of a problem for the currently available unitized geonets. Considerably more concern should be focused on the interface friction behavior with respect to the materials above and beneath the geonet. This is particularly the case for geonet composites with covering geotextile(s). Due to the generally low shear strength of a geotextile to a geonet, it is common practice to bond the geotextile(s) to the geonet at the manufacturer's facility. Bonding has been done in the past using adhesives, but most manufacturers currently use thermal methods (i.e., hot wedge, infrared, etc.). The test method commonly used as a quality control test to determine the tensile peel strength of the geotextile to the geonet is ASTM D413. The appropriate interface shear test method is ASTM D5321 and it is obviously a product-specific and site-specific test that must be performed for each set of conditions that arise; see Lydick and Zagorski [4] in this regard. There is no known relationship between the tension peel and interface shear tests. It is an interesting research area.

4.1.3 Hydraulic Properties

The in-plane hydraulic test to determine planar flow rate, or transmissivity, of geonets should be performed using ASTM D4716 or ISO/DIS 12958. Both test methods use a planar transmissivity device and not the radial transmissivity device that was described and illustrated in Section 2.3.4 on geotextiles. This is necessary because the flow regime in a geonet is surely turbulent (consisting of irregular flow paths and eddies) whereas in a geotextile it is probably laminar (thus allowing for radial flow and integration over the flow path length). Some relevant points regarding the ASTM D4716 test method are as follows.

- The specimen size must be 150 by 150 mm, or larger. Kolbasuk et al. [5] have shown that sample size and aspect ratio (i.e., length-to-width ratio) can have significant effect on the resulting measured flow rates.
- As an index test, the geonet test specimen is sandwiched between rigid plates. However, the cross-section can be varied as desired to approach becoming a site-specific performance test.
- Normal stresses are applied for a 15 min duration, and flow rates are measured at different hydraulic gradients during a subsequent 15 min duration.
- Stress is then increased to the next level and the cycle is repeated.
- De-aired water at 5 ppm, or lower, of dissolved oxygen is not explicitly specified, although for critical situations it might be necessary.
- A standard laboratory temperature of 20°C is required.

Following this procedure and using a 200 mm square test specimen with a 6.3 mm thick solid-rib extruded biaxial geonet in the test device shown in Figure 4.6, the flow rate

(a) Photograph of flow rate device

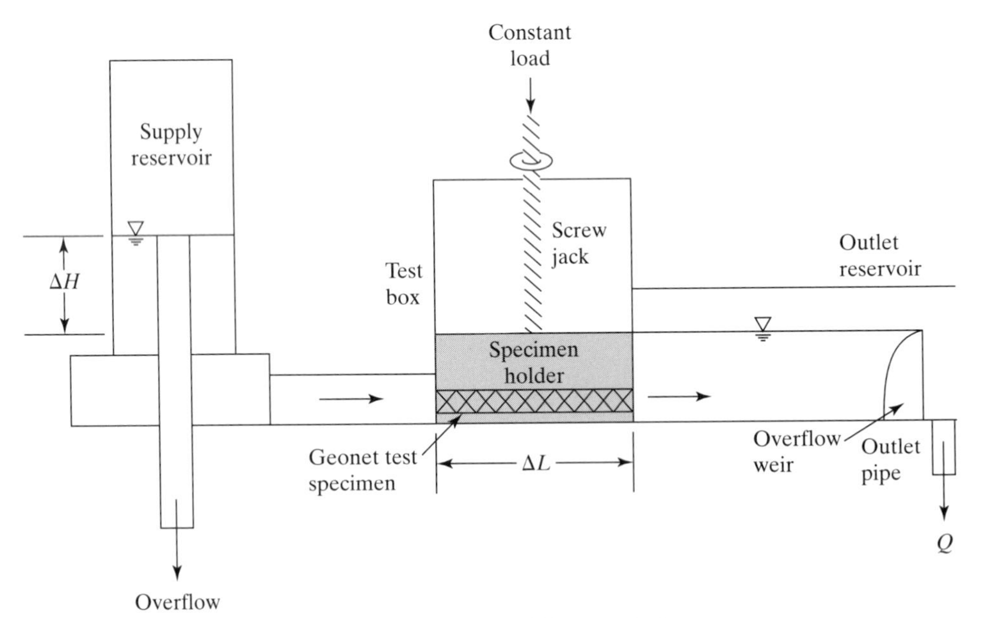

(b) Schematic diagram of flow rate device

Figure 4.6 Permeability device for measuring transmissivity (parallel in-plane flow) of a geonet.

curves of Figure 4.7 resulted. Here a decreasing flow rate can be observed at each hydraulic gradient evaluated. Obviously, the higher the hydraulic gradient, the higher the flow rate. From these data we can easily calculate a transmissivity value, the assumption being that the system is saturated at all times. When we consider that these flow rates

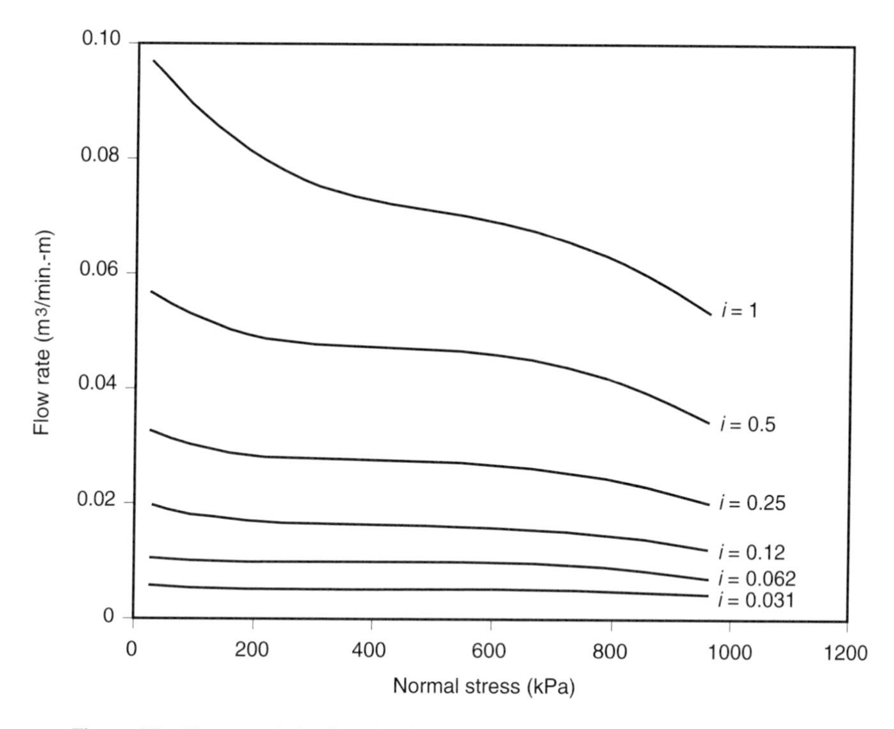

Figure 4.7 Flow-rate behavior of a 6.3 mm thick geonet sandwiched between two 1.5 mm HDPE geomembranes.

are extremely high in comparison to the flow rates in soil, it is clear that caution should be used when relying on transmissivity. For example, 300 mm of sand, having a coefficient of permeability of 0.1 cm/s at a hydraulic gradient of 1.0, can carry only 0.02 m^3/min-m. Thus geonets can handle large flow rates compared to soil due to the higher velocity of flow within the significantly large voids. Furthermore, the preferred flow rate in triaxial geonets is 5 to 10 times higher than in biaxial geonets. Clearly, the flow regime in geonets is turbulent in its behavior. Thus the discussion in this chapter will generally be using flow rate values rather than transmissivity values.

The data shown in Figure 4.7 are of the *index* test type. Site-specific situations, however, can be included in the test procedure so as to make the test results more *performance* oriented. To do so we must have the representative conditions above and below the test specimen, and use liquids of the type (and sometimes at the temperature) to be conveyed in the actual system. Figure 4.8 illustrates the impact of one such variation. Here a cross-sectional profile consisting of a layer of clay (kaolinite clay at 15% water content), a 540 gm/m^2 nonwoven needle-punched polyester geotextile, a geonet, and a 1.5 mm HDPE geomembrane is used. Since the geonet is the same type as that producing the data in Figure 4.7, the results can be compared directly, the only difference being the clay soil/geotextile separator placed over the geonet. The marked decrease in the flow rates of Figure 4.8 as compared to Figure 4.7 comes from intrusion of the geotextile into the core space of the geonet via the pressure applied to the clay. Numerically the comparison results in Table 4.1, in which the reduction in flow rates are

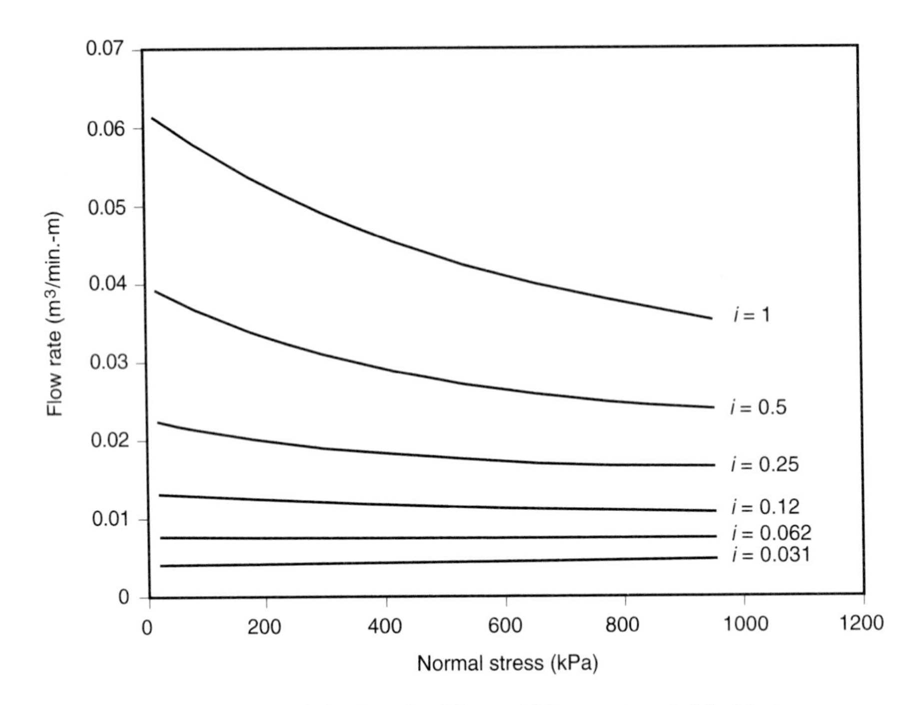

Figure 4.8 Flow-rate behavior of a 6.3 mm thick geonet sandwiched between a 540 g/m^2 nonwoven needle-punched geotextile with clay above and a 1.5 mm HDPE geomembrane below.

TABLE 4.1 FLOW RATE (m^3/min-m) AND REDUCTIONS (%) BETWEEN CURVES OF FIGURES 4.7 AND 4.8

Normal Stress (kPa)	Cross Section	Hydraulic Gradient (i)					
		0.03	0.06	0.12	0.25	0.50	1.00
50	HDPE (both sides)	0.005	0.011	0.019	0.032	0.055	0.095
	GT/Clay (one side)	0.004	0.008	0.013	0.022	0.038	0.059
	Difference	0.001	0.003	0.006	0.010	0.017	0.036
	Reduction	20%	27%	31%	31%	31%	38%
250	HDPE (both sides)	0.005	0.011	0.017	0.028	0.048	0.077
	GT/Clay (one side)	0.004	0.008	0.012	0.019	0.032	0.052
	Difference	0.001	0.003	0.005	0.009	0.016	0.025
	Reduction	20%	27%	29%	32%	33%	32%
500	HDPE (both sides)	0.005	0.010	0.017	0.027	0.047	0.072
	GT/Clay (one side)	0.004	0.007	0.012	0.018	0.028	0.043
	Difference	0.001	0.003	0.005	0.009	0.019	0.029
	Reduction	20%	30%	29%	30%	40%	40%
950*	HDPE (both sides)	0.005	0.008	0.013	0.021	0.035	0.054
	GT/Clay (one side)	0.004	0.007	0.011	0.017	0.024	0.035
	Difference	0.001	0.001	0.002	0.004	0.011	0.019
	Reduction	20%	12%	15%	19%	31%	35%

*Note that roll-over of the geonet structure has occurred at this normal stress (recall Figure 4.5). Thus the reduction percentages are not as large as they are with the other normal stress values.

seen to be as high as 40% from the index test values. Additionally, the geotextile must be capable of sustaining the applied stress, suggesting that long-term tests are required to adequately assess such situations. Sustained load deformation, or creep, will be discussed in Section 4.1.4.

It should be emphasized that *flow rates per unit width* values are not *transmissivity* values. To convert flow rate per unit width to transmissivity we use Darcy's formula, which tacitly assumes saturated conditions and laminar flow, neither of which are met with the typical flow regime in a geonet. Yet, current U.S. EPA leak detection regulations [6] state the following:

- For landfills and waste piles, the geonet's *transmissivity* must be

$$\theta \geqslant 3 \times 10^{-5}\ \mathrm{m^2/s}$$

- For surface impoundments, the geonet's *transmissivity* must be

$$\theta \geqslant 3 \times 10^{-4}\ \mathrm{m^2/s}$$

We convert from flow rate per unit width to transmissivity as follows.

$$q = ki\,A \tag{4.1}$$

$$q = ki(t \times W)$$

$$\frac{q}{W} = i(k \times t)$$

$$\frac{q}{W} = i\theta \tag{4.2}$$

Thus it is seen in Eq. (4.2) that the units of q/W and θ are identical, but only at $i = 1.0$ are the numeric values the same. A hydraulic gradient of 1.0 is when the geonet is placed vertically, as in the lining of a tank wall, not on the sloping surfaces of a typical landfill or surface impoundment. Example 4.1 illustrates the numeric conversion.

Example 4.1

Given a geonet tested in the laboratory under site-specific conditions resulting in a flow rate per unit width of $0.65 \times 10^{-4}\ \mathrm{m^2/s}$ at a hydraulic gradient 0.040. (**a**) What is the equivalent transmissivity in units of $\mathrm{m^2/s}$ and (**b**) does this value meet the EPA criteria for landfills and surface impoundments?

Solution: Using Eq. (4.2) and converting units we have the following:

(a) The equivalent transmissivity is

$$\theta = \left(\frac{q}{W}\right)\left(\frac{1}{i}\right)$$

$$= 0.65 \times 10^{-4}\left(\frac{1}{0.040}\right)$$

$$\theta = 16.2 \times 10^{-4}\ \text{m}^2/\text{s}$$

(b) Comparing the above to the EPA criteria,

$162 \times 10^{-5} > 3.0 \times 10^{-5}\ \text{m}^2/\text{s}$ which is easily acceptable for landfills

$16.2 \times 10^{-4} > 3.0 \times 10^{-4}\ \text{m}^2/\text{s}$ which is also acceptable for surface impoundments

4.1.4 Endurance Properties

The major endurance properties of concern when using geonets have to do with the long-term sustained deformation of the material and its ability to continue to transmit the required in-plane flow rate. Four issues are relevant: *type* of polyethylene resin, *creep* of the geonet structure, *intrusion* of adjacent materials into the geonet's apertures, and the possible *extrusion* of clay through a covering geotextile.

Type of Polyethylene Resin. Depending upon the type of polyethylene resin, primarily characterized by its density, the geonet will have different mechanical and endurance properties. The high-density resins (e.g., greater than 0.950 mg/l) will result in relatively high modulus, high strength, and high creep resistance. Conversely, lower-density resins (e.g, less than 0.945 mg/l) will be more flexible and can deform under high compressive stresses more easily. The lower-density resins, however, often have the advantage of having better stress-crack resistance properties. (Stress-crack resistance is also critical for geomembranes, as will be discussed in Chapter 5.) The importance of stress-crack resistance for a geonet is a design issue and related to the site-specific situation. Work is on-going in this regard.

Creep Behavior. Sustained load (or creep) is the reduction in thickness of a geonet under an applied compressive stress. Here the density of the resin (as described above), type of structure, and composition of the rib junctions are all significant. Figure 4.9 presents geonet creep data at 480 kPa for 1000 hours. Note that even this is too short a time for conventional practice because the extrapolation of trends beyond one order of magnitude is questionable. Thus data for 10,000 hr, extrapolated to 100,000 hr ($\cong$ 11 years), just begins to get into the frame of the life expectancy of many engineered systems using geonets. To shorten the testing time, other techniques such as time-temperature superposition may be appropriate. It is a good research area.

Intrusion of Adjacent Materials. All geonets will necessarily be covered on the upper and lower surfaces with a geotextile(s), geomembrane, concrete wall surface, or some other material. The geonet's surfaces must be covered or else soil will invade its apertures, rendering flow impossible. *Intrusion* refers to the deformation of the flexible covering materials, primarily geotextiles, occupying some of the open space, as illustrated by the flow-rate reductions in Table 4.1. As shown by the data, intrusion into the core space is a very real phenomenon causing flow-rate decrease. The photographs

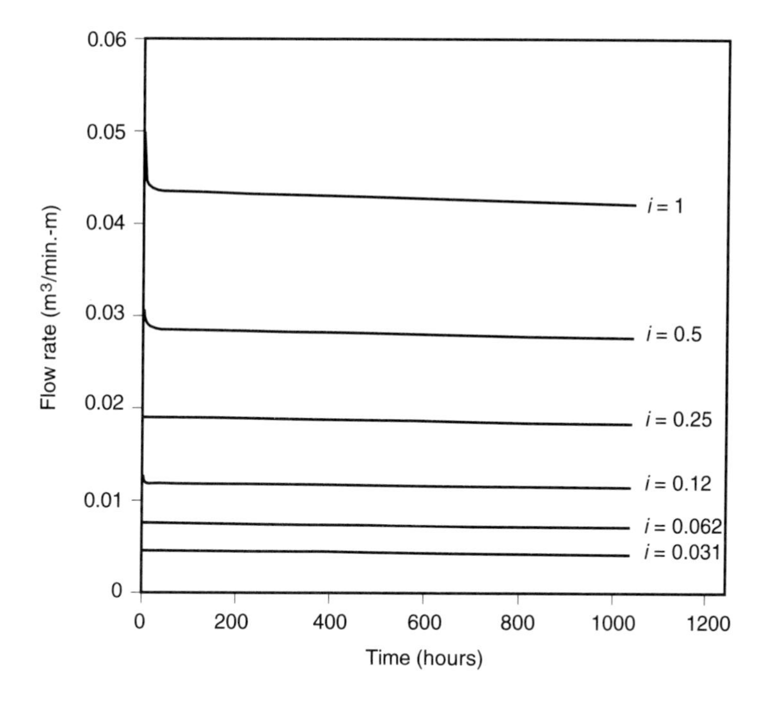

Figure 4.9 Long-term creep test results of a cross section that is the same as Figure 4.8, except sustained load at 480 kPa.

in Figure 4.10 illustrate the phenomenon. They were made by using a compressible foam (representing soil) above the geomembrane or geotextile covering the geonet. Under load, the geomembranes have no observable intrusion. The stiff heat-bonded nonwoven geotextile on the upper surface of the geonet produces some intrusion. The flexible needle-punched nonwoven geotextile, however, has considerable intrusion. See Hwu et al. [7] and Eith et al. [8] for data in this regard. It should be mentioned that thermal bonding of the geotextile to the geonet has a tendency to decrease some of the intrusion.

Now superimpose on these short-term reductions, sustained compressive stresses for extended times. Figure 4.9 presents the results of such a test series, where each test has load maintained for 1000 hours. Note that the response curves are essentially horizontal. Thus creep intrusion is not an issue. The geotextile in this case is a nonwoven needle-punched polyester of 550 gm/m^2 mass per unit area with clay above. Clearly the selection of the geotextile is important in sustaining the applied compressive stresses. The initial amount of intrusion, not necessarily the long-term behavior, is primarily a function of the geotextile's initial modulus. All other things being equal, those nonwoven geotextiles with high modulus will have the minimum amount of initial intrusion and hence higher flow rates. The long-term creep behavior depends upon the polymer type, stress level, distance between geonet ribs, and so on, and is best quantified by simulated laboratory testing.

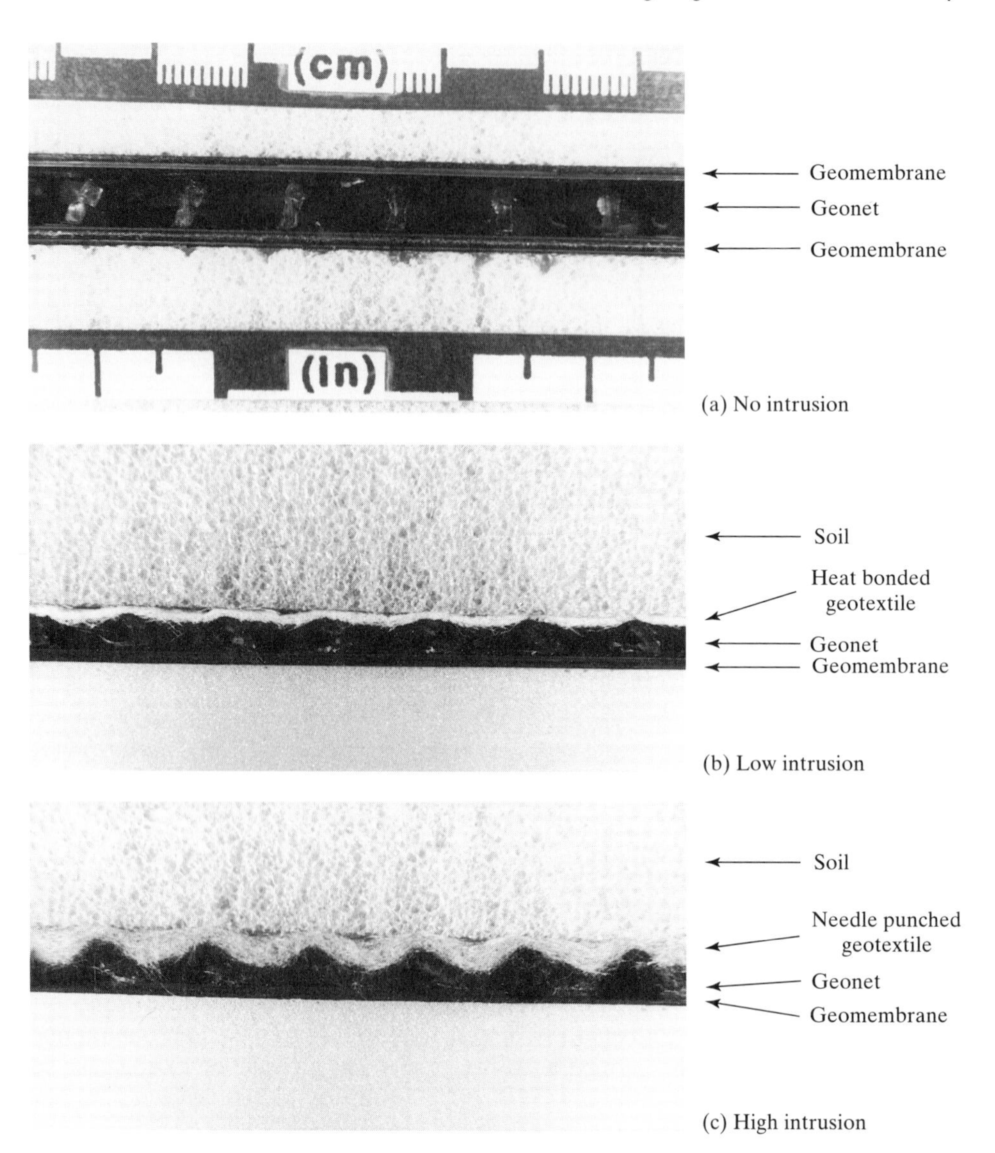

Figure 4.10 Intrusion of geomembranes and geotextiles into geonet's apertures.

Extrusion of Clay Materials. If a compacted clay liner or a geosynthetic clay liner is placed adjacent to a geonet composite, there is a possibility of the hydrated clay particles *extruding* through the geotextile's voids into the geonet. This is a disaster for the flow-rate capability of the geonet—it must be avoided under all circumstances. Such a situation has happened in laboratory tests with woven monofilament geotextiles on the geonet and could happen with woven slit-film geotextiles as well. Conversely, nonwoven geotextiles have generally been effective in preventing soil extrusion. The minimum mass per unit area is a site-specific issue. Further studies are warranted in this regard.

4.1.5 Environmental Properties

A series of environmentally related issues can have impact upon the flow-rate performance of geonets. The first that comes to mind is temperature. Under high temperatures, flow rates increase over standardized laboratory test conditions. The converse is true for cold temperatures. These are usually minor effects and can be calculated on the basis of viscosity corrections (recall Section 2.3.4). Perhaps more important is that creep of the polymers (geonet and adjacent geotextile or geomembranes) increases under increasing temperature. Simulated testing under these conditions is possible, but costly. In lieu of such testing, a conservative design approach regarding creep is warranted.

The second environmental consideration focuses on the nature of the liquid being transmitted. If chemicals or leachate are being transmitted, a number of questions arise. One of these is the chemical resistance of the polymers being used for the geonet and the covering geotextiles and/or geomembranes to the site-specific liquid. Here the choice of polyethylene for geonets is fortunate, since it is very resistant to most aggressive leachates. Again, laboratory testing can be performed using the actual or simulated leachate, but this can create concerns in a laboratory that is not equipped to handle contaminated liquids. The turbidity and viscosity effects of leachate (versus water) used in testing are another consideration, but these are often of second-order importance compared to some of the other issues being raised. Furthermore, they can be corrected by density and viscosity corrections (recall Section 2.3.4).

The third environmental consideration has to do with biological growth within the geonet and/or on the geotextiles that allow liquid to enter the geonet. In most transportation-related systems, such as roads and walls, the problem does not appear to be too serious. In leachate-related systems (e.g., landfill leachate collection systems), the issue should be addressed. At the bottom of a landfill, temperatures can be high, ample carbon (as a biological food source) is available, and bacteria and fungi could indeed thrive. Whether oxygen is available or not only dictates whether aerobic or anaerobic conditions prevail. No data regarding microorganism clogging of geonets is presently available. Procedurally, we must use a high flow-rate factor of safety or have systems designed so that flushing is possible. This area simply begs for future inquiry. Research should also focus on the filtration and drainage of agricultural wastewater systems as well. The biological growth on geotextiles has been addressed and a design procedure is available (see Koerner et al. [9]).

The fourth environmental consideration, resistance to light and weather, is not felt to be a serious concern for most situations in which geonets are used. Polyethylene is quite resistant to weather-related degradation, and carbon black is included in all of the known products. Nevertheless geonets should be covered as soon as possible after placement.

4.1.6 Allowable Flow Rate

As described previously, the very essence of the design-by-function concept is the establishment of an adequate factor of safety. For geonets, where flow rate is the primary function, this takes the following form.

$$FS = \frac{q_{\text{allow}}}{q_{\text{reqd}}} \tag{4.3}$$

where

FS = factor of safety (to handle unknown loading conditions or uncertainties in the design method, etc.),

q_{allow} = allowable flow rate as obtained from laboratory testing, and

q_{reqd} = required flow rate as obtained from design of the actual system.

Alternatively, we could work from transmissivity to obtain the equivalent relationship.

$$FS = \frac{\theta_{\text{allow}}}{\theta_{\text{reqd}}} \tag{4.4}$$

where θ is the transmissivity, under definitions as above. As discussed previously, however, it is preferable to design with flow rate rather than with transmissivity because of nonlaminar flow conditions in geonets.

Concerning the allowable flow rate or transmissivity value, which comes from hydraulic testing of the type described in Section 4.1.3, we must assess the realism of the test setup in contrast to the actual field system. If the test setup does not model site-specific conditions adequately, then adjustments to the laboratory value must be made. This is usually the case. Thus the laboratory-generated value is an ultimate value that must be reduced before use in design; that is,

$$q_{\text{allow}} < q_{\text{ult}}$$

One way of doing this is to ascribe reduction factors on each of the items not adequately assessed in the laboratory test. For example,

$$q_{\text{allow}} = q_{\text{ult}}\left[\frac{1}{RF_{IN} \times RF_{CR} \times RF_{CC} \times RF_{BC}}\right] \tag{4.5}$$

or if all of the reduction factors are considered together.

$$q_{\text{allow}} = q_{\text{ult}}\left[\frac{1}{\Pi RF}\right] \tag{4.6}$$

where

q_{ult} = flow rate determined using ASTM D4716 or ISO/DIS 12958 for short-term tests between solid platens using water as the transported liquid under laboratory test temperatures,

q_{allow} = allowable flow rate to be used in Eq. (4.3) for final design purposes,
RF_{IN} = reduction factor for elastic deformation, or intrusion, of the adjacent geosynthetics into the geonet's core space,
RF_{CR} = reduction factor for creep deformation of the geonet and/or adjacent geosynthetics into the geonet's core space,
RF_{CC} = reduction factor for chemical clogging and/or precipitation of chemicals in the geonet's core space,
RF_{BC} = reduction factor for biological clogging in the geonet's core space, and
ΠRF = product of all reduction factors for the site-specific conditions.

Some guidelines for the various reduction factors to be used in different situations are given in Table 4.2. Please note that some of these values are based on relatively sparse information. Other reduction factors, such as installation damage, temperature effects, and liquid turbidity, could also be included. If needed, they can be included on a site-specific basis. On the other hand, if the actual laboratory test procedure has included the particular item, it would appear in the above formulation as a value of unity. Examples 4.2 and 4.3 illustrate the use of geonets and serve to point out that high reduction factors are warranted in critical situations.

Example 4.2

What is the allowable geonet flow rate to be used in the design of a capillary break beneath a roadway to prevent frost heave? Assume that laboratory testing was done at the proper design load and hydraulic gradient and that this testing yielded a short-term between-rigid-plates value of 2.5×10^{-4} m²/s.

Solution: Since better information is not known, average values from Table 4.2 are used in Eq. (4.5).

TABLE 4.2 RECOMMENDED PRELIMINARY REDUCTION FACTOR VALUES FOR EQ. (4.5) FOR DETERMINING ALLOWABLE FLOW RATE OR TRANSMISSIVITY OF GEONETS

Application Area	RF_{IN}	RF_{CR}*	RF_{CC}	RF_{BC}
Sport fields	1.0 to 1.2	1.0 to 1.5	1.0 to 1.2	1.1 to 1.3
Capillary breaks	1.1 to 1.3	1.0 to 1.2	1.1 to 1.5	1.1 to 1.3
Roof and plaza decks	1.2 to 1.4	1.0 to 1.2	1.0 to 1.2	1.1 to 1.3
Retaining walls, seeping rock, and soil slopes	1.3 to 1.5	1.2 to 1.4	1.1 to 1.5	1.0 to 1.5
Drainage blankets	1.3 to 1.5	1.2 to 1.4	1.0 to 1.2	1.0 to 1.2
Surface water drains for landfill covers	1.3 to 1.5	1.1 to 1.4	1.0 to 1.2	1.2 to 1.5
Secondary leachate collection (landfills)	1.5 to 2.0	1.4 to 2.0	1.5 to 2.0	1.5 to 2.0
Primary leachate collection (landfills)	1.5 to 2.0	1.4 to 2.0	1.5 to 2.0	1.5 to 2.0

*These values are sensitive to the density of the resin used in the geonet's manufacture. The higher the density, the lower the reduction factor. Creep of the covering geotextile(s) is a product-specific issue.

$$q_{\text{allow}} = q_{\text{ult}}\left[\frac{1}{RF_{IN} \times RF_{CR} \times RF_{CC} \times RF_{BC}}\right]$$

$$= 2.5 \times 10^{-4}\left[\frac{1}{1.2 \times 1.1 \times 1.3 \times 1.2}\right]$$

$$= 2.5 \times 10^{-4}\left[\frac{1}{2.06}\right]$$

$$= 1.21 \times 10^{-4}\ \text{m}^2/\text{s}$$

Example 4.3

What is the allowable geonet flow rate to be used in the design of a secondary leachate collection system? Assume that laboratory testing at the proper design load and proper hydraulic gradient gave a short-term between-rigid-plates value of 2.5×10^{-4} m²/s.

Solution: Average values from Table 4.2 are used in Eq. (4.5) (however, note the large reduction).

$$q_{\text{allow}} = q_{\text{ult}}\left[\frac{1}{RF_{IN} \times RF_{CR} \times RF_{CC} \times RF_{BC}}\right]$$

$$= 2.5 \times 10^{-4}\left[\frac{1}{1.75 \times 1.7 \times 1.75 \times 1.75}\right]$$

$$= 2.5 \times 10^{-4}\left[\frac{1}{9.11}\right]$$

$$= 0.27 \times 10^{-4}\ \text{m}^2/\text{s}$$

4.2 DESIGNING FOR GEONET DRAINAGE

This section will be subdivided into a discussion of required theory (which somewhat repeats previously described issues, due to its importance), drainage examples in the transportation-related field, and drainage examples in the environmental-related field.

4.2.1 Theoretical Concepts

Design-by-function requires the formulation of a factor of safety as follows.

$$\text{FS} = \frac{\text{allowable (test) value}}{\text{required (design) value}}$$

For geonets serving as a drainage medium, the targeted value is flow rate and the above concept becomes the following equation.

$$\text{FS} = \frac{q_{\text{allow}}}{q_{\text{reqd}}} \tag{4.3}$$

where

q_{allow} = allowable flow rate (as discussed in Section 4.1), and
q_{reqd} = required flow rate (to be discussed here).

As stated previously, calculations can be based on Darcy's formula (assuming saturated conditions and laminar flow) if we desire, as an alternative to the flow rate, the transmissivity θ. This important concept is repeated from Eq. (4.1) and (4.2).

$$q = kiA \tag{4.1}$$

$$q = ki(W \times t)$$

$$q = (kt)iW$$

$$kt \equiv \theta = \frac{q}{iW} \tag{4.2}$$

where

q = volumetric flow rate (m^3/s),
k = coefficient of permeability (m/s),
i = hydraulic gradient (dimensionless),
A = flow cross-sectional area (m^2),
θ = transmissivity (m^2/s),
W = width (m), and
t = thickness (m).

In Eq. (4.2), it is seen that q/W and θ carry the same units and are directly related to one another by means of the hydraulic gradient i. At a hydraulic gradient of 1.0, they are numerically identical. At other values of hydraulic gradient they are not equal. Also note that the system should be saturated and flow must be laminar in order to use transmissivity. When in doubt, it is usually best to use flow rate per unit width.

4.2.2 Transportation-Related Applications

While geonets have been introduced as an alternative drainage material to granular soils in the environmental field, there is no reason why they should not be used in transportation-related applications as well. This section addresses two such applications, both of which are very relevant.

Example 4.4

Given an area that is to be retrofitted with a new pavement because of past problems with frost heave, the new scheme is intended to have a geonet with thermally bonded geotextiles on both sides and will be placed immediately beneath the depth of maximum frost penetration (see the sketch below). Based upon the rising capillary water, the required flow rate to be conveyed is estimated to be 0.17×10^{-4} m^2/s. A candidate geonet has been selected, and tests performed at a gradient of 0.05 have resulted in a flow rate of 0.83×10^{-4} m^2/s. What is the factor of safety?

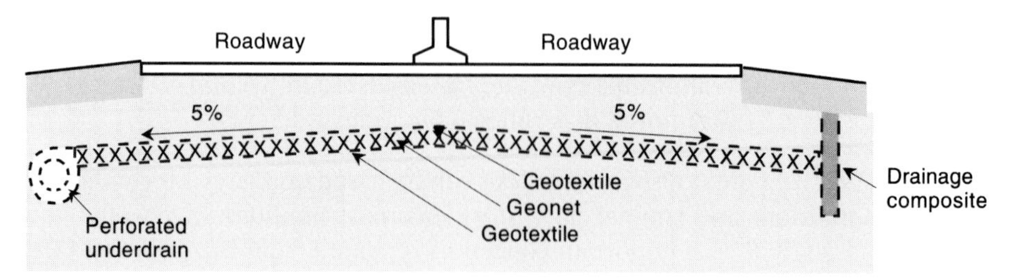

Solution:

(a) Assuming that the laboratory tests were short-term and between rigid plates, the value of 0.83×10^{-4} m²/s must be reduced according to Table 4.2, as per the earlier example problems. Thus,

$$q_{\text{allow}} = 0.83 \times 10^{-4}\left[\frac{1}{1.2 \times 1.1 \times 1.3 \times 1.2}\right]$$

$$q_{\text{allow}} = 0.83 \times 10^{-4}\left[\frac{1}{2.06}\right]$$

$$= 0.403 \times 10^{-4} \text{ m}^2/\text{s}$$

(b) Now the actual flow rate factor of safety can be determined.

$$\text{FS} = \frac{q_{\text{allow}}}{q_{\text{reqd}}}$$

$$= \frac{0.403 \times 10^{-4}}{0.17 \times 10^{-4}}$$

$$= 2.4 \qquad \text{which is adequate}$$

Example 4.5

Determine the factor of safety that the geonet response curve shown in Figure 4.7 has for the 8 m high cantilever retaining wall shown below. The soil backfill is a silty sand (ML-SW) with $k = 5.0 \times 10^{-5}$ m/s. Note that this is the same problem that was attempted using a geotextile in Section 2.9.3 (Example 2.26), where the factor of safety was found to be 0.0062.

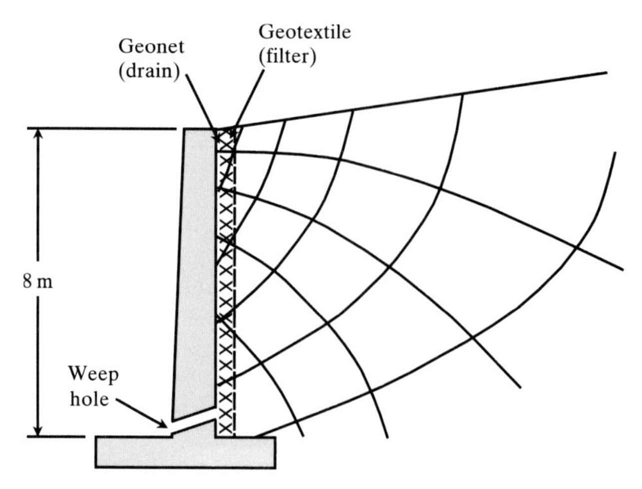

Solution:

(a) Calculate the maximum flow rate coming to the geonet. From the flow rate above, we have

$$q = kh\left(\frac{F}{N}\right)$$

$$= (5.0 \times 10^{-5})(8)\left(\frac{5}{5}\right)$$

$$= 4.0 \times 10^{-4}\ \text{m}^2/\text{s}$$

(b) Determine the flow gradient within the geotextile.

$$i = \sin 90$$

$$= 1.0$$

(c) Calculate the required transmissivity. Note that transmissivity is used so that the results can be directly compared to Example 2.26, using geotextiles.

$$q = kiA$$

$$= ki(t \times W)$$

$$= (kt)(i \times W)$$

$$kt = \frac{q}{i \times W}$$

$$= \frac{4.0 \times 10^{-4}}{1.0 \times 1.0}$$

$$\theta_{\text{reqd}} = 4.0 \times 10^{-4}\ \text{m}^2/\text{s}$$

(d) From the laboratory data in Figure 4.7, obtain the ultimate flow rate, convert it to a transmissivity value, and then reduce it to an allowable transmissivity.

$$\sigma_n \cong 0.5(8)(18) \qquad \text{(from Fig. 4.7)}$$

$$\sigma_n = 72\ \text{kPa}$$

$$q = 0.094\ \text{m}^2/\text{min} = 15.6 \times 10^{-4}\ \text{m}^2/\text{s}$$

For a vertical wall, where $i = 1.0$, the flow rate per unit width (q/W) is identical to the transmissivity θ [recall Eq. (4.2)], so

$$\theta_{\text{ult}} = 15.6 \times 10^{-4}\ \text{m}^2/\text{s}$$

Reducing the value according to Table 4.2,

$$\theta_{\text{allow}} = 1.56 \times 10^{-3}\left[\frac{1}{1.4 \times 1.3 \times 1.2 \times 1.2}\right]$$

$$= 1.56 \times 10^{-3}\left[\frac{1}{2.62}\right]$$

$$= 0.595 \times 10^{-3}\ \text{m}^2/\text{s}$$

(e) Knowing both allowable and required values of transmissivity, the global factor of safety can be calculated.

$$\begin{aligned}FS &= \frac{\theta_{\text{allow}}}{\theta_{\text{reqd}}}\\ &= \frac{0.595 \times 10^{-3}}{0.40 \times 10^{-3}}\\ &= 1.48 \qquad \text{which is marginally acceptable}\end{aligned}$$

Thus, the geonet characterized by the Figure 4.7 data is only marginally adequate to drain the wall, whereas multiple layers of geotextiles were not adequate. However, even this factor of safety is somewhat questionable and a higher flow capacity drainage composite should be investigated, for example, a triaxial flow geonet or other type of drainage geocomposite. The problem will be repeated a third time in Chapter 8.

4.2.3 Environmental-Related Applications

Geonets are widely used in landfill-liner systems as the primary leachate collection systems on side slopes and as leak detection and collection systems between the primary and secondary liners (see Bonaparte et al. [10] and Lundell and Menoff [11]). They are placed with either a geomembrane on both sides, or a geomembrane beneath and a geotextile above. In this latter case, the geotextile has either granular soil or clay above it. Note that the inverse cross section is also possible, that is, a geomembrane above and a geotextile with soil below. The flow-rate results illustrated in Figures 4.7 and 4.8 indicate some of these alternatives.

Example 4.6

Assuming that there are no regulations governing the situation, is the geonet whose response is shown in Figure 4.7 adequate for the following specific leak detection and collection system? The geonet lies between two HDPE geomembranes, and the design flow is 100 times *de minimis** leakage. (De minimus leakage is approximately 10 l/ha-day). The minimum slope of the bottom of the 300 m long landfill is 6%, and the landfill when completed will be 35 m high with a unit weight of waste being 13 kN/m^3.

Solution:

(a) The required flow rate converted to comparable units is

$$q_{\text{reqd}} = \frac{(100)(10)(0.001)}{(10{,}000)(24 \times 60 \times 60)} \times 300$$

*U.S. EPA Regulatory Note: The primary (or upper) geomembrane of a hazardous waste facility should allow no more than *de minimis* leakage of all polluting species through the liner itself. The concept *de minimis* comes from the legal principle *de minimis non curat lex* (the law does not concern itself with trifles). De minimis leakage is considered to be the amout that is of no threat to human health or the environment. It is recognized that geomembranes, since they are not impermeable, will allow some transmission of waste constituents by such means as vapor permeation or via very small imperfections. The actual level of de minimis leakage of a constituent is specific to the site, the constituent's toxicity, and the mobility and biodegradability of the constituent. Specific levels for individual constituents have not been set, although the U.S. EPA believes that total de minimis leakage should be no more than 1 gallon per acre per day (gpad). This is approximately 10 l/ha-day. Current legislation [12], however, avoids setting a general action leakage rate (ALR) and instead requires each permit application to set its own site-specific ALR value.

$$= 3.5 \times 10^{-7} \text{ m}^2/\text{s}$$

(b) The ultimate flow rate is taken from Figure 4.7 (at $\sigma_n = 35 \times 13 = 455$ kPa and $i = 0.06$).

$$q_{\text{ult}} = 0.01 \text{ m}^2/\text{mm} = 1.66 \times 10^{-4} \text{ m}^2/\text{s}$$

and from Table 4.2 it is reduced to obtain an allowable value

$$q_{\text{allow}} = q_{\text{ult}}\left[\frac{1}{1.75 \times 1.7 \times 1.75 \times 1.75}\right]$$

$$= 1.66 \times 10^{-4}\left[\frac{1}{9.11}\right]$$

$$= 0.182 \times 10^{-4} \text{ m}^2/\text{s}$$

(c) Therefore, the factor of safety is

$$\text{FS} = \frac{q_{\text{allow}}}{q_{\text{reqd}}}$$

$$= \frac{0.182 \times 10^{-4}}{3.5 \times 10^{-7}}$$

$$\text{FS} = 52 \qquad \text{which is more than adequate}$$

(d) Note that for the data of Figure 4.8, for the geotextile and clay over the geonet, the flow rate is approximately half of the above and the factor of safety is reduced accordingly.

Example 4.7

What is the factor of safety of a geonet placed above a geomembrane and beneath a geotextile in the final cover of a completed solid-waste facility (landfill or waste pile). As shown in the sketch, the slope is 10% and the cover soil fill height is 1.25 m. Assume that the slope length is 120 m long. Short-term in-plane flow tests show that the flow rate is 2.5×10^{-4} m²/s at a hydraulic gradient of 0.10 and at a normal pressure of 50 kN/m² (which includes soil weight plus equipment loads).

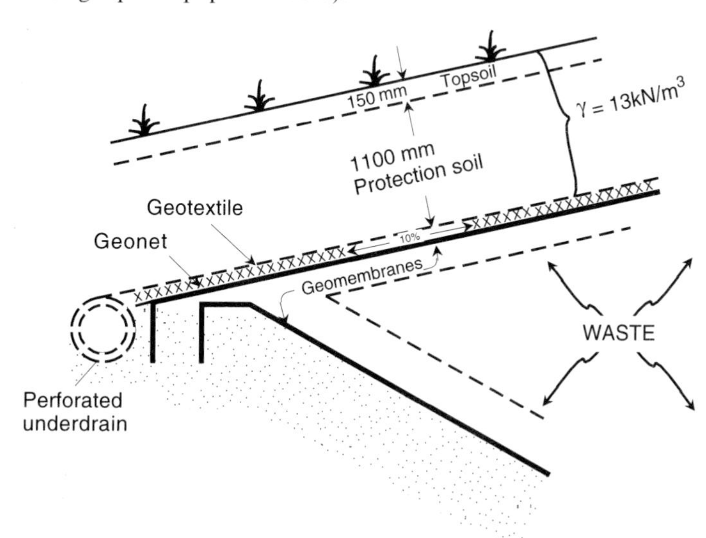

Solution:

(a) The allowable flow rate is again taken from Eq. (4.5) and Table 4.2, using average values

$$q_{\text{allow}} = 2.5 \times 10^{-4}\left[\frac{1}{1.4 \times 1.25 \times 1.1 \times 1.35}\right]$$

$$= 2.5 \times 10^{-4}\left[\frac{1}{2.60}\right]$$

$$= 0.96 \times 10^{-4}\text{m}^2/\text{s}$$

(b) The required flow rate either must be approximated or must be determined by using a liquid-mass balance, including local hydrological data, soil storage, leakage, and so on. This was done in this case, following an hourly rainfall modeling procedure (see Koerner and Daniel [13]), giving a required flow rate 0.17×10^{-4} m^2/s at the end of a 120 m long section of closure. Note that a computer model entitled Hydrologic Evaluation of Landfill Performance or HELP (Schroeder et al. [14]) is widely used by regulatory personnel and consultants working in the solid-waste field in the U.S., but it can only track seepage on a daily basis, and is not felt to adequately model intense storms.

(c) The final factor of safety becomes

$$\text{FS} = \frac{q_{\text{allow}}}{q_{\text{reqd}}}$$

$$= \frac{0.96 \times 10^{-4}}{0.17 \times 10^{-4}}$$

$$\text{FS} = 5.6$$

This value of factor of safety, being well above 1.0, is acceptable. However, seepage pressures in landfill final covers have been so troublesome that very high FS values are required. The final decision on acceptability is site-specific.

4.3 DESIGN CRITIQUE

The design examples just presented focused entirely on FS values based on either flow rate or transmissivity. This was done to reinforce the concept of in-plane drainage as the primary and unique function of geonets. In this regard, the geonet must be properly specified. At least three items are necessary to make a proper flow rate specification: the flow rate (which is preferred to transmissivity), the normal stress, and the hydraulic gradient.

A few words about the normal stress are in order. To avoid rib lay-down and/or creep deformation, the normal strength capability of the geonet must be higher than the design value. This value should be approximately 1.5 times (for short-service lifetimes) to 2, or more, times (for long-service lifetimes in critical situations). Thus the structural stability of the geonet must be ensured against creep deformation or collapse. The actual flow-rate value used for design purposes, however, can be taken from the curves at the design load at which the system will be operating.

When soil is adjacent to the geonet, the type of geotextile covering it is of great significance. While for most situations the geotextile is usually designed as a filter, it must also span the apertures of the geonet without excessively intruding or collapsing into the core space. There will always be some intrusion, and just how much is allowable depends on the site-specific situation. This can be evaluated experimentally, and values given in Table 4.2 reflect a series of such experiments. A possible method of minimizing intrusion could be the use a high-modulus woven monofilament geotextile. For environmental-related facilities, however, this might not be appropriate. If hydrated clay is above the geotextile and is under high pressure from the weight of the landfill above it, the clay will be extruded through the open spaces in the woven geotextile directly into the geonet openings. This is completely unacceptable. Thus, a nonwoven needle-punched geotextile with a labyrinth of overlapping fibers is necessary. However, some amount of intrusion must be anticipated and adequately accounted for. A nonwoven heat-bonded geotextile might be a compromise geotextile, with both high modulus (to prevent excessive intrusion) and high fiber overlapping (to prevent extrusion). Depending upon the actual stress level, 200 g/m^2 should be the minimum mass per unit area. The major problem in using this type of geotextile appears to be thermal bonding the geotextile to the geonet.

These same considerations must be expressed when a geosynthetic clay liner (GCL) is placed over a geonet. The lower geotextile of the GCL must be viewed in the same light as other geotextiles in this discussion. Depending upon the type of geotextile on the GCL facing the geonet, it is quite likely that an additional geotextile may be required between the GCL and the geonet.

4.4 CONSTRUCTION METHODS

Geonets are supplied in rolls from 2.3 to 6.7 m wide. They should be placed and covered in a timely manner. While UV and heat effects are not as severe in geonets as they are in geotextiles (because of thicker ribs, in contrast to thin yarns and fibers), it is good practice not to leave the material exposed and subjected to accidental damage or contamination of any variety. Contamination can occur from soil, miscellaneous sediment, construction debris, ingrowing roots, and so on.

The rolls are usually placed with their roll directions oriented up-and-down slopes, rather than along (or parallel to) them. There are two reasons for this: First, the machine direction has the greatest strength (recall Figure 4.4); second, it eliminates seams along the flow direction. If triaxial geonets are being used for their high flow in a unique direction, the proper orientation is critical during placement.

The seaming or joining of geonets is difficult. Assuming stress does not have to be transferred from one roll to the next, plastic electrical ties, threaded loops, and wires have all been used with a relatively small overlap of 50 to 100 mm. Overall, there is room for improvement in this regard (see Zagorski and Wayne [15]). Metal hog rings should never be used when geonets are used adjacent to geomembranes. There are questions as to what influence overlapping has on the geonet's flow rate.

Notwithstanding the above concerns, geonets are very impressive with respect to their flow rate capability, ease of construction, savings in airspace, and overall economy in many facilities where drainage must be accommodated. Some aspects of geonets will be revisited in Chapter 8 when we discuss drainage geocomposites.

REFERENCES

1. Austin, R. A., "The Manufacture of Geonets and Composite Products," *Proc GRI-8 on Geosynthetic Resins, Formulations and Manufacturing,* St. Paul, MN: IFAI, 1995, pp. 127–138.
2. Williams N., Giroud, J-P., and Bonaparte, R., "Properties of Plastic Nets for Liquid and Gas Drainage Associated with Geomembranes," *Proc. Intl. Conf. Geomembranes,* St. Paul, MN: IFAI, 1984, pp. 399–404.
3. Corcoran, G. T., Cheng, S.-C. J., and Spear, A. D., "High Normal Stress Compression of Geosynthetic Lining Systems," *Proc. 5th IGS Conf.,* Rotterdam: A.A. Balkema, 1994, pp. 837–840.
4. Lydick, L. D., and Zagorski, G. A., "Interface Friction of Geonets: A Literature Survey," *J. Geotextiles and Geomembranes,* Vol. 10, Nos. 5–6, 1991, pp. 167–176.
5. Kolbasuk, G. M., Lydick, L. D., and Reed, L. S., "Effects of Test Procedures on Geonet Transmissivity Results," *J. Geotextiles and Geomembranes,* Vol. 11, Nos. 4–6, 1992, pp. 153–166.
6. EPA 40 CFR Parts 260, 264, 265, 270 and 271, *Federal Register,* Vol. 57, No. 19, *Rules and Regulations,* January 29, 1992, p. 3463.
7. Hwu, B.-L., Sprague, C. J., and Koerner, R. M., "Geotextile Intrusion into Geonets, *Proc. 4th Intl. Conf. on Geotextiles, Geomembranes and Related Products,* Rotterdam: A.A. Balkema, 1990, pp. 351–356.
8. Eith, A. W., and Koerner, R. M., "Field Evaluation of Geonet Flow Rate (Transmissivity) under Increasing Load," *J. Geotextiles and Geomembranes,* Vol. 11, Nos. 5–6, 1992, pp. 153–166.
9. Koerner, G. R., Koerner, R. M., and Martin, J. P., "Geotextile Filters Used for Leachate Collection Systems: Testing, Design and Field Behavior," *J. Geotechnical Eng. Div., ASCE,* Vol. 120, No. 10, 1994, pp. 1792–1803.
10. Bonaparte, R., Williams, N., and Giroud, J-P, "Innovative Leachate Collection Systems for Hazardous Waste Containment Systems," *Proc. Geotechnical Fabrics Conf. '85,* St. Paul, MN: IFAI, 1985, pp. 10–34.
11. Lundell, C. M., and Menoff, S. D., "The Use of Geosynthetics as Drainage Media at Solid Waste Landfills," *Proc. Geosynthetics '89,* St. Paul, MN: IFAI, 1989, pp. 10–17.
12. Matrecon, Inc., *Lining of Waste Containment and Other Impoundment Facilities,* 600/2-88/052, U.S. EPA, Risk Reduction Eng. Lab., Cincinnati, OH, Sept. 1988.
13. Koerner, R. M., and Daniel, D. E., *Final Covers for Solid Waste Landfills and Abandoned Dumps,* New York: ASCE Press, 1997.
14. Schroeder, P. R., Dizier, T. S., Zappi, P. A., McEnroe, B. M., Sjostrom, J. W., and Peyton, R. L., "The Hydrologic Evaluation of Landfill Performance (HELP) Model: Engineering Documentation for Version 3," EPA/600/R-94/168b, U.S. E.P.A., Risk Reduction Eng. Lab., Cincinnati, OH, 1994.
15. Zagorski, G. A., and Wayne, M. H., "Geonet Seams," *J. of Geotextiles and Geomembranes,* Vol. 9, Nos. 4–6, 1990, pp. 207–220.

PROBLEMS

4.1. Geonets are used specifically for their in-plane drainage capability. Give the reasons they are not used for the following:

(a) separation

(b) reinforcement
(c) filtration
(d) containment (moisture barrier)

4.2. In their use for the drainage function, what keeps the adjacent soil from getting in their apertures and blocking the flow?

4.3. If a geotextile is placed adjacent to a geonet, what function(s) does the geotextile provide? How does the combination of geotextile and geonet accommodate flow?

4.4. All of the geonets described in this chapter are made of polyethylene. Can they be made from other polymers? Why do you suppose they are made from polyethylene?

4.5. It is noted in the chapter that the aperture size varies from product to product. What effect does aperture size have on flow and intrusion?

4.6. In the typical extruded geonets, the vertical axes of the intersecting ribs are not quite perpendicular to one another. What implications does this have for the compressive load-carrying capacity of the geonet?

4.7. For the foamed extruded geonets shown in Figure 4.2, answer the following questions.
(a) What are the foamed pores within the ribs filled with?
(b) What are the long-term implications of this?
(c) How would Fick's law (of diffusion) enter into this discussion?

4.8. For triaxial geonets, the flow rate is significantly higher than for biaxial geonets.
(a) Where in a typical landfill configuration can these geonets be used?
(b) Why is knowledge of the slope direction critical to know?
(c) Why is the flow in the cross-machine direction not particularly important?

4.9. The shear strength between a geotextile and a geonet can be quite low and troublesome when used in side-slope design. Describe two methods by which the geotextile can be attached to the geonet to avoid the potential problem. Include the advantages and disadvantages of each method.

4.10. The flow-rate reduction between Figures 4.7 and 4.8, as tabulated in Table 4.1, are up to 40% for up to 500 kPa. Beyond this stress level, the reductions are lower. Why are they lower at the higher stress levels?

4.11. Regarding the placement and use of a compacted clay liner over a geotextile placed on a geonet:
(a) What are the implications of using a lighter-weight geotextile for the results shown in Figure 4.8?
(b) What would happen if a woven monofilament geotextile of 6% open area were used?
(c) What would happen if a nonwoven heat-bonded geotextile were used?

4.12. For a geotextile covering a geonet that has a clay liner placed above it, discuss the difference between extrusion and intrusion.

4.13. Using a geonet beneath an artificial surface in a tennis court requires an allowable flow rate of 1.6×10^{-4} m²/s. What would be the necessary ultimate flow rate using the average values from Table 4.2?

4.14. The ultimate flow rate of a geonet being considered for the primary leachate collection system on a landfill side slope is 7.3×10^{-4} m²/s. Using the maximum values in Table 4.2, what is the allowable flow rate?

4.15. Recalculate Example 4.6 (Section 4.2.3) concerning secondary leachate collection systems for design flows from 1 to 4000 times de minimus. Plot the resulting factors of safety against required flow rate.

4.16. A geonet is being considered for primary leachate collection on the side slopes of a landfill. Using the data in Figure 4.8, interpolating at a normal stress of 700 kPa and a hydraulic gradient of 0.184, what is the factor of safety for a flow rate of 25,000 l/ha-day? The cumulative reduction factors should be 8.0. (Note that this leachate flow rate is typical of primary leachate flow rates in the state of New York).

4.17. Recalculate Example 4.7 (Section 4.2.3) concerning a geonet in a landfill cover, where the required flow rate varies from 1×10^{-3} to 1×10^{-5} m^2/s. Plot the resulting factors of safety against slope angle.

4.18. What are the long-term normal stress implications for a geonet's flow-rate capability?

4.19. What are the long-term environmental implications for a geonet's flow-rate capability?

5 Designing with Geomembranes

5.0 INTRODUCTION

According to ASTM D4439, a geomembrane is defined as follows:

> **geomembrane,** *n.* Very low permeability synthetic membrane liner or barrier used with any geotechnical engineering related material so as to control fluid migration in a human-made project, structure, or system.

Geomembranes are made from relatively thin continuous polymeric sheets, but they can also be made from the impregnation of geotextiles with asphalt or elastomer sprays or as multilayered bitumen geocomposites. In this chapter, we will focus on continuous polymeric sheet geomembranes. Impregnated geotextiles are covered in various parts of Chapter 2, and bituminous geomembranes are covered in Chapter 8.

Polymeric geomembranes are not absolutely impermeable (actually nothing is), but are relatively impermeable when compared to geotextiles or soils, even to clay soils. Typical values of geomembrane permeability as measured by water-vapor transmission tests are in the range 1×10^{-12} to 1×10^{-15} m/s. Thus the primary function is always as a barrier to liquids or vapors. As noted in Section 1.6.3, the current market for geomembranes is extremely strong. New applications are regularly being developed, and this is directly reflected in sales volume; geomembranes are currently the largest segment of geosynthetics, as far as product sales are concerned.

Since the primary function of geomembranes is always containment, as a barrier to a liquid and/or gas, this chapter is organized on the basis of different application areas. Liquid containment is treated first, and then solid waste containment, followed by numerous geotechnical applications. As a counterpoint for the design calculations to follow, however, the next section describes geomembrane properties and the test methods used to obtain these properties. This information allows the design of the major geomembrane-related systems that are currently in use. These designs then form the subsequent parts of the chapter.

5.1 GEOMEMBRANE PROPERTIES AND TEST METHODS

To design-by-function (the theme of this book) is to make a conscious decision about the adequacy of the ratio of the allowable property of a geomembrane to the required property, that is, the factor of safety value. This section on properties is devoted to providing the test methods for the numerator of this ratio. The required, or design, value will come later.

5.1.1 Overview

The vast majority of geomembranes are relatively thin sheets of flexible thermoplastic polymeric materials (recall Figure 1.1). These sheets are prefabricated in a factory and transported to the job site, where placement and field-seaming are performed to complete the job. In this category are the materials listed in Table 5.1, the principal ones currently in use; these are the geomembranes that will be focused upon. Section 1.2 gives an overview of the various polymers listed and describes a number of identification or fingerprinting tests used to quantify their composition and formulation. Hybrid systems, many of which are made on-site by in situ methods, are quite different in their properties. For example, bitumen-impregnated geotextiles possess the fundamental properties of the base geotextile with the obvious exception of their modified hydraulic properties. In fact, impregnating a geotextile with bitumen essentially renders it impermeable, as does impregnating it with bitumen/rubber or elastomeric materials (recall Sections 2.6.2 and 2.10.2). These modifications are each unique to the individual process, and as a result the performance properties of the resulting product must be individually assessed. Some of the tests do carry over (e.g., water-vapor transmission) to hybrid systems.

Many geomembrane test methods and standards are available or are being developed by standards-setting organizations around the world. Where appropriate, the ASTM and ISO standards will be referenced. The individual test methods will be grouped into the following categories: physical properties, mechanical properties, chemical properties, biological properties, and thermal properties.

TABLE 5.1 GEOMEMBRANES IN CURRENT USE

Most widely used	Somewhat less widely used
High density polyethylene (HDPE)	Flexible polypropylene (fPP)
Very flexible polyethylene (VFPE)	Flexible polypropylene-reinforced (fPP-R)
Very low density polyethylene (VLDPE)	Chlorosulfonated polyethylene-reinforced
Linear low density polyethylene (LLDPE)	
Low density linear polyethylene (LDLPE)	(CSPE-R)
Polyvinyl chloride (PVC)	Ethylene interpolymer alloy-reinforced (EIA-R)

5.1.2 Physical Properties

Physical properties have to do with the geomembrane in an as-received and relaxed state. They are important for quality control, quality assurance, and proper identification.

Thickness. Depending on the type of geomembrane, there are three types of thickness to be considered: the thickness of smooth sheet, the core thickness of textured sheet, and the thickness (or height) of the asperities of textured sheet.

Smooth Sheet. The determination of the thickness of a smooth geomembrane is performed by a straightforward measurement. The test uses an enlarged-area micrometer under a specified pressure, resulting in the desired value. ASTM D5199 is the test method generally used for measuring geomembranes. The pressure exerted by the micrometer is specified at 20 kPa. A number of measurements are taken and an average value is obtained. When measuring the thickness of a geomembrane, there is little ambiguity in the procedure. Polyethylenes, polyvinyl chlorides, and nonreinforced polypropylene are made in thicknesses from 0.5 to 3 mm. When measuring multiply materials with scrim reinforcement or aged membranes that have swelled, extreme care must be exercised, particularly in the preparation of the test specimen and in the application of pressure. Test conditions and applied pressures should always be given together with the actual values. Scrim-reinforced geomembranes are manufactured from single plies in the range 0.25 to 0.38 mm, which when laminated together result in geomembranes of considerable thickness (recall Figure 1.26).

Textured Sheet. The roughened surface of a textured geomembrane results in a significant increase in interface friction with adjacent materials versus the same geomembrane with a smooth surface. The thickness of such textured sheets is measured as the minimum core thickness between the roughened peaks or *asperities*. To measure the core thickness, a tapered-point micrometer for measuring machine screw threads is recommended. The tapered-point dimensions per ASTM D5994 are a 60° angle with the extreme tip at 0.08 mm diameter. The normal load on the tapered point is 0.56 N. For a single-sided textured sheet, only one tapered point is needed, while a double-sided textured sheet requires a micrometer with two opposing tapered tips. Within a limited area, the local minimum core thickness is obtained. Typically, 10 measurements across the roll width are taken and an average core thickness value is calculated and compared to the specification value. The test is quite controversial.

Asperity Height. For textured geomembranes the roughness pattern is of interest insofar as it relates to mobilizing the maximum amount of interface shear strength with the opposing surface. Optimized texturing is a daunting task and a topic of interest to both the manufacturing and user communities. Profilometry has been attempted (Dove and Frost [1]), as well as an effort extended toward the use of fractals (Vallejo and Zhou [2]). Less involved, but still useful as a quality control and quality assurance method, is to merely measure the height of the asperities. To do so, a depth gage micrometer with a 1.3 mm diameter pointed stylus is recommended. The gage is zeroed on a flat surface and then is placed on the peaks of the textured sheet with the stylus falling into the valley created by the texturing. The localized maximum depth is the asperity height. A number of measurements are taken across the roll width and an average asperity height is obtained.

Density. The density or specific gravity of a geomembrane is dependent on the base material from which it is made. There are distinct differences, however, even in the same generic polymer. For example, polyethylene comes in very-low-density, low-density, linear low-density, medium-density, and high-density varieties. The range for all geomembrane polymers falls within the general limits of 0.85 to 1.5 mg/l. A relevant ASTM test method is D792. This test method is based on the fundamental Archimedean principle of specific gravity as the weight of the object in air divided by its weight in water.

A more accurate method is ASTM D1505, "Density Determination by the Density Column." Here a long glass column containing liquid varying from high-density at the bottom to low-density at the top is used. For example, isopropanol with water is often used for measuring densities less than 1.0, while sodium bromide with water is used for densities greater than 1.0. Upon setup, spheres of known densities are immersed in the column to generate a calibration curve. Small pieces of the polymer test specimen are then dropped into the column. Their equilibrium level within the column is used with the calibration curve to find the specimen's density. Accuracy is very good, within 0.002 mg/l, when proper care is taken.

A comment on the density of HDPE geomembranes should be made. The ASTM classification for HDPE resin requires a density $\geq$ 0.941 mg/l. However, all commercially available HDPE geomembranes use polyethylene resin from 0.934 to 0.938 mg/l; the resin, itself, is actually in the medium-density range (MDPE). Only by adding carbon black and additives to the formulation is its gross density raised to 0.941 mg/l or slightly higher. Thus what is called HDPE by the geomembrane industry is actually MDPE resin to the polymer producer.

Moreover, included in the very flexible polyethylene (VFPE) classification are polyethylenes of lower formulated density such as: linear low density polyethylene (LLDPE): 0.928 to 0.936 mg/l, low density linear polyethylene (LDLPE): 0.934 to 0.938 mg/l, and very low density polyethylene (VLDPE): 0.920 to 0.930 mg/l.

Melt (Flow) Index. The melt flow index or melt index (MI) test is used routinely by geomembrane manufacturers as a method of controlling polymer uniformity. It relates to the flowability of the polymer in its molten state. It is used for both the incoming resin and the final geomembrane sheet. The test method often used for geomembrane polymers is ASTM D1238. Here, a given amount of the polymer is heated in a furnace until it melts. A constant weight forces the fluid mass through an orifice and out of the bottom of the test device. The MI value is the weight of extruded material in grams for 10 min duration. The higher the value of melt flow index, the lower the density of the polymer, all other things being equal. High MI values suggest a lower molecular weight, and vice versa, albeit by a relatively crude method in comparison to the techniques discussed in Section 1.2.2.

The test is also performed using two different weights forcing the molten polymer out of the orifice; for example, the test is first performed at 2.16 kg, and then at 5.0 kg. The resulting MI values are then made into a ratio as follows.

$$FRR = \frac{MI_{5.0}}{MI_{2.16}} \tag{5.1}$$

where

FRR = flow-rate ratio,
$MI_{5.0}$ = melt flow index under 5.0 kg weight, and
$MI_{2.16}$ = melt flow index under 2.16 kg weight.

High values of FRR indicate broad molecular weight distributions and various empirical relationships have been proposed.

Both MI and FRR tests are very important in the quality control and quality assurance of polyethylene resins and geomembranes.

Mass per Unit Area (Weight). The weight of a geomembrane (actually its *mass per unit area* but invariably called simply *weight*) can be determined using a carefully measured area of a representative specimen and accurately measuring its mass. It is measured in g/m^2. The test is straightforward to perform and usually follows ASTM D1910 procedures.

Water-Vapor Transmission. Since nothing is absolutely impermeable, the assessment of the relative impermeability of geomembranes is an often-discussed issue. The discussion is placed with the physical properties for want of a better location. The test itself could use an adapted form of a geotechnical engineering test using water as the permeant; however, this would be impractical. In such a case, the hydraulic heads required are so great that leaks or failed specimens invariably result. At lower heads, long test times leading to evaporation problems become a major obstacle. Instead, a completely different approach is taken, whereby water *vapor* is used as the permeant and diffusion is the fundamental mechanism of permeation. In the water-vapor transmission (WVT) test, a test specimen is sealed over an aluminum cup with either water or a desiccant in it and a controlled relative-humidity difference across the geomembrane boundary is maintained. The ASTM test method is covered under E96. With water in the cup (i.e., 100% relative humidity) and a lower relative humidity outside of it, a weight loss over time can be monitored (see Figure 5.1). With a desiccant in the cup (i.e., 0% relative humidity) and a higher relative humidity outside of it, a weight gain over time can be seen and appropriately monitored. The required test time varies, but it is usually from 3 to 30 days. Water vapor transmission, permeance, and (diffusion) permeability are then calculated, as shown in Example 5.1.

Example 5.1

Calculate the WVT, permeance, and (diffusion) permeability of a 0.75 mm thick PVC geomembrane of area 0.003 m^2, which produced the test data in Figure 5.1 at an 80% relative-humidity difference at a temperature of 30°C.

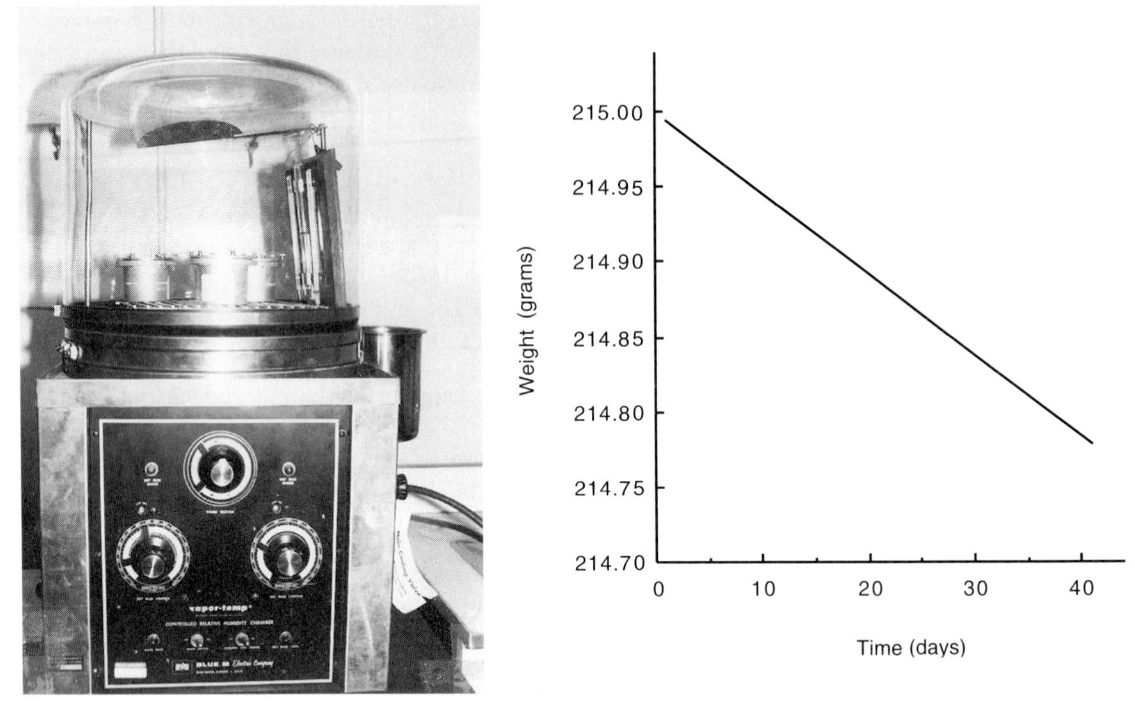

Figure 5.1 Photograph of water vapor transmission test setup and resulting data for 0.75 mm thick PVC geomembrane.

Solution: Calculations proceed in stages using the slope of the curve in Figure 5.1.

(a) Find the water vapor transmission.

$$\text{WVT} = \frac{g \times 24}{t \times a} \tag{5.2}$$

where

g = weight change (g),
t = time interval (h), and
a = area of specimen (m^2).

$$\text{WVT} = \frac{(0.216)(24)}{(40)(24)(0.003)} = 1.80 \text{ g/m}^2\text{-day}$$

(b) The permeance is given as

$$\text{permeance} = \frac{\text{WVT}}{\Delta P} = \frac{\text{WVT}}{S(R_1 - R_2)}$$

where

ΔP = vapor pressure difference across membrane (mmHg),
S = saturation vapor pressure at test temperature (mmHg),

R_1 = relative humidity within cup, and
R_2 = relative humidity outside cup (in environmental chamber).

$$\text{permeance} = \frac{1.80}{32(1.00 - 0.20)} = 0.0703 \text{ metric perm}$$

(c) (Diffusion) permeability = permeance × thickness

$$= (0.0703)(0.75) = 0.0527 \text{ metric perm-mm}$$

Note that this is a vapor-diffusion permeability following Fickian diffusion and not the customary Darcian permeability.

Example 5.2

Using the information and data from Example 5.1, calculate an equivalent hydraulic permeability (i.e., a Darcian permeability, or hydraulic conductivity) of the geomembrane as is customarily measured in a geotechnical engineering test on clay soils.

Solution: The parallel theories are Darcy's formula for hydraulic permeability, $q = kiA$,

$$q\left(\frac{\text{cm}^3}{\text{s}}\right) = k\left(\frac{\text{cm}}{\text{s}}\right)\frac{\Delta h}{\Delta l}\left(\frac{\text{cm H}_2\text{O}}{\text{cm soil}}\right)a(\text{cm}^2)$$

and the WVT test for diffusion permeability,

$$\text{flow}\left(\frac{\text{cm}^3}{\text{s}}\right) = k\left(\frac{\text{cm}^3}{\text{cm}^2\text{-s-cm H}_2\text{O/cm liner}}\right)\text{pressure}\left(\frac{\text{cm H}_2\text{O}}{\text{cm liner}}\right)A(\text{cm}^2)$$

Thus we must now modify the data in Example 5.1 into the proper units.

$$\text{WVT} = 1.80\frac{\text{g}}{\text{m}^2\text{-day}}\frac{1}{(10^4)(24)(60)(60)}$$

$$= 2.08 \times 10^{-9}\frac{\text{g}}{\text{cm}^2\text{-s}}$$

$$\text{permeance} = \frac{\text{WVT}}{\Delta P} = \frac{\text{WVT}}{S(R_1 - R_2)}$$

$$= \frac{2.08 \times 10^{-9}}{(32)(1.00 - 0.20)}$$

$$= 0.812 \times 10^{-10}\frac{\text{g}}{\text{cm}^2\text{-s-mm Hg}}$$

$$\text{permeability} = \text{permeance} \times \text{liner thickness}$$

$$= 0.812 \times 10^{-10}(0.075)$$

$$= 0.609 \times 10^{-11}\frac{\text{g}}{\text{cm}^2\text{-s-mm Hg/cm liner}}$$

$$= 6.09 \times 10^{-11}\frac{\text{g}}{\text{cm}^2\text{-s-cm Hg/cm liner}}$$

In terms of water pressure,

$$\text{hydraulic conductivity} = 6.09 \times 10^{-11} \frac{\text{g}}{\text{cm}^2\text{-s-}\dfrac{\text{cm Hg}}{\text{cm liner}} 13.6 \dfrac{\text{water}}{\text{mercury}}}$$

$$= 0.448 \times 10^{-11} \frac{\text{g}}{\text{cm}^2\text{-s-cm Hg/cm liner}}$$

Now using the density of water,

$$\text{hydraulic conductivity} = 0.448 \times 10^{-11} \frac{\text{g}}{\text{cm}^2\text{-s-}\dfrac{\text{cm water}}{\text{cm liner}} 1.0 \dfrac{\text{g}}{\text{cm}^3}}$$

and canceling the units out, we get a comparable k value for the geomembrane of

$$k \cong 0.5 \times 10^{-11} \text{ cm/s} \quad \text{or} \quad 0.5 \times 10^{-13} \text{ m/s}$$

The WVT values for a number of common geomembranes of different thicknesses are given in Table 5.2. It should be mentioned, however, that the above-described test method is extremely difficult to conduct for thick geomembranes and particularly for of HDPE since its WVT values are so low. The least amount of leakage around the test specimen-to-cup seal will greatly distort the resulting test results. As such the test is not recommended for general use and an entirely different configuration may be necessary, although the concept and theory will be the same.

Of particular interest is the conversion of 1.0 g/m^2-day, approximately equal to 10 l/ha-day, which is the leakage sometimes associated with a flawlessly placed geomembrane. It has been referred to in various regulations as de-minimus leakage (see the footnote in Section 4.2.3). Note that if such a low value is used, it automatically

TABLE 5.2 WATER-VAPOR TRANSMISSION VALUES

Geomembrane Polymer	Thickness (mm)	WVT (g/m^2-day)	WVT (perm-cm)
PVC	0.28	4.4	1.2×10^{-2}
	0.52	2.9	1.4×10^{-2}
	0.76	1.8	1.3×10^{-2}
CPE	0.53	0.64	0.32×10^{-2}
	0.79	0.32	0.24×10^{-2}
	0.97	0.56	0.51×10^{-2}
CSPE	0.89	0.44	0.84×10^{-2}
EPDM	0.51	0.27	0.13×10^{-2}
	1.23	0.31	0.37×10^{-2}
HDPE	0.80	0.017	0.013×10^{-2}
	2.44	0.006	0.014×10^{-2}

1.0 g/m^2-day = 10.0 l/ha-day
Source: After Haxo et al. [3]

eliminates many geomembrane materials, even without a single leak! It also suggests that materials with extremely low WVT values are the best for all liquid containments. But, as we will see in the next section, this is not necessarily the case.

Solvent-Vapor Transmission. When containing liquids other than water, the concept of *permselectivity* must be considered. Here the molecular size and attraction of the liquid vis-a-vis the polymeric liner material might result in very different vapor diffusion values than when using water. Organic solvents are in this category.

The test itself is a parallel to E96, the water-vapor transmission test, except now the solvent of interest is placed within the cup. Obviously, care and proper laboratory procedures must be exercised when using such hazardous or sometimes radioactive materials. As with the WVT test, proper sealing to prevent leakage is extremely difficult to achieve. However, some solvent-vapor transmission data is available and is reproduced in Table 5.3. Notice the tremendous range compared to the values given in Table 5.2 for water-vapor transmission. Clearly, if solvents are to be contained by a geomembrane, the site-specific solvent-vapor transmission test should be used to assess the geomembrane's capability in this regard.

The area of solvent-vapor transmission has been extended by Rowe et al. [5] who use synthesized leachates to measure diffusion through different thicknesses of HDPE geomembranes. A number of organic compounds are evaluated to obtain their diffusion coefficients. It is shown that the geomembrane provides an excellent barrier to acetic acid and chloride. Conversely, organic compounds [such as dichloromethane, 1,1-dichloroethane, 1,2-dichloroethane, and 2-butanone (MEK)] can diffuse though much more rapidly. Preventative options in this regard are to use relatively thick geomembranes; to back up the geomembrane with a clay liner (CCL or GCL) for attenuation,

TABLE 5.3 SOLVENT VAPOR TRANSMISSION VALUES

		Geomembrane Polymer Type (Avg Thickness)			
Property	Solvent	CSPE-R (1.10 mm)	HDPE (0.80 mm)	HDPE (2.6 mm)	LDPE (0.75 mm)
SVT (g/m^2-day)	Methyl alcohol	—	0.16	—	0.74
	Acetone	221	0.56	—	2.83
	Cyclohexane	—	11.7	—	161
	Xylene	—	21.6	6.86	116
	Chloroform	—	54.8	15.8	570
Solvent vapor permeance (10^{-2} metric perms) (SVT/mm Hg)	Acetone	104	0.26	—	1.33
	Methyl alcohol	—	0.14	—	0.66
	Cyclohexane	—	13.1	—	181
	Xylene	—	308	97.9	1650
	Chloroform	—	30.8	8.88	320
Solvent vapor permeability (10^{-2} metric perms-cm)	Methyl alcohol	—	0.01	—	0.05
	Acetone	11.4	0.02	—	0.10
	Cyclohexane	—	1.05	—	13.6
	Xylene	—	24.6	25.6	124
	Chloroform	—	2.46	2.32	24.0

Source: After Matrecon [4]

thereby creating a composite liner; or to use double-liner systems with an intermediate drainage layer. Note that diffusion tests of this type have the liquids stationary on the geomembrane, while in a typical landfill the sloped surface causes movement of the liquids to a sump area where they are removed.

5.1.3 Mechanical Properties

There are a number of mechanical tests that have been developed to determine the strength of polymeric sheet materials. Many have been adopted for use in evaluating geomembranes. This section attempts to sort out those having applicability to design and to distinguish between index and performance tests.

Tensile Behavior (Index). The various tensile tests performed on geomembrane specimens are quite small in size and are used routinely for the quality control and quality assurance of the manufactured geomembranes. The test procedures generally used are covered in ASTM D638, D882, and D751. Table 5.4 gives the currently recommended tests for commonly used geomembranes. (Also given are the currently used tests for seamed samples, a topic which will be discussed in Section 5.10.)

The results for several of these geomembranes are given in Figure 5.2. Here is seen that the scrim-reinforced geomembrane CSPE-R resulted in the highest strength, but failed abruptly when the scrim broke. The response, however, does not drop to zero because the geomembrane plies on both sides of the scrim remained intact until ultimate failure occurred. The HDPE geomembrane responded in a characteristic fashion by showing a pronounced yield point, dropping slightly and then extending in strain to approximately 1000% when failure actually occurred. The PVC geomembrane gave a relatively smooth response, gradually increasing in stress until its failure at about 480% strain. The VLDPE geomembrane also gave a relatively smooth, but lowest strength, response until it failed at approximately 700% strain.

The curves were generated using the specimen's original width and thickness to calculate stress and the original length to calculate strain. Thus the axes are engineering stress and strain, rather than true stress and strain. Quantitative data gained from these curves are focused around:

- maximum stress (at ultimate for PVC and VLDPE, at scrim break for CSPE-R, and at yield for HDPE),
- maximum strain (usually called *elongation* in the geomembrane literature),
- modulus (the slope of the initial portion of the stress-strain curve),
- ultimate stress at failure (or strength), and
- ultimate strain (or elongation) at failure.

Table 5.5a gives these values for the four materials shown in Figure 5.2. While all of the listed values of strength are significant, attention is often focused on the maximum stress. It must be recognized, however, that polymers are viscoelastic materials and strain invariably plays an important role.

TABLE 5.4 RECOMMENDED TEST METHOD DETAILS FOR GEOMEMBRANES AND GEOMEMBRANE SEAMS IN SHEAR AND IN PEEL

Test	HDPE	VFPE; PP	PVC	CSPE-R; PP-R	EIA-R
Tensile Test on Sheet					
ASTM Test Method	D638	D638	D882	D751	D751
Specimen Shape	Dumbbell	Dumbbell	Strip	Grab	Grab
Specimen Width (mm)	6.3	6.3	25	100 (25 Grab)	100 (25 Grab)
Speciment Length (mm)	115	115	150	150	150
Gage Length (mm)	33	33	50	75	75
Strain Rate (mm/min)	50	500	500	300	300
Strength	Force/($w \times t$)	Force/($w \times t$)	Force/($w \times t$)	Force	Force
Strain (mm/mm)	Elong/33	Elong/33	Elong/50	Elong/75	Elong/75
Modulus	From graph	From graph	From graph	n/a	n/a
Shear Test on Seams					
ASTM Test Method	D4437	D4437	D3083	D751	D751
Specimen Shape	Strip	Strip	Strip	Grab	Grab
Specimen Width (mm)	25	25	25	100 (25 grab)	100 (25 grab)
Specimen Length (mm)	150 + seam	150 + seam	150 + seam	225 + seam	225 + seam
Gage Length (mm)	100 + seam	100 + seam	100 + seam	150 + seam	150 + seam
Strain Rate (mm/min)	50	500	500	300	300
Strength	Force/($w \times t$)	Force/($w \times t$)	Force/($w \times t$)	Force	Force
Peel Test on Seams					
ASTM Test Method	D4437	D4437	D413	D413	D413
Specimen Shape	Strip	Strip	Strip	Strip	Strip
Specimen Width (mm)	25	25	25	25	25
Specimen Length (mm)	100	100	100	100	100
Gage Length (mm)	n/a	n/a	n/a	n/a	n/a
Strain Rate (mm/min)	50	500	50	50	50
Strength	Force/($w \times t$)	Force/($w \times t$)	Force/w	Force/w	Force/w

Abbreviations: n/a = not applicable
w = specimen width (mm)
t = specimen thickness (mm)
Force = maximum force attained at specimen failure (ultimate or break)

Source: Modified and extended from [4]

Tensile Behavior (Wide-Width). A major criticism of the previously described index test specimens is their contraction within the central region, giving a one-dimensional behavior not experienced with wide sheets in field situations. Thus uniform width and wider test specimens are desirable. Just how wide is a matter of debate. A width of 200 mm has been used for testing geotextiles and has been adopted for testing geomembranes; see ASTM D4885. The strain rate for testing geomembranes is, however, different from geotextiles. D4885 recommends using 1.0 mm/min. For a 100 mm long specimen with 200% strain at failure, the test would require 3.3 hours to complete. For a geomembrane with 1000% strain at failure, the test would require 16.7 hours. Clearly such tests are not of the index or quality-control variety and should be considered performance-oriented.

Figure 5.3 presents tensile stress-versus-strain curves on the same four geomembranes that are shown in Figure 5.2, but now for a uniform 200-mm width. While the

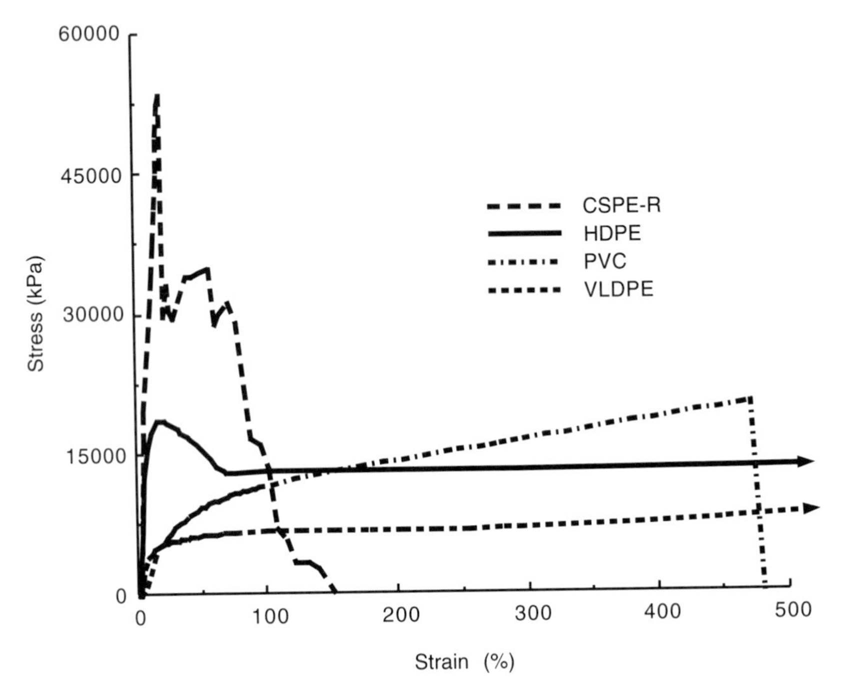

Figure 5.2 Index tensile test results of commonly used geomembranes using criteria given in Table 5.4.

general shape of each material is the same, the results of the various properties of interest are quite different. These results are tabulated in Table 5.5b. It is felt that the use of a 200 mm width test specimen results in a much more design-oriented value than do test results from dumbbell or narrow-width specimens. This is particularly the case when plane-strain conditions are assumed in the design process (e.g., in side-slope stability calculations).

Tensile Behavior (Axi-Symmetric). There are situations that call for a geomembrane's tensile behavior when mobilized by out-of-plane stresses. Localized de-

TABLE 5.5a TENSILE BEHAVIOR PROPERTIES OF HDPE, VLDPE, PVC, AND CSPE-R

		Index Tension Tests (Figure 5.2)			
Test Property	Unit	HDPE	VLDPE	PVC	CSPE-R
Maximum stress and	(kPa)	18,600	8,300	21,000	54,500
corresponding strain	(%)	17	500+	480	19
Modulus	(MPa)	330	76	31	330
Ultimate stress and	(kPa)	13,800	8,300	20,700	5,700
corresponding strain	(%)	500+	500+	480	110

Nom. thicknesses are: HDPE 1.5 mm, VLDPE 1.0 mm, PVC 0.75 mm, CSPE-R 0.91 mm.

Abbreviations: + = did not fail

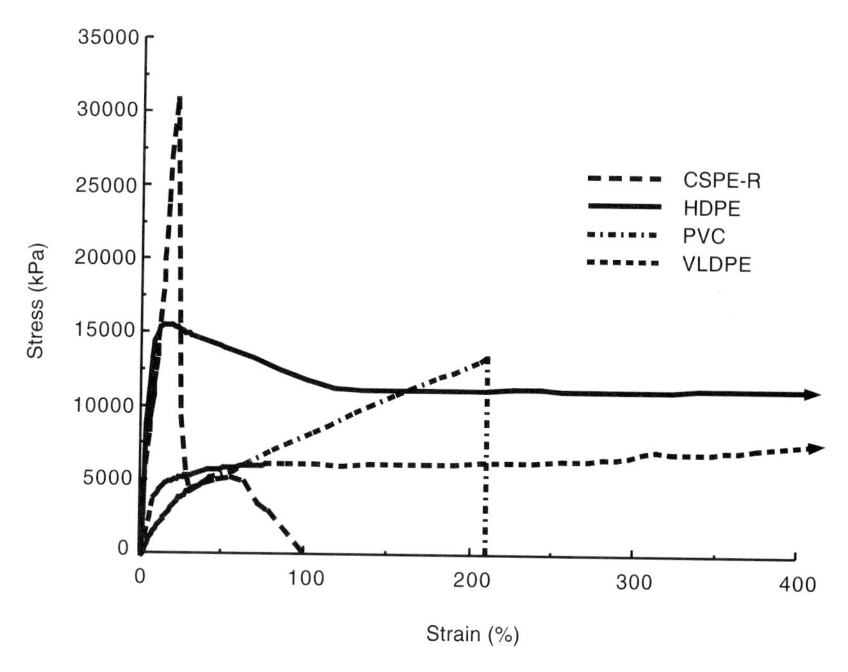

Figure 5.3 Tensile test results on 200 mm wide-width specimens of commonly used geomembranes using ASTM D4885 test method.

formation beneath a geomembrane is such a case. This type of behavior could well be anticipated for a geomembrane used in a landfill cover placed over differentially subsiding solid-waste material. The situation can be modeled by placing the geomembrane in an empty container, as shown in Figure 5.4. An appropriate seal is made with the cover section and water is introduced above the geomembrane. Pressure is mobilized until the failure of the test specimen occurs. Beginning with Stefan [6], a number of variations of this test have been made. It is currently formalized as ASTM D5716.

TABLE 5.5b TENSILE BEHAVIOR PROPERTIES OF HDPE, VLDPE, PVC, AND CSPE-R

		Wide-Width Tension Tests (Figure 5.3)			
Test Property	Unit	HDPE	VLDPE	PVC	CSPE-R
Maximum stress and	(kPa)	15,900	7,600	13,800	31,000
corresponding strain	(%)	15	400$^+$	210	23
Modulus	(MPa)	450	69	20	300
Ultimate stress and	(kPa)	11,000	7,600	13,800	2,800
corresponding strain	(%)	400$^+$	400$^+$	210	79

Nom. thicknesses are: HDPE 1.5 mm, VLDPE 1.0 mm, PVC 0.75 mm, CSPE-R 0.91 mm.
Abbreviations: $^+$ = did not fail

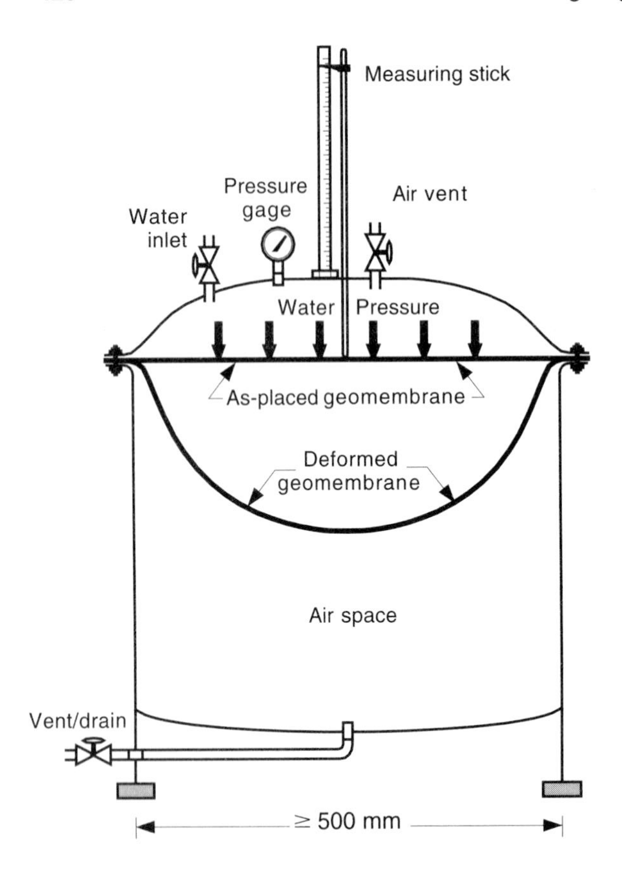

Figure 5.4 Schematic diagram and photograph of three-dimensional axi-symmetric geomembrane tension test apparatus.

The data generated by the test is pressure versus centerpoint deflection, which can be plotted and used directly. This is customary in Germany. To obtain stress-versus-strain (which is more desirable), certain assumptions must be made. The original analysis by Koerner et al. [7], has been modified by Merry and Bray [8]. Both are based on the assumption that the deflected shape is being deformed as a portion of a sphere with a gradually moving center point along the centerline axis. The latter is preferred, however, since it is based on a constant geomembrane-volume hypothesis. The resulting equations for stress and strain are as follows [8].

$$\sigma = \frac{(L^2 + 4\delta^2)^2 p}{16\delta L^2 t} \qquad \text{for all } \delta\text{-values} \tag{5.3}$$

and

$$\varepsilon(\%) = \left[\frac{\tan^{-1}\left(\dfrac{4L\delta}{L^2 - 4\delta^2}\right)\left(\dfrac{L^2 + 4\delta^2}{4\delta}\right) - L}{L} \right] \times 100 \qquad \text{for } \delta < \frac{L}{2} \tag{5.4}$$

$$\varepsilon(\%) = \left\{ \frac{\left[\dfrac{L^2 + 4\delta^2}{4\delta}\right]\left[\pi - \sin^{-1}\left(\dfrac{4L\delta}{L^2 + 4\delta^2}\right)\right] - L}{L} \right\} \times 100 \qquad \text{for } \delta \geqslant \frac{L}{2} \tag{5.5}$$

where

L = diameter of test specimen, that is, the container (mm)
δ = centerpoint deflection (mm)
p = pressure on test specimen (kPa)
t = original thickness to geomembrane (mm)
σ = geomembrane tensile stress (kPa)
ε = geomembrane tensile strain (%)

In conducting the test, it is observed that HDPE and reinforced geomembranes like CSPE-R and fPP-R fail at relatively low deflections ($\delta < L/2$) and extensible geomembranes like VFPE, fPP and PVC fail at relatively high deflections ($\delta \geqslant L/2$).

Using a 600 mm diameter test vessel, hydrostatically pressurized at a rate of 7.0 kPa per minute, the data produce the curves of Figure 5.5. Note that both the HDPE and CSPE-R geomembranes fail at relatively low strains (but high stresses), whereas the VLDPE and PVC fail at significantly greater strains. These test results are also in Table 5.5c, which taken with Tables 5.5a and 5.5b provide a comparison of the different tensile tests. Note the relatively large differences between this test and the other tension tests presented earlier. Clearly, the lesson here is that appropriate modeling of a field situation is absolutely necessary if a design-by-function approach is to be used. Additional insight into the behavior of the test results under varying conditions is found in Nobert [9].

Tensile Behavior (Seams). The joining of geomembrane rolls and panels results in a seam that can be weaker than the geomembrane itself. This is particularly true of seams made in the field versus those made in the fabrication factory, where quality control can be exercised more rigorously. (Various types of seaming methods will be discussed in Section 5.10.) To determine the strength of a geomembrane seam, a number of tests are available: typical shear tests are ASTM D4437, D3083, and D751; typical peel tests are ASTM D4437 and D413. Table 5.4 identifies them with respect to the

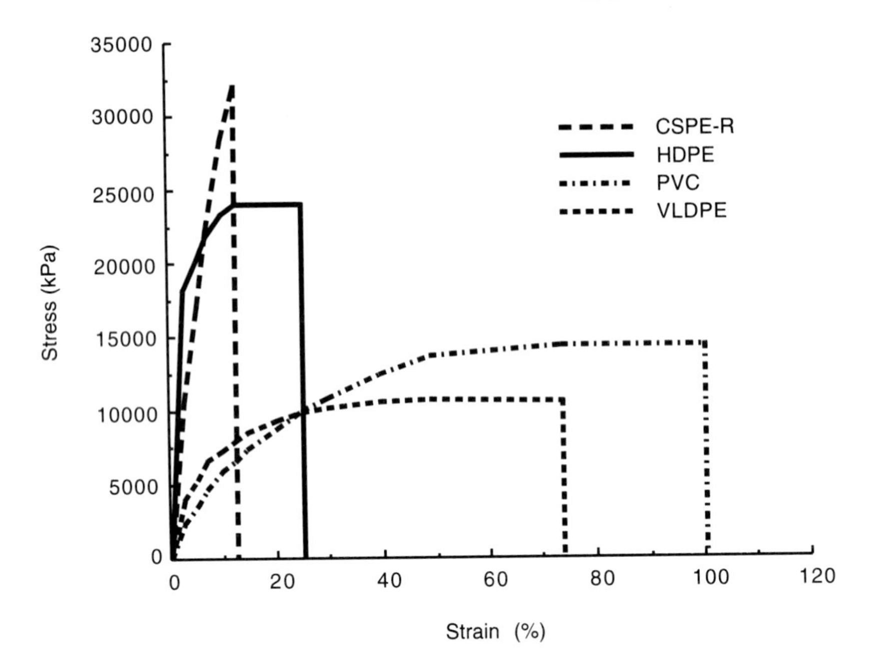

Figure 5.5 Stress-versus-strain response curves of various types of geomembranes under axi-symmetric hydrostatic pressure. (After Koerner et al. [7])

type of geomembranes. In both shear and peel tests a representative specimen (usually 25 mm wide) is taken across the seam and the unseamed ends are placed in the grips of a tensile testing machine. For the shear test, the two separate pieces of the seam are pulled apart, placing the joined or seamed portion between the geomembrane sheets in shear. For the peel test, one end of the geomembrane and the closest end of the adjacent piece are gripped, placing the seamed portion between them in a tensile mode. Fig-

TABLE 5.5c TENSILE BEHAVIOR PROPERTIES OF HDPE, VLDPE, PVC, AND CSPE-R

		Axi-Symmetric Tension Tests (Figure 5.5)			
Test Property	Unit	HDPE	VLDPE	PVC	CSPE-R
Maximum stress and	(kPa)	23,500	10,300	14,500	31,000
corresponding strain	(%)	12	75	100	13
Modulus	(MPa)	720^-	170^-	100^-	350^-
Ultimate stress and	(kPa)	23,500	10,300	14,500	31,000
corresponding strain	(%)	25	75	100	13

Nom. thicknesses are: HDPE 1.5 mm, VLDPE 1.0 mm, PVC 0.75 mm, CSPE-R 0.91 mm.

Abbreviations: $^-$ = values felt to be high

ure 5.6 shows sketches of the two configurations and also the results of tests on solvent-bonded PVC geomembrane seams compared to the sheet material itself with no seams and extrusion fillet-welded HDPE geomembrane compared to the sheet itself. It is clearly seen that for these situations the peel test is a great deal more critical than the shear test and that in all cases the seams have lower strength than the parent material. It is sometimes said that the shear test simulates a performance mode, whereas the peel test is more of an index test. It is important to perform both in order to fully evaluate the quality of the seam. It should be cautioned, however, that these results vary greatly with the type of geomembrane and the type of seam being evaluated. Of particular concern are the following questions.

- Which type of seam test (shear, peel, or both) should be used for construction quality control and quality assurance?
- Has the failure occurred in the seamed region or in the parent material adjacent to the seam?
- What percentage of parent material strength should the seam itself support?
- Is elongation at failure of significance in assessing seam quality?
- Should seams be evaluated in wide-width or axi-symmetric modes that better simulate performance tests?

Haxo [10] has written extensively on the subject, as have Rollin and Fayoux [11] and Peggs [12], but the final decision rests with the specification writer and/or owner-permitting agency.

Tear Resistance. The measurement of tear resistance of a geomembrane can be done in a number of ways. ASTM D2263, D1004, D751, D1424, D2261, and D1938 all cover the general topic. The first method, D2263, also called *trapezoidal tear,* is sometimes recommended. In this test a geomembrane specimen is cut in the shape of a trapezoid of dimensions 100 mm on one side, 25 mm on the other, and 75 mm long. An initiating cut of 12.5 mm is made in the center of the 25 mm side. This specimen is then mounted in the grips of a tensile testing machine in such a way that the 25 mm side (with the cut) is taut and the 100 mm side has 75 mm of slack in it. As the test machine elongates, the specimen tears from the 25 mm side to the 100 mm side, beginning at the initiating cut. The maximum load is reported as the tear resistance.

The above test yields results similar to D1004, which uses a template to form a test specimen shaped such as to have a 90° angle where tear can begin to propagate. The two ends on each side of the angle are gripped in a tensile testing machine and tearing proceeds across the specimen perpendicular to the application of load. The major difference in these two tests (D2263 and D1004) is the length of test specimen for the tear to propagate across. Here the D2263 is greater, 75 mm versus 12.5 mm, and a more accurate trend of behavior can be assessed. However, the required sample size is larger, which may be a limitation for incubated materials.

The tear resistance of many thin nonreinforced geomembranes is quite low, from 18 to 130 N. The implication of this is important during geomembrane handling and installation. Extreme care must be exercised during construction when moving rolls or panels into place and during periods of high wind. The placement of a scrim within the

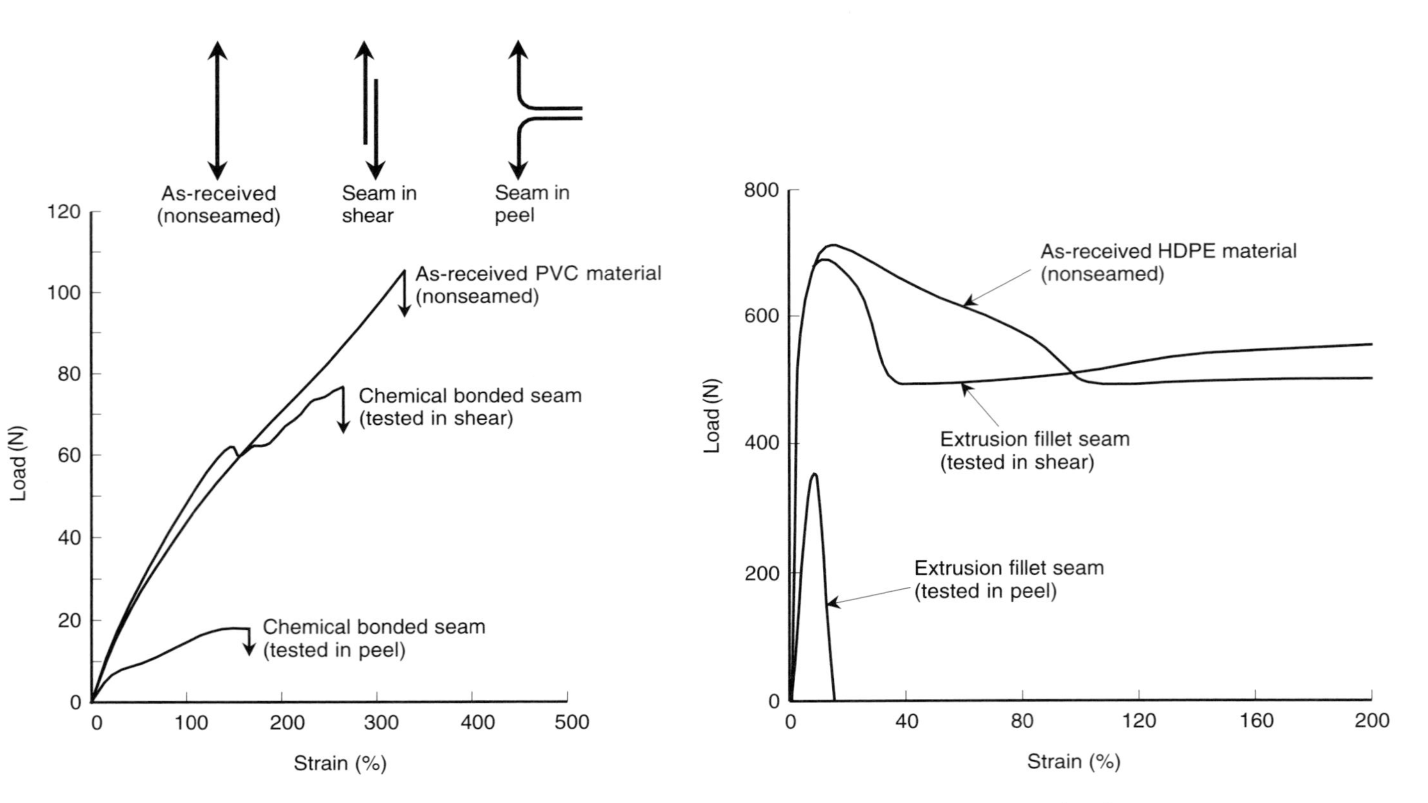

Figure 5.6 Tensile test results on 25 mm wide 0.5 mm PVC geomembrane compared to chemical bonded seams on the same material tested in shear and in peel; and 25 mm wide 1.5 mm HDPE geomembrane compared to extrusion fillet welded seams on the same material tested in shear and in peel.

geomembrane greatly helps this situation. The tear test often recommended for scrim-reinforced geomembranes is the tongue tear test, D2261. Here a 75 mm wide specimen with a 25 mm initiating cut is put in the grips of a tension testing machine and pulled against itself. The test specimen is 200 mm long and the maximum value of resistance is reported as the tear strength. Values of tear resistance for scrim-reinforced geomembranes increase significantly, to the range 90 to 450 N.

As the geomembrane becomes thicker, tear during installation becomes less of an issue and the design-related tear stresses become the critical values. These are site-specific situations that must be individually assessed.

Impact Resistance. Falling objects, including cover soils, can penetrate geomembranes, either causing leaks themselves or acting as initiating points for tear propagation. Thus an assessment of geomembrane impact resistance is relevant. There are a number of ASTM options available; among them are: ASTM D1709 (free-falling dart), ASTM D3029 (falling weight), and ASTM D1822, D746 and D3998 (pendulum types).

Rather than using a separate test device, it is sometimes convenient to use the Spencer impact adaptation of the Elmendorf tear test, ASTM D1424. The Elmendorf tear apparatus is a pendulum-type device that results in an energy-to-failure value in joules. The Spencer impact attachment is merely a specimen-holding device for the penetration of a point at the end of the pendulum swing.

The results in Table 5.6 were obtained using such a device. It is clearly seen that the thicker geomembranes have a greater impact resistance than do the thinner ones. The effect of scrim-reinforcement is not significant. Note that there is a significant difference in impact resistance between different types of geomembranes. The maximum value listed in the table is 21 J, which is typical of many geomembranes when a blunt penetrating object is used. It is also the maximum capacity of the test apparatus just described. For geomembranes with greater impact resistance, or for geotextiles underlying and/or overlying the geomembrane, significantly higher impact resistances will result. Alternatively, a notched dumbbell-shaped test specimen can be used in which the pendulum imparts a tensile impact stress on the specimen. This is used with poly-

TABLE 5.6 IMPACT RESISTANCE (J) OF VARIOUS GEOMEMBRANES, ASTM D1424 SPENCER IMPACT METHOD

	Point Geometry Angle*				
Geomembrane	15°	30°	45°	60°	90°
PVC (0.50 mm)	6.5	8.9	14.9	> 21	> 21
PVC (0.75 mm)	9.2	13.6	18.3	> 21	> 21
HDPE (1.0 mm)	7.6	9.4	11.3	11.3	8.7
EPDM-R (0.91 mm) 10 × 10 scrim	11.7	11.9	14.6	19.5	> 21
CSPE-R (0.91 mm) 10 × 10 scrim	12.2	12.7	14.0	19.3	> 21

*These are conical points where the angle is measured from the central axis of the cone to its surface. In each case the total surface area of the points is kept constant.

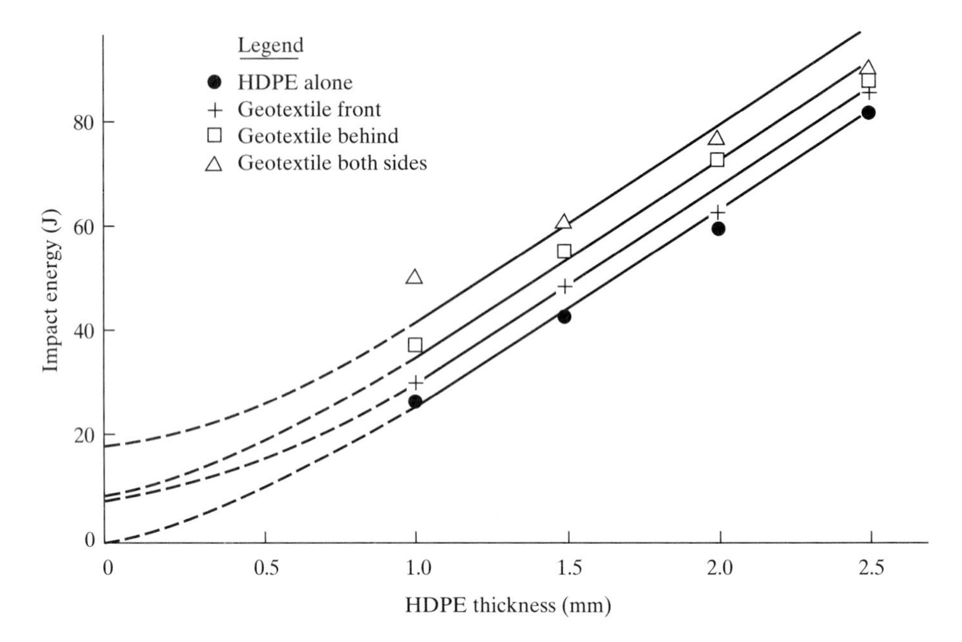

Figure 5.7 Falling pendulum impact test results on varying thicknesses of HDPE geomembranes and different combinations of geomembranes with 400 g/m2 nonwoven needle-punched geotextiles. ([13])

ethylene geomembranes and can be done in cold temperatures; see ASTM D746 in this regard. For the testing of geocomposite systems a larger floor-mounted pendulum device can be used. ASTM D3998 describes such a device; the data from such a test produce the curves shown in Figure 5.7. Here varying thicknesses of HDPE are evaluated with an approximate linear response as the result. When using a 400 g/m^2 nonwoven needle-punched geotextile on the front or back of the geomembrane, an improvement in impact resistance is noted. Furthermore, a geotextile on both sides of the geomembrane improves the impact resistance even further. The response curves can be used to determine the economic efficiency of a thicker geomembrane versus a geosynthetic composite.

Puncture Resistance. Geomembranes placed on, or backfilled with, soil containing stones, sticks, or other debris are vulnerable to puncture during and after loads are placed on them. Such puncture is an important consideration because it occurs after the geomembrane is covered and cannot be detected until a leak from the completed system becomes obvious. Repair costs at that time are often enormous.

The closest ASTM test modeling this situation is D5494, the pyramid puncture test. Alternatively D4833 can be used, since it is the test method used by manufacturers for quality control purposes. Here a geomembrane is clamped over an empty mold of 45-mm diameter. The assembly is placed in a compression testing machine fitted with a 8 mm diameter rod with a flat, but edge-beveled, bottom. The rod is pressed into the geomembrane until it punches through. The recommended load rate is 300 mm/min. The value reported as puncture resistance is the maximum load registered on the test machine. Typical values of geomembrane puncture resistance are 50 to 500 N for thin nonreinforced geomembranes and 200 to 2000 N for reinforced geomembranes. Here again the influence of the scrim reinforcement is seen. Note that the placement of a geotextile below and/or above a nonreinforced geomembrane greatly increases the puncture resistance of the geomembrane and essentially takes all the load before the geomembrane absorbs any of it. The results of the test of resistance of varying thicknesses of HDPE geomembrane to puncture are shown in Figure 5.8. As expected, the thicker the geomembrane, the higher the puncture resistance. The response is approximately linear. With a 400 g/m^2 nonwoven needle-punched geotextile on front or on back, however, the improvement in puncture resistance is quite impressive. Clearly a strong composite action is occurring. This is further evidenced by the upper curve, which has a similar geotextile on front and on back of the geomembrane.

Recognizing that the tests above are index tests and the importance of puncture resistance when large stone aggregate is used for leachate collection layers in landfills, waste piles and heap leach pads, the need for a field-simulated performance test becomes obvious. Activity in this regard is ongoing. Most efforts use a large-diameter pressure vessel with the subgrade beneath the geomembrane test specimen. Several variations are being evaluated: the actual subgrade (sand, gravel, stone, etc.) at the targeted density; the actual subgrade set in an epoxy cast (so-called "rock pizza"), so as to

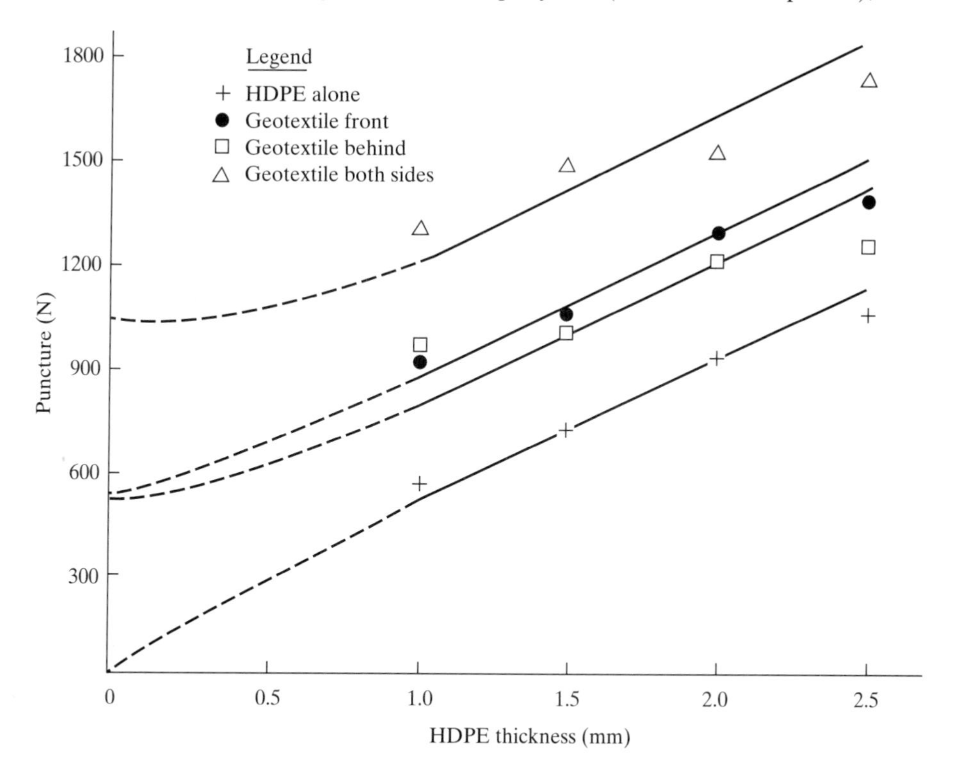

Figure 5.8 Beveled flat-tip puncture test results on varying thicknesses of HDPE geomembranes and different combinations of geomembranes with 400 g/m2 nonwoven needle-punched geotextiles. ([13])

have the particles in the same configuration for each test; and truncated cones in a triangular array to simulate a worst-case subgrade condition. A test method is currently available, ASTM D5514, and a paper on the truncated cone test has evaluated a number of common geomembranes (HDPE, CSPE-R, PVC, and VLDPE), both with and without geotextile protection layers [14]. It has also been extended into creep testing to assess the viscoelastic properties of the geomembranes and protection layers [15].

Interface Shear. Critically important for the proper design of geomembrane-lined side slopes of landfills, reservoirs, and canals, is the soil-to-geomembrane shear strength. As pointed out by Boschuk [16], numerous side slope failures have occurred. Often cover soils slide over geomembranes, but sometimes the geomembrane fails (or pulls out of the anchor trench) moving on a lower-friction surface beneath. The test method used to access the situation is an adapted form of a geotechnical engineering direct shear test for determining soil-to-soil friction.

The experimental setup to evaluate soil-to-geomembrane friction uses the geomembrane in one half of a direct shear box with the opposing soil surface in the other half; see Figure 5.9a. Generally the soil is compacted to its intended density and moisture content in the lower half and the geomembrane is firmly bonded to a wooden platen in the upper half. A number of site-specific conditions must be addressed in order to have realistic results, for example: the type and gradation of soil to be used, the type of liquid, the density and moisture content of soil to be placed, the moisture condition during test (i.e., dry, moist or saturated), the normal stress(es) to apply, the time for saturation and/or consolidation, the strain rate to use during shear, and the deformation required to attain residual strength. In short, every geotechnical engineering question we might have in designing a soil system must be addressed when designing a geomembrane system.

Additionally, the size of the shear box must be considered. For geomembranes against sands, silts or clays, I favor a 100 × 100 mm square shear box. Many commercial soil-testing devices are available of this type. Only if we are evaluating gravel or other large particle-sized material against a geomembrane must the shear box be larger. This decision on shear box size should really be left to the design engineer, since the test is always performance-oriented. Unfortunately, ASTM D5321 on direct shear evaluation of geosynthetic-to-soil, or geosynthetic-to-geosynthetic, recommends a 300 × 300 mm square shear box for all situations (unless it can be proven that some other size is needed). While such a large shear box is appropriate for geogrid or geonet shear testing, it is felt to be excessive by the author. The paramount reasons are the very long times for saturation, the nonuniform normal stresses (particularly during the test), and the insufficient travel to attain true residual strength.

Irrespective of box size, conducting a direct shear test is straightforward. Upon deciding the above mentioned site-specific issues, a series of at least three separate tests (each time with new test specimens) is performed at different normal stresses (σ_n) centered around the site-specific normal stress. The data should be similar to that in Figure 5.9b. The peak and residual strengths are sometimes similar and sometimes very different.

Using the peak and residual shear stress values (i.e., the shear strengths) from these three graphs allows for a new graph to be developed; see Figure 5.9c. This is the Mohr-Coulomb failure envelope, consisting of three failure points at the corresponding

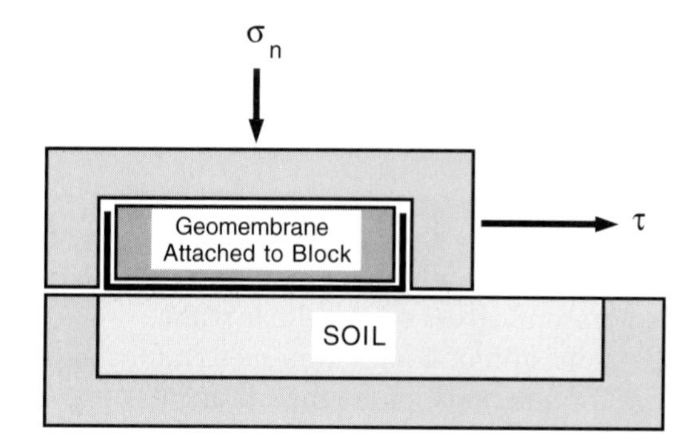

(a) Direct shear test device

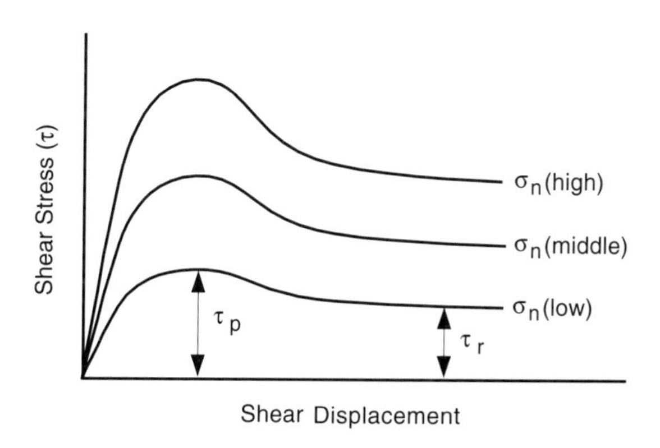

(b) Direct shear test data

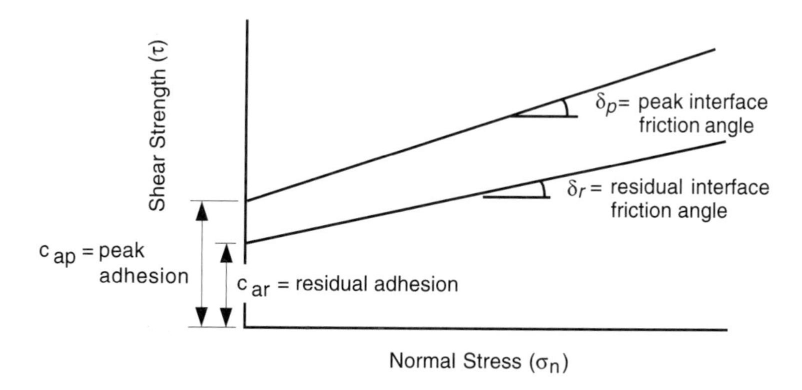

(c) Mohr-Coulomb stress space

Figure 5.9 Direct shear testing concept and resulting shear strength parameters.

normal stresses for peak and (if different) residual strengths. The straight-line response results in the following equation

$$\tau = c_a + \sigma_n \tan \delta \tag{5.6}$$

where

τ = shear strength of geomembrane to the opposing surface,
σ_n = normal stress on the shear plane,
c_a = adhesion of geomembrane to opposing surface, and
δ = friction angle of geomembrane to opposing surface.

If soil is the opposing surface, the tests can be repeated with soil in both halves of the shear box. Treating the data in an identical manner results in another Mohr-Coulomb failure envelope, which results in the following equation.

$$\tau = c + \sigma_n \tan \phi \tag{5.7}$$

where

τ = shear strength of soil,
c = cohesion of soil, and
ϕ = friction angle of soil.

As for other geosynthetics, efficiencies can be calculated in the standard manner.

$$E_c = \left(\frac{c_a}{c}\right)100 \tag{5.8}$$

$$E_\phi = \left(\frac{\tan \delta}{\tan \phi}\right)100 \tag{5.9}$$

There are many papers available on the interface friction between geomembranes and numerous other surfaces (soils and geosynthetics). Results from an early effort focusing on peak friction values (Martin et al. [17]) are given in Table 5.7.

As shown in Table 5.7, the peak friction angles of soil-to-geomembrane are always less than soil-to-soil, with the smoother, harder geomembranes being the lowest (e.g., HDPE); conversely, the softer and rougher geomembranes (PVC and CSPE-R) have relatively high friction values. For geotextile-to-geomembrane friction values for situations using a geotextile underliner and/or overliner, great variations are shown in the table, with the smooth, hard geomembranes on smooth geotextiles giving the lowest friction values. For reference purposes, Table 5.7 also gives the soil-to-geotextile friction values that are necessary for a slope design of lined slopes with geotextiles under or over the liner. Note that this study predates the development of textured HDPE and VLDPE geomembranes; as described in Section 1.6.2, these very roughened surfaces can develop the full shear strength of many types of soils. This is particularly the case if the soil particle sizes are equal to or smaller than the surface pattern of the geomembrane's texturing.

TABLE 5.7 PEAK FRICTION VALUES AND EFFICIENCIES*

Soil-to-Geomembrane Friction Angles

Geomembrane	Soil type: Concrete sand ($\phi = 30°$)		Ottawa sand ($\phi = 28°$)		Mica schist sand ($\phi = 26°$)	
HDPE (smooth)	18°	(0.56)	18°	(0.61)	17°	(0.63)
PVC						
rough	27°	(0.88)	—	—	25°	(0.96)
smooth	25°	(0.81)	—	—	21°	(0.79)
CSPE-R	25°	(0.81)	21°	(0.72)	23°	(0.87)

Geomembrane-to-Geotextile Friction Angles

Geotextile	Geomembrane: HDPE (smooth)	PVC: Rough	PVC: Smooth	CSPE-R
Nonwoven needle-punched	8°	23°	21°	15°
Nonwoven heat-bonded	11°	20°	18°	21°
Woven monofilament	6°	11°	10°	9°
Woven slit-film	10°	28°	24°	13°

Soil-to-Geotextile Friction Angles

Geomembrane	Soil type: Concrete sand ($\phi = 30°$)		Ottawa sand ($\phi = 28°$)		Mica schist sand ($\phi = 26°$)	
Nonwoven needle-punched	30°	(1.00)	26°	(0.92)	25°	(0.96)
Nonwoven heat-bonded	26°	(0.84)	—	—	—	—
Woven monofilament	26°	(0.84)	—	—	—	—
Woven slit-film	24°	(0.77)	24°	(0.84)	23°	(0.87)

*Efficiency values (in parentheses) are based on the relationship $E = (\tan \delta)/(\tan \phi)$.

Source: After Martin et al. [17]

The frictional behavior of geomembranes placed on clay soils is of considerable importance for composite liners containing solid or liquid wastes. The current requirements are for the clay to have a hydraulic conductivity equal to or less than 1×10^{-7} cm/s and for the geomembrane to be placed directly upon the clay. While an indication of the shear strength parameters has been investigated (e.g., [18]), the data are so sensitive to the variables discussed previously that site-specific and material-specific tests should always be performed. *In such cases, literature values should never be used for final design purposes.*

Much of the direct shear literature data is for peak shear strengths (e.g., the data in Table 5.7). Stark and Poeppel [19] have challenged this situation by testing various geosynthetic interfaces in a ring-shear device. In using such a device, significantly larger deformations can be mobilized than in conventional direct shear testing. In so doing, they found that residual strengths are often considerably lower than peak strengths. Among other findings, they identified a geomembrane *polishing action* that could occur at large deformations, decreasing peak friction angles by 50 to 100%. Clearly, shear deformation tests must be continued further than has been done in past practice to see if, and how much, shear strength decreases beyond the peak values.

Anchorage. In certain problem situations, a geomembrane might be sandwiched between two materials and then tensioned by an external force. The termination of a geomembrane liner within an anchor trench is such a situation. To simulate this behavior in a laboratory environment we can use a 200 mm wide geomembrane embedded between back-to-back channels. Here the channels are placed under pressure using a hydraulic jack, and the exposed end is held fixed in the grips of a tension testing machine. The channel surfaces are fitted or faced with the actual or simulated (e.g., sandpaper) adjacent materials. Tension is mobilized from the fixed end of the geomembrane to the opposite end within the anchored zone. For design purposes we are searching for the anchorage depth necessary to mobilize the geomembrane's strength (recall Figure 5.3). The target value could be the tensile strength at yield, at scrim break, or at an allowable strain. Figure 5.10 shows the embedment depth required to mobilize full anchorage strength for HDPE, CSPE-R, and PVC geomembranes. Depending upon the applied normal stress, the anchorage distance varies from approximately 50 to 300 mm; that is, it is very small. Other curves for different geomembranes and confinement conditions can be similarly generated.

Stress-Cracking (Bent Strip). Called *environmental stress-cracking* in ASTM D1693, this test is only applicable to semicrystalline materials like HDPE. Furthermore, the higher the density (hence, the crystallinity), the more significant the test. Small test specimens of 38 × 13 mm are prepared with a controlled imperfection on one surface; a notch about one-half of the thickness running centrally along the long dimension. The specimens are bent into a U-shape and placed within the flanges of a channel holder. This assembly is then immersed in a surface-wetting agent at an elevated temperature, usually 50°C. Since stress-cracking is defined as "an external or internal rupture in a plastic caused by tensile stress less than its short-time mechanical strength," the test records the proportion of the total number of specimens that crack in a given time. Older geomembrane specifications that call for this test require that there be no stress-cracked specimens within 1500 hours. The test is not a good challenge for currently used HDPE resins and is not recommended by the author for further use. It should be discontinued.

Stress-Cracking (Constant Load). A different type of stress-cracking test for polyethylene geomembranes has been developed and adopted by ASTM. It is called the notched constant tension load (NCTL) test, designated D5397. It places centrally notched dumbbell-shaped test specimens under a constant load (at a known percentage of their yield stress) in a surface-wetting agent at an elevated temperature. Igepal is the usual wetting agent and 50°C is the recommended temperature. When a series of test specimens are evaluated at different percentages of their yield stress, a ductile-to-

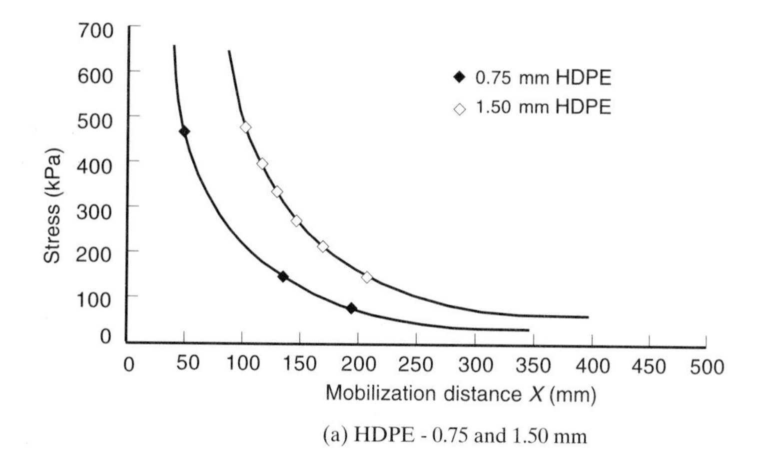

(a) HDPE - 0.75 and 1.50 mm

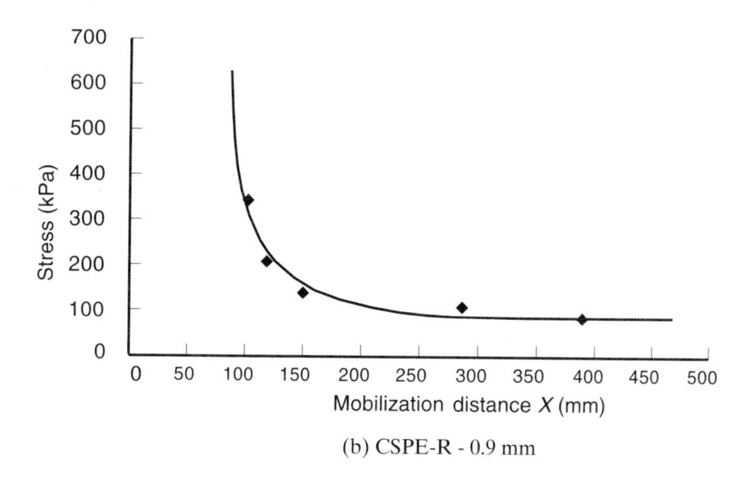

(b) CSPE-R - 0.9 mm

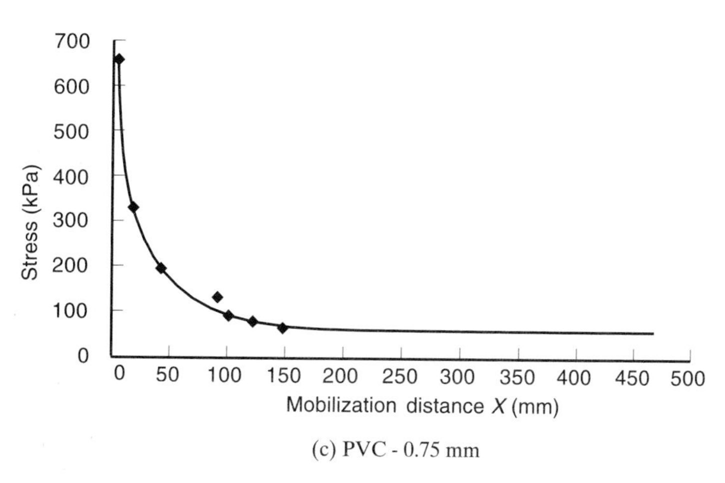

(c) PVC - 0.75 mm

Figure 5.10 Embedment depth curves versus applied normal stress for various geomembranes.

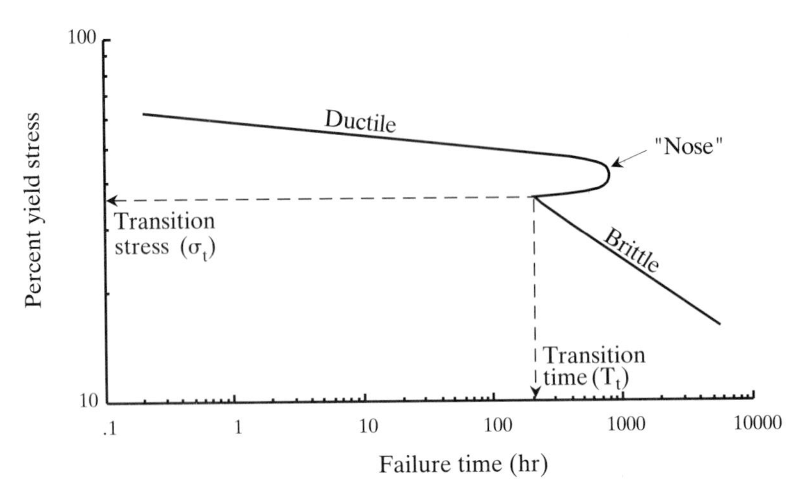

(a) A typical overshoot (or nose) response curve for a complete NCTL test

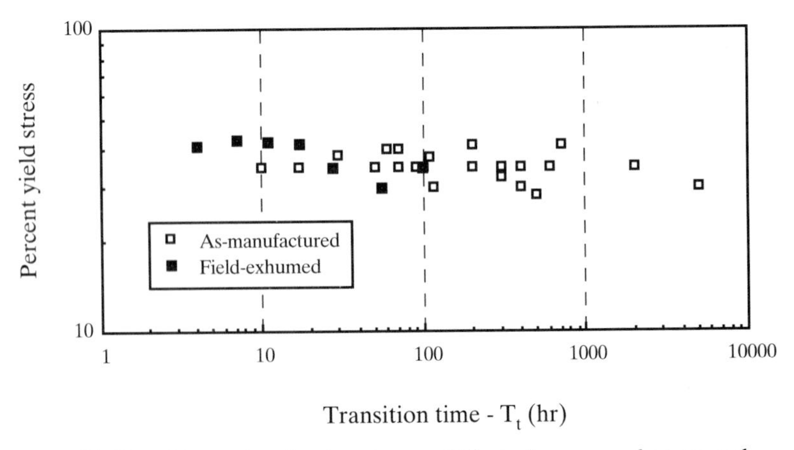

(b) Transition points for twenty-one (21) virgin geomembranes and seven (7) field retrieved geomembranes

Figure 5.11 Typical response of NCTL test to evaluate stress-crack resistance of HDPE geomembranes and resins and summary of transition point variations. (After Hsuan et al. [20])

brittle behavior is indicated (see Figure 5.11a). Evaluating 21 commercially available virgin HDPE geomembranes, it is seen that the transition time (T_t) varies from 10 to 5000 hours. Additionally seven field-exhumed HDPE geomembranes that had evidenced stress-cracking problems are evaluated. Their transition times range from 4 to 97 hr. These 25 data points are shown on Figure 5.11b. The current recommendation for an acceptable stress-crack resistant HDPE geomembrane is a transition time equal to or greater than 100 hr. Since stress-crack resistance is largely a resin-dependent mechanism, the NCTL test can be performed on samples made as plaques from the base resin, as well as on the finished geomembrane sheet. See ASTM D1928 for the preparation of plaques.

Stress-Cracking (Single-Point). While the NCTL test just described is the premier test for evaluating the stress-cracking behavior of HDPE geomembranes, it has the disadvantage of requiring many tests and being quite lengthy in developing the full curve shown in Figure 5.11a. To make the procedure into more of a quality control test, a single-point version is available; called a SP-NCTL test, it is outlined in the appendix to ASTM D5397.

Procedurally we use the same type of test specimens and load device, but select a specific value of stress, in this case 30% of yield stress. If the specimen does not fail within 200 hr, it signifies that the transition time for the full curve is at least 100 hr and would thus fulfill the specified value mentioned in the previous section. The SP-NCTL test is further described by Hsuan [21] and is recommended for use in HDPE geomembrane specifications.

5.1.4 Endurance Properties

Any phenomenon that causes polymeric chain scission or bond breaking within the polymer structure of a geomembrane must be considered as being detrimental to its long-term performance. There are a number of potential concerns in this regard. While each is site-specific, the general behavioral trend is to cause the polymer to become brittle in its stress-strain behavior over time. The mechanical properties to track in monitoring most types of long-term polymer degradation are: the decrease in elongation at failure, the increase in modulus of elasticity, the increase (then decrease) in stress at failure (i.e., strength), and the general loss of ductility. Obviously many of the physical and mechanical properties discussed in this section could be used to monitor the polymeric degradation process.

Ultraviolet. As described in Section 2.3.6 for geotextiles, ultraviolet degradation of geomembranes occurs. Short-wavelength energy for sunlight can penetrate the polymer structure causing chain scission and bond breaking. Figure 2.26 presents the wavelength spectrum of natural sunlight and indicates that the UV-B range is the most sensitive region. By virtue of the specific surface area differences between geotextiles and geomembranes, however, the relative degree of ultraviolet degradation of geomembranes is much less than with geotextiles. Thus the temporary covering of geomembranes before placement is not generally necessary from an ultraviolet degradation perspective. Note in Table 1.5 that all polymers used for geosynthetics have carbon black or pigments (2% to 25%) included to act as screening or blocking agents to minimize ultraviolet degradation. Furthermore, many have ultraviolet chemical stabilizers as part of their additive package. Thus the timely cover of geomembranes, written into many specifications, is often included to eliminate accidental or intentional damage of the geomembrane rather than because of ultraviolet degradation.

If we wish to estimate the UV-exposed lifetime of a particular geomembrane, a number of accelerated laboratory exposure tests are available, for example, the ultraviolet fluorescent tube method, ASTM G26, or the xenon arc method, ASTM D4355.

Since it is difficult to estimate the lifetime from these tests, we can also perform outdoor weathering tests such as ASTM D1435, D3334, or D5970. These tests will generally require long exposure times and are sometimes accelerated by means of rotating mirrors that intensify the sunlight's energy. The general area of estimating lifetime from any of these tests (laboratory or field-exposed), needs considerable research.

The current trend for permanently exposed geomembranes (e.g., floating covers in reservoirs or exposed side-slope liners of landfills and surface impoundments) is to use the manufacturers' warranties. Warranties of up to 20 years can be obtained in some situations for CSPE-R and HDPE geomembranes. Other commonly used geomembranes have lower exposed lifetimes and must be covered with soil or with a sacrificial material, e.g., a replaceable geotextile. For geomembranes that are covered in a timely manner, ultraviolet degradation is a nonissue and material warranties are irrelevant and unnecessary.

Radioactive Degradation. It is quite possible that radioactivity higher than 10^6 to 10^7 rads will cause polymer degradation via chain scission. Thus the containment of high-level radioactive waste would not be likely to use geomembranes. Low-level radioactive waste, however, is much lower in its activity and can quite possibly be disposed of in containment systems that include geomembranes. Kane and Widmayer [22] describe a number of HDPE lined radioactive waste containment scenarios including: landfill-liner systems, landfill-cover systems, below-ground vaults, uranium mill tailings disposal cells, and high-integrity containers. The last have been in use at low-level waste disposal facilities since 1979.

Other than internal government reports, however, there are very few references in the open literature on radioactive-degradation of geomembranes. Since many countries are being required to site low-level radioactive waste landfills, additional research should be undertaken.

Biological. There are a tremendous number of living organisms in the soil (see Figure 5.12). The area of biological hazards to geomembranes indeed is a vast one; we will only take a superficial glance here. Our discussion will focus only on areas where there is a perceived concern.

Animals. A major concern for soil-buried geomembranes is animals burrowing through them. Tests in Germany [24] have focused on mice and rats. Technically, only those substances harder than the burrower's tooth enamel or claws can avoid an attack if the animal is persistent enough. Thus geomembranes are indeed vulnerable to burrowing animals, but to what degree is largely unknown. Unfortunately, there are no established test procedures available; only intuitively can it be assumed that the stronger, harder, and thicker the geomembrane, the better its resistance to animal attack. The exceptions appear to be those geomembranes that use large amounts of plasticizer in their compounds. It is possible that the animals develop an addiction to plasticizers, although it is not known whether the addiction is of a chemical or tactile nature [24].

Fungi. Fungi include yeasts, molds, and mushrooms. They depend on organic matter for carbon, nitrogen, and other elements. Their numbers can be very large, as much as 10 to 20 million per gram of dry soil, and their population is constantly chang-

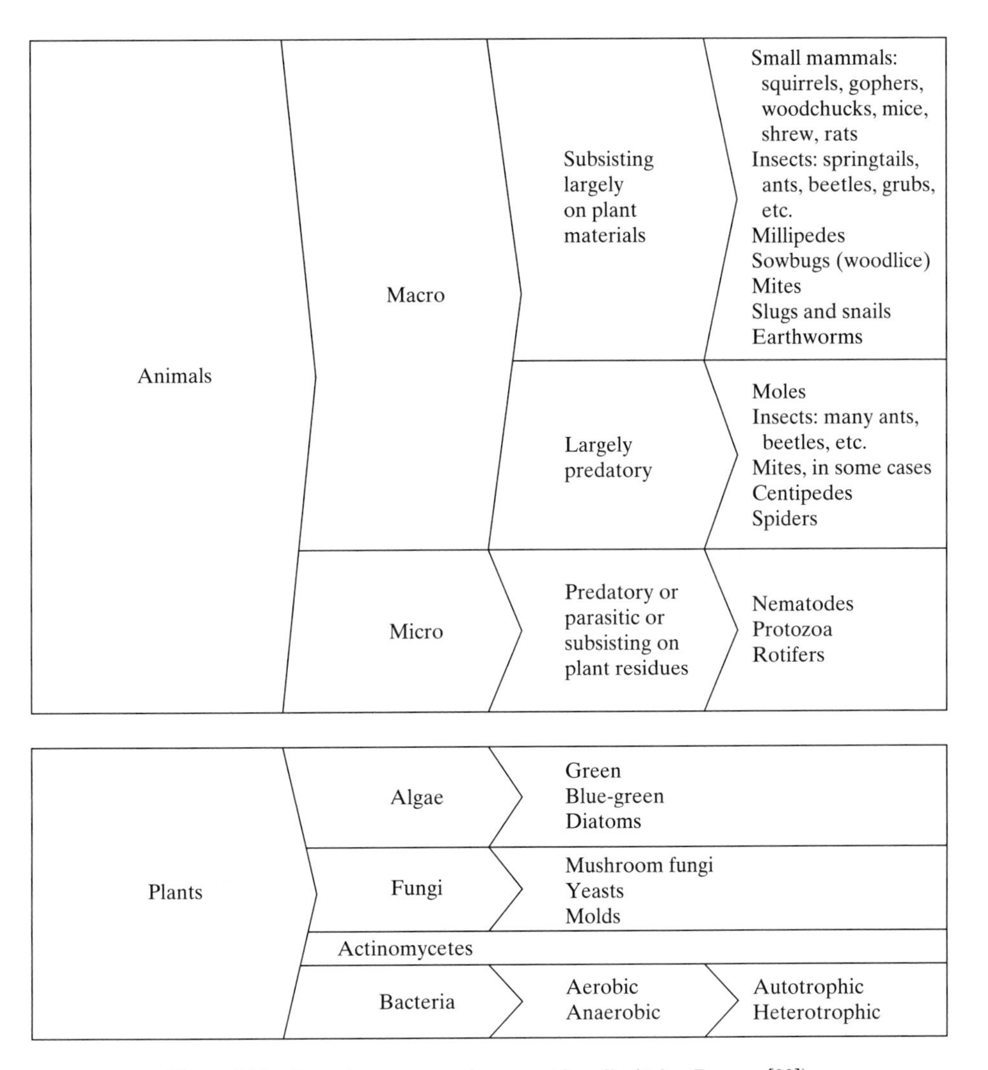

Figure 5.12 Organisms commonly present in soils. (After Dragun [23])

ing. Placing geomembranes in decomposing organic residue often causes concern about degradation. However, the high-molecular-weight polymers generally used for geomembranes seem very insensitive to such degradation. ASTM G21 deals with the resistance of plastics to fungi.

Of very real concern, however, is the possibility of fungal deposits clogging and binding flow through or within the geotextiles, geonets, and drainage geocomposites that are often associated with geomembranes.

Bacteria. Bacteria are single-cell organisms, among the simplest and smallest known forms of life. They rarely exceed 5 μm in length and are usually round, rodlike, or spiral in shape. Their numbers are enormous: more than 1 billion per gram of soil. They participate in all organic transformations, and thus the discussion on fungi could essentially be repeated here. The test method for evaluation of the resistance of plastics to bacteria is ASTM G22.

As with fungi, the greatest concern about bacteria regarding geomembranes is not polymeric degradation, but the fouling and clogging of the drainage systems often constructed in conjunction with the liner.

Chemical. The chemical resistance of a geomembrane vis-a-vis the substance(s) it is meant to contain is always important, and often it is the foremost aspect of the design process. For example, in domestic-waste or hazardous-waste containment, the pollutant will interface directly with the geomembrane. Thus the geomembrane's resistance must be assured for the life of the facility. This situation has long been recognized, and manufacturers and fabricators have evaluated many situations. This has resulted in various chemical resistance charts, such as Table 5.8, which lists generic chemicals against many common geomembranes on a relative or ranked basis. These charts and their tests are sometimes incorrectly called chemical compatibility charts or tests. To a chemist, compatibility is when two substances properly mix with one another; this is exactly the opposite of the trend we are considering in this section. Although such tables are generally reliable, there are many circumstances where geomembrane-specific testing is required.

- When the chemical is not a single-component material and possible synergistic effects are unknown
- When the composition of the resulting chemical is simply not known, as in landfill leachates before the facility is constructed

TABLE 5.8 GENERAL CHEMICAL RESISTANCE GUIDELINES OF SOME COMMONLY USED GEOMEMBRANES*

	Geomembrane Type							
	PE		PVC		CSPE		EPDM	
Chemical	38°C	70°C	38°C	70°C	38°C	70°C	38°C	70°C
General:								
Allphatic hydrocarbons	x	x						
Aromatic hydrocarbons	x	x						
Chlorinated solvents	x	x					x	
Oxygenated solvents	x	x					x	x
Crude petroleum solvents	x	x						
Alcohols	x	x	x	x			x	x
Acids:								
Organic	x	x	x	x	x		x	x
Inorganic	x	x	x	x	x		x	x
Heavy Metals	x	x	x	x	x		x	x
Salts	x	x	x	x	x		x	x

*x = generally good resistance.

Source: After Vandervoort [25].

- When the geomembrane is not a single-component material but is made from a blend of materials
- When the geomembrane is modified at the seams or is seamed with material that is different from that of the geomembrane sheets
- When the containment must function over a very long period and the leachate many change over time during the course of the service lifetime
- When untested circumstances, such as extreme heat or cold conditions, exist at the particular site
- When the chart or table does not list new types and/or formulations of geomembranes; for example, Table 5.8 does not list VLDPE, LLDPE, LDLPE, PP, and PP-R

Thus there exists a need for a specific test procedure. Chemical resistance tests on geomembranes for the specific conditions mentioned above require four important decisions to be made: the selection of the particular liquid to be used, the precise details of the coupon incubation (temperature, atmosphere, orientation, and removal), the manner and type of specimen testing, and the assessment of the results of the testing.

The selection of the liquid is surely site-specific. A large database is given in [4], but the range of ingredients is enormous. Clearly, there is no typical leachate. What liquid is selected is a matter of agreement among the various parties involved. Often it is a difficult decision, and when the situation is critical it is decided on a worst-case basis. Thus the most aggressive liquid chemicals envisioned (e.g., various organic solvents) in the highest possible concentrations are often used for the incubating liquid.

The coupon incubation can sometimes be done in open containers or tubs, but generally it is being done in closed containers of the type shown in Figure 5.13. Here the container is sealed with the liquid circulating and being constantly monitored as to its consistency and temperature. There is no available headspace in the container, so organic solvents cannot escape from the completely filled chamber. Individual coupons are removed at 30, 60, 90, and 120 days, according to ASTM D5322 and D5496, and then cut into test specimens for evaluation.

There are many types of test(s) used to quantify the geomembrane's performance after chemical incubation. The following are most common.

Physical property tests. These are for thickness, mass, length, width, and hardness and are the easiest and most straightforward to perform.

Mechanical property tests. The tensile test properties of strength at yield and/or break, elongation at yield and/or break, and modulus along with tear, puncture, and impact are the usual values measured. These are done as previously described.

Transport property tests. Perhaps the most sensitive tests to perform (and undoubtedly the most difficult) are tests for water- or solvent-vapor transmission through the incubated geomembranes.

If a specific procedure, such as ASTM D5747, is followed, the test methods to be used will be specified.

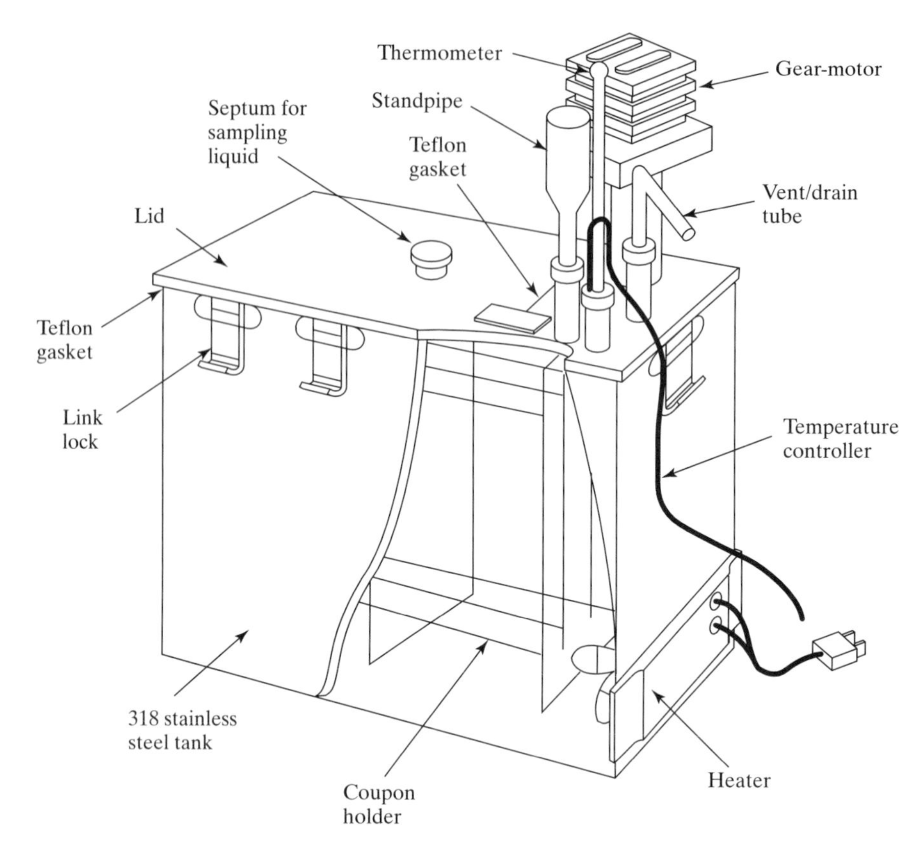

Figure 5.13 Schematic of incubation container for evaluating chemical resistance behavior of geomembranes. (After Metracon [4])

For the assessment of the test results, the response curves for the above-mentioned tests should be plotted as the percent change in the measured property from the original versus the duration of incubation. Figure 5.14 shows such response curves for HDPE in a municipal solid-waste leachate at 50°C. The curves presented are the type often seen, in that the changes in the physical properties are significantly less than the changes in the mechanical properties and no consistent trend is established, either a uniform increase or decrease. In this particular set of tests, the 23°C incubation data behaved similarly (see [26]). If the geomembrane is reactive to the leachate, we expect uniform behavioral changes and the changes at the higher temperature to be greater than those at the lower temperature. With no discernible trend to indicate a reaction, and hence degradation, of the geomembrane, it may be concluded that the scatter results from inherent variation in the materials and the test methods themselves. Furthermore, the property that changed the greatest amount, the modulus values in Figure 5.14b, is subject to the greatest amount of judgmental error of all the values presented. While there are no established rules on the allowable variation from the original test properties (see Table 5.9 for suggested values), it is clear that polyethylene will be more resistant to most organic solvents and aggressive chemicals than will the other common

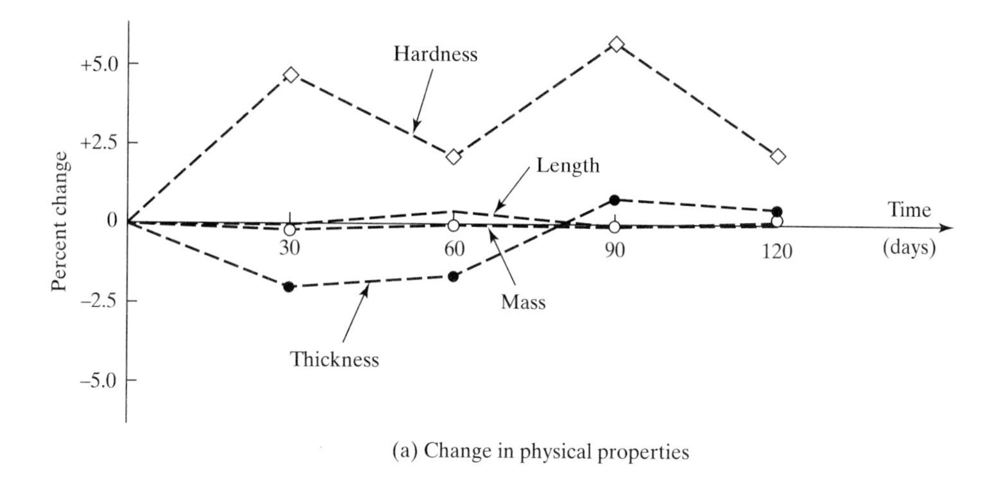

(a) Change in physical properties

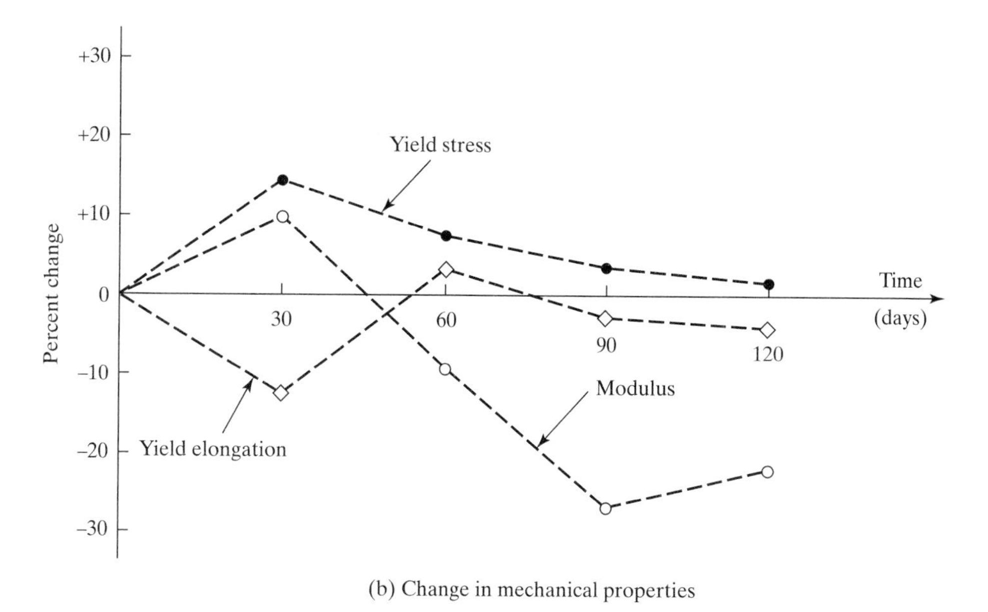

(b) Change in mechanical properties

Figure 5.14 Immersion behavior of HDPE samples to landfill leachate at 50°C up to 120 days. (After Tisinger [26])

geomembrane polymers. Furthermore, the higher the density, the better the chemical resistance. Thus high-density polyethylene (HDPE) geomembranes are widely used for many landfill liners.

Thermal. Various properties of geomembranes, as they are made from polymers, are sensitive to changes in temperature. Both warm and cold temperatures have their own unique effects.

Warm temperatures. Geomembrane materials exposed to heat can be subjected to changes in physical, mechanical, or chemical properties. The magnitude and duration

TABLE 5.9 SUGGESTED LIMITS OF DIFFERENT TEST VALUES FOR INCUBATED GEOMEMBRANES*

Thermoset and Thermoplastic Polymers (e.g., PVC, CSPE-R, EPDM-R)

Property	Resistant	Not Resistant
Permeation rate (g/m²/hr)	< 0.9	> 0.9
Change in weight (%)	< 10	> 10
Change in volume (%)	< 10	> 10
Change in tensile strength (%)	< 20	> 20
Change in elongation at break (%)	< 30	> 30
Change in 100% or 200% modulus (%)	< 30	> 30
Change in hardness (points)	< 10	> 10

Semicrystalline Polymers (e.g., HDPE)

	O'Toole [28]		Little [27]		Koerner	
Property	Resistant	Not Resistant	Resistant	Not Resistant	Resistant	Not Resistant
Permeation rate (g/m²-hr)	—	—	< 0.9	⩾ 0.9	< 0.9	⩾ 0.9
Change in weight (%)	< 0.5	> 1.0	< 3	⩾ 3	< 2	⩾ 2
Change in volume (%)	< 0.2	> 0.5	< 1	⩾ 1	< 1	⩾ 1
Change in yield strength (%)	< 10	> 20	< 20	⩾ 20	< 20	⩾ 20
Change in yield elongation (%)	—	—	< 20	⩾ 20	< 30	⩾ 30
Change in modulus (%)	—	—	—	—	< 30	⩾ 30
Change in tear strength (%)	—	—	—	—	< 20	⩾ 20
Change in puncture strength (%)	—	—	—	—	< 30	⩾ 30

*For low crystallinity polymers (e.g., VFPE, fPP) the properties should be the same as those for thermoset and thermoplastic polymers and the values for resistance should be slightly more restrictive. No specific recommendations are available.

Source: Thermoset and thermoplastic polymer data after Little [27].

of exposure determine the extent of this change. ASTM D794 covers the recommended procedure for determining permanent effects of heat on plastics—a tubular oven method (ASTM D1870), which consists of an oven with a coupon rack to allow for air circulation. Failure due to heat is defined as "a change in appearance, weight, dimensions, or other properties that alter the material to a degree that it is no longer acceptable for the service in question." This statement is of a qualitative nature and seems to suggest that comparison testing of candidate geomembranes for critical situations be done or that new samples be used for each incubation time and tensile tests be performed for comparison purposes.

Cold temperatures. Testing to evaluate the effect of cold on geomembranes follows along the same general lines as testing the effects of heat, but the behavior of the material is, of course, completely opposite. Cold will generally not degrade the geomembrane in any appreciable way, at least under the temperatures normally encountered. Furthermore, tests on a variety of geomembranes and different seam types have shown no adverse effects to cyclic cold temperatures for 500 cycles (see Hsuan et al. [29]). (The problems of ice leading to puncturing and tearing problems have been discussed earlier.) The only meaningful effect that cold has on the constructability of the system is that flexibility is decreased and seams are more difficult to make. The latter point is perhaps the most significant aspect of cold conditions. The proposed seaming method should be attempted at site-installation temperatures on test specimens on simulated subgrades and evaluated to see that a satisfactory seam strength will indeed result.

Thermal expansion and contraction. There are a number of procedures that can be used to determine the coefficient of thermal contraction or expansion of a material, for example, ASTM D2102 and D2259 for contraction, and D1042 and D1204 for expansion and dimensional changes. All of them subject the test specimen to a constant source of cold (or heat) and carefully measure the separation distance between two given initial locations. Some typical data are presented in Table 5.10. Example 5.3 uses this data in adding slack during the installation of a geomembrane.

Example 5.3

Calculate the amount of slack to be added during the installation of a HDPE liner for a surface impoundment anticipating a 40°C temperature change. Base the calculations on a 30 m distance and a coefficient of thermal linear expansion in Table 5.10.

Solution: Minimum slack: $11 \times 10^{-5}(40)(30)(1000) = 132$ mm

Maximum slack: $13 \times 10^{-5}(40)(30)(1000) = 156$ mm

It is easily seen that the calculated amounts in Example 5.3 are quite significant and that adding slack for temperature compensation is an important field-placement consideration.

Oxidation. Whenever a free radical is created (e.g., on a carbon atom in the polyethylene chain), oxygen can create progressive long-term degradation. Oxygen combines with the free radical to form a hydroperoxy radical, which is passed around

TABLE 5.10 COEFFICIENTS OF LINEAR THERMAL EXPANSION

Polymer Type	Thermal Linear Expansivity ($\times 10^{-5}$/°C)
Polyethylene	
high-density	11–13
medium-density	14–16
low-density	10–12
very-low-density	15–25
Polypropylene	5–9
Polyvinyl chloride	
unplasticized	5–10
35% plasticizer	7–25
Polyamide	
nylon 6	7–9
nylon 66	7–9
Polystyrene	3–7
Polyester	5–9

within the molecular structure. It eventually reacts with another polymer chain, creating a new free radical and causing chain scission. The reaction generally accelerates once it is triggered, as shown in the following equations.

$$R\cdot + O_2 \rightarrow ROO\cdot \tag{5.10}$$

$$ROO\cdot + RH \rightarrow ROOH + R\cdot \tag{5.11}$$

where

$R\cdot$ = free radical
$ROO\cdot$ = hydroperoxy free radical
RH = polymer chain
$ROOH$ = oxidized polymer chain

Antioxidation additives (antioxidants) are added to the compound to scavenge these free radicals in order to halt, or at least to interfere with, the process. These additives, or stabilizers, are specific to each type of resin (recall Table 1.5). This area is very sophisticated and quite advanced with all the resin manufacturers being involved in a meaningful and positive way. The specific antioxidants that are used are usually proprietary (see Hsuan et al. [30] for a review of the topic). The removal of oxygen from the geomembrane's surface, of course, eliminates the concern. Thus once placed and covered with waste or liquid, degradation by oxidation should be greatly retarded. Conversely, exposed geomembranes or those covered by nonsaturated soil will be proportionately more susceptible to the phenomenon. It should be recognized that oxidation of polymers will eventually, perhaps after hundreds of years, cause degradation even in the absence of other types of degradation phenomena.

There are two related test methods that are used to track the amount and/or depletion of antioxidants. They are called *oxidative induction time* (OIT) tests and are performed with a DSC device as described in Section 1.2.2.

Standard OIT. (ASTM D3895); The oxidation is conducted at 35 kPa and 200°C. This test appears to misrepresent antioxidant packages containing thiosynergists and/or hindered amines due to the relatively high test temperature.

High Pressure OIT. (ASTM D5885); The oxidation is conducted at 3500 kPa and 150°C. This test can be used for all types of antioxidant packages and is the preferred test.

By conducting a series of simulated incubations at elevated temperatures, OIT testing can be conducted on retrieved specimens to monitor the antioxidant depletion rate. As will be seen in Section 5.1.5, this leads to lifetime prediction via Arrhenius modeling.

Synergistic Effects. Each of the previous degradation phenomena has been described individually and separately. In practice, however, it is likely that two or more mechanisms are acting simultaneously. For example, a waste-containment geomembrane may have anaerobic leachate above it and a partially saturated leak-detection network containing oxygen below it. Thus chemical degradation from above and oxidation degradation from below will be acting on the liner. Additionally, elevated temperature from decomposing solid waste and the local stress situation may complicate the situation further. Evaluation of these various phenomena is the essence of geomembrane lifetime prediction.

5.1.5 Lifetime Prediction Techniques

Clearly, the long time frames involved in evaluating individual degradation mechanisms at field-related temperatures and stresses, compounded by synergistic effects, are not providing answers regarding geomembrane behavior fast enough for the decision-making practices of today. Thus accelerated testing, either by high stress, elevated temperatures and/or aggressive liquids, is very compelling. As will be seen, all of the methods use these ways of accelerating the test, time-temperature superposition being of fundamental importance.

Stress-Limit Testing. Focusing almost exclusively on HDPE pipe for natural gas transmission, many institutions are active in various aspects of plastic pipe research and development. Stress-limit testing in the plastic pipes has proceeded to a point where there are generally accepted testing methods and standards. ASTM D1598 describes a standard experimental procedure and ASTM D2837 gives guidance on the interpretation of the results of the D1598 test method.

In these experiments, long pieces of unnotched pipe are capped and placed in a constant temperature environment; a room temperature of 23°C is usually used. The pipes are placed under various internal pressures, which mobilize different values of hoop stress in the pipe walls. The pipes are monitored until failure occurs, which is indicated by a sudden loss of pressure. Then the values of hoop stress are plotted versus failure times on a log-log scale (see Figure 5.15). If the graph is reasonably linear, a straight line is extrapolated to the desired, or design, lifetime, which is often 10^5 hours or 11.4 years. The stress at this failure time multiplied by an appropriate factor is called the *hydrostatic design basis stress.* While this is of interest for pipelines, the stress state of geomembranes is essentially unknown and is extremely difficult to model. Thus the technique is not of direct value for geomembrane design. It leads, however, to the next method.

Rate Process Method (RPM) for Pipes. Research at the Gas Institute of The Netherlands (Wolters [31]), uses the method of pipe aging that is most prevalent in Europe. The experiments are again performed using long pieces of unnotched pipe that are capped, but now they are placed in various constant-temperature environments. So as to accelerate the process, elevated temperature baths up to 80°C are used. The pipes are internally pressurized so that hoop stress occurs in the pipe walls. The pipes are monitored until failure occurs, resulting in sudden loss of pressure. Two distinct types of failures are found: ductile and brittle. The failure times corresponding to each applied pressure are recorded. A response curve is presented by plotting hoop stress against failure time on a log-log scale.

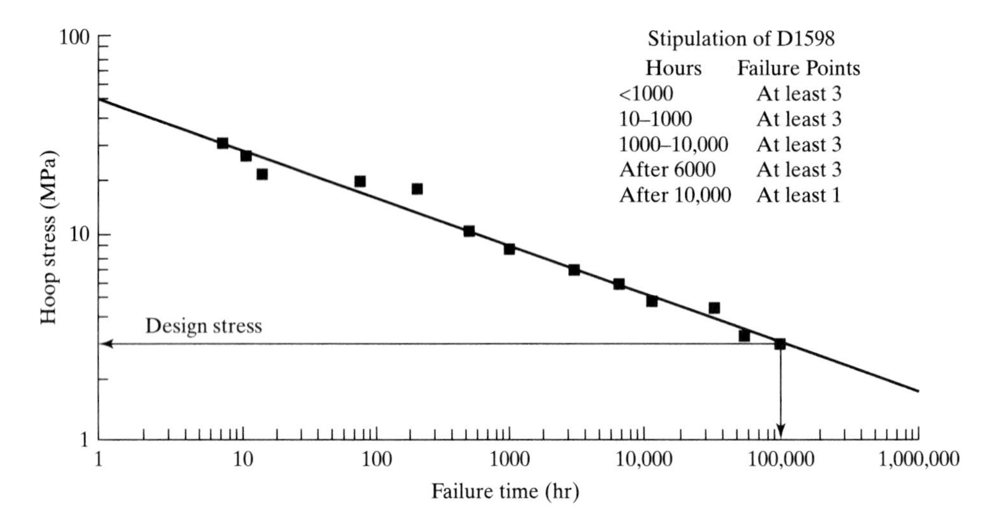

Figure 5.15 Schematic plot of time of failure versus pipe hoop stress for burst testing of unnotched MDPE pipe.

The rate process method (RPM) is then used to predict a failure curve at some temperature other than the high temperatures tested, that is, at a lower (field-related) temperature. This method is based on an absolute reaction rate theory and is explained in [32]. The relationships between the failure time and stress are expressed in the form of one of the following equations:

$$\log t_f = A_0T^{-1} + A_1T^{-1}\sigma \tag{5.12}$$

$$\log t_f = A_0 + A_1T^{-1} + A_2T^{-1}P \tag{5.13}$$

$$\log t_f = A_0 + A_1T^{-1} + A_2 \log P \tag{5.14}$$

$$\log t_f = A_0 + A_1T^{-1} + A_2T^{-1} \log P \tag{5.15}$$

where

t_f = time to failure,
T = temperature,
σ = tensile stress,
A_0 and A_1 = constants, and
P = internal pipe pressure proportional to the hoop stress in the pipe.

The application of RPM requires a minimum of two experimental failure curves at different elevated temperatures, generally above 40°C. The equation that yields the best correlation to these curves is then used in the prediction procedure for a response curve at a field-related temperature (e.g., 10 to 25°C). Two separate extrapolations are required, one for the ductile response and one for the brittle response. Three representative points are chosen on the ductile regions of the two experimental curves. One curve will be selected for two points, and the other, for the remaining point. These data are substituted into the chosen equation to obtain the prediction equation for the ductile response of the curve at the desired (lower) temperature. The process is now repeated for the predicted brittle response curve at the same desired temperature. The intersection of these two lines defines the transition time.

Figure 5.16 shows two experimental failure curves that were conducted at temperatures of 80°C and 60°C along with the predicted curves at 20°C. The intersection of the linear portions of the 20°C curve represents the anticipated time for transition in the HDPE pipe from a ductile to a brittle behavior of the material. For pipe design, however, the intersection of the desired service lifetime, say 50 years, with the brittle curve as the focal point. A factor of safety is then placed on this value, for example, note that it is lowered in Figure 5.16. This value of stress is used as a limiting value for the internal pressure in the pipe.

Rate Process Method (RPM) for Geomembranes. A similar RPM method to that just described for HDPE pipes can be applied to HDPE geomembranes. The major difference is the method of stressing the material. The geomembrane tests are performed using a notched constant load test (NCTL) (recall Section 5.1.3). Figure 5.17 shows typical experimental curves at 50°C and 40°C, which are very similar to the behavior of MDPE pipe. Here distinct ductile and brittle regions can be seen along with a clearly defined transition time.

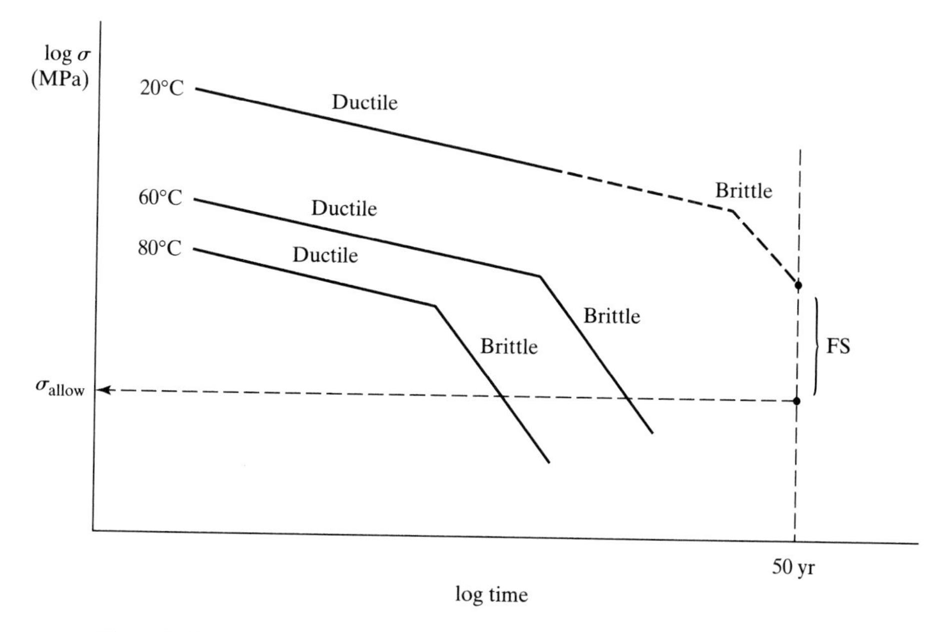

Figure 5.16 Burst test data for unnotched MDPE pipe in tap water. The intersection of the ductile portion of the 20°C line and 50 years has been lowered by the appropriate factor of safety.

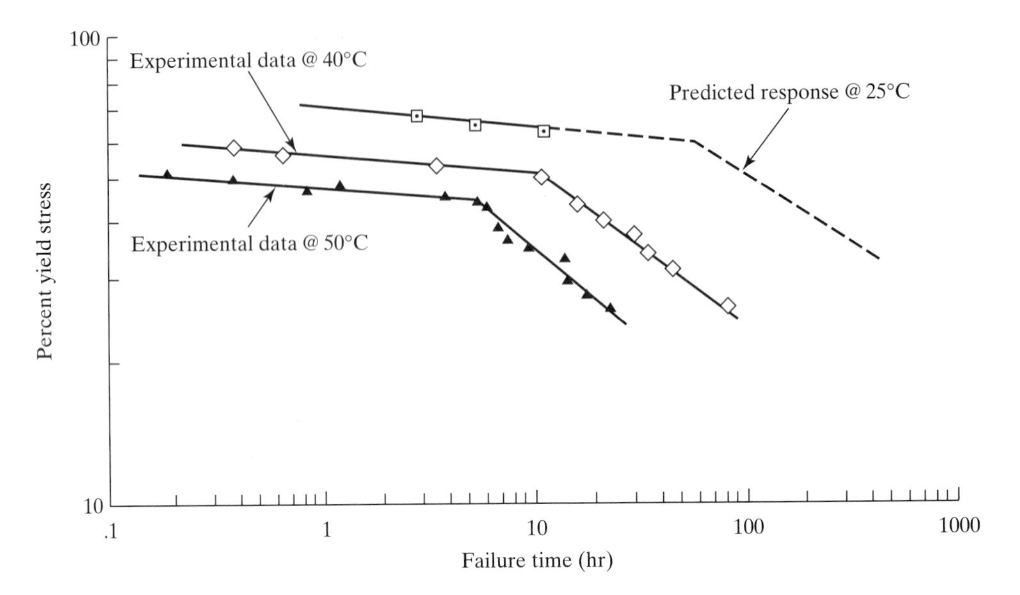

Figure 5.17 Notched constant load tests on HDPE geomembrane samples immersed in 10% Igepal/90% tap water solution.

In order to use these elevated temperature curves to obtain the transition time for a realistic temperature of a geomembrane beneath solid-waste or liquid impoundments (e.g., 20 to 25°C), only Eqs. (5.14) or (5.15) can be used, due to the data being plotted in a log-log scale. How the curves are shifted to the site-specific lower temperature is explained in detail in [32].

The process of predicting lifetime can now follow that outlined in Figure 5.16. A design lifetime can be assumed and a percent allowable stress can be determined. Conversely, a maximum allowable design stress can be given from which the unknown lifetime can be determined. Work is ongoing using this approach.

Hoechst Multiparameter Approach. The Hoechst research laboratory in Germany has been active in the long-term testing of polyethylene pipe since the 1950s. They have also applied their expertise and experience to the long-term behavior of HDPE geomembranes (Koch et al. [33]). The Hoechst long-term testing for a geomembrane sheet consists of the following procedure:

1. Perform a modified burst testing of pipe (of the same material as the geomembrane) with additional longitudinal stress to produce an isotropic biaxial stress state. Note that the site-specific liquid should be used.
2. Assume a given subsidence strain-versus-time profile.
3. Measure the stress relaxation curves in sheets that have been stressed biaxially at strain values encountered in field.
4. Use Steps 2 and 3 to predict the stress as a function of time.
5. See how these maximum stresses compare with the stress-lifetime curves determined in the normal constant stress-lifetime pipe measurements of Step 1. The

constant stress-lifetime curves are modified (as in the normal pipe testing) to accommodate the effects of various chemicals and for seams.

6. Use the variable stress curves to predict failure from the constant stress-lifetime curves if linear degradation is assumed.

This approach should certainly be considered seriously. One of the main impediments to its viability is that pipe may have different stress conditions than geomembranes. Furthermore, the residual stresses could be quite different. Other studies of a related nature can be found in Hessell and John [34] and Gaube et al. [35].

Elevated Temperature and Arrhenius Modeling. Using an experimental chamber, as shown in Figure 5.18, Mitchell and Spanner [36] have superimposed compressive stress, chemical exposure, elevated temperature, and long testing time into a single experimental device. At the Geosynthetic Research Institute, twenty of these columns have been constructed with five each at 85, 75, 65, and 55°C constant temperatures (see Figure 5.18). Each is under a normal stress of 260 kPa and is under 300 mm of liquid head on its upper surface. The subgrade sand is dry and vented to the atmosphere. The test coupons are 1.5 mm thick HDPE geomembranes.

Coupons are removed periodically and evaluated for changes in numerous physical, mechanical, and chemical test properties. The anticipated behavior is shown in Figure 5.19a; however, at this time, only the A and B stages have been quantified.

For stage A, the antioxidant depletion time, the HDPE geomembrane selected results in approximately 200 years (Hsuan and Koerner [37]). On the basis of exhumed HDPE milk containers at the bottom of a landfill, the measured induction time (stage B) is from 20 to 30 years. Thus in a buried environment there is a time span of approximately 220 years with essentially no engineering-property degradation. This leads directly to stage C. Deciding on a maximum property change to establish stage C (e.g., a 50% reduction or *half-life* at each temperature) allows the plotting of another curve (see Figure 5.19b). This graph is inverse temperature versus reaction rate (actually the inverse time from Figure 5.19a) and is called the *Arrhenius curve*. The slope of the line is the activation energy divided by the gas constant.

We can now extrapolate graphically to a lower site-specific temperature, as shown by the dashed line on Figure 5.19b, or extend the curve analytically. Examples 5.4 and 5.5 illustrate how this is accomplished, using literature values for the activation energy (see [38] for additional details). The essential equation for the extrapolation is

$$\frac{r_{T\text{-test}}}{r_{T\text{-site}}} = e^{-\frac{E_{act}}{R}\left[\frac{1}{T\text{-test}} - \frac{1}{T\text{-site}}\right]} \tag{5.16}$$

where

E_{act}/R = slope of Arrhenius plot,
T-test = incubated (high) temperature, and
T-site = site-specific (lower) temperature.

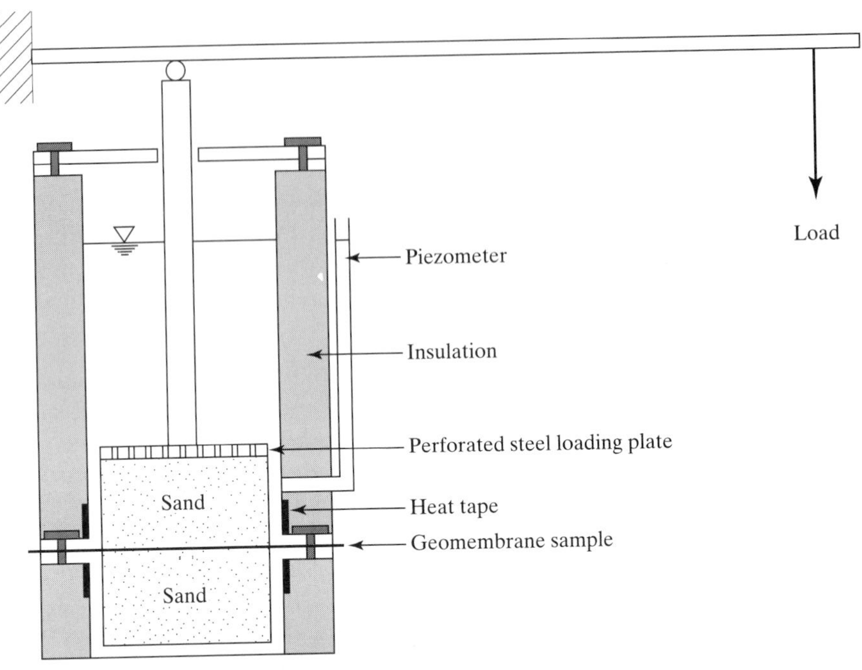

Figure 5.18 Incubation unit for accelerated aging and photograph of a number of similar units used at the Geosynthetic Research Institute.

Example 5.4

Using experimental data from Martin and Gardner [39] for the half-life of the tensile strength of a PBT plastic, the E_{act}/R value is $-12{,}800$ K. Determine the estimated life, extrapolating from the 93°C actual incubation temperature (which took 300 hours to complete) to a site-specific temperature of 20°C.

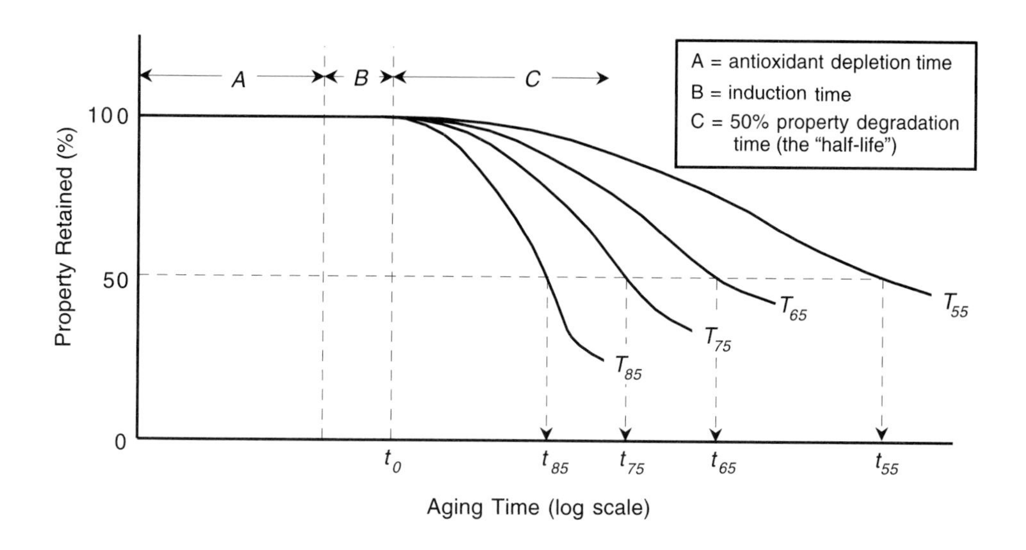

(a) Incubated property behavior

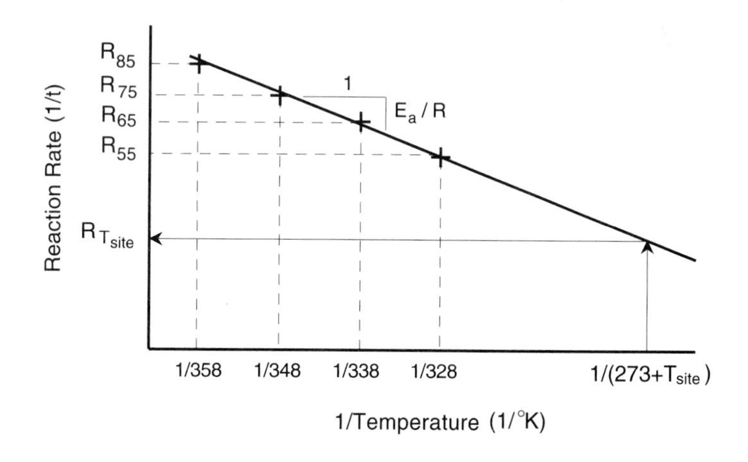

(b) Arrhenius plot for half-life property

Figure 5.19 Arrhenius modeling for lifetime prediction via elevated temperature aging.

Solution: After converting from centigrade to kelvin

$$\frac{r_{93°C}}{r_{20°C}} = e^{-\frac{E_{act}}{R}\left[\frac{1}{93+273} - \frac{1}{20+273}\right]}$$

$$= e^{-12,800\left[\frac{1}{366} - \frac{1}{293}\right]}$$

$$= 6083$$

If the 93°C reaction takes 300 hours to complete, the comparable 20°C reaction will take

$$r_{20°C} = 6083(300)$$

$$= 1,825,000 \text{ hr}$$

$$= 208 \text{ yr}$$

Thus the predicted time for this particular polymer to reach 50% of its original strength at 20°C is approximately 200 years, its predicted lifetime for stage C.

Example 5.5

Using Underwriters Laboratory Standard [40] data for HDPE cable shielding, the E_{act}/R value is −14,000 K. This comes from the half-life of impact strength tests. One of the high-temperature tests was at 196°C and it took 1000 hours to obtain these data. What is the life expectancy of this material at 90°C?

Solution. After converting from centigrade to kelvin

$$\frac{r_{196°C}}{r_{90°C}} = e^{-\frac{E_{act}}{R}\left[\frac{1}{196+273} - \frac{1}{90+273}\right]}$$

$$= e^{-14,000\left[\frac{1}{469} - \frac{1}{363}\right]}$$

$$= 6104$$

If the 196°C reaction takes 1000 hours to complete, the comparable 90°C reaction will take

$$r_{90°C} = 6104(1000)$$

$$= 6,104,000 \text{ hr}$$

$$= 697 \text{ yr}$$

Thus the predicted time for this particular polymer to reach 50% of its original impact strength at 90°C is approximately 700 years, its predicted lifetime for stage C.

5.1.6 Summary

This relatively long section on properties and test methods has, hopefully, served to illustrate the wealth of test methods available for the characterization and design considerations of geomembranes. Many of the established tests and standardized test methods have come by way of the plastics and rubber industries for nongeotechnical-related uses. This is fortunate, for it gives a base or reference plane to work from. However, for many some variation is required before they can be used in below-ground construction. Still others demand completely new tests and test methods. In standards-setting institutes the world over there is an awareness of the problems and vibrant activity to develop such test methods and procedures. Until they are available, however, we must act on intuition and develop methods that model the required design information as closely as possible. Many of the tests and information presented in this section are done in that light. It should also be obvious that the complexity of the tests have progressed from the quite simple thickness test on smooth geomembranes to the very complex degradation tests. Indeed, a very wide range of test methods are available.

Finally, a rather lengthy discussion of durability and aging gives insight into the potential service lifetime of geomembranes. In a buried environment, the lifetimes promise to be very long—for example, with stage A of Figure 5.19a being 200 years, stage B being 20 years and stage C being hundreds of years, the HDPE geomembrane being evaluated promises to far outlast other engineering materials in comparable situations. In my experience with geosynthetics over the past 20 years, my original conclusion was that geosynthetics were easy to place but wouldn't last very long; this has shifted dramatically to where I sense that geosynthetics have extremely long service lifetimes, but I have very real concerns as to the proper installation of geosynthetics. Clearly, the geosynthetic material must survive its initial placement if these long predicted lifetimes are to be achieved.

5.2 SURVIVABILITY REQUIREMENTS

For any of the design methods presented in this chapter to function properly, it is necessary that the geomembrane survive the packaging, transportation, handling, and installation demands that are placed on it. This aspect of design cannot be taken lightly or assumed simply to take care of itself. Yet there is a decided problem in formulating a generalized survivability design for every application, since each situation is unique. Some of the major variables affecting a given situation are the following:

- Storage at the manufacturing facility
- Handling at the manufacturing facility
- Transportation from the factory to the construction site
- Offloading at the site
- Storage conditions at the site
- Temperature extremes at the site
- Subgrade conditions at the site
- Deployment at the approximate location
- Movement into the final seaming location
- Treatment at the site during seaming

- Exposure at the site after seaming
- Placement of the cover material or soil backfill on the completed geomembrane

Note that each of these topics is largely out of the hands of the designer. Only by rigid specifications, a complimentary construction quality assurance (CQA) document, competent *full-time* inspection by CQA personnel, and cooperation of the installation contractor can the geomembrane survive to the point of beginning to function as designed (see [10]). Remembering that each situation is surely different, some empirical guidelines are necessary and some properties and their minimum values are offered in Table 5.11.

Geomembranes are most often vulnerable to tear, puncture, and impact while being stored, transported, handled, and installed. Such events often come about accidentally, due to vandalism, or due to poor workmanship. Typical situations are the dropping of tools on the geomembrane, the driving of autos or pickup trucks on the unprotected liner, high winds getting beneath the geomembrane during placement, the awkwardness of moving large sheets of the geomembrane into position, and so on. The geomembrane property most involved with resistance or susceptibility to tear, puncture, and impact damage is thickness. At least a linear, and sometimes an exponential, increase in resistance to the above actions is seen as thickness increases. For this reason many agencies require a minimum thickness under any circumstance. For example, the U.S. Bureau of Reclamation requires a minimum thickness of 0.50 mm for canal liners, while the U.S. Environmental Protection Agency requires a minimum thickness for geomembranes for solid-waste liners of 0.75 mm. For similar applications in Germany, the use of a 2.0 mm thick geomembrane is required. Rather than use a single regulated value for all conditions, however, the minimum thickness and its subsequent properties should be related to site-specific conditions. Using a concept similar to the placement of geotextiles, Table 5.11 shows four required survivability levels. Note that these val-

TABLE 5.11 RECOMMENDED MINIMUM PROPERTIES FOR GENERAL GEOMEMBRANE INSTALLATION SURVIVABILITY

	Required Degree of Installation Survivability*			
Property and Test Method	Low	Medium	High	Very High
Thickness (D1593) (mm)	0.63	0.75	0.88	1.00
Tensile D882 (25 mm strip) (kN/m)	7.0	9.0	11	13
Tear (D1004 Die C) (N)	33	45	67	90
Puncture (D4833) (N)	110	140	170	200
Impact (D3998 mod.) (J)	10	12	15	20

**Low* refers to careful hand placement on a very uniform well-graded subgrade with light loads of a static nature, typical of vapor barriers beneath building floor slabs.

Medium refers to hand or machine placement on a machine-graded subgrade with medium loads, typical of canal liners.

High refers to hand or machine placement on a machine-graded subgrade of poor texture with high loads, typical of landfill liners and covers.

Very high refers to hand or machine placement on machine-graded subgrade of very poor texture with high loads, typical of liners for heap leach pads and reservoir covers.

ues are not to be used in place of design but as a check on design, to see that installation can be properly assured.

5.3 LIQUID CONTAINMENT (POND) LINERS

The U.S. EPA estimates that there are over 200,000 surface impoundments storing hazardous and nonhazardous liquids in the United States, the vast majority of which are unlined. This total does not include potable water and nonregulated reservoirs and impoundments. Worldwide, the number is unknown, but it is obviously enormous. Certainly there is a major need for and use of geomembranes in the area of liquid containment of surface impoundments. In fact, the name *geomembrane* is actually one that supersedes the name *pond liner,* reflecting the original use of the polymeric materials to which this section is devoted. In addition to the containment of the above types of liquids, the agriculture industry has a pressing need to store water, and hence both the U.S. Department of Agriculture and the U.S. Bureau of Reclamation were involved in early research into synthetic pond liners. While thermoset (rubber) liners may have been used prior to the 1930s, the use of polyvinyl chloride sheeting for liners began in the 1940s. When covered with a minimum of 300 mm of soil, these PVC liners apparently performed well. Uncovered, however, there was a tendency for progressive brittleness and cracking. Other thermoplastic liner materials, less susceptible to this problem, followed in rapid succession (e.g., CPE and CSPE). Today all of the geomembrane materials listed in Table 5.1 are used for the containment of all types of liquids.

5.3.1 Geometric Considerations

Before selecting the geomembrane type the desired liquid volume to be contained versus the available land area must be considered. Such calculations are geometric by nature and result in a required depth on the basis of assumed side-slope angles. For a square or rectangular section with uniform side slopes, the general equation for volume is

$$V = HLW - SH^2L - SH^2W + 2S^2H^3 \tag{5.17}$$

where

V = volume of reservoir,
H = height (i.e., depth of reservoir at the center),
W = width at ground surface,
L = length at the ground surface, and
S = slope ratio (horizontal to vertical).

Eq. (5.17) can be solved in a variety of ways and various design curves can be generated. Such design curves are given in Figure 5.20 for a side-slope angle of 18.4°, which is 3 to 1 (horizontal to vertical), written as 3(H) to 1(V), and a square configuration. Example 5.6 uses this concept.

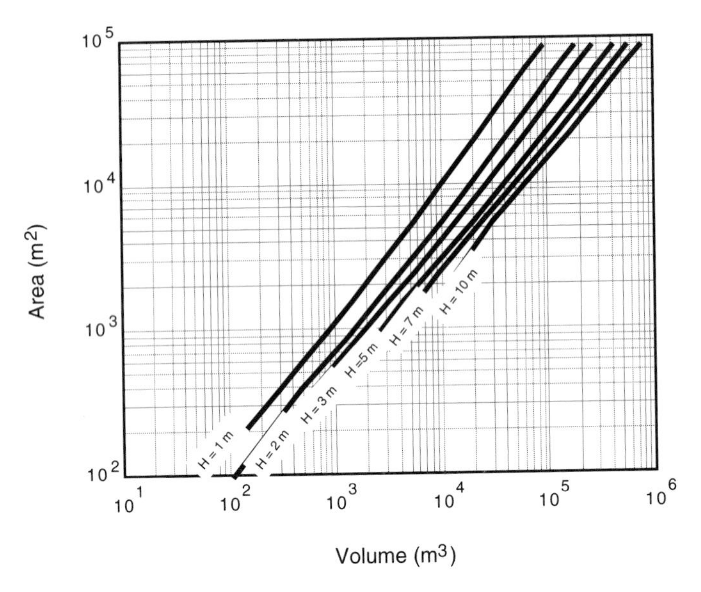

Figure 5.20 Volume-versus-area design chart for liquid-containment ponds with side slopes of 3(H) to 1(V).

Example 5.6

A square area 125 × 125 m is available for constructing a reservoir for storage of 60,000,000 liters of industrial process water. At estimated side slopes of 3(H) to 1(V), what is the required height (i.e., depth of the pit)?

Solution:

$$\text{Volume} = 60{,}000{,}000 \text{ liters}$$

$$= 60{,}000 \text{ m}^3$$

Using Eq. (5.17),

$$60{,}000 = H(125)(125) - (3)H^2(125) - (3)H^2(125) + 2(9)H^3$$

$$60{,}000 = 15{,}625H - 750H^2 + 18H^3$$

$$H = 4.83 \text{ m}$$

Note that the above result agrees with the curves given in Figure 5.20; however, the lined impoundment must be somewhat deeper to allow for freeboard against overfilling, wave action, and so on.

From Example 5.6, it can be seen that to contain large volumes of liquids we will require massive land areas and/or deep containment pits. If such a land area is not available, the required depths often lead to additional problems, such as: interception of the water table, difficulty in stabilizing the bottom and sides of the excavation, interception of unsuitable soil conditions, interception of bedrock, high excavation costs, and excavated soil disposal problems.

These problems are compounded in areas with soft cohesive soils or granular cohesionless soils in which very flat side slopes or geosynthetically reinforced slopes are required, as described in Section 3.2.7.

5.3.2 Typical Cross Sections

Upon first consideration, digging a hole, putting a liner in it, and then filling it with the liquid to be contained is simplicity itself. Indeed, for an ideal site, with proper liner material, proper construction techniques, and maintenance during its service lifetime, it is a straightforward task. Note the sketch of Figure 5.21a, which shows such a liner on a prepared soil subgrade anchored in trenches around the perimeter of the site. Unfortunately, such ideal conditions and situations are seldom encountered.

The first complication has to do with atmospheric exposure and damage to the geomembrane. To shield the liner from ozone, ultraviolet light, temperature extremes, ice damage, wind stresses, accidental damage, and vandalism, a soil cover of at least 300

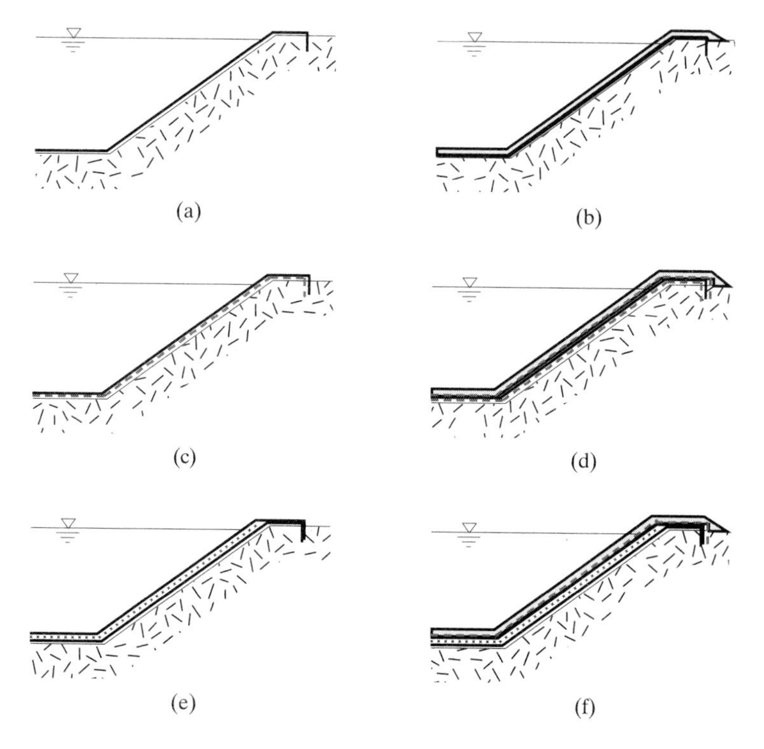

Figure 5.21 Various types of geomembrane liner systems for liquid-containment: (a) single unprotected liner system; (b) liner with soil covering; (c) liner with geotextile underliner; (d) liner with geotextile underliner and overliner with soil covering; (e) double-liner with geonet leak detection between liners; (f) double-liner with geonet leak detection between liners and soil covering, which may or may not contain a geotextile or geogrid as veneer reinforcement.

mm thickness is usually required. Vandalism is particularly troublesome in areas of noncontrolled site access. Figure 5.21b shows that the soil cover extends up out of the pit and over the liner anchorage areas. This soil cover is particularly troublesome on the side slopes, where gravitational sloughing of the cover soil compounded by liquid drawdown is often a problem. Friction between the liner and cover soil must be evaluated and the appropriate design procedures followed. This aspect is discussed later in this section.

The use of a geotextile beneath the geomembrane (Figures 5.21c,d), placed directly on the prepared soil subgrade before liner placement, is considered proper design for a number of reasons [41].

- It provides a clean working area for making field seams.
- It provides added puncture resistance when loads (either during construction or from the cover soil) are applied.
- It can add frictional resistance to the geomembrane-to-soil interface, thereby preventing excessive stresses on the geomembrane as it enters the anchor trench, or allowing for steepened side slopes.
- If properly selected, the geotextile will allow for the lateral and upward escape of subsurface water and gases that rise up beneath the geomembrane during its service life (see Figure 5.22). Upward-moving water is caused by high groundwater

Figure 5.22 Subsurface-generated gases pushing up a geomembrane, creating very high stresses on liner and seams.

levels and flooding in nearby water courses. Upward-moving gases are caused by biodegradation of organic material in the subsurface soils and from rising water-table levels that expel the air from the soil voids. In such cases nonwoven needle-punched geotextiles, geonets, or drainage geocomposites with sufficient transmissivity to handle the estimated flows are required; Example 5.7 illustrates the procedure.

Example 5.7

Consider a 7 m deep geomembrane-lined pond that will create a barrier to rising gases from the biodegradation of the organic silt layer, as shown in the sketch below. The width of the impoundment is 200 m, with the grade rising up from the center as shown. A high estimate of gas generation is 0.001 m^3/day − m^2 at a pressure of 7.0 kPa. The proposed underliner to be used is an 550 g/m^2 nonwoven needle-punched geotextile. What is the factor of safety of this geotextile's transmissivity for this set of conditions?

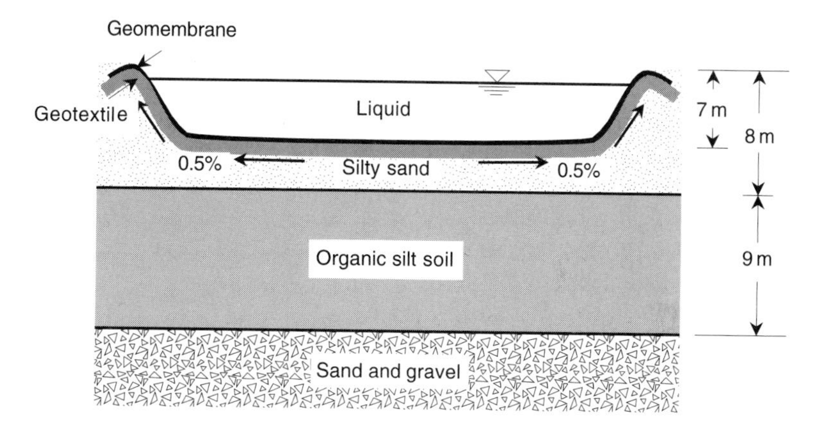

Solution:

(a) The flow rate is

$$q = 0.001\left(\frac{200}{2} \times 1\right)$$

$$q = 0.10 \text{ m}^3/\text{day} \qquad \text{at the high elevation}$$

$$= \frac{0.10}{(24)(60)}$$

$$= 6.94 \times 10^{-5} \text{ m}^3/\text{min}$$

(b) The critical slope is along the bottom of the reservoir.

$$0.5\% \text{ slope} = 0.005$$

(c) Although this problem is probably not one of laminar-flow conditions, use Darcy's formula since it is a conservative approach and air-flow transmissivity data are available in the literature.

$$q = kiA$$

$$= ki(t \times W)$$

$$kt = \theta_{\text{reqd}} = \frac{q}{i \times W}$$

$$\theta_{\text{reqd}} = \frac{6.94 \times 10^{-5}}{(0.005)(1.0)}$$

$$= 0.0139 \text{ m}^3/\text{min-m}$$

(d) The actual transmissivity of geotextiles of the type proposed is given in Figure 2.18a. Here we obtain a θ_{allow} of 0.192 m^3/min-m at a stress of 7 (9.81) $\cong$ 70 kPa and an air pressure of 7 kPa.

(e) The actual factor of safety is

$$\text{FS} = \frac{\theta_{\text{allow}}}{\theta_{\text{reqd}}}$$

$$= \frac{0.192}{0.0139}$$

$$= 14 \qquad \text{acceptable}$$

For significantly larger air flows, the solution calls for a geonet or drainage geocomposite of the type described in Chapters 4 and 8.

Figure 5.21d illustrates the ultimate in protection of the geomembrane where it is sandwiched between two geotextiles. The underlying geotextile serves the same purposes just discussed. The overlying one is useful to maintain stability of the side-slope cover soil and to prevent stones in the cover soil from puncturing the geomembrane. The latter point is a consideration only if properly graded cover soil is not available and poor-quality soil must be used. This type of composite design is also becoming customary in the secondary containment of underground storage tanks for groundwater protection.

Figure 5.21e illustrates a double-lined surface impoundment with a geonet leak-detection system between them. The first use of a geonet in 1984 was of the type shown in Figure 5.21e, which is still in use for storing hazardous liquid at a private site in Virginia. The variation shown in Figure 5.21f has soil covering of the upper (primary) geomembrane for its protection and to shield it from ultraviolet degradation. Oftentimes, such coverings need reinforcement by means of a geotextile or geogrid inclusion. This topic is covered in Section 3.2.7 under veneer reinforcement.

Not shown in Figure 5.21 are various alternatives using geosynthetic clay liners and/or compacted clay liners. The variations are enormous and are clearly site-specific vis-a-vis the liquid to be contained.

It should also be mentioned that federal and state regulations could very well prescribe a cross section that is considered to be minimum technology guidance (MTG). Generally, a double-liner cross section with leak detection is necessary if the stored liquid is hazardous and a single composite liner (a geomembrane and compacted clay or a geosynthetic clay liner) if the liquid is nonhazardous. The variations in regulations are significant among the different regulatory agencies.

5.3.3 Geomembrane Material Selection

Concerning the selection of the type of material to be used in the manufacture of the liner itself, chemical resistance to the contained liquid is of utmost importance. The entire design process becomes ludicrous in the absence of such chemical resistance. Furthermore, this resistance must be considered for the entire service life of the particular installation.

For potable water storage, service lifetimes of approximately 20 years must be considered. This is similar to general water storage for agricultural use. Of the liner types noted in Table 5.1, PVC has been widely used, due in large part to its ease of installation compared with that of other materials. As noted earlier, it must be covered with soil to prevent excessive degradation, and this tends to offset somewhat its lower installation cost when compared to other liner materials that are not soil-covered (e.g., CSPE-R and HDPE). Indeed, any of the material types listed in Table 5.1 are candidates for potable or storage water containment, due to the relative inertness of this type of liquid.

For the storage of liquids containing known acids, bases, heavy metals, salts, or commonly stored chemicals, the chemical resistance chart in Table 5.8 should be consulted. Note that most manufacturers have similar charts and that these too should be reviewed. One consideration in this regard that is often overlooked is the resistance of the seams to the liquid being contained. This is particularly important for adhesive-bonded seams and less so for solvent, thermal, and extrusion seams. (Seams are described in Section 5.10.)

For the storage of liquids that are combinations of industrial process effluents, the most aggressive of the individual liquids to polymeric materials should be used for the selection process. This assumes that there are no synergistic effects occurring within the different liquids that may be placed in the reservoir. For the majority of these situations, chemical resistance charts are available (as in Table 5.8) for proper material selection.

For the storage of liquids that are unidentifiable or of an unknown variety (e.g., from industrial processes that are in the design stage and not yet on-stream) or for leachates of a very heterogeneous nature, extreme conservatism must be used. Because of its relative inertness with chemicals, HDPE will often be the material of choice. Seaming is done by thermal fusion or extrusion welding, with no foreign material additives used. It is prudent, however, and oftentimes required to incubate coupons in the laboratory using the synthesized liquid to see if reactions are occurring. This procedure is described in Section 5.1.4.

5.3.4 Thickness Considerations

There are several empirical relationships between geomembrane thickness and the depth of the contained liquid. While geomembrane thickness is indeed related to the pressures exerted upon it, such empirical guides are completely unfounded and certainly not in keeping with the mechanics-based design that will follow. This design is based on the subsurface deformation that the liner might experience during its service lifetime. Such deformations can come about in a number of ways: by randomly located differential settlement of subgrade soils, by settlement of backfilled zones beneath the geomembrane (e.g., in pipe trenches), by localized settlements around soft areas beneath the geomembrane, by seismic disturbances that may modify the subgrade conditions, and by any kind of anomalous conditions that place the liner in tension.

The basic model we will work from requires a deformation-mobilized tensile force to occur and is shown in Figure 5.23. Here deformation is induced by one of the settlement mechanisms mentioned above, thereby defining the value of β. This induces

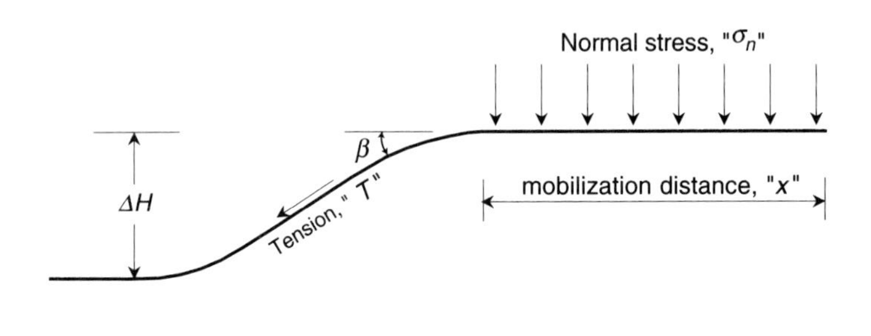

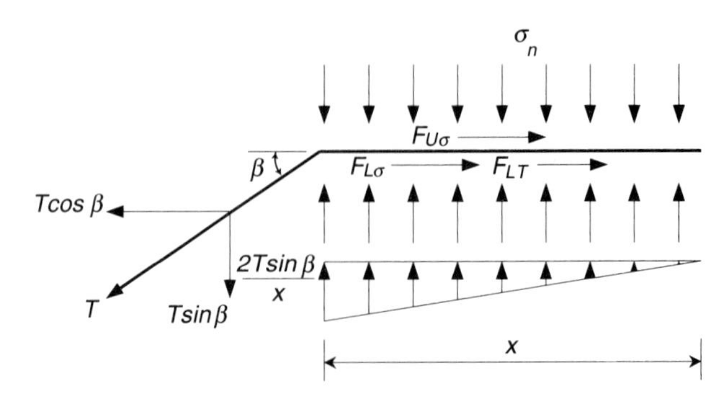

Figure 5.23 Design model and related forces used to calculate geomembrane thickness.

tension in the geomembrane, which is equal to the allowable stress times the unknown thickness.

$$T = \sigma_{allow} t$$

where

T = tension mobilized in the geomembrane,
σ_{allow} = allowable geomembrane stress, and
t = thickness of the geomembrane.

The tension is now resolved into its horizontal component, which must be resisted by the shear forces shown in Figure 5.23. Note that the vertical component (which is assumed to be dissipated along the mobilization distance x) must be added to the normal stress imposed by the overlying liquid (and soil, if applicable). Thus,

$$\Sigma F_x = 0$$

$$T \cos \beta = F_{U\sigma} + F_{L\sigma} + F_{LT}$$

$$T \cos \beta = \sigma_n \tan \delta_U(x) + \sigma_n \tan \delta_L(x) + 0.5\left(\frac{2T \sin \beta}{x}\right)(x) \tan \delta_L$$

$$T = \frac{\sigma_n x(\tan \delta_U + \tan \delta_L)}{\cos \beta - \sin \beta \tan \delta_L}$$

But $T = \sigma_{allow} t$, so

$$t = \frac{\sigma_n x(\tan \delta_U + \tan \delta_L)}{\sigma_{allow}(\cos \beta - \sin \beta \tan \delta_L)} \tag{5.18}$$

where

β = settlement angle mobilizing the geomembrane tension,
$F_{U\sigma}$ = shear force above geomembrane due to liquid pressure (note that if liquid is being contained, the shear stress is zero; this is essentially the same even if a thin soil cover is above the geomembrane),
$F_{L\sigma}$ = shear force below geomembrane due to the overlying liquid pressure (and soil if applicable),
F_{LT} = shear force below geomembrane due to the vertical component of T,
σ_n = applied stress from reservoir contents,
δ = angle of shearing resistance between geomembrane and the adjacent material (i.e., soil or geotextile), and
x = distance of mobilized geomembrane deformation.

and where the general ranges of the variables above are

σ_n = 20 to 100 kPa ($\cong$ 2 to 10 m of liquid),
β = 0 to 45°,

x = 15 to 100 mm (determined from the laboratory test described in Section 5.1.3, see Figure 5.10,

σ_{allow} = 6000 to 30,000 kPa (determined from laboratory tests described in Section 5.1.3), see Figure 5.3,

δ_U = 0° for liquid containment and 10 to 40° for landfill containment (determined from laboratory tests described in Section 5.1.3), and

δ_L = 10 to 40° (determined from laboratory tests described in Section 5.1.3).

Example 5.8 shows the procedure to be used for a specific case.

Example 5.8

Determine the thickness of a geomembrane to be used in the containment of a 7 m deep water reservoir where the settlement over a backfilled collector pipe could result in a 45° settlement angle. The geomembrane is a textured fPP of 8,000 kPa allowable stress. There is only a thin soil cover over the geomembrane (i.e., $\delta_U \cong 0$) and a nonwoven needle-punched geotextile is to be placed beneath it ($\delta_L = 25°$) The estimated mobilized distance for liner deformation is 50 mm.

Solution: Using Eq. (5.18), we obtain the required thickness.

$$t = \frac{\sigma_n x(\tan \delta_U + \tan \delta_L)}{\sigma_{\text{allow}}(\cos \beta - \sin \beta \tan \delta_L)}$$

$$= \frac{(68.7)(0.050)[\tan 0 + \tan 25]}{(8000)[\cos 45 - (\sin 45)(\tan 25)]}$$

$$= \frac{1.60}{3018}$$

$$= 0.00053 \text{ m}$$

$$t = 0.53 \text{ mm}$$

Figure 5.24 carries the procedure further into a set of design curves for the conditions cited therein. Note that the result is in terms of a thickness coefficient, which when multiplied by the height of liquid in meters (water is assumed) gives the required geomembrane thickness in mm. Other types of design charts can also be generated but all are based on the premise that a subsidence occurs beneath the geomembrane, giving rise to the analysis. If no subsidence occurs, this type of analysis is not appropriate and thickness is based on installation and/or regulatory minimum values.

Note that adequate geomembrane thickness cannot be addressed solely on the basis of the analysis above. Other factors, such as construction equipment driving on the geomembrane during liner installation or reservoir cleaning operations during its service lifetime can impose severe stresses on the liner. Since all mechanical properties of the liner increase with increasing thickness, there is usually a minimum thickness that is recommended irrespective of design calculations. As shown in Table 5.11, this value is 0.63 mm. Some regulatory agencies, however, have their own minimum thickness standards that, if greater than the above value, would take precedence in those specific instances.

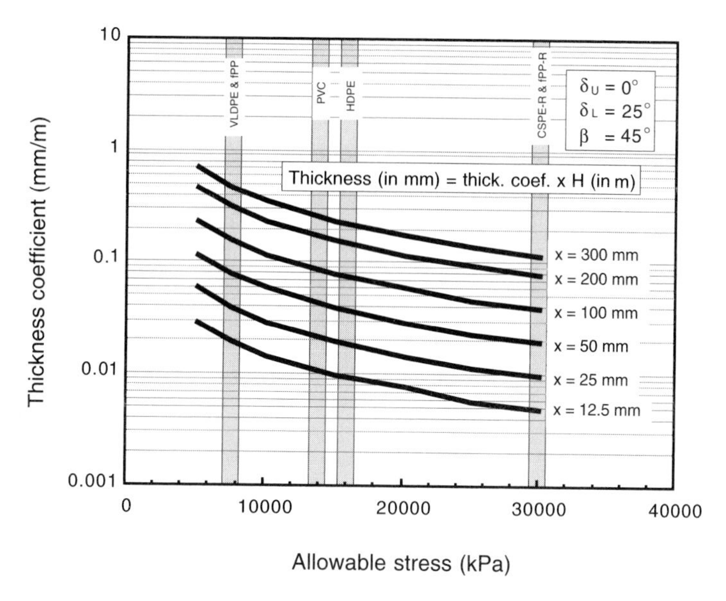

Figure 5.24 Design curves for geomembrane thickness based on unit height of water.

5.3.5 Side-Slope Considerations

The design of side slopes for liquid retention ponds falls within the scope of geotechnical engineering with some minor modifications. The analyses involved can be as simple or as detailed as the particular situation warrants. For the purposes of this particular section, separate aspects of the problem considering both the general slope stability (with and without a geomembrane) and the stability of cover soils placed over the geomembrane will be treated.

General (Global) Slope Stability. In considering the general slope stability of the soil mass beneath a geomembrane, a circular failure arc is generally assumed to be the mode of likely failure. In keeping with this assumption, several classes of failures can occur. As presented in Figure 5.25, these failures are a base failure, toe failures (either beyond or within the anchor trench), and a slope failure.

The usual design procedure involves the slope height, the soil properties, and the soil shear strength parameters, and has as its unknown the slope angle β. Furthermore, a total stress analysis is customary, since the entire site is generally above the watertable and in an equilibrium state. Proceeding on the basis of an assumed center of rotation and radius, the procedure for the circles labeled (a) and (b) in Figure 5.25 is to subdivide the mass involved into vertical slices and to take moment equilibrium about the center of rotation. This yields the following factor of safety equation.

$$\text{FS} = \sum_{i=1}^{n} \frac{[(W_i \cos \theta_i) \tan \phi + \Delta l_i c]R}{(W_i \sin \theta_i)R} \tag{5.19}$$

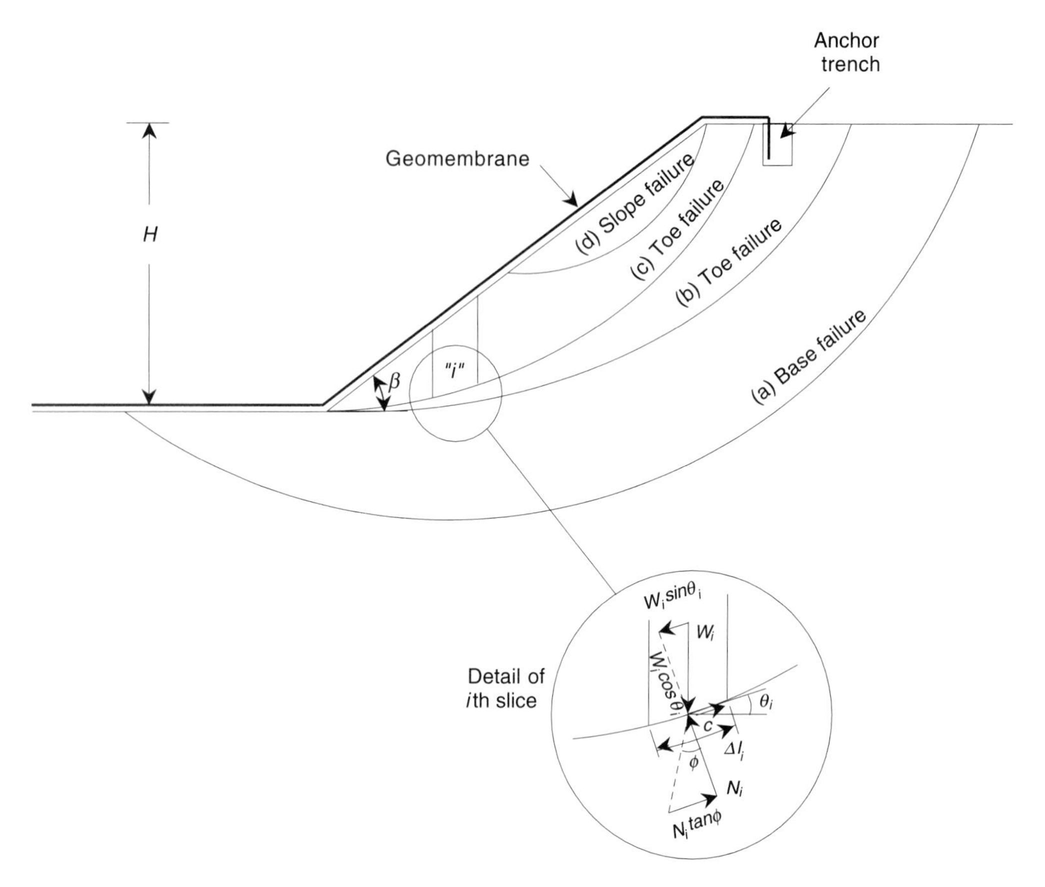

Figure 5.25 Various types of geomembrane-covered soil slope stability failures.

where

W_i = weight of the ith slice,
θ_i = angle the ith slice makes with the horizontal,
Δl_i = arc length of the ith slice,
ϕ = angle of shearing resistance of the soil,
c = cohesion of the soil,
R = radius of the failure circle, and
n = the number of slices utilized.

Note that in Eq. (5.19) R factors out from both numerator and denominator and cancels out. This is not the case if other terms, such as seismic forces and live loads, are involved. Once the factor of safety has been calculated for the arbitrarily selected center and radius of the assumed arc, a search is conducted to determine what particular arc gives the lowest factor of safety. When found, this value is assessed under the criteria that $FS < 1.0$ is unacceptable, $FS \cong 1.0$ is incipient failure, and $FS > 1.0$ is stable, with higher values being even safer. For example, a $FS \geqslant 1.5$ is often the targeted value. If

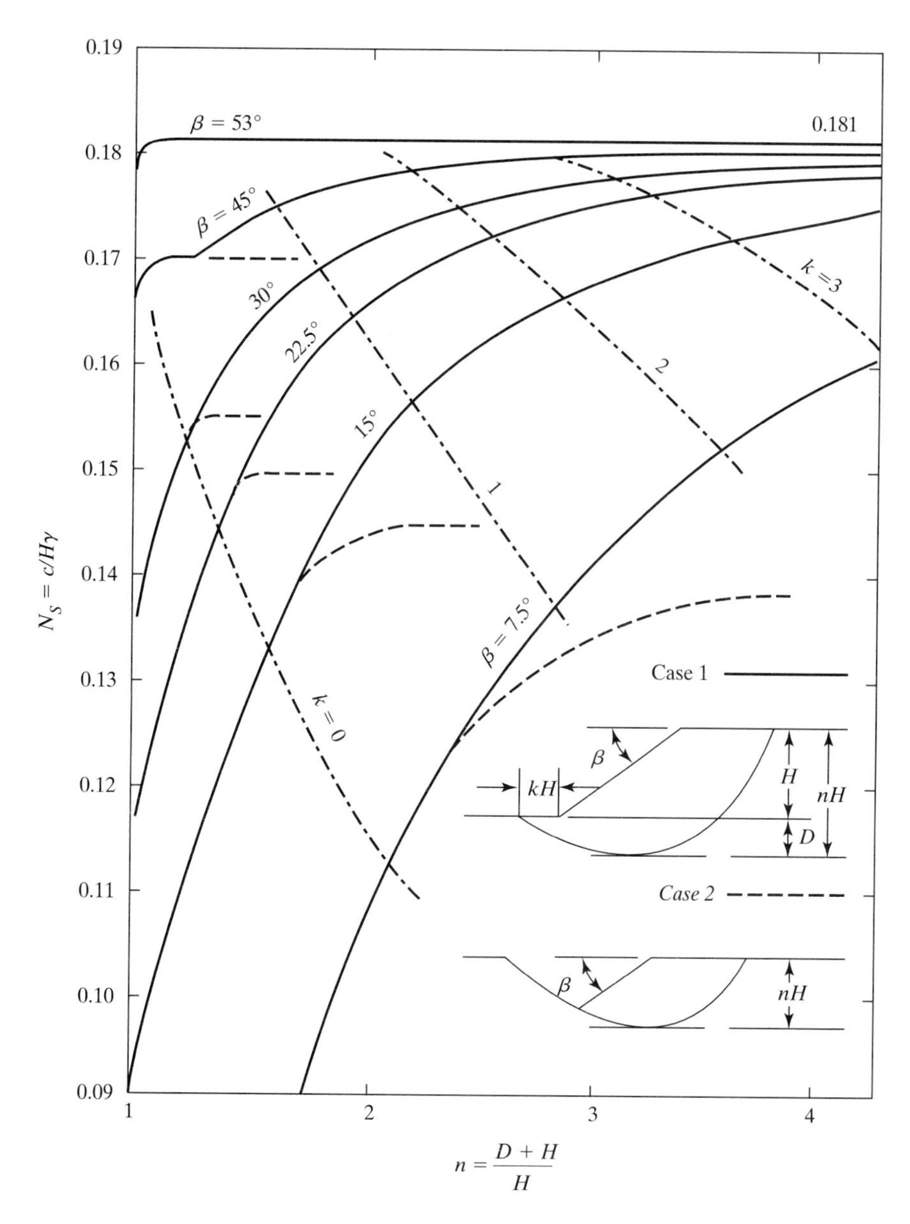

Figure 5.26 Stability curves for soils whose strength can be approximated by undrained conditions. (After Taylor [42])

the FS value is too low, the slope angle β is decreased or horizontal benches are set in the slope until the FS value is adequate.

The process described above is very time-consuming, and numerous design charts have been developed over the years to obtain approximate solutions. Figures

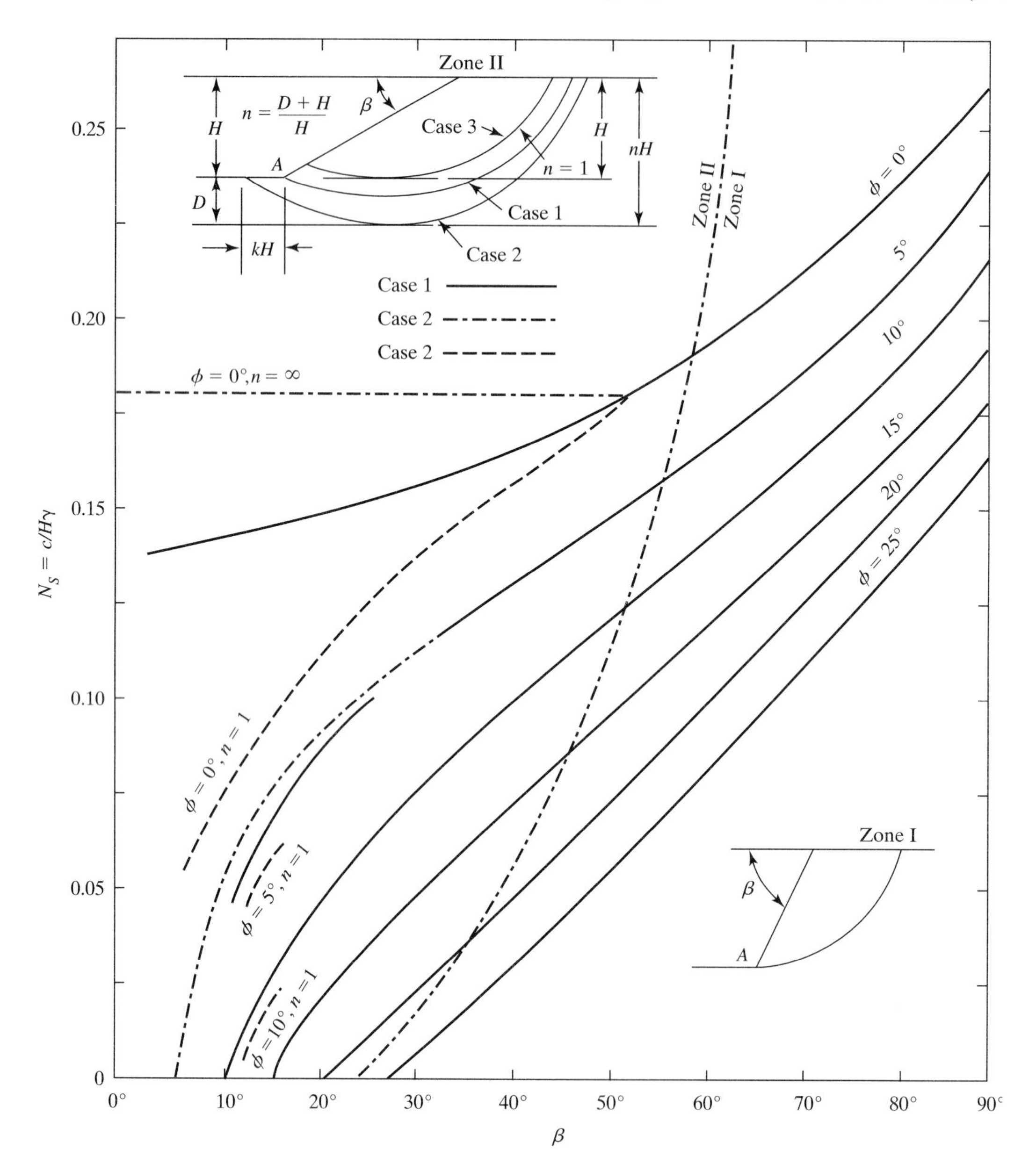

Figure 5.27 Stability curves for soils whose strength comes about from cohesion and friction. (After Taylor [42])

5.26 and 5.27 are of this type. In the use of these curves, the factor of safety is calculated as follows.

$$\text{FS} = \frac{c}{N_s \gamma H} \tag{5.20}$$

where

FS = factor of safety,
c = undrained strength of the soil (kN/m^2),
γ = total unit weight (kN/m^3),
H = vertical height of slope (m), and
N_s = stability number taken from Figures 5.26 or 5.27.

The curves can also be used to back-calculate the height or slope angle on the basis of a given factor of safety by solving for the appropriate term. These curves, however, should be used with a considerable degree of caution and for situations that are not considered to be critical (e.g., shallow ponds, where a failure would not cause loss of life or serious property damage). Example 5.9 demonstrates the use of one of the charts.

Example 5.9

A 4 m deep geomembrane-lined reservoir is to be placed in an area where the soil has an undrained shear strength of 14 kN/m^2 and a total unit weight of 15 kN/m^3. There is a hard layer of sand and gravel 3 m beneath the bottom of the proposed reservoir; thus a base failure to this depth is envisioned. What slope angle will be required on the basis of FS = 1.5?

Solution: Working from Eq. (5.20) for FS and Figure 5.26,

$$\text{FS} = \frac{c}{N_s \gamma H}$$

$$1.5 = \frac{14}{N_s(15)(4)}$$

$$N_s = 0.155$$

and

$$n = \frac{D + H}{H}$$

$$= \frac{3 + 4}{4}$$

$$= 1.75$$

Using Figure 5.26, the required slope angle is 20°, which is approximately 2.7(H) to 1.0(V).

Regarding the failure circles labeled (c) and (d) in Figure 5.25, a slight variation can be considered. If the liner is covered with a soil layer and if the liner is in tension as it comes up the slope and enters the anchor trench (as it usually is), a tensile force can be included in the analysis. The factor-of-safety equation then becomes

$$\text{FS} = \sum_{i=1}^{n} \frac{[(W_i \cos \theta_i) \tan \phi + \Delta l_i c]R + Ta}{(W_i \sin \theta_i)R} \tag{5.21}$$

where (in addition to the terms previously defined)

$T = \sigma_{\text{allow}} t$, in which
σ_{allow} = allowable strength of the geomembrane, and
t = thickness of the geomembrane; and
a = the moment arm equal to R as its maximum.

If a geotextile underliner and/or overliner is used in conjunction with the geomembrane, it is to be included in a similar manner. Whatever the case, the net effect is to increase the factor of safety for a given circle location and radius. If we omit the term entirely, the error is on the conservative side. Regarding such a tensile force at the bottom of the failure arc, it is felt to be rarely of benefit in arresting a potential failure. Certainly, if the liner is not covered with soil, there is no normal stress to mobilize resistance and, even if covered, the net effect is minimal.

Due to their tedious and repetitious nature, slope-stability calculations are well-suited for computer adaptation. Many such programs exist and can readily be adapted for inclusion of the tensile forces just described.

Stability of Cover Soil—Infinite Slope. In general, geomembranes should be covered. The reasons for this are numerous and include the following: protection against oxidation, protection against ultraviolet degradation, minimization of elevated temperature that increases degradation, protection against ice puncture in cold climates, protection against puncturing or tearing by sharp objects, elimination of wind uplift, protection against accidental damage, and protection against intentional damage. The usual covering is a relatively thin layer of soil, which has the unfortunate tendency when placed on side slopes to slide gravitationally downward. For example, when compacted clay liners are placed on side slopes, they sometimes fail by slumping. The accumulated soil gathers at the toe of the slope and part of the denuded subsoil is thereby exposed. This same situation is even more apt to occur for cover soils placed on geomembranes, which are invariably lower in frictional resistance than the subgrade soils from which the slope itself is formed. The design method to prevent this unraveling of cover soils from occurring is straightforward and based on limit equilibrium conditions. Figure 5.28a shows a segment of subsoil, geomembrane, and cover soil having a uniform thickness. For this set of conditions a force summation equation along the slope angle β can be written, resulting in a factor of safety against failure.

$$\text{FS} = \frac{\text{resisting forces}}{\text{driving forces}}$$

$$= \frac{F}{W \sin \beta}$$

$$= \frac{N \tan \delta}{W \sin \beta}$$

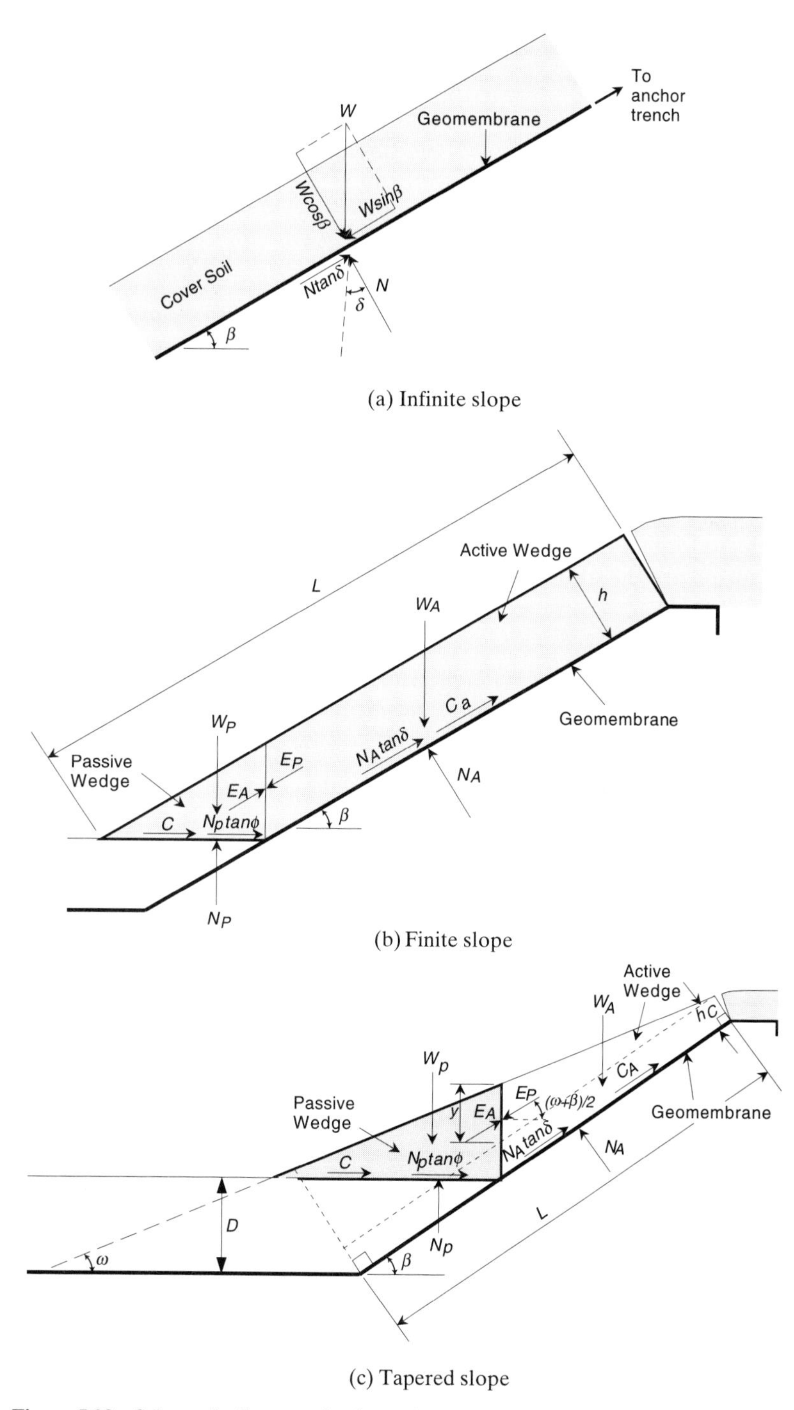

Figure 5.28 Schematic diagrams for forces involved with cover soils on geomembrane-lined slopes: (a) infinite slope; (b) finite slope with uniform cover soil thickness; (c) finite slope with tapered cover soil thickness.

$$= \frac{W \cos \beta \tan \delta}{W \sin \beta}$$

$$FS = \frac{\tan \delta}{\tan \beta} \tag{5.22}$$

where

β = slope angle, and
δ = friction angle between the geomembrane and its cover soil.

(Note that failure will occur at the cover-soil interface because the geomembrane is held in place at the upper ground surface by the horizontal runout and anchor trench.) The design process for this situation is usually one where the slope angle β is known and a factor of safety is selected, leaving the friction angle between the geomembrane and the cover soil unknown. Since the type of geomembrane is probably already selected, it is seen that the quality of the cover soil or the possibility of geomembrane texturing are the ultimate variables.

Example 5.10

What type of soil is required for covering an CSPE-R liner on a $3(H)$ to $1(V)$ slope using a (relatively low) factor of safety of 1.3?

Solution: A 3-to-1 slope is a slope angle of 18.4°. Eq. (5.22) gives the required friction angle between the geomembrane and the cover soil.

$$FS = \frac{\tan \delta}{\tan \beta}$$

$$1.3 = \frac{\tan \delta}{\tan 18.4}$$

$$\tan \delta = 0.433$$

$$\delta = 23.4°$$

Going to the data in Table 5.7a for CSPE-R, we see that both concrete sand and decomposed mica schist have δ-values that are acceptable and either one would be adequate in this case. Also notice that the Ottawa sand would not be suitable.

The critical issue in Example 5.10 is not the analysis (which unrealistically assumes an infinite slope) but the critical importance of determining an accurate value of interface shear strength, that is, the δ-value. The direct shear test discussed in Section 5.1.3 must be performed on the site-specific materials and under proper simulation conditions.

Stability of Cover Soil—Finite Slope. The slopes of geomembrane-lined facilities are not infinite in their length and are usually limited to about 30 m. If they are longer, horizontal benches are used to break up the continuous length into smaller increments. For finite-length slopes, there exists a small passive wedge at the toe and the analysis can be generalized to include other variations of the problem. Figure 5.28b illustrates the free-body diagram, which is illustrated in Chapter 3 in order to set up the veneer-reinforcement application using geogrids. The analysis will not be repeated, only the equations necessary to illustrate the technique and its extension into a tapered cover soil scenario will be presented. The necessary equations, repeated here from Section 3.2.7, are illustrated in Example 5.11.

$$W_A = \gamma h^2\left(\frac{L}{h} - \frac{1}{\sin\beta} - \frac{\tan\beta}{2}\right) \tag{3.14}$$

$$N_A = W_A \cos\beta \tag{3.15}$$

$$W_p = \frac{\gamma h^2}{\sin 2\beta} \tag{3.17}$$

The resulting FS value is then obtained from the following equation.

$$\text{FS} = \frac{-b + \sqrt{b^2 - 4ac}}{2a} \tag{3.22}$$

where

W_A = total weight of the active wedge,
W_P = total weight of the passive wedge,
N_A = effective force normal to the failure plane of the active wedge,
N_P = effective force normal to the failure plane of the passive wedge,
γ = unit weight of the cover soil,
h = thickness of the cover soil,
L = length of slope measured along the geomembrane,
β = soil slope angle beneath the geomembrane,
ϕ = friction angle of the cover soil,
δ = interface friction angle between cover soil and geomembrane,
C_a = adhesive force between cover soil of the active wedge and the geomembrane,
c_a = adhesion between cover soil of the active wedge and the geomembrane,
$a = (W_A - N_A \cos\beta)\cos\beta$,
$b = -[(W_A - N_A\cos\beta)\sin\beta\tan\phi + (N_A\tan\delta + C_a)\sin\beta\cos\beta + \sin\beta(C + W_P\tan\phi)]$, and
$c = (N_A\tan\delta + C_a)\sin^2\beta\tan\phi$.

Example 5.11

Given a 30 m long slope with a uniformly thick cover soil of 300 mm at a unit weight of 18 kN/m^3, the soil has a friction angle of 30° and zero cohesion (i.e., it is a sand). The cover soil is on a geomembrane, as shown in Figure 5.28b. Direct shear testing has resulted in an interface friction angle between the cover soil and geomembrane of 22° and zero adhesion. What is the FS value at a slope angle of 3(H) to 1(V), that is, 18.4°?

Solution: Using the formulae just presented:

$$W_A = \gamma h^2\left(\frac{L}{h} - \frac{1}{\sin\beta} - \frac{\tan\beta}{2}\right)$$

$$= (18)(0.30)^2\left[\frac{30}{0.30} - \frac{1}{\sin 18.4} - \frac{\tan 18.4}{2}\right]$$

$$= 157 \text{ kN/m}$$

$$N_A = W_A \cos\beta$$

$$= (157)(\cos 18.4)$$

$$= 149 \text{ kN/m}$$

$$W_p = \frac{\gamma h^2}{\sin 2\beta}$$

$$= \frac{(18)(0.30)^2}{\sin 36.8}$$

$$= 2.70 \text{ kN/m}$$

$$a = (W_A - N_A\cos\beta)\cos\beta$$

$$= (157 - 149\cos 18.4)\cos 18.4$$

$$= 14.8 \text{ kN/m}$$

$$b = -[(W_A - N_A\cos\beta)\sin\beta\tan\phi + (N_A\tan\delta + C_a)\sin\beta\cos\beta + \sin\beta(C + W_p\tan\phi)]$$

$$= -[(157 - 149\cos 18.4)\sin 18.4\tan 30$$

$$+ (149\tan 22 + 0)\sin 18.4\cos 18.4$$

$$+ \sin 18.4(0 + 2.70\tan 30)]$$

$$= -21.4 \text{ kN/m}$$

$$c = (N_A\tan\delta + C_a)\sin^2\beta\tan\phi$$

$$= (149\tan 22 + 0)\sin^2 18.4\tan 30$$

$$= 3.46 \text{ kN/m}$$

$$\text{FS} = \frac{-b + \sqrt{b^2 - 4ac}}{2a}$$

$$= \frac{21.4 + \sqrt{(-21.4)^2 - 4(14.8)(3.46)}}{2(14.8)}$$

$$\text{FS} = 1.25$$

Example 5.11 has been extended into a set of design curves in Figure 5.29a. As expected the FS value decreases for increasing slope angles and increases for increasing soil-to-geomembrane friction angles. Since the curves have been generated for the same conditions as the example problem, the resulting FS value is easily verified.

It should be noted that there can be a number of destabilizing forces which can reduce the FS value. These are equipment forces when moving down the slope, seepage forces within the cover soil, and seismic forces. They are addressed in the manner paralleling this section in Koerner and Soong [43].

Note that there are two commonly used methods to increase the FS value. One is the inclusion of a geogrid or geotextile as veneer reinforcement. This is covered in Section 3.2.7. Less common due to the additional toe space required, but certainly possible, is to use a tapered cover soil thickness.

Stability of Cover Soil Tapered Thickness. From Example 5.11 it is easily seen that cover soils over geomembranes can become unstable quite easily, even under static conditions. To alleviate this situation, it is possible to taper the cover soil, placing it so that it is thicker at the bottom and gradually thinner going toward the top (see Figure 5.28c). Note that the slope of the top of the cover soil is at an angle ω, where $\omega < \beta$. The formulation follows that of the previous section.

Considering the active wedge,

$$W_A = \gamma\left[\left(L - \frac{D}{\sin\beta} - h_c \tan\beta\right)\left(\frac{y\cos\beta}{2} + h_c\right) + \frac{h_c^2 \tan\beta}{2}\right]$$

$$N_A = W_A \cos\beta$$

$$C_a = c_a\left(L - \frac{D}{\sin\beta}\right)$$

Balancing the forces in the vertical direction, the following formulation results.

$$E_A \sin\left(\frac{\omega + \beta}{2}\right) = W_A - N_A\cos\beta - \frac{N_A \tan\delta + C_a}{\text{FS}}(\sin\beta)$$

Hence the interwedge force acting on the active wedge is

$$E_A = \frac{(\text{FS})(W_A - N_A\cos\beta) - (N_A\tan\delta + C_a)\sin\beta}{\sin\left(\frac{\omega+\beta}{2}\right)(\text{FS})}$$

The passive wedge can be considered in a similar manner.

$$W_P = \frac{\gamma}{2\tan\omega}\left[\left(L - \frac{D}{\sin\beta} - h_c\tan\beta\right)(\sin\beta - \cos\beta\tan\omega) + \frac{h_c}{\cos\beta}\right]^2$$

$$N_p = W_P + E_P \sin\left(\frac{\omega+\beta}{2}\right)$$

$$C = \frac{\gamma}{\tan\omega}\left[\left(L - \frac{D}{\sin\beta} - h_c\tan\beta\right)(\sin\beta - \cos\beta\tan\omega) + \frac{h_c}{\cos\beta}\right]$$

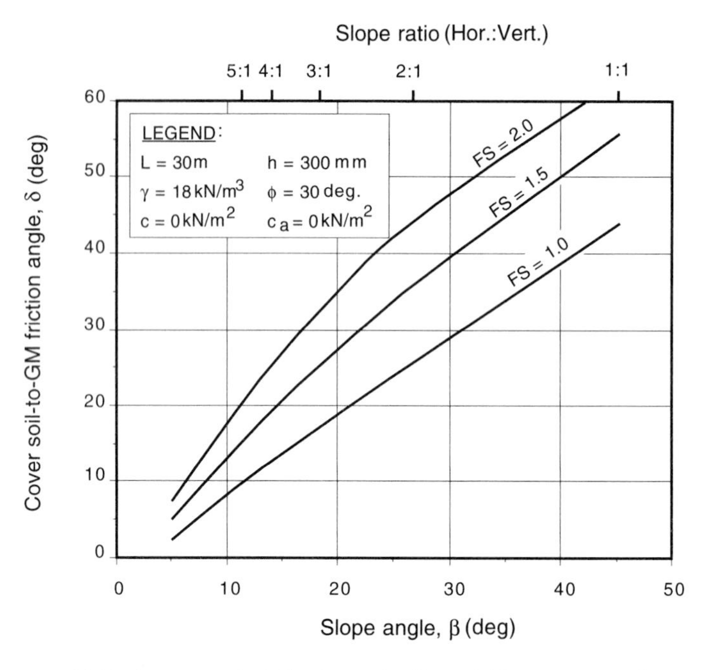

(a) Design curves for FS-values of uniform thickness cover soils

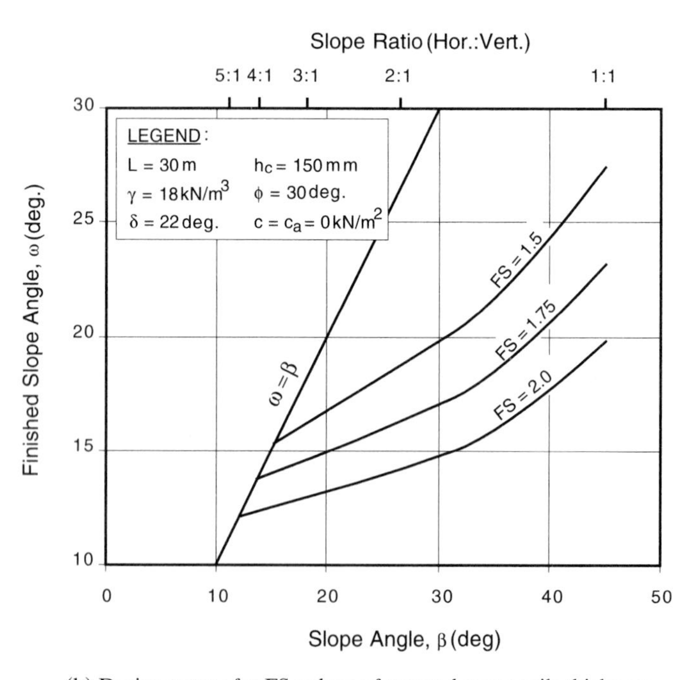

(b) Design curves for FS-values of tapered cover soils thickness

Figure 5.29 Design curves for cover soils on geomembrane-lined slopes.

Balancing the forces in the horizontal direction, the following formulation results.

$$E_P \cos\left(\frac{\omega + \beta}{2}\right) = \frac{\mathrm{C} + \mathrm{N_P} \tan \phi}{\mathrm{FS}}$$

Hence, the interwedge force acting on the passive wedge is

$$E_P = \frac{C + W_P \tan \phi}{\cos\left(\frac{\omega + \beta}{2}\right)(\mathrm{FS}) - \sin\left(\frac{\omega + \beta}{2}\right)\tan \phi}$$

Again, by setting $E_A = E_P$, the following equation can be arranged in the form of $ax^2 + bx + c = 0$ which in our case is

$$a(\mathrm{FS})^2 + b(\mathrm{FS}) + c = 0 \tag{5.23}$$

where

$$a = (W_A - N_A \cos \beta) \cos\left(\frac{\omega + \beta}{2}\right),$$

$$b = -\left[(W_A - N_A \cos \beta) \sin\left(\frac{\omega + \beta}{2}\right) \tan \phi + (N_A \tan \delta + C_a) \sin \beta \cos\left(\frac{\omega + \beta}{2}\right)\right.$$

$$\left. + \sin\left(\frac{\omega + \beta}{2}\right)(\mathrm{C} + \mathrm{W_P} \tan \phi)\right], \text{ and}$$

$$c = (N_A \tan \delta + C_a) \sin \beta \sin\left(\frac{\omega + \beta}{2}\right) \tan \phi.$$

Again, the resulting FS value can then be obtained as before.

$$\mathrm{FS} = \frac{-b + \sqrt{b^2 - 4ac}}{2a} \tag{5.24}$$

where (see also Figure 2.28c and the preceding section)

D = thickness of cover soil at bottom of the slope, measured vertically,
h_c = thickness of cover soil at crest of the slope, measured perpendicular to the slope,

$y = \left(L - \frac{D}{\sin \beta} - h_c \tan \beta\right)(\sin \beta - \cos \beta \tan \omega)$, see Figure 5.28c, and

ω = finished cover soil slope angle, note that $\omega < \beta$.

Example 5.12

Given a 30 m long slope with a tapered thickness cover soil of 150 mm at the crest extending at an angle ω of 16° to the intersection of the cover soil at the toe. The soil thickness at the bottom is 300 mm. The unit weight of the cover soil is 18 kN/m^3. The soil has a friction angle of 30° and zero cohesion (i.e., it is a sand). The interface friction angle with the underlying geomembrane is 22° and zero adhesion. What is the FS value at an underlying soil slope angle β of 3(H) to 1(V) which is equal to 18.4°?

Solution:

$$y = \left(L - \frac{D}{\sin\beta} - h_c \tan\beta\right)(\sin\beta - \cos\beta\tan\omega)$$

$$= \left(30 - \frac{0.30}{\sin 18.4} - (0.15)\tan 18.4\right)(\sin 18.4 - (\cos 18.4)(\tan 16))$$

$$= 1.28 \text{ m}$$

$$W_A = \gamma\left[\left(L - \frac{D}{\sin\beta} - h_c\tan\beta\right)\left(\frac{y\cos\beta}{2} + h_c\right) + \frac{h_c^2\tan\beta}{2}\right]$$

$$= 18\left[\left(30 - \frac{0.30}{\sin 18.4} - (1.28)\tan 18.4\right)\left(\frac{1.28\cos 18.4}{2} + 0.15\right) + \frac{(0.15)^2\tan 18.4}{2}\right]$$

$$= 390 \text{ kN/m}$$

$$N_A = W_A\cos\beta$$

$$= 390\cos 18.4$$

$$= 370 \text{ kN/m}$$

$$W_P = \frac{\gamma}{2\tan\omega}\left[\left(L - \frac{D}{\sin\beta} - h_c\tan\beta\right)(\sin\beta - \cos\beta\tan\omega) + \frac{h_c}{\cos\beta}\right]^2$$

$$= \frac{18}{2\tan 16}\left[\left(30 - \frac{0.30}{\sin 18.4} - 1.28\tan 18.4\right)(\sin 18.4 - \cos 18.4\tan 16) + \frac{0.15}{\cos 18.4}\right]^2$$

$$= 65.1 \text{ kN/m}$$

$$a = (W_A - N_A\cos\beta)\cos\left(\frac{\omega + \beta}{2}\right)$$

$$= (390 - 370\cos 18.4)\cos\left(\frac{16 + 18.4}{2}\right)$$

$$= 37.2 \text{ kN/m}$$

$$b = -\left[(W_A - N_A\cos\beta)\sin\left(\frac{\omega + \beta}{2}\right)\tan\phi + (N_A\tan\delta + C_a)\sin\beta\cos\left(\frac{\omega + \beta}{2}\right)\right.$$

$$\left. + \sin\left(\frac{\omega + \beta}{2}\right)(\mathrm{C} + \mathrm{W_P}\tan\phi)\right]$$

$$= -\left[(390 - 370\cos 18.4)\sin\left(\frac{16 + 18.4}{2}\right)\tan 30 + (370\tan 22 + 0)\sin 18.4\cos\left(\frac{16 + 18.4}{2}\right)\right.$$

$$\left. + \sin\left(\frac{16 + 18.4}{2}\right)(0 + 65.1\tan 30)\right]$$

$$= -62.8 \text{ kN/m}$$

$$c = (N_A \tan \delta + C_a) \sin \beta \sin \left(\frac{\omega + \beta}{2}\right) \tan \phi$$

$$= (370 \tan 22 + 0) \sin 18.4 \sin \left(\frac{16 + 18.4}{2}\right) \tan 30$$

$$= 8.07 \text{ kN/m}$$

$$\text{FS} = \frac{-b + \sqrt{b^2 - 4ac}}{2a}$$

$$= \frac{62.8 + \sqrt{(-62.8)^2 - 4(37.2)(8.07)}}{2(37.2)}$$

$$\text{FS} = 1.55$$

Example 5.12 has also been extended to a set of design curves as seen in Figure 5.29b. The anticipated trends are again noted, as is the agreement with the worked-out example. Clearly, this type of stabilizing solution can be used if space at the toe of the slope is available (often it is not or it occupies valuable air space) and if ample soil of the desired type is relatively inexpensive.

5.3.6 Runout and Anchor Trench Design

As shown in Figure 5.21 and the subsequent profile sections of geomembrane-lined reservoirs the liner comes up from the bottom of the excavation, covers the side slopes, and then runs over the top a short distance. It often terminates vertically down into an anchor trench. This anchor trench is typically dug by a small backhoe or trenching machine; the liner is draped over the edge, and then the trench is backfilled with the same soil that was there originally. The backfilled soil should be compacted in layers as the backfilling proceeds. Although concrete has been used as an anchorage block, it is rarely justified, at least on the basis of calculations, as will be seen in this section.

Regarding design, two separate cases will be analyzed: one with geomembrane runout only and no anchor trench at all (as is often used with canal liners), and the other as described above, with both runout and anchor trench considerations (as with reservoirs and landfills). Figure 5.30 defines the first situation, together with the forces and stresses involved. Note that the cover soil applies normal stress due to its weight, but does not contribute frictional resistance above the geomembrane. This is due to the fact that the soil moves along with the geomembrane as it deforms and undoubtedly cracks, thereby losing its integrity.

From Figure 5.30, the following horizontal force summation results, which leads to the appropriate design equation.

$$\Sigma F_x = 0$$

$$T_{\text{allow}} \cos \beta = F_{U\sigma} + F_{L\sigma} + F_{LT}$$

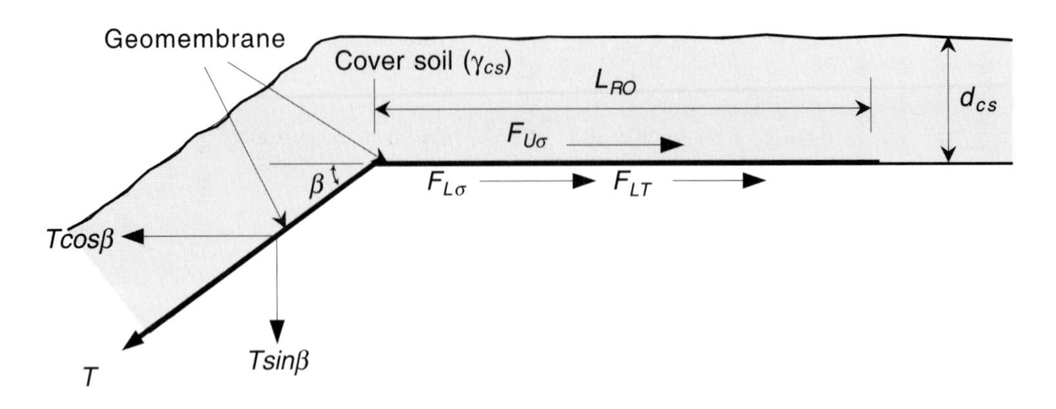

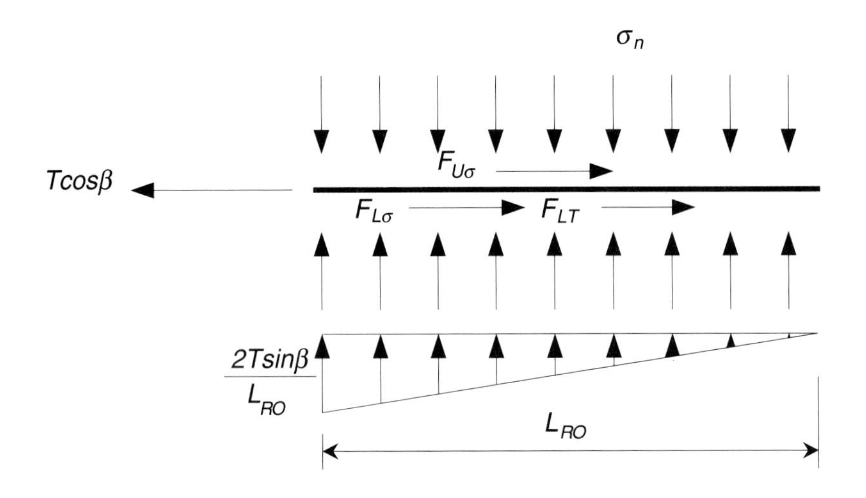

Figure 5.30 Cross section of geomembrane runout section and related stresses and forces involved.

$$= \sigma_n \tan \delta_U(L_{RO}) + \sigma_n \tan \delta_L(L_{RO}) + 0.5\left(\frac{2T_{\text{allow}} \sin \beta}{L_{RO}}\right)(L_{RO}) \tan \delta_L$$

$$L_{RO} = \frac{T_{\text{allow}}(\cos \beta - \sin \beta \tan \delta_L)}{\sigma_n(\tan \delta_U + \tan \delta_L)} \tag{5.25}$$

where

T_{allow} = allowable force in geomembrane stress = $\sigma_{\text{allow}} t$, where
σ_{allow} = allowable stress in geomembrane, and
t = thickness of geomembrane;
β = side slope angle;
$F_{U\sigma}$ = shear force above geomembrane due to cover soil (note that for thin cover soils tensile cracking will occur and this value will be negligible);

$F_{L\sigma}$ = shear force below geomembrane due to cover soil;
F_{LT} = shear force below geomembrane due to vertical component of T_{allow};
σ_n = applied normal stress from cover soil;
δ = angle of shearing resistance between geomembrane and adjacent material (i.e., soil or geotextile); and
L_{RO} = length of geomembrane runout.

Example 5.13 illustrates the use of the concept and the equations just developed.

Example 5.13

Consider a 1.0 mm thick VLDPE geomembrane with a mobilized allowable stress of 7000 kPa, which is on a $3(H)$ to $1(V)$ side slope. Determine the required runout length to resist this stress without use of a vertical anchor trench. In this analysis use 300 mm of cover soil weighing 16.5 kN/m^3 and a friction angle of 30° with the geomembrane.

Solution: From the design equations just presented,

$$T_{allow} = \sigma_{allow} t$$
$$= (7000)(0.001)$$
$$T_{allow} = 7.0 \text{ kN/m}$$

and

$$L_{RO} = \frac{T_{allow}(\cos\beta - \sin\beta\tan\delta_L)}{\sigma_n(\tan\delta_U + \tan\delta_L)}$$
$$= \frac{(7.0)[\cos 18.4 - (\sin 18.4)(\tan 30)]}{(16.5)(0.30)[\tan 0 + \tan 30]}$$
$$= \frac{5.37}{2.86}$$
$$L_{RO} = 1.9 \text{ m}$$

Note that this value is strongly dependent on the value of mobilized allowable stress used in the analysis. To mobilize the failure strength of the geomembrane would require a longer runout length or embedment in an anchor trench. This, however, might not be desirable. Pullout without geomembrane failure might be a preferable phenomenon. It is a site-specific situation.

The situation with an anchor trench at the end of the runout section is illustrated in Figure 5.31. The configuration requires some important assumptions regarding the state of stress within the anchor trench and its resistance mechanism. In order to provide lateral resistance, the vertical distance within the anchor trench has lateral forces acting upon it. More specifically, an active earth pressure (P_A) is tending to destabilize the situation, whereas a passive earth pressure (P_P) is tending to resist pullout. As will

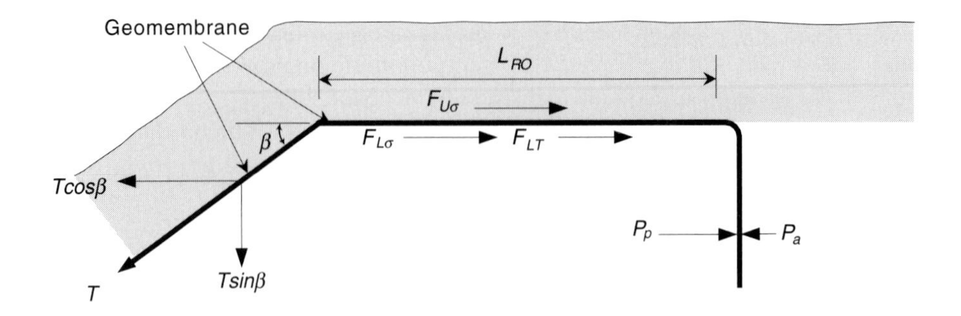

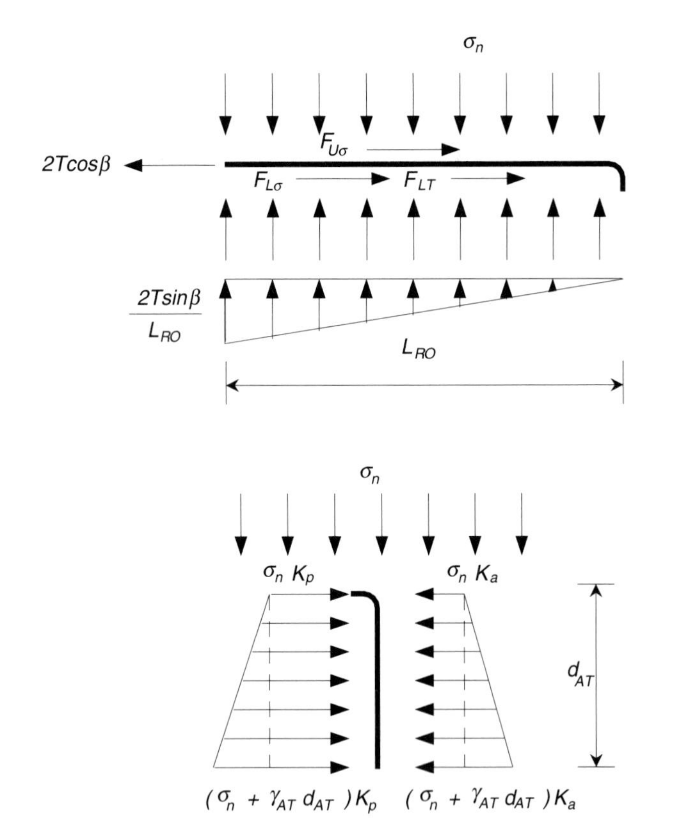

Figure 5.31 Cross section of geomembrane runout section with anchor trench and related stresses and forces involved.

be shown, this passive earth pressure is very effective in providing a resisting force (see Holtz and Kovacs [44]). Using the free-body diagram in Figure 5.31,

$$\Sigma F_x = 0$$

$$T_{\text{allow}} \cos \beta = F_{U\sigma} + F_{L\sigma} + F_{LT} - P_A + P_P \tag{5.26}$$

where

T_{allow} = allowable force in geomembrane = $\sigma_{allow} t$, where
σ_{allow} = allowable stress in geomembrane, and
t = thickness of geomembrane;
β = side slope angle;
$F_{U\sigma}$ = shear force above geomembrane due to cover soil (note that for thin cover soils, tensile cracking will occur, and this value will be negligible);
$F_{L\sigma}$ = shear force below geomembrane due to cover soil;
F_{LT} = shear force below geomembrane due to vertical component of T_{allow};
P_A = active each pressure against the backfill side of the anchor trench; and
P_P = passive earth pressure against the in-situ side of the anchor trench.

The values of $F_{U\sigma}$, $F_{L\sigma}$, and F_{LT} have been defined previously. The values of P_A and P_P require the use of lateral earth pressure theory.

$$P_A = \tfrac{1}{2}(\gamma_{AT} d_{AT}) K_A d_{AT} + (\sigma_n) K_A d_{AT}$$

$$P_A = (0.5\gamma_{AT} d_{AT} + \sigma_n) K_A d_{AT} \quad (5.27)$$

$$P_P = (0.5\gamma_{AT} d_{AT} + \sigma_n) K_P d_{AT} \quad (5.28)$$

where

γ_{AT} = unit weight of soil in anchor trench,
d_{AT} = depth of the anchor trench,
σ_n = applied normal stress from cover soil,
K_A = coefficient of active earth pressure = $\tan^2(45 - \phi/2)$,
K_P = coefficient of passive earth pressure = $\tan^2(45 + \phi/2)$, and
ϕ = angle of shearing resistance of respective soil.

This situation results in one equation with two unknowns; thus a choice of either L_{RO} or d_{AT} is necessary to calculate the other. As with the previous situation, the factor of safety is placed on the geomembrane force T, which is used as an allowable value, T_{allow}. Example 5.14 illustrates the procedure.

Example 5.14

Consider a 1.5 mm thick HDPE geomembrane extending out of a facility as shown in Figure 5.31. What depth anchor trench is needed if the runout distance is constrained to 1.0 m? In the solution, use a geomembrane allowable stress of 16,000 kPa on a 3(H) to 1(V) side slope. There are 300 mm of cover soil at 16.5 kN/m^3 placed over the geomembrane runout and anchor trench (this is also the unit weight of the anchor trench soil). The friction angle of the geomembrane to the soil is 30° (although assume 0° for the top of the geomembrane under a soil-cracking assumption) and the soil itself is 35°.

Solution: Using the previously developed design equations based on Figure 5.31:

$$\begin{aligned} T_{\text{allow}} &= \sigma_{\text{allow}} t \\ &= 16000(0.0015) \\ &= 24.0 \text{ kN/m} \end{aligned}$$

and

$$\begin{aligned} F_{U\sigma} &= \sigma_n \tan \delta_U (L_{RO}) \\ &= (0.3)(16.5) \tan 0 (L_{RO}) \\ &= 0 \\ F_{L\sigma} &= \sigma_n \tan \delta_L (L_{RO}) \\ &= (0.3)(16.5) \tan 30 (L_{RO}) \\ &= 2.86 L_{RO} \\ F_{LT} &= T_{\text{allow}} \sin \beta \tan \delta_L \\ &= (24.0) \sin 18.4 \tan 30 \\ &= 4.37 \text{ kN/m} \\ P_A &= (0.5\gamma_{AT} d_{AT} + \sigma_n) K_A d_{AT} \\ &= [(0.5)(16.5) d_{AT} + (0.3)(16.5)] \tan^2 (45 - 35/2)\, d_{AT} \\ &= [8.25 d_{AT} + 4.95](0.271) d_{AT} \\ &= 2.24 d_{AT}^2 + 1.34 d_{AT} \\ P_P &= (0.5\gamma_{AT} d_{AT} + \sigma_n) K_P d_{AT} \\ &= [(0.5)(16.5) d_{AT} + (0.3)(16.5)] \tan^2 (45 - 35/2)\, d_{AT} \\ &= [8.25 d_{AT} + 4.95](3.69) d_{AT} \\ &= 30.4 d_{AT}^2 + 18.3 d_{AT} \end{aligned}$$

This is substituted into the general force equation [Eq. (5.26)] to arrive at the solution in terms of the two variables L_{RO} and d_{AT}.

$$\begin{aligned} T_{\text{allow}} \cos \beta &= F_{U\sigma} + F_{L\sigma} + F_{LT} - P_A + P_P \\ (24.0) \cos 18.4 &= 0 + 2.86 L_{RO} + 4.37 - 2.24 d_{AT}^2 - 1.34 d_{AT} + 30.4 d_{AT}^2 + 18.3 d_{AT} \\ 18.4 &= 2.86 L_{RO} + 17.0 d_{AT} + 28.2 d_{AT}^2 \end{aligned}$$

Since $L_{RO} = 1.0$ m, the equation can be solved for the unknown d_{AT}.

$$d_{AT} = 0.50 \text{ m}$$

Using this formulation we can develop a design chart for a wide range of geomembranes and thicknesses as characterized by different values of T_{allow}. For the specific conditions of Example 5.14,

$$\beta = 18.4°, \text{ which is } 3(H) \text{ to } 1(V)$$

$$\sigma_n = d_{cs}\gamma_{cs}$$

$$= (0.30)(16.5)$$

$$= 4.95 \text{ kN/m}^2$$

$$\delta_U = 0°$$

$$\delta_L = 30°$$

$$\phi = 35°$$

$$\gamma_{AT} = 16.5 \text{ kN/m}^3$$

$$\delta_{AT} = 30°$$

the response in terms of the two unknowns L_{RO} and d_{AT} is given in the following figure. Using this figure, Example 5.14 with the 1.5 mm thick HDPE at 24.0 kN/m gives an anchor trench depth of 0.50 m for an assumed runout length of 1.0 m. Other values can be readily selected.

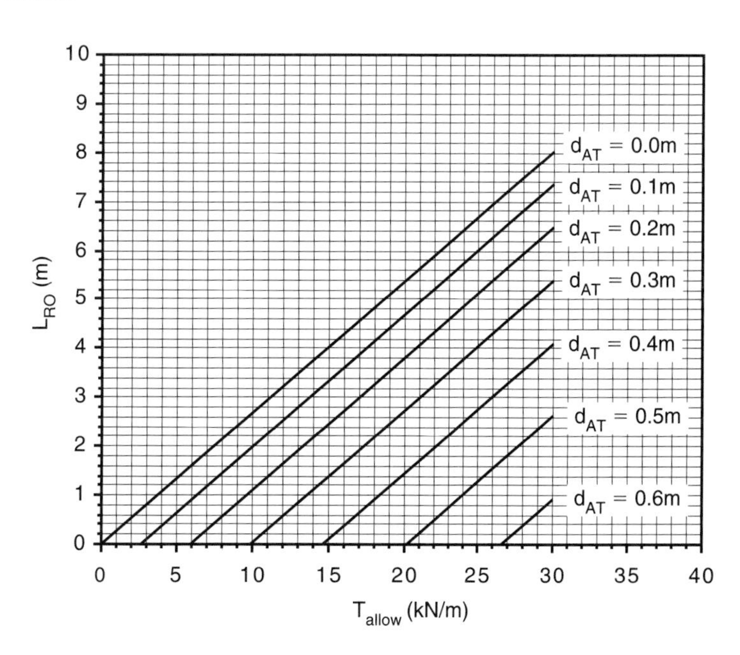

It should be noted that many manufacturers specify 500 mm deep anchor trenches and 1000 mm long runout sections. As seen above, this is very simplistic, for each membrane type and thickness requires its own analysis. By using a model as presented here, any set of conditions can be used to arrive at a solution. Even situations in which geotextiles and/or geonets are used in conjunction with the geomembrane (under, over, or both) and brought into the anchor trench can be analyzed in a similar manner.

5.3.7 Summary

Projects involving liquid containment using geomembranes are often extremely large. With large size come some inherent advantages over smaller projects. Foremost of these advantages is that most parties involved take the project seriously and approve of and enter into a planned and sequential design procedure. This section was laid out with this in mind, so that the design process proceeded step by step. Each element of design that is made leads to a new issue, which is followed by a new design element. Eventually, the quantitative process is concluded and details, often qualitative by nature, must be attended to. These details, such as seam type, seam layout, piping layout, and appurtenance details, are extremely important. They are, however, common to all geomembrane projects and therefore will be handled in Sections 5.10 and 5.11.

Although such large projects obviously warrant a careful design procedure, it does not follow that smaller projects do not deserve the same attention. Indeed, failures of small liner systems can be significant. Many warrant a design effort comparable to that of large projects, as illustrated in this section.

With this section behind us, we can now focus on other applications involving geomembranes. Where a similar analysis is called for, reference will be made back to this section. Thus only new and/or unique features of geomembrane projects will form the basis of the sections to follow.

5.4 COVERS FOR RESERVOIRS

Geomembrane covers are often used above the liquid surface of storage reservoirs. They are of fixed, floating, or suspended types.

5.4.1 Overview

There are a number of important reasons why liquid containment structures should be covered. These include: losses due to evaporation (up to 84% per year; see Cooley [45]), savings on chlorine treatment (for water reservoirs), savings on algae control chemicals (for water reservoirs), reduced air pollution (for reservoirs holding chemicals), reduced need for drainage and cleaning, increased safety against accidental drowning, protection from natural pollution entering the reservoir (e.g., animal excretion), and protection from intentional pollution (i.e., sabotage).

Obviously, a rigid roof structure could be constructed over the reservoir, but the costs involved are usually prohibitively high. At a far lower cost, both during initial construction or in a retrofitted system, is the use of an impermeable liner. All the materials listed in Table 5.1 are candidate covers for this purpose; however, those geomembranes with superior ultraviolet and exposed weathering resistance have a decided advantage. For smaller structures (where the span length from side to side is less than approximately 5 m) the cover can be fixed and remain stationary. For larger span lengths, the use of a floating cover that resides directly on the liquid's surface as it varies in elevation will be considered. Finally, the use of a totally encapsulated enclosure around the liquid will be presented.

5.4.2 Fixed Covers

Fixed covers are usually used in conjunction with small-diameter tanks whose sides are made of wood, concrete, or steel. The geomembrane is fixed at the upper edge of the tank and takes a catenary shape toward the center. Positive fixity is required at the edges, since stress concentrations are very high when the tank's liquid level is beneath the elevation of the lower point of the cover. Small holes are put in the cover for rainwater drainage, but snow and wind loads can create a problem. The following example illustrates this situation.

Example 5.15

(a) Calculate the edge stresses in a 0.91 mm geomembrane cover over a 6 m diameter wooden water storage tank, as shown in the sketch below. The combined loading to be used is 1.0 kPa, which will deflect the cover into a 15 m radius deformed shape. **(b)** If a CSPE cover is used with an ultimate strength of 4.5 kN/m when nonreinforced or 10.5 kN/m when reinforced with a 10 × 10 polyester scrim, what are the resulting factors of safety?

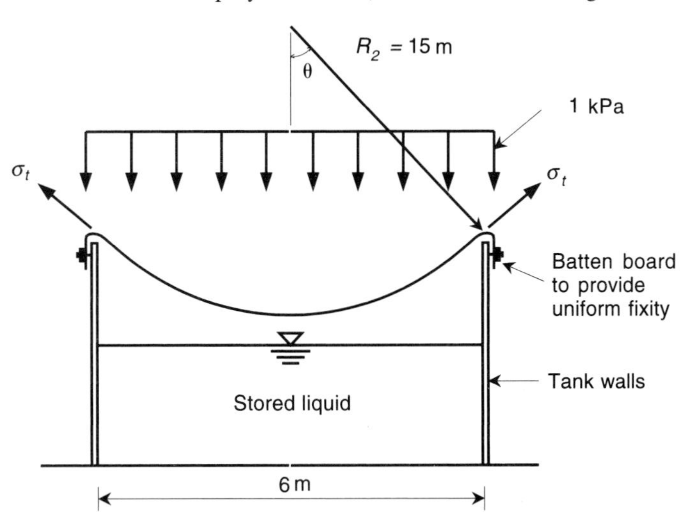

Solution:

(a) For a uniform loading on a horizontal projected area with a tangential top edge support with $\theta \leq 90°$, the maximum stress is

$$\sigma_t = \frac{wR_2}{2t} \tag{5.29}$$

where

σ_t = required stress (kPa),
w = uniformly distributed load (kPa),
R_2 = deformed radius (m), and
t = thickness of membrane (m).

$$\sigma_t = \frac{(1.0)(15)}{2(0.91/1000)}$$

$$= 8240 \text{ kPa}$$

(b) For the factors of safety, the stress and resulting FS of the nonreinforced cover are

$$\sigma_{\text{ult}} = \frac{4.5}{(0.91/1000)}$$

$$= 4950 \text{ kPa}$$

$$\text{FS} = \frac{\sigma_{\text{ult}}}{\sigma_t}$$

$$= \frac{4950}{8240}$$

$$= 0.60 \qquad \text{not acceptable}$$

For the reinforced cover, the stress and resulting FS are,

$$\sigma_{\text{ult}} = \frac{10.5}{(0.91/1000)}$$

$$= 11{,}500 \text{ kPa}$$

$$\text{FS} = \frac{\sigma_{\text{ult}}}{\sigma_t}$$

$$= \frac{11{,}500}{8240}$$

$$= 1.40 \qquad \text{acceptable}$$

It is easily seen in Example 5.15 that very high edge stresses occur, which gives a distinct advantage to scrim(fabric)-reinforced membranes. It should also be obvious that a very carefully planned and executed method of fixing the lining to the tank must be made. Such details will be addressed in Section 5.11. Since the edge stresses become extremely high for large-diameter tanks when the cover is fixed in position, the concept of floating covers has great appeal and current use.

5.4.3 Floating Covers

Floating covers of polymeric materials have made a strong impact on reservoir and liquid-holding facilities. All the materials listed in Table 5.1 are candidate cover materials, but superior ultraviolet and weathering resistance are an obvious advantage. Due to the desirability of decreasing the dead load, however, some additional materials have also been developed for this application. Foamed polymers (e.g., foamed EPDM rubber) in the form of a closed-cell structure of density as low as 0.16 mg/l have been successfully used. These lightweight foam liners have been used on tanks up to 15 m diameter, and the design and construction procedure is somewhat unusual [46]. Small holes are cut into the liner to allow the cover to drain rainwater and snowmelt accumulating on top of it. The holes also help during installation, allowing any air trapped under the cover to escape. The edge is not attached to the tank but is stiffened by the addition of bonded foam-rubber strips, which prevent wind from getting beneath the cover. Wires are sometimes stretched across the top of the tank to prevent the loss of cover material when it is accidentally overfilled. A freeboard of 300 mm should be maintained.

However, for large water storage reservoirs of the type discussed in Section 5.3, the liner materials listed in Table 5.1 are generally used, particularly CSPE-R and HDPE. As can be intuitively appreciated, tensile strength, tear, puncture, and impact resistance are critical mechanical properties in this application. The covers are fixed to the sidewall of the tank or soil side-slope anchor trench in these cases. The dimensions of the cover must be greatest when the reservoir is empty. This creates a problem during filling, when the slack accumulates on the leeward sides. This results in the stressing of the cover on the windward sides and does not allow for rainwater or snowmelt collection and disposal in other areas. Thus accommodation of the slack is a major design consideration.

Gerber [47] presents a number of slack-accommodating designs using combinations of floats and sump weights. The floats are made of lightweight foamed or beaded polymer materials and are usually arranged beneath the cover and attached to its underside. Weights are similarly constructed, but the pockets are filled with sand or gravel and constructed on the top side of the cover. The designs are intricate and some are shown in Figures 5.32 and 5.33. In Figure 5.32 the central float grid is sufficiently stiff to maintain its integrity as the reservoir fills. All slack is forced to the sides, where rainwater and snowmelt collects and is pumped away when it becomes excessive. The drawbacks of these designs are: (1) walking on the cover in the peripheral area is very unsafe; (2) wind pushes the central float section to the leeward side of the reservoir, leaving little or no slack on the windward side; and (3) rain puddles, ice, and debris tend to collect in certain areas of the central float section. Thus the concept of a *defined sump reservoir cover* is preferred. Figure 5.33 shows additional designs that essentially eliminate the previous problems.

Important in all floating cover designs are the access hatches (which are normally provided), projecting structures (which are very troublesome), and strategies for dealing with ice (which can cause puncture and tear). Attachment to the edge anchorage is very important, since wind-generated stresses can exert large tear and shear forces. For this reason the anchorage is usually a concrete footing with anchor bolts to which the cover, along with batten strips and nuts, is fastened for positive fixity. Figure 5.34 shows

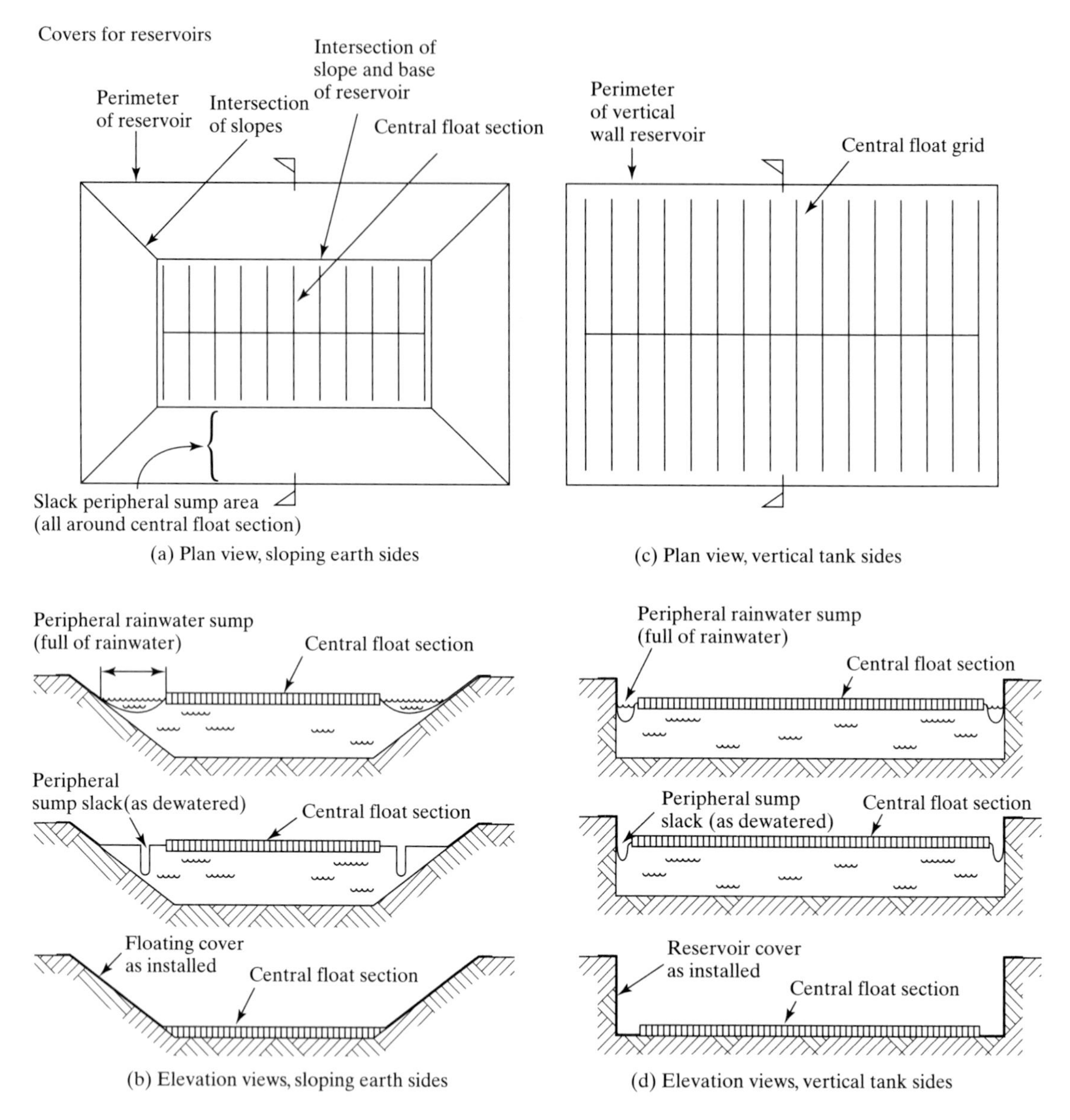

Figure 5.32 Designs for untensioned, centrally floated, peripheral sump reservoir covers. (After Gerber [47])

some details of typical cover attachments, together with the geomembrane liner or in its own separate anchor trench. The geomembrane cover should be folded back over itself with a 12 mm nylon or polypropylene rope placed within the fold. The rope butts up against the back of the batten strip(s), thereby relieving some of the concentrated

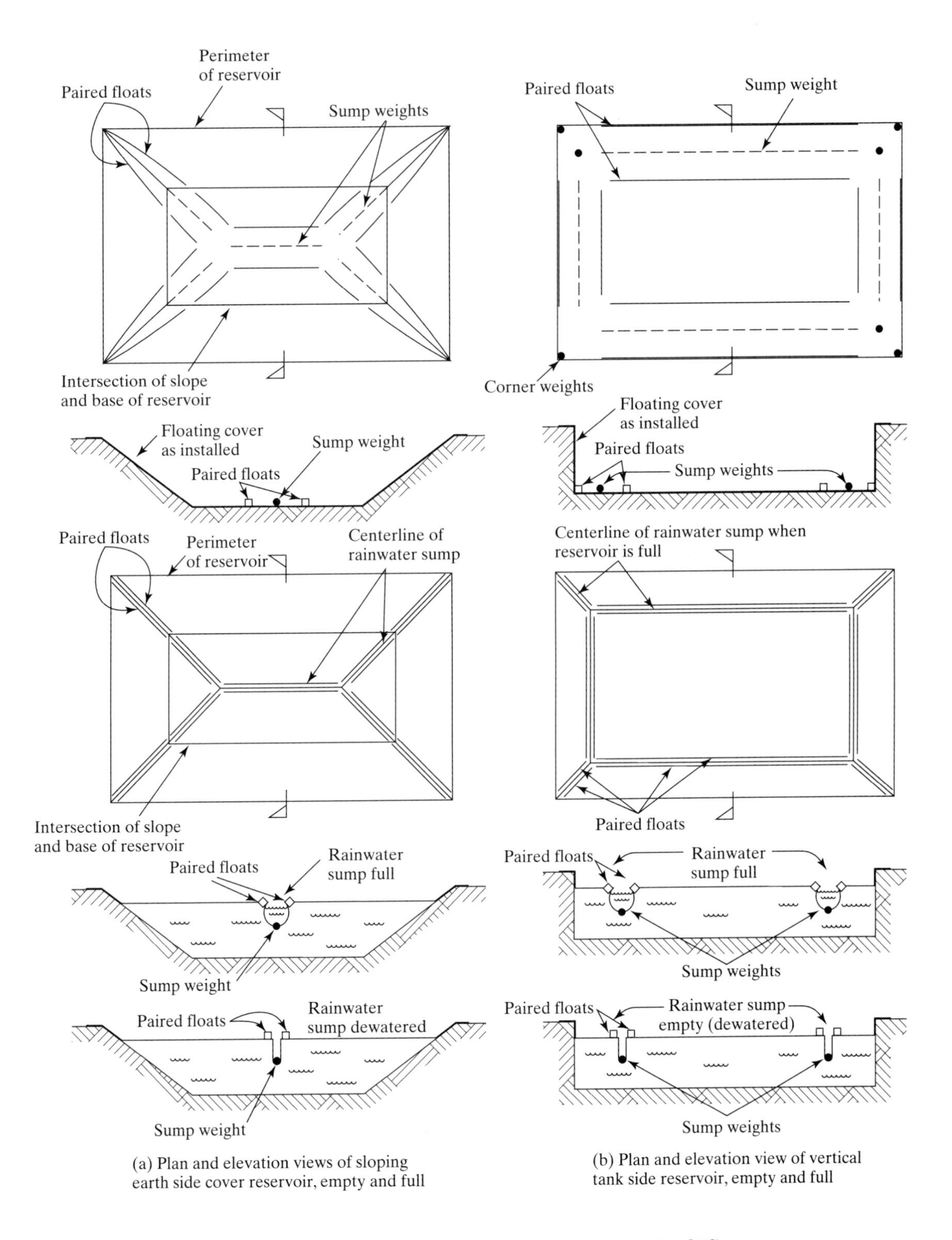

(a) Plan and elevation views of sloping earth side cover reservoir, empty and full

(b) Plan and elevation view of vertical tank side reservoir, empty and full

Figure 5.33 Designs for defined sump reservoir covers. (After Gerber [47])

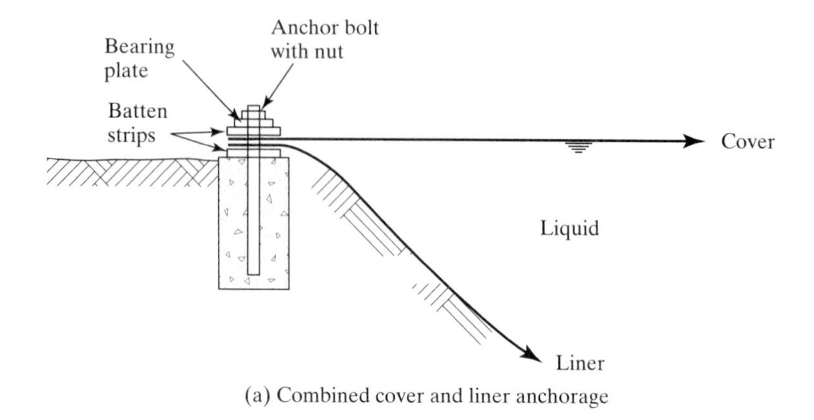

(a) Combined cover and liner anchorage

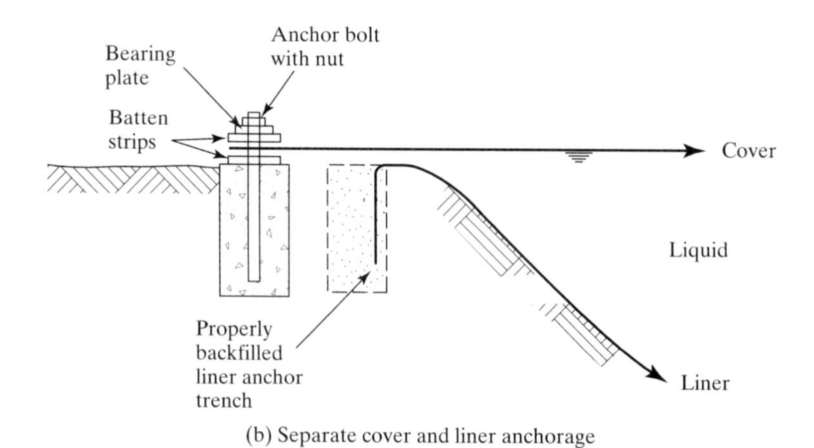

(b) Separate cover and liner anchorage

Figure 5.34 Various configurations for anchoring geomembrane floating covers and their edges.

stresses at the anchor bolt penetrations. The anchor trench for the cover is almost always made from cast-in-place concrete.

The design of a floating cover's anchor trench is difficult because of the uncertainty in selecting the required stresses. In the absence of wind-tunnel design values (which, incidentally, is an excellent research topic), we could focus on the tensile strength or on the tear strength of the proposed cover material. Using Figure 5.35, the relevant equations are developed as follows.

Using $\Sigma F_x = 0$,

$$T_{\text{reqd}} = F_L + P_p - P_A \tag{5.30}$$

where

T_{reqd} = required tensile strength = $\sigma_{\text{reqd}}\, t$, in which
σ_{reqd} = required tensile stress

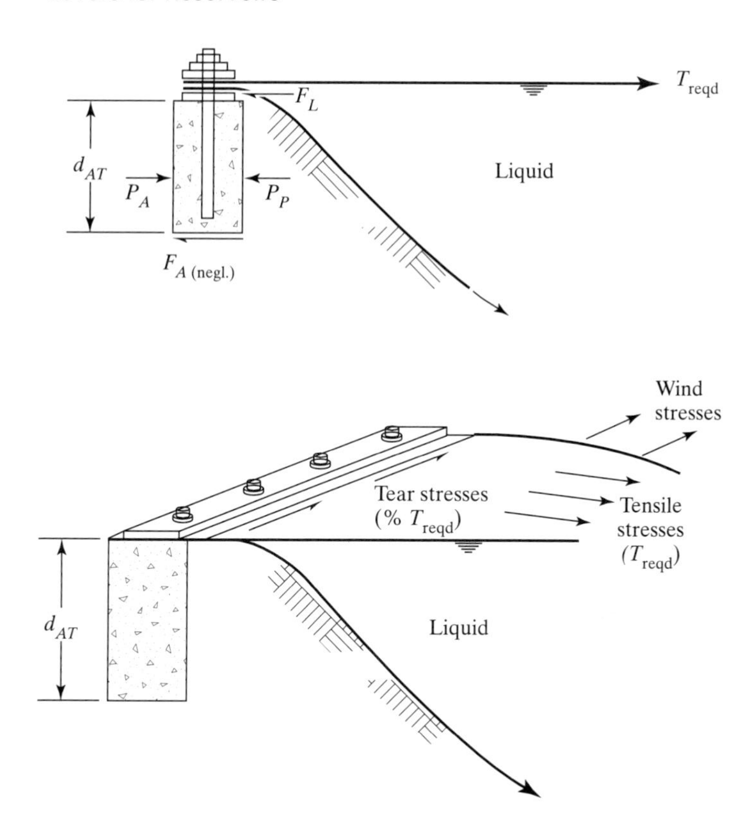

Figure 5.35 Design schemes for geomembrane floating covers.

t = geomembrane thickness
F_L = frictional resistance along upper berm material (ignored as a worst-case assumption when the reservoir is full and the cover is horizontal);
P_A = active earth pressure = $\frac{1}{2}\gamma d_{AT}^2 K_a$, in which
γ = unit weight of backfill soil,
d_{AT} = depth of anchor trench,
K_a = $\tan^2(45 - \phi/2)$, where
ϕ = angle of shearing resistance of soil; and
P_p = passive earth pressure = $\frac{1}{2}\gamma d_{AT}^2 K_p$, in which
K_p = $\tan^2(45 + \phi/2)$.

Example 5.16 illustrates the importance of this aspect of floating geomembrane covers.

Example 5.16

Calculate the factor of safety of a 1.15 mm thick reinforced CSPE-R geomembrane floating reservoir cover of allowable stress of 3500 kN/m^2 when it is connected to an 500 mm deep concrete anchor trench, as shown in Figure 5.35. The friction angle of the backfill soil is 35° and its unit weight is 19 kN/m^3.

Solution: Using the previously developed design equations with $F_L = 0$,

$$K_A = \tan^2\left(45 - \frac{\phi}{2}\right)$$

$$= \tan^2\left(45 - \frac{35}{2}\right)$$

$$= 0.27$$

$$K_p = \tan^2\left(45 + \frac{\phi}{2}\right)$$

$$= \tan^2\left(45 + \frac{35}{2}\right)$$

$$= 3.69$$

$$T_{\text{reqd}} = F_L + P_P - P_A$$

$$= 0 + (0.5)(19)(0.5)^2(3.69) - (0.5)(19)(0.5)^2(0.27)$$

$$= 0 + 8.76 - 0.64$$

$$= 8.12 \text{ kN/m}$$

Additionally,

$$T_{\text{reqd}} = \sigma_{\text{reqd}} t$$

$$\sigma_{\text{reqd}} = \frac{8.12}{(1.15/1000)}$$

$$= 7060 \text{ kN/m}^2$$

$$\text{FS} = \frac{\sigma_{\text{allow}}}{\sigma_{\text{reqd}}}$$

$$= \frac{3500}{7060}$$

$$\text{FS} = 0.50 \qquad \text{which is not acceptable}$$

Thus the geomembrane cover will fail before the anchor trench is pulled out of its position.

The problem of estimating the actual tear stresses mobilized by wind acting over (and on) the central section is a very difficult one. To be sure, the stresses are high, and the survivability data of Table 5.11 in the very high category represents the absolute minimum values that should be used.

5.4.4 Quasisolid Covers

We have concentrated in this section on reservoirs containing liquids. Sometimes it is necessary to cover quasisolid (or semiliquid) substances. One application area is the covering of odorous substances, such as manure and other biodegradable viscous wastes. Here the key material property is the geomembrane's vapor transmission as determined in Section 5.1.2. The actual gas vapor that is to be contained should be used in the test. Not only is the vapor permeability of interest, but also the possibility of an adverse chemical reaction to the geomembrane should be investigated.

The second area in which quasisolids need to be covered is in the treatment of sewage sludge to increase the efficiency of anaerobic digestion. In addition to the geomembrane cover's low gas permeability, the black color increases the temperature beneath the cover, which further increases the bacterial activity. The generated methane and hydrogen sulfide gases are collected at the high point of the site and flared off or appropriately used.

The key to both of these application areas is just how liquid-like or solid-like the material is. If it is more liquid, the material must be placed in an excavation or container and the cover designed per the details of this section. If it is more solid, the cover will be designed as with solid-waste landfill closures, which will be described later in Section 5.7.

5.4.5 Complete Encapsulation

By lining the bottom and sides of reservoirs (as in Section 5.3) and now covering the contents with similar geomembrane materials (as in Section 5.4.3), it seems only natural that a completely fabricated enclosure should be considered. Indeed, such *superbags* are available, and in standard sizes up to 5 million liters! Even larger sizes can be fabricated by special order. The concept is straightforward in that the entire fabrication occurs at the factory, where the complete bag is transported to a prepared site (a 5 Ml nitrile rubber bag weighs only 55 kN) and placed accordingly. Filling with the liquid to be contained can begin immediately. Figure 5.36 shows such a bag inflated with air (to give visual impact) and installed in a prepared earth substructure after filling.

5.5 WATER CONVEYANCE (CANAL) LINERS

This section covers the use of geomembranes to line canals. The usual liquid is water, but many other liquids, including industrial chemicals and wastes, can also be conveyed.

5.5.1 Overview

Often the source of water is located at a considerable distance from the intended user. As a result, many and varied attempts have been made to convey this valuable resource, sometimes requiring herculean feats. Consider, for example, the Romans, whose aqueducts are among the premier engineering achievements of recorded history. An important element in the economic functioning of such water-conveyance

(a)

(b)

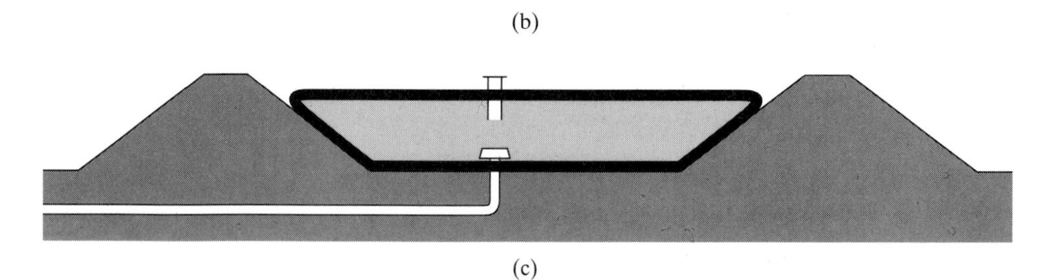

(c)

Figure 5.36 Inflatable reservoirs. (Compliments of Firestone Coated Fabrics Co. and Fabritank Co.)

canals is that they hold the water placed in them during the journey from source to user. Excessive leakage is obviously unacceptable, making the liner of the canal a key element in a successful system. With this in mind, engineers have tried almost everything to line their water conveyance canals at one time or another. These include soil liners (mainly clay soil), nonflexible liners (bricks, paving blocks, concrete, shotcrete, gunite, etc.), and flexible liners (bituminous panels, spray-on chemicals, and geomembranes). The emphasis in this chapter, naturally, will be on polymeric geomembranes. When properly designed, constructed, and maintained, geomembrane materials should be capable of having a significant impact on the canal lining industry. As will be seen later in this section, this impact on the canal lining market could easily extend to the rehabilitation of old canals and their linings, as well. The scope of the problem is enormous—indeed, it is worldwide.

Most countries have national committees and specific agencies studying canal linings. These organizations are specifying geomembranes regularly. In the U.S., the American Society of Agricultural Engineers, the U.S. Bureau of Reclamation, and the U.S. Army Corps of Engineers have been very active in this area. The activity is international: For instance, a joint U.S.-Russian commission has made a series of tests and has reported on the subject. Many standards are being developed for geomembranes used specifically as canal linings. When searching the literature, however, the topic will usually come under a heading other than geomembranes, such as: canal liners, synthetic canal liners, plastic canal liners, rubber canal liners, or polymeric canal liners.

5.5.2 Basic Considerations

With potable water shortages looming in many parts of the world, for example, Asia (particularly the Near East and China), the arid regions of Africa, and even in some areas of South America, the efficient transportation of water is absolutely necessary. Geomembrane liners represent an economical and realistic seepage-control material in almost every instance.

Geometry. The design of canal geometry for uniform flow is a well established branch of hydraulics within the general category of *open channel flow.* Many textbooks are available on the subject and it is a required course in most undergraduate civil engineering curricula. It is well known that the preferred hydraulic cross sections are: trapezoid (half of a hexagon), rectangle (half of a square), triangle (half of a square), semicircle, parabola, and catenary (hydrostatic).

Due to various layout, excavation, and compaction problems that are encountered, curved surfaces are not as widely used as those consisting of linear segments. Furthermore, rectangular sections must be supported by a separate structure of wood, concrete, or steel. The most common cross sections are therefore trapezoid (for large flows) and triangle (for small flows). Regarding the side-slope angles of these sections, the slope-stability considerations of Section 5.3.5 are applicable.

Once the shape and side slopes are selected, the depth of the section is calculated from a *section factor* as follows.

$$AR^{2/3} = \frac{nq}{\sqrt{S}} \tag{5.31}$$

where

$AR^{2/3}$ = section factor,
= area (m^2),
R = hydraulic radius (m),
n = Manning coefficient (the typical range of which is 0.020 to 0.035 depending on the flow; $n = 0.028d_{50}^{0.1667}$ is sometimes used where d_{50} is the average size of the cover soil in meters),
q = flow rate (m^3/s), and
S = slope of water surface (dimensionless).

The A and R values are functions of the depth (in meters) and can be solved for explicitly or taken from design charts that are available [48].

Cross Sections. When a geomembrane is used as the liner material, it is placed either directly on the prepared soil subgrade or on a previously installed geotextile. A uniform or tapered-thickness soil cover is commonly placed over the geomembrane. Thus the completed cross section is typically like that shown in Figure 5.21b,d. The difference in this case, however, is that the liquid is flowing, and thus the possibility of scour of the cover soil must be addressed. Many studies (mostly empirical) have been directed at predicting a maximum permissible velocity of the liquid as a function of the type of cover soil. The values seem to range from 30 to 100 m/min, depending on cover soil type and the turbidity of the flow water (see Table 5.12). This table also includes the tractive force mobilized by the flowing water. For a trapezoidal section, these forces are distributed as shown in Figure 5.37. If these forces are such that the soil cover is eroded (or if the geomembrane has no soil cover to begin with), they will act directly on the geomembrane. Particularly vulnerable are the seams, which have been shown to be low in strength in peel or tension mode (recall Figure 5.6). Because of such problems, it is not uncommon to cover the liner with a nonerodable cover of asphalt, shotcrete or concrete, although problems occur here too, including the oxidation of asphalts and the thermal shrinkage and cracking of cementatious materials. Steel reinforcement is often used to control thermal stresses.

Since the quantity of liquid moving in the canal does not remain constant, a freeboard consideration must be addressed. Two definitions are needed: one of the height of the top of bank above the water surface (F_B), the other of the height of the geomembrane liner above the water surface (F_L). Figure 5.38 gives the Bureau of Reclamation's experience in this regard. Note that at high flow capacity generous amounts of freeboard are required in all cases, since overtopping would cause scour beneath the geomembrane and rapid failure of the system. Also given is the height of a clay lining (F_E) above the water surface, which is less than that of a geomembrane liner. This probably reflects a caution in that scour is more serious in undermining geomembrane liners than it is in clay linings.

Material Selection. For the conveyance of potable or agricultural water, all of the geomembranes listed in Table 5.1 are candidate materials. Due to its historical use and ease of construction, however, PVC is the most widely-used liner material for wa-

TABLE 5.12 MAXIMUM PERMISSIBLE VELOCITIES AND THE CORRESPONDING UNIT-TRACTIVE-FORCE VALUES

		Clear water		Turbid water	
Material	Manning coefficient	V (m/s)	T_0 (Pa)	V (m/s)	T_0 (Pa)
Fine sand, colloidal	0.020	0.45	1.29	0.76	3.59
Sandy loam, noncolloidal	0.020	0.54	1.77	0.76	3.59
Silt loam, noncolloidal	0.020	0.61	2.30	0.91	5.26
Alluvial silts, noncolloidal	0.020	0.61	2.30	1.07	7.18
Ordinary firm loam	0.020	0.76	3.59	1.07	7.18
Volcanic ash	0.020	0.76	3.59	1.07	7.18
Stiff clay, very colloidal	0.025	1.14	12.4	1.52	22.0
Alluvial silts, colloidal	0.025	1.14	12.4	1.52	22.0
Shales and hardpans	0.025	1.83	32.0	1.83	32.0
Fine gravel	0.020	0.76	3.59	1.52	15.3
Graded loam to cobbles when noncolloidal	0.030	1.14	18.2	1.52	31.0
Graded silts to cobbles when colloidal	0.030	1.22	20.6	1.68	38.0
Coarse gravel, colloidal	0.025	1.22	14.3	1.83	32.0
Cobbles and shingles	0.035	1.52	44.0	1.68	53.0

For straight channels of small slope, after aging.

Source:

V values from Fortier and Scobey [49]. The Fortier and Scobey values were recommended for use in 1926 by a Special Committee on Irrigation Research of the American Society of Civil Engineers.

T_0 values converted by the U.S. Bureau of Reclamation.

ter conveyance canals. Of course, certain situations may favor the choice of another material, but this choice would be on an individual basis. One situation of particular concern is in arid, desert areas, where consistently high temperatures are encountered. Here heat aging and weathering tests are critical in proper geomembrane selection.

For the conveyance of liquids such as chemicals, the chemical resistance chart in Table 5.8 (or its equivalent) should be used. When a mixed-waste stream or complex effluent is being transported, it may be necessary to run chemical-resistance tests as described in Section 5.1.4. Where leak detection is important, a double liner with a drainage layer (sand, geonet, or drainage geocomposite) between the primary and secondary liners can be used. Such double liners are quite common in solid-waste landfill liners and will be described later.

Thickness. A thickness design like that presented in Section 5.3.4 is certainly appropriate. However, the usual care in subgrade preparation, low normal stress, low hydraulic heads, moving water, and so on, make the design such that calculated thicknesses are quite low. For this application it is best to use a minimum allowable liner based on experience or the survivability chart in Table 5.11. The U.S. Bureau of Reclamation recommends a minimum 0.50 mm thickness for water-conveyance canals and sometimes uses 1.0 mm thick geomembranes.

Side Slopes. The design of canal side slopes follows exactly the procedures described in Section 5.3.5.

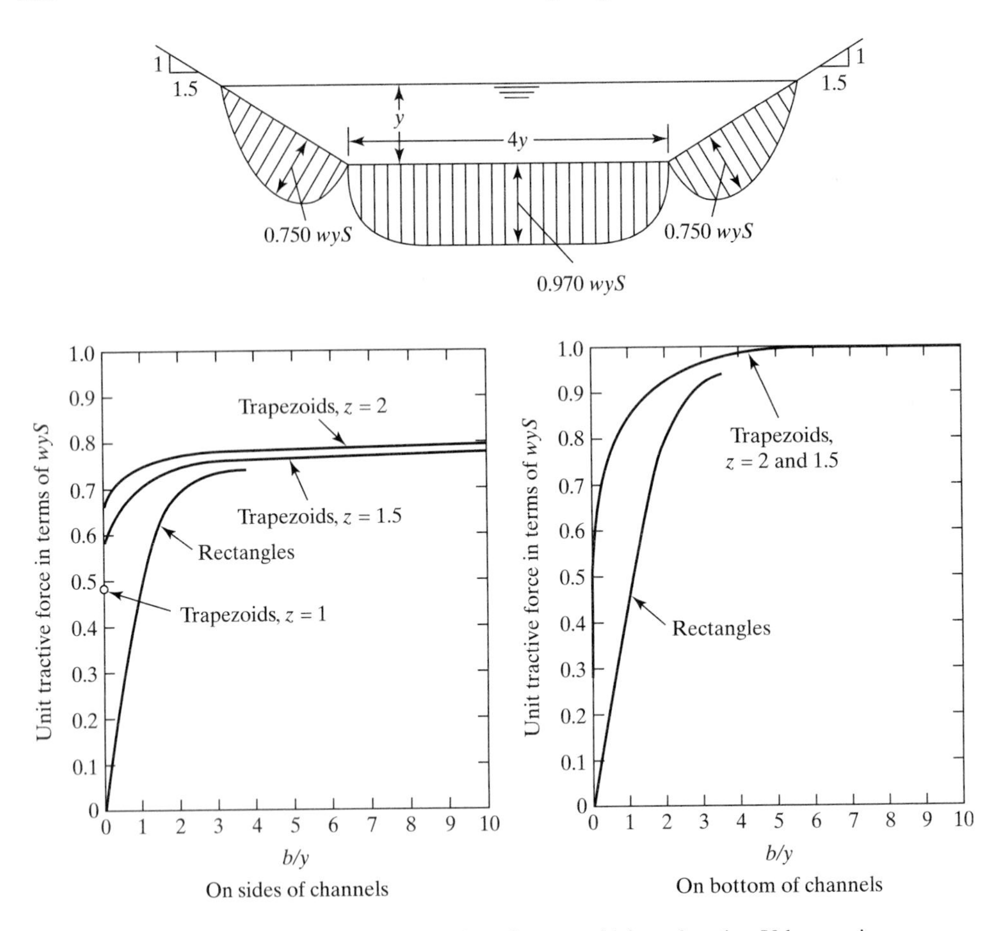

Figure 5.37 Distribution of tractive force in trapezoidal canal section. Values are in terms of wyS, where w = unit weight of fluid, y = depth, S = slope, and b = base width. (After Chow [48])

Runout Length. The design of geomembrane runout coming over the side slopes follows exactly the procedures described in Section 5.3.6. Quite often there is no anchor trench, per se, involved. Usual runout lengths are approximately 1.5 m. The design follows exactly the procedure described in Section 5.3.6. When the canal is of a rectangular section, however, the liner must be anchored to a rigid structural member. Typical details for joining at connections are given in Section 5.11.1.

5.5.3 Unique Features

Due to the empirical nature of the use of geomembranes as canal liners, there are many specialty features that play a role in the success of a particular project.

Cover Soil. Cover soils from 300 to 600 mm in thickness are needed on most geomembrane-lined canals for a number of reasons: to resist erosion, particularly at the

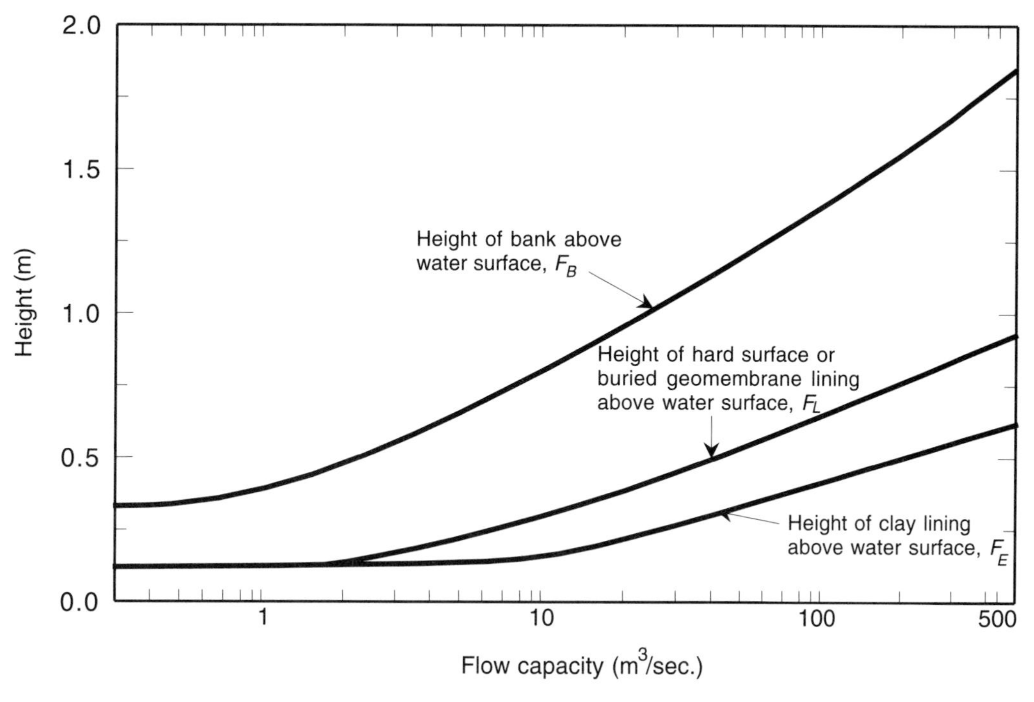

Figure 5.38 Bank heights for canals; and freeboard for hard surface, buried geomembrane, and clay linings. (After Morrison and Starbuck [50])

air-water interface; to hold the liner in place and to dissipate the tractive forces; to protect the liner from exposure from UV light, ozone, wind, and so on; and to protect the liner from damage from water action, plant growth, animals, vandalism, and canal maintenance equipment. However, due to the moving liquid in canal sections, the likelihood of cover soil scour is very high. Therefore, carefully selected cover soil particle sizes and shapes must be considered. The U.S. Bureau of Reclamation recommendations are given in Figure 5.39, where it is seen that the required cover soil is a well-graded sandy gravel. The material's particle shape should be angular or subangular, so as to provide for a high in situ density with correspondingly high shear strength. Compaction to at least 95% standard Proctor compaction is necessary. Because of the angular nature of the cover soil, it is sometimes prudent to place at least a thin (say, 200 g/m^2) geotextile over the geomembrane or to use a two-layer cover soil approach with finer-sized soil particles on the bottom layer next to the geomembrane. The thickness of each layer is usually 150 to 300 mm.

Seam-Joint Overlap. Although geomembrane seams (joints) have not been specifically addressed (see Section 5.10), it should be intuitive that overlap should be placed downstream and should be relatively long. Thus 250 to 300 mm overlaps are recommended, versus the usual 75 to 100 mm overlaps in other applications. If water is the transported liquid, seams are not as important as in other geomembrane applications

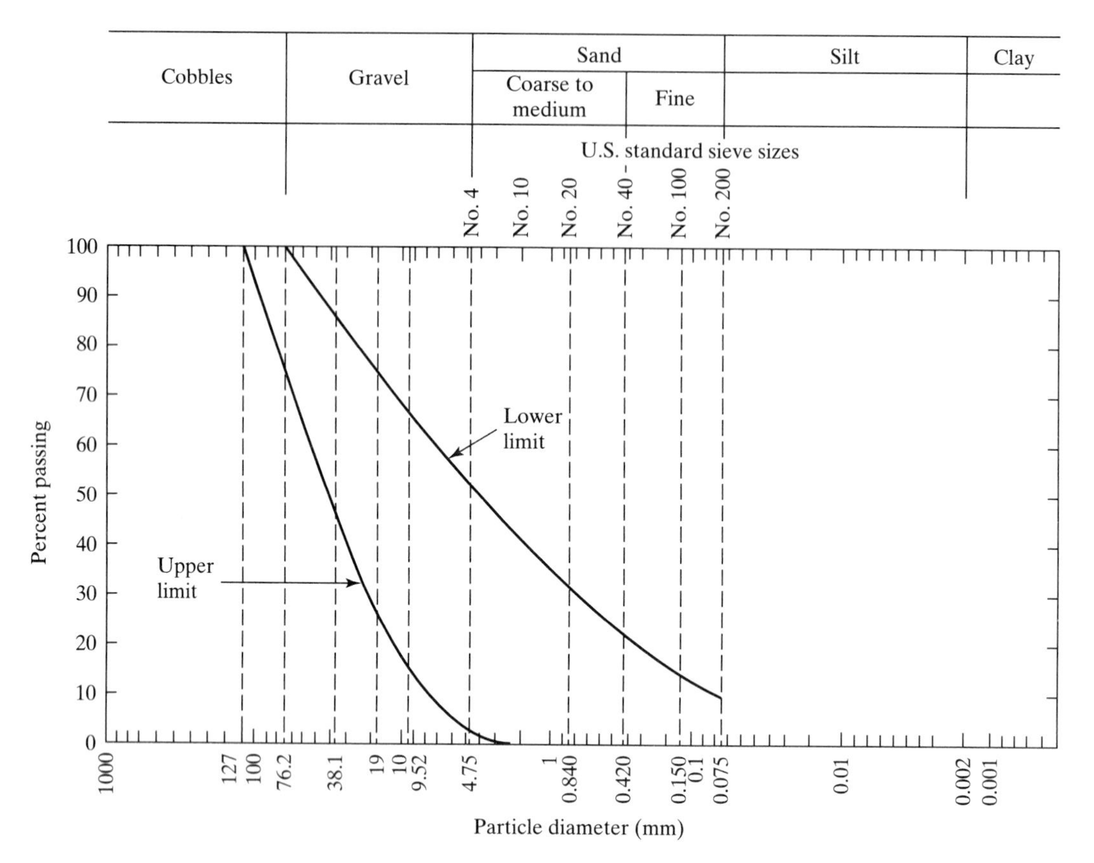

Figure 5.39 Limiting gradation curves for cover soil in canals. (After Morrison and Starbuck [50])

since some water leakage can usually be tolerated. Often a long unseamed overlap is adequate. It obviously is a site-specific situation.

Remediation Work. Since many canals in the past have been made from or lined with nonflexible linings (asphalt, concrete, etc.) and have subsequently deteriorated, remediation projects are plentiful. They occur from a number of situations, such as: settlement-induced cracking; thermal cyclic fatigue cracking; deteriorated expansion joints; opened construction joints; deterioration due to chemical attack, oxidation, and natural aging; and so on. When a crack occurs in a concrete lining, the amount of leakage is alarming. While crack-filling with bitumen is a normal maintenance item, it is a temporary measure and becomes unwieldy as the situation grows progressively worse.

Geomembranes have served nicely as remediation liners. The canal surface must be cleaned, and loose sections removed and repaired. No loose sections of concrete can remain in place. Depending on the surface conditions, a thick nonwoven needle-punched geotextile protection layer is sometimes placed first, then the geomembrane.

Bonding to the concrete is generally not necessary. Edge details, however, are very important, and positive fixity is required (see Section 5.11).

In the absence of a geotextile protection layer, the use of a thick geomembrane is required. Hammer et al. [51] report on the use of a 5.0 mm polypropylene liner placed in a live conveyance canal carrying paper mill effluent. The 3.2 km long project used a mechanical seal, since conventional seaming was not possible while working in live flowing water conditions.

Concrete Cover. Rather than put the liner over the concrete (analogous to putting the cart before the donkey), new construction often justifies putting the geomembrane on the prepared soil subgrade and then concrete on top of it. The water coming through the concrete joints or cracks should be removed; this can be done by placing the concrete over a nonwoven needle-punched geotextile that is directly above the geomembrane. Figure 5.40 gives the transmissivity results of two lightweight geotextiles when fresh concrete is placed upon them. It is easily seen that the in-plane flow rates of the geotextiles are drastically reduced because of intrusion of the wet concrete. If greater flow rates are required, geotextiles heavier than 350 g/m^2 will be required. The use of composite fabrics with a tight pore structure beneath the concrete, and then a high transmissivity portion, would also be possible.

The concrete used to pave over liners for canals is usually reinforced with welded wire mesh and is generally 150 to 300 mm thick. Standard road-paving techniques are used in most cases. In larger projects, special paving equipment can be developed and slip-form paving techniques have been used successfully even under conditions of retrofitting live canals (Comer et al. [53]).

Low Cost Seepage Control. In many western states of the U.S., irrigation canals run over exposed rock surfaces that are very porous and/or highly fractured. Obviously, the seepage losses are significant; at times up to 75%! The U.S. Bureau of Reclamation is evaluating a number of seepage-control schemes, many of which involve geosynthetics.

Embarking on a 10-year feasibility study, the Bureau is evaluating approximately twenty 300 m long sections of water irrigation canals in central Oregon (Swihart [54]). The subgrade is a highly fractured volcanic basalt. The various liner trial sections consist of the following:

- Shotcrete placed directly on the rock subgrade
- Shotcrete over a 0.75 mm PVC geomembrane
- Shotcrete with steel fibers over a 0.75 mm PVC geomembrane
- Shotcrete with polypropylene fibers over a 0.75 mm PVC geomembrane
- Shotcrete with polyethylene microgrids over a 0.75 mm PVC geomembrane
- Grout-filled mattress placed directly on the rock subgrade
- Grout-filled mattress over a 0.75 mm PVC geomembrane
- Geotextile/0.75 mm PVC geomembrane/geotextile composite
- Geotextile/0.91 mm CSPE-R geomembrane composite

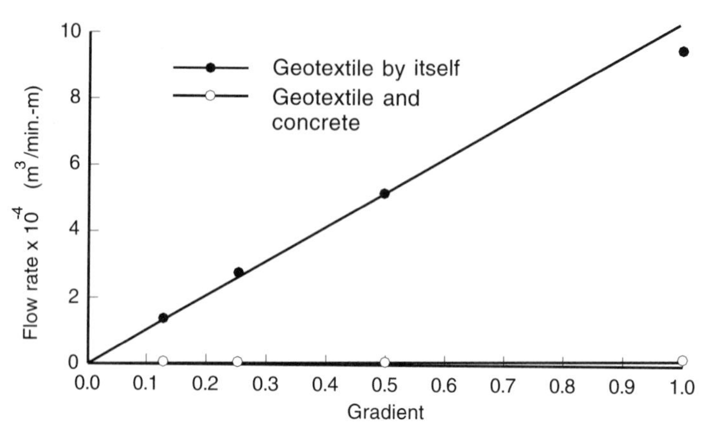

(a) Needle punched nonwoven geotextile of 120 g/m^2

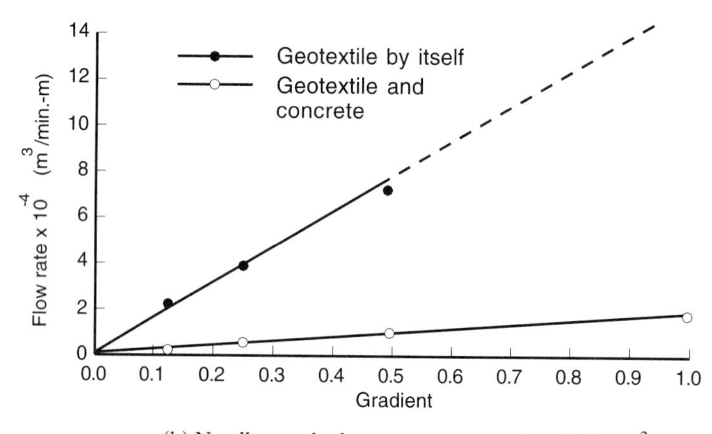

(b) Needle punched nonwoven geotextile of 250 g/m^2

Figure 5.40 In-plane flow rate of geotextile before and after placement of fresh concrete at 2.4 kPa normal pressure on their upper surface. (After [52])

- Geotextile/1.15 mm CSPE-R geomembrane composite
- Geotextile/1.0 mm textured HDPE geomembrane
- Textured 2.0 mm HDPE geomembrane
- Textured 2.5 mm HDPE geomembrane
- Various other combinations

The efficiency of the various sections is being evaluated by means of ponding tests at 2, 5, and 10 year intervals. These values will be contrasted to the initial cost of the installation plus maintenance costs, which are being kept by the local Water District.

5.5.4 Summary

Geomembrane use in canal linings is a rapidly growing field for both new and remediation work. While most liners are covered with soil or concrete, uncovered geomembranes can also be used. As with any meaningful installation, design must be carefully considered using a logical procedure. In this section the design followed directly from the earlier section on reservoir liners, with some notable exceptions. As shown, the tools are available for a rational design.

5.6 SOLID-MATERIAL (LANDFILL) LINERS

The amount of solid waste generated is enormous by any standard of measure. Table 5.13 categorizes waste volume in the United States by type of waste and method of disposal. The projection of waste volume in the year 2000 is 463 million metric tons per year! In the method of disposal portion of the table, it should be recognized that both landfill/surface impoundment and treatment/stabilization of wastes require lined facilities, as does the ash from incinerated wastes. Thus by the year 2000, the projection indicates that $(150 + 136 + (\cong)\ 30) = 316/463$ or 68% of the total waste will be in lined facilities of the type to be described in this section.

Table 5.13 continues the waste volume study by associating with it the products and services that will be necessary. An astounding $43 billion per year by the year 2000 is projected for waste disposal. Within the services grouping, the consulting/engineering, analytical testing, and landfill disposal items should be of keen interest to the readers of this book.

Unfortunately for society, there are still further quantities of solid waste *not* included in Table 5.13. Some of these materials are:

- Bottom and fly ash from incinerators, such as municipal incinerators, hazardous waste incinerators, and trash-to-steam facilities, where the concern is over concentrated heavy metals in the residual ash.
- Dredged river and harbor sediments, which may be contaminated to various levels with a wide range of possible pollutants.
- Waste residue from sewage treatment facilities, where the same concern of heavy metals is often expressed.
- Bottom and fly ash from coal-burning power plants, which represents a tremendous quantity of material.
- Residue from precious metal extraction of previously mined rock, that is, heap leach operations.
- Solidified radioactive wastes, which are sometimes in a vitrified form.

Adequate and safe storage of all of the above material must be ensured. Furthermore, the situation is similar for all the industrialized nations and even looms as a problem for developing nations (see Mackey [56]). Currently geomembranes are being used as a

TABLE 5.13 VOLUME AND SALES IN THE WASTE MARKET IN THE U.S.

Waste Volume (millions of metric tons)

	1977	1988	1993	2000
By type of waste				
Heavy metals	46	103	136	178
Organic chemicals	38	91	120	163
Petroleum derived	15	30	40	55
Inorganic chemicals	16	32	39	50
Other hazardous waste	5	9	12	17
By method of disposal				
Landfill/surface impoundment	11	181	204	150
Treatment/stabilization	2	12	45	136
Incineration	neg	14	32	86
Resource recovery	2	11	28	68
Deep-well injection	5	13	14	9
Illegal disposal	100	32	19	5
Other methods	neg	2	5	9
Total	120	265	347	463

Products and Services (millions of U.S. dollars)

	1977	1988	1993	2000
Products				
Treatment chemicals	40	520	2,550	9,000
Incineration equipment	40	1,600	4,000	8,500
Processing equipment	10	340	1,020	2,700
Analytical instruments	25	390	700	1,100
Other products	5	650	1,030	1,700
(Subtotal)	120	3,500	9,300	23,000
Services				
Transportation	185	2,560	3,770	4,700
Consulting/engineering	5	700	1,400	3,500
Remediation/cleanup	neg	645	1,290	3,400
Incineration services	10	370	920	2,500
Analytical testing	35	610	1,070	2,300
Landfill disposal	15	605	1,370	2,000
Other services	135	710	1,080	1,600
(Subtotal)	385	6,200	10,900	20,000
Total	505	9,700	20,200	43,000

Values are estimated taking into account reduction efforts. Neg means negligible.
Source: After Hanson [55].

primary strategy for such containment. Their proper design and construction is at the heart of this important section.

5.6.1 Overview

As a groundwater pollution-control mechanism, the use of some type of liner on the bottom and sides of a landfill has been considered necessary for many years. This necessity is created by the moisture in the incoming materials augmented by rainfall and snowmelt interacting with the already placed solid waste forming a liquid called *leachate.* This leachate flows gravitationally downward and, if not for a liner, would continue to flow until it encountered groundwater, posing the threat of pollution. Although both the quantity and quality of leachate are of concern, it is the quality that can have horrendous characteristics while at the same time being extremely variable in its composition. Table 5.14 gives the range of leachate values at 18 municipal solid waste (MSW) sites. This table is in contrast to Table 5.15, which gives the range of leachate values at selected *hazardous* waste sites. In particular, note the high levels of several types of organic solvents.

The types of liners that have been used for leachate containment are indeed numerous (see Kays [59] for an excellent review), but the predominant liner material until the early 1980s was compacted clay. When of the proper type, clay liners can achieve hydraulic conductivity (or permeability) values in the range 1×10^{-6} to 5×10^{-8} cm/s and perform very satisfactorily. There are two drawbacks to compacted clay liners (CCLs), however, both of which have given impetus to the use of geomembranes for landfill liners.

- Clay liners must be typically 600 to 1500 mm thick, which takes up significant landfill volume that could be used to house the waste itself.
- Clay liners have been shown to be subject to chemical reactions and subsequent shrinkage when evaluated in fixed-wall permeameters and exposed to full concentrations of organic solvent leachates (e.g., xylene, methanol, aniline, and acetic acid) [60].

TABLE 5.14 RANGE OF LEACHATE CHARACTERISTICS FROM MUNICIPAL SOLID WASTE LANDFILLS*

Chemical oxygen demand (COD)	40–89,520
Biological oxygen demand (BOD)	81–33,360
Total organic carbon (TOC)	256–28,000
pH	3.7–8.5
Total solids (TS)	0–59,200
Total dissolved solids (TDS)	584–44,900
Total suspended solids (TSS)	10–700
Specific conductance	2810–16,800
Alkalinity ($CaCO_3$)	0–20,800
Hardness ($CaCO_3$)	0–22,800
Total phosphorus (P)	0–130
Ortho-phosphorus (P)	6.5–85
NH_4–N	0–1106
$NO_3 + NO_2$–N	0.2–10.29
Calcium (Ca^{2+})	60–7200
Chlorine (Cl^-)	4.7–2467
Sodium (Na^+)	0–7700
Sulfate $(-SO_3)^{2+}$	1–1558
Manganese (Mn)	0.09–125
Magnesium (Mg)	17–15,600

*All values in mg/l, except specific conductance, which is in μS/cm, and pH, which is in standard units.

Source: Chian and deWalle [57].

TABLE 5.15 LEACHATE CHARACTERISTICS FROM A HAZARDOUS SOLID WASTE LANDFILL

Parameter	Unit	Value
Alkalinity	mg/l as $CaCO_3$	11600
BOD_5	mg/l	19500
COD	mg/l	37200
Conductivity	μMhos	38575
Oil & Grease	mg/l	210
pH	St. units	9.1
Silica	mg/l as SiO_2	141
Total Dissolved Solids	mg/l	50100
Total Organic Carbon	mg/l	11500
Total Suspended Solids	mg/l	212
Turbidity	NTU	156
Calcium	mg/l	30
Iron	mg/l	19
Magnesium	mg/l	16
Nickel	mg/l	28
Potassium	mg/l	2715
Sodium	mg/l	13250
Chloride	mg/l	12100
Nitrate	mg/l as N	51
Phosphorus	mg/l as P	21
Sulfate	mg/l as SO_4	3850
Benzene	μg/l	6500
Chloroform	μg/l	1330
1,1-Dichlorethane	μg/l	2900
Ethyl Benzene	μg/l	35
Phenol	μg/l	14100
Styrene	μg/l	95
Toluene	μg/l	21100
m–Xylene	μg/l	284
o–Xylene	μg/l	93

Source: Dudzik and Tisinger [58]

This second feature of clay liners caused the U.S. Environmental Protection Agency (EPA) to promulgate the following regulations on July 6, 1982:

> Prevention (via geomembranes), rather than minimization (via compacted clay liners), of leachate migration produces better environmental results in the case of landfills used to dispose of hazardous wastes. A liner that prevents rather than minimizes leachate migration provides added assurance that environmental contamination will not occur.

The above series of events has ushered in increased awareness of, interest in, and demand for geomembranes made from polymeric materials. This section and the following one are directed at geomembrane designs for landfill liners beneath the waste and cover systems above the waste.

Note, however, that all landfills are not toxic, hazardous, or radioactive, and some are as harmless as uncontaminated building demolition debris. A suggested ranking of environmental and health concerns from lowest to highest is as follows:

- Building demolition
- Power plant ash
- Dredged river and harbor sediments
- Sewage-treatment sludge
- Treated or incinerated waste ash
- Nontreated nontoxic waste
- Untreated municipal waste
- Untreated biological (hospital) waste
- Heap leach residual waste
- Near-hazardous waste
- Hazardous waste
- Radioactive waste: low level, transuranic, high level

Municipal waste is the highest in quantity, followed by industrial waste and power plant ash. Hazardous and radioactive wastes are much less in quantity, but significantly more dangerous in quality. For this reason the U.S. Environmental Protection Agency has expended considerable effort in making a distinction between hazardous and nonhazardous waste. For *hazardous waste* (as classified as having any one of 800+ priority pollutants above legislated acceptable limits) federal regulations fall under 40 CFR 264.221 (1986). Such Subtitle C hazardous-waste landfills, surface impoundments, and waste piles must have

> . . . two or more liners and a leak detection system between such liners. The liners and leak detection system must protect human health and the environment . . . The requirement for the installation of two or more liners . . . may be satisfied by the installation of a primary liner designed, operated, and constructed of materials to prevent the migration of any constituent into such liner during the period such facility remains in operation (including any post-closure monitoring period), and a secondary liner designed, operated, and constructed to prevent the migration of any constituent through such liner during such period.

Furthermore, the leachate collection and removal system regulations for double-lined waste piles and landfills specifically require that the system be designed and operated to ensure that the leachate depth over the primary liner does not exceed 300 mm. The system must also be chemically resistant to wastes and leachate, sufficiently strong to withstand landfill loadings, and protected from excessive clogging. Minimum technology guidance for double liners provides specific design criteria for the leachate-collection and leak-detection systems. These criteria are as follows:

- The leachate collection system should be capable of maintaining a leachate head of less than 300 mm above the primary liner.
- Both leachate collection and leak detection systems should have at least a 300 mm granular drainage layers that are chemically resistant to the waste and

leachate, with a hydraulic conductivity not less than 0.01 cm/s or an equivalent geosynthetic drainage material (e.g., a geonet).

- The minimum bottom slope of the facility should be two percent.
- The leachate-collection system should have a granular soil filter or geotextile above the drainage layer to prevent excessive clogging.
- Both systems when made of natural soils should have a drainage system of interconnected pipes to efficiently collect leachate; the pipes should have sufficient strength and chemical resistance to perform under anticipated landfill loadings.
- By virtue of the leak-detection rules 40 CFR 260, 264, 265, 270, 271 (1992), a site-specific action leakage rate (ALR) must be set for each facility.
- A construction quality assurance (CQA) program must be developed to see that the constructed facility meets or exceeds all design criteria, plans, and specifications.

Note, that most federal and many state regulations refer to the leachate-collection systems above the primary liner as the primary leachate collection and removal systems (PLCRS), and to the leak-detection systems between the two liner systems as the secondary leachate collection and removal systems (SLCRS). For simplicity, we will refer to these two drainage systems as *leachate collection* (above the primary liner) and *leak detection* (between the primary and secondary liners).

For *nonhazardous waste* (those wastes not containing priority pollutants, or at least not higher than prescribed levels) federal regulations fall under 40 CFR Parts 257 and 258 Subtitle D for solid-waste disposal. These regulations clearly identify MSW as being the focus material of the regulations. However, it is presumed that nonhazardous industrial waste could fall under these Subtitle D regulations, rather than under those of Subtitle C. Some salient points regarding Subtitle D liner systems are as follows.

- A leachate collection system should be located above the liner system.
- The leachate collection system should be capable of maintaining a leachate head of less than 300 mm.
- The liner system could be a single composite liner (i.e., it is not required to have a double-liner system with leak detection capability).
- The single composite liner must be a geomembrane placed over a compacted clay liner.
- The geomembrane must be at least 0.75 mm thick, unless it is HDPE. A HDPE geomembrane must be at least 1.50 mm thick.
- The geomembrane must have "direct and uniform contact with the underlying compacted soil component." Furthermore the phrase "intimate contact" is used in some state regulations.
- The compacted clay liner must be at least 600 mm thick and of a permeability of 1×10^{-7} cm/s, or less.

Thus it is seen that by regulatory mandate there is an extremely large use of geomembranes and associated drainage systems for liner systems in the U.S. In both hazardous- and nonhazardous-waste legislation, geosynthetic materials can be substituted for natural soil materials, if technical equivalency can be shown. Thus geonets can often be used to replace drainage soils, geotextiles can often be used to replace filter soils, and geosynthetic clay liners can sometimes be used to replace compacted clay liners. It is quite clear that liner and drainage systems of this type are currently driving the geosynthetics industry in North America.

The regulations just mentioned are minimum technology guidance (MTG) and individual states can, and often do, exceed these requirements. For example, in New York state, all solid waste (both hazardous and nonhazardous) goes into a landfill that consists of a double-composite (geomembrane and compacted clay) liner system. Various other states also have regulations that are distinctly more restrictive than the federal standards. Indeed, a consulting engineering firm under contract to an owner/operator developing a landfill liner system must be fully cognizant of the state regulations where the facility will be located. It is usually the state that must issue the permit to proceed with construction.

While the preceding text on landfill liner regulations was focused on the U.S., it must be noted the German regulations are also fully developed. Regulatory personnel from the two countries have long interacted with each other, resulting in largely parallel systems. Some differences contained in German regulations are:

- The only geomembrane resin type that is permitted is HDPE.
- Its minimum thickness must be 2.0 mm.
- The compacted clay liner beneath the geomembrane is highly engineered, thicker, and of lower hydraulic conductivity than in the U.S.
- The drainage stone above the geomembrane is prescribed and must be 16/32 mm diameter rounded stone.
- The protection layer beneath the drainage stone must be such that no more than 0.25% strain is imposed to the underlying geomembrane.
- Intimate contact, translated directly as "press fit," must exist between the geomembrane and the underlying compacted clay.
- The seaming of the geomembranes is done under highly regulated circumstances.

These differences, and many similarities, between the two countries have been the subject of a 1996 workshop (Corbet [61]). This is an important report because most countries appear to look to either the U.S. or Germany for guidance in formulating emerging landfill regulations.

5.6.2 Siting Considerations and Geometry

Due largely to nontechnical considerations (i.e., social, political, and legal), the siting of solid-waste landfills of any type is very difficult. This difficulty is increased even further when the waste contains hazardous or radioactive materials. Nowhere is the NIMBY (*N*ot *I*n *M*y *B*ackyard) syndrome more obvious. An even higher (more politically oriented) level of difficulty is expressed by the acronym NIMTO (*N*ot *I*n *M*y *T*erm of *O*ffice). Yet when properly sited, designed, constructed, and maintained, landfills can be made secure for as long as they generate leachate—even in the case of landfills containing low-level radioactive waste. (High-level radioactive wastes have completely different containment strategies than those to be discussed in this book.) In siting, the following list of items are important to consider: the stratigraphy and geology of site, the depth to the watertable, the quality and significance of subsurface water, the use of down-gradient water, the population density, the weather conditions (particularly precipitation), the seismicity of the region, and any other concerns unique to the particular site.

Regarding the geometry for such landfills, the general recommendations and specific designs discussed in Section 5.3.1 have applicability here as well. A major difference, however, is the manner of placement of the contained materials, which are now solid rather than liquid. Solid waste landfills are of the following configurations: in an excavation below grade, as a fill above grade, as a combination of below and above grade, and within a canyon between two hillsides. The waste depths and/or elevations have no technical limits and the tendency is toward large regionalized *megafills,* versus small localized sites.

The planning of the landfill must be done in the design stage with particular emphasis on the leachate collection (and leak detection, if any) system and the leachate removal. Separate cells within a landfill are often made, each being an internal containment zone partitioned off by a small soil embankment or *berm.* The external embankment dikes surrounding the site are usually quite steep and sometimes reinforced with geogrids or geotextiles. When below grade, a haul road is made at the top and used for access during construction and filling. Accepted solid-waste practices must be used (see Tchobanoglous et al. [62]). As an aside, it is suggested that future cells may be enclosed within a geomembrane itself (a super trash bag, if you will) or be placed beneath a temporary enclosure or canopy made as a tension structure or air-supported structure. Such actions would drastically reduce leachate generation, but are very expensive from an initial cost perspective.

5.6.3 Typical Cross Sections

A critical element in the proper functioning of a landfill is the containment system. This is often referred to as a *liner system* and thus includes geomembranes, compacted clay liners (CCLs) and geosynthetic clay liners (GCLs). Note that a geomembrane placed directly over a CCL or GCL is a single liner, albeit one of a composite nature. For solid-waste landfills, a leachate collection (and removal) system must be integrated into the system, and in cases of a double liner, a leak detection system is needed as well. The leachate collection and removal system is located above the uppermost, or primary, liner. For single-liner systems, the only way to monitor for a leak through the liner is when the leachate become fugitive. This has traditionally necessitated downstream monitoring wells and, for comparative purposes, upstream wells. If the wells are numerous enough and properly sited, the difference in water quality between downstream and upstream wells is indicative of the functioning of the landfill liner. If the quality is the same, the lined landfill is functioning as intended. If not, a leak is suggested. Considerably better than such a hit-or-miss leak detection approach is to construct a double-lined landfill liner with a gravel and perforated pipe system or geonet drainage system between them. When graded to a low spot beneath the landfill, any leachate getting through the primary (upper) liner indicates a leak.

The leakage can come from a number of different sources. Gross et al. [63] have identified the following possible sources: primary liner leakage, expulsion of construction water, expulsion of compression water, compacted clay liner consolidation water, and infiltration water from other areas. Particularly troublesome is the consolidation water from a CCL as the lower component of a primary composite liner. When such a CCL is placed at wet of optimum water content, as it usually is, each layer of waste placed in the facility will expel consolidation pore water, which must necessarily be captured and transmitted in the leak detection network. The quantities can be very large and can easily mask the true leakage that may be coming through the primary liner. The situation can become quite contentious if an ALR has been set for the site and it is exceeded because of consolidation water from the overlying CCL.

If, indeed, leachate leakage through the primary liner is noted above the ALR, corrective measures must be instituted. The prescribed corrective measures will be delineated in the response action plan (RAP), which necessarily accompanies the regulatory permit and sets the site specific ALR value. Such actions might be: continuous monitoring and tracking of the leakage rate, chemical analysis of both the liquid in the leak detection system and the leachate in the primary collection system for comparative purposes, placement of downstream monitoring wells (or additional ones), secession of waste placement in the facility, and removal of waste from the facility to find and repair the leak(s).

Using these concepts, Figure 5.41 shows the genesis of liner systems that have been used in the U.S. since the enactment of the 1982 legislation mentioned earlier. Table 5.16 contains details for each of the parts of the figure. Figure 5.41a shows the best that we could hope for in a solid waste liner system prior to 1982. There were no regulations on the clay thickness or on its permeability. (Recall that throughout this book, permeability is used rather than hydraulic conductivity.) The enactment of the 1982 regulations gave rise to the cross section of Figure 5.41b. Shortly thereafter, in 1983, it was recognized that redundancy of geomembrane liners was desirable which allowed for a leak detection layer to be placed between the two geomembranes; see Figure 5.41c.

CCLs were not to be denied. The study in [60] was flawed to the extent that neat chemicals were used and evaluated in rigid-wall permeameters. With additional research, CCLs re-entered the cross section as part of a composite liner; see Figure 5.41d. Work by Giroud and Bonaparte [65] confirmed the low leakage rates that could be accomplished through composite liners that have intimate contact with one another.

It was soon recognized that these layered geosynthetic and natural soil systems were difficult to construct, particularly due to the problem of the drainage soil's stability when placed on geomembranes on side slopes. Geonets entered as the leak-detection network (no perforated pipes were required), as shown in Figure 5.41e. Thus stability was assured, as well as a considerable savings in volume.

The effectiveness of a composite liner is not to be forgotten. Furthermore, if it is the best strategy for the secondary liner, why not for the primary liner as well? Hence we have the cross section in Figure 5.41f. The CCL above the geonet leak-detection system, however, is extremely difficult to properly place and compact. Furthermore, the

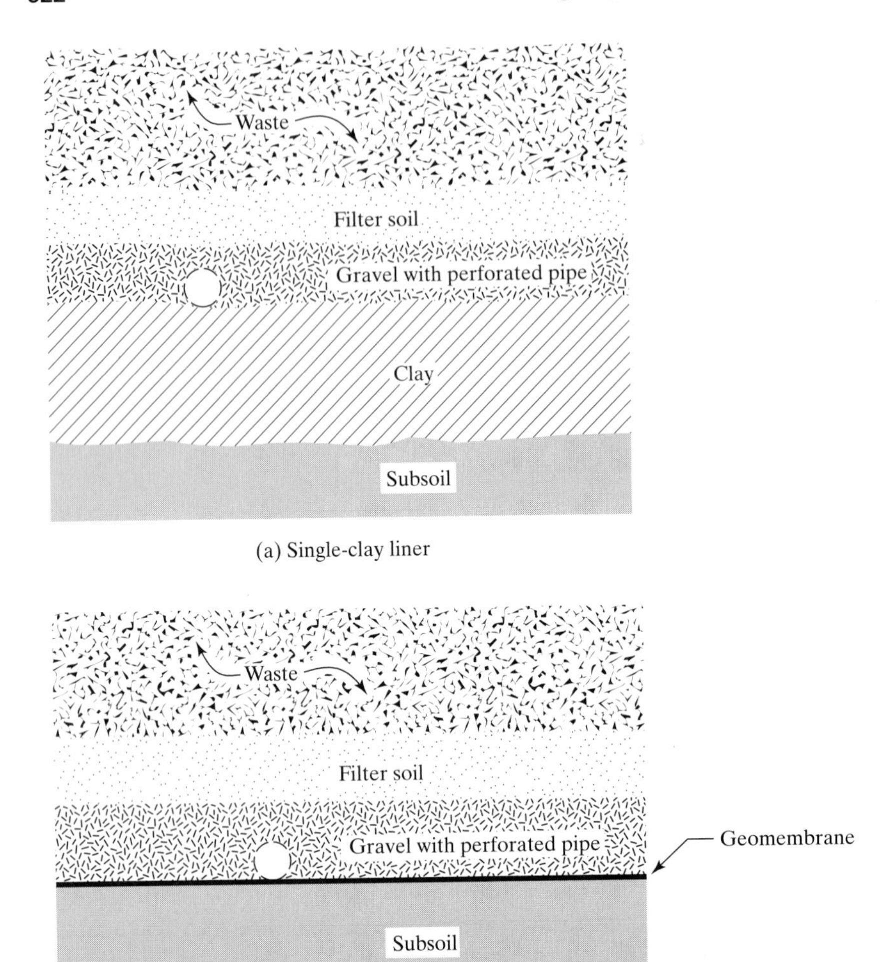

Figure 5.41 Genesis of liner cross sections used to contain solid waste in the U.S. since 1982.

consolidation water expelled during waste placement is so troublesome that an alternate scheme is very attractive. GCLs as the lower component of the primary liner nicely solve this situation, with the added attraction of saving volume; see Figure 5.41g.

Finally, the use of either geonets (biaxial or triaxial) or high compressive strength geocomposites for leachate collection above the primary liner is being used in some facilities, particularly on side slopes; see Figure 5.41h. Thus it is seen that geosynthetics have replaced natural soils in the entire cross section, with the exception of the lower component of the secondary liner. In this location, directly on top of the soil subgrade, a CCL can be placed and compacted with no danger of damaging the associated geosynthetic materials.

It is readily seen in the cross section of Figure 5.41h that great demands are being placed on geosynthetics in solid-waste liner systems. The designs to follow, as well as

TABLE 5.16 GENESIS OF LINER SYSTEMS USED IN THE U.S.

Fig. 5.41 Part	Type of Liner System	Approx. Date in Use	Leachate Collection System	Primary Liner	Leak Detection System	Secondary Liner
(a)	single CCL	pre-1982	soil/pipe	CCL	none	none
(b)	single GM	1982	soil/pipe	GM	none	none
(c)	double GM	1983	soil/pipe	GM	soil/pipe	GM
(d)	single GM, single composite	1984	soil/pipe	GM	soil/pipe	GM/CCL
(e)	single GM, single composite	1985	soil/pipe	GM	GN	GM/CCL
(f)	double composite	1987	soil/pipe	GM/CCL	GT/GN	GM/CCL
(g)	double composite	1989	soil/pipe	GM/GCL	GT/GN	GM/CCL
(h)	double composite	1993	GT/GC	GM/GCL	GT/GN	GM/CCL

Abbreviations: GM = geomembrane
GN = geonet
GT = geotextile
GC = geocomposite
CCL = compacted clay liner
GCL = geosynthetic clay liner

others that have been already presented in this book, focus on many of these details. In this discussion on liner systems we mentioned savings in volume, or air space, several times. Example 5.17 illustrates the financial impact of air space to a facility's owner/operator.

Example 5.17

In a 5 ha landfill cell, a designer is considering using a 5 mm thick drainage geonet to replace a 300 mm thick sand layer as a leak-detection layer. Technical equivalency has been shown numerically in Section 4.2.3 and has been corroborated in the field [66]. How much will this replacement save if the tipping fee of the solid waste is \$80 per cubic meter, as it is currently for municipal solid waste in the Philadelphia area.

Solution: The air space saved is first calculated.

$$\Delta H = 300 - 5$$
$$= 295 \text{ mm}$$

For a 5 ha cell at \$80/m^3,

$$\text{Savings} = (80)(10{,}000)\left(\frac{295}{1000}\right)$$
$$= \$236{,}000/\text{ha}$$
$$\text{Savings} = \$1{,}180{,}000/5 \text{ ha} \ldots \text{clearly a significant amount!}$$

5.6.4 Grading and Leachate Removal

The profile and configuration of the bottom of a landfill must be such that gravitational flow to a low point (a sump) always exists. This must be true for both the leachate collection and, when present, for the leak detection system, as well. Thus accurate grading

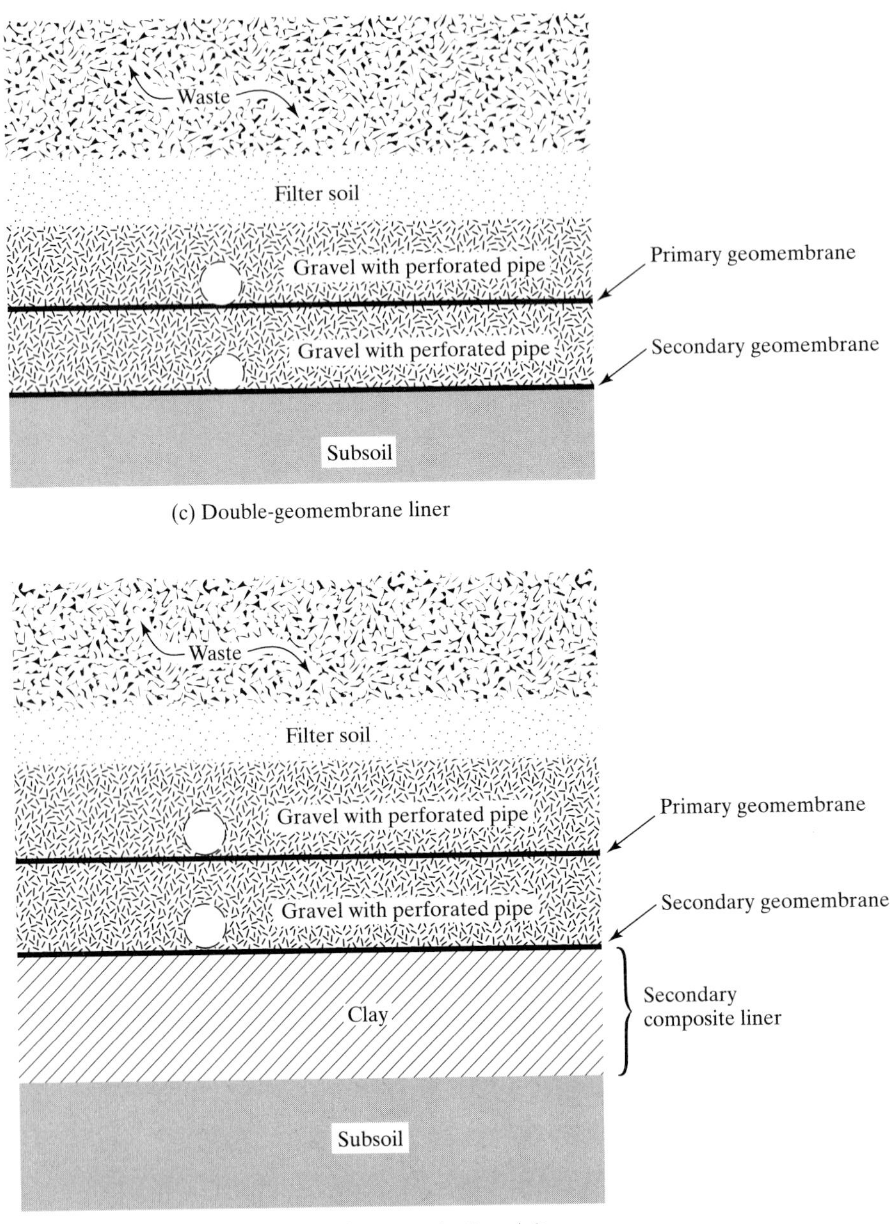

(c) Double-geomembrane liner

(d) Single-geomembrane, single-composite liner (after Burarek and Pacey [64])

Figure 5.41 (*continued*)

of the bottom of the landfill (or cell within a landfill) is very important. The consequence of improper design (localized low points, subsidence of subsoil, poor construction quality control and assurance, etc.) is that leachate will pond above the geomembrane and eventually diffuse through it, rather than being properly removed and

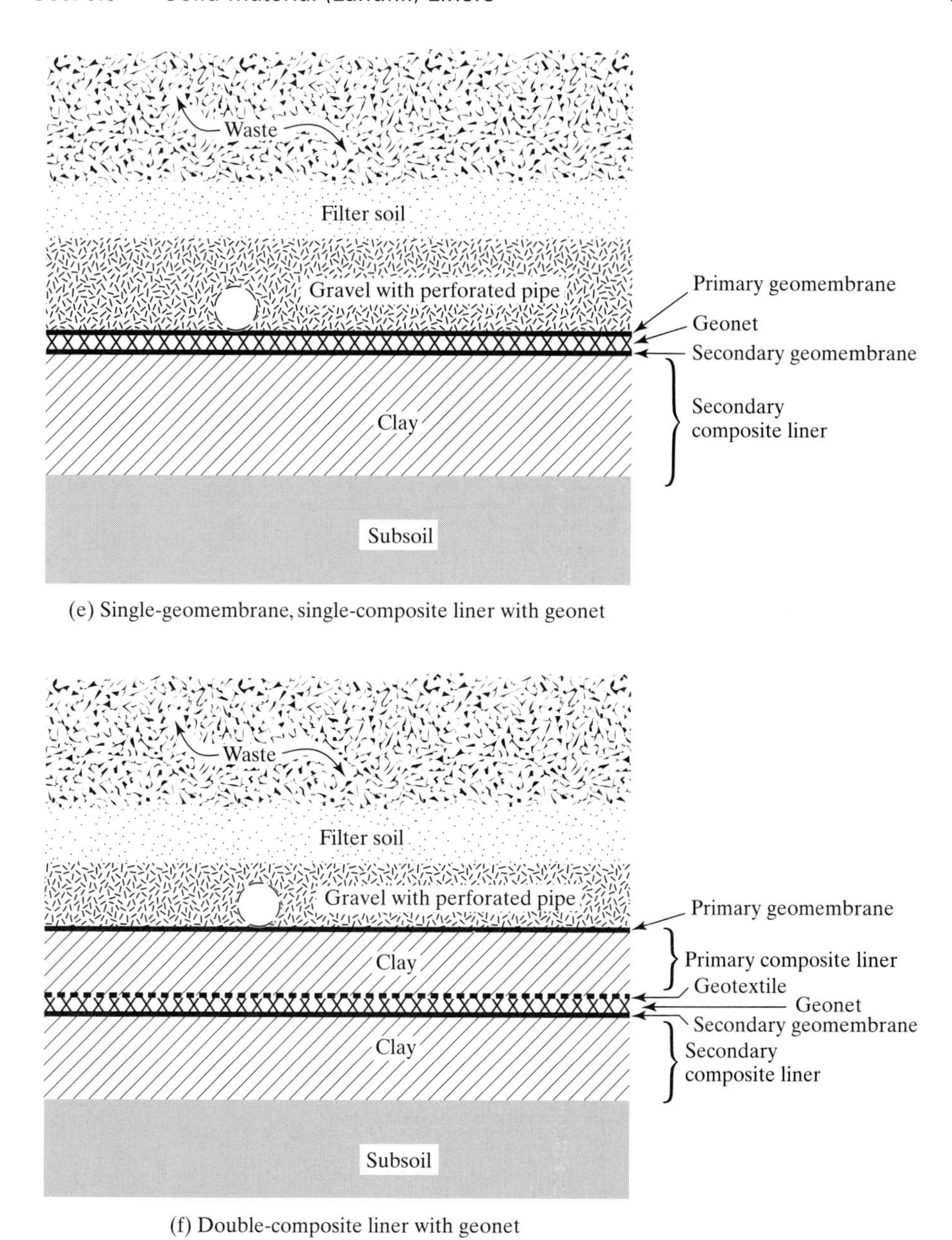

(e) Single-geomembrane, single-composite liner with geonet

(f) Double-composite liner with geonet

Figure 5.41 (*continued*)

treated. Grading of the site for gravity flow leachate collection and leak detection is most important and can be most difficult as well.

For large sites with no watertable restrictions, grades of 2% or higher can be designed and constructed with relative ease. Note, however, that such grades in a large

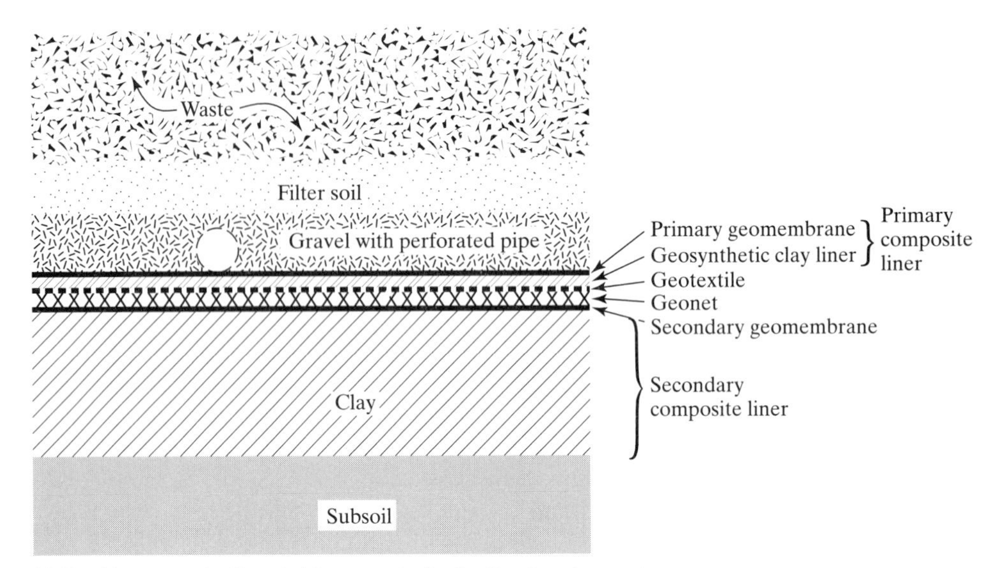

(g) Double-composite liner (with geosynthetic clay liner) and geonet

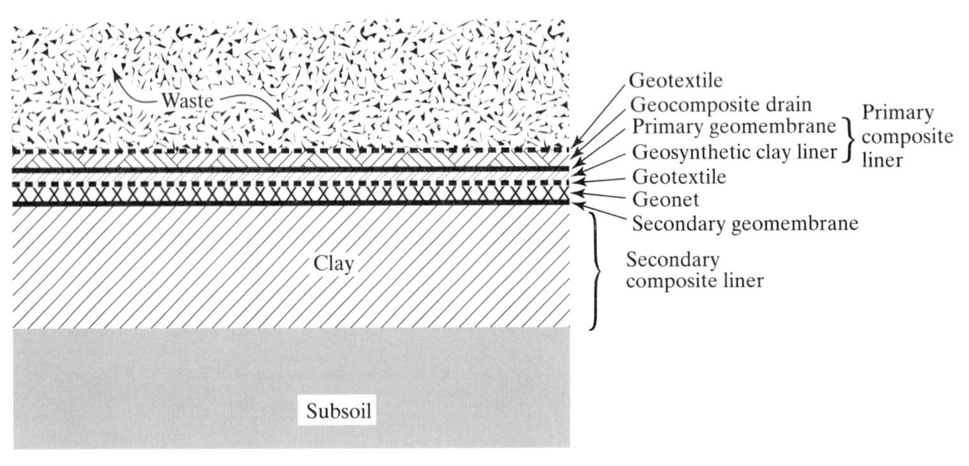

(h) Double-composite liner (with geosynthetic clay liner), geonet, and geocomposite

Figure 5.41 *(continued)*

landfill take considerable air space from the facility and alternate designs (e.g., accordian-shaped profiles) become necessary. For smaller sites and/or high watertables, however, the design is usually on the basis of 0.5 to 1% slopes, which is very difficult and costly. Figure 5.42 gives some actual case histories of contours for gravity flow drainage. The low point of the leachate collection system must terminate at a sump with an outlet stemming from this location to beyond the landfill or cell.

For sites where removal pipe systems within the leachate collection soil must be periodically inspected, cleaned, and flushed, both access and egress must be available.

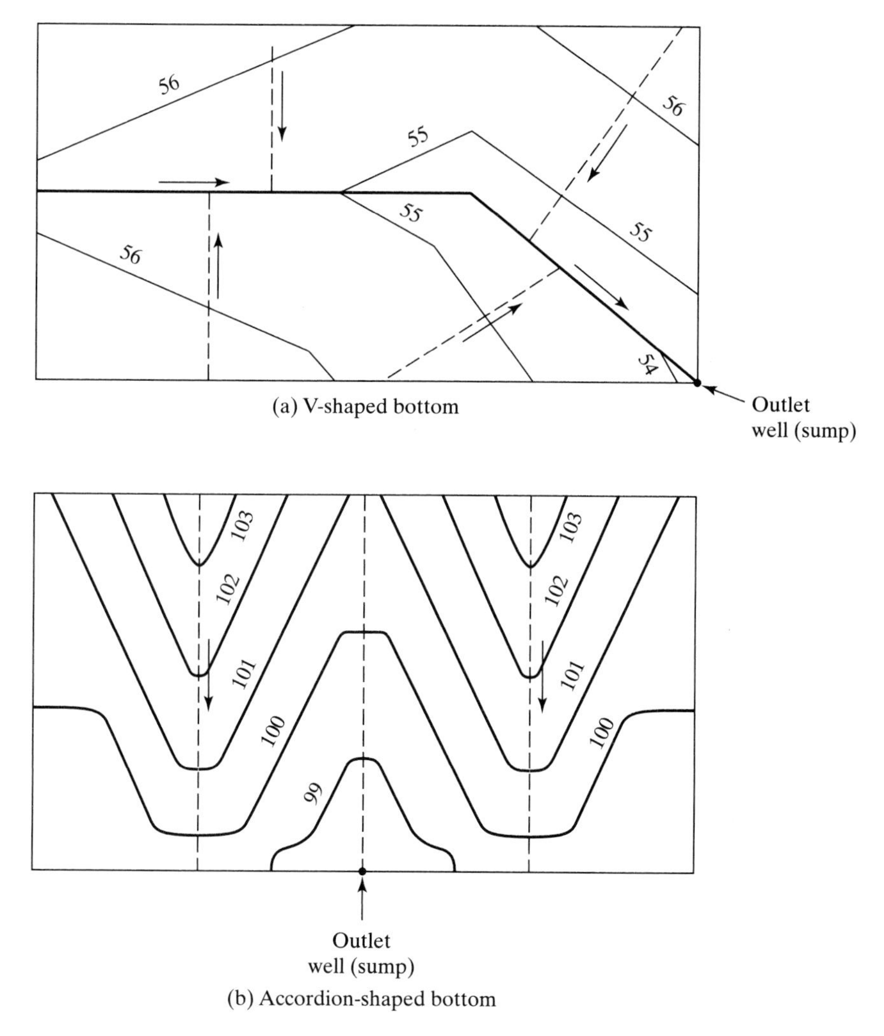

Figure 5.42 Various shapes for bottom of landfills for proper leachate collection and/or leak detection.

This will require careful planning and can completely dictate the nature of the grading plan. Some permits require only the header pipe to be cleaned; thus feeder liners can be laid out in a herringbone fashion and V-shaped contours over the entire cell are acceptable. Figure 5.42a shows such a scheme, but sharp bends, both horizontal and vertical (i.e., up side slopes), must be made with wide-angle fittings. If all the pipes must be inspected, cleaned, and flushed, the accordion profile shown in Figure 5.42b must be considered. Within each trough will be a perforated pipe having access at the top of the slope and egress at the bottom of the slope. This latter case is very difficult to handle when constructing individual cells within a larger permitted facility. Careful detail and planning must be considered and it must be designed accordingly.

Figures 5.43a,b show details of low- and high-volume leachate-removal sumps and manholes [67, 68]. The manhole risers extend vertically through the waste as it is being placed and eventually penetrate the landfill cover when the site is closed. Leachate is removed using submersible pumps until the leachate is no longer generated. Note that the post-closure period for hazardous waste landfills according to the U.S. EPA is 30 years. A critical aspect of this type of leachate removal manhole is the downward pressure exerted on the outside of the pipe risers by the subsiding waste mass. Called *downdrag* or *negative skin friction* by geotechnical engineers in dealing with end bearing piles, piers, and caissons, it can result in very large downward pressures, hence, the careful detail beneath the footing in both sketches of Figure 5.43. Numeric examples from Richardson and Koerner [67] illustrate this feature. To relieve the pressures somewhat, the outside of the concrete riser sections extending through the waste is sometimes wrapped in a low-friction material, such as HDPE. Other downdrag-reducing methods such as bitumen slip layers are also possible [69]. Vertical manhole risers of this type through the waste, however, are generally *not recommended* for leachate removal.

The downdrag issue is essentially eliminated by using a side-wall riser scheme for leachate removal, as shown in Figure 5.43c. In so doing, the vertical risers in the waste mass itself, which result in many operational problems, are avoided. For side-wall risers, a large-diameter HDPE pipe, typically 300 mm, terminates with a T-section that has numerous perforations in it. A HDPE rub sheet is placed under the T-section. The riser is brought up the side slope and into a shed. Here a submersible pump is lowered into the pipe for the removal of leachate on demand. The pump can be withdrawn for maintenance or if problems arise. The photographs in Figure 5.44 show the type of sump area along with the riser pipes. In the lower photograph, the two large pipes are for primary leachate removal, the smaller central pipe is for header cleanout, and the smaller pipe on the right is for leak detection monitoring. An additional favorable aspect of side-wall risers of the type shown in Figure 5.43c is that the sump can be quite large in area, for example, 5 to 10 m in length and width. By so doing, the depth of leachate in the sump can be limited to the usually prescribed value of 300 mm. Thus regulatory constraints are met and the placement and seaming of the geomembrane is much simpler than with the deep sump area of Figure 5.43b.

The monitoring and removal of liquid from the leak detection system between the primary and secondary liners in a double-lined system is also necessary. To do so, a HDPE pipe of 100 to 150 mm diameter is placed between the primary and secondary liners from a small sump in the secondary liner up the side slope, as shown in Figure 5.45. It is unfortunately necessary to penetrate the primary liner at the upper slope, but pipe boots can be carefully fitted and seamed for this detail. A small diameter bailer or pump is generally used to extract and monitor the liquid within this leak detection piping system (see the right-side pipe in the lower photograph of Figure 5.44).

Liquid in leak detection sumps can be quantitatively measured by a number of techniques. Some of the following have been used at sites and mentioned in the literature: monitoring the change in liquid depth in the sump or riser pipe, using a flow meter with a mechanical or automatic accumulator coming from the sump, use of a tipping bucket for gravity systems with a mechanical or automatic counter, and adapting a weir to the tipping bucket for gravity systems.

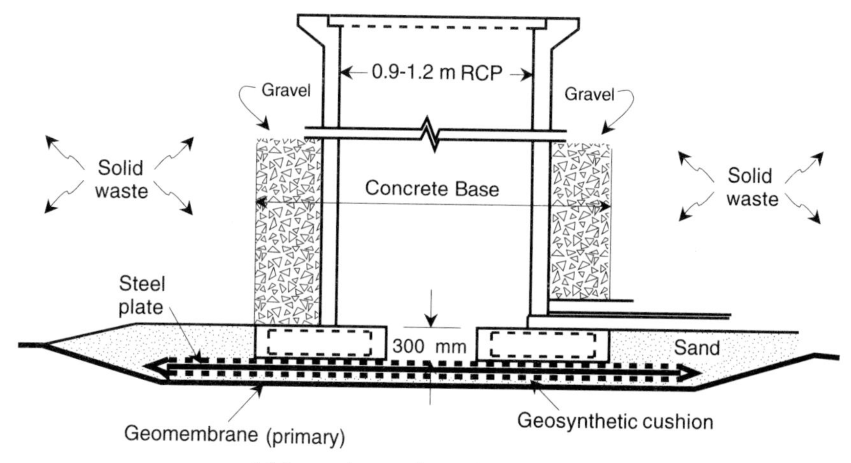

(a) Low-volume primary leachate collection

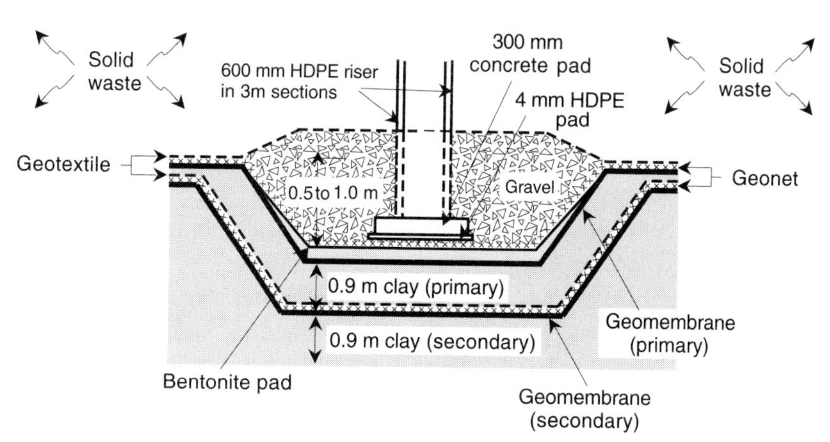

(b) High-volume primary leachate collection manhole (after Chemical Waste Management, Inc.)

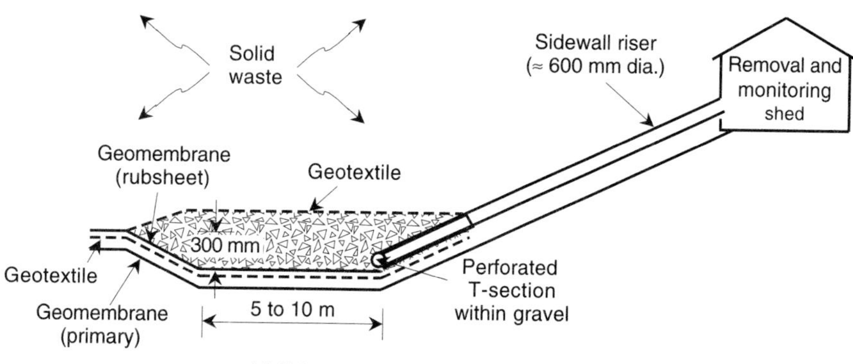

(c) Side wall primary leachate collection riser

Figure 5.43 Various removal designs for primary leachate collection systems.

Figure 5.44 Primary leachate collection sump side-wall riser pipes leading from sump into a collection and monitoring shed.

The quantity of the liquid gathered in the leak detection sump must, of course, be compared to the ALR, as prescribed in the site-specific response action plan (RAP). Recall the discussion in Section 5.6.3. The quantities of liquids found in the leak detection sumps of different facilities vary widely. The actual quantity appears to depend upon: the attitude of the installation contractor, the workmanship of the contractors' personnel, solid waste landfills versus surface impoundments, appropriate CQA versus

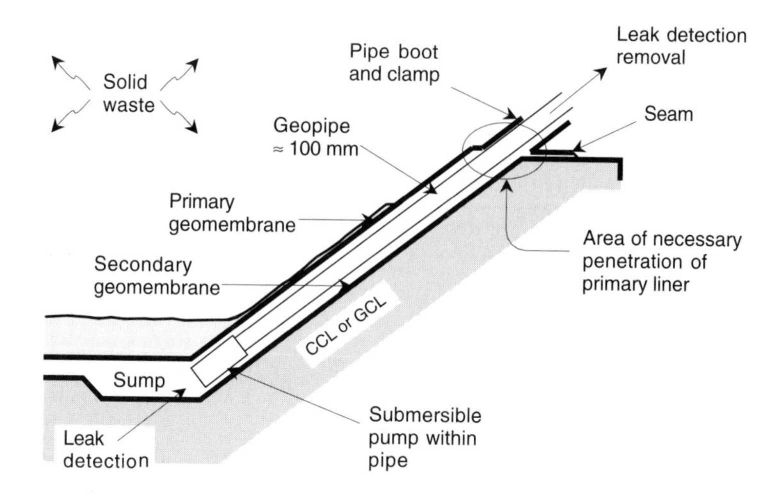

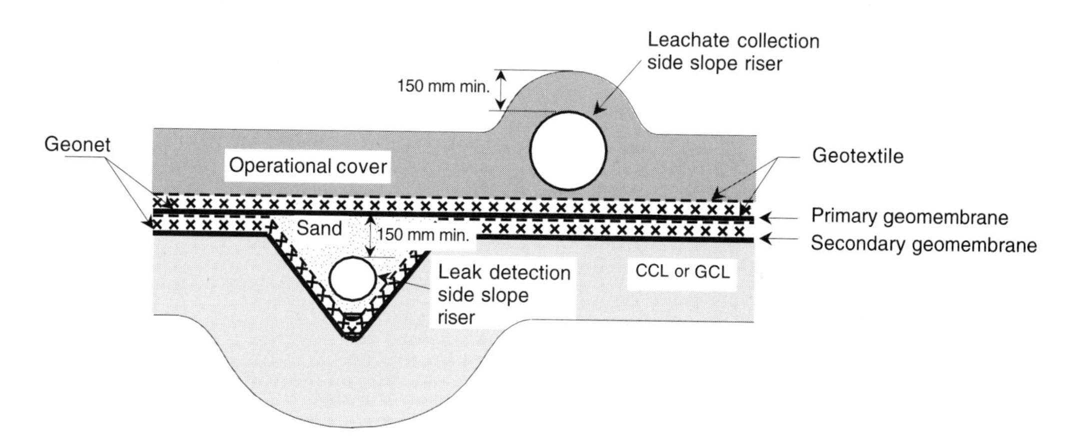

Figure 5.45 Leak detection and removal system (upper) and side wall details when using a geonet (lower).

inappropriate (or no) CQA, the type of primary liner (GM, GM/CCL or GM/GCL), the site operations (e.g., type of waste and daily cover), the site location (e.g., hydrology and precipitation), and other site-specific circumstances (e.g., consolidation water from CCLs).

A study by Othman et al. [70] surveys 93 double lined landfills with leak detection compatibility. The survey data appears in Table 5.17. The data is grouped according to the type of primary liner and the type of leak detection layer material. The liner grouping is either a geomembrane, geomembrane/CCL or geomembrane/GCL. The leak detection layer is either sand or a geonet. Three stages of operations have been assessed: during construction, active landfilling, and post closure. Some relevant conclusions from this interesting study follow:

TABLE 5.17 FLOW RATES FROM THE LEAK DETECTION SYSTEMS (LDS) OF MODERN DOUBLE-LINED LANDFILLS

	Liner/LDS Type								
	Type I (GM–Sand)			Type II (GM–GN)			Type III (GM/CCL–Sand)		
Lifecycle Stage	1	2	3	1	2	3	1	2	3
Average Flow	380	170	64	94	103	ND	220	140	64
Minimum Flow	7.6	0	0.19	4.8	1.4	ND	1.2	22	0
Maximum Flow	2140	1480	245	374	355	ND	1180	664	271
Number of points	30	32	8	7	11	ND	31	41	15
Number of landfills	11	11	4	4	6	ND	11	11	4
	Type IV (GM/CCL–GN)			Type V (GM/GCL–Sand)			Type VI (GM/GCL–GN)		
Lifecycle Stage	1	2	3	1	2	3	1	2	3
Average Flow	170	83	65	130	22	0.28	6.5	2.6	ND
Minimum Flow	0	0	0	0	0	0	0	0	ND
Maximum Flow	692	505	131	972	280	0.93	34	9.3	ND
Number of points	21	27	12	19	19	4	6	4	ND
Number of landfills	6	9	3	3	3	1	2	2	ND

Notes: All Flow Rates in l/ha–day (lphd).

Number of points: number of measuring points (i.e., outlets of single or multiple cells)

Lifecycle stages: 1. Construction life
2. Active life
3. Post closure

Abbreviations: GM = geomembrane
GN = geonet
CCL = compacted clay liner
GCL = geosynthetic clay liner
ND = no data

Source: After Othman et al. [70]

- The highest flow rates are from the GM/CCL composite liner, with consolidation water undoubtedly being a major contributor.
- The flow rates from the geomembrane by itself are also quite high with the sand leak detection being higher than the geonet leak detection.
- Geomembranes are difficult (but not impossible) to place without flaws.
- The reason for the sand leak detection layer being higher than the geonet may be that construction water was in the sand and was expelled over time.
- The lowest leakage rates were for the GM/GCL composite primary liner. Clearly, this data set makes a compelling argument for this type of liner. Interestingly, it is precisely where the original GCLs were placed when they were first offered on the market in the late 1980s. (GCLs are the topic of Chapter 6.)

5.6.5 Material Selection

The serious consequences of leaks caused by leachate reactions with the geomembrane or its premature degradation makes solid waste landfill liner selection more critical than for any other application. Making the selection process more difficult is the extreme variety of solid waste leachates (recall Tables 5.14 and 5.15). Thus candidate liner testing (generally via ASTM D5322) with the actual or synthesized leachate is often necessary to select the proper liner material. This incubation process is then followed by a series of physical and mechanical tests (generally via ASTM D5747) over varying time periods to determine if the original geomembrane properties have changed during the incubation period. If several geomembrane materials are being considered, the one with no change or the least change is the obvious choice. Recall the details of this procedure from Section 5.1.4.

The importance of the selection of the incubating leachate cannot be overemphasized. If a worst-case leachate is selected, for example, one containing organic solvents and similarly aggressive chemicals as noted in Table 5.15, the choice geomembrane will probably be some form of PE. The more concentrated and aggressive the leachate, the higher the required density of the polyethylene. This worst-case leachate selection has indeed been the trend the past and has resulted in the common use of HDPE for solid waste landfill liners. In Germany, HDPE is the only type of polymer that can be used for geomembranes in waste containment applications.

While the chemical resistance of HDPE is indeed a desirable and necessary feature of the polymer, it comes along with some less than desirable characteristics.

Sensitivity to stress cracking due to its high crystallinity. It is hoped that the NCTL test described in Section 5.1.3 along with its 100 hr minimum required transition time will prove effective in this regard (see [20]). Alternatively, the SP-NCTL test is more of a quality control test and has a comparable test time of 200 hr (see Section 5.1.3 and [21]).

Poor conformance to subgrade materials due to its stiffness and relatively high coefficient of thermal expansion (recall Table 5.10). This leads to relatively large waves that challenge the requirement that a composite liner have intimate contact between the geomembrane and underlying CCL or GCL.

A low friction coefficient leading to stability concerns. These concerns are eliminated, however, when using textured HDPE. Although the processes by which texturing is accomplished differ among manufacturers, all result in a major improvement in interface friction.

Poor axi-symmetric tensile elongation (recall Section 5.1.3). This is only a concern for those sites with poor subgrade stability, like landfill covers above degrading waste. In general, it should not be a problem beneath the base of a properly sited and designed landfill. If questionable subgrades are anticipated, out-of-plane deformation can be offset by using coextruded HDPE/VLDPE/HDPE geomembrane. As shown in Figure 5.5, the core of VLDPE is excellent in this particular stress mode.

In spite of the above limitations, it is felt that, with proper selection of the resin, an awareness of proper design methods, and careful installation CQC and CQA, HDPE should be used for landfill liners that contain aggressive leachates. This is not to say that other existing geomembranes cannot be used, or that other new formulations will not appear in the future. It is only meant to explain to the reader the current widespread use of HDPE as landfill liners.

5.6.6 Thickness

According to U.S. EPA regulations, the required minimum thickness of a geomembrane liner for hazardous materials containment is 0.75 mm (with timely cover) and 1.5 mm if it is made from HDPE. Recall that German regulations require a minimum thickness for HDPE of 2.0 mm and, furthermore, that HDPE is the only polymer that can be used. Whatever the regulatory situation, the technical design should proceed along the same lines as that of any liner, as described in Section 5.3.4. As with thickness design dealing with reservoir liners, solid-waste geomembrane thickness can be calculated and then compared to the above minimum values if regulations apply, or to the minimum survivability values of Table 5.11 if there are no regulations. When the secondary liner is also a geomembrane, it should also be the same thickness and type as the primary liner.

The design uses the same formulation as that developed in Section 5.3.4.

$$t = \frac{\sigma_n x(\tan \delta_U + \tan \delta_L)}{\sigma_{\text{allow}}(\cos \beta - \sin \beta \tan \delta_L)} \tag{5.18}$$

where

t = thickness of the geomembrane,
σ_n = applied stress from the landfill contents,
x = distance of mobilized geomembrane deformation,
δ_U = angle of shearing resistance between geomembrane and the upper material,
δ_L = angle of shearing resistance between geomembrane and the lower material,
σ_{allow} = allowable geomembrane stress, and
β = settlement angle mobilizing the geomembrane tension.

Example 5.18 illustrates the procedure.

Example 5.18

Obtain the required thickness of a HDPE geomembrane beneath a 50 m high landfill containing solid waste of unit weight 12.5 kN/m^3. The localized subsoil settlement is estimated to result in a liner deformation angle of 20°. Drainage sand is above the geomembrane and a geonet is below it.

Solution: The necessary information for solving the design equation is

(a) For out-of-plane tension testing, the yield-stress of HDPE (from Table 5.5c) is conservatively estimated as 20,000 kPa.
(b) The mobilization distance for HDPE at $50 \times 12.5 = 625$ kPa (from Fig. 5.10) is approximately 80 mm.
(c) The friction angle (from Table 5.7) for HDPE against Ottawa sand (δ_U) is 18°.
(d) The friction angle for HDPE against a geonet (separate test results) (δ_L) is 10°.
(e) These values give the required geomembrane thickness.

$$t = \frac{(625)(0.080)[\tan 18 + \tan 10]}{(20{,}000)[\cos 20 - (\sin 20)(\tan 10)]}$$

$$= \frac{25.1}{17600}$$

$$= 0.00143 \text{ m}$$

$$t = 1.43 \text{ mm}$$

Thus the regulated values of 1.5 mm in the U.S. or 2.0 mm in German regulations would control in this situation.

5.6.7 Puncture Protection

There are many circumstances where geomembranes are placed on or beneath soils containing relatively large-sized stones, for example poorly prepared soil subgrades with stones protruding from the surface or resting on the surface, soil subgrades over which geomembranes (particularly textured) have been dragged dislodging near-surface stones, and cases where crushed-stone drainage layers are to be placed above the geomembrane. All of these situations, particularly the last (which is unavoidable since it is a design situation), could use a protective geotextile to avoid puncturing the geomembrane. Note that if the soil subgrade is a CCL, a geotextile cannot be used and the isolated stones must be physically removed. For the drainage layer case, which is common to all landfills, a nonwoven needle-punched geotextile can provide significant puncture protection (recall Figure 5.8). The issue of required mass per unit area of the geotextile becomes critical.

In a series of papers, Wilson-Fahmy, Narejo, and Koerner [71, 72, 73] have presented a design method that focuses on the protection of 1.5 mm thick HDPE geomembranes. The method uses the conventional factor of safety equation.

$$FS = \frac{p_{allow}}{p_{act}} \tag{5.32}$$

where

FS = factor of safety (against geomembrane puncture),
p_{act} = actual pressure due to the landfill contents (or surface impoundment), and

TABLE 5.18 MODIFICATION FACTORS AND REDUCTION FACTORS FOR GEOMEMBRANE PROTECTION DESIGN USING NONWOVEN NEEDLE-PUNCHED GEOTEXTILES

Modification Factors

MF_S		MF_{PD}		MF_A	
Angular	1.0	Isolated	1.0	Hydrostatic	1.0
Subrounded	0.5	Dense, 38 mm	0.83	Geostatic, shallow	0.75
Rounded	0.25	Dense, 25 mm	0.67	Geostatic, mod.	0.50
		Dense, 12 mm	0.50	Geostatic, deep	0.25

Reduction Factors

		RF_{CR}			
			Protrusion (mm)		
RF_{CBD}		Mass per unit area (g/m^2)	38	25	12
Mild leachate	1.1	Geomembrane alone	N/R	N/R	N/R
Moderate leachate	1.3	270	N/R	N/R	> 1.5
Harsh leachate	1.5	550	N/R	1.5	1.3
		1100	1.3	1.2	1.1
		> 1100	$\cong 1.2$	$\cong 1.1$	$\cong 1.0$

N/R = Not recommended

p_{allow} = allowable pressure using different types of geotextiles and site-specific conditions.

Based on a large number of ASTM 5514 experiments, an empirical relationship for p_{allow} has been obtained, Eq. (5.33). It requires the set of modification factors and reduction factors given in Table 5.18.

$$p_{\text{allow}} = \left(50 + 0.00045\,\frac{M}{H^2}\right)\left[\frac{1}{MF_S \times MF_{PD} \times MF_A}\right]\left[\frac{1}{RF_{CR} \times RF_{CBD}}\right] \tag{5.33}$$

where

p_{allow} = allowable pressure (kPa),
M = geotextile mass per unit area (g/m^2),
H = protrusion height (m),
MF_S = modification factor for protrusion shape,
MF_{PD} = modification factor for packing density,
MF_A = modification factor for arching in solids,
RF_{CR} = reduction factor for long-term creep, and
RF_{CBD} = reduction factor for long-term chemical/biological degradation.

Note that in the above all MF values $\leqslant 1.0$ and all RF values $\geqslant 1.0$.

The situation can be approached from a given mass per unit area geotextile to determine the unknown FS value, or from an unknown mass per unit area geotextile and a given FS value. Example 5.19 uses the latter approach.

Example 5.19

Given a coarse-gravel ($d_{50} = 38$ mm) leachate collection layer to be placed on a 1.5 mm HDPE geomembrane under a 50 m high landfill, what geotextile mass per unit area is necessary for a FS value of 3.0? Assume that the solid waste weighs 12 kN/m^3.

Solution: Use H = 25 mm = 0.025 m, which is an estimate since the gravel particles are not isolated, but are adjacent to one another, $MF_S = 1.0$ for shape, $MF_{PD} = 0.20$ for packing density, $MF_A = 0.50$ for arching, $RF_{CR} = 1.5$ for creep, and $RF_{CBD} = 1.5$ for long-term degradation.

Now calculate the value of p_{allow} using Eq. (5.32).

$$FS = \frac{p_{allow}}{p_{act}}$$

$$3.0 = \frac{p_{allow}}{(50)(12)}$$

$$p_{allow} = 1800 \text{ kN/m}^2$$

Then calculate the required mass per unit area of the geotextile using Eq. (5.33).

$$p_{allow} = \left(50 + 0.00045\frac{M}{H^2}\right)\left[\frac{1}{MF_S \times MF_{PD} \times MF_A}\right]\left[\frac{1}{FS_{CR} \times FS_{CBD}}\right]$$

$$1800 = \left[50 + 0.00045\frac{M}{(0.025)^2}\right]\left[\frac{1}{1.0 \times 0.20 \times 0.50}\right]\left[\frac{1}{1.5 \times 1.5}\right]$$

$M = 493 \text{ g/m}^2$ use a 500 g/m^2 geotextile

5.6.8 Runout and Anchor Trenches

The terminus of geomembranes (and geonets if they are also involved) is a short horizontal runout at the top of the slope (recall Figure 5.30), and then (usually) a short drop into an anchor trench (recall Figure 5.31). The anchor trench is backfilled with soil and suitably compacted. Concrete anchor trenches with full fixity to the liner should generally not be used since geomembrane pullout is probably more desirable than geomembrane failure, although both should obviously be avoided.

The design method is explained and illustrated in Section 5.3.6 and will not be repeated here. Both analyses (runout alone and runout plus anchor trench) are applicable, with the latter being the most common.

For termination of double liner systems, the designer is faced with a number of possible choices (see Figure 5.46). The major considerations are to protect the integrity of both geomembranes and to keep surface water out of the leak detection system (shown as a geonet in Figure 5.46).

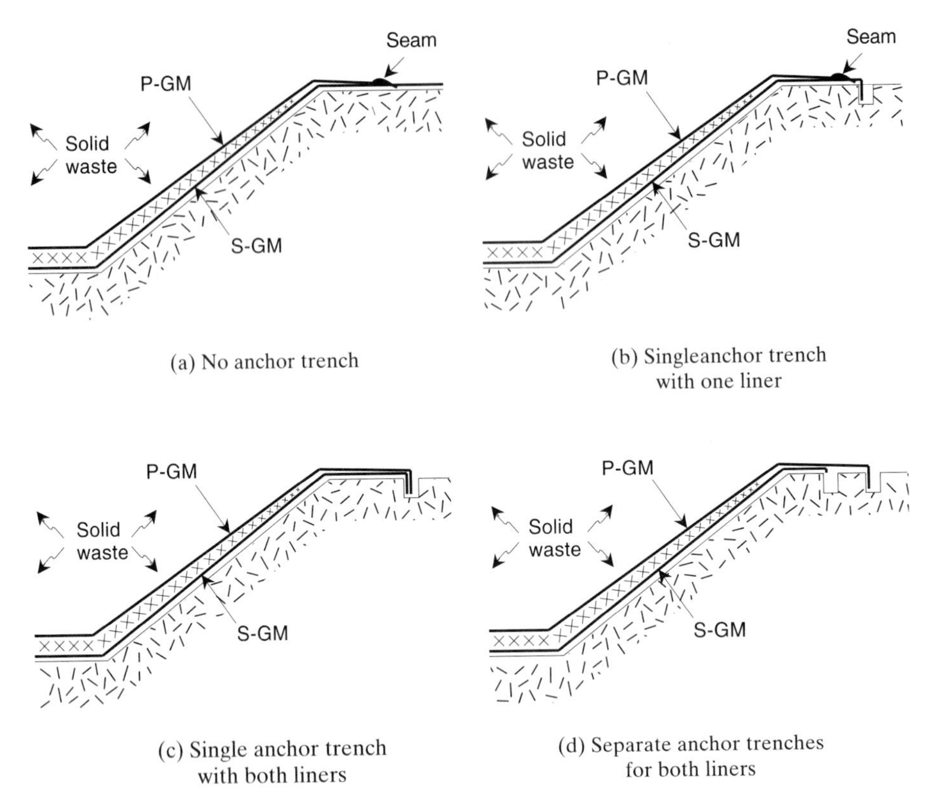

Figure 5.46 Various methods of terminating double liner systems (where P-GM = primary geomembrane and S-GM = secondary geomembrane).

The terminus of the liner of a completed internal cell within a zoned landfill, with its eventual continuation into an adjacent cell, is usually done by overlapping and seaming along the horizontal runout length. When waste fills the second cell, it is then continued from cell to cell over the seamed area, resulting in an encapsulated berm. Shear stresses on the geomembranes in both cells over this berm have been evaluated by large-scale laboratory models and found to be generally small and geomembrane-dependent (see Koerner and Wayne [74]). In such cases where higher stresses are felt to be generated, an auxiliary (or sacrificial) geomembrane rub-sheet over the crest of the berm should effectively dissipate the stresses before they propagate down to the underlying primary geomembrane.

5.6.9 Side Slope Soil Stability

The design of the stability of the soil mass beneath the liner system of a solid waste landfill is carried out in exactly the same manner as was discussed for liquid containment (reservoir) slopes and berms (recall Section 5.3.5). The process can include the strength of the covering liner materials, but if they are not included in the analysis, the error is on the conservative side. Interior berms, with or without geosynthetic inclusions, are also handled in the same manner as previously described.

5.6.10 Side Slope Liner Stability

The situation of a liner and its leachate collection cover soil's stability, or slumping, becomes quite complicated for multilayered liner and leachate collection systems of the types shown in Figure 5.41. Consider such a system, as shown in Figure 5.41e. The leachate collection system soil gravitationally induces shear stress through the system, thereby challenging each of the interface layers that are in the cross section. If all of the interface shear strengths are greater than the slope angle, stability is achieved and the only deformation involved is a small amount to achieve elastic equilibrium (Wilson-Fahmy and Koerner [75]). However, if any interface shear strengths are lower than the slope angle, wide width tensile stresses are induced into the overlying geosynthetics. This can cause the failure of the geosynthetic, pullout from the anchor trench, or result in quasistability via tensile reinforcement. If the last is the case, we can refer to the overlying geosynthetics as acting as nonintentional veneer reinforcement.

If the situation consists of a double liner system as shown in Figure 5.47, all of the interface surfaces can be made quite stable by the proper selection of the geosynthetics. For example, textured geomembranes could be selected and this together with nonwoven needle punched geotextiles will usually result in friction angles in excess of 20°. Furthermore, by thermally bonding the geotextiles in the leak detection system to the geonet, these surfaces are also stable at relatively high slope angles. This leaves the critical interfaces at the upper (cover soil) surface and the lower (clay liner) surface. The upper surface is analyzed exactly as described in Section 3.2.7 for the case without geogrid reinforcement. The proper selection of the cover soil, however, against a nonwoven needle punched geotextile (acting as a protection material, recall Section 5.6.7) should also result in a friction angle in excess of 20°. This leaves the lower surface of the secondary geomembrane against the clay liner as being the potentially low-interface

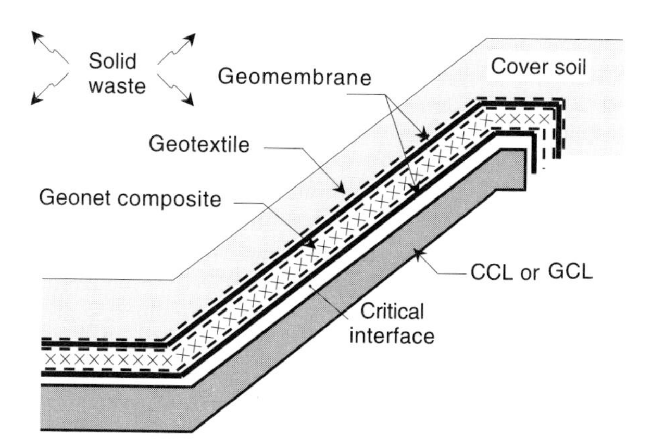

Figure 5.47 Geotextile/geomembrane/geonet composite/geomembrane above a CCL or GCL.

surface. If the clay liner is a CCL, the concern is with the expelled consolidation water lubricating the interface. This surface has been involved in a major failure of a hazardous waste liner system, as reported by Byrne et al. [76]. If the clay liner is a GCL, the concern is the hydrated bentonite being extruded out of the upper geotextile and lubricating the interface. This surface was involved in two slides in full-scale field tests, as reported by Koerner et al. [77].

The analysis of multilined slopes of the type being discussed is a direct extension of the veneer reinforcement model presented in Section 3.2.7 on geogrids. Recalling Figure 3.23, the analysis results in the following equation.

$$a(\text{FS})^2 + b(\text{FS}) + c = 0 \tag{3.21}$$

where

$a = (W_A - N_A \cos\beta - T\sin\beta)\cos\beta,$
$b = -[(W_A - N_A \cos\beta - T\sin\beta)\sin\beta\tan\phi + (N_A\tan\delta + C_a)\sin\beta\cos\beta$
$\quad + \sin\beta(C + W_P\tan\phi),$ and
$c = (N_A\tan\delta + C_a)\sin^2\beta\tan\phi.$

The resulting FS value is then obtained from Eq. (3.22).

$$\text{FS} = \frac{-b + \sqrt{b^2 - 4ac}}{2a} \tag{3.22}$$

The variables and values of W_A, N_A, T, W_P are defined in Sections 3.2.7 and 5.3.5. The critical parameter in the above equation is T, the allowable wide width tension strength of the geogrid reinforcement. For the cross section shown in Figure 5.47, T represents the allowable (?) strength of all of the geosynthetic materials above the critical interface. Not only is the issue of reduction factors difficult to assess for the liner materials, per se, but the issue of strain compatibility is also unwieldy. In this latter regard, the wide width tensile strength of each geosynthetic material must be determined, plotted on the same axes, and assessed at a specific value of strain. That is, the liner system components cannot act individually and must act as an equally strained unit. Example 5.20 illustrates the situation.

Example 5.20

For a 30 m long slope at 3(H) to 1(V) (18.4°), lined with a double liner system consisting of GT/GM/GC/GM/CCL or GCL (as in Figure 5.47), the lowest friction angle is the GM/CCL or GM/GCL interface, which is 10°. All other interface friction angles are in excess of 18.4°. The wide width tensile behavior of the various candidate geosynthetics is given in the curves on the graph below. The cover soil is 450 mm thick with a unit weight of 18.0 kN/m^3 and a friction angle of 30°. What is the factor of safety of the slope based on a cumulative reduction factor of 2.0?

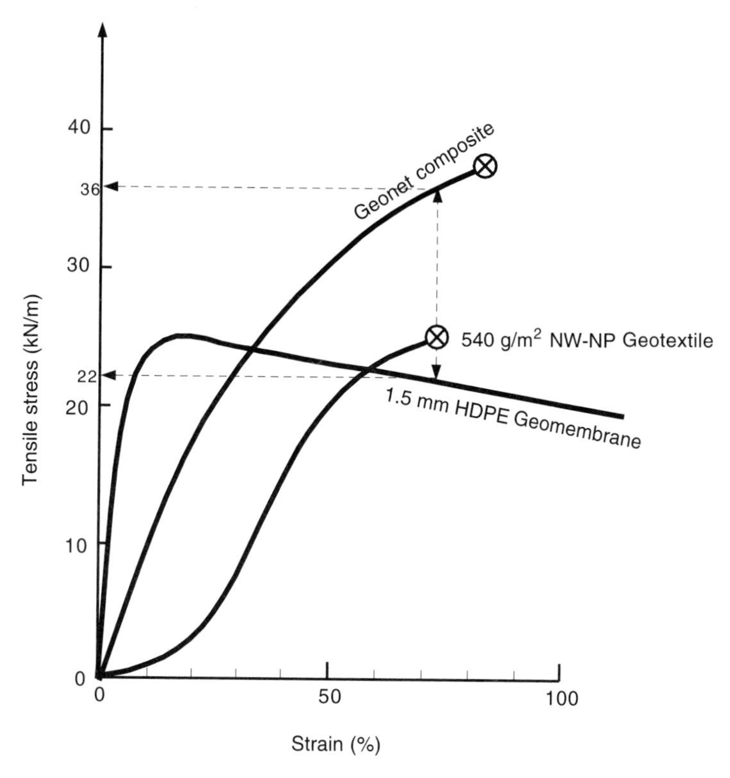

Solution:

$$W_A = \gamma h^2\left(\frac{L}{h} - \frac{1}{\sin\beta} - \frac{\tan\beta}{2}\right)$$

$$= (18.0)(0.45)^2\left[\frac{30}{0.45} - \frac{1}{\sin 18.4} - \frac{\tan 18.4}{2}\right]$$

$$= 3.65[63.3]$$

$$= 231 \text{ kN/m}$$

$$N_A = W_A \cos\beta$$

$$= 231 \cos 18.4$$

$$= 219 \text{ kN/m}$$

$$W_p = \frac{\gamma h^2}{\sin 2\beta}$$

$$= \frac{(18.0)(0.45)^2}{\sin 36.8}$$

$$= 6.08 \text{ kN/m}$$

T_{ult}, taken at the first geosynthetic failure, which is the nonwoven needle punched geotextile at 25kN/m, is

$$T_{ult} = 25 + 2(22) + 36$$
$$= 105 \text{ kN/m}$$

For a reduction factor of 2.0,

$$T_{allow} = \frac{105}{2.0}$$
$$= 52.5 \text{ kN/m}$$

$$a = (W_A - N_A \cos\beta - T\sin\beta)\cos\beta$$
$$= (231 - 219\cos 18.4 - 52.5\sin 18.4)\cos 18.4$$
$$= 6.1 \text{ kN/m}$$

$$b = -[(W_A - N_A\cos\beta - T\sin\beta)\sin\beta\tan\phi$$
$$+ (N_A\tan\delta + C_a)\sin\beta\cos\beta + \sin\beta(C + W_p\tan\phi)]$$
$$= -[(231 - 219\cos 18.4 - 52.5\sin 18.4)\sin 18.4\tan 30$$
$$+ (21.9\tan 10 + 0)\sin 18.4\cos 18.4 + \sin 18.4(0 + 6.08\tan 30)]$$
$$= -[1.17 + 11.57 + 1.11]$$
$$= -13.8 \text{ kN/m}$$

$$c = (N_A\tan\delta + C_a)\sin^2\beta\tan\phi$$
$$= (219\tan 10 + 0)\sin^2 18.4\tan 30$$
$$= 2.22 \text{ kN/m}$$

$$\text{FS} = \frac{-b + \sqrt{b^2 - 4ac}}{2a}$$
$$= \frac{13.8 + \sqrt{(-13.8)^2 - 4(6.1)(2.22)}}{2(6.1))}$$
$$\text{FS} = 2.10$$

While the value appears to be acceptable, it is nevertheless disconcerting that the liner system, per se, is being used as the veneer reinforcement mechanism. Had higher reduction factors been used, the resulting FS value would be proportionately decreased.

5.6.11 Access Ramps

For below-grade landfills it is necessary to prepared the subgrade to accommodate the necessary access ramp(s), line the entire facility, and then construct a road above the liner cross section. A typical geometry is shown in Figure 5.48a. Particularly trouble-

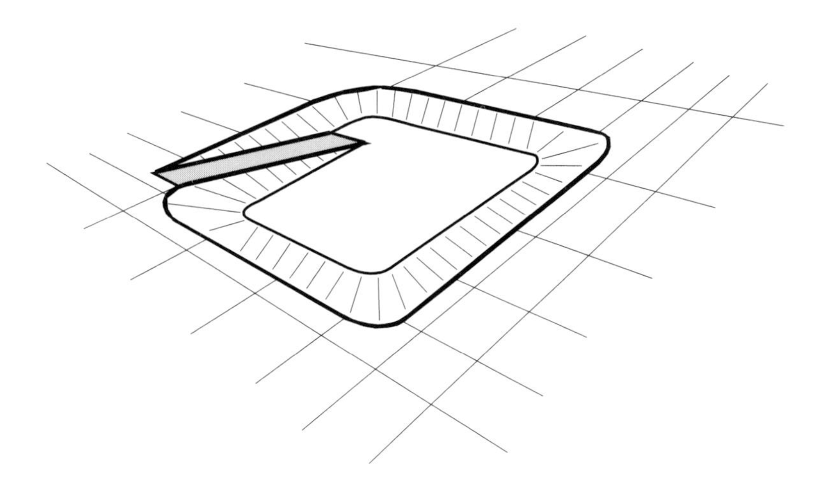

(a) Geometry of typical ramp-grades from nil to 25% (14.0°)

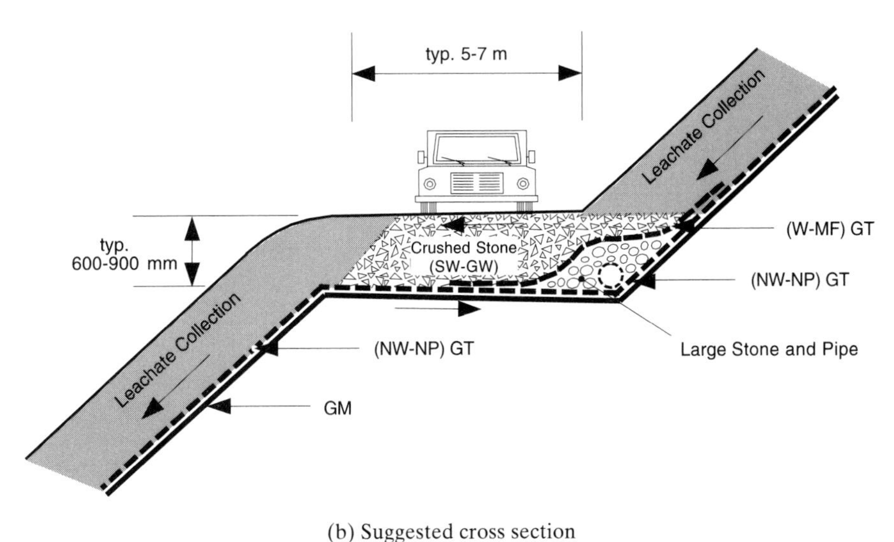

(b) Suggested cross section

Figure 5.48 Typical geometry and cross section of a below-grade landfill access ramp.

some is that the road must be built above the completed liner system. Past problems have been:

- Inadequate drainage where the ramp meets the upper slope, with subsequent erosion and scour of the roadway itself.
- Inadequate roadway thickness above the liner system with ramp soil sliding off the upper geomembrane due to truck traffic.

- Inadequate roadway thickness above the liner system with the upper geomembrane failing in tension along the slope due to truck traffic.
- Inadequate roadway thickness above the liner system with an underlying hydrated GCL creating slippage of the overlying geomembrane and entire roadway.

Clearly, a conservative design is required, with Figure 5.48b presenting some recommendations. While a 600 to 900 mm thickness might seen excessive, the dynamic stresses caused by braking trucks are high and, furthermore, the soil can be removed in whole or in part as the waste elevation rises during filling operations.

5.6.12 Stability of Solid Waste

Upon first consideration, the stability of solid waste failing within itself should present no particular concern since its shear strength characteristics should be quite high. Singh and Murphy [78] present shear strength parameters of waste transitioning from high in friction (24 to 36°) to being high in cohesion (80 to 120kPa). Obviously, the aging of the waste is an issue, but at all times the shear strength is quite high. Paradoxically, there have been some massive failures of solid waste. They are shown in Figure 5.49 and have all resulted from different mechanisms. None involved geosynthetics, although if a geomembrane were present it would not have altered the situation.

Figure 5.49a was a failure wherein the face of the solid waste was too steep, causing a bearing capacity failure of the weak foundation soil. As the foundation soil deformed, the waste was unable to maintain stability within itself. The failure was quite slow in its mobilization.

Figure 5.49b was caused by the inadvertent excavation of a stiff clay layer at the toe of the waste. The underlying clay was very weak. Augmented by a heavy rainfall, the waste failed in a rapid succession of circular arcs, as seen in Figure 5.50. The case study is analyzed in Reynolds [79].

While the previous two failures were clearly rotational in their behavior, Figure 5.49c represents a translational failure. The waste slid on, or within, a thin leachate saturated remolded colluvium layer over shale rock on a 8 to 9% slope. The initiating action was probably the toe excavation for additional waste capacity.

5.6.13 Piggyback Landfills

In closing this section on waste related geosynthetic systems, the concept of *piggybacking* a new landfill on an existing one should be mentioned. Many existing landfills are filled and there is nowhere else to go but up. Thus a new landfilling operation above an existing one sometimes becomes necessary. Some precautions regarding this type of vertical expansion are as follows:

- Total settlement of the existing landfill must be anticipated and estimated. Thus the slopes of the leachate collection system must reflect this requirement and will probably be quite high, as much as 10 to 15%.
- Estimation of differential settlements within the existing landfill may require a high-strength geogrid or geotextile network to be placed over all or a portion of the site (recall Section 3.2.6 and Example 3.11.)

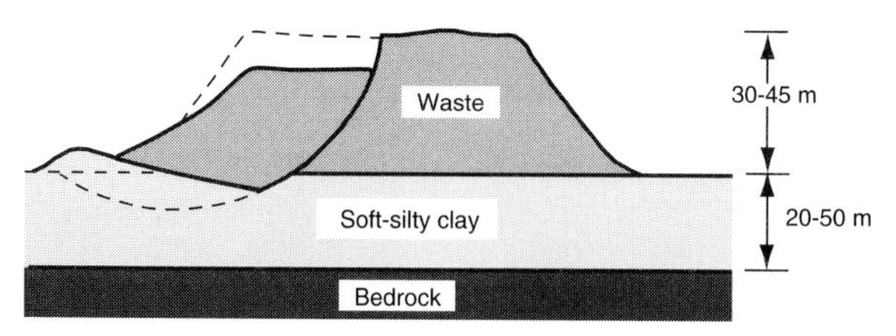

(a) Single rotational failure

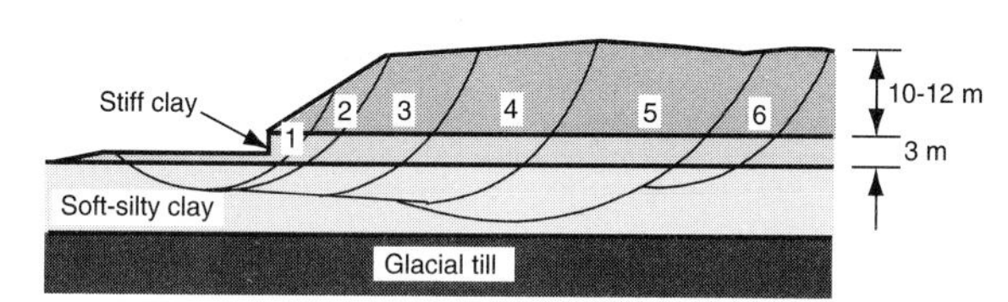

(b) Multiple rotational failures

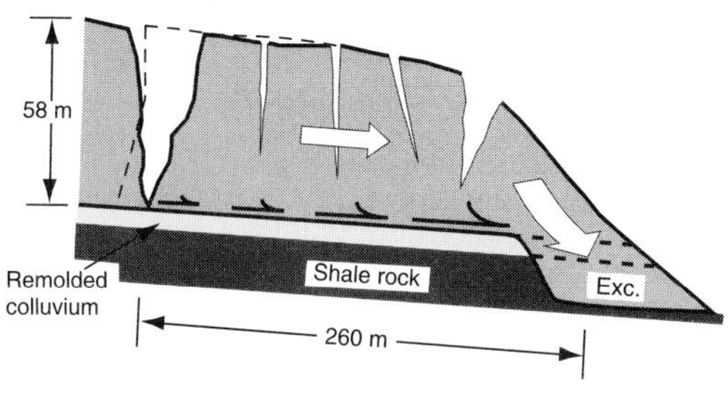

(c) Translational failure

Figure 5.49 Failure of municipal solid waste landfills within the mass itself.

- Waste placement in the new landfill must be carefully sequenced to balance stress on the existing landfill. The stability of the waste situation just discussed is exacerbated greatly by the addition of a large surcharge stress, which is what the piggybacked landfill represents to the underlying waste.
- Methane gas (if generated) migrating from the existing landfill must be carried laterally under the new landfill liner to side-slope venting and/or collection locations. Active gas collection systems may be required.
- Leachate collection from the existing landfill should be considered. If required, directionally drilled withdrawal wells at the perimeter of the facility may be a consideration.
- Access to the site via haul roads must be carefully considered so as not to cause damage to, or instability of, the underlying liner system.

Figure 5.50 Failure of a municipal solid waste landfill within the waste mass itself.

5.6.14 Heap Leach Pads

Heap leach pads consist of a geomembrane with an overlying drainage system, and then a metal (gold, silver or copper) bearing ore heaped above. A cyanide or sulfuric acid solution is sprayed on top of the ore, leaches through it reacting with the metals, and carries the solution to the drainage system where it is collected. Separation of the ore from the leachate occurs in an on-site processing plant. The leaching solution is renewed and the process is repeated until it is no longer economical. Figure 5.51 illustrates the general configuration.

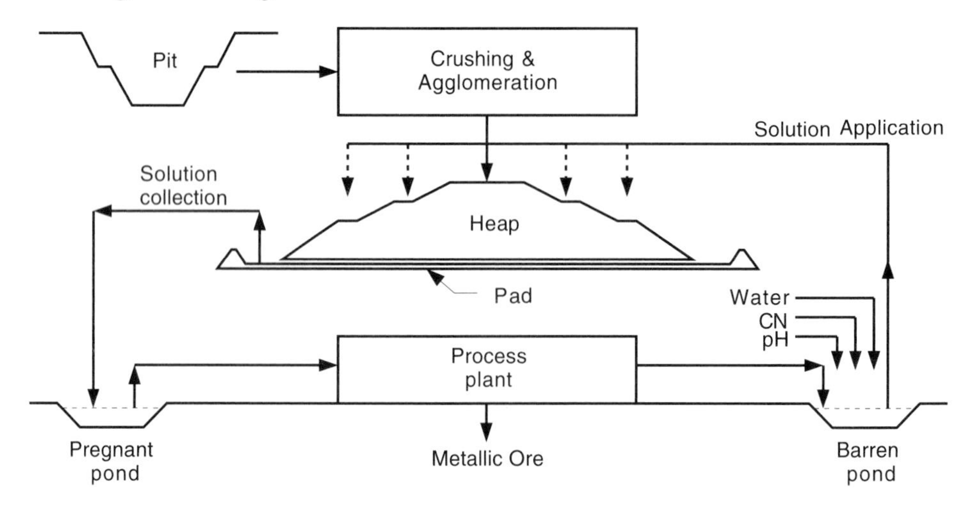

Figure 5.51 General configuration of a heap leach operation for extraction of metallic ores. (After Leach et al. [80])

The heap itself is often enormous in its proportions. Ores of 22 kN/m^3 unit weight at heights up to 70 m produce large stresses on the drainage system and geomembrane. The concept is used widely in the western U.S. and Canada and in many South American countries (see Smith and Welkner [81]).

Regarding the design of the geomembrane, its thickness and type are very subjective and all resin types have been used to varying degrees. The drainage system is usually coarse gravel, allowing for rapid and efficient removal of the ore-bearing solution from beneath the heap. This situation requires the consideration of a very thick protection geotextile between the geomembrane and drainage/collection gravel. The design method presented in Section 5.6.7 should be considered, with the reminder that it is developed on the basis that different geomembrane thicknesses and types will behave differently.

5.6.15 Solar Ponds

There are a number of solid material liner systems that have not yet been mentioned. A small but growing segment of these systems is solar ponds [82]. Here the geomembrane is placed in an excavation and then it is filled with salt. Solar energy is collected and stored as heat. A salt gradient effect is created, whereby zones are set up constantly replenishing new heat as it is gradually withdrawn from the lower storage zone for useful heat. The main consideration insofar as the geomembrane is concerned is that temperatures up to 93°C must be withstood without loss of strength that is required for containment. The design, as with other situations not mentioned, should be possible using the guidelines set up in this section.

5.6.16 Summary

This section on the design of solid material containment liners followed closely the concepts developed in Section 5.3 on liquid containment (pond) liners. The notable exceptions are that leachate collection systems above the primary geomembrane are always necessary; cover soils (e.g., the leachate collection system) develop side-wall shear stresses that can induce stresses in the liner system; the shear stresses can result in tensile stresses in individual geosynthetic components that can be very high; leak detection systems, and hence double liners, are oftentimes necessary or required by regulations; and the cross sections can become very complex, requiring numerous design details to be considered. These extra design considerations require the determination of many physical and mechanical properties of the various geosynthetic and natural soils components and of the solid material itself. The criticality of proper laboratory testing of the physical and mechanical properties for a rational design approach becomes obvious. Tables 5.19 and 5.20 summarize the specific design problems for geomembranes and drainage geosynthetics, respectively. Also indicated is the large number of required properties of the geosynthetics and of the contained solid material. Keep in mind, however, that all of the these values are obtainable, and a design-by-function approach is indeed possible for solid waste containment facilities.

With all of the geosynthetic components discussed in this section, and in the cover to be discussed in the next section, it should come as no surprise that the solid waste

TABLE 5.19 VARIOUS DESIGN MODELS FOR GEOMEMBRANES IN WASTE DISPOSAL SITUATIONS

Problem	Liner Stress	Free-Body Diagram	Required Properties		Typical Factor of Safety
			Geomembrane	Landfill	
Liner self-weight	tensile	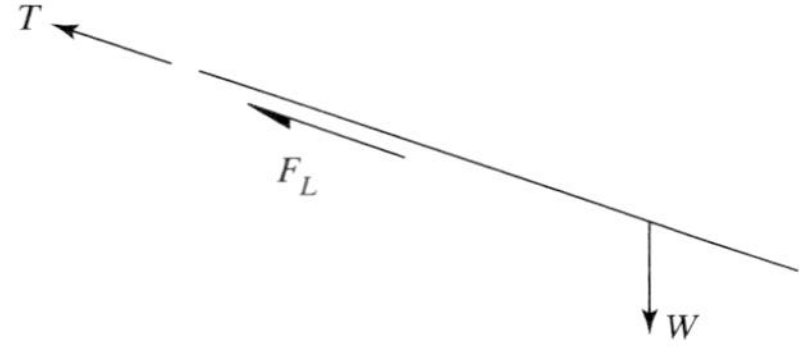 T, F_L, W	$G, t, \sigma_{allow}, \delta_L$	β, H	≥ 10
Weight of cover soil	tensile	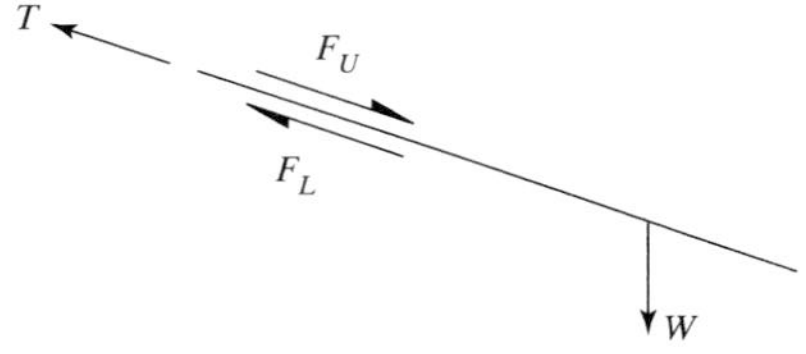 T, F_U, F_L, W	$t, \sigma_{allow}, \delta_U, \delta_L$	β, h, γ, H	0.5 to 2
Impact during construction	impact	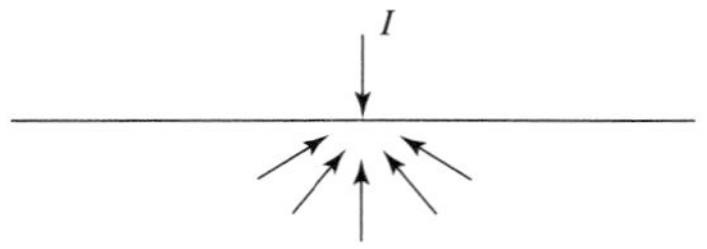 I	I	d, w	0.1 to 5
Weight of landfill	compression	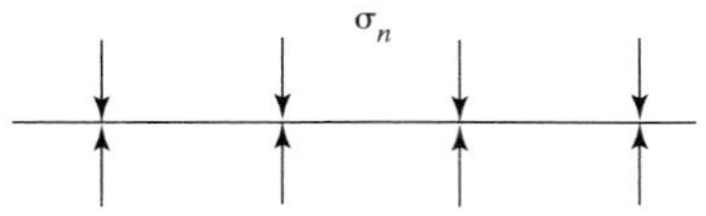 σ_n	σ_{allow}	γ, H	≥ 10
Puncture	puncture	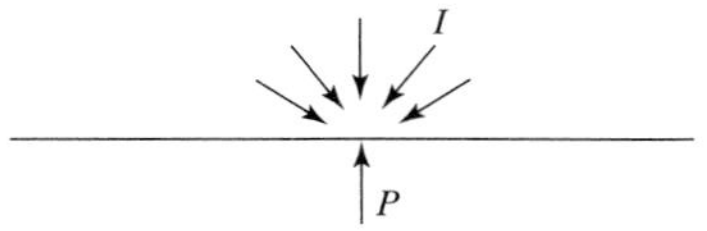 I, P	σ_p	γ, H, P, A_p	0.5 to 3

TABLE 5.19 VARIOUS DESIGN MODELS FOR GEOMEMBRANES IN WASTE DISPOSAL SITUATIONS

Problem	Liner Stress	Free-Body Diagram	Required Properties		Typical Factor of Safety
			Geomembrane	Landfill	
Anchorage	tensile	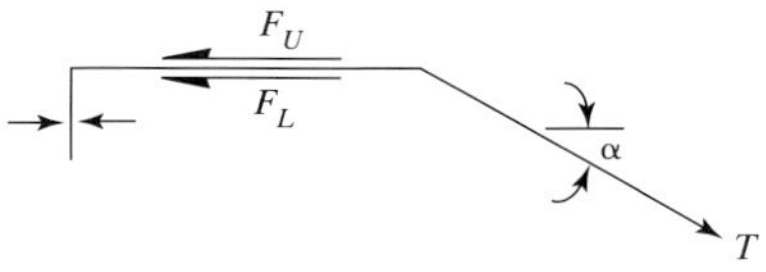 F_U, F_L, α, T	$t, \sigma_{\text{allow}}, \delta_U, \delta_L$	β, γ, ϕ	0.7 to 5
Settlement of landfill	shear	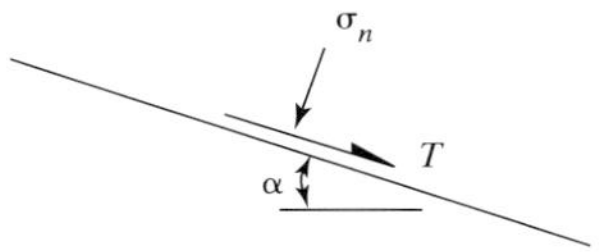 σ_n, T, α	τ, δ_U	β, γ, H	$\geqslant 10$
Subsidence under landfill	tensile	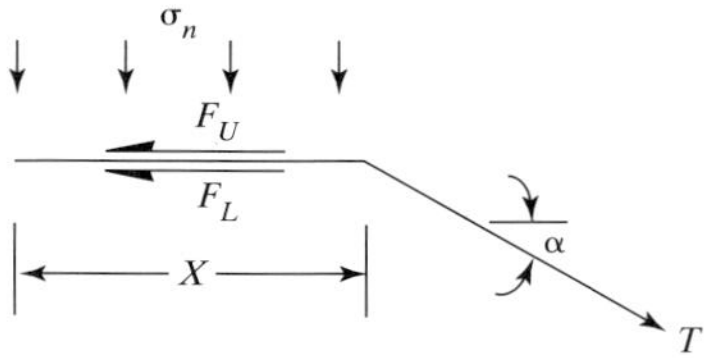 σ_n, F_U, F_L, X, α, T	$t, \sigma_{\text{allow}}, \delta_U, \delta_L, \chi$	α, γ, H	0.3 to 10

Abbreviations: Geomembrane properties:

G = specific gravity
t = thickness
σ_{allow} = allowable stress (yield, break or allow)
τ = shear stress
I = impact resistance
σ_p = puncture stress
δ_U = friction with material above
δ_L = friction with materal below
X = mobilization distance

Landfill properties:

β = slope angle
H = height
γ = unit weight
h = lift height
α = subsidence angle
ϕ = friction angle
d = drop height
W = weight
p = puncture force
A_p = puncture area

Source: Koerner and Richardson [83].

TABLE 5.20 DESIGN CONSIDERATIONS FOR DRAINAGE GEOCOMPOSITES (USUALLY GEONETS) IN WASTE DISPOSAL SITUATIONS

Problem	Reason	Approach	Required Properties: Geosynthetic	Required Properties: Landfill	Status of Problem
Stability of core	avoid crushing of core	FS $= \sigma_{ult}/\sigma_{max}$	σ_{ult}	γ, H	designable
Flow in core	first approximation	FS $= q_{allow}/q_{reqd}$	q_{allow}	γ, H, i, q_{reqd}	designable
Creep of core	first reduction	FS $= q'_{allow}/q_{reqd}$	q'_{allow}	γ, H, q_{reqd}	designable
Elastic intrusion of geomembrane	second reduction	elastic plate theory	E, μ, x, y	γ, H, q_{reqd}	designable
Elastic intrusion of geotextile	second reducton	elastic plate theory	E, μ, x, y	γ, H, q_{reqd}	designable
Creep intrusion of geomembrane	third reduction	creep theory	$\varepsilon(\sigma,t), x, y$	γ,H,t	unknown
Creep intrusion of geotextile	third reduction	creep theory	$\varepsilon(\sigma,t), x, y$	γ,H,t	unknown

Abbreviations: Geocomposite properties:
σ_{ult} = ultimate compression strength
q_{allow} = allowable flow rate
q'_{allow} = reduced allowable flow rate
t = time
E = modulus of elasticity
μ = Poisson's ratio
x, y = core dimensions
$\varepsilon(\sigma,t)$ = strain rate

Landfill Properties:
γ = unit weight
H = height
i = hydraulic gradient
q_{reqd} = actual (design) flow rate
t = time
σ_{max} = maximum stress
σ = applied stress

Source: Koerner and Richardson [83].

(landfill) area is pushing geosynthetics to new levels. Design models, testing methods, installation practices, inspection techniques, and geosynthetic product sales are all evident when we consider the landfill liner and closure sketch of Figure 5.52. Here we see in a single cross section the awesome use of geosynthetics in the solid waste containment application area. Yet, every component must be analyzed via a technically sound equivalency argument and justified via a reasonable benefit/cost basis. In the liner system beneath the waste, each component should make sense. Now let's fill the site with solid waste and consider the cover system to be placed above the waste.

5.7 LANDFILL COVERS AND CLOSURES

In order to minimize or entirely eliminate leachate generation after solid waste filling is complete, a *cover* is required over a landfill, waste pile, or other mass of solid waste. Covers are also referred to in the literature as *landfill caps* and *landfill closures.*

5.7.1 Overview

There comes a time in the life of any landfill when it cannot accept additional material. When this occurs it is necessary to construct a cover above the waste. Depending on its inherent stability, stabilization work on the disposed waste materials may be necessary

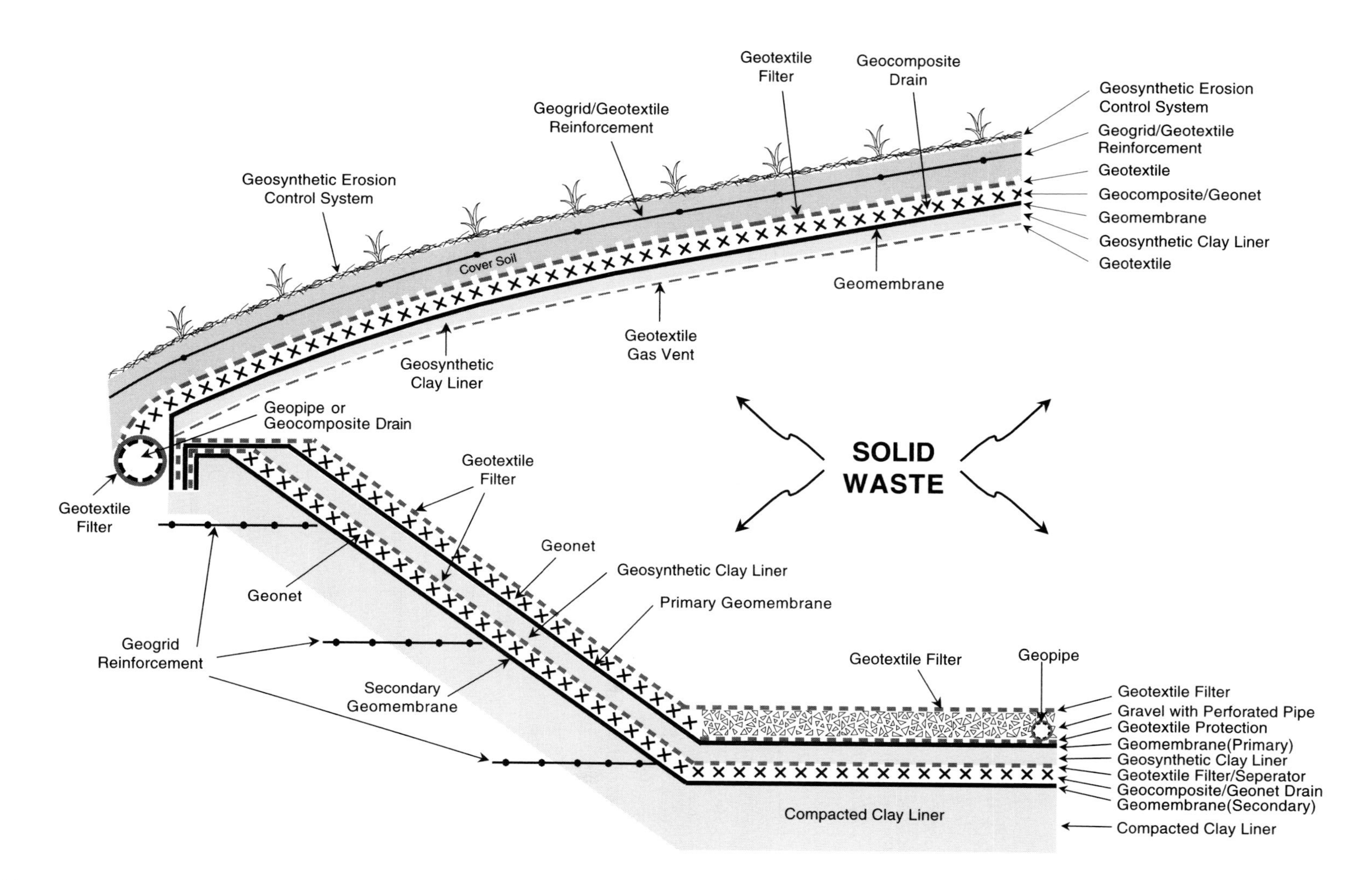

Figure 5.52 Solid waste containment system with high geosynthetic utilization.

before closure begins. Recall Table 5.13; the category treatment/stabilization applies to this topic. Take, for example, sludge lagoons of viscous fluids or spent hydrocarbon wastes or suspensions of materials exhibiting Brownian movement, each of which requires some stabilization before a permanent closure can be attempted. If it is not stabilized, the viscous waste will gradually work its way up through the cover soil and emerge at the ground surface. It is a form of hydrofracturing. An example is buried car tires that have worked themselves up through 5 m of fill in a period of 10 years. Stabilization of viscous liquid materials is usually done by mixing them with soil, cement, fly ash, lime, or other matrix or reagent material. The exact composition depends on the desired mechanical stability and fixity of any toxic, hazardous, or radioactive pollutants, plus the availability of local and/or inexpensive materials. Mixing is best done outside the site, where proportioning is controllable and backfilling can be done in a systematic manner. Often, however, this is not possible, due to the unavailability of space or the chance of contamination exposure. In such cases, in situ mixing is required. In situ mixing with stabilizing agents is difficult to control and generally results in a randomly stabilized landfill. This must be taken into consideration during the closure design. The literature is abundant in this regard.

Certainly a municipal solid waste landfill represents random stabilization at best. The compaction during placement, type of waste, type and thickness of cover soil, and so on, all interact to result in a very uncertain post-closure subsidence pattern. Landfill subsidence strains of 5% to 30% over a 20-year period have been measured (see Figure 5.53). Analytic modeling of landfill subsidence has also been attempted using column models [85] and centrifuge modeling [86]. However, for a site-specific situation, quantitative values are essentially unavailable (which, incidentally, makes it an area that clamors for research).

Whatever the situation, the landfill must eventually be covered and the following five components (from the waste to the ground surface) must be considered: the gas collection layer, the hydraulic/gas barrier layer, the drainage layer, the protection layer, and the surface layer. Each of these components can be seen in the cover portion of the cross section on Figure 5.52. They are shown in the context of geosynthetic materials, but it should be noted that most regulations call for the use of natural soil materials. These same regulations also state that geosynthetics can be used as alternatives, if technical equivalency can be shown. In this book, of course, geosynthetics will be emphasized.

5.7.2 Various Cross Sections

There are a large number of variations, using both geosynthetics and natural soils, that can be selected in designing the final cover for a landfill, or (equally important) an abandoned dump site. Koerner and Daniel [87] use classifications for the existing waste mass as hazardous, nonhazardous, and abandoned, the last being essentially unknown in its classification. Equally (if not more) important is a knowledge of the regulations that apply for the site under consideration.

For final covers above *hazardous waste,* the U.S. EPA [88] requires the following technical details.

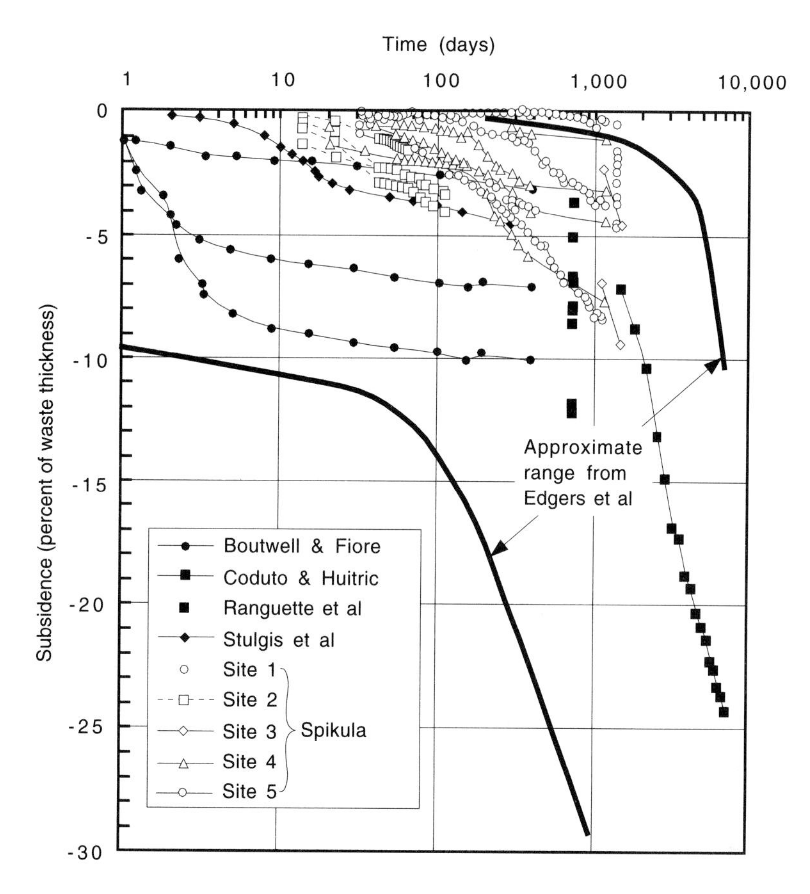

Figure 5.53 Municipal solid waste landfill subsidence. (After Spikula [84])

- The low-permeability soil layer, or CCL, should have a minimum thickness of 600 mm and a maximum in-place saturated hydraulic conductivity of 1×10^{-7} cm/s.
- The geomembrane barrier above the compacted clay should have a minimum thickness of 0.75 mm.
- There should be adequate bedding above and below the geomembrane.
- The drainage layer above the geomembrane should have a minimum hydraulic conductivity of 0.01 cm/s and a final bottom slope of 2% after settlement and subsidence.
- The cover topsoil and protection soil layer must have a minimum thickness of 600 mm.

In considering geosynthetic alternatives to the above regulations, as seen in Figure 5.52, the following is possible.

- The CCL could be replaced by a GCL.
- The drainage layer could be replaced by a geocomposite or geonet drain.

- The filter layer above the drain could be replaced by a geotextile filter.
- The methane gas-collection layer could be replaced by a thick geotextile.
- For steep slopes, the cover soil may need geogrid or geotextile acting as veneer reinforcement.
- The top soil may need some type of geosynthetic erosion control system.

For final covers above *nonhazardous waste* landfills, the EPA appears to be in a state of flux. The federal regulations [89] are quite loosely written and state that the cover must have a permeability as low as or lower than the liner system. Lending credibility to the importance of the cover system vis-a-vis the liner system beneath the waste, it also states that a "bathtub" effect is to be avoided. On June 26, 1992, a "clarification" was issued that has created considerable confusion in that it relaxes the regulations about the barrier layer of a landfill's composite liner to one requiring only a thin geomembrane of 0.5 mm thickness (1.5 mm if HDPE) placed over a soil of 1×10^{-5} cm/s permeability. In my opinion, such a regulation can never be interpreted as being comparable in its impermeability to a thicker geomembrane and a clay soil 100 times lower in permeability!

Lastly, as far as regulations are concerned, the U.S. Army Corps of Engineers guidance on *abandoned dump* covers should be mentioned. In their documents they give equal credibility to 1.0 mm thick PVC and VFPE geomembranes for covers. This is based on the usual conditions of only having a surface water interface with the geomembrane and the possibility for geomembrane conformance to out-of-plane deformations beneath the cover. Recall Figure 5.5, which shows that these particular geomembranes perform very well in this type of stress state. The situation is further heightened in that the U.S. EPA requires a landfill gas emission control system for landfills of more than 2.5 million tonnes capacity or one that emits more than 50 tonnes of nonmethane organic compounds annually, that is, most large landfills.

Instead of focusing so much on regulations, however, it is preferable to proceed from basics, particularly as the previous regulations are only for the U.S. The five essential layers of a final cover for an engineered landfill or abandoned dump are the following: gas collection layer, barrier layer, drainage layer, protection layer, and surface layer. Each will be described in the order stated; for additional details see [87].

5.7.3 Gas Collection Layer

Municipal solid waste (MSW) can generate tremendous quantities of gas during its decomposition. The two primary constituents are methane (CH_4) and carbon dioxide (CO_2). To give an idea of the mechanisms and quantities, Baron et al. [90] cite the following series of events:

1. After closure, the aerobic phase of microorganism growth is relatively short, since oxygen supplies are rapidly (but not completely) depleted.
2. The anaerobic acid-forming microorganisms begin to appear.
3. The bacteriological organisms (aerobic and anaerobic) break down the long-chain organic compounds in the waste (mainly carbohydrates) to form organic acids, mainly CO_2.

4. This phase produces as much as 90% of the CO_2 and peaks 11 to 40 days after closure. It depletes the remainder of the available oxygen.
5. The methane-forming microorganisms become dominant.
6. The methane-forming anaerobic bacteria use the acids to form CH_4, some additional CO_2, and water.
7. Over time, the CH_4 increases and the CO_2 decreases. This takes about 180 to 500 days after closure. Thus one to two years is required to initiate a continuous flow of CH_4.
8. In the following two years, approximately 30% of the CH_4 will be generated, and in five years approximately 50% will be generated. Thereafter, CH_4 generation continues but at a diminished rate. For example, 90% will have been generated after 80 years and 99% after 160 years. Note, however, that these values are very site-specific and are only meant to illustrate the trends and implications of methane gas generation in MSW landfills.
9. The range of quantities of CH_4 that are generated in a MSW landfill is 0.13 to 0.64 m^3 of CH_4 per kN of municipal solid waste per year. For a large landfill of 3 million metric tons per year, this results in 8,500 to 43,500 m^3 of CH_4 produced per day!

It is obviously necessary to provide a gas-collection layer and then a suitable venting and capturing system so as to avoid air pollution. Most large landfills use the gas for energy production at the landfill site or sell the energy over the local energy utility system. In the absence of such a gas-collection system, *blow-outs* of the geomembrane barrier, shown in Figure 5.54, are becoming more frequent. Note that the gas pressure generated from the decomposing waste has completely displaced the 1.5 m of cover soil at this particular site. Methane gas vents, typically on 15 to 50 m centers must unfortunately penetrate and pass through the cover system. Figure 5.55 presents some sketches of these details. These vents, however, are of no meaning unless the gas can arrive at the lower entry level. To accomplish this, a sand layer or a nonwoven needle punched geotextile is placed beneath the barrier system shown in Figure 5.52. The design is based on air transmissivity and is similar to Example 5.7 in Section 5.3.2 for the relief of air pressure beneath a liner system. The comments in that section apply here as well.

While the above discussion focuses on gases generated for municipal solid waste (MSW) typical of the U.S., waste in other countries may be more degradable and produce even greater quantities. Conversely, wastes other than MSW will generally produce significantly lower quantities of gas (e.g., hazardous waste, ash, or building demolition waste). The gas generation rates are clearly waste-specific. Be cautioned, however, that if a geomembrane is located in the barrier layer, even small quantities of gases cannot be released and the situation shown in Figure 5.54 can easily result.

5.7.4 Barrier Layer

In designing the barrier layer for a landfill closure one could consider several possible options: a single compacted clay liner (CCL), a single geomembrane (GM), a single geosynthetic clay liner (GCL), a two-component composite (GM/CCL), a two-component composite (GM/GCL), a three-component composite liner (GM/CCL/GM), and

Figure 5.54 Geomembrane expanded by methane gas from a closed landfill pushing aside ≃1.5 m of cover soil and topsoil.

a three-component composite liner (GM/GCL/GM). The last two options are sometimes attractive when the moisture content in the clay or bentonite components are particularly sensitive to environmental conditions (i.e., desiccation when dry or low shear strength when saturated).

The critical factors that affect the selection of a specific barrier layer are climate, the amount of differential settlement, the vulnerability of the cover soil to erosion or puncture, the amount of water percolation that can be tolerated through the cover system, the need for collecting waste generated gas, and the slope steepness.

In assessing the seven barrier layer alternatives, Daniel and Koerner [91] have scored each of those factors from 1 (not recommended) to 5 (recommended). For example, in cases involving large differential settlement, they scored a CCL as 1 and a geomembrane 5. The authors suggest that designers consider these assessments, given in Table 5.21, for general guidance only. Unique conditions at a particular site will call for special design precautions.

The assessed scores can then be extended using a cost estimate and a general benefit/cost ratio can be computed for each of the seven alternate barrier layers. The following installed cost estimates were used:

- CCLs vary from about 5 to \$20 per m^2, depending on thickness, availability, size, and type of facility. In extreme conditions such as lack of locally available clay, the

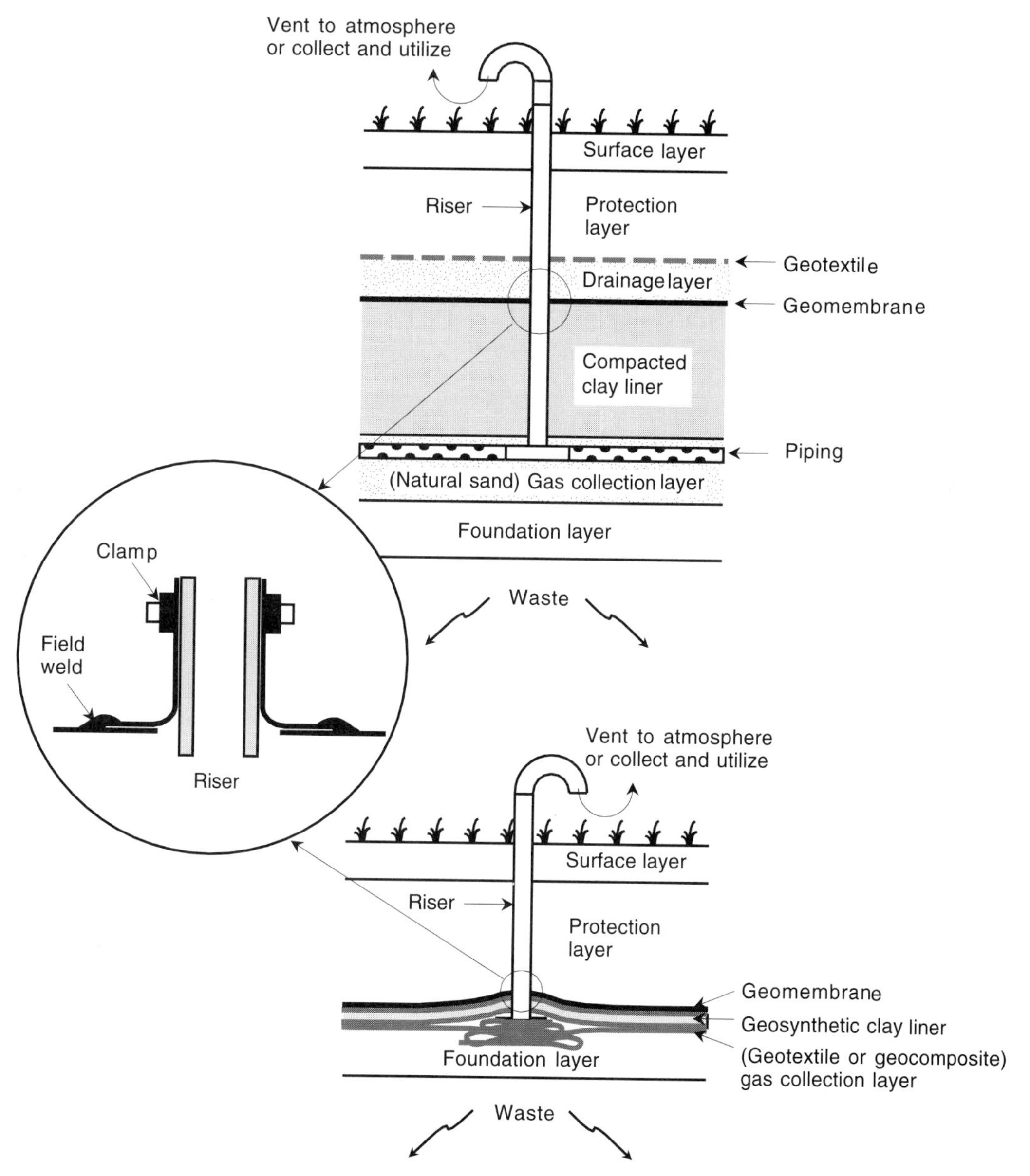

Figure 5.55 Selected venting systems to transmit landfill gases from the gas collection layer (natural soil and geosynthetics) to the ground surface.

cost can be much higher. A value of \$7.50 per m^2 was used as the estimated cost for a 600 mm thick CCL of 1×10^{-7} cm/s hydraulic conductivity.

- The cost range for GMs is narrower. A value of \$7.50 per m^2 for a 1.0 mm thick geomembrane with good out-of-plane deformation properties was used.

TABLE 5.21 PERFORMANCE OF BARRIER LAYER MATERIALS

		Climate			Differential settlement		
Alternate	Liner component	Arid	Cyclic	Humid	Major	Moderate	Nominal
A	CCL	1	1	3	1	1	3
B	GM	5	4	4	4	5	5
C	GCL	3	3	4	2	3	4
D	GM/CCL	2	3	4	2	3	4
E	GM/GCL	5	4	5	3	4	5
F	GM/CCL/GM	4	4	5	3	4	5
G	GM/GCL/GM	5	5	5	4	5	5

		Cover erosion/ puncture vulnerability			Allowable percolation		
Alternate	Liner component	Major	Moderate	Low	Essentially none	Very little	Moderate
A	CCL	1	2	3	1	2	3
B	GM	1	1	3	1	3	5
C	GCL	1	1	3	1	2	3
D	GM/CCL	3	4	4	3	4	4
E	GM/GCL	2	3	4	3	4	5
F	GM/CCL/GM	4	5	5	5	5	5
G	GM/GCL/GM	4	5	5	5	5	5

		Gas collection		Slope inclination		
Alternate	Liner component	Gas	No Gas	Less than 9°	9–18°	Greater than 18°
A	CCL	1	1	5	4	3
B	GM	5	5	5	5	3
C	GCL	1	5	4	3	3
D	GM/CCL	3	5	5	3	2
E	GM/GCL	4	5	5	3	2
F	GM/CCL/GM	4	5	5	2	1
G	GM/GCL/GM	4	5	5	3	2

Note: 1 = Not recommended; 2 = Marginal; 3 = Possibly acceptable (depends on specific conditions); 4 = Acceptable; and 5 = Recommended.

Source: After Daniel and Koerner [91]

- GCLs vary from 5 to \$10 per m², depending on site conditions. Again \$7.50 per m² was used.

Thus, with each of the three barrier layer components costing \$7.50 per m², schemes with two elements cost twice as much as a single-element layer, and schemes with three elements, triple. For the benefit/cost ratio of each alternative, the rows in Table 5.21 were added to obtain an overall benefit, which was then divided by the cost. This arbitrary procedure gives an overall ranking of how the alternatives perform over a broad range of site conditions; see Table 5.22.

TABLE 5.22 BENEFIT/COST RATIOS, BARRIER LAYER OPTIONS

Design alternate	Description	Overall benefit	Estimated cost (dollars/m^2)	Benefit/cost ratio	Ranking in group
One-Layer Barrier					
A	CCL	36	7.50	4.8	3
B	GM	64	7.50	8.5	1
C	GCL	46	7.50	6.1	2
Two-Layer Barrier					
D	GM/CCL	58	15.00	3.9	2
E	GM/GCL	66	15.00	4.4	1
Three-Layer Barrier					
F	GM/CCL/GM	71	22.50	3.2	2
G	GM/GCL/GM	77	22.50	3.4	1

Source: After Daniel and Koerner [91]

Among the single-layer systems, the single geomembrane (GM) outperforms the GCL in benefit/cost terms. Both of these materials outperform a CCL, which is the poorest overall technical choice as well as benefit/cost choice of any single-layer system. Paradoxically, a CCL is the most widely used and permitted single-layer material currently in use for landfill covers.

Of the composites, the GM/GCL outperforms the GM/CCL in both of the situations that were evaluated. This is important since most commonly constructed composite liners are a GM/CCL combination. For triple systems, the differences are relatively small.

On the basis of this relatively subjective analysis, the authors recommend either a single geomembrane liner (GM), or a geomembrane over a geosynthetic clay liner (GM/GCL). CCLs may have a place in cover systems, but only for limited situations with very little total and differential settlements and adequate protection from both desiccation and freeze/thaw.

5.7.5 Surface Water Drainage Layer

Since the normal stresses on a landfill cover are quite low (construction equipment is probably the largest), a wide range of geonets or geocomposites can be used for infiltration water drainage. Such geosynthetics would then be an alternative to a sand or gravel drainage layer and would appear as shown in Figure 5.52. All drainage geocomposites require a geotextile as a filter and separator above them, but this design element is a straightforward one and is covered in Chapter 2. In addition to geonets, a wide variety of available drainage geocomposite products will be presented in Chapter 8. Note that the geotextile filter/separator on all geonets and most geocomposites is thermally bonded to the drainage core so as to avoid a potentially weak interface layer. The design follows along the traditional lines, with the formulation of a flow rate factor of safety, that is,

$$FS = \frac{q_{\text{allow}}}{q_{\text{reqd}}} \tag{5.34}$$

where

$$FS = \text{factor of safety,}$$
$$q_{\text{allow}} = \text{allowable (test) flow rate, and}$$
$$q_{\text{reqd}} = \text{required (design) flow rate.}$$

The last term is usually estimated using the computer model HELP, developed by Schroeder et al. [92]. HELP contains hydrologic data from 200+ cities and has a great deal of design flexibility. Its limitation for this problem is that it calculates flow rate on a daily basis, which underestimates intense rainstorms. Alternatively, hand calculation of the infiltration quantity on an hourly basis is possible and is detailed in [87]. The allowable flow rate is obtained directly from laboratory testing via the ASTM D4716 test method, which is described in Chapter 3 and further described in Chapter 8. The simulated cross section at design pressures (or greater, if creep is of concern) should be used.

5.7.6 Protection Layer

While closures of hazardous waste landfills are usually regulated insofar as their minimum thickness of protection soil is concerned, other types of closures are not. Many abandoned landfills are being temporarily or permanently closed under a variety of legislation where the designer has considerable flexibility. Generally the thickness of a soil protection layer should be greater than the greatest frost-penetration depth in order that the surface water drainage system is constantly operative and that clay soils in the barrier layer do not freeze. Beyond this restriction, the soil cover thickness should vary in accordance with the protection needed against infiltration and intrusion into the landfill. In general, the following guide can be used:

municipal and ash landfill caps:	300 to 600 mm
industrial landfill caps:	450 to 900 mm
hazardous material landfill caps:	750 to 1200 mm
low-level-radioactive-material caps:	1200 to 2000 mm

The thickness of a protection layer is always at issue in northern climates involving frost depth. This issue must be viewed in light of the type of barrier system and, to a lesser extent, the drainage system. For CCLs, the upper surface of the clay must be beneath the depth of maximum frost penetration. As shown in Table 5.22, however, geomembranes and GCLs are the preferred barrier materials. For geomembranes and their seams, frost is essentially a nonissue [29]. For GCLs, work in this area is still ongoing, but clearly they are less susceptible to frost than CCLs (see Chapter 6 for additional detail). Regarding the drainage layer, both natural soils and geosynthetics must contend with the same situation, thus no decided advantage is given to either design strategy.

5.7.7 Surface Layer

Since the final cover of a landfill or abandoned dump is a long-term structural system, its integrity must be assured for many years. Just how long is a hot debate. Legislatively, the post-closure care period for hazardous waste landfills in the U.S. is 30 years. But to many, your author included, this is far too short a time. What happens after 30 years? Can rainwater and snowmelt enter the waste, generating a new leachate for an unknown someone to remove and properly treat? Consider also radioactive waste with hundreds or thousands of years of active lifetime. Indeed, many questions arise, with too few answers.

With the above discussion in mind, it is necessary to anticipate the various mechanisms that might intrude or disturb the buried waste, thereby negating the cover system. Of major concern are the following mechanisms, which are offered with some selected comments. See [87, 88, 89, 91] for additional insight into this important area.

Erosion by water. With proper cover soil and vegetation, it is possible to design against water erosion. The local agricultural station, Soil Conservation Service, or local highway department can be consulted for the proper type of indigenous plants and shrubs. Gradients are very important. For above-grade landfills they can be quite steep. In such cases, the use of erosion control geocomposites becomes important. Both temporary and permanent types are available and they are discussed in Chapter 8.

Erosion by wind. Wind erosion is quite difficult to design against because of the wind's widely varying velocity and direction. It is not a problem if vegetation is present over the entire cover area, hence we again see the importance of proper plants and shrubs. In arid regions it might be necessary to use some type of erosion control geocomposite or even hard armor treatment. Chapter 8 presents these materials and their alternatives.

Root penetration. With some types of vegetation, deep root penetration into the cover soil can occur. The barrier layer should be chosen to prevent penetration, but even so, the surface water drainage system might become clogged. Proper plant and shrub selection is again important, as it was with both types of above-described erosion.

Burrowing animals. We have seen (Figure 5.12) that a large number of macro-animals exist in soil. When they are burrowing, it is possible that the barrier layer could be encountered and penetrated (recall Section 5.1.4). If this action poses a problem, the use of a rock layer above the drainage system might be necessary. References [91, 93] discuss the use of large stone *bio-barriers* of this type.

Accidental intrusion. Of particular concern is accidental intrusion by drillers, pipeline excavators, site developers, and others who have interest in relatively remote sites. The proper posting of signs and maintained fencing should be adequate for preventing accidental intrusion by construction workers into buried and capped landfill sites.

Intentional intrusion. Why anyone would want to intrude through the final cover of an old landfill is beyond comprehension of the author. I suppose that it is always possible and for some sites essentially impossible to prevent. Periodic inspections, with appropriate maintenance, is required for this and most of the other intrusion mechanisms as well.

5.7.8 Aesthetic Considerations

To date, the final covers over completed landfills and abandoned dumps have been ominous zones, usually buffered from the public by fencing. The effect on the region is much like that obtained by quarantined areas: vast open spaces that appear to be permanently lost for public use. Recently some landfill owners have begun to explore alternatives for such culturally dead zones. Mackey [94] reports on closed landfills used for sport fields, walking and jogging paths, and other nonpermanent uses. It is important to note that with a geomembrane in the barrier layer, the offensive odor of methane is eliminated. A different approach is the recent commissioning of a graphic artist by the Hackensack Meadowland Development Commission [95] to transform a New Jersey municipal waste landfill final cover into an environmental art form. The 23 ha cover will be transformed into a Sky Mound that includes earth mounds up to 30 m in height. These mounds will frame sunrises and sunsets when viewed from the center of the cover. The astronomy theme is carried onto an interior lunar zone that is surrounded by a circular moat that serves as part of the surface water collection system and the looping arches of the methane recovery system. Pipe tunnels through selected mounds are aligned with the stellar helical settings of the stars Sirius and Vega. These extraordinary features are shown on Figure 5.56. Land surrounding the landfill cover will be converted to a wild bird refuge.

It is clear that the negative impact of a landfill closure can be minimized. However, it should be cautioned that features such as earth mounds or surface impoundments within the cover must be carefully engineered to prevent damage to the underlying drainage and barrier systems. The long-term performance of the cover must not be compromised by surface structures, regardless of their function or intent.

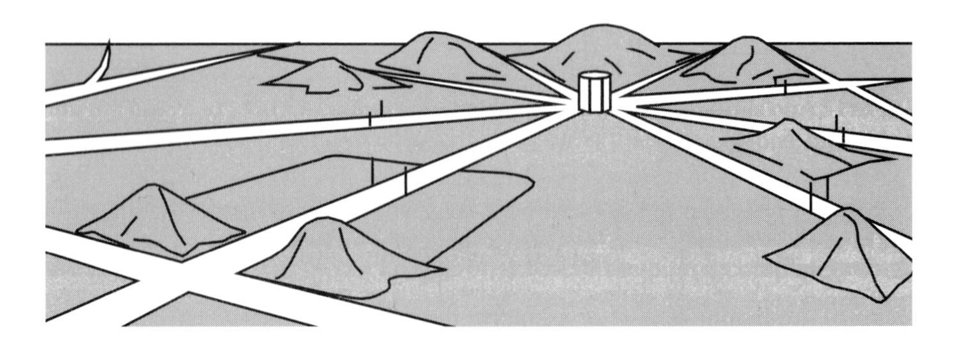

Figure 5.56 Artist's rendering of the topography of final cover. (After Meadowlands Redevelopment Authority, Pinyon [95])

5.8 UNDERGROUND STORAGE TANKS

Leakage from underground storage tanks (many of which contain hydrocarbon products) represents a very serious threat to downgradient water supplies. Such leakage has necessitated various containment strategies in the form of secondary containment, which are described in this section.

5.8.1 Overview

Depending on the study selected, there are as many as six million underground storage tanks in the U.S. containing hydrocarbon products. Of these, anywhere from 10 to 30% are thought to be leaking. To realize the seriousness of this number, consider that a 6 mm diameter hole in a standard 75,000 liter gasoline storage tank will pollute the drinking water supply of a 100,000-person community beyond acceptable background levels. Thus many states have enacted legislation requiring secondary containment of underground storage tanks. Two different strategies using geomembranes will be described.

5.8.2 Low-Volume Systems

As the secondary containment around a steel or fiberglass storage tank, we can wrap a geonet directly around the tank with a geomembrane around the geonet (see Figure 5.57a). The geonet (also called a *stand-off mesh*) becomes the leak detection network, while the geomembrane acts as the secondary liner. Both the geonet and geomembrane must be chemically resistant against the liquid in the tank, and HDPE is often used in this particular approach. A leak monitoring and removal pipe is placed at the low point of the geonet to monitor for primary liner (i.e., the tank itself) leaks.

5.8.3 High-Volume Systems

For high-volume systems, rather than fit each tank individually as above, an excavation for a number of underground tanks can be made and the entire excavation lined with a geotextile-geomembrane-geotextile composite. Figure 5.57b illustrates such a procedure. The leak detection media is drainage stone, which acts as bedding for the tanks. A pipe monitor is placed in this drainage system to check for tank leaks.

An interesting feature of this system is that the piping system leading to the gasoline station pumps can be handled in the same manner. Leaks often occur in or near the connections and fittings. The geomembrane composite actually encases the entire pipe network and travels with it wherever it goes. Hydrocarbon resistance of all the geosynthetic components is obviously necessary and EIA-R is often used in this particular approach, due to its greater flexibility over HDPE.

Underground storage tank owners who have sites underlain by granular soils with high seasonal watertables should be cognizant of these ongoing environmental activities.

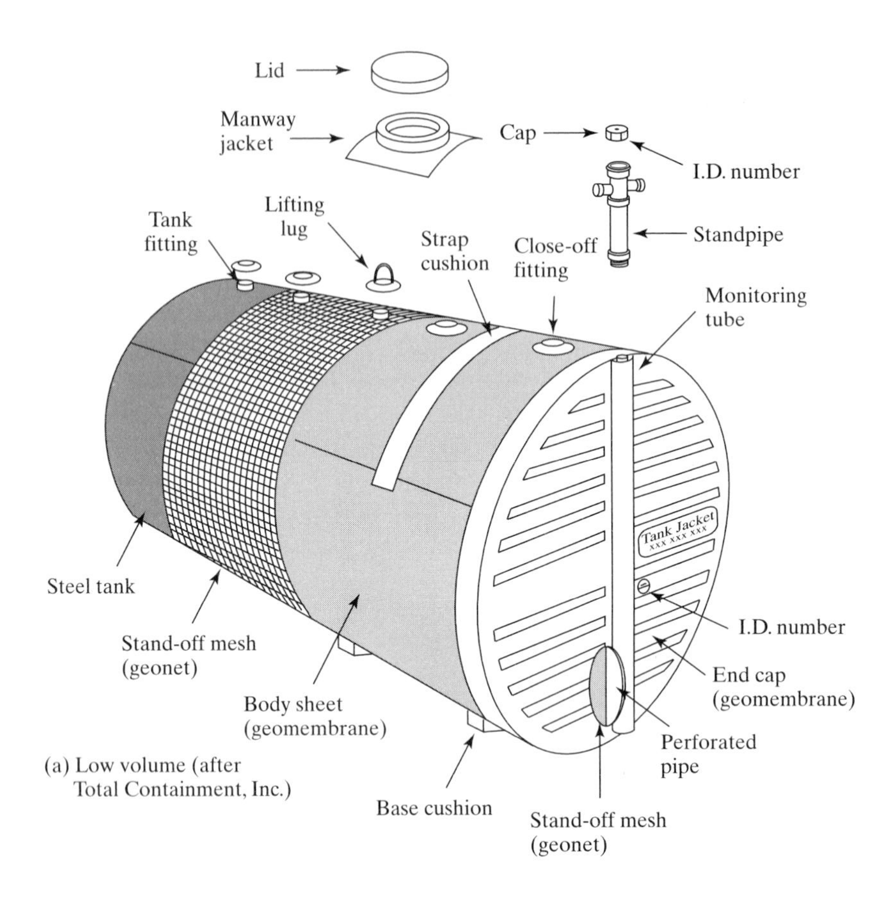

(a) Low volume (after Total Containment, Inc.)

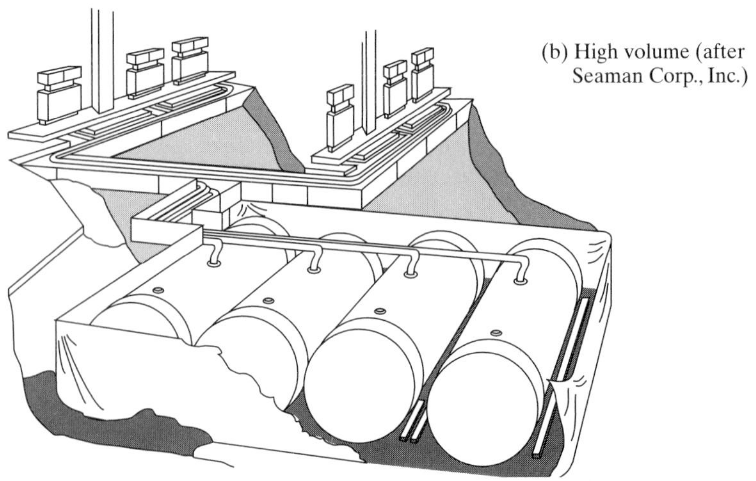

(b) High volume (after Seaman Corp., Inc.)

Figure 5.57 Schematic diagram of secondary leak detection and liner systems for underground storage tanks.

5.8.4 Tank Farms

Large holding tanks, some containing 40 Ml of liquid, require earth embankment containment, called *fire walls,* in case of leaks or rupture. These same embankments place the tanks in an underground category, even though they are located at the ground surface. Nevertheless, the need for secondary containment is required in many states. The resulting configuration appears exactly like any one of the sketches shown in Figure 5.21, with the exception that a steel or concrete tank is sitting in the center of the surrounding embankments. Thus any one of the geomembrane schemes illustrated can be used. For a long service lifetime, the geomembrane should be soil-buried and it obviously must be resistant to the liquid being contained in the tank. For hydrocarbon storage tanks, EIA-R and HDPE geomembranes are generally used. In this same type of application, GCLs have also been used, but they first must be exposed to water in order to hydrate the bentonite clay and mobilize its low permeability.

5.9 GEOTECHNICAL APPLICATIONS FOR GEOMEMBRANES

Geomembranes have been used in many innovative ways in connection with traditional geotechnical engineering structures. Often these are in connection with dams, tunnels, and seepage control.

5.9.1 Earth and Earth/Rock Dams

In most situations where low-permeability materials are desired to inhibit high pore water pressure and/or excessive seepage, geomembranes offer a logical and competitive alternative to the use of clays. Zoned earth and earth/rock dams, as well as roller-compacted concrete dams, require an impervious core, which has traditionally been of low-permeability silts, clays, or their mixtures. It seems natural to use a geomembrane as an alternative for cases where such fine-grained soils are difficult to obtain or to place.

Eigenbrod et al. [96] report on the use of an impervious upstream geomembrane blanket for controlling seepage in a tailings dam. Shown in Figure 5.58a is a cross section of the completed structure, in which it is seen that a geomembrane is on the upstream face of the compacted tailings immediately beneath a crushed rock layer. A high-strength 300 g/m^2 woven polypropylene geotextile was placed between the geomembrane and the crushed rock to protect the geomembrane from puncture. The geomembrane was a 0.75 mm nonreinforced CSPE liner. It was selected on the basis of the following criteria: elongation sufficient to withstand 900 mm of settlement in a 3 m diameter void, hence it was nonreinforced; resistance to the chemicals in the tailings materials being contained; satisfactory performance at -40°C; seams joinable at $+30$°C; and cost-competitiveness with other geomembranes.

Figure 5.58b shows such a geomembrane being placed on the upstream face of a roller-compacted concrete dam. When completed, a thick nonwoven needle punched protection geotextile will be placed over the geomembrane and then crushed rock

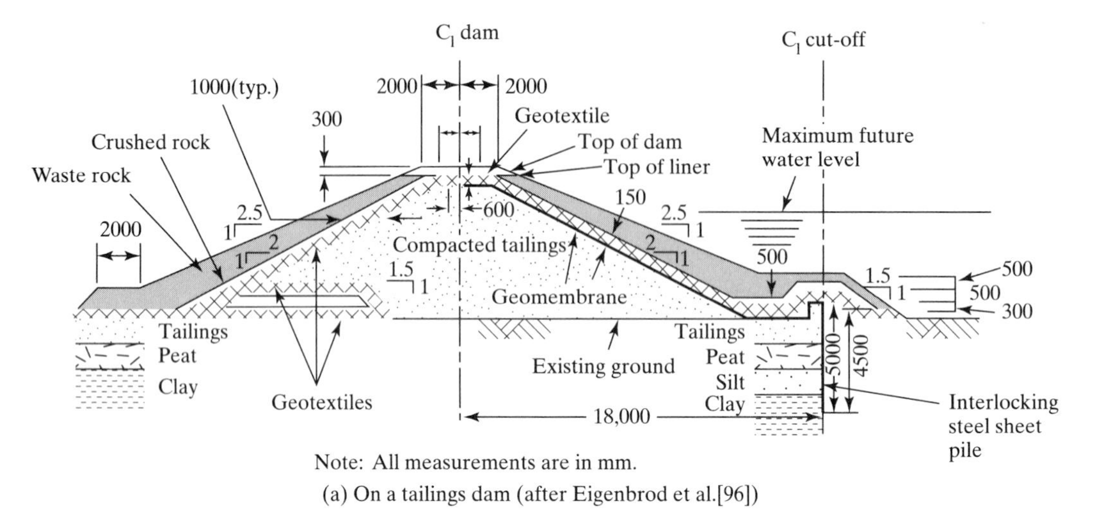

(a) On a tailings dam (after Eigenbrod et al.[96])

(b) On a roller compacted concrete dam in Italy (compliments of CARPI, Inc.)

Figure 5.58 Geomembrane upstream blanket for seepage control.

placed, similar to Figure 5.58a. Sembenelli and Rodriguez [97] illustrate many additional uses of geosynthetics in earth and masonry dams.

5.9.2 Concrete and Masonry Dams

Many existing concrete dams and spillways are showing signs of deterioration due to old age. Leakage from cracks in such structures can be very large, hence remediation is often necessary. Monari [98] illustrates how a geomembrane was used to cover the upstream side of a 37 m high concrete dam. The liner was 2.0 mm PVC with 300% elongation. Fixity was achieved by a clever series of steel ribs that were fastened to the concrete prior to the geomembrane installation. In this way future repairs and/or replacement of the geomembrane can be made easily. This case history represents one of many ways in which geomembranes can be used to control seepage. Cazzuffi [99] shows how widespread the technique is on a series of Italian dams; see Table 5.23. Scuero and Vaschetti [100] show how the same methods are adaptable to masonry

TABLE 5.23 DAM AND GEOMEMBRANE CHARACTERISTICS FOR WATERPROOFING SEVERAL ITALIAN DAMS

	Dam							
	Contrada Sabetta	Lago Baitone	Lago Miller	Lago Nero	Locone	Castreccioni	Cignana	Piano Barbellino
Owner	ENEL	ENEL	ENEL	ENEL	CBAL	CBM	ENEL	ENEL
Type	R	M	M	C	Coff	C	C	C
Height (m)	32	37	11	40	13	67	58	69
Construction date	1957–59	1927–30	1925–26	1924–29	1982	1981–86	1925–28	1926–31
Geomembrane								
Location	UF	UF	UF	UF	UF	RA	UF	UF
Slope	1/1	V	V	V	1/2.5	1/2.5–1/3.0	V	V
Surface (m^2)	2600	3500	1500	4000	28,000	46,000	10,000	5500
Support	DC	—	—	GT	—	GT	GT	GT
Protection	CS	—	—	—	—	GT + RR	—	—
Application	1959	1969–71	1976	1980–81	1982	1984–85	1986–87	1986–87
Type	EG	PIB	PVC	PVC	IIR	PVC	PVC	PRC
Thickness (mm)	2.0	2.0	1.8	1.9	1.5	1.2	2.5	2.5

Abbreviations:

R: Rockfill dam
M: Masonry dam
C: Concrete dam
Coff: Cofferdam
ENEL: Ente Nazionale Energia Elettrica
CBAL: Consorzio Bonifica Apulo-Lucano
CBM: Consorzio Bonifica Musone
UF: Upstream face
RA: Reservoir area
V: Vertical
DC: Draining concrete
CS: Concrete slabs
RR: Riprap
GT: Geotextile
EG: Elastomeric geomembrane
PIB: Polyisobutylene
PVC: Polyvinyl chloride
IIR: Isoprene–isobutylene rubber

Source: After Cazzuffi [99]

dams of all types. Probably the most serious design consideration is that of puncture resistance, particularly in areas where ice will form. Several clever air bubbling systems have been devised to reduce the danger of this problem.

5.9.3 Geomembrane Dams

The use of water-inflated tubes to block off streams and create reservoirs or to control downstream water levels has been attempted. The tubes, made from geomembrane materials, have been used to contain water levels up to about 2.5 m in height. Various systems have been used, consisting of different materials. All have been of a reinforced variety (e.g, three- and five-ply CSPE-R, fPP-R, etc.). The seams are obviously critical, as are connections to the bottom and sides of the stream banks.

The most ambitious of all the schemes of this type (it was ultimately rejected) was the proposed damming of the three shipping channels in the Po River valley leading to Venice, Italy. Figure 5.59 shows schematic drawings of the dams in their deflated and inflated positions. The intention was for them to remain deflated for most of the time, but when high waters in the Adriatic Sea occurred, the dams would be inflated, cutting off shipping but also foiling the destructive *aqua alta,* which is doing damage to Venice itself.

5.9.4 Tunnels

The waterproofing of tunnels has successfully deployed geomembranes, particularly in connection with the New Austrian Tunneling Method [102]. Here the tunnel is excavated and immediately shotcreted to prevent inward movement. The following series of steps are taken, leading to the completed section shown schematically in Figure 5.60.

1. Rock (usually) or soil (occasionally) is excavated.
2. The exposed surface is shotcreted.
3. A thick nonwoven needle-punched geotextile of 400 g/m^2 minimum mass per unit area is attached by means of an oversized washer.
4. The geotextile is fixed to underdrains on each side of the tunnel base.
5. A geomembrane (either PVC or VFPE) is placed over the geotextile, which is heat bonded to the previously placed washers.
6. The concrete liner segments or slip-formed concrete are placed against the geomembrane, completing the system.

5.9.5 Vertical Cutoff Walls

The use of vertically deployed geomembranes to control seepage in a number of systems as they are constructed can be extended to their use in vertical cutoff trenches for remediation work. This type of cutoff can be placed in a number of positions depending on actual circumstances: at or near the upstream toe (as shown in Figure 5.58a in place of the steel sheet piling); at or near the downstream toe (which is much less desirable because of boiling considerations, but is sometimes necessary where dewatering cannot occur); vertically through the entire dam itself, from its crest down to the top of

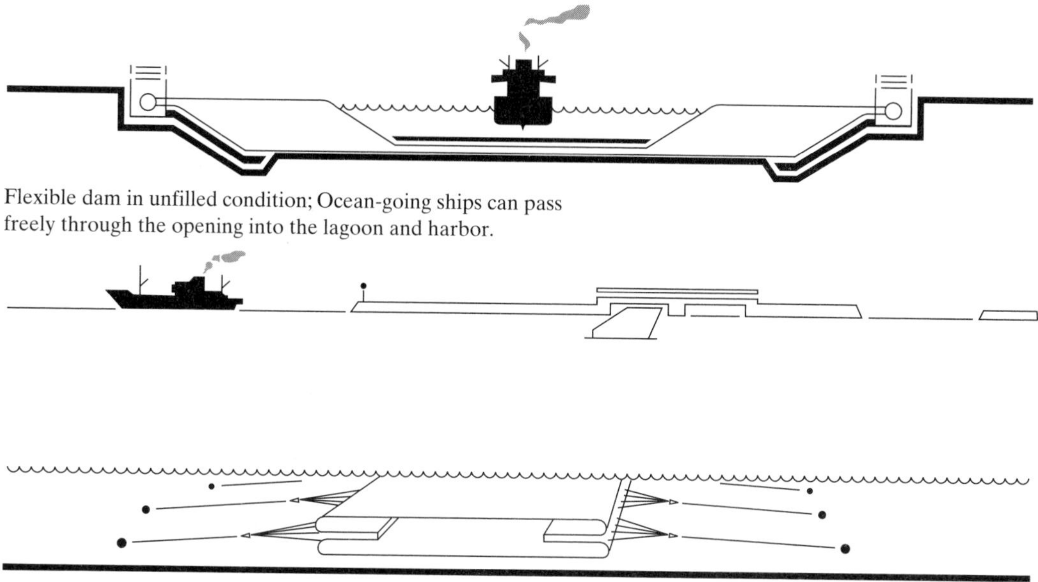

The sinkable high-water protection dam is designed as 30 m wide and 15 m high (when filled). It will be 470 m long for the Malamocco Canal, 550 m long for the Chioppia and 900 m long for the Lido Canal.

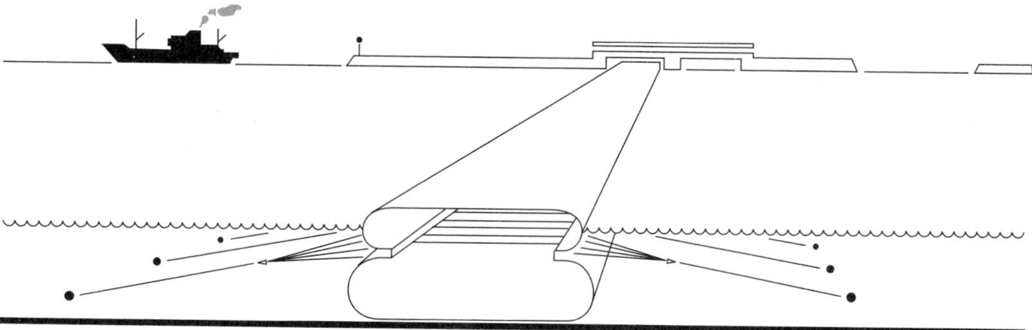

At flood stage the dam is pumped to its maximum height.

Figure 5.59 Proposed scheme for damming off channels leading to Venice using inflatable fabric tubes filled with seawater. (After Koerner and Welsh [101])

the foundation or even into the foundation itself; as single or double seepage cutoff rings around the site (e.g., to contain seepage coming from landfills or hazardous waste sites; see Koerner and Gugliemetti [103]).

The construction process calls for excavating a trench and placing the seamed geomembrane in it, and then backfilling it, thereby pushing the liner to one side of the trench. For deep trenches, this is usually not possible due to soil collapse, so the use of a slurry-supported trench is necessary. Here a mixture of water and bentonite (in approximate proportions of 20 to 1) is used to balance the pressures exerted by the in situ soil and groundwater so as to retain stability. Trenches 1 m wide and 20 m deep have

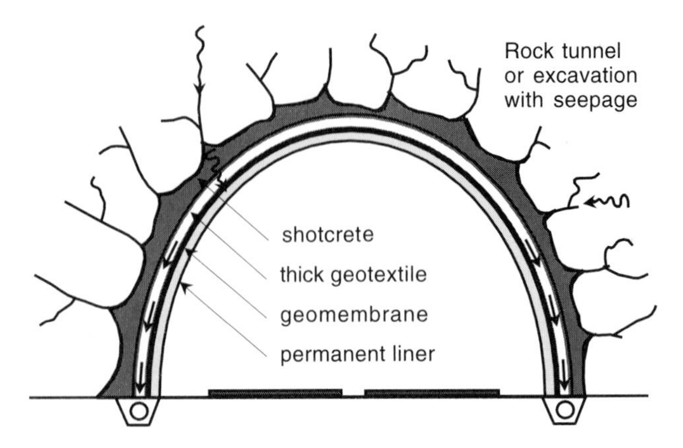

Figure 5.60 Procedure involved in tunnel waterproofing using geomembranes.

been constructed in this manner. Once the trench is dug to its intended depth, the fabricated geomembrane is placed in the slurry. Since most geomembrane materials have a specific gravity near unity, it is necessary to weight the bottom of the liner so that it sinks properly. Weights consisting of steel rods or metal pipes attached to the bottom edge of the geomembrane can be used. When the geomembrane is properly in place, the backfill is introduced, which displaces the slurry, forcing the geomembrane to the side of the trench. One representation of this concept is shown in schematic form in Figure 5.61.

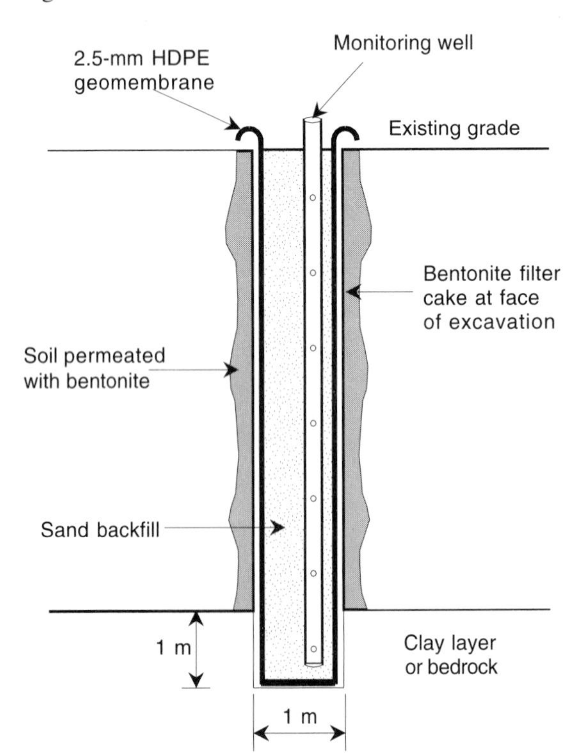

Figure 5.61 Envirowall concept of placing a geomembrane in a slurry-constructed trench to form a seepage cut-off wall. (After ICOS of America, Inc.)

Since installation of the above scheme is very difficult, other competing systems on this same theme have recently become available. These usually center around thick HDPE or nonplasticized PVC, in the form of tongue-and-groove sheeting. It is exactly the same as with interlocking steel sheeting, except now with stiff (noncorroding) polymer materials. For seepage control within the interlocks, a water-expandable gasket or a polymer-filled tube is used. Initial trials began with narrow sheets, but now wide sheets attached to an insertion plate are placed within a slurry supported trench (see Figure 5.62). The sheets are folded around the bottom of the insertion plate and held by pins until the proper depth is reached. The insertion plate is then removed, leaving the geomembrane cut-off wall in place and ready for the slurry displacing backfilling. The same system has been deployed in soft soil with no predug trench, using a vibratory pile hammer attached to the insertion plate. Several variations of the connections of one sheet to the next are shown in Koerner and Guglielmetti [103].

5.10 GEOMEMBRANE SEAMS

If a class were asked to make a list of the most important aspects of construction with geomembranes, *seams* should be at the top of everyone's paper. Indeed, without proper seaming the whole concept of using a geomembrane as a liner or vapor barrier is foolish. The topic can further be viewed from the aspect of factory versus field seams. The individual geomembrane sheets are sometimes made into larger sheets by factory-seaming them together (e.g., PVC and CSPE-R). These seams are generally very good, having been made in a controlled and clean environment with good quality control. The resulting panels are then brought to the project site and field seamed in their final configuration. Geomembranes supplied in roll form (e.g., HDPE and VFPE) come directly to the site for field seaming. It is the field seams that are particularly vulnerable to problems. When quality control is poor, leaks invariably arise. This important section addresses the type and manner of both factory-and field-seaming of geomembranes.

5.10.1 Seaming Methods

The field seaming of deployed geomembrane rolls or panels is a critical aspect of their successful functioning as a barrier to liquid and/or gas flow. This section describes the various seaming methods in current use and describes the concept and importance of test strips (or trial seams). It draws heavily from an EPA Technical Guidance Document [104].

Overview. The fundamental mechanism of seaming polymeric geomembrane sheets together is to temporarily reorganize the polymer structure (by melting or softening) of the two opposing surfaces to be joined in a controlled manner that, after the application of pressure and after the passage of a certain amount of time, results in the two sheets being bonded together. This reorganization results from an input of energy that originates from either *thermal* or *chemical* processes. These processes may involve the addition of an extra polymer in the bonded area.

Figure 5.62 Gundlok vertical cut-off walls using HDPE interlocking piles. (After GSE, Inc., Houston, TX)

Ideally, seaming two geomembrane sheets should result in no net loss of tensile strength across the two sheets, and the joined sheets should perform as one single geomembrane sheet. However, due to stress concentrations resulting from the seam geometry, current seaming techniques may result in minor tensile strength loss relative to the parent geomembrane sheet. The characteristics of the seamed area are a function of the type of geomembrane and the seaming technique used. Various factors, such as residual strength, geomembrane type, and seaming type, should be recognized by the designer when applying appropriate design factors of safety for the overall geomembrane function and facility performance.

The methods of seaming the geomembranes listed in Table 5.1 are given in Table 5.24 and shown schematically in Figure 5.63.

Seam Details. Within the entire group of thermoplastic geomembranes that will be discussed (note that thermoset geomembrane seams are not included, due to the nonuse of these types of geomembranes at this time), there are four general categories of seaming methods *extrusion welding, thermal fusion* or *melt bonding, chemical fusion*, and *adhesive seaming.* Each will be explained along with its specific variations, so as to give an overview of field seaming technology.

Extrusion welding is presently used exclusively on HDPE and VFPE (VLDPE, LLDPE, LDLPE) geomembranes. A ribbon of molten polymer is extruded over the edge of, or in between, the two slightly roughened surfaces to be joined. The molten extrudate causes the surfaces of the sheets to become hot and melt, after which the entire mass cools and bonds together. The technique is called *extrusion fillet* seaming when the extrudate is placed over the leading edge of the seam, and *extrusion flat* seaming when the extrudate is placed between the two sheets to be joined. The latter technique is no longer used in North America. It should be noted that extrusion fillet seaming is essentially the only method for seaming polyethylene geomembrane patches, for use in poorly accessible areas such as sump bottoms and around pipes, and for extremely short seam lengths. Temperature and seaming rate both play important roles in obtaining an acceptable bond; too much melting weakens the geomembrane and too little melting results in inadequate extrudate flow across the seam interface and in poor seam strength.

There are two thermal fusion or melt bonding methods that can be used on all of the thermoplastic geomembranes listed in Table 5.1. In both of them, the surface portions of the opposing surfaces are truly melted. This being the case, temperature, pressure, and seaming rate all play important roles in that excessive melting weakens the

TABLE 5.24 FUNDAMENTAL METHODS OF JOINING POLYMERIC GEOMEMBRANES

Thermal Processes	Chemical Processes
Extrusion:	Chemical:
Fillet	Chemical Fusion
Flat	Bodied Chemical Fusion
Fusion:	Adhesive:
Hot Wedge	Chemical Adhesive
Hot Air	Contact Adhesive

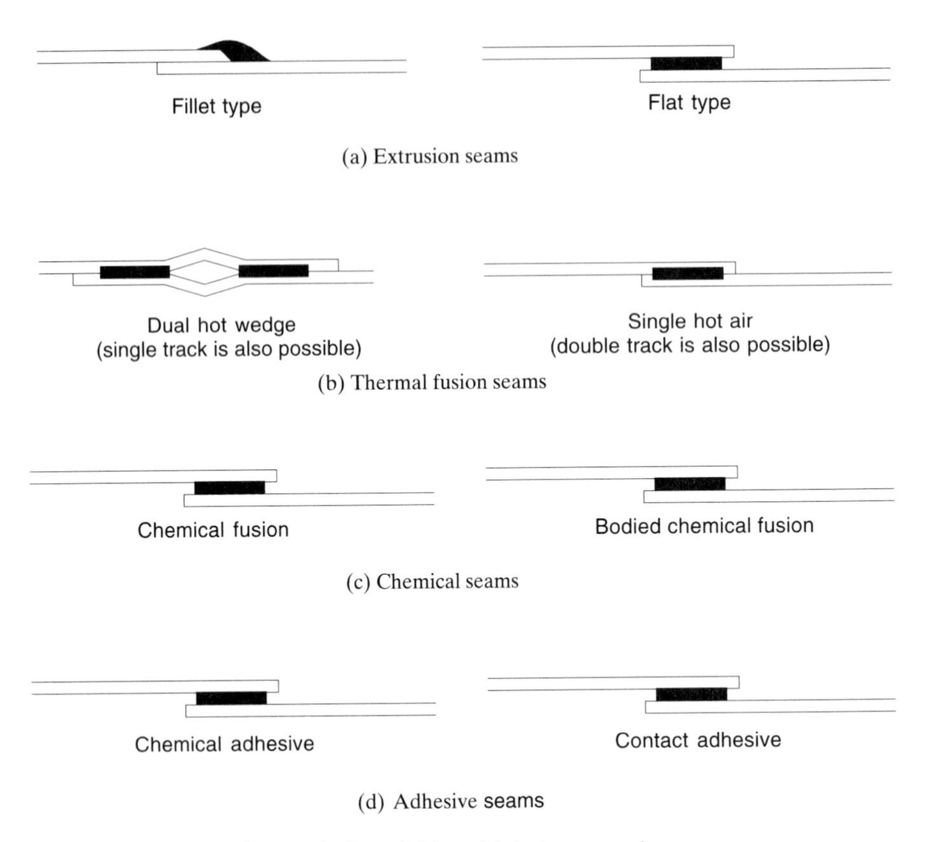

Figure 5.63 Various methods available to fabricate geomembrane seams.

geomembrane and inadequate melting results in poor seam strength. The *hot wedge* method consists of using an electrically heated resistance element in the shape of a wedge that travels between the two sheets to be seamed. As it melts the surface of the two sheets being seamed, a shear flow occurs across the upper and lower surfaces of the wedge. Roller pressure is applied as the two sheets converge at the tip of the wedge to form the final seam. Hot wedge units are automated as far as temperature, amount of pressure applied, and travel rate. A standard hot wedge creates a single uniform-width seam, while a dual (or *split*) hot wedge forms two parallel seams with a uniform unbonded space between them. This space can be used to evaluate seam quality and the continuity of the seam by pressurizing the unbonded space with air and monitoring any drop in pressure that may signify a leak in the seam. The technique can be adapted to data acquisition welders (as is routinely done in Germany) and even for computer-controlled systems (see [105]). The *hot air* method makes use of a device consisting of a resistance heater, a blower, and temperature controls to force hot air between two sheets to melt the opposing surfaces. Immediately following the melting of the surfaces, pressure is applied to the seamed area to bond the two sheets. As with the hot wedge method, both single and dual seams can be produced. In selected situations, this technique will be used to temporarily tack weld two sheets together until the final seam or weld is made and accepted.

There are two chemical fusion seam types. *Chemical* seams make use of a liquid solvent applied between the two geomembrane sheets to be joined. After a few seconds to soften the surfaces, pressure is applied to make complete contact and bond the sheets together. As with any of the chemical seaming processes to be described, a portion of the two adjacent materials to be bonded is truly transformed into a viscous phase. The technique is only used for those geomembranes that can be dissolved by the applied solvent. Methyl ethyl ketone (MEK) is the solvent usually used. Excessive solvent will weaken the adjoining sheets, and inadequate solvent will result in a weak seam. *Bodied chemical* seams are similar to chemical seams except that 1 to 20% of the parent lining resin or compound is dissolved in the MEK solvent and then is used to make the seam. The purpose of adding the resin or compound is to increase the viscosity of the liquid for slope work and/or to adjust the evaporation rate of the solvent. This viscous liquid is applied between the two opposing surfaces to be bonded. After a few seconds, pressure is applied to make complete contact. *Chemical adhesive* seams make use of a dissolved bonding agent (an adherent) in the chemical or bodied chemical, which is left after the seam has been completed and cured. The adherent thus becomes an additional element in the system. *Contact adhesives* are bonding agents applied to both mating surfaces. After reaching the proper degree of tackiness, the two sheets are placed on top of one another, followed by roller pressure. The adhesive forms the bond and is an additional element in the system.

In order to gain an overview as to which seaming methods are customarily used for which of the thermoplastic geomembranes in Table 5.1, Table 5.25 is offered. It is generalized, and meant to be used to introduce the primary seaming methods.

Test Strips (or Trial Seams). Test strips (also called trial seams or qualifying seams) are an important aspect of field seaming procedures. They are meant to serve as

TABLE 5.25 POSSIBLE FIELD-SEAMING METHODS FOR VARIOUS GEOMEMBRANE TYPES

	Type of Geomembrane						
Seaming Method	HDPE	VFPE	PP	PP–R	PVC	CSPE–R	EIA–R
Extrusion (fillet and flat)	A	A	A	A	n/a	n/a	n/a
Thermal fusion (hot wedge and hot air)	A	A	A	A	A	A	A
Chemical fusion (chemical and bodied chemical)	n/a	n/a	n/a	n/a	A	A	A
Adhesive (chemical and contact)	n/a	n/a	n/a	n/a	A	A	A

A = method is applicable

n/a = method is not applicable

VFPE includes VLDPE, LLDPE, and LDLPE geomembranes

Figure 5.64 Fabrication of a geomembrane test strip.

a prequalifying experience for personnel, equipment, and procedures for making seams on the identical geomembrane material under the same climatic conditions as will be the actual field production seams. The test strips are usually made on two narrow pieces of excess geomembrane, varying in length from 1.0 to 3.0 m (see Figure 5.64). The test strips should be made in sufficient lengths, preferably as a single continuous seam, for all required testing purposes.

The goal of these test strips is to imitate all aspects of the actual production field seaming activities intended to be performed in the immediately upcoming work session, so as to determine equipment and operator proficiency. Ideally, test strips can estimate the quality of the production seams while minimizing the field sampling of the installed geomembrane through destructive mechanical testing. Test strips are typically made every four hours (for example, at the beginning of the work day and after the lunch break). They are also made whenever personnel or equipment are changed and when climatic conditions reflect wide changes in geomembrane temperature or other conditions that could affect seam quality.

The destructive testing of the test strips should be done as soon as the installation contractor feels that the strength requirements can be met. Thus it behooves the contractor to have all aspects of the test strip seam fabrication in complete working order, just as would be done in fabricating production field seams. For extrusion and thermal fusion seams, destructive testing can be done as soon as the seam cools. For chemical fusion and adhesive seams, testing must wait for curing, possibly several days, and the use of a field oven to accelerate the curing of the seam is possible.

From two to six test specimens are cut from the test strip using a 25 mm wide die. The specimens are then tested in both peel and shear, using a field tensiometer (see Figure 5.65). If any of the test specimens fail, a new test strip is fabricated. If additional specimens fail, the seaming apparatus and seamer should not be accepted and should not be used for seaming until the deficiencies are corrected and successful trial welds are achieved. If the specimens pass, seaming operations can move directly to production seams in the field.

The flow chart in Figure 5.66 gives an idea of the various decisions that can be reached depending upon the outcome of destructive tests on test strip specimens. Here

Figure 5.65 A field tensiometer performing a geomembrane seam test. (Compliments of Wegener Co.)

it is seen that failed test strips are linked to an increased frequency of destructive tests done on production field seams that are made during the time interval between making the test strip and its testing. Furthermore, it is seen that there are only two chances to make adequate test strips before production field seaming is stopped and repairs are initiated. These details should be covered in either the contract specification or the quality assurance work plan.

5.10.2 Destructive Seam Tests

After a field seaming crew has made a series of production seams, it is important to evaluate their performance. The obvious procedure is to cut out a sample, send it to the laboratory, and pull it until failure in either shear or peel modes (recall Section 5.1.3). Another option is to test it directly at the field site. But considering a geomembrane sheet layout as shown in Figure 5.67, the questions become where and how many? Remember that each seam sample becomes a hole, which must be appropriately patched and then retested. For this reason it is common to reduce the number of field seam samples to a bare minimum, and then to assess only the method of seaming, not its continuity. By *method* we mean installation type, temperature, dwell time, pressure, and other operational details affecting seam quality. Continuity is handled by nondestructive testing, which will be described later.

Sampling Protocol. Destructive test sampling can be done on a random basis or on a periodic basis. Metracon [4] recommends a frequency of six samples per kilometer of seam on a random basis, or one sample per 150 m of seam on a uniform basis. Recognize that even within a uniform interval the samples can be selected randomly (see [104]). Furthermore, there are sampling strategies that reward good seaming by requiring fewer samples and penalize poor seaming by requiring additional samples. See

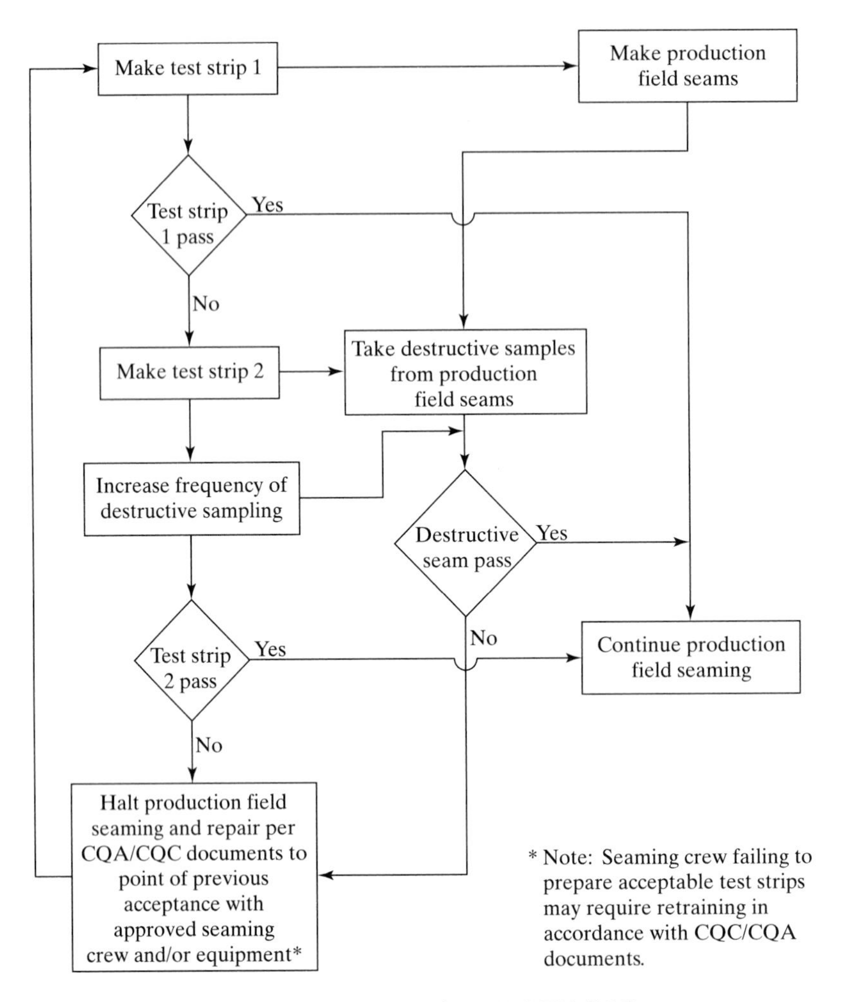

Figure 5.66 Test strip process flow chart. (After U.S. EPA [104])

Richardson [106] for two such strategies, the method of attributes and the use of control charts.

Sample Size. The size of the destructive test sample depends on the specification and quality assurance plan at the site. It can be as small as 300 mm along the seam length or up to 1500 mm. After taking a sample, it is further subdivided among one or all of the following organizations: the owner/operator (for archiving), the construction quality control firm (for testing), the construction quality assurance firm (for testing), the general contractor (for archiving), and the regulatory agency (for inspection or archiving). Additional detail is given in Reference 104.

Shear and Peel Testing. Shown in Figures 5.68a,b are examples of shear and peel testing of geomembrane seam test specimens. While such tests appear straightfor-

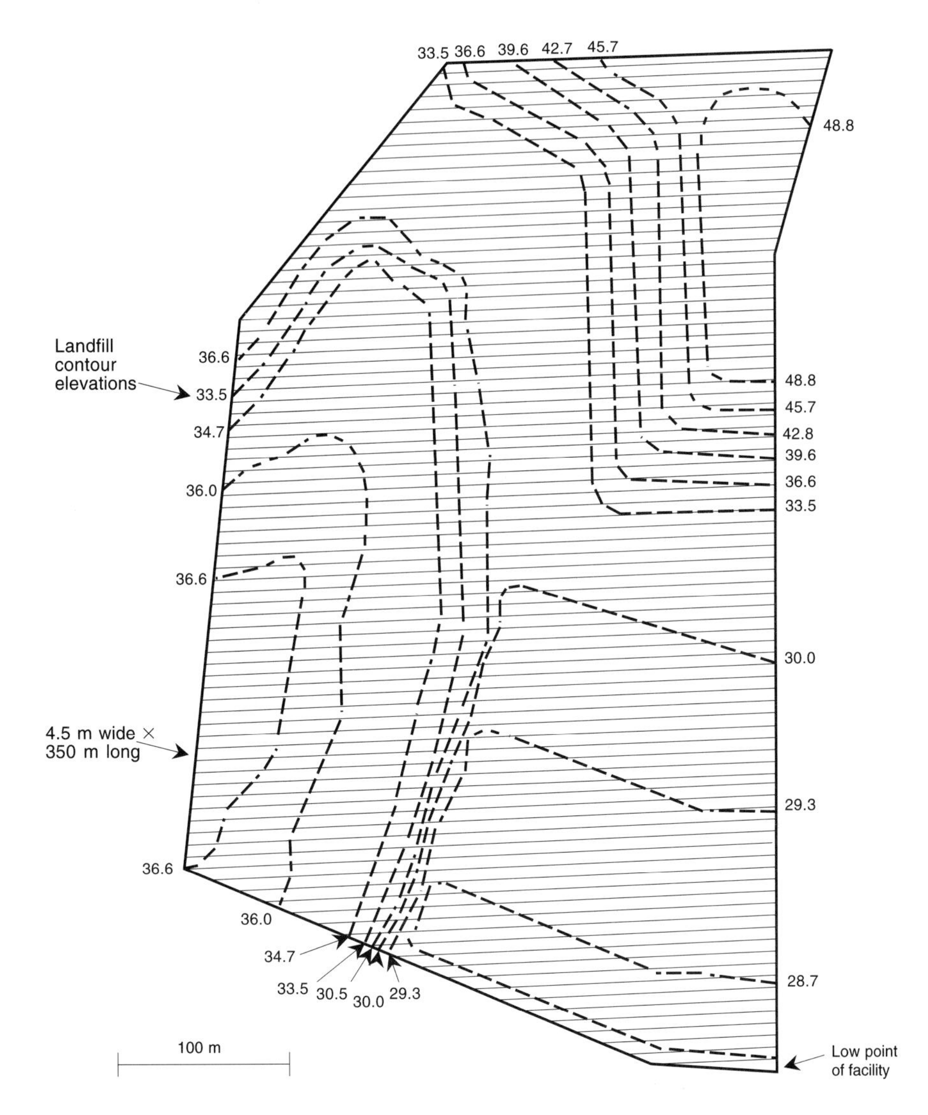

Figure 5.67 Geomembrane sheet layout for a solid waste landfill; lined area equals approximately 76,800 m^2. (After National Seal Co.)

ward, there are many nuances depending on the type of geomembrane being evaluated. Table 5.4 presents the current status of evaluating the various geomembranes mentioned in Table 5.1. Given is the type of ASTM shear test, the peel test, and the comparison test on the nonseamed sheet, unless the strength of the seam test is targeted to a limiting value. Insofar as a passing test is concerned, we are focusing on two issues.

(a) Typical shear test

(b) Typical peel test

Figure 5.68 Basic types of geomembrane seam tests illustrated for HDPE.

- The sheets on either side of the seam must fail, that is, the seam cannot delaminate. This is called a *film tear bond* failure as per Metracon [4].
- The magnitude of the force required for failure should meet or exceed a specified value. For seams tested in a shear mode, failure forces of 90 to 100% of the unseamed sheet failure are usually specified. For seams tested in a peel mode, failure forces of 50 to 80% of the unseamed sheet are often specified for thermally bonded seams. Other strategies are used for chemical and adhesive bonded seams. These percentages underscore the severity of peel tests as compared to shear tests. As seen in the data in Figure 5.6, this is indeed the case. For assessing seam quality, the peel test is indeed the target test that should be focused on.

5.10.3 Nondestructive Seam Tests

Although it is important to properly assess the method of seaming, shear and peel tests tell nothing of the continuity and completeness of the entire seam. It does little good if one section of a seam comes up to be 100% of the strength of the parent material, only to have the section next to it missed completely by the field seaming crew. Thus continuous methods of a nondestructive testing (NDT) nature will be discussed here (see [67, 104]). In each of these methods, the goal is to check 100% of the seams. The methods are listed in Table 5.26 in the order that they will be discussed. Note that the primary user has also been identified. Construction quality control (CQC) refers to the firm ac-

TABLE 5.26 NONDESTRUCTIVE GEOMEMBRANE SEAM TESTING METHODS

Nondestructive Test Method	Primary User		Cost of Equipment	Speed of Tests	Cost of Tests	Type of Result	Recording Method	Operator Dependency
	CQC	CQA						
Air lance	yes	—	$200	fast	nil	yes–no	manual	very high
Mechanical point (pick) stress	yes	—	nil	fast	nil	yes–no	manual	very high
Dual seam (positive pressure)	yes	—	$200	fast	moderate	yes–no	manual	low
Vacuum chamber (negative pressure)	yes	yes	$1000	slow	very high	yes–no	manual	high
Electric wire	yes	yes	$500	fast	nil	yes–no	manual	high
Electric field	yes	yes	$20,000	slow	high	yes–no	manual and automatic	low
Ultrasonic pulse echo	—	yes	$5000	moderate	high	yes–no	automatic	moderate
Ultrasonic impedance	—	yes	$7000	moderate	high	qualitative	automatic	unknown
Ultrasonic shadow	—	yes	$5000	moderate	high	qualitative	automatic	moderate

CQC = Construction Quality Control (geomembrane manufacturer or installer)

CQA = Construction Quality Assurance via design engineer or inspection organization

Source: Modified from Richardson and Koerner [67]

tually doing the seam fabrication; construction quality assurance (CQA) refers to a separate organization working on behalf of the regulatory agency, but paid by and reporting to the facility's owner and/or operator.

The *air lance* method uses a jet of air at approximately 350 kPa pressure coming through an orifice of 5 mm diameter. It is directed beneath the upper edge of the overlapped seam to detect unbonded areas. When such an area is located, the air passes through, causing inflation and fluttering in the localized area. The audible sound also changes when unbonded areas are encountered. The method works best on relatively thin, less than 1.0 mm, flexible geomembranes, but it works only if the defect is open at the front edge of the seam, where the air jet is directed. It is essentially a contractor/installer's tool to be used in a CQC manner.

The *mechanical point stress* or *"pick"* test uses a dull tool (such as a blunt screwdriver) under the top edge of a seam. With care, an installer can detect an unbonded area, which is easier to separate than a properly bonded area. It is a rapid test that obviously depends completely on the care and sensitivity of the person doing it. Detectability is similar to that using the air lance, and both are very operator-dependent. Again, this test is to be performed only by the installation contractor and/or geomembrane manufacturer. Design or inspection engineers have no business poking objects into seamed regions and should use one or more of the following techniques.

The *pressurized dual seam* method was mentioned earlier in connection with the double-wedge thermal seaming method. The air channel that results between the double seam is inflated using a hypodermic needle and is pressurized to approximately 200 kPa. If no drop on a pressure gage occurs over a given time period, the seam is acceptable. The test method for polyethylene geomembranes is ASTM D5820. If an excessive drop in pressure occurs, a number of actions can be taken: the distance can be systematically halved until the leak is located, the section can be tested by some other leak detection method, or a cap strip can be seamed over the entire edge. The test is excellent for long straight seam runs. It is generally performed by the installation contractor, but often with the CQA personnel viewing the procedure and assessing the results.

Vacuum chambers (boxes) have been used where a 1 m long box with a transparent top is placed over the seam and a vacuum of approximately 15 kPa applied. When a leak is encountered, the soapy solution previously placed over the seam shows bubbles, thereby reducing the vacuum. This is due to air entering from beneath the liner and passing through the unbonded zone. The test is slow to perform and it is often difficult to make a vacuum-tight joint at the bottom of the box where it passes over the seam edges. Due to upward deformations of the geomembrane into the vacuum box, only geomembrane thicknesses greater than 1.0 mm should be tested in this manner. For thin flexible geomembranes, the bottom of the box can be fitted with a steel mesh to avoid excessive deformations. It should also be noted that vacuum boxes are a common form of nondestructive test used by design engineers and CQA inspectors. If 100% of the field seams are to be inspected by this method, it will take a large number of field personnel who will get bored out of their minds, and the test still will not cover the area around sumps, anchor trenches, and patches with any degree of assurance. The method is also essentially impossible to use on side slopes, since the adequate downward pressure required to make a good seal cannot be mobilized (as this is usually done by standing on top of the box).

Electric sparking (not listed in Table 5.26) is an old technique used to detect pinholes in thermoplastic liners. The method uses a high-voltage (15 to 30 kV) current, and any leakage to ground (through an unbonded area) will result in sparking. The method is not very sensitive to overlapped seams of the type used in liners and is used only rarely for this purpose. The technique has been revived in a somewhat different form by manufacturing a high carbon black coextruded polyethylene geomembrane on the lower surface. By applying a suitable voltage, the entire geomembrane system (sheets and seams) can be monitored for leaks by electric spark testing.

The *electric wire* method places a copper or stainless steel wire between the overlapped geomembrane regions and actually embeds it into the completed seam. After seaming, a charged probe of about 20,000 volts is connected to one end of the wire and slowly moved over the length of the seam. A seam defect between the probe and the embedded wire results in an audible alarm from the unit. The method is strongly advocated by some installation firms, giving rise to extremely low seam failure instances.

The *electric field* test was developed by Schultz et al. [107] and utilizes a liquid-covered geomembrane to contain an electric field. Note that the entire bottom of the lined facility must be covered with liquid, usually water. The depth can be nominal (e.g., 150 mm). A current source is used to impose current across the boundary of the liner. When a current is applied between the source and the remote current return electrodes, current flows either around the entire site (if no leak is present) or bypasses the longer travel path through the leak itself (when one is present). Potentials measured on the surface are affected by the distributions and can be used to locate the source of the leak. These potentials are measured by *walking* a probe in the water. The operator walks on a predetermined grid layout and marks where anomalies exist. These can be rechecked after the survey is completed by other methods, such as a vacuum box. Note that the technique must be modified where water does not cover the geomembrane, as on side slopes. The technique is currently commercially available; see Laine et al. [108].

The last group of nondestructive test methods can collectively be called *ultrasonic methods.* A number of ultrasonic methods are available for seam testing and evaluation. (See [109] for greater detail than presented here.) The *ultrasonic pulse echo* technique is basically a thickness measurement technique and is only for use with nonreinforced geomembranes. Here a high-frequency pulse is sent into the upper geomembrane and (in the case of a good seam) reflects off the bottom of the lower one. If, however, an unbonded area is present, the reflection will occur at the unbonded interface. The use of two transducers, a pulse generator, and a CRT monitor are required. It cannot be used for extrusion fillet seams, because of their nonuniform thickness. The *ultrasonic impedance plane* method works on the principle of acoustic impedance. A continuous wave of 160 to 185 kHz is sent through the seamed liner, and a characteristic dot pattern is displayed on a CRT screen. Calibration of the dot pattern is required to signify a good seam; otherwise, it is not. The method has potential for all types of geomembranes but still needs additional development work. The *ultrasonic shadow method* uses two roller transducers: one sends a signal into the upper geomembrane and the other receives the signal from the lower geomembrane on the other side of the

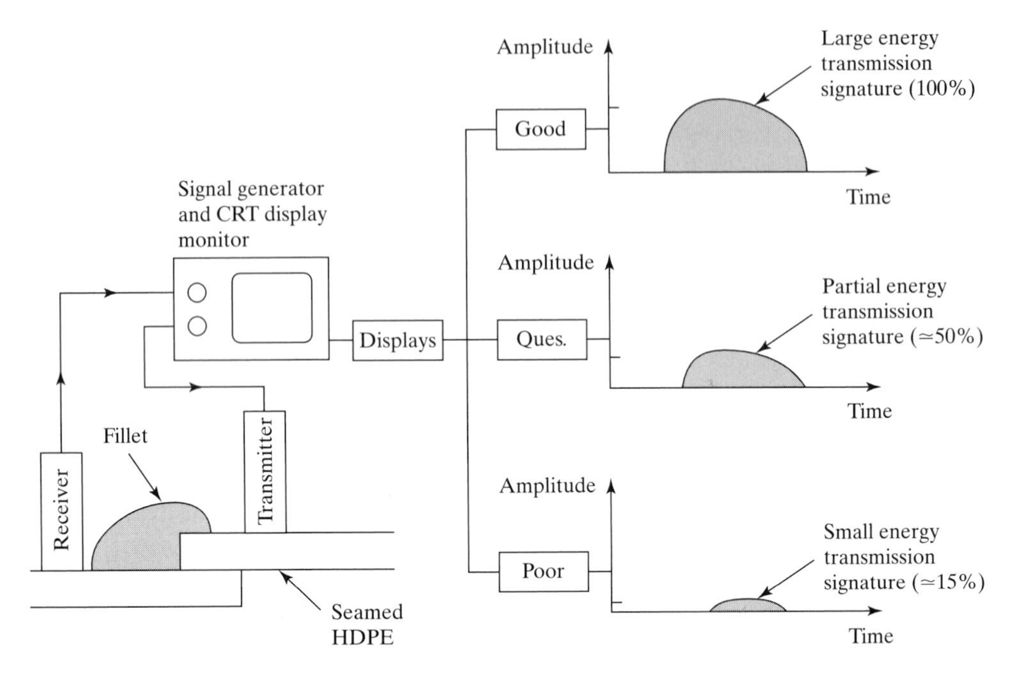

Figure 5.69 Schematic diagram of operational concepts of ultrasonic shadow method. (After Koerner et al. [110])

seam. Figure 5.69 illustrates the technique, showing the type of traces received by the display monitor. It has been found [110] that for received signals greater than 50% full-scale height, HDPE seams subsequently tested destructively were all acceptable. Received signals less than 20% full-scale height indicated correctly that the seams were unacceptable. The 50% to 20% range had mixed results. The technique can be used for all types of seams, even those in difficult locations, such as around manholes, sumps, and appurtenances. It is best suited to semicrystalline geomembranes, including HDPE, and will not work for scrim reinforced liners.

5.10.4 Summary

It is generally recognized that the geomembrane industry's ability to manufacture near-flawless sheets far surpasses its ability to seam separate sheets together. Furthermore, it is also recognized that the ability to make factory seams is generally considered to be better than the ability to make field seams. Field seams owe this difficulty to a number of sources: horizontal (sloped) preparation surfaces; nonuniform (or yielding) preparation surfaces; nonconforming sheets to the subsurface (air pockets); slippery liners made of low friction materials; wind-blown dirt in the areas to be seamed; moisture and dampness in the subgrade beneath the seam; frost in the subgrade beneath the seam; moisture on the upper surface of the geomembrane; penetrations, connections, and appurtenances; wind fluttering the sheets out of position; ambient temperature variations during seaming; uncomfortably high temperatures for careful working; uncomfortably low temperatures for careful working; and expansion and/or contraction of sheets during seaming.

With so many potential problems,* it is natural that emphasis on high-quality field seams and on subsequent seam inspection is commonly referred to in the literature. This need grows progressively more important depending upon the implications of the contained material (usually liquid) escaping. Thus hazardous and radioactive waste facilities have the highest priority, while recreational reservoirs and aesthetic ponds have a much lower priority.

Of the different geomembrane seaming methods described, the option is probably best left to the manufacturer and/or installing contractor. This is not to say that the status of the methods might not change in the future. If, indeed, one seaming method shows itself to be superior to others, future specifications might mandate its use. It is a fruitful area for additional research and development.

In this regard the hot wedge fusion system deserves further commentary. At the outset, recognize that *all* thermoplastic geomembranes can be seamed by this method. This includes every type of geomembrane listed in Table 5.1. The method has three controllable features: wedge temperature, nip roller pressure, and travel rate (speed). Currently these controls are set manually, based in part on the outcome of trial seams, as previously discussed. Since trial seams are typically made at four-hour intervals, weather conditions can change, and the operator must adjust the device accordingly. To avoid subjective modifications, current efforts are being made at data acquisition, on-line sensing, and computer-controlled feedback and adjustment [105]. Typically, the speed will be increased if the geomembrane temperature warms, and be decreased if it cools. A number of new and exciting systems are currently operational are available in Europe.

Of equal importance to the type of seam are seam testing methods. While destructive tests are invariably required, they are self-defeating at the outset. The worst-looking locations of a lined facility is at every location where a sample has been cut out for testing, patched, retested, and sometimes patched again. When samples must be taken by or distributed to the regulatory agency, the owner, the contractor, the designer, and the CQA organization, the situation can become ludicrous. It begs for a nondestructive test that assesses both quality and continuity. At this point in time, the vacuum box method is heavily relied upon. In the author's opinion, such reliance is foolish. One-hundred percent seam inspection by vacuum box testing simply cannot be done. Usually those locations where the vacuum box cannot be used are where problems arise, namely, on slopes, in corners, at sumps and at penetrations. In this light, ultrasonic methods—particularly the shadow method—show some potential. A major thrust to investigate its capabilities and limitations is warranted, yet such an investigation must be in addition to the constant search for test and seaming methods that are ever more accurate and efficient.

*Note that in Germany, geomembranes are required to have thin plastic tear strips (150 mm wide) on the top and bottom edges of the rolls, which are removed immediately before seaming.

5.11 DETAILS AND MISCELLANEOUS ITEMS

As mentioned in Section 5.10 on seams, difficulties often arise where details are required. Where the space is limited and automated equipment cannot be used, hand labor and experience are all important.

5.11.1 Connections

The primary guidelines that a designer should follow regarding geomembrane connections are to maintain as smooth a transition as possible and to use materials with the least possible change in stiffness. In this context, stiffness can be assessed by modulus, where the following ranking of materials (from highest to lowest modulus) is well known: steel (and other metals), concrete, wood, stiff polymers, soft polymers. Thus geomembranes connected to metal and concrete structures must be very carefully designed.

Design in this case is really a matter of detailing and visualizing how settlements, deformations, and other stress-and-strain-mobilizing phenomena will influence the connection. Experience is certainly important in this regard. Most manufacturers and installers of liners have details showing proper procedures. Some of these details are presented in Figure 5.70, where thick polyethylene pieces or woodstrips are used to make the transition from liner to concrete structure. Metal structures can be treated in a similar manner. Note that there is the possibility of making a pressure seal to concrete, as shown on the one detail. Such designs have been used to force subsurface groundwater to drainage sumps where gravitational flow was not possible.

Although these details are straightforward to visualize and easy to show, their proper construction is not so simple. Care and true craftsmanlike work are required for trouble-free and leak-free performance.

5.11.2 Appurtenances

Appurtenances are any adjunct item necessary for proper functioning of the total system. When dealing with geomembrane-related systems, this refers to inlet and outlet systems together with pipe racks, vents, sumps, structural monitoring frames, and the like.

As for pipes penetrating the geomembrane, prefabricated *boots* are commonly used, which fit snugly over the pipe and are then sealed to the liner (see Figure 5.71). Mastic, O-rings, or fillet extrusion welds are then used to complete the seal. The boot should be made of the same material as the liner, and the seal should be the same type as used in the rest of the system. Direct connections to flanges and base outlet pipes are even more difficult to construct, as the details of Figure 5.71 illustrate. Problems can arise here, not so much from the initial installation but because such outlets represent a separate structure. These structures often have settlement profiles very different from the rest of the impoundment; hence, differential settlements should be anticipated. An important example is the leachate collection sump of a landfill liner. This sump must be connected to the ground surface by pipe sections during filling and poses a severe puncturing problem for the liner. The degree of severity, however, is very much site-specific.

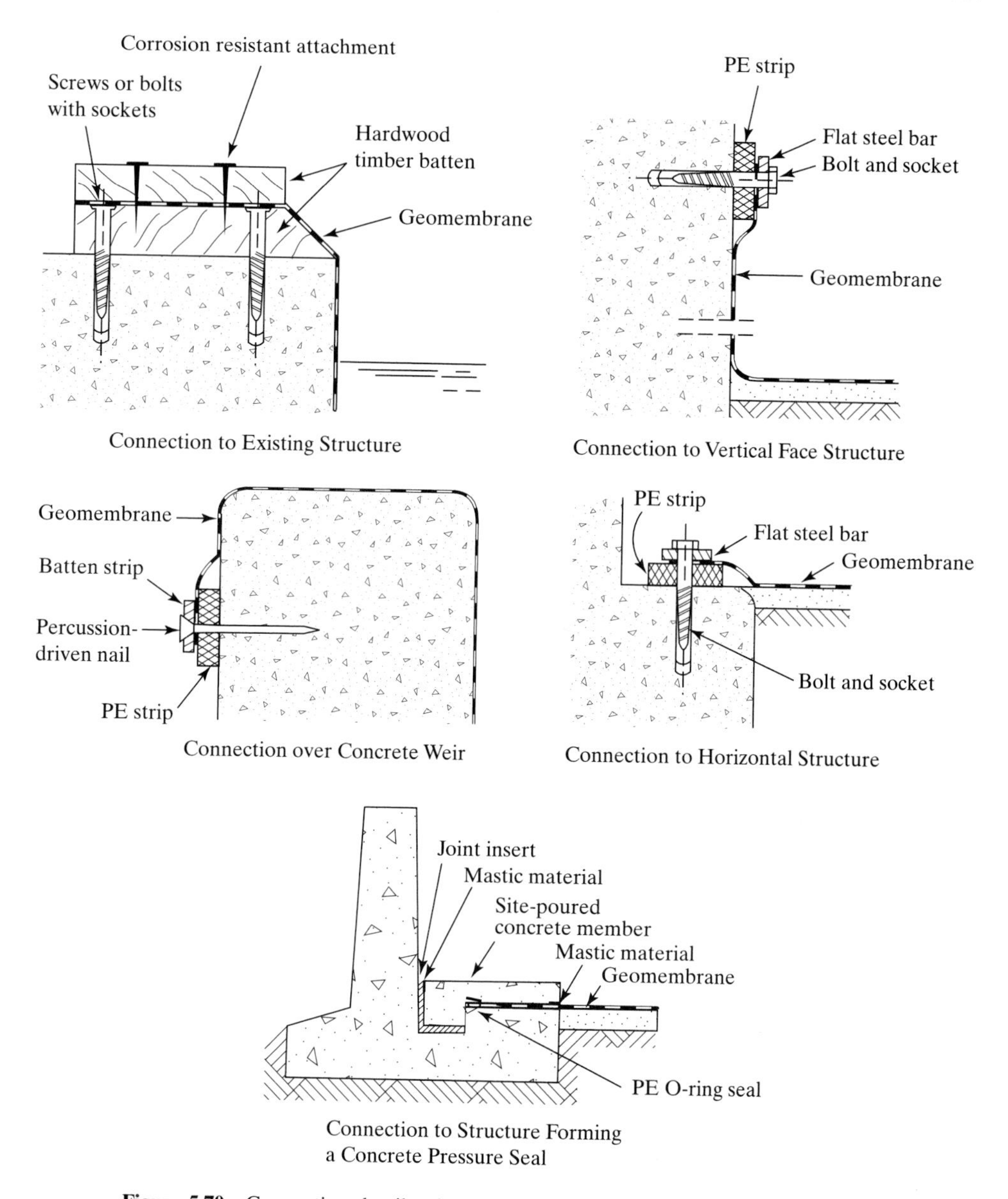

Figure 5.70 Connection details of geomembranes to different structures. (After GSE Lining Technology, Inc.)

Great care in detailing, construction, and inspection should be exercised in these appurtenance items.

For gas generating subsurface conditions, a geotextile underliner is recommended for collection and transmission, but eventually this gas must be released to a collection system or vented to the atmosphere. Vents are often made at the top of the side slope berm or along the runout length. They are either open cutout areas with flap valves

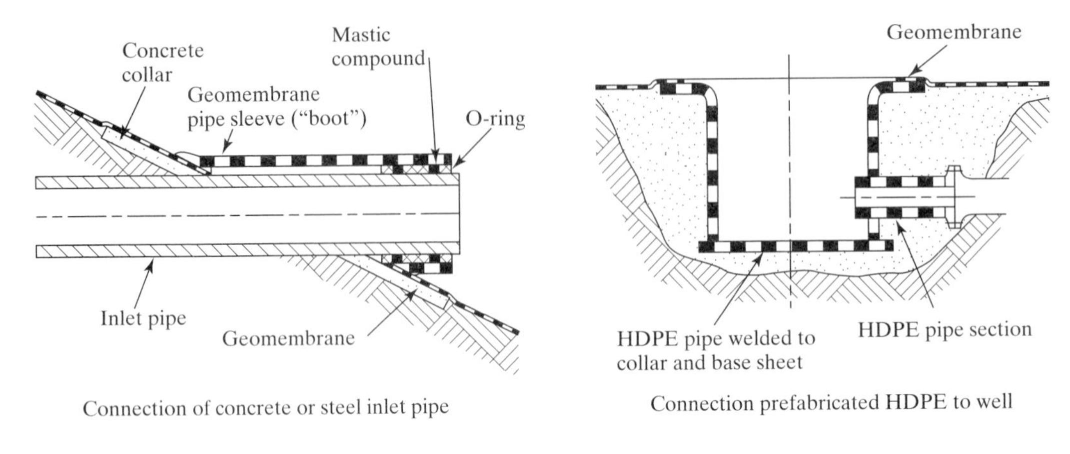

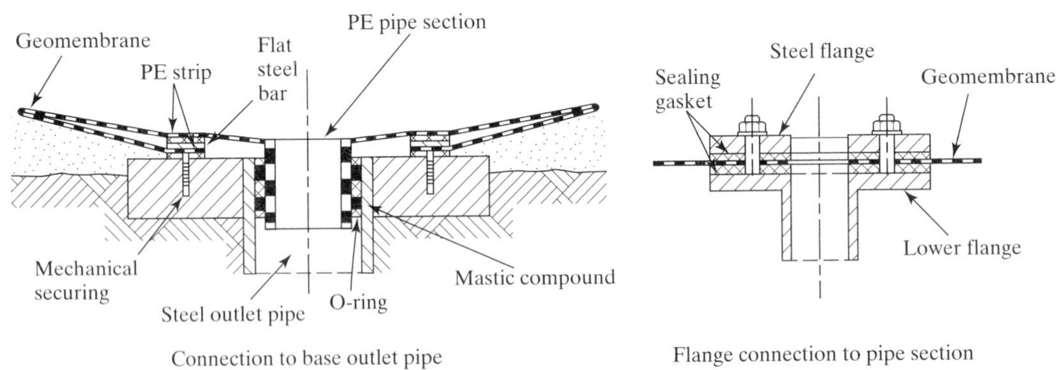

Figure 5.71 Appurtenance details of geomembranes to various pipe systems. (After GSE Lining Technology, Inc.)

(generally not recommended) or stack vents (preferred) at approximately 10 to 30 m centers (see Figure 5.72). An alternative is a rotating wind cowl assembly that always points downwind, thereby venting the system and at the same time pulling a slight vacuum that holds the liner snugly to the ground surface.

5.11.3 Leak Location Techniques

Once a leak occurs in a lined impoundment it is often too late to initiate corrective action. However, we cannot be so cavalier when the leak is from a hazardous- or radioactive waste site, and downstream water contamination can result. Thus numerous attempts have been made at solving the problem of leak location (versus leak detection as described in Section 5.6.3). At the outset it should be mentioned that all the techniques to be reviewed are in various stages of research and development. None have been used with unconditional success and some are experimental. They are mentioned only to emphasize the importance of leak location and to illustrate some the efforts are currently ongoing.

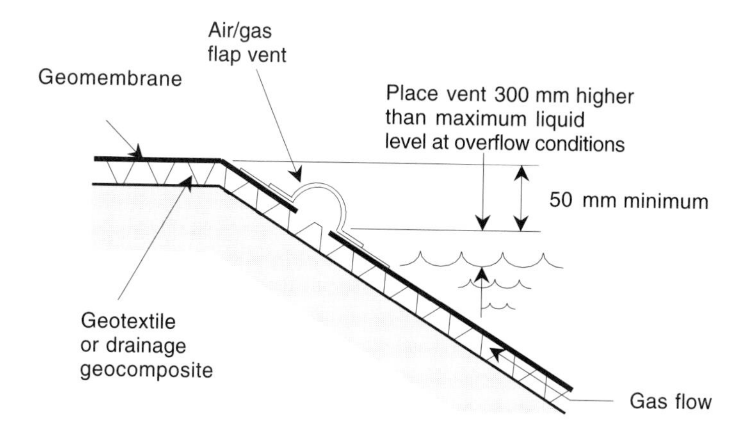

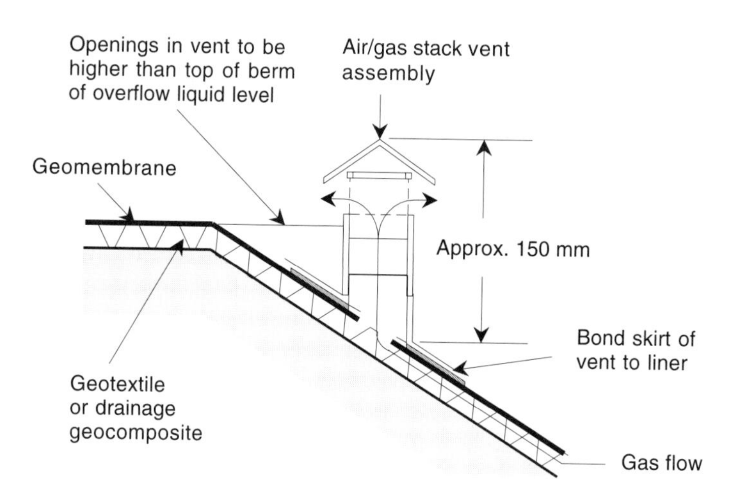

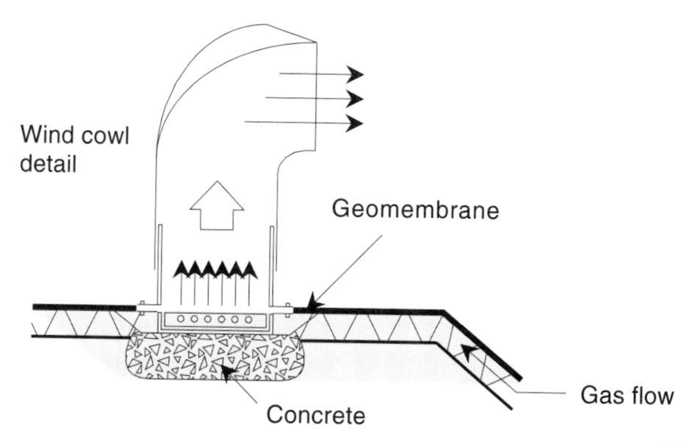

Figure 5.72 Typical gas vent details. (After Richardson and Koerner [67])

Downstream well monitoring is generally held out as a possible approach to the problem. In this approach, pollutants carried by the groundwater can possibly be detected in a well by proper sampling and testing methods. If this same pollutant is detected in other downstream wells and enough wells are available, pollution concentration gradients can be drawn. When back-extrapolated to beneath the landfill, some idea of the leak location might be possible. It is a long shot, however, and the number of wells required for accurate contouring is quite large, and hence very costly.

Rather than using discrete data, we could possibly obtain continuous data from a nondestructive testing (NDT) technique. The geophysical method of *electrical resistivity* is a candidate method in this regard. Electrical resistivity (as with all electromagnetic methods) is very sensitive to high ion concentrations. If a contaminated seepage plume (usually high in ionic content) forms downstream from a landfill, electrical resistivity traces can be made, detecting both the location of the plume and the concentration within the plume. By using back-extrapolated contours an approximate leak location can be determined. Problems arise in that many other subsurface features also influence electrical resistivity, such as stratigraphy, density, buried objects, and so on. Even under ideal conditions, leak location under the landfill is approximate at best. Furthermore, the classical techniques are quite labor intensive. An alternative, based on the same principle of electrical resistivity, is a portable electromagnetic induction system. Using it makes the survey much more efficient and as accurate as older resistivity methods, at least for shallow-depth tracing.

The use of *tracers*, vegetable or chemical dyes, injected into various locations of the leachate collection system has been attempted. An estimate based on the time it takes for the tracer to reach a leak detection monitor (for double lined facilities) or a downstream monitoring well (for single lined facilities) serves to approximate where the leak is located. It is not known how successful this technique is or how tracer dilution or multiple leaks might complicate the process. It appears, however, that only very large leaks can be identified using such tracers.

Other leak location methods used within the boundaries of the facility itself must be designed before construction and installed accordingly. For example, when wires are placed beneath the facility (e.g., in a geotextile underliner) during its construction, three different NDT methods can possibly be used to locate leaks. *Acoustic emission monitoring* senses the sounds that the leaks make as they pass over or near to the wires. By having a grid of wires, the emissions can be monitored at the edges of the impoundment. These pulses collected over a timespan of a few minutes can be plotted in the x and y directions, and contours of equal emission-count rate can be obtained. The convergence of these contours signifies the leak location. Feasibility and laboratory demonstrations have been attempted [111, 112]. In a similar manner, *time-domain reflectometry* uses transmission-line wires placed beneath the geomembrane itself during construction. These are placed in sets, and depending on their response to questioning, signify leaks and, by implication, the location of these leaks. The technique has been attempted on a prototype landfill with success [111], but suffers from the same drawback as acoustic emission, in that the conducting leads (wires) must be placed during liner construction. Long term corrosion of the wires is a concern for both techniques. Thus there must be a conscious effort by the designer before construction to include such a provision for potential leak location. Such a technique is available and has been used in a few final cover installations in Germany. Rödel [113] describes a set of electrodes placed under the geomembrane and another set placed perpendicularly above the geomembrane. When voltage is induced across the two sets of electrodes the geomembrane acts as an insulator unless it has defects (holes). The resistance then drops on these electrodes near the defect and it can be located by the electrode grid arrangement. Computer software can be used for a graphic display.

5.11.4 Wind Uplift

Geomembranes, when exposed during installation or permanently, can be greatly affected by wind. Wind traveling over the ground surface is influenced by friction with the ground and turbulence within the flowing air mass. Uplift forces develop as a result of wind flow separation, which occurs when the air mass decelerates or when irregular boundary shapes are encountered. Downwind from the air-flow separation, a wake of turbulent eddies is formed and the air flow reverses. This results in uplift forces being exerted on the surface of the geomembrane. If forces are excessive with respect to weight of the geomembrane and its anchorage (if any), it will be uplifted and unceremoniously pulled out of position in a very random manner. As seen in Figure 5.73, the geomembrane can easily be torn and severely damaged.

The obvious solution to this situation is to use sandbags to hold the deployed geomembrane in position until final cover is placed or suitable anchorage is provided. As seen in Table 5.27, however, the number of sandbags becomes unreasonably high as wind speeds become severe. While no easy solution is offered, the possibility of wind-displaced geomembranes must be discussed by all parties *before* construction of the geomembrane begins. The proper time is at the preconstruction meeting when all parties are involved. Possible remedies are to merely reposition the disturbed geomembrane, reseam or cap strip the torn locations, test the damaged geomembrane at creases and severe distortions, or (in a worst-case situation) reject the roll(s) or panel(s) involved. It must be mentioned, however, that if rejection of roll(s) or panel(s) is decided upon, the installer may not have replacement material readily available and the project will probably be delayed. Furthermore, the important issue of payment for the replaced geomembrane must be openly and carefully discussed at the preconstruction meeting.

5.11.5 Quality Control and Quality Assurance

Of all the geosynthetic materials described in this book, none are as unforgiving as geomembranes. The smallest leak when placed under hydrostatic pressure will produce alarmingly high flow rates; see Giroud et al. [117]. If the facility is a regulated landfill or surface impoundment, the leak rates can readily exceed the ALR of a waste containment facility. Thus inspection is clearly warranted in almost all applications. Such inspection comes under the dual headings of quality control (QC) and quality assurance (QA). For geosynthetics that are manufactured and constructed (usually by different organizations), a further subdivision is necessary.

It is important to define and understand the differences between MQC and MQA, and between CQC and CQA, and to counterpoint where the different activities contrast and/or compliment one another. These four definitions follow [104].

Figure 5.73 Examples of wind-damaged geomembranes on two projects.

Manufacturing Quality Control (MQC). A planned system of inspections that is used to directly monitor and control the manufacture of a material that is factory-originated. MQC is normally performed by the manufacturer (or fabricator) of geosynthetic materials and is necessary to ensure minimum, or maximum, specified values in the manufactured product. MQC refers to measures taken by the manufacturer to determine compliance with the requirements for materials and workmanship as stated in certification documents and contract plans and specifications.

Manufacturing Quality Assurance (MQA). A planned system of activities that provide assurance that the materials were manufactured as specified in the certification documents and contract plans and specifications. MQA includes manufacturing and fabrication facility inspections, verifications, audits, and evaluation of the raw materials and geosynthetic products to assess the quality of the manufactured materials. MQA refers to measures taken by the MQA organization to determine if the manufacturer or fabricator is in compliance with the product certification and contract plans and specifications for the project.

TABLE 5.27 TRACTIVE FORCES AND 300 N SANDBAG SPACING TO RESIST WIND UPLIFT OF GEOMEMBRANES AT VARIOUS WIND SPEEDS AND GEOMETRIC ORIENTATIONS

Tractive (Uplift) Forces (Pa)

	Wind Speed				
C_p Values	40 km/hr	80 km/hr	120 km/hr	160 km/hr	200 km/hr
−0.2	15	60	140	240	380
−0.4	31	120	270	490	770
−0.6	46	180	410	730	1100
−0.8	61	240	550	980	1500
−1.0	77	310	690	1200	1900

Sandbag Spacing Requirements (m^2) to Compensate for Uplift Forces

	Wind Speed				
C_p Values	40 km/hr	80 km/hr	120 km/hr	160 km/hr	200 km/hr
−0.2	20	4.9	2.1	1.2	0.79
−0.4	9.7	2.5	1.1	0.61	0.39
−0.6	6.5	1.7	0.73	0.41	0.27
−0.8	4.9	1.2	0.55	0.31	0.20
−1.0	3.9	0.97	0.43	0.25	0.16

C_p values are geometrically related to the orientation of the geomembrane lined facility; see Dedrick [115, 116].

Source: After Wayne and Koerner [114].

Construction Quality Control (CQC). A planned system of inspections that are used to directly monitor and control the quality of a construction project. Construction quality control is normally performed by the geosynthetics installer to achieve the highest quality in the constructed or installed system. CQC refers to measures taken by the installer or contractor to determine compliance with the requirements for materials and workmanship as stated in the plans and specifications for the project.

Construction Quality Assurance (CQA). A planned system of activities that provide assurance that the facility was constructed as specified in the design. Construction quality assurance includes inspections, verifications, audits, and evaluations of materials and workmanship necessary to determine and document the quality of the constructed facility. CQA refers to measures taken by the CQA organization to assess if the installer or contractor is in compliance with the plans and specifications for the project.

MQA and CQA are performed independently of MQC and CQC. Although MQA/CQA and MQC/CQC are separate activities, they have similar objectives and, in a smoothly running construction project, the processes will complement one another. An effective MQA/CQA program can lead to the identification of deficiencies in the MQC/CQC process, but a MQA/CQA program by itself (in complete absence of a MQC/CQC program) is unlikely to lead to acceptable quality management. Quality is best ensured with effective MQC/CQC *and* MQA/CQA programs. See Figure 5.74 for the usual interaction of the various organizations in a total inspection program. Note that the flow chart includes both geosynthetic and natural soil materials, since both require similar concern and care. Of particular importance is the qualifications of the various parties involved. The current recommendations [104] are given in Table 5.28.

The proper and intended functioning of a geomembrane or other geosynthetic system in an engineered facility is strongly dependent on the MQC/MQA of its manufacture and the CQC/CQA of its installation. This section has defined the scope and definition of those activities with emphasis on their interrelationship to one another. While the level of effort will differ from project to project, the concepts outlined should be present for all situations. Geosynthetics are relatively new engineered materials in comparison to steel, concrete, timber, and so on, and every detail must be considered in order to avert failures. Such failures are simply unacceptable if they occur any time up to the intended design lifetime of the facility. Proper consideration of MQC/MQA and CQC/CQA will serve well to position geomembranes (and all types of geosynthetics) as bona fide engineering materials for the future.

5.12 SUMMARY

Throughout this chapter on geomembranes functioning as liquid or vapor barriers, the emphasis has been on a step-by-step design procedure. These steps generally are taken in the following order.

1. Site selection
2. Geometric layout (width and depth)
3. Geotechnical considerations
4. Cross-section determination
5. Geomembrane material selection
6. Thickness determination
7. Side slope and cover soil details
8. Anchor trench details
9. Solid waste stability (if applicable)
10. Final cover design and details (if applicable)
11. Seam type decision
12. Seam testing strategy
13. Design of connections and appurtenances
14. Leak scenarios and corrective measures
15. Proper MQC and CQC
16. Proper MQA and CQA

Within the context of unifying a variety of geomembrane types certain generalities have been made, but most of the elements above must be handled on a site-specific basis insofar as design is considered.

As mentioned in Section 5.11 the details and installation concerns cannot be denied. A recent conference [118] has been devoted to these concerns and the complimentary nature of the included papers to design should be apparent.

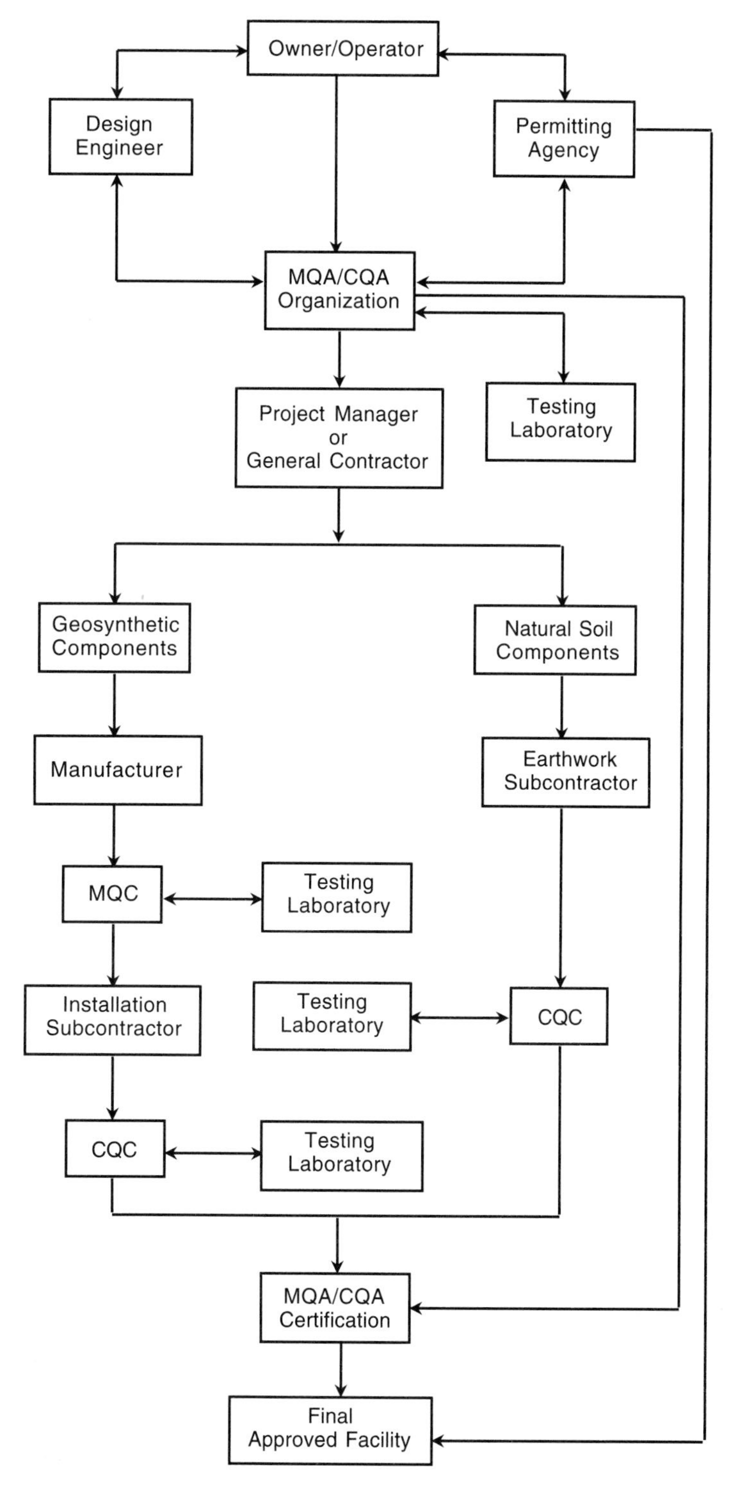

Figure 5.74 Organizational structure of MQC/MQA and CQC/CQA inspection activities. (After EPA [104])

TABLE 5.28 RECOMMENDED PERSONNEL QUALIFICATIONS

Individual	Minimum Recommended Qualifications
Design engineer	Registered Professional Engineer
Owner's representative	Registered Professional Engineer or Engineer-in-Training
Manufacturer/fabricator	Experience in manufacturing, or fabricating, at least 1,000,000 m^2
MQC personnel	Manufacturer, or fabricator, trained personnel in charge of quality control of the geosynthetic materials to be used in the specific facility
MQC officer	The specific designated individual by a manufacturer or fabricator, in charge of geosynthetic material quality control
Geosynthetic installer's representative	Experience installing at least 1,000,000 m^2
CQC personnel	Employed by the general contractor, installation contractor or earthwork contractor involved in similar facilities; certified by NICET* to the extent shown in [104]
CQA personnel	Employed by an organization that operates separately from the contractor and the owner/operator; certified by NICET* to the extent shown in [104]
MQA/CQA engineer	Employed by an organization that operates separately from the contractor and owner/operator; registered Professional Engineer and approved by permitting agency
MQA/CQA certifying engineer	Employed by an organization that operates separately from the contractor and owner/operator; registered Professional Engineer in the state in which the facility is constructed and approved by permitting agency

*NICET = National Institute for Certification in Engineering Technologies, 1420 King Street, Alexandria, VA 22314, U.S.
Source:U.S. EPA [104]

REFERENCES

1. Dove, J. E., and Frost, J. J., "A Method for Measuring Geomembrane Surface Roughness," *Geosynthetics Int.,* Vol. 3, No. 3, 1996, pp. 369–392.
2. Vallejo, L. E., and Zhou, Y., "Fractal Approach to Measuring Roughness on Geomembranes," *J. of Geotechncal Eng.,* ASCE, Vol. 121, No. 5, pp. 442–447.
3. Haxo, H. E., Jr., Miedema, J. A., and Nelson, N. A., "Permeability of Polymeric Membrane Lining Materials," *Proc. Intl. Conf. Geomembranes,* St. Paul, MN: IFAI, 1984, pp. 151–156.
4. Matrecon, Inc., *Lining of Waste Containment and Other Impoundment Facilities,* U.S. EPA/600/2-88/052, Cincinnati: OH, 1988.
5. Rowe, R. K., Hrapovic, L., and Armstrong, M. D., "Diffusion of Organic Pollutants through HDPE Geomembranes and Composite Liners and Its Influence on Groundwater Quality," *Proc. Geosynthetics: Applications, Design and Construction,* Rotterdam: A.A. Balkema, 1996, pp. 737–742.
6. Steffen, H., "Report on Two Dimensional Strain-Stress Behavior of Geomembranes with and without Friction," *Proc. Intl. Conf. Geomembranes,* St. Paul, MN: IFAI, 1982, pp. 181–185.
7. Koerner, R. M., Koerner, G. R., and Hwu, B.-L., "Three Dimensional, Axi-Symmetric Geomembrane Tension Test," *Proc. Geosynthetic Testing for Waste Containment Applications, ASTM STP 1081,* ed. R. M. Koener, Philadelphia, PA: ASTM, 1990, pp. 170–184.

8. Merry, S. M., and Bray, J. D., "Time-Dependent Mechanical Reponse of HDPE Geomembranes," *J. Geotechnical and Geoenvironmental Eng.*, ASCE, Vol. 123, No. 1, 1957, pp. 57–68.
9. Nobert, J., "The Use of Multi-Axial Burst Test to Assess Field Performance of Geomembranes," *Proc. Geosynthetics '93*, St. Paul, MN: IFAI, 1993, pp. 685–702.
10. Haxo, H. E., Jr., "Quality Assurance of Geomembranes Used as Linings for Hazardous Waste Containment," *J. Geotextiles and Geomembranes,* London: Elsevier Appl. Sci., 1986, pp. 225–247.
11. Rollin, A. L., and Fayoux, D., "Geomembrane Seaming Techniques," in *Geomembranes: Identification and Performance Testing,* RILEM Report 4, ed. A. Rollin and J.-M. Rigo, London: Chapman and Hall, 1991, pp. 59–79.
12. Peggs, I. D., "Assessment of HDPE Geomembrane Seam Specifications," *Proc. Geosynthetics: Applications, Design and Construction,* Rotterdam: A.A. Balkema, 1996, pp. 693–695.
13. Koerner, R. M., Monteleone, M. J., Schmidt, J. R., and Roethe, A. T., "Puncture and Impact Resistance of Geosynthetics," *Proc. 3d Intl. Conf. Geotextiles,* 1986, Vienna: Austrian Society of Engineers, pp. 677–682.
14. Hullings, D. E., and Koerner, R. M., "Puncture Resistance of Geomembranes Using a Truncated Cone Test," *Proc. Geosynthetics '91,* St. Paul, MN: IFAI, 1991, pp. 273–286.
15. Narejo, D. B., Wilson-Fahmy, R., and Koerner, R. M., "Geomembrane Puncture Evaluation and Use of Geotextile Protection Layers," *Proc. PennDOT/ASCE Conf. Geotechnical Eng.,* Harrisburg, PA: Central Pennsylvania Section, ASCE, 1993, pp. 1–16.
16. Boschuk, J., Jr., "Landfill Covers—An Engineering Perspective," *Geotechnical Fabrics Rpt,* Vol. 9, No. 2, 1991, pp. 23–24.
17. Martin, J. P., Koerner, R. M., and Whitty, J. E., "Experimental Friction Evaluation of Slippage between Geomembranes, Geotextiles and Soils," *Proc. Intl. Conf. Geomembranes,* St. Paul, MN: IFAI, 1984, pp. 191–196.
18. Koerner, R. M., Martin, J. P., and Koerner, G. R., "Shear Strength Parameters between Geomembranes and Cohesive Soils," *J. Geotextiles and Geomembranes,* Vol. 4, No. 1, 1986, pp. 21–30.
19. Stark, T. D., and Poeppel, A. R., "Landfill Liner Interface Strengths from Torsional-Ring-Stress Tests," *J. Geotechnical Eng.,* ASCE, Vol. 120, No. 3, pp. 597–617.
20. Hsuan, Y. G., Koerner, R. M., and Lord, A. E., Jr., "Stress Crack Resistance of High Density Polyethylene Geomembranes," *J. Geotechnical Eng. Div., ASCE,* Vol. 119, No. 11, 1993, pp. 1840–1855.
21. Hsuan, Y. G., and Koerner, R. M., "The Single Point—Notched Constant Tension Load Test: A Quality Control Test for Assessing Stress Crack Resistance," *Geosynthetics Int.,* Vol. 2, No. 5, 1995, pp. 831–843.
22. Kane, J. D., and Widmayer, D. A., "Considerations for the Long-Term Performance of Geosynthetics at Radioactive Waste Disposal Facilities," *Durability and Aging of Geosynthetics,* ed. R. M. Koerner, London: Elsevier, 1989, pp. 13–27.
23. Dragun, J., "The Fate of Hazardous Materials in Soils," *Hazardous Materials Control Magazine,* March/April 1988, pp. 31–77.
24. Steiniger, F., "The Effect of Burrower Attack on Dike Liners," *Wasser und Boden,* Berlin: Ernst and Son, Inc., 1968, pp. 16–24.
25. Vandervoort, J., *The Use of Extruded Polymers in the Containment of Hazardous Wastes,* The Woodlands, TX: Schlegel Lining Technology, Inc.
26. Tisinger, L. G., "Chemical Compatibility Testing: A Typical Program," *Geotechnical Fabrics Rpt,* Vol. 7, No. 3, 1989, pp. 22–25.
27. Little, A. D. Inc., "Resistance of Flexible Membrane Liners to Chemicals and Wastes," U.S. EPA Report PB86-119955, Cincinnati, OH, 1985.

28. O'Toole, J. L., *Design Guide, Modern Plastics Encyclopedia,* New York: McGraw-Hill, 1985–1986, pp. 398–446.
29. Hsuan, Y. G., Sculli, M. L., Guan, Z. C., and Comer, A. I., "Effects of Freeze-Thaw Cycling on Geomembrane Sheets and Their Seams," *Proc. Geosynthetics '97*, St. Paul, MN: IFAI, 1997, pp. 201–216.
30. Hsuan, Y. G., Koerner, R. M., and Lord, A. E., Jr., "A Review of the Degradation of Geosynthetic Reinforcing Materials and Various Polymer Stabilization Methods," *Geosynthetic Soil Reinforcement Testing Procedure, ASTM STP 1190,* ed. S. C. Jonathan Cheng, Philadelphia, PA: ASTM, 1993, pp. 228–244.
31. Wolters, M., "Prediction of Long-Term Strength of Plastic Pipe System," *Proc. 10th Plastic Fuel Gas Pipe Symp.,* Columbus, OH: AGA, 1987, pp. 164–174.
32. Koerner, R. M., Halse, Y-H., and Lord, A. E., Jr., "Long-Term Durability and Aging of Geomembranes," *Proc. Waste Containment Systems Geotech.* Spec. Publ. #26, ASCE, ed. R. Bonaparte, Philadelphia, PA: ASCE, 1990, pp. 106–134.
33. Koch, R. et al., "Long Term Creep Resistance of Sheets of Polyethylene Geomembrane," Report TR-88-0054, Hoechst A. G., Frankfurt, Germany, 1987.
34. Hessel, J., and John, P., "Long Term Strength of Welded Joints in Polyethylene Sealing Sheets," *Werkstofftechnik,* Vol. 18, 1987, pp. 228–231.
35. Gaube, E., Diedrick, G., and Muller, W., "Pipes of Thermoplastics; Experience of 20 Years of Pipe Testing," *Kunstoffe,* Vol. 66, 1976, pp. 2–8.
36. Mitchell, D. H., and Spanner, G. E., "Field Performance Assessment of Synthetic Liners for Uranium Tailings Ponds," Status Report, Battelle PNL, U.S. NRC, NUREG/CR-4023, PNL-5005, Washington, D.C., 1985.
37. Hsuan, Y. G., and Koerner, R. M., "Antioxidant Depletion Lifetime for HDPE Geomembranes," *J. Geotechnical and Geoenvironmental Eng.,* under review.
38. Koerner, R. M., Lord, A. E., Jr., and Hsuan, Y. H., "Arrhenius Modeling to Predict Geosynthetic Degradation," *J. Geotextiles and Geomembranes,* Vol. 11, No. 2, 1992, pp. 151–183.
39. Martin, J. R., and Gardner, R. J., "Use of Plastics in Corrosion Resistance Instrumentation," paper presented at 1983 Plastic Seminar, NACE, 1993.
40. Underwriters Laboratory Standard UL 746 B, "Polymeric Materials—Long Term Property Evaluation," Northbrook, IL, 1987.
41. Koerner, R. M., Bove, J. A., and Martin, J. P., "Water and Air Transmissivity of Geotextiles," *J. Geotextiles and Geomembranes,* Vol. 1, No. 1, 1984, pp. 57–73.
42. Taylor, D. W., *Fundamentals of Soil Mechanics,* New York: John Wiley and Sons, 1948.
43. Koerner, R. M., and Soong, T.-Y., "Analysis and Design of Veneer Cover Soils," *Proc. 6th IGS Conf.,* St. Paul, MN: IFAI, 1998, to appear.
44. Holtz, R. D., and Kovacs, W. D., *An Introduction to Geotechnical Engineering*, Englewood Cliffs, NJ: Prentice-Hall, 1981.
45. Cooley, K. R., "Evaporation Reduction: Summary of Long-Term Tank Studies," *J. Irrigation Drainage Div., ASCE,* Vol. 109, No. 1, 1983, pp. 89–98.
46. Dedrick, A. R., "Foam-Rubber Covers for Evaporation Control," *Proc. Intl. Conf. Geomembranes,* St. Paul, MN: IFAI, 1984, pp. 89–91.
47. Gerber, D. H., "Floating Reservoir Cover Designs," *Proc. Intl. Conf. Geomembranes,* St. Paul, MN: IFAI, 1984, pp. 79–84.
48. Chow, V. T., *Open Channel Hydraulics,* New York: McGraw-Hill, 1959.
49. Fortier, S., and Scobey, F. C., "Permissible Canal Velocities," *Trans. ASCE,* Vol. 89, 1926, pp. 940–956.
50. Morrison, W. R., and Starbuck, J. G., *Performance of Plastic Canal Linings,* USDI, Bureau of Reclamation, REC-ERC-84-1, 1984.

51. Hammer, H., Ainsworth, J. B., and Beckham, R., "Case Study of an In-Situ, Uninterrupted Flow Repair of a Concrete Sluce Channel," *Proc. Intl. Conf. Geomembranes,* St. Paul, MN: IFAI, 1984, pp. 343–345.
52. Koerner, R. M., and Lawrence, C. A., *Transmissivity of Geotextiles After Placement of Fresh Concrete,* Internal Report to U.S.B.R., W. R. Morrison, Denver, CO, 1988.
53. Comer, A. I., Kube, M., and Sayer, M., "Remediation of Existing Canal Linings," *J. Geotextiles and Geomembranes,* Vol. 14, Nos. 5–6, 1996, pp. 313–326.
54. Swihart, J. J. "Deschutes Canal Lining Demonstration—Construction Report," *Proc. 5th Intl. Geosynthetic Conf.,* Singapore: Southeast Asia Chapter, IGS 1994, pp. 553–556.
55. Hanson, D. J., "Hazardous Waste Management: Planning to Avoid Future Problems," tables by Freedonia Group, *Chemical and Engineering News,* 1989, pp. 60–63.
56. Mackey, R. E., "Is Thailand Ready for Lined Landfills?" *Geotechnical Fabrics Rpt,* Vol. 14, No. 7, 1996, pp. 20–25.
57. Chian, E. S. K., and deWalle, F. B., "Sanitary Landfill Leachates and Their Treatment," *J. Environmental Eng. Div., ASCE,* Vol. 102, No. EE2, 1976, pp. 411–431.
58. Dudzik, B. E., and Tisinger, L. G., "An Evaluation of Chemical Compatibility Test Results of HDPE Geomembrane Exposed to Industrial Waste Leachate," *Proc. Geosynthetic Testing for Waste Containment Applications, ASTM STP 1081,* ed. R. M. Koerner, Philadelphia, PA: ASTM, 1990, pp. 37–56.
59. Kays, W. B., *Construction of Linings for Resevoirs, Tanks, and Pollution Control Facilities,* 2d ed., New York: John Wiley and Sons, 1986.
60. Anderson, D. C., Brown, K. W., and Green, J., "Organic Leachate Effects on the Permeabilities of Clay Liners," *Proc. Natl. Conf. Management of Uncontrolled Hazardous Waste Substances,* Washington, D.C.: HMCRI, 1981, pp. 223–229.
61. Corbet, S., Workshop Report on "USA-Germany Landfill Liner Practices," *J. Geotextiles and Geomembranes*, to appear.
62. Tchobanoglous, G., Theisen, H., and Eliassen, R., *Solid Wastes,* New York: McGraw-Hill, 1997.
63. Gross, B. A., Bonaparte, R., and Giroud, J. P., "Evaluation of Flow from Landfill Leakage Detection Layers," *Proc. 4th Intl. Conf. on Geotextiles, Geomembranes and Related Products,* Rotterdam: A.A. Balkema, 1990, pp. 481–486.
64. Buranek, D., and J. Pacey, "Geomembrane-Soil Composite Lining Systems Design, Construction, Problems and Solutions," *Proc. of Geosynthetics '87,* Vol. 2, St. Paul, MN: IFAI, 1987, pp. 375–384.
65. Giroud, J. P., and Bonaparte, R., "Leakage Through Liners Constructed with Geomembranes. Part II, Composite Liners," *J. Geotextiles and Geomembranes,* Vol. 8, No. 2, 1989, pp. 71–112.
66. Eith, A. E., and Koerner, R. M., "Field Evaluation of Geonet Flow Rate (Transmissivity) under Increasing Load," *J. Geotextiles and Geomembranes,* Vol. 11, Nos. 4–6, 1992, pp. 489–502.
67. Richardson, G. N., and Koerner, R. M., *Geosynthetic Design Guidance for Hazardous Waste Landfill Cells and Surface Impoundments,* Final Report U.S. EPA Contract No. 68-03-3338, Philadelphia, PA: GRI, 1987.
68. Matrecon, Inc., *Lining of Waste Impoundment and Disposal Facilities,* Final Report U.S. EPA, SW-870, 1988.
69. Koerner, R. M., and Mukhopadhyay, C., *The Behavior of Negative Skin Friction on Model Piles in Medium Plasticity Silt,* H. R. B. Record #405, 1972, pp. 34–44.
70. Othman, M. A., Bonaparte, R., and Gross, B. A., "Preliminary Results of Study of Composite Liner Field Performance," *J. of Geotextiles and Geomembranes,* to appear.
71. Wilson-Fahmy, R. F., Narejo, D., and Koerner, R. M., "Puncture Protection of Geomembranes. Part I: Theory," *Geosynthetics Intl.,* Vol. 3, No. 5, 1996, pp. 605–628.

72. Narejo, D., Koerner, R. M., and Wilson-Fahmy, R. F., "Puncture Protection of Geomembranes. Part II: Experimental," *Geosynthetics Intl.,* Vol. 3, No. 5, 1996, pp. 629–653.
73. Koerner, R. M., Wilson-Fahmy, R. F., and Narejo, D., "Puncture Protection of Geomembranes. Part III: Examples," *Geosynthetics Intl.,* Vol. 3, No. 5, 1996, pp. 655–676.
74. Koerner, R. M., and Wayne, M. H., "Geomembrane Anchorage Behavior Using a Large Scale Pullout Device," in *Geomembranes, Identification and Performance Testing,* ed. A. Rollin and J.-M Rigo, RILEM, London: Chapman and Hall, 1991, pp. 204–218.
75. Wilson-Fahmy, R. F., and Koerner, R. M., "Finite Element Analysis of Stability of Cover Soil on Geomembrane Lined Slopes," *Proc. Geosynthetics '93,* St. Paul, MN: IFAI, 1993, pp. 1425–1438.
76. Byrne, R. J., Kendall, J., and Brown, S., "Cause and Mechanism of Failure of Kettleman Hills Landfill," *Proc. ASCE Conf. Stability and Performance of Slopes and Embankments II,* ASCE, 1992, pp. 1–23.
77. Koerner, R. M., Carson, D. A., Daniel, D. E., and Bonaparte, R., "Current Status of the Cincinnati GCL Test Plots," *J. of Geotextiles and Geomembranes,* to appear.
78. Singh, S., and Murphy, B., "Evaluation of the Stability of Sanitary Landfills," *Geotechnics of Waste Fills—Theory and Practice, ASTM STP 1070,* ed. Arvid Landva and G. David Knowles, Philadelphia, PA: ASTM, 1990, pp. 240–258.
79. Reynolds, R. T., "Geotechnical Field Techniques Used in Monitoring Slope Stability at a Landfill," *Proc. Field Measurements in Geotechnics,* ed. G. Sorum, Rotterdam: A.A. Balkema, 1991, pp. 883–891.
80. Leach, J. A., Harper, T. G., and Tape, R. T., "Current Practice in the Use of Geosynthetics in the Heap Leach Industry," *Proc. Geosynthetics '87,* St. Paul, MN: IFAI, 1987, pp. 365–374.
81. Smith, M. E., and Welkner, P. M., "Liner Systems in Chilean Cooper and Gold Heap Leaching," *Proc. 5th IGS Conferences,* Singapore: Southeast Asia Chapter, IGS, 1994, pp. 1063–1068.
82. Attaway, D. C., "New Applications for Geomembranes: Lining Solar Ponds," *Proc. Intl. Conf. Geomembranes,* St. Paul, MN: IFAI, 1984, pp. 55–59.
83. Koerner, R. M., and Richardson, G. N., "Design of Geosynthetic Systems for Waste Disposal," *Proc. Conf. Geotechnical Practice for Waste Disposal '87,* Philadelphia, PA: ASCE, 1987, pp. 65–86.
84. Spikula, D. R., "Subsidence Performance of Landfills," *J. Geotextiles and Geomembranes,* to appear.
85. Murphy, W. L., and Gilbert, P. A., "Estimation of Maximum Cover Subsidence Expected in Hazardous Waste Landfills," EPA Proc. on Land Disposal of Hazardous Waste, EPA 600/9-84-007, 1984, pp. 222–229.
86. Sterling, H. J., and Ronayne, M. C., "Simulating Landfill Cover Subsidence," *Proc. 11th Symp. Land Disposal of Hazardous Waste,* U.S. EPA 600/9-85/013, Washington, D.C., 1985, pp. 236–244.
87. Koerner, R. M., and Daniel, D. E., *Final Covers for Engineered Landfills and Abandoned Dumps,* New York: ASCE, 1997.
88. U.S. Environmental Protection Agency, *Covers for Uncontrolled Hazardous Waste Sites,* EPA-540/2-85-002, Cincinnati, OH, 1986.
89. U.S. Environmental Protection Agency, *Design and Construction of Covers for Solid Waste Landfills,* EPA-600/2-79-165, Cincinnati, OH, 1979.
90. Baron, J. L. et al., *Landfill Methane Utilization Technology Workbook,* U.S. DOE, CPE-810, Contract 31-109-38-5686, Argonne National Laboratory, 1981.
91. Daniel, D. E., and Koerner, R. M., *Final Cover Systems,* in *Geotechnical Aspects of Waste Disposal,* ed. D. E. Daniel, London: Chapman and Hall, 1992, pp. 455–496.

92. Schroeder, P. R., Dizier, T. S., Zappi, P. A., McEnroe, B. M., Sjostrom, J. W., and Peyton, R. L., "The Hydrologic Evaluation of Landfill Performance (HELP) Model: Engineering Documentation for Version 3," EPA/600/R-94/168b, U. S. EPA, Risk Reduction Engineering Laboratory, Cincinnati, OH.
93. Cline, J. F., *Biobarriers Used in Shallow-Buried Ground Stabilization,* Battelle Pacific Northwest Laboratory Report PNL-2918, Richland, WA, 1979.
94. Mackey, R. E., "Three End Uses for Closed Landfills and Their Impacts to the Geosynthetic Design," *J. Geotextiles and Geomembranes,* Vol. 14, Nos. 7–8, 1996, pp. 409–424.
95. Pinyan, C., "Sky Mound to Raise from Dump," *ENR,* June 11, 1987, pp. 28–29.
96. Eigenbrod, K. D., Irwin, W. W., and Roggensack, W. D., "Upstream Geomembrane Liner for a Dam on a Compressible Foundation," *Proc. Intl. Conf. Geomembranes,* St. Paul, MN: IFAI, 1984, pp. 99–103.
97. Sembenelli, P., and Rodriguez, E. A., "Geomembranes for Earth and Earth-Rock Dams: State-of-the-Art Report," *Proc. Geosynthetics: Applications, Design and Construction,* ed. M. B. de Groot, G. den Hoedt and R. J. Termaat, Rotterdam: A. A. Balkema, 1996, pp. 877–888.
98. Monari, F., "Waterproofing Covering for the Upstream of the Lago Nera Dam," *Proc. Intl. Conf. Geomembranes,* St. Paul, MN: IFAI, 1984, pp. 105–110.
99. Cazzuffi, D., "The Use of Geomembranes in Italian Dams," *Intl. J. Water Power and Dam Construction,* Vol. 26, No. 2, 1987, pp. 44–52.
100. Scuero, A. M., and Vaschetti, G. L., "Geomembranes for Masonry and Concrete Dams: State-of-the-Art Report," *Proc. Geosynthetics: Applications, Design and Construction,* ed. M. B. de Groot, G. den Hoedt and R. J. Termaat, Rotterdam: A. A. Balkema, 1996, pp. 889–898.
101. Koerner, R. M., and Welsh, J. P., *Construction and Geotechnical Engineering Using Synthetic Fabrics,* New York: John Wiley and Sons, 1980.
102. Frobel, R. K., "Geosynthetics in the NATM Tunnel Design," *Proc. Geosynthetics for Soil Improvement,* Geotech. Spec. Publ. 18, ASCE, 1988, pp. 51–67.
103. Koerner, R. M., and Guglielmetti, J., "Vertical Barriers; Geomembranes," in *Assessment of Barrier Technologies,* ed. R. R. Rumer and J. K. Mitchell, NTIS, PB96-180583, 1995, pp. 95–118.
104. Daniel, D. E., and Koerner, R. M., "MQC/MQA and CQC/CQA of Waste Containment Liner and Cover Systems," Also available as U.S. EPA/600/R-93/182, Technical Resource Document, from ASCE, 1993.
105. U. S. Environmental Protection Agency, *Proc. Workshop on Geomembrane Seaming, Data Acquisition and Control,* EPA/600/R-93/112, 1993.
106. Richardson, G. N., "Construction Quality Management for Remedial Action and Remedial Design of Waste Containment Systems," Technical guidance document, EPA/540/R-92/073, Cincinnati, OH: U.S. EPA, 1992.
107. Schultz, D. W., Duff, B. M., and Peters, W. R., "Performance of an Electrical Resistivity Technique for Detecting and Locating Geomembrane Failures," *Proc. Intl. Conf. Geomembranes,* St. Paul, MN: IFAI, 1984, pp. 445–449.
108. Laine, D. L., Binley, A.M., and Darilek, G. T., "Locating Geomembrane Liner Leaks under Waste in a Landfill," *Proc. Geosynthetics '97,* St. Paul, MN: IFAI, 1997, pp. 407–411.
109. Lord, A. E., Jr., Koerner, R. M., and Crawford, R. B., "Nondestructive Testing Techniques to Assess Geomembrane Seam Quality," *Proc. Management of Uncontrolled Hazardous Waste Sites,* Washington, DC: HMCRI, 1986, pp. 272–276.
110. Koerner, R. M., Lord, A. E., Jr., Crawford, R. B., and Cadwallader, M., "Geomembrane Seam Inspection Using the Ultrasonic Shadow Method," *Proc. Geosynthetics '87,* St. Paul, MN: IFAI, 1987, pp. 493–504.

111. Waller, M. J., and Singh, R., "Leak Detection Techniques and Repairability for Lined Waste Impoundment Sites," *Proc. Management of Uncontained Hazardous Waste Sites,* Washington, DC: HMCRI, 1983, pp. 147–153.
112. Koerner, R. M., Lord, A. E., Jr., and Luciani, V. A., "A Detection and Monitoring Technique for Location of Geomembrane Leaks," *Proc. Intl. Conf. Geomembranes,* St. Paul, MN: IFAI, 1984, pp. 379–384.
113. Rödel, A., "Geologger—A New Type of Monitoring System for the Total Area of Geomembranes on Landfill Sites," *Proc. Geosynthetics: Applications, Design and Construction,* ed. M. B. de Groot, G. den Hoedt and R. J. Termaat, Rotterdam: A. A. Balkema, 1996, pp. 625–626.
114. Wayne, M. H., and Koerner, R. M., "Effect of Wind Uplift on Liner Systems," *Geotechnical Fabrics Rpt,* Vol. 6, No. 4, 1988, pp. 26–29.
115. Dedrick, A. R., "Air Pressure Over Surfaces Exposed to Wind. I. Water Harvesting Catchments," *Trans. ASAE,* Vol. 17, No. 5, 1974, pp. 140–147.
116. Dedrick, A. R., "Air Pressures over Surfaces Exposed to Wind. II. Reservoirs," *Trans. ASAE,* Vol. 18, No. 3, 1975, pp. 78–82.
117. Giroud, J. P., Badu-Tweneboah, K., and Bonaparte, R., "Rate of Leakage Through a Composite Liner due to Geomembrane Defects," *J. Geotextiles and Geomembranes,* Vol. 11, No. 1, 1992, pp. 1–28.
118. Koerner, R. M., ed., *Proc. 6th GRI Sem. MQC/MQA and CQC/CQA of Geosynthetics,* St. Paul, MN: IFAI, 1993.

PROBLEMS

5.1. What is the difference between thermoplastic and thermoset geomembranes? Why are thermoset geomembranes rarely used (and not even listed in Table 5.1)?

5.2. Describe the differences between noncrystalline and semicrystalline thermoplastic geomembranes with regard to their anticipated behavior. Draw a sketch of each type, showing proper mixtures within noncrystalline formulations like PVC and CSPE and the tie molecules bonding together crystalline regions in semicrystalline geomembranes like HDPE.

5.3. Describe the differences between geomembranes made from thermoplastic materials versus polymer-impregnated geotextiles, as discussed in Section 2.6.2.

5.4. Regarding geomembrane test methods and standards:
- **(a)** In what standards area would you expect to find relevant test methods for the plastics from which geomembranes are made?
- **(b)** Why are there so few existing standards under a geomembrane category?

5.5. Regarding common scrims used in reinforced geomembranes such as CSPE-R and PP-R:
- **(a)** Briefly describe how scrim-reinforced geomembranes are manufactured.
- **(b)** What do 20×20, 10×10, and 6×6 designations for the scrim mean?
- **(c)** Why are scrims less than 6×6 not available?
- **(d)** Why are scrims greater than 20×20 not available?
- **(e)** What kind of polymer are the scrim yarns generally made from?
- **(f)** Why is it a woven fabric rather than a nonwoven one?
- **(g)** What advantages and disadvantages do reinforced geomembranes have over nonreinforced?

5.6. Describe the difference in behavior between a scrim-reinforced geomembrane and a spread-coated geomembrane if the scrim is a 10 × 10 polyester yarn and the spread-coated substrate is a 300 g/m^2 nonwoven needle punched continuous-filament polyester fabric.

5.7. If a HDPE formulation is 97% resin at a density of 0.936 g/cc, 2.5% carbon black at a density of 1.85 g/cc, and 0.5% antioxidant at a density of 2.05 g/cc, what is its overall density?

5.8. Calculate the WVT, permeance, and permeability of a 1.5 mm HDPE geomembrane that shows a weight loss of 0.0045 g in 14 days at a relative humidity difference of 70% at a constant temperature of 20°C. The area of the test specimen is 0.0032 m^2.

5.9. For the 25 mm wide tensile test results shown in the graphs below, fill out the following table.

	fPP		
	0.75 mm	0.90 mm (6 × 6)	0.90 mm (10 × 10)
Strength at Break (kN/m^2)			
Strain at Break (%)			
Strength at Failure (kN/m^2)			
Strain at Failure (%)			
Modulus (kN/m^2)			

	CSPE		
	0.50 mm	0.75 mm	0.90 mm (10 × 10)
Strength at Break (kN/m^2)			
Strain at Break (%)			
Strength at Failure (kN/m^2)			
Strain at Failure (%)			
Modulus (kN/m^2)			

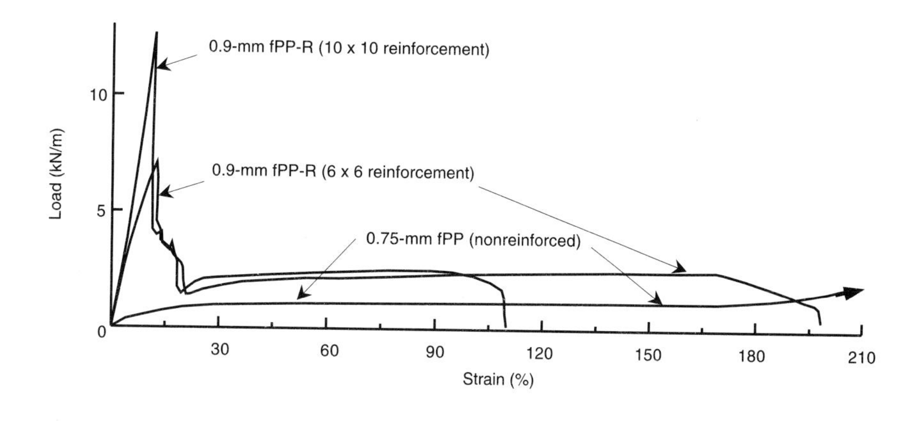

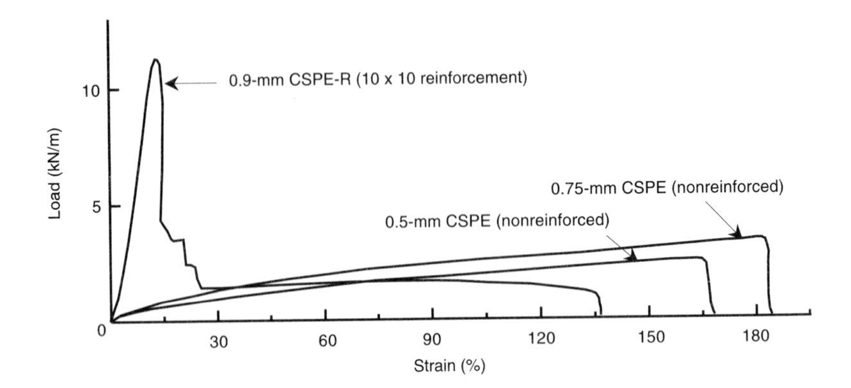

5.10. When could tensile test results from dumbbell-shaped specimens be used for design-simulation purposes? What is the basic purpose of such tests?

5.11. What type of tensile test would you use for the following design situations, and why?

(a) A geomembrane coming up out of an excavation and over a berm into an anchor trench

(b) Localized subsidence of a geomembrane beneath a reservoir due to gradual decay of organic subgrade material

(c) Differential settlement of a geomembrane in the final cover (closure) of a municipal solid waste landfill

5.12. What proportion of full-sheet strength is the shear and peel seam test data in Figure 5.6? Give answers for both PVC and HDPE, and comment on the adequacy of these seam strengths.

5.13. Plot the impact resistance of the five geomembranes listed in Table 5.6 against the point geometry angle listed. Comment on the various types of response curves that result.

5.14. Regarding interface friction tests:

(a) Given the following set of data from direct shear tests of a CSPE-R geomembrane on Ottawa sand, calculate the resulting peak friction angle and its efficiency, based on the friction of the sand as shown in Table 5.7.

Normal stress (kN/m^2)	35	70	105	140
Shear strength (kN/m^2)	12.5	25.0	40.7	53.0

(b) For a VLDPE geomembrane on mica schist soil, the following data result. Calculate its peak friction angle and its efficiency based on the soil's friction angle of 26°.

Normal stress (kN/m^2)	35	70	105	140
Shear strength (kN/m^2)	15.1	31.7	47.5	63.2

5.15. The stress-versus-displacement direct shear curves in Figure 5.9b show that the residual strength is less than the peak strength. List the types of geomembranes where this is likely to occur.

5.16. Regarding Section 5.1.3 on stress cracking of geomembranes:
(a) Stress cracking of geomembranes is usually considered to be possible in semicrystalline geomembranes like HDPE. What are the conditions that can bring about this phenomenon?
(b) Where are the most likely locations for stress cracking in field-deployed HDPE geomembranes?
(c) Can stress cracking occur in other geomembranes? If so, describe the phenomenon for VFPE, PVC, fPP, and CSPE-R.

5.17. The currently recommended acceptance level for HDPE geomembranes with respect to stress cracking via the NCTL test is a minimum transition time of 100 hours (recall Figure 5.11b). How was this determined? How could this value change in the future? In which direction would it change?

5.18. The current recommendation of the SP-NCTL test presented in Section 5.1.3 is 200 hours. As a quality control test, this is too long. List the ways and procedures by which the time might be shortened to a more practical 24 hours.

5.19. Regarding ultraviolet degradation of geomembranes:
(a) How does UV degrade the various polymers used to manufacture geomembranes?
(b) How do manufacturers limit the phenomenon?
(c) Rank the geomembranes listed in Table 5.1 according to their resistance to ultraviolet light degradation.

5.20. List a series of situations in which chemical resistance charts of the type shown in Table 5.8 are *not* of use.

5.21. As a consultant or regulator, describe how you would select a leachate for chemical immersion tests like that described in Section 5.1.4 for a facility that is not built (i.e., it is in the proposal stage) for the following situations.
(a) hazardous waste landfill
(b) municipal solid waste landfill
(c) surface impoundment of a chemical plant
(d) surface impoundment for sewage sludge

5.22. Given the chemical resistance immersion data in Figure 5.14, does this particular HDPE geomembrane qualify as being "resistant" in terms of O'Toole, Little and Koerner, in Table 5.9?

5.23. What properties of a geomembrane would you require to be high when working in cold regions with liquid containment systems?

5.24. Immersing a geomembrane in water generally increases its weight (i.e., it swells via water absorption). In some cases, however, there is a weight loss. How can this occur?

5.25. Using average data from Table 5.10, calculate the change in length for 10, 20, 30, and 40°C variations of a 100 m long geomembrane roll if it were made from:
(a) HDPE
(b) VLDPE
(c) plasticized PVC

5.26. In the various lifetime prediction techniques discussed in Section 5.1.5, elevated temperatures are always involved. Why is this the case, that is, describe the concept of time-temperature superposition?

5.27. Consider the aging response curves in Figure 5.19a.
(a) Describe how antioxidants function in producing Stage A of the curve.
(b) Describe the mechanism involved in Stage B.
(c) Describe the mechanism(s) involved in Stage C.

5.28. The discussion in Section 5.1.5 centered around HDPE pipe and geomembranes. What type of aging response would you expect from PVC geomembranes? Illustrate your answer via a set of curves similar to those in Figure 5.19a.

5.29. Regarding the minimum thickness of a geomembrane, three considerations are necessary: survivability, regulations, and estimated stresses.

(a) From a survivability perspective, what is the minimum geomembrane thickness for the closure of a landfill (see Table 5.11)?

(b) A 1992 EPA clarification document on this topic calls for a minimum thickness of 0.50 mm. Please comment in light of your answer from a survivability perspective in part (a).

5.30. What liquid volume in liters will a site hold that is 35 by 35 m at the ground surface, has 3(H)-to-1(V) side slopes, and is 3.5 m deep at its center?

5.31. A 60 million liter water reservoir is to be constructed on a site measuring 160 by 115 m. The reservoir is to have uniform side slopes but requires a minimum 10 m buffer zone around each side. Determine the required depth of the basin for the following:

(a) 1-to-4 side slopes

(b) 1-to-3 side slopes

(c) 1-to-2 side slopes

5.32. Regarding a geotextile placed under a geomembrane as per Section 5.3.2:

(a) What is the required air transmissivity of a geotextile underliner beneath a geomembrane in the case of a rising watertable of 1.2 m in a three-day period? The soil porosity is 0.32 and the covered site measures 15 by 45 m. The slope of the geotextile is 1.0%.

(b) If the pond is filled with 7.5 m of water, what pressure is generated on the bottom of the liner?

(c) If this pressure is a potential problem, what are some possible design alternatives?

5.33. Regarding composite geomembranes:

(a) What would be the benefits and limitations of having a geotextile bonded directly to the geomembrane on its lower side?

(b) Answer the question in part (a) for two geotextiles that are bonded to both the lower and upper sides.

5.34. How thick should an unreinforced fPP geomembrane be if it subsides locally to a 45° angle under 5.0 m of water and is mobilized over a 50 mm zone? The liner is not soil-covered and has a friction angle of 25° with the soil beneath it. The allowable strength is 7000 kN/m^2. Check your answer with the design chart in Figure 5.24.

5.35. In the thickness design model presented in Section 5.3.4, which of the variables are experimentally obtained? Which are design assumptions? Which are dictated by the problem under consideration?

5.36. What is the factor of safety of a slope behind a geomembrane lined pond when it is empty if the soil fails under undrained conditions? The slope is 2(H) to 1(V), 7.5 m deep, and will be of the toe failure type (i.e., $n = 1.0$). The soil's unconfined compression strength is $q_u =$ 35 kN/m^2 (note $q_u = 2c$) and its unit weight is 18 kN/m^3.

5.37. Repeat Problem 5.36 for side slopes of 3(H) to 1(V).

5.38. Using a factor of safety of 1.75, what is the maximum slope angle (and H to V values) for a 9.0 m high slope of 16.5 kN/m^3 unit weight with a cohesion of 10 kN/m^2 and friction angle of 15°?

5.39. Recalculate the stability of the 300 mm uniform thickness cover soil in Example 5.11 (Section 5.3.5) for 600, 900, and 1200 mm thickness and plot the FS response curve.

5.40. Recalculate the stability of the 22° friction angle soil in Example 5.11 (Section 5.3.5) for soil-to-geomembrane friction angles of 10, 14, 18, 22 (the example), and 26° and plot the FS response curve.

5.41. Recalculate the stability of the tapered thickness cover soil in Example 5.12 (Section 5.3.5) for ω values of 14, 15, 16 (the example), 17 and 18.4° and plot the FS response curve. (Does the 18.4° result check with the analysis given in Section 5.3.5 for uniformly thick cover soils?)

5.42. Regarding geomembranes in anchor trenches with runout only:

(a) Recalculate Example 5.13 (Section 5.3.6) on runout length for various thicknesses of VLDPE geomembrane. Use 1.0 mm (the example), 1.5, 2.0, and 2.5 mm values and plot the results.

(b) Recalculate Example 5.13, assuming the geomembrane is fPP-R with an allowable stress of 30,000 kPa. Plot runout length versus thickness, where thickness varies from 1.0, 1.5, 2.0, and 2.5 mm.

5.43. Regarding geomembranes in anchor trenches with runout and vertical depth:

(a) Recalculate Example 5.14 (Section 5.3.6) on runout length plus anchor trench depth for various thickness of HDPE geomembranes. Use 1.0, 1.5 (the example), 2.0, and 2.5 mm values and plot the results.

(b) Recalculate Example 5.14, assuming the geomembrane is PVC with an allowable stress of 13,000 kPa. Plot runout length, and hold the anchor trench depth constant; then reverse by holding runout length constant and vary the anchor trench depth.

5.44. What would be the maximum stresses mobilized in a geomembrane covering a 10 m diameter wooden tank if the deformed radius conforms to a 25 m arc? The loading is 0.75 kN/m^2 and the thickness varies from 0.5, 1.0, 1.5, and 2.0 mm. Since some recent covers are made from foamed polyurethane up to 7.0 mm thick, include this value in your calculations and comment on such an approach.

5.45. In designing liquid containment systems the term *freeboard* is used. What does this term mean?

5.46. In the design of floating covers for reservoirs, the location of floats and weights are critical.

(a) How are floats made on a geomembrane?

(b) How are weights made on a geomembrane?

5.47. Geomembranes that are used for floating covers must be exposed to the atmosphere indefinitely.

(a) Which of the types of geomembranes listed in Table 5.1 are most suitable for long-term exposure?

(b) How would you go about estimating the lifetime of such *exposed* geomembranes?

(c) What physical and mechanical properties should be emphasized in writing a specification for a floating cover?

5.48. For large *superbags* as shown in Figure 5.36, what type of lateral containment is necessary to prevent the rupture of the enclosure?

5.49. Calculate the maximum tractive force on the bottom of a trapezoidal geomembrane lined water canal of 1.5 m depth and 5.0 m width at a slope of 2%, and compare it to the values given in Table 5.12.

5.50. Rank the low-cost canal liners mentioned in Section 5.5.3 when placed over highly fractured volcanic basalt rock foundation in the order of

(a) their perceived seepage control (i.e., a benefit ranking)

(b) a cost ranking on the basis of perceived cost for the various systems

(c) the ranking of subsequent benefit/cost ratio

5.51. Regarding the siting of a lined landfill (a *greenfield* site).

(a) What are some of the major technical features to be considered?

(b) What are some of the major nontechnical issues to be considered?

5.52. Describe the chemical interaction process by which organic solvents decrease the hydraulic conductivity (or coefficient of permeability) of clay soils.

5.53. When speaking of natural soil clay liners, the choices are either compacted clay liners or amended clay liners. Describe each of these types of soil liners.

5.54. The cost of an in-place natural soil liner can vary tremendously, while geomembranes (and geosynthetic clay liners) vary very little. Why is this the case and what are some of the major variables involved in natural clay soil liners?

5.55. Comment on the advantages and disadvantages of a composite geomembrane/natural clay secondary liner as shown in Figures 5.41d,e.

5.56. Comment on the advantages and disadvantages of a composite geomembrane/natural clay primary liner as shown in Figures 5.41f.

5.57. Comment on the advantages and disadvantages of a composite geomembrane/GCL primary liner as shown in Figure 5.41g,h.

5.58. For composite action to occur, do the geomembrane and clay have to be directly in contact? Can a geotextile, for puncture resistance, be placed between them? What is the effect of waves or wrinkles left in the geomembrane after backfilling?

5.59. For the outlet of a leachate collection pipe system beneath a landfill, a collection well at its lowest elevation is necessary. Should this collection be located within or outside the liner,

(a) for the primary removal system?

(b) for the leak-detection removal system?

5.60. Prepare a specification for the gravel soil to be used around leachate collection and leak detection pipe systems as shown in Figures 5.41a–g.

5.61. What are some technical equivalency issues that must be addressed to replace a gravel soil drain in either leachate collection or leak detection with the following:

(a) geonet, as in Figures 5.41e–h

(b) geocomposite, as in Figure 5.41h

5.62. Why is density control of subgrade soils beneath the lowest layer of a geosynthetic important when constructing a lined landfill facility? What density requirement should be specified?

5.63. Regarding geomembrane thickness:

(a) What is the minimum thickness allowed by the U.S. EPA for a hazardous material geomembrane landfill liner?

(b) Why are geomembrane thicknesses for hazardous material landfills liners greater than the thicknesses required for other situations?

(c) Is geomembrane thickness the key issue in leak prevention in lined landfills?

5.64. What is the thickness required for a VFPE liner containing a landfill of 13 kN/m^3 material 22 m deep under the following conditions: $\sigma_{\text{allow}} = 8000$ kN/m^2, $\beta = 30°$, $x = 38$ mm, $\delta_U = 20°$, and $\delta_L = 35°$?

5.65. Repeat Problem 5.64 for HDPE at $\sigma_y = 16{,}000$ kN/m^2.

5.66. Repeat Problem 5.64 for CSPE-R at $\sigma_b = 30{,}000$ kN/m^2.

5.67. The geomembrane puncture-protection design in Section 5.6.7 is based on 1.5 mm thick HDPE.

(a) How would the design vary for different thicknesses of HDPE?
(b) How would the design vary for the different types of geomembranes given in Table 5.1?
(c) How would the design vary for different type of geotextiles (i.e., other than nonwoven needle punched geotextiles)?

5.68. Repeat Example 5.19 (Section 5.6.7) on geomembrane puncture protection for the following. Plot the response curve for each variation.
(a) Different FS values, that is, FS = 1.0 to 10.0
(b) Different landfill heights, that is, 20 to 100 m
(c) Different protrusion heights, that is, 0.005 to 0.050 mm

5.69. Repeat Example 5.20 (Section 5.6.10), for the multilined side-slope stability using cumulative reduction factors 3.0 and 4.0, and plot the result against 2.0 (the example).

5.70. Calculate the required thickness of an access ramp as shown in Figure 5.48b based on trucks of 80 kN wheel loads at tire inflation pressures of 480 kPa. Assume that the liner system beneath the ramp has a cross section as shown in Figure 5.47 with a grade of 14° where the critical issues are the tensile strength of the 1.5 mm thick HDPE geomembranes and the roll-over compressive strength of the biaxially extruded geonet. The critical shear interface is beneath the secondary geomembrane and its interface friction angle is 22°.

5.71. Repeat Problem 5.70 assuming a dynamic impact factor of 3.0 times the static weight.

5.72. Describe and illustrate the concept of heap leach mining and the removal, treatment, and recirculation of the chemicals used in the process.

5.73. Describe and illustrate the concept of a solar pond together with the *salt gradient effect.*

5.74. List each of the geosynthetic materials shown in Figure 5.52 against its primary function in table below (either a yes or no is adequate).

Location	Type of Geosynthetic	Separation	Reinforcement	Filtration	Drainage	Barrier
Liner System						
Cover System						

5.75. Stabilization of liquid waste and quasi-solid waste materials is required before final capping and closure. What are the two basic functions of the stabilization process?

5.76. Incomplete in situ stabilization sometimes leaves unmixed zones in the area to be capped, which eventually results in hydrofracturing. What is meant by this term?

5.77. What are *bathtubs* with reference to stabilized landfill closures, and what problems do they create?

5.78. Regarding the geotextile gas-transmission layer in a landfill closure:
(a) Determine the required transmissivity of a nonwoven needle punched geotextile beneath the barrier layer of a closure system as shown in Figure 5.52. The infiltration rate based on decomposition modeling is 10,000 m^3/day and the grading of the system is 5%.
(b) If the candidate geotextile has a measured transmissivity of 0.0020 m^3/min-m under a normal stress of 50 kPa, what is its resulting factor of safety?

5.79. Subsidence prediction of randomly placed municipal solid waste landfills represents a difficult challenge insofar as cap design is concerned. For a deep municipal landfill consisting of domestic refuse, what steps would you take to quantify subsidence versus time? What are the basic mechanisms? How long do you suspect they will take to mobilize?

5.80. The thickness of the cover soil-protection layer of a landfill closure discussed in Section 5.7.6 is generally required to be greater than the maximum depth of frost penetration. Is this same requirement justified for all barrier materials? Compare and contrast the situation for
(a) compacted clay liners (CCLs)
(b) geosynthetic clay liners (GCLs)
(c) geomembranes (GMs)

5.81. List some possible uses for closed and properly capped landfill sites.

5.82. Corewalls in earth dams (whether of clay or geomembrane) can be centrally located, located near the upstream side, or located near the downstream side. When using a geomembrane, where is the favored location and why?

5.83. List the durability issues of exposed geomembranes when placed on concrete and masonry dams as described in Section 5.9.2.

5.84. Geomembranes can be used as cut-off walls in many situations, as described in Section 5.9.5. When placed in a slurry filled trench, how are the seams made?

5.85. Thinner, more flexible, geomembranes like PVC and CSPE-R are generally seamed into large panels in a fabrication facility. The seams are called factory seams, versus the subsequent joining of the panels together at the job site, called field seaming. List some advantages and disadvantages of factory seams versus field seams for the same type of geomembrane.

5.86. Regarding factory versus field seams:
(a) What percentage of factory seams should be inspected via destructive tests? Via nondestructive tests?
(b) What percentage of field seams should be inspected via destructive tests? Via nondestructive tests?

5.87. Why can't either thermal or solvent seam methods be used on thermoset elastomeric geomembranes?

5.88. The two fundamentally different seaming methods used to join thermoplastic geomembranes are either thermal or chemical methods (recall Table 5.24). List the advantages and disadvantages of each of these different seaming methods.

5.89. For thermal extrusion seaming methods the sheets to be joined require surface grinding.
(a) Why is this required?
(b) How deep should the grinding be?
(c) What is the preferred grinding direction, or orientation, with respect to the direction of the seam?
(d) What is the extent of the grinding with respect to the width of the extrudate?
(e) How long before seaming should the surfaces be ground?

5.90. Describe the impacts of the following environmental situations on geomembrane seam quality:
(a) moisture
(b) soil particles
(c) extreme heat
(d) extreme cold

5.91. In thermal wedge welding of geomembrane seams, you often consider a two-dimensional window or even a three-dimensional bubble. Plot the seaming variables to describe both of these situations.

5.92. Regarding automated hot wedge welding devices:
- **(a)** What is meant by a data acquisition welder? What type of data would be acquired?
- **(b)** What is meant by a process-control welder? How would an on-board computer help in this regard?

5.93. What is the fundamental problem with test strips made via solvent or adhesive seaming methods when it comes to following the flow chart in Figure 5.66?

5.94. Using 4.5 m wide geomembranes from the site illustrated in Figure 5.67:
- **(a)** How many linear meters of field seams are required?
- **(b)** At one destructive test per 150 m, how many destructive tests are required?
- **(c)** For the CQC or CQA organization (which typically performs 5 shear and 5 peel tests for each sample), how many tests are required assuming that all the samples pass and that no resampling is required?

5.95. What is the most commonly used nondestructive test for
- **(a)** PVC geomembranes?
- **(b)** CSPE-R and fPP-R geomembranes?
- **(c)** extrusion welded PE geomembranes?
- **(d)** hot wedge welded PE geomembranes?

5.96. If a second nondestructive test is required beyond your answer to Problem 5.95, what would be your choice for each geomembrane type?

5.97. Regarding leak location methods, which of those mentioned in Section 5.11.3 are field tested and functioning systems? Rate them as to advantages and disadvantages.

5.98. What are some test methods you would use to evaluate severely folded and/or creased geomembranes that are subjected to wind uplift and displacement (recall Section 5.11.4)?

5.99. Regarding quality control and quality assurance:
- **(a)** What technical skills are required to perform MQA?
- **(b)** What technical skills are required to perform CQA?
- **(c)** Can (or should) the MQA organization be the same as the CQA organization?

5.100. How early in the process of manufacturing geomembranes should the MQA organization become involved?

5.101. Regarding MQA/CQA:
- **(a)** Who contracts for (i.e., pays for) these services?
- **(b)** What is the QA plan or QA document all about?
- **(c)** When should a QA plan be developed?
- **(d)** What is the relationship between the QA organization and the site designer?
- **(e)** What is the relationship between the QA organization and the QC organization?
- **(f)** What is the relationship between QA organization and the regulatory agency?
- **(g)** What do you estimate are costs of MQA/CQA as a percentage of the cost of the entire liner system?

5.102. Why is there no MQC/MQA involved with natural soil components in the flow chart in Figure 5.74?

5.103. List some advantages and disadvantages of requiring 1,000,000 m^2 of experience for a geosynthetic installer for minimum qualifications (Table 5.28).

5.104. What is the necessity and/or value of having NICET, or equal, inspectors doing QC/QA work on geosynthetic systems?

5.105. In your opinion, is there a necessity for the certification of laboratories that do geosynthetic testing? Why?

5.106. List your ideas on the most common causes of failure in geomembrane systems.

5.107. What research and development areas do you feel are most important for the future development of geomembranes in civil engineering applications? Frame your comments in the following groupings:

(a) transportation-related
(b) environmental-related
(c) geotechnical-related
(d) hydraulics-related

6 Geosynthetic Clay Liners

6.0 INTRODUCTION

Factory-fabricated clay products to use as a barrier to migrating liquids have been available to the building construction industry for many years. Such *waterproofing barriers* in the form of semirigid panels sandwiched between cardboard sheets have been used for over 25 years. What is new, and the focus of this chapter, are large *flexible* rolls of factory-fabricated clay barrier materials that can be used to great advantage in pollution control facilities as landfill liners, reservoirs liners, landfill covers, and containment of underground storage tanks, not to mention the more traditional uses in geotechnical and transportation applications. A definition of geosynthetic clay liners or GCLs is

> **Geosynthetic clay liners** are factory-manufactured hydraulic barriers consisting of a layer of bentonite or other very low permeability material supported by geotextiles and/or geomembranes, mechanically held together by needling, stitching, or chemical adhesives.

Section 1.7 describes the manufacturing of the currently available GCLs. Also mentioned are the general application areas in which GCLs are currently used.

Note that GCLs were not known at the time of writing of the first edition of this book (1986), were only mentioned briefly in the second edition (1990), had a chapter of their own in the third edition (1994), and now have an expanded chapter devoted to their properties, equivalency (to other soil hydraulic barriers), design, and construction details. Their development and usage is certainly a welcomed addition to the field of geosynthetics since they offer a bentonitic clay liner material in a factory-manufactured form. As such, they form a hydraulic barrier material somewhere between thick field-placed CCLs and the polymeric geomembranes described in Chapter 5.

(a) Claymax of CETCO

(b) Bentofix of BTI, Naue Fasertechnik, and Terrafix

(c) Bentomat of CETCO

(d) Gundseal of GSE Lining Technology, Inc.

(e) NaBento of Huesker, Inc.

Figure 6.1 Photographs of commercially available GCLs with as-received product on right side and hydrated product on left side.

There are currently five GCL product types available in North America, and numerous others worldwide. Each type has various styles and properties. The photographs in Figure 6.1 show the as-received GCL on the right side and the GCL in a free-swell hydrated condition on the left side.

Claymax is manufactured by the Colloid Environmental Technology Company (CETCO), a subsidiary of the American Colloid Company. Claymax was the first GCL product to be designed and introduced into the market as a waste containment hydraulic barrier. Several types are manufactured. Claymax 200R is the original product, which consists of approximately 5.0 kg/m^2 of adhesive bonded granular sodium bentonite sandwiched between an upper woven geotextile and a lower woven geotextile. Claymax 600SP is similar, but has high internal shear and tensile strength properties that are achieved by stitch bonding the upper and lower geotextiles together (see Figure 6.1a). Current stitch spacing is at 100 mm, running continuously in the machine direction of the product. Figure 6.2a illustrates the cross section of the original product and Figure 6.2b the stitch bonded type. A new type of Claymax with a film-laminated fabric is also available.

The needle punching method of producing thick nonwoven geotextiles has been adapted to the manufacture of GCLs by Naue Fasertechnik GmbH of Germany. In North America, the products are produced by Bentofix Technologies Inc. Bentofix products are marketed in Canada by Terrafix Geosynthetics Inc. and in the rest of North America as well as in South America by the National Seal Company and Fluid Systems Inc. Bentofix products are manufactured with granular sodium bentonite. The mechanical integrity between the covering geotextiles (see Figure 6.1b) is provided by needle punching. Bentofix also uses a heat bonding process to burnish the needle punched fibers and lock them into place. Figure 6.2c shows a cross section; however, numerous different styles are available in North America, based upon two primary configurations:

Bentofix Thermal Lock NS. 3.5 to 5.0 kg/m^2 of granular sodium bentonite needle punched between a nonwoven and a woven geotextile.

Bentofix Thermal Lock NW. 3.5 to 5.0 kg/m^2 of granular sodium bentonite needle punched between two nonwoven geotextiles, one nonwoven being scrim reinforced.

In Europe, additional variations on the aforementioned products are available through Naue Fasertechnik, including GCLs made with HDPE geotextiles as well as geotextile/bentonite/geotextile/bentonite/geotextile composites.

Bentomat is a second type of GCL manufactured by CETCO. It consists of approximately 5.0 kg/m^2 of bentonite granules placed between upper and lower geotextiles and needle punched throughout to provide mechanical stability. See the photograph in Figure 6.1c, and also the cross section of Figure 6.2c. There are currently two types of Bentomat that are distinguished by their geotextile types:

Bentomat ST. A woven slit film geotextile on its upper surface and a nonwoven needle punched geotextile on the lower surface.

Bentomat DN. Nonwoven needle punched geotextiles on both upper and lower surfaces.

Variations in Bentomat can also come about from the use of different types of modified bentonite for specialty applications, for example, water containment, mild leachate, and high levels of salts, acids, alkalis, and organic and inorganic contaminants.

Gundseal is manufactured by GSE Lining Technology Inc. Gundseal consists of approximately 5.0 kg/m^2 of adhesive bonded bentonite adhered to a geomembrane. Depending on the types of fluids that may come in contact with the GCL, the bentonite

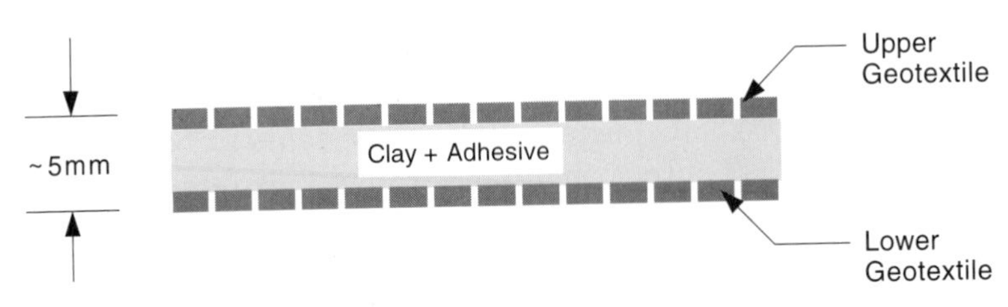

(a) Adhesive bound clay to upper and lower geotextiles

(b) Stitch bonded clay between upper and lower geotextiles

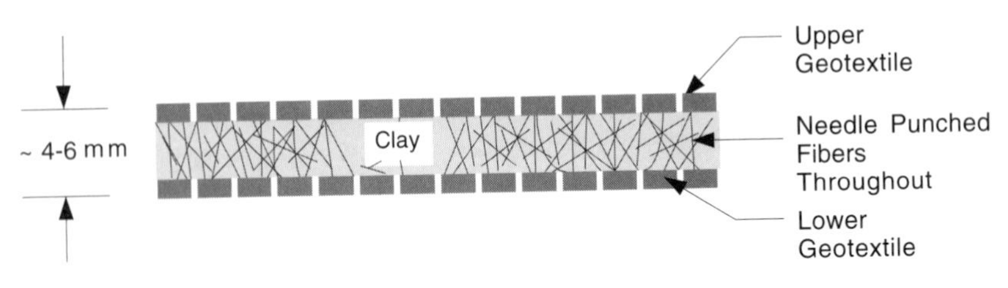

(c) Needle punched clay through upper and lower geotextiles

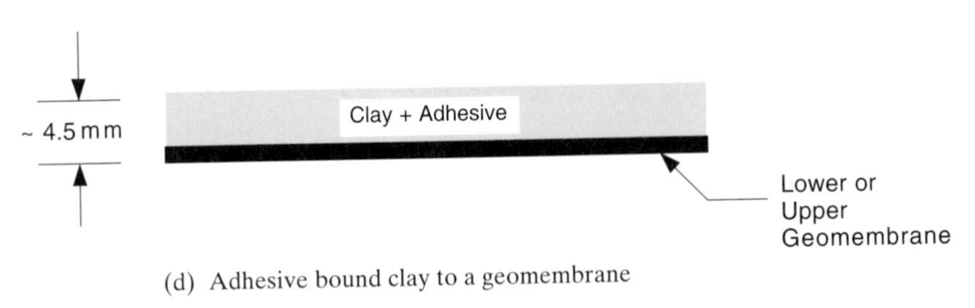

(d) Adhesive bound clay to a geomembrane

Figure 6.2 Cross-section sketches of currently available GCLs.

can either be a Wyoming sodium bentonite or a treated contaminant-resistant bentonite. The geomembrane can be either HDPE or VFPE with thickness ranging from 0.5 to 2.0 mm. Textured geomembranes can be used, as well. The photograph in Figure 6.1d and the cross section in Figure 6.2d illustrate the product. Gundseal can be installed in two ways. The first configuration is with the bentonite side facing downward against the subgrade forming a two-layer composite liner within itself. The second configuration is with the geomembrane side facing downward against the subgrade and with the bentonite side facing upward against an overlying field deployed geomembrane, forming a three-layer composite system.

NaBento is manufactured by Huesker Synthetic Co. and Nabento Vliesstoff GmbH in Germany. It is marketed by the Huesker organization in Germany and the U.S. The product consists of approximately 5.0 kg/m^2 of sodium-activated bentonite clay between two geotextiles that are stitch bonded together. See the photograph in Figure 6.1e and also the cross section in Figure 6.2b. The rows of stitch bonding are on approximately 20 mm centers and are continuous in the machine direction of the product.

It should be recognized that changes are occasionally made to existing GCLs and that new GCLs appear on a regular basis. In this latter category are the following.

Equiva-Seal from Geosynthetics, Inc. in the U.S., which consists of a geonet thermally bonded to a carrier geotextile, filled with bentonite in its apertures and then sealed with a covering geotextile that is thermally bonded to the upper surface of the geonet. The main purpose of the geonet is to contain the bentonite from lateral squeezing after it hydrates.

Nodulo Geobent from Laviosa Technology S.r.L. in Italy, which is a calcium bentonite adhesively bonded to a woven slit film geotextile on the lower surface and a spun laced geotextile on the upper surface.

An unnamed product from SBS in Austria, which is a calcium bentonite entangled via a fleece and needle punched between upper and lower geotextiles.

An unnamed product from Rawell Co. in the U.K., which is a prehydrated calcium bentonite that is then dewatered to a thickness of approximately 7 to 9 mm. It can be manufactured with any type of carrier geotextile.

Additional details are available in Koerner [1] and in the various manufacturers' brochures and commercial literature for their respective products. The various manufacturers should be queried as to their current line of products, and their properties and availability.

6.1 GCL PROPERTIES AND TEST METHODS

Inasmuch as GCLs are relatively new, there is considerable ongoing activity. A number of ASTM methods have been approved and still others are in various stages of development.

6.1.1 Physical Properties

A number of physical properties of GCLs are of interest: clay type, thickness, mass per unit area, adhesives, coverings, and moisture content.

Clay Type. X-ray diffraction (XRD) is a precise method of determining the composition of clays, however, the test is costly and relatively few testing laboratories are capable of performing it. Although not as accurate, The American Petroleum Institute's methylene blue analysis is easy to perform and is thought to give conservative results. Methylene blue dye is added to a bentonite pyrophosphate solution in 1 ml increments. Dye is added to the solution until a spot of the solution forms a blue halo when placed on a filter paper. The volume of dye added then relates to the cation exchange capacity (CEC), which relates to the montmorillonite content. Using such a methylene blue test, a montmorillonite content of at least 70% is felt to be required to yield adequate swell and permeability values. This value is approximately equivalent to an X-ray diffraction value of 90%, Heerten et al. [2].

Thickness. While seemingly easy to measure, thickness presents a measurement problem for all types of GCLs. It is essentially impossible to measure the thickness of the bentonite component of a GCL with its associated geotextiles or geomembrane. Even if it were possible to measure, the moisture content of the bentonite would have to be accurately stipulated. As a result, the thickness of a GCL usually refers to the composite material. Three items play into variations in thickness measurements.

Moisture content of the bentonite, which can be controlled by stipulating oven-dry test specimens.

Geotextile thickness variation under pressure, which can be controlled by stipulating a precise pressure.

Variation across the specimen width due to needle punching (which is minor and relatively uniform) or stitch bonding (which is major, depending on the particular product).

Thickness in the as-manufactured state (or dry, as an index value) is really not a major property and can be considered at best to be a quality control item for manufacturing. Where thickness is relevant as a performance role is in permeability testing to convert a flow rate (or flux) value to a hydraulic conductivity or permeability value. Here the thickness of the hydrated test specimen is required and the issue is quite controversial.

Mass Per Unit Area. The measurement of mass per unit area of a composite GCL is ASTM D5993. It is somewhat subjective for reasons similar to those just discussed in the thickness section. In addition, cutting out a GCL test specimen, having powdered or granular bentonite, without losing material is very difficult. Moisture has been added around the edges before cutting with little success and the mass of the adsorbed water must be deducted, which is difficult if a spray is used.

A measurement that can be made with assurity is the roll average mass per unit area of the complete GCL roll. This is a valuable property but it is more of a manufacturing quality control check than a field or conformance value since the rolls generally weigh about 1.5 tonnes.

Concerning the mass per unit area of the bentonite (without the associated geotextiles or geomembranes) difficulties arise with respect to sampling, the removal of the geotextiles or geomembrane, and the deduction for the amount of adhesive (if present). Unfortunately, the GCLs that are easiest to sample are those that contain adhesives, which is a difficult situation if we desire a separate mass per unit area for the bentonite component. Note that most GCLs are targeted to have 5.0 kg/m^2 of bentonite.

Clearly work needs to be done in evaluating the mass per unit area of GCLs, particularly from a field-conformance perspective.

Adhesives. The adhesives used to bond the bentonite powder or granules to themselves and to their adjacent geotextiles (as in Figure 6.2a) or to a geomembrane (as in Figure 6.2d) are proprietary materials. While the chemical identification methods described in Table 1.4 could identify the type of adhesive, it is rarely done. Instead, other more performance oriented tests, such as permeability or swelling, are used to see that the adhesive is not detrimental to the performance of the final product.

Coverings. As shown in Figure 6.2a,b,c, geotextiles cover the upper and lower surfaces of most types of GCLs; a geomembrane covers the surface of one of them (see Figure 6.2d). All of the geotextiles and geomembranes used for GCLs are described in Chapters 2 and 5, respectively. The reader is referred to these chapters for details. This comment, however, should not be taken as meaning that the geotextiles or geomembranes are not an important element of GCLs. Many properties are directly related to the covering materials: the uniformity of bentonite distribution, the containment of the hydrated bentonite during installation and service lifetime, the shear strength of the geocomposite at its two surfaces and internally, the puncture resistance of the geocomposite, the cross-plane permeability, and the overlap-seam permeability. It should be cautioned, however, that the original properties of the geotextiles (or geomembrane) will be significantly altered by virtue of the GCL-manufacturing process. For example, the grab tensile strength of the original geotextile will be decreased considerably after needle- or stitch-bonding during the production process. Thus if the geotextile is removed from the GCL, its physical, mechanical, and hydraulic properties will be significantly changed from the original geotextile properties as received by the GCL manufacturer before fabrication.

Moisture Content. Bentonite is a very hydrophilic mineral. As such it will generally have a measurable moisture content at all times. The value can be as high as 20% in humid areas, yet this is still considered to be the as-received, or dry, condition. The situation is further complicated by those GCLs that contain adhesives in the bentonite. Generally, some adhesive in liquid form remains after oven heating. This, along with the adsorption of the bentonite, can lead to a total moisture content of up to 30% as the product leaves the manufacturing facility.

The measurement of moisture content is straightforward via ASTM D4643 and is defined as the moisture content divided by the oven-dry weight of the specimen, expressed as a percentage. Some manufacturers base the moisture content on the wet weight of the test specimen, which results in slightly lower values.

Alternatively, we could use a thermogravimetric analysis (TGA) to determine both the free water and adsorbed water (see Section 1.2.2). In a TGA test the amount of adhesive used, for those GCLs containing adhesives, is also quantified.

6.1.2 Hydraulic Properties

Since GCLs are used in their primary function as hydraulic barriers this section is critically important. The hydraulic properties considered here are: hydration liquid, free swell, moisture absorption, fluid loss, and permeability.

Hydration Liquid. Bentonite, the essential low permeability component of all GCLs currently available, is known to hydrate differently depending upon the nature of the hydrating liquid. It is also known to hydrate differently as a function of the applied normal stress. Figure 6.3 illustrates the hydration response of four GCLs to the following five different liquids: distilled water, Philadelphia tap water, mild landfill leachate, harsh landfill leachate, and automotive diesel fuel. In all cases, the distilled water hydrated the GCLs the greatest. In contrast, the diesel fuel resulted in no hydration. Obviously, with diesel fuel the adsorbed water layer on the bentonite particles never developed and no swelling occurred. This is an important and well-known finding, in that GCLs must be pre-hydrated with water if they are to be used to contain hydrocarbons and nonpolar fluids. The two types of landfill leachates and tap water fall intermediate between these two extremes. All three fluids have anions and cations within them, which diminish the hydration potential of the bentonite from the ideal case of distilled water. It is somewhat disconcerting to see that local Philadelphia tap water was found to be quite close in its response to leachate, but at least it was the mild leachate!

Swell Index. The amount of swelling of bentonite under zero normal stress has been formalized in a test known as swell index, and designated as ASTM D5890. In this test, a graduated cylinder is filled with 100 ml of water and 2.0 g of bentonite is added. The bentonite is milled to a powder and added to the water slowly so as to allow the clay to flocculate and settle to the bottom of the cylinder. After leaving the cylinder undisturbed for 24 hours, the volume occupied by the clay is measured and a recommendation is given. Heerten et al. [2] recommend a minimum swell index value of 25 ml.

A similar test, albeit under a very low seating load but one that can be readily performed in the field as a conformance test, is GRI test method GCL-1. In this procedure a CBR swelling test device is used wherein 100 g of the GCL clay component (along with its adhesive, if present) is removed from the product and placed in the mold. A light seating load of 0.68 kPa is placed on the test specimen with a dial gage attached. The test specimen is saturated and readings are taken for 24 hours. The hydration behavior is recorded (see Figure 6.4) and if the swelling meets or exceeds the manufacturer's value, the clay component is acceptable.

Moisture Absorption. The fact that the bentonite in GCLs can readily absorb water from the adjacent soil has been shown by Daniel et al. [4]. They placed samples of GCLs on sand soils of varying water contents from 1 to 17% and measured the uptake of water in the GCL. Figure 6.5 shows the resulting curves. Two important messages stem from this data: soils as dry as 1% can result in GCL hydration to 50% and the time for hydration is extremely rapid (e.g., within 5 to 15 days).

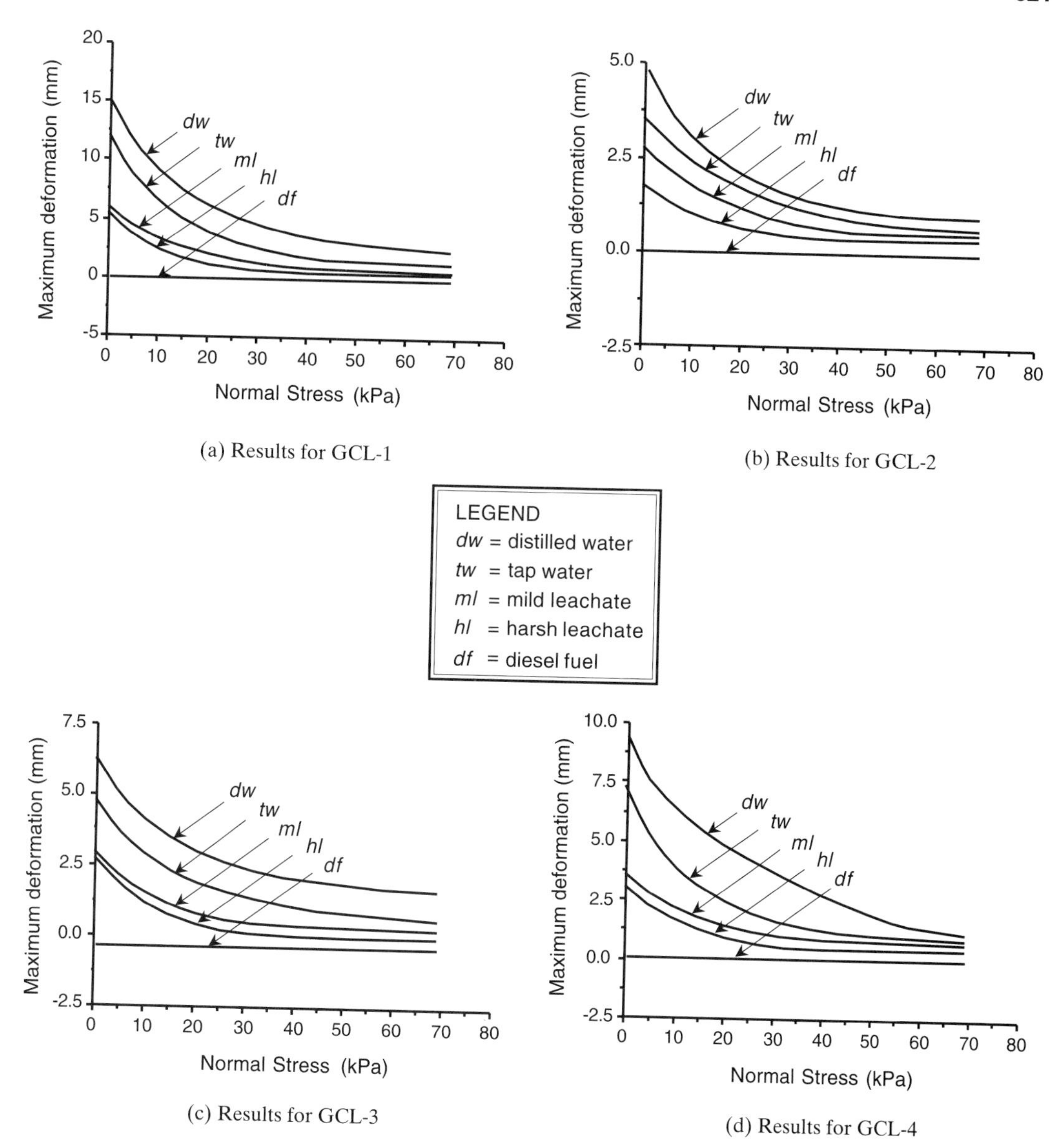

Figure 6.3 Hydration of GCLs using different liquids. (After Leisher [3])

For a laboratory determination of absorption, some GCL manufacturers report a plate water absorption test, performed in accordance with ASTM E946, to determine the volumetric increase of a clay sample as it draws water from an underlying saturated porous stone. Alternatively, Heerten et al. [2] recommend the Enslin-Neff test, which uses 0.4 g of bentonite on a glass filter within a cylinder. The cylinder funnels into a graduated capillary tube filled with water. The bentonite draws the water through the

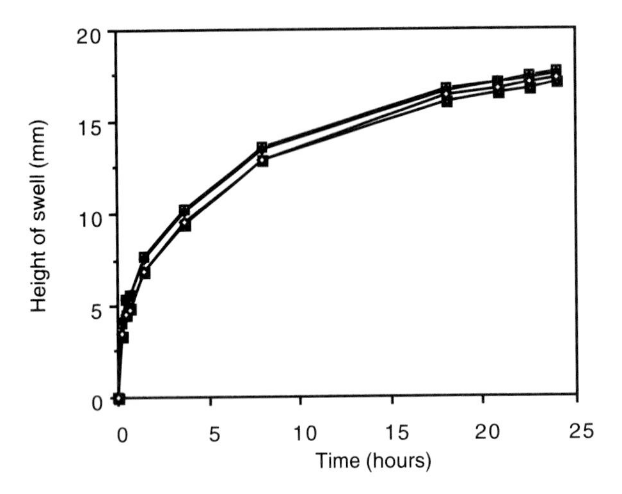

(a) Type of swell behavior from Product A

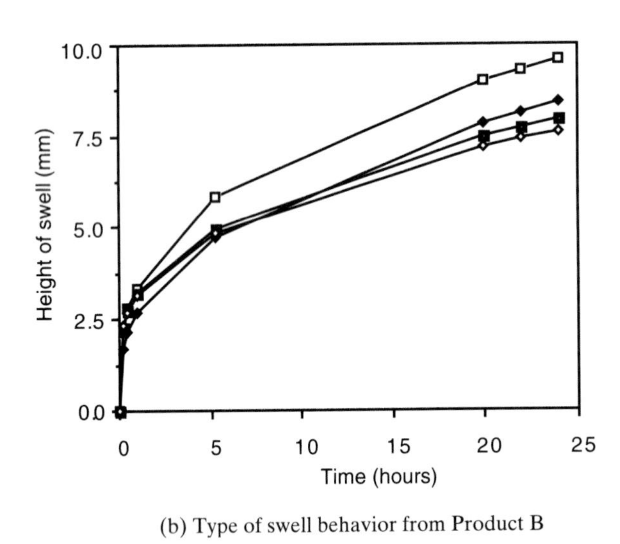

(b) Type of swell behavior from Product B

Figure 6.4 Typical hydration response curves for 100 g of clay component removed from two commercially available GCLs (each graph shows the reproducibility of four replicate tests).

filter, causing a reduction in the water level of the tube. After 24 hours, the change in volume and the corresponding weight of water is recorded as a percentage of the original weight of bentonite.

Fluid Loss. Another index test now focused on the fluid loss of the bentonite tested under pressure is ASTM D5891. It is an indirect measure of adhesive characteristics of the pore water to the clay particles.

Permeability (Hydraulic Conductivity) and Flux. Although the proper term is *hydraulic conductivity*, we will continue to use the word *permeability* since it is embedded in the GCL literature in this field at this time. As with CCLs, the permeability

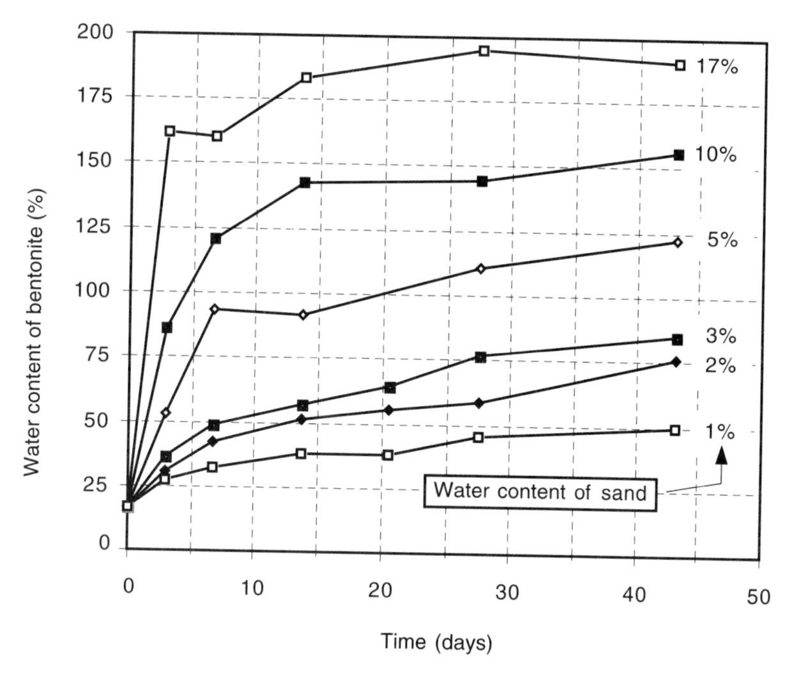

Figure 6.5 Water content versus time for GCL samples placed in contact with sand at various water contents. (After Daniel et al. [4])

of a GCL should be evaluated under field-simulated conditions in a flexible wall permeameter. The general performance test for CCLs is ASTM D5084. Since numerous modifications are necessary for GCLs, an alternate test is available as GRI test method GCL-2. For the index test permeability of a GCL, the following conditions are recommended in the GRI GCL-2 test procedure.

A 100 mm diameter GCL test specimen is used and placed in a triaxial permeameter. It is subjected to a total stress of 350 kPa and then back pressure saturated at 275 kPa with tap water for 2 days. Permeation is initiated by raising the pressure on the influent side of the test specimen to 310 kPa. Permeation is continued until inflow and outflow are equal to ± 25%, or until the permeability is sufficiently low to ensure conformance to a required value. The final value of permeability is then reported. The test can also be conducted using conditions set forth by the parties involved.

To assess the accuracy of the above procedure, Daniel et al. [5] conducted an interlaboratory testing program among 18 commercial laboratories and found that the permeability of a specific GCL ranged from 2×10^{-11} to 2×10^{-12} m/s. This is the general range for geotextile-related GCLs made from sodium bentonite. If calcium bentonite is used, the value of the permeability will be somewhat higher. Conversely, if the calcium bentonite is treated with sodium, the permeability is reduced to approximately the value for naturally occurring sodium bentonite. A considerable body of literature is available on bentonite modification and its subsequent engineering properties.

In the above described experimental procedure, a flow rate per unit area through the test specimen is actually measured. This value is also called the *flux*. It, along with the hydraulic gradient, is plotted for different values of total head to produce the permeability. Darcy's formula in terms of flow rate and flux illustrates the numeric procedure.

$$q = kiA \tag{6.1}$$

$$\frac{q}{A} = k\left(\frac{\Delta h}{t}\right) \tag{6.2}$$

where

q = flow rate (m^3/s),
A = area (m^2),
q/A = flux (m/s),
k = permeability (m/s),
Δh = total head (m), and
t = thickness (m).

All of the values, with the notable exception of the thickness, are easily measured. The thickness of the hydrated test specimen at the conclusion of the text is very troublesome to measure. The edges of the test specimen are often thinner or thicker than the center. For stitch bonded GCLs, a waved surface is observed. What measurement to use is elusive and consequently ASTM D5887 calls for only the flux to be measured and reported. While this is completely appropriate and can be done accurately, it leaves both the designer and regulator at a loss as to how to compare a GCL's permeability to another clay barrier such as a CCL.

Permeability of Overlap Seam. Since GCL roll edges and ends are sealed in the field by an overlap configuration, it is important that flow does not occur between the upper and lower GCL sheets.

Using a large laboratory test tank measuring 2.4 m long × 1.2 m wide × 0.9 m high with an overlap seal along the long direction of the tank, the permeability in the overlap region was measured by Estornell and Daniel [6]. The amount of overlap was 150 mm. Above the GCLs, 300 mm of gravel supplied the applied normal stress. Within experimental error, the overlap permeability was as low as with the control sample having no overlap seam. This study, in general, confirms the manufacturers' recommendations of a minimum overlap requirement of 150 mm. A smaller version of this test is currently under investigation.

It is important to mention that those GCLs with nonwoven needle punched geotextiles on both upper and lower surfaces *must* have bentonite powder or paste placed within the overlap area. The amount is typically recommended to be 0.4 kg/m, but it does depend on the type of geotextiles that are involved. The manufacturers' recommendations must be followed in this regard.

Permeability under Deformation. Recognizing that GCLs are being used in landfill caps and closures and that differential subsidence is likely in such applications, the issue of GCL permeability in an out-of-plane deformation mode must be addressed. LaGatta [7] has evaluated a number of conditions in a large-scale laboratory tank. The tank bottom was fitted with a centrally located rubber bladder filled with water, and the GCL test specimen was placed above it. Both full sections of GCLs and overlap seamed GCLs were evaluated. After 300 mm of gravel was placed above the GCL and the system was saturated, the bladder was sequentially emptied, producing an out-of-plane deformation in the GCL. The deformation was characterized by a Δ/L ratio, where Δ is the settlement and L is the radius of the deformation. Although the results are clearly product specific, it was found that:

- The integrity of the different overlapped seams is compromised at Δ/L values of 0.12 to 0.81. (Note, however, that the tests were conducted at very low normal stresses.)
- Vertical separation of the overlapping sheets can occur at very low normal stresses.
- There is a potential for resealing after an initial movement in the overlapped region has occurred.
- In cases of anticipated differential settlement and low normal stresses, LaGatta recommends a larger overlap distance, for example, 225 mm instead of the customary 150 mm.

6.1.3 Mechanical Properties

GCLs placed on side slopes, under high shear stresses, adjacent to rough or yielding subgrades, under thermal stresses, and so on, can readily challenge the individual product's mechanical properties thereby affecting its functionality as a hydraulic barrier. Invariably, some aspect of tensile stress will be involved.

Wide Width Tension. Using ASTM D4595, a GCL can be evaluated for its wide width tensile behavior. Since the clay component has little tensile strength, either dry or saturated (in comparison to the geosynthetics), the recommended manner of testing is the dry state. The resulting strength will be essentially that of the geotextiles or geomembrane involved. The GCL should be tested as a composite, however, and not as individual geosynthetics or assessed by using the published values of the individual original geosynthetic materials. Thus we can anticipate getting clay all over the testing machine. It is possible to seal the test specimen edges with hot glue, but if this is done carelessly it could affect the tensile strength response.

Two GCLs were tested in their as-received dry state with results given in Figure 6.6. The first, referred to as GCL A, consisted of bentonite clay sandwiched between a needle punched geotextile on one side and a composite geotextile on the other side. The composite was a woven slit film geotextile incorporated into a needle punched geotextile. GCL B consisted of bentonite clay sandwiched between a woven slit film geotextile and a nonwoven needle punched geotextile. All the geotextiles used in the manufacture of GCLs A and B were made from polypropylene yarns. Note that the tests were first conducted at zero normal stress (the customary manner) and then at

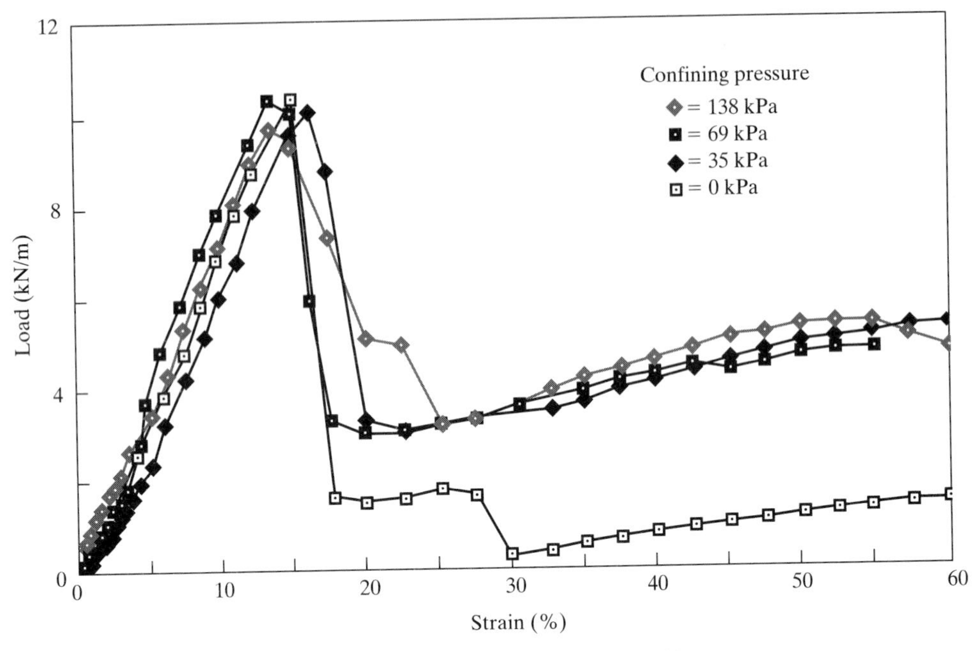

(a) Load-extension behavior of GCL

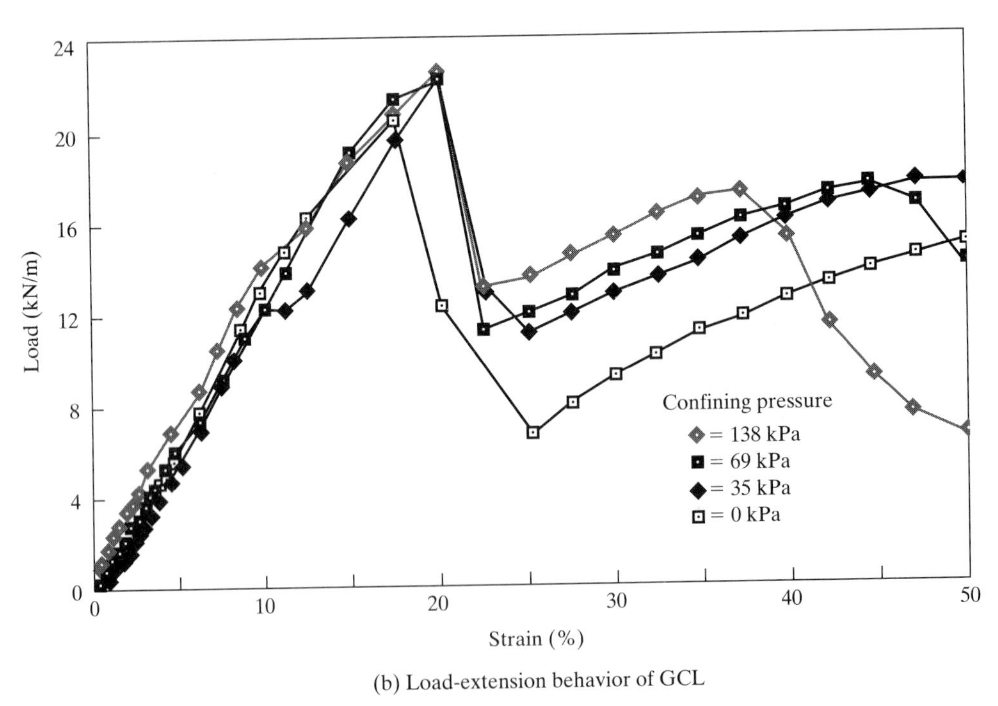

(b) Load-extension behavior of GCL

Figure 6.6 Wide width tensile behavior of two GCLs per ASTM test method D4595 (mod.). (Wilson-Fahmy et al. [8])

varying amounts of confined normal stress. The following comments focus only on the unconfined behavior.

The initial response of both GCLs was strongly influenced by the woven slit film geotextiles, which took the load uniformly until this component failed. The loss of strength is seen to be quite pronounced. Thereafter, the nonwoven component took the load until its ultimate failure at 50 to 60% strain. The modulus, strain at initial failure, and peak strength are the targeted values from such tests, but clearly, they are product-specific and dominated by the type of geotextile(s).

For the geomembrane associated GCL of Figure 6.2d, the tensile behavior is anticipated to be quite close to that of the geomembrane by itself; see [8] for selected data from geomembranes in this same mode of testing.

Confined Wide Width Tension. Using a confined wide width tension device (recall Section 2.3.3), the same two GCLs as in the previous section were evaluated, but now at 35, 69, and 138 kPa normal pressure [8].

Figure 6.6a shows the load-extension response of GCL A at these pressures and compares the responses to the zero confining pressure. It can be seen that up to peak load, there is practically no effect of confinement. This is not surprising since most of the load is carried by the woven slit film geotextile, which has a much higher modulus compared to the needle punched geotextiles making up the remainder of the product. In fact, the peak load is always associated with the rupture of the woven geotextile. The same behavior is noticed in Figure 6.6b for GCL B, where the load up to peak is again carried by the woven slit film geotextile. After peak is reached in both types of GCLs, the stress drops off significantly and very erratic behavior is observed. This behavior is quite complex in that the nonwoven needle punched geotextiles, the needling process, and the interaction of the clay particles are all involved in some way. For the purposes of reinforcement, however, this behavior is academic since the modulus, peak strength and associated peak strain are the focal points for any design process.

The conclusion of confined wide width test appears to be that *if* a woven slit film geotextile (possibly also a woven monofilament) is in the upper or lower geotextile, the effect of lateral confinement is negligible and the unconfined tension test is adequate. Use of ASTM D4595 in isolation is recommended. It also is much easier to set up and is considerably faster to perform. This probably holds true for GCLs with geomembranes as well. If, however, the GCL has only nonwoven needle punched geotextiles associated with it, the effect of confinement is measurable and tests with lateral confinement, as illustrated in this section, are warranted.

Axi-symmetric Tension. The question frequently arises as to the axi-symmetric tensile behavior of GCLs, particularly when used in landfill closure applications. Koerner et al. [9] have used an axi-symmetric tension test setup for geomembranes and modified it for GCLs. A test setup as shown in Figure 5.4 is used with a VFPE geomembrane over the GCL test specimen. Hydrostatic pressure is applied until failure occurs, which is always in the GCL before the VFPE geomembrane. The load taken by the VFPE geomembrane is deducted (via a separate test on the geomembrane by itself), and the stress-versus-strain of the GCL is obtained. In general, the failure strain for the geotextile related GCLs was from 10 to 19%, and for the geomembrane-related

GCLs 15 to 22%. Such values are orders of magnitudes higher than CCLs, and in keeping with stiffer geomembranes like HDPE and fabric-reinforced geomembranes (e.g., fPP-R and CSPE-R). Geomembranes with the highest flexibility, like VFPE, fPP, and PVC, strain considerably further (50 to 100%) in this type of test (recall Figure 5.5).

Direct Shear. Using a 100 mm × 100 mm shear box with the center of the GCL test specimens located at the split in the upper and lower shear boxes, a series of internal shear tests have been performed. The strain rate in all cases was 1.0 mm/min. Note that this is a relatively rapid strain rate and creep shear tests are currently ongoing for comparison purposes. Normal stresses varied from 0.7 to 140 kPa. The same four GCLs hydrated in the same five liquids described in Section 6.1.2 were used for these internal shear tests. The tests were performed in three different states: as-received condition (dry), hydration while under zero normal stress (free swell), and hydration under the same normal stress as the respective shear tests (constrained swell). Table 6.1 presents the resulting shear strength parameters. Please refer to Sections 2.3.3 and 5.1.3 for details of the direct shear test. Some observations regarding these internal shear test results follow.

- The GCLs were strongest in the dry condition and weakest in the free swell condition. The results from constrained swell conditions are intermediate between the two extremes.
- The type of hydrating liquid affects shear strength, but to a lesser extent than other factors. Hydration with distilled water is the worst-case condition in this regard.
- GCLs fabricated by needle punching between the two geotextiles required much larger displacements than unreinforced GCLs to reach their limiting shear strength stress.
- Needle punching significantly increases the shear strength under all conditions. (Note that stitch bonded GCLs were not evaluated in this series of tests, but similar and even greater improvements have been measured in separate tests.)

Depending upon the nature of the upper and lower surfaces of the GCL and on the adjacent soil or other geosynthetic materials, separate direct shear tests will be needed for these interfaces. These interfaces must be evaluated with respect to the site-specific materials: geotextiles (Section 2.3.3), geogrids (Section 3.1.2), geonets (Section 4.1.2), or geomembranes (Section 5.1.3). Also note that the interface surface may be considerably changed from the as-received geosynthetic materials, due to hydrated bentonite intruding into nonwoven needle punched geotextiles or extruding out of the woven geotextiles into the interface of concern. Slit film, spun laced, and monofilament geotextiles with even the slightest open area between fibers or yarns are all of concern in this regard. Product-specific simulated testing is called for in most circumstances.

Puncture Resistance. Due to the relative thinness of GCLs compared to CCLs, puncture-resistance concerns are often heard. There are a number of puncture tests that can be used with GCLs, including: ASTM D4833, which uses a 8.0 mm punc-

TABLE 6.1 SUMMARY OF GCL (INTERNAL) DIRECT SHEAR TEST RESULTS*

		Hydration with Distilled Water			Hydration with Tap Water		
		Dry	Constrained Swell	Free Swell	Dry	Constrained Swell	Free Swell
GCL-1	ϕ (degrees)	37°	16°	0°	37°	18°	0°
	C (kPa)	6.9	2.8	4.1	6.9	2.8	3.4
GCL-2	ϕ (degrees)	36°	31°	10°	36°	34°	15°
	C (kPa)	68	6.9	9.0	68	6.9	6.9
GCL-3	ϕ (degrees)	42°	37°	23°	42°	43°	26°
	C (kPa)	14	8.5	4.8	14	5.5	10
GCL-4	ϕ (degrees)	26°	19°	0°	26°	18°	0°
	C (kPa)	50	4.8	2.8	50	4.8	3.4

		Hydration with Mild Leachate			Hydration with Harsh Leachate		
		Dry	Constrained Swell	Free Swell	Dry	Constrained Swell	Free Swell
GCL-1	ϕ (degrees)	37°	24°	4°	37°	19°	0°
	C (kPa)	6.9	6.2	3.4	6.9	5.5	2.8
GCL-2	ϕ (degrees)	36°	43°	20°	36°	39°	30°
	C (kPa)	68	4.8	12	68	4.1	8.3
GCL-3	ϕ (degrees)	42°	39°	25°	42°	45°	32°
	C (kPa)	14	8.3	14	14	4.8	12
GCL-4	ϕ (degrees)	26°	18°	13°	26°	13°	0°
	C (kPa)	50	4.8	3.4	50	7.6	3.4

Hydration with Diesel Fuel									
		Dry	Constrained Swell	Free Swell			Dry	Constrained Swell	Free Swell
GCL-1	ϕ (degrees)	37°	44°	38°	GCL-3	ϕ (degrees)	42°	42°	40°
	C (kPa)	6.9	4.1	6.2		C (kPa)	14	6.2	4.8
GCL-2	ϕ (degrees)	36°	51°	46°	GCL-4	ϕ (degrees)	26°	24°	29°
	C (kPa)	68	4.1	4.8		C (kPa)	50	4.1	6.2

**Dry* refers to the GCL as received, placed under desired normal stress, then sheared at midplane.
Constrained swell refers to GCL hydrated under the desired normal stress, then sheared at the midplane.
Free swell refers to GCL hydrated under zero normal stress, then placed under the desired normal stress, and immediately sheared at midplane.
Source: After Leisher [3].

turing probe; GRI GS-1, which uses a CBR puncturing probe of 50 mm diameter probe; and ISO/DIS 12236, which also uses a 50 mm diameter probe. While all of these are straightforward to perform, it is important to recognize the self-healing characteristics of the bentonite-comprising GCLs. The photograph in Figure 6.7 illustrates this feature for hole made by a bolt that penetrated the GCL, which upon hydration appears to have sealed itself quite nicely. No puncture test by itself can reproduce this self-sealing mechanism, since the GCL is being used as a hydraulic barrier and puncture, per se, may not be a defeating, or even limiting, phenomenon.

Figure 6.7 Bolt puncture of a GCL, illustrating self-healing quality of bentonite clay. (Photograph courtesy of CETCO)

6.1.4 Endurance Properties

Since the solid component of the barrier material in a GCL is clay, its long-term integrity is generally assured. However, the liquid that activates the bentonite, resulting in its low permeability, is certainly an issue insofar as moisture barrier endurance is concerned.

Freeze-Thaw. The central property of a hydrated GCL insofar as freeze-thaw behavior is concerned is its permeability. Daniel et al. [10] used a rectangular laboratory flow box and subjected the entire assembly to 10 freeze-thaw cycles. The permeability showed a slight increase from 1.5×10^{-9} to 5.5×10^{-9} cm/s. Kraus et al. [11] report no change in flexible wall permeability tests of the specimens evaluated after 20 freeze-thaw cycles.

While the moisture in the bentonite of the GCL can freeze, causing disruption of the soil structure, upon thawing the bentonite is very self-healing and apparently returns to its original state. In this regard, it is fortunate that most GCLs have geotextile or geomembrane coverings so that fugative soil particles cannot invade the bentonite structure during the expansion cycle.

Shrink-Swell. The behavior of alternate wet and dry cycles on a GCL's permeability is important in many circumstances, particularly so when the duration and intensity of the dry cycle is sufficient to cause the desiccation of the clay component of the GCL. Boardman and Daniel [12] evaluated a single, albeit severe, wet-dry cycle on a number of GCLs and found essentially no change in the permeability. The results are encouraging and mimic the freeze-thaw results, but the results of numerous wet-versus-dry cycles awaits further investigation.

Perhaps more significant than the change in permeability is that shrinkage can cause loss of overlap at the roll edges or ends. If this occurs in the field, friction with the underlying surface will prevent expansion back to the original overlapped condition. Thus cover soil, placed in a timely manner and sufficiently thick to resist shrinkage, is necessary.

Adsorption. The adsorptive capacity of GCLs is important when they are used for landfill liners and interface with the various leachates that they are meant to contain. Both organic and inorganic solutes are of concern. The situation is described in [13], particularly in comparison to CCLs and addressing the issue of making an equivalency assessment. The cation exchange capacity of the bentonite clay must be determined and, along with its thickness, such a comparison can be made. It is in this particular instance that GCLs usually are not considered to be equivalent to the much thicker CCLs. For this reason there is a tendency to use a three-component composite liner (i.e., GM/GCL/CCL). The CCL component, however, can be significantly higher in its permeability than the usual regulated value. The hydraulic conductivity of the system has been analytically investigated by Giroud et al. [14].

Water Breakout Time. Particularly for GCLs used in landfill closures, the water breakout time is of interest. It is at this point that steady-state seepage will occur through the GCL and into the underlying solid waste. The data can be obtained from a permeability test, as described in Section 6.1.2, but now starting with the as-received dry GCL instead of starting with a fully saturated test specimen.

Solute Breakout Time. For a GCL placed beneath a landfill or surface impoundment, it is the solute breakout time (rather than water) that is of concern. The test method is again the permeability test (see Section 6.1.2), but now with the liquid of concern (e.g, with the leachate), as the permeant. This is an area where research seems to be warranted, particularly in light of showing the equivalency of GCLs to CCLs.

6.2 EQUIVALENCY ISSUES

Since CCLs (both natural soil and amended soil types) have been used historically as liquid barriers, it is only fitting that GCLs should have to compare favorably with, *or be better than,* CCLs in order to be used as replacement barrier materials. They may have to be better than CCLs, since GCLs are the replacement material and concerns are often voiced when the use of new materials is contemplated. The obvious issues are due to the fundamental differences listed in Table 6.2.

As first glance, we would assume that a technical equivalency argument could be based on the flow rate or flux through the competitive materials. Such a calculation is straightforward (it is illustrated in Section 6.3.1) and is routinely used for such purposes. However, this particular calculation is only the beginning of a complete equivalency comparison since numerous hydraulic, physical/mechanical, and construction issues need evaluation. Within each issue there are specific questions that can be raised in order to arrive at a complete equivalency assessment. Furthermore, for waste containment systems, we can identify functional differences between a barrier material beneath a waste facility (e.g., landfills, surface impoundments, and waste piles) and a barrier material placed above a waste facility (e.g., landfill covers and closure situations). In addition, the comparison may differ depending on whether the GCL is compared to a CCL when each is used by itself (as with a single barrier) or when they are used in a composite barrier, as with a GM/GCL compared to a GM/CCL.

TABLE 6.2 DIFFERENCES BETWEEN GEOSYNTHETIC CLAY LINERS AND COMPACTED CLAY LINERS

Characteristic	GCL	CCL
Material	Bentonite clay, adhesives, geotextiles and/or geomembranes	Native soils or blends of soil and bentonite clay
Construction	Factory manufactured and then installed in the field	Constructed and/or amended in the field
Thickness	$\cong$ 6 mm	300 to 900 mm
Permeability of clay	10^{-10} to 10^{-12} m/s	10^{-9} to 10^{-10} m/s
Speed and ease of construction	Rapid, simple installation	Slow, complicated construction
Installed Cost	\$0.05 to \$0.10 per m^2	Highly variable (estimated range \$0.07 to \$0.30 per m^2)
Experience level	Limited due to newness	Has been used for decades

The aforementioned contrasts can be arranged via a comparison that includes the various issues for liners versus covers. See Tables 6.3 and 6.4 for a relatively complete set of equivalency issues that often require an analysis. The tables can serve best as a guide or checklist for a site-specific comparison to be made by the user.

In both Table 6.3 (for liners) and Table 6.4 (for covers) it is seen that regarding the *hydraulic issues,* the chemical adsorptive capacity of a GCL compared to the typical CCL is generally not equivalent. It is site-specific just how dominant an issue this is. If it is significant, the use of a combined GCL/CCL composite is an alternative (see Giroud et al. [14]). Similarly the water and solute breakout times for the geotextile-covered GCLs shown in Figures 6.2a,b are probably not equivalent to CCLs, but the geomembrane-backed GCL probably is. Again the relevancy of breakout time must be assessed in light of site-specific considerations. Intimate contact of geomembranes with both GCLs and CCLs is an area in need of additional research.

Regarding *physical/mechanical issues,* GCLs are generally equivalent to or better than CCLs, with the exception of bearing capacity when the GCLs are hydrated and trafficked without sufficient soil cover. This issue must be avoided by proper specifications and follow-through in the CQC and CQA activities. Shear strength (internal and interface) is a site-specific and/or product-specific issue.

Regarding *construction issues,* it appears that only the puncture resistance and need for very careful subgrade preparation of GCLs are lacking. The self-healing characteristics of bentonite clay, however, must be considered in regard to puncture—recall the bolt photograph of Figure 6.7. Regarding subgrade conditions Scheu et al. [15] describe GCLs placed over very rough subgrades. Even further, GCLs are used as protection mats in Germany placed *over* geomembranes and beneath coarse drainage stone in leachate collection layers. Lastly, a key issue, as with all geosynthetics and natural soil materials, is proper CQC and CQA insofar as their installation is concerned.

Thus it is felt that in most cases a GCL can replace a CCL on the basis of technical equivalency. One important issue not addressed in Tables 6.3 and 6.4 is cost. In areas where proper natural clay soils are plentiful, CCLs will be competitive to GCLs. In

TABLE 6.3 GENERALIZED TECHNICAL EQUIVALENCY ASSESSMENT FOR GCLS BENEATH LANDFILLS AND SURFACE IMPOUNDMENTS

Category	Criterion for evaluation	Probably superior	Probably equivalent	Probably not equivalent	Equivalency dependent on site or product
Hydraulic Issues	Steady flux of water		X		
	Steady solute flux		X		
	Chemical adsorption capacity			X	
	Breakout time				
	Water				X
	Solute				X
	Horizontal flow in seams or lifts		X		
	Horizontal flow beneath geomembrane		X		
	Generation of consolidation water	X			
Physical/ Mechanical Issues	Freeze-thaw behavior	X			
	Total settlement		X		
	Differential settlement	X			
	Slope stability				X
	Bearing stability			X	
Construction Issues	Puncture resistance			X	
	Subgrade conditions			X	
	Ease of placement	X			
	Speed of construction	X			
	Availability of materials	X			
	Requirements for water	X			
	Air pollution concerns	X			
	Weather constraints				X
	Quality assurance considerations		X		

Source: After Koerner and Daniel [13].

areas where they are not, and blending of native soils with admixed bentonite clay is necessary, the GCLs will usually be very cost effective. This is obviously a site-specific consideration.

6.3 DESIGNING WITH GCLS

The single, and obviously primary, function of a GCL is as a liquid barrier. Thus the examples given in this section will illustrate this function for different applications.

6.3.1 GCLs as Single Liners

GCLs have been used as single liners, that is, by themselves with no composite or backup geomembranes, in a number of notable cases. Two will be described in this section: canal liners and underground storage tank liners.

TABLE 6.4 GENERALIZED TECHNICAL EQUIVALENCY ASSESSMENT FOR GCL COVERS ABOVE LANDFILLS AND ABANDONED DUMPS

Category	Criterion for evaluation	Probably superior	Probably equivalent	Probably not equivalent	Equivalency dependent on site or product
Hydraulic Issues	Steady flux of water		X		
	Breakout time of water				X
	Horizontal flow in seams or lifts		X		
	Horizontal flow beneath geomembrane		X		
	Generation of consolidation water	X			
	Permeability to gases		X		X
Physical/ Mechanical Issues	Freeze-thaw behavior	X			
	Shrink-swell behavior	X			
	Total settlement		X		
	Differential settlement	X			
	Slope stability				X
	Vulnerability to erosion				X
	Bearing stability			X	
Construction Issues	Puncture resistance			X	
	Subgrade conditions			X	
	Ease of placement	X			
	Speed of construction	X			
	Availability of materials	X			
	Requirements for water	X			
	Air pollution concerns	X			
	Weather constraints				X
	Quality assurance considerations		X		

Source: Koerner and Daniel [13].

Heerten and List [16] report on GCLs being used to rehabilitate old clay liners in German canals. The study cites a canal that was dewatered, properly regraded, and had the GCL placed directly on the soil subgrade. The particular GCL was of the needle-punched type shown in Figure 6.2c with a 800 g/m^2 nonwoven geotextile as a substrate and a 300 g/m nonwoven geotextile a superstrate. The permeability of the calcium bentonite was 1×10^{-10} m/s. (Note that all North American manufactured GCLs described in Section 6.1 have sodium bentonite clay as their primary mineral, which is approximately one order of magnitude lower in its permeability than calcium bentonite.) A 300 mm thick gravel layer was placed over the GCL (see Figure 6.8a). The side slopes varied segmentally from 4.8 to 18.4 to 26.6 to 29.7° as the cross section rose to the top of the slope. Internal shear tests on this particular product resulted in friction angles of approximately 34°, according to the data in Table 6.1. Thus the needling process and upper and lower nonwoven geotextiles provide the assurity of adequate slope stability for all sections along the side slopes.

There are a huge number of steel storage tanks that require secondary containment liners for environmental safety in case of a failure. The usual deployment is shown

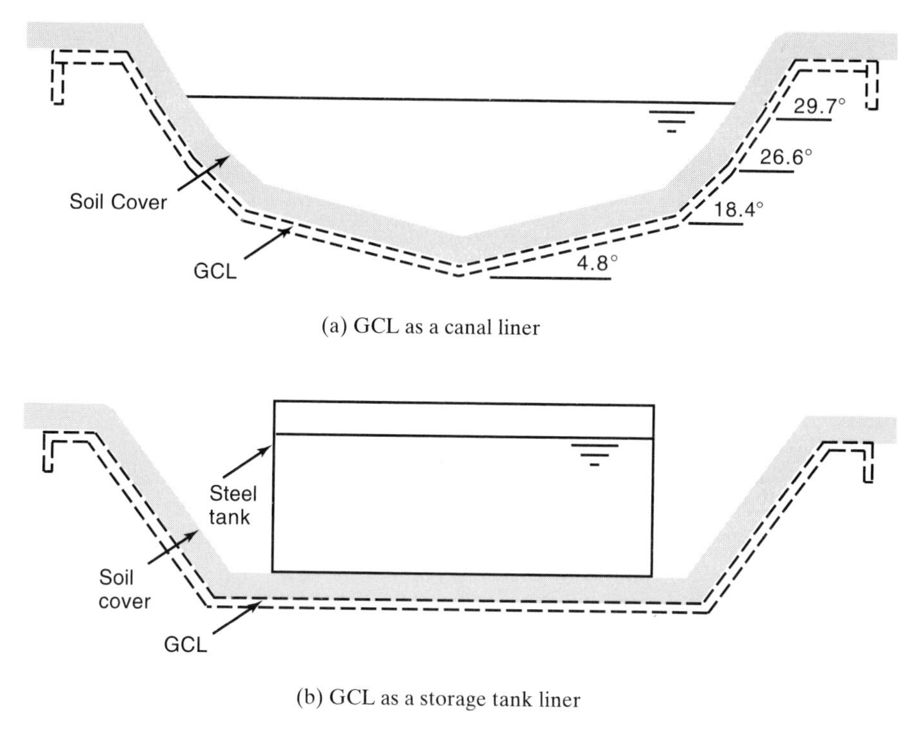

(a) GCL as a canal liner

(b) GCL as a storage tank liner

Figure 6.8 Cross section of GCLs used on single liners. (After Heerten and List [16])

in Figure 6.8b. The purpose of the GCL is to contain the liquid in the storage tank in the event of a pipe burst or tank leak. It is the same application as described in Section 5.8, where the solution was to use a geomembrane. Since side slope angles are usually quite small, most GCLs are candidates for this application. However, one caution must be raised. If the soil cover, typically a gravel, placed over the GCL is limestone, the leaching of calcium and magnesium from the stone into the sodium bentonite clay may cause a base exchange to occur, thereby increasing the permeability of the bentonite clay in the GCL. Thus it is important to prehydrate GCLs whenever they will not be water-saturated during their initial liquid immersion (recall the diesel fuel hydration tests in Figure 6.3).

The general issues in the design for a GCL used as a single liner are as follows.

1. Calculate the flow rate in the context of alternative materials (see Example 6.1), adsorption, and breakout time for both water and the solute in the containment application.
2. Calculate the shear strength for side slopes under all possible internal and interface conditions.
3. Assess the possible implications of puncture, tear, and loss of bentonite, considering both the materials above and below the GCL.

4. Carefully consider installation survivability of the GCL, considering both the subgrade and the backfill materials.

Example 6.1

Calculate the water flow rate coming from a GCL in a 4.0 m deep canal if the permeability of the GCL is 5×10^{-11} m/s and it is 5 mm thick. Compare this value to a CCL that is 450 mm thick with a permeability of 1×10^{-9} m/s, and compare the results.

Solution: Using Darcy's formula, Eq. (6.1), for each liner material where i is the hydraulic gradient and is equal to the total head divided by the liner thickness and based on a unit area we have the following.

for the GCL

$$\begin{aligned} q &= kiA \\ &= \left(5 \times 10^{-11}\right)\left(\frac{4.005}{0.005}\right)(1.0 \times 1.0) \\ &= 40.1 \times 10^{-9}\ \text{m}^3/\text{s} \end{aligned}$$

for the CCL

$$\begin{aligned} q &= kiA \\ &= \left(1 \times 10^{-9}\right)\left(\frac{4.450}{0.450}\right)(1.0 \times 1.0) \\ &= 9.9 \times 10^{-9}\ \text{m}^3/\text{s} \end{aligned}$$

Thus flow rate through the GCL is 4 times higher than through the CCL. This is due to the large total head causing the flow. Even so, for a water canal the GCL's flow rate may be quite acceptable. It should be noted that the same type of calculation for the typical 300 mm total head loss in a landfill will highly favor the GCL solution.

6.3.2 GCLs as Composite Liners

GCLs have seen their greatest use to date as the lower component of a composite liner for landfills and surface impoundments. Thus a geomembrane will be the upper component and the GCL the lower component. Such a composite liner has been used to great advantage as the primary liner of a double-lined landfill facility (recall Figures 5.41g,h). The reduction in leakage rates for facilities constructed with a GM/GCL composite, versus a GM alone *or* a GM/CCL composite, is remarkable. In the case of the GM by itself, the occurrence of a hole brings leachate directly into the leak detection system, while consolidation water from a GM/CCL is very large and difficult to distinguish from actual leakage. The GCL, being placed dry, attenuates any leakage through holes or flaws in the geomembrane giving in many cases near-zero leakage rates (recall Table 5.17 after Othman et al. [17]).

Such a composite GM/GCL can also be used as a secondary liner system, but regulations sometime require a GM/CCL secondary liner. Thus we would have to show complete equivalency of the GCL to the CCL. While this can be done (recall Section 6.2), some regulators are reluctant to give changes for traditionally thick elements of the cross section, such as a CCL. An alternative composite liner that has been used is a three-component GM/GCL/CCL; although the CCL component is somewhat higher in permeability than the regulated value, the GCL/CCL together will give the regulated value or lower (Giroud et al. [14]).

In a GM/GCL composite liner application, the upper geotextile covering of the GCLs is controversial (Figures 6.2 a,b,c). This raises the issue of intimate contact. The original concept of composite liner behavior is shown by comparing Figures 6.9a,b. Here a potential hole in the geomembrane directly meets the underlying clay, where it

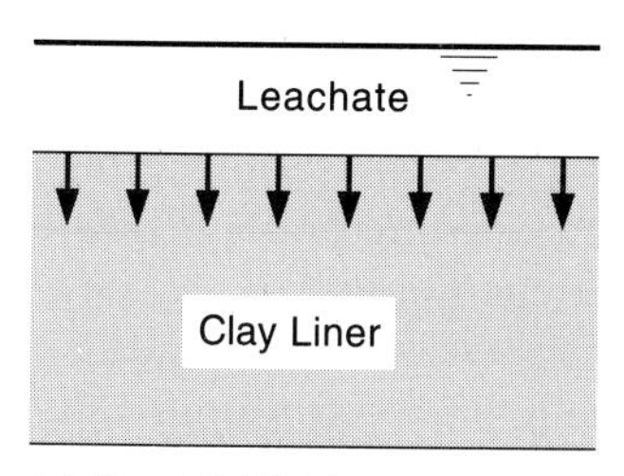

(a) Clay (CCL) Liner
(CCL by itself)

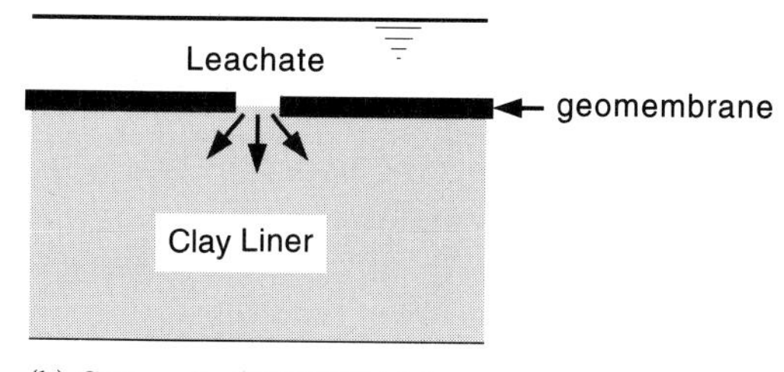

(b) Composite (GM/CCL) Liner
(geomembrane and CCL
with intimate contact)

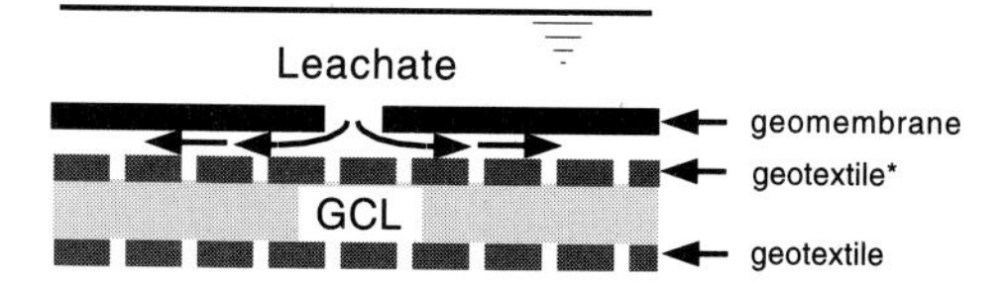

* Does this geotextile's transmissivity compromise the composite liner concept?

(c) Composite (GM/GCL) Liner
(geomembrane and GCL with
questionable intimate contact)

Figure 6.9 Composite liner concept, illustrating the issue of intimate contact.

is forced to radially propagate through the clay soil. No drainage layer (sand or geotextile) is generally allowed between the two materials. Thus it is reasonable to express a concern when a GCL is used instead of a CCL and the GCL has a geotextile as its upper surface. Figure 6.9c illustrates the essence of this concern. Here the lateral transmission of the leachate in the plane of the geotextile, allowing for the attack of the clay component over an area larger than the hole itself can be envisioned. While this is a reasonable concern, we must realistically question the transmissivity of the geotextile in light of the *quantity* of liquid being transmitted. Table 2.7 presents the transmissivity of various geotextiles. Generally, woven slit film and woven monofilament geotextiles are not of concern and furthermore, the extrusion of bentonite through the open voids of such woven geotextiles will probably lower the transmissivity value to that of the bentonite itself. Thus the concern in this situation is the lowering of the shear strength of the interface between the bottom of the geomembrane and the top surface of the GCL. This situation has resulted in two slides of full-scale test plots at 2(H)-to-1(V) slopes (Koerner et al. [18]). As a result, when GCLs are on relatively steep side slopes, the upper geotextile (beneath the covering geomembrane) should be a nonwoven needle punched type. The minimum mass per unit area becomes a trade-off between being sufficiently thick to avoid the extrusion of the hydrated bentonite and sufficiently thin so that composite action is not an issue. To assess the situation, the radial transmissivity test described by Wilson-Fahmy et al. [19] can be considered.

The general issues involved in the design for a composite GM/GCL liner are as follows.

1. Calculate the composite liner flow rate for water containment applications, and the flow rate adsorption and breakout time for leachate (solute) containment applications. See Examples 6.2 and 6.3.
2. Assess the internal and interface shear strengths for the side slopes and intermediate berms under short- and long-term loading conditions.
3. Consider the GCL wide width strength design and anchor trench reaction. If the interface friction angle below the GCL is lower than those above it and also lower than the side slope angle, the difference must be carried by the GCL in tension.
4. Avoid puncture, tear, and loss of bentonite into an underlying leak detection system by using carefully worded specifications followed by strict CQC and CQA. Both static and dynamic conditions must be addressed (the latter being of concern for access ramps).
5. Avoid lateral migration, or squeezing, and loss of thickness of the hydrated bentonite under heavy long-term static loads. This is listed as bearing stability in Table 6.3. A suitably thick soil cover layer is necessary before trafficking.
6. Avoid the contamination of overlying geomembrane seam areas from loss of bentonite. This has been a problem particularly with textured geomembranes.
7. Consider both the subgrade and the backfill materials in the GCLs' survivability during installation.

Example 6.2

Calculate the leakage from a slit in the geomembrane overlying a GCL. That is, what is the upper-bound flow rate through a 2.0 mm wide slit in a geomembrane overlying a 10 mm thick GCL having a permeability of 7.0×10^{-12} m/s? The slit is long with respect to its width. The composite liner is under a constant total head of 300 mm. Use the formulation presented by Giroud and Bonaparte [20].

Solution: Assuming radial flow through the underlying GCL, the following formula gives the flow rate per unit length of slit in the geomembrane.

$$q = \frac{\pi k_s (h_w + t)}{\ln (2t/b)} \tag{6.3}$$

where

q = flow rate,
k_s = GCL permeability,
h_w = total head loss,
t = GCL thickness, and
b = length of slit in geomembrane.

$$q = \frac{\pi(7.0 \times 10^{-12})(0.300 + 0.010)}{\ln (0.020/0.002)}$$

$$q = 3.0 \times 10^{-12} \text{ m}^3/\text{s-m of slit length}$$

Example 6.3

Calculate the leakage from a hole in the geomembrane overlying a GCL. That is, what is the upper-bound flow rate through a 2.0 mm circular hole in a geomembrane overlying a 10 mm thick GCL having a permeability of 7.0×10^{-12} m/s? The composite liner is under a constant total head of 300 mm. Again, use the formulation presented by Giroud and Bonaparte [20].

Solution: Assuming radial flow through the underlying GCL, the following formula gives the estimated leakage rate.

$$q = \frac{\pi k_s (h_w + t) d}{(1 - 0.5d/t)} \tag{6.4}$$

where d = hole diameter.

$$q = \frac{\pi(7.0 \times 10^{-12})(0.300 + 0.010)(0.002)}{(1 - 0.001/0.010)}$$

$$q = 0.015 \times 10^{-12} \text{m}^3/\text{s}$$

6.3.3 GCLs as Composite Covers

In exactly the same way as they are used for liners beneath solid waste, a composite GM/GCL can be used in a cover above the solid waste (see Koerner and Daniel [21]). In describing such barrier strategy for landfill covers (also called closures or caps), the typical benefit/cost ratios favor a GM/GCL over a GM/CCL (see Koerner and Daniel [22]).

The fundamental difference between the two applications of a composite liner above the waste and below the waste is that differential settlement will likely occur in cover situations. The GM/GCL application has been developed and used in a number of covers for abandoned dumps, many with no liner of any type beneath the waste.

The general issues involved in the design for a composite GM/GCL cover placed above a solid waste landfill are as follows.

1. Calculate the composite cover flow rate for water, as illustrated in Ex. 6.2. and 6.3.
2. Assess the internal and interface shear strengths for cover slopes under short- and long-term loading conditions, including live loadings.
3. Assess the GCL strength design and factor of safety. If the interface friction angle below the GCL is lower than those above it and also lower than the slope angle, the difference must be carried by the GCL in tension.
4. Carry the tensile stresses via anchor trenches or, if symmetry of the cover exists, via equal and opposite reactions on each side of the crest.
5. Evaluate the retention of the GCL's low permeability in the event of out-of-plane deformation due to subsidence of the underlying solid waste material (i.e., due to differential settlement). See Example 6.4.
6. Avoid puncture, tear and loss of bentonite, considering both the materials above and below the GCL.
7. Avoid the contamination of geomembrane seam areas from loss of bentonite. This has been a problem, particularly with textured geomembranes.
8. Consider both the subgrade and the backfill materials in the GCLs' survivability during installation.

Example 6.4

For the out-of-plane deformation configuration shown in the following sketch, calculate the approximate tensile strain in the GCL and in the geomembrane. What is the factor of safety of each material if the GCL loses its hydraulic barrier integrity at 14% tensile strain and the geomembrane is VFPE (recall Figure 5.5 and Eq. 5.4)?

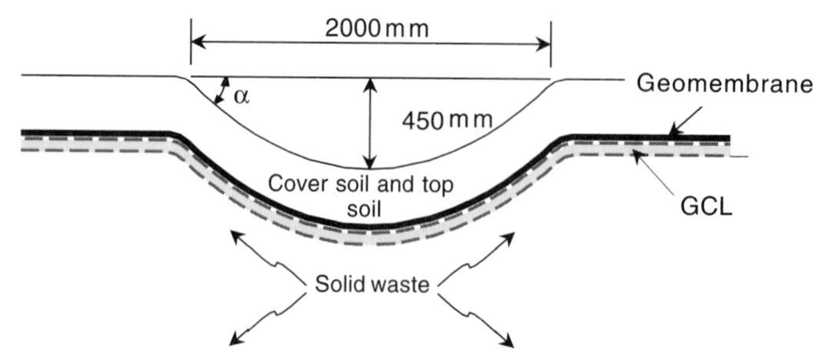

Solution: The deflected shape is used for calculations as follows.

$$\varepsilon(\%) = \left[\frac{\tan^{-1}\left(\dfrac{4L\delta}{L^2 - 4\delta^2}\right)\left(\dfrac{L^2 + 4\delta^2}{4\delta}\right) - L}{L} \right] \times 100 \qquad \text{for } \delta < \frac{L}{2} \tag{5.4}$$

$$\varepsilon(\%) = \left\{ \frac{\tan^{-1}\left[\frac{4\,(2.0)\,(0.45)}{(2.0)^2 - 4\,(0.45)^2}\right]\left[\frac{(2.0)^2 + 4\,(0.45)^2}{4\,(0.45)}\right] - 2.0}{2.0} \right\}(100)$$

$$\varepsilon(\%) = 13.0\%$$

Thus the GCL is satisfactory, having FS = 14.0/13.0 = 1.1; while the GM is also satisfactory, having FS = 100/13.0 = 7.7.

6.3.4 GCLs on Slopes

The 14 test plots [9 on 2(H)-to-1(V) slopes and 5 on 3(H)-to-1(V) slopes] described in [18] are focused on assessing the long term internal shear strength of various types of GCLs. It is a worthwhile and on-going effort, since the shear strength of hydrated bentonite is low. How low is a matter of the site-specific conditions as assessed by simulated direct shear tests (recall Table 6.1).

The issue, insofar as GCLs on slopes is concerned, is that the bentonite must either remain dry (e.g., protected between two geomembranes) or be internally reinforced. The most common methods of internal reinforcement are by needle-punching or stitch-bonding. Examples 6.5 and 6.6 illustrate the magnitude of the required internal strength; first with no support from above, then with overlying geosynthetics providing additional reinforcement.

Example 6.5

What is the required long term strength of the internal reinforcement of a hydrated GCL (i.e., the needle punched fibers or stitch bonded yarns) to achieve a FS = 1.5 when it is placed on a 3(H)-to-1(V) slope that is 30 m long and is covered with 450 mm of well-graded sand ($\phi = 35°$) weighing 17 kN/m^3? Assume that no tensile strength is afforded by the upper geotextile (i.e., worst-case assumption). Also assume that the upper and lower surfaces of the GCL have sufficient interface shear strength so as to force the potential failure plane within the internal structure of the GCL. Use $\delta = 6°$ for the bentonite.

Solution: Using the formulation given in Section 3.2.7, on geogrid veneer reinforcement, the required reinforcement strength can be found from back-calculation, using a given value of factor of safety. (Actually a computer program was written for this solution).

$$W_A = \gamma h^2\left(\frac{L}{h} - \frac{1}{\sin\beta} - \frac{\tan\beta}{2}\right) \tag{3.14}$$

$$= (17.0)(0.45)^2\left[\frac{30}{0.45} - \frac{1}{\sin 18.4} - \frac{\tan 18.4}{2}\right]$$

$$= 218 \text{ kN/m}$$

$$N_A = W_A \cos\beta \tag{3.15}$$
$$= (218)\cos 18.4$$
$$= 207 \text{ kN/m}$$

$$W_P = \frac{\gamma h^2}{\sin 2\beta} \tag{3.17}$$
$$= \frac{(17.0)(0.45)^2}{\sin 36.8}$$
$$= 5.75 \text{ kN/m}$$

$$\text{FS} = \frac{-b + \sqrt{b^2 - 4ac}}{2a} \tag{3.22}$$

where

$$\begin{aligned}
a &= (W_A - N_A\cos\beta - T\sin\beta)\cos\beta \\
&= (218 - 207\cos 18.4 - T\sin 18.4)\cos 18.4 \\
&= 21 - 0.299T \\
b &= -[(W_A - N_A\cos\beta - T\sin\beta)\sin\beta\tan\phi + (N_A\tan\delta + C_a)\sin\beta\cos\beta \\
&\quad + \sin\beta(C + W_p\tan\phi)] \\
&= -[(218 - 207\cos 18.4 - T\sin 18.4)\sin 18.4\tan 35 + (207\tan 6 + 0)\sin 18.4 \\
&\quad \cos 18.4 + \sin 18.4(0 + 5.75\tan 35)] \\
&= -[(39.5 - 35.6 - 0.057T) + 8.4] \\
c &= (N_A\tan\delta + C_a)\sin^2\beta\tan\phi \\
&= (207\tan 6 + 0)\sin^2 18.4\tan 35 \\
&= 1.52
\end{aligned}$$

When Eq. (3.22) is set equal to FS = 1.5 this results in the required strength of the internal reinforcement.

$$T = 51.0 \text{ kN/m}$$

Example 6.5 could certainly be modified to account for the tensile strength of the covering geotextile and perhaps other overlying geosynthetics as well. Example 6.6 illustrates this more realistic situation.

Example 6.6

Continue Example 6.5, taking into account that the upper geotextile of the GCL is a nonwoven needle punched fabric with a wide width strength of 16 kN/m. Furthermore, it is overlain by a 1.5 mm thick HDPE geomembrane with a wide width strength of 13 kN/m. Assume that strain compatibility exists and that both the upper geotextile and geomembrane are held firmly in the anchor trench at the crest of the slope.

Solution: The required strength of the internal GCL reinforcement is reduced in direct proportion to the overlying geosynthetics, since they are acting as nonintentional veneer reinforcement. Thus,

$$T = 51.0 - 16 - 13$$
$$= 22 \text{ kN/m}$$

Alternatively, a FS = 1.5 (it could be considered as a reduction factor on the overlying geosynthetics) could be put on both the geotextile and geomembrane, resulting in a balanced factor of safety for each component of the system. Thus,

$$T = 51.0 - \left(\frac{16}{1.5}\right) - \left(\frac{13}{1.5}\right)$$

$$= 31.7 \text{ kN/m}$$

What remains at this point is to calculate the actual internal resisting strengths of the various needle punched and stitch bonded GCLs. This is a very interesting and difficult textile engineering problem, which is being investigated. Until such time as quantified answers are available, long term test plots are providing the confidence needed to use GCLs on relatively steep slopes.

6.4 DESIGN CRITIQUE

When designing with GCLs we must first recognize that a liquid barrier is the focal point of attention. Thus the flow rate for water containment problems and the flow rate, adsorption, and breakout time for solute problems are involved. For water, the situation is reasonable due to bentonite's long history as a waterproofing material. The solute aspects are more difficult due to the complex nature of leachate and its many possible constituents.

Secondly, shear strength considerations (generally the bentonite is considered to be hydrated) are very important when GCLs are placed on side slopes. Direct shear testing (of interfaces and internally) is necessary and site-specific conditions should be simulated in every way possible. Bentonite is a known material to geotechnical engineers and the superposition of geosynthetic considerations (e.g., needle punching or stitch bonding) should not be overwhelming. Limit equilibrium has been illustrated in Section 5.6.10 for multilined slopes and the inclusion of GCLs into the cross section is straightforward. Perhaps the greatest uncertainty in a strength design with respect to side slopes are the long term considerations. Long term direct shear and wide width creep tension tests are both required if the situation warrants this feature. This is clearly the case for landfill closures, but generally *not* the case for landfill liners. This is because solid waste will be placed against the liner system during the filling of the landfill, thus providing a passive and stabilizing force.

Thirdly, we must consider the possibility of puncture. In the geotextile chapter, a puncture analysis was provided (recall Section 2.5.4). Using this model and the puncture resistance of the GCL, a factor of safety could be formulated. However, the calculation may not be relevant. If an object punctures the GCL and the bentonite provides a seal against it, the liquid barrier function might still be adequate (recall Figure 6.7). On the other hand, if we have a GM/GCL composite liner such a puncturing situation could be very significant. This leads directly to the importance of construction methods, that is, CQC and CQA, when using GCLs.

Finally, we must consider that the GCL will hydrate quickly and its bearing capacity against lateral squeezing is quite low (Fox et al. [23]). To avoid this situation, an adequately thick soil fill must be placed over the GCL (and covering geosynthetics, if they are involved) before trafficking the site with construction equipment. This also leads directly to the importance of CQC and CQA when using GCLs.

6.5 CONSTRUCTION METHODS

The contract plans and specifications involving GCLs should be very specific as to their construction details. Panel layouts, as with geomembranes, should be addressed. The orientation of the overlap-seam shingling and the length of overlap must be clearly stated. It is considered good practice in GCL manufacturing to have an overlap line marked on the products for guidance in this regard. If additional bentonite (dry or paste) is to be placed in the overlapped region, it must be stated accordingly and constructed in the recommended manner. Generally, those GCLs with nonwoven needle punched geotextiles on both sides should be treated in this manner.

The manner of placement should also be mentioned. Heavy vehicles should never ride directly on geosynthetics of any type, including GCLs. Even though puncture might not occur, the thinning of the material will play havoc with the flow rate calculations. This is the bearing capacity issue mentioned previously. Once the first geosynthetic of any type, including GCLs, is placed, only lightweight vehicles, such as all-terrain vehicles (ATVs), can be permitted. Figure 6.10 shows teams of laborers deploying a GCL, which typically weighs 1.5 tonnes. Alternatively, low-tire pressure, lightweight equipment can be used.

The premature wetting of GCLs before they are covered or backfilled is generally a problem. The contract documents must be clear as to the disposition of fully hydrated GCLs before covering. In a completely opposite manner to the above, the drying of GCLs has been a problem. If the GCL, at its as-received moisture content of 10 to 20%, dries, it will shrink and a loss of the overlap distance will occur, even to the point of completely separating. The situation may not be noticed if a geomembrane is placed over the GCL and then left exposed to summer sunlight, particularly on side slopes. Temperatures beneath the geomembrane will rise dramatically and if shrinkage occurs, the overlap distance will be reduced or even lost completely since there is essentially no normal stress keeping the adjacent GCLs in position.

There are two aspects to be considered in quality control and assurance: the manufacturing and the field construction. These are referred to as MQC/MQA and CQC/CQA, respectively. (See the definitions in Section 5.11.5.) For the MQC tests per ASTM D5889, see Table 6.5. Also included is the recommended frequency of each test. If the MQA document requires additional tests or greater testing frequency, the manufacturer should be notified accordingly.

For field CQC/CQA, many of the tests in Table 6.5 cannot be readily performed. Sampling is most difficult for GCLs and the seriousness of edge disturbance and redistribution of bentonite is yet to be resolved for the various products. Thus the search is currently for field-oriented test methods (i.e., for conformance tests), that can be done to assure that the intended product is delivered and properly installed. Some detail in this regard is found in Daniel and Koerner [24]. A possible field permeability test has

Figure 6.10 Photographs of GCLs being installed.

been proposed by Didier et al. [25]. A guide for installation is available as ASTM D6102.

In summary, GCLs are indeed viable and true geosynthetic materials. They deserve to be included in a book such as this, and to have a separate chapter devoted to them. Many applications are being developed and implemented on a regular basis, not only in the environmental containment area, but also in transportation areas (highways, airfields, etc.); see Koerner et al. [26]. As with all geosynthetic materials, the GCL market is quite mobile with new styles and products being developed on a regular basis.

TABLE 6.5 MINIMUM TYPES OF TESTS AND THEIR FREQUENCIES FOR THE MQC OF GCLS, ASTM D5889

Test Designation	Test Method	Frequency of Testing	Report Value
Clay:[A]			
Free swell	D5890	One per truck or railcar but min, every 50 tonnes	Minimum average[E]
Fluid loss	D5891	One per truck or railcar but min, every 60 tonnes	Minimum average[E]
Geosynthetic Materials:[B]			
Geotextiles			
Mass per unit area	D5261	20,000 m^2	Typical and MARV[E]
Grab tensile strength (MD and CD)	D4632	20,000 m^2	MARV[E]
Geomembrane[B]			
Mass per unit area	D5261	20,000 m^2	Typical and MARV
Thickness	D5199	20,000 m^2	MARV
Tensile strength at break and yield (MD and CD)	D638	20,000 m^2	MARV
Finished GCL:			
Clay mass per unit area (dried)[C]		4,000 m^2	MARV
Clay moisture constant	D4643	4,000 m^2	Average value[F]
Grab tensile strength (MD and CD)[D]	D4632	20,000 m^2	MARV
Index flux	D5887	Once weekly with the last 20 values reported[G]	Maximum value

[A]These tests on the bentonite are to be performed on the as-received material before fabrication into the GCL product.

[B]Components from finished GCL product should not be separated and tested, because the process may alter the properties of the components.

[C]Dried bentonite should be defined as 0% moisture content.

[D]This test may not be applicable for geomembrane-backed GCLs.

[E]Certification letter from bentonite manufacturer or QA from GCL manufacturer, or both. Certification letters must arrive and be checked before the bentonite is used for the GCL production.

[F]Noncritical value and only for information.

[G]The last 20 values to be reported should end at the production date of the supplied GCL. If the manufacturer has more production facilities or production lines, or both, tests must be performed and reported for each line.

This vitality is considered to be a welcomed asset and will, hopefully, be sustained into the future.

REFERENCES

1. Koerner, R. M., "Perspectives on Geosynthetic Clay Liners," in *Testing and Acceptance Criteria for Geosynthetic Clay Liners, ASTM STP 1308,* ed. Larry W. Well, Philadelphia, PA: ASTM, 1997, pp. 3–20.
2. Heerten, G., von Maubeuge, K., Simpson, M., and Mills, C., "Manufacturing Quality Control of Geosynthetic Clay Liners—A Manufacturers Perspective," *Proc. 6th GRI Seminar, MQC/MQA and CQC/CQA of Geosynthetics,* St. Paul, MN: IFAI, 1993, pp. 86–95.
3. Leisher, P. J., "Hydration and Shear Strength Behavior of Geosynthetic Clay Liners," MSCE Thesis, Drexel University, Philadelphia, PA, 1992.

4. Daniel, D. E., Shan, H.-Y., and Anderson, J. D., "Effects of Partial Wetting on the Performance of the Bentonite Component of a Geosynthetic Clay Liner," *Proc. Geosynthetics '93,* St. Paul, MN: IFAI, 1993, pp. 1483–1496.
5. Daniel, D. E., Bowders, J. J., and Gilbert, R. B., "Laboratory Hydraulic Conductivity Testing of GCLs in Flexible-Wall Permeameters," in *Testing and Acceptance Criteria for Geosynthetic Clay Liners, ASTM STP 1308,* ed. Larry W. Well, Philadelphia, PA: ASTM, 1997, pp. 208–228.
6. Estornell, P., and Daniel, D. E., "Hydraulic Conductivity of Three Geosynthetic Clay Liners," *J. Geotechnical Eng.,* ASCE, Vol. 118, No. 10, 1992, pp. 1592–1606.
7. LaGatta, M. D., "Hydraulic Conductivity Tests on Geosynthetic Clay Liners Subjected to Differential Settlement," MSCE Thesis, University of Texas, Austin, TX, 1992.
8. Wilson-Fahmy, R. G., Koerner, R. M., and Fleck, J. A., "Unconfined and Confined Wide Width Testing of Geosynthetics," *ASTM STP 1190,* ed. S. J. Cheng, Philadelphia, PA: ASTM, 1993, pp. 44–63.
9. Koerner, R. M., Koerner, G. R., and Eberlé, M. A., "Out-of-Plane Tensile Behavior of Geosynthetic Clay Liners," *Geosynthetics Int.,* Vol. 3, No. 2, 1996, pp. 277–296.
10. Daniel, D. E., Trautwein, S. J., and Goswami, P. K., "Measurement of Hydraulic Properties of Geosynthetic Clay Liners Using a Flow Box," in *Testing and Acceptance Criteria for Geosynthetic Clay Liners, ASTM STP 1308,* ed. Larry W. Well, Philadelphia, PA: ASTM, 1997, pp. 196–207.
11. Kraus, J. B., Benson, C. H., Erickson, A. E., and Chamberlain, E. J., "Freeze-Thaw Cycling and Hydraulic Conductivity of Bentonite Barriers," *J. Geotechnical and Geoenvironmental Eng.,* Vol. 123, No. 3, 1997, pp. 229–238.
12. Boardman, B. T., and Daniel, D. E., "Hydraulic Conductivity of Desiccated Geosynthetic Clay Liners," *J. Geotechnical Eng., ASCE,* Vol. 122, No. 3, 1996, pp. 204–208.
13. Koerner, R. M., and Daniel, D. E., "A Suggested Methodology for Assessing the Technical Equivalency of GCLs to CCLs," in *Geosynthetic Clay Liners,* ed. R. M. Koerner, E. Gartung, and H. Zanzinger, Rotterdam: A. A. Balkema, 1995, pp. 73–100.
14. Giroud, J.-P., Badu-Tweneboah, K., and Soderman, K. L., "Comparison of Leachate Flow through Compacted Clay Liners and Geosynthetic Clay Liners in Landfill Liner Systems," *Geosynthetics Int.,* in press.
15. Scheu, C., Johannssen, K., and Soatloff, F., "Nonwoven Bentonite Fabrics—A New Fiber Reinforced Mineral Liner System," *4th Intl. Conf. Geotextiles, Geomembranes and Related Products,* ed. Den Hoedt, Rotterdam: A. A. Balkema, 1990, pp. 467–472.
16. Heerten, G., and List, F., "Rehabilitation of Old Liner Systems in Canals," *4th Intl. Conf. Geotextiles, Geomembranes and Related Products,* ed. Den Hoedt, Rotterdam: A. A. Balkema, 1990, pp. 453–456.
17. Othman, M. A., Bonaparte, R., and Gross, B. A., "Preliminary Results of Study of Composite Liner Field Performance," *J. Geotextiles and Geomembranes,* 1997, to appear.
18. Koerner, R. M., Carson, D. A., Daniel, D. E., and Bonaparte, R., "Current Status of the Cincinnati GCL Test Plots," *J. Geotextiles and Geomembranes,* 1997, to appear.
19. Wilson-Fahmy, R. F., and Koerner, R. M., "Leakage Rates through Holes in Geomembranes Overlying Geosynthetic Clay Liners," *Proc. Geosynthetics '95,* St. Paul, MN: IFAI, 1995, pp. 655–668.
20. Giroud, J.-P., and Bonaparte, R., "Leakage through Liners Constructed with Geomembranes. Part II. Composite Liners," *J. Geotextiles and Geomembranes,* Vol. 8, No. 2, 1989, pp. 71–112.
21. Koerner, R. M., and Daniel, D. E., *Final Covers for Solid Waste Landfills and Abandoned Dumps,* New York: ASCE Press, 1997.
22. Koerner, R. M., and Daniel, D. E., "Better Cover-Ups," in *Civil Engineering,* New York: ASCE, 1992, pp. 55–57.

23. Fox, P. J., DeBattista, D. J., and Chen, S.-H., "A Study of the CBR Bearing Capacity Test for Hydrated Geosynthetic Clay Liners," in *Testing and Acceptance Criteria for Geosynthetic Clay Liners, ASTM STP 1308,* ed. L. W. Well, Philadelphia, PA: ASTM, 1997, pp. 251–264.
24. Daniel, D. E., and Koerner, R. M., "MQC/MQA and CQC/CQA of Waste Containment Liner and Cover Systems," U.S. EPA Technical Resource Document, EPA/600/R-93/182, Cincinnati, OH, 1993.
25. Didier, G., and Cazaux, D., "Field Permeability Measurement of Geosynthetic Clay Liners," *Proc. Geosynthetics: Applications, Design and Construction,* ed. M. B. deGroot, G. den Hoedt and R. J. Tremaat, Rotterdam: A. A. Balkema, 1996, pp. 837–843.
26. Koerner, R. M., Gartung, E., and Zanzinger, H., eds., *Geosynthetic Clay Liners,* Rotterdam: A. A. Balkema, 1995.

PROBLEMS

6.1. Regarding clay mineral soils (from a mineralogical perspective):
- **(a)** Sketch the chemical structure of montmorillonite clay.
- **(b)** Compare this structure to kaolinite and illite clays.
- **(c)** How does bentonite clay relate to the clay soils in parts (a) and (b) and what is its background and past usage?

6.2. Most bentonite clay deposits in Wyoming and North Dakota, U.S., are sodium bentonites. Elsewhere in the world, there are large deposits of calcium bentonite clay.
- **(a)** What is the difference between sodium and calcium bentonite clays?
- **(b)** How does one sodium activate a calcium clay?
- **(c)** What is the permeance of such an activation process?

6.3. Some GCLs are made from bentonite powder and others from bentonite granules. What are the pros and cons of each?

6.4. If the moisture content of an as-received GCL is 18.5% based on measured values $W_w =$ 28.1 g and $W_s = 152$ g, using dry weight (as in standard geotechnical engineering practice), what is its moisture content based on wet weight?

6.5. Describe how anions and cations in the hydrating liquid of a GCL might affect hydration behavior. How do you think this would affect its internal shear behavior?

6.6. The wide width tension behavior of both GCLs shown in Figure 6.6 give little increase due to the effect of lateral confinement. What is the main contributing component in these GCLs that resulted in this lack of response?.

6.7. For the two wide-width tensile response curves shown in Figure 6.6, there is a post-peak response that continues, albeit at lower strength, out to a 50 to 60% strain. What is the main contributing component in these GCLs that resulted in this response?

6.8. Regarding the internal direct shear tests shown in Table 6.1:
- **(a)** The distilled water free swell tests in GCLs 1 and 4 give a zero friction angle. Why is this the case?
- **(b)** In the same tests, GCLs 2 and 3, which are both needle punched products, give higher but different values (10 and 23°, respectively). Why is this the case and what are the possible reasons for the differences in the needle punched products?
- **(c)** Constrained swell tests in this same series give increases of 16, 21, 14, and 19°, respectively (over the free swell friction angle results). Why is this the case?
- **(d)** In comparison to parts (a)–(c), the diesel fuel tests give tremendously high friction angles in both free and constrained swell conditions, 24 to 51°. Why is this the case?

6.9. With respect to the GCL to CCL equivalency summary in Table 6.4, discuss the following issues. Illustrate your logic by using sketches wherever possible.

(a) Intimate contact is product-specific. Why? Which are the preferred GCLs in this situation?

(b) Water breakout time is product-specific. Which is the preferred GCL in this situation?

(c) Slope stability is product-specific. Why?

(d) Erosion potential is product-specific and site-specific. Why?

(e) Bearing capacity is product-specific. What is bearing capacity and why is it important?

(f) Subgrade condition is site-specific. Why?

(g) Why are CQC/CQA procedures significant in such an equivalency comparison?

6.10. The puncture resistance of a GCL in Table 6.4 is noted as being probably not equivalent to a CCL. Why is this the case and why may it not be significant?

6.11. Calculate the water flux ratio of a GCL to CCL, as per Example 6.1 (Section 6.3.1) for total hydraulic heads of 8 m, 4 m (the example), 2 m, 1 m, 0.5 m, 0.30 m (the common regulatory limit), and 0.15 m and graph the result of each calculation.

6.12. Calculate the tensile strain in a GCL as it deforms in an out-of-plane mode (as per Example 6.4 (Section 6.3.3) for deformations of 100, 300, 450 (the sample), and 1000 mm and graph the resulting values. The radius of the depression remains constant at 1.0 m.

6.13. Recalculate the required long term shear strength of the internal reinforcement of a GCL presented in Example 6.5 (Section 6.3.4) for the following cover soil thicknesses: 300, 600, and 900 mm and graph the results along with that for 450 mm (the example).

6.14. What long term internal reinforcement concerns of GCLs might be expressed for

(a) needle punched types

(b) stitch bonded types

(c) nonreinforced types when placed between two geomembranes

6.15. Why should construction equipment be prohibited from traveling directly on a deployed GCL? Is the situation more critical with certain GCLs? What is the effect on a equivalency calculation for those GCLs that may have thinned?

6.16. What is the flow rate (water flux) through an intact GCL having a permeability of 5×10^{-12} m/s under 300 mm of total head difference if it is originally 10 mm thick? What is it if it has been thinned during construction and placement to: 8 mm, then 5 mm, then 2 mm? Plot the resulting response curve.

6.17. What are the implications for a GCL that is deployed and then hydrates before it can be covered?

6.18. The loss of seam overlap in the field after deployment (e.g., by shrinkage) may be a concern in certain unique situations. How can GCL seams be given structural integrity? In other words, list and describe some possible GCL seam joining methods or some newer GCLs that have internal structural stability.

7 Designing with Geopipes

7.0 INTRODUCTION

The traditional materials used for the underground pipeline transmission of water, gas, oil, and various other liquids have been steel, cast iron, concrete, and clay. These pipe materials are classified as *rigid* and are strength-related as far as their material behavior is concerned. Polymers, in the form of plastic pipe, however, are making significant inroads into these markets and they certainly deserve a separate chapter in a book devoted to geosynthetic materials. Pipes made from polymeric materials are classified as *flexible* and are deflection-related as far as their material behavior is concerned. This chapter deals with geopipes, plastic pipe placed beneath the ground surface and subsequently backfilled.

There are a number of polymer resins used in the fabrication of plastic pipe. Currently they are high density polyethylene (HDPE), polyvinyl chloride (PVC), polypropylene (PP), polybutylene (PB), acrylonitrile butadiene styrene (ABS), and cellulose acetate butyrate (CAB). They all entered the market as solid wall constant-thickness pipes of relatively small diameter. For example, every hardware store handles small diameter PVC pipe and fittings for household water and drainage systems for do-it-yourselfers. Today's plastic pipes, however, can be very large in diameter and very thick in their wall dimension (see Figure 7.1). Advanced extrusion and seaming processes have also led to differing types of wall sections, consisting of ribs, cores, and corrugated profiles of a wide variety of cross-sectional shapes and sizes. These pipes are generally referred to as profiled, ribbed, or corrugated wall pipes. Additionally, many applications such as agriculture drains and leachate collection systems require holes, slots, or other types of perforations through the wall section to allow for the inflow of fluid. Geopipe of the types just mentioned are being used at a tremendously increasing rate. They offer the user many significant advantages. They are low in initial cost, are

Figure 7.1 Photographs of plastic pipe used in below ground construction. (Compliments of Phillips Driscopipe, Carlon, and ADS Cos.)

lightweight and easy to install, are easy to join together, have numerous prefabricated appurtenances, have an excellent flow regime, and have excellent durability. A number of informative conference proceedings on plastic pipe are available (see Jeyapalan [1]).

As with all pipe materials, there are a number of potential failure modes of geopipe that must be assessed by the design engineer. They include the following, each of which has had attention drawn to it due to past problems and current concerns.

- Excessive pipe deflection due to improper backfilling, leading to excessive deflection or localized material overstressing
- Seam separation of joined pipe ends due to ring compression stress
- Wall crushing due to high overburden stress
- Wall buckling due to external pressure and/or internal vacuum
- Impact cracking of pipe in extremely cold environments
- Brittle stress cracking of pipe and pipe connections
- Polymer material degradation due to the conveyance of aggressive chemicals

Within this list of concerns, there are three areas in need of a focused discussion: properties, design, and construction. These are the topics in this chapter.

Finally, it should be clearly recognized that plastic pipe is an extremely large and well-developed area with respect to the other types of geosynthetic materials. There are large numbers of books, reports, test methods, manufacturers' guides, and so on, that are available. In particular, note that the Plastic Pipe Institute (PPI) is very active in providing a wealth of information through its own literature and that of its member companies. To make the chapter manageable in size, we will focus only on gravity flow situations involving either profiled or solid wall pipe. This arbitrary selection eliminates water and gas pipelines, which function under positive pressure and are the focus of many other publications.

7.1 GEOPIPE PROPERTIES AND TEST METHODS

Various test methods for evaluating the properties of plastic pipe are available from standardization organizations throughout the world. Among the most active is the ASTM, which has several specific committees on the subject. Within ASTM, the main committees are Committee F-17 on Plastic Piping Systems and Committee D20 on Plastics. Other organizations have plastic pipe standards that are often oriented toward specific application areas. For example, the American Water Works Association (AWWA) has standards oriented toward drinking water pipe conveyance, the Underwriters Laboratory (UL) has standards for cable shielding, the National Sanitation Foundation (NSF) has standards for waste-water pipe conveyance, and AASHTO has standards for highway drainage applications. Also note that PPI has general plastic pipe standards for various applications.

For the sake of uniformity with other chapters in this book, we will subdivide this section on properties and test methods not by application (since geopipes always act in a drainage function), but by the category that best fits the various tests into a design mode: physical, mechanical, chemical, biological, and thermal properties.

7.1.1 Physical Properties

Physical properties have to do with the pipe in the as-received and nonstressed state. These properties are important for proper identification, for comparing one style or product to another, and for conformance and quality assurance purposes.

Wall Thickness. The thickness of smooth wall constant-thickness pipe can be measured according to ASTM D2122. It is a straightforward measurement that uses a caliper accurate to 0.02 mm. The value reported is the minimum wall thickness of the pipe at any cross section. Wall thicknesses and tolerances for HDPE pipe are given in ASTM D2447.

When dealing with profiled pipe (either corrugated or spiral wound) the situation is more difficult. The number of variations in this regard are enormous; see Figure 7.2, which shows some of the most common types. At minimum, the inner liner thickness can be measured. The undulating external surface of the pipe is fabricated for the purpose of stiffening the pipe. The effects of this profiled surface will be clearly evidenced in Section 7.1.2, on mechanical testing.

Diameter. Plastic pipes are generally measured by their outside diameter and the test procedure is detailed in ASTM D2122. The average outside diameter and its tolerance (for average values and for out-of-roundness) is based on the nominal (inside) size of the pipe and is given by "schedule" in ASTM D2447. Note that schedule refers to a pipe sizing system (outside diameters and wall thicknesses) which was originated by the iron pipe industry.

As with the thickness measurements, geopipes of the profiled (corrugated or spiral wound) types present difficulties with diameter measurements, recall Figure 7.2. The inside pipe diameter is the relevant measurement in this regard.

Standard Dimension Ratio. Of importance in comparing uniform-wall pipe diameters and thicknesses to one another is the standard dimension ratio (SDR). It is defined as follows.

$$\text{SDR} = \frac{D}{t} \tag{7.1}$$

where

SDR = standard dimension ratio,
D = outside pipe diameter, and
t = minimum pipe wall thickness.

SDR values have an extremely wide range, varying from a minimum of about 8 to a maximum of 40. As will be seen later, SDR is related to a number of meaningful parameters, including internal-pressure capability and external-strength considerations.

Density. The density or specific gravity of a plastic pipe is dependent upon the base polymer from which it is manufactured. Recognize, however, that there are distinct ranges of density within each polymer type. The range for polymers used to

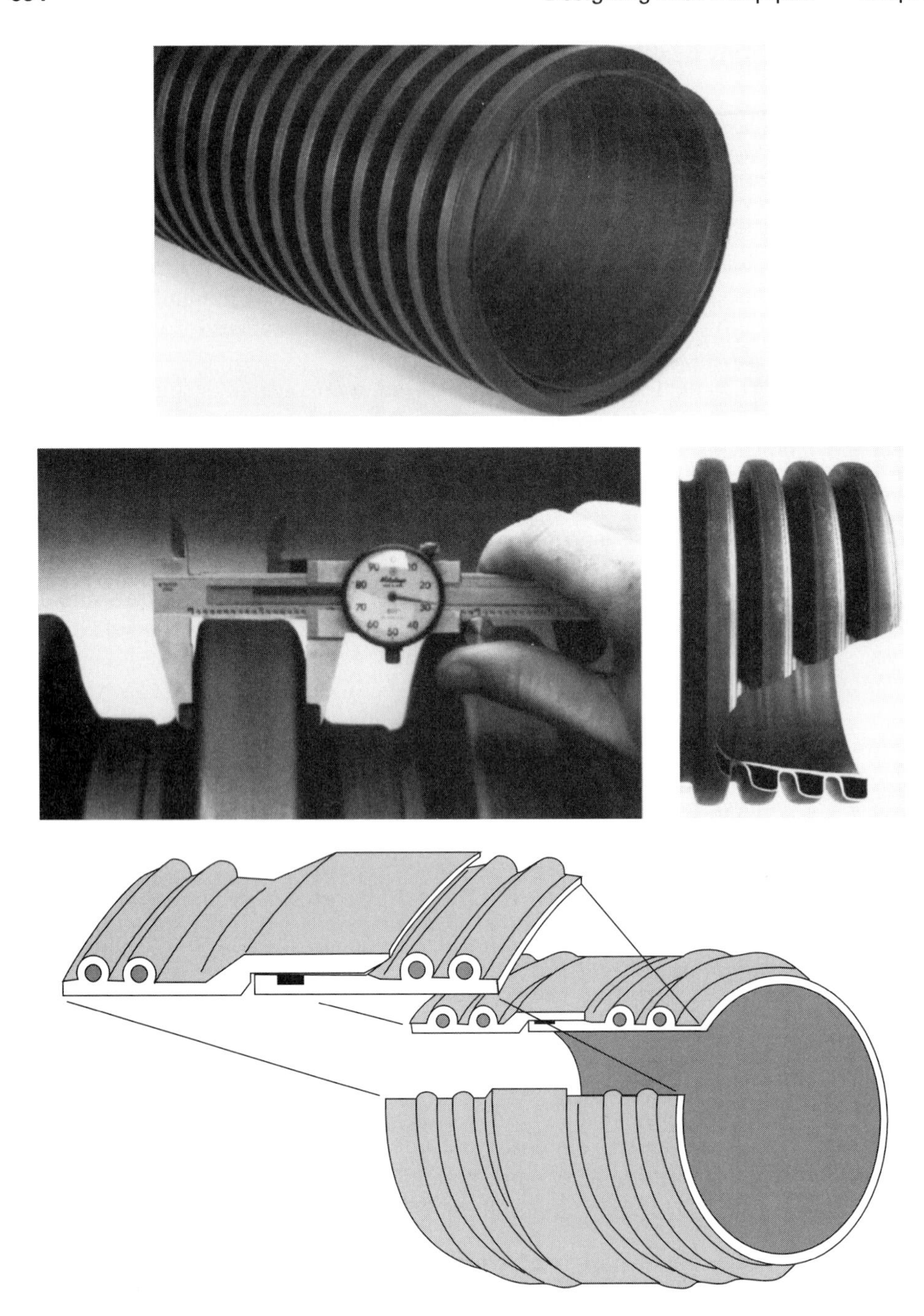

Figure 7.2 Various cross sections of profiled geopipe used for a variety of drainage applications. (Compliments of ADS Co., Spirolite Corp., and Hancor, Inc.)

manufacture plastic pipe falls within the general limits of 0.85 to 1.5. A relevant ASTM test method is D792, based on the fundamental Archimedean principle of the specific gravity as the weight of the object in air divided by its weight in water.

A more accurate method is ASTM D1505. Here a long glass column containing liquid, varying from high density at the bottom to low density at the top, is used. For example, isopropanol and water is often used for measuring densities less than 1.0, while sodium bromide and water is used for densities greater than 1.0. Upon setup, glass spheres of known density are immersed in the column to form a calibration curve. Small pieces of the test specimen polymer weighing about 20 to 30 mg are then dropped into the column after they have been properly cleaned of surface impurities. Their equilibrium position within the column is located and used with a calibration curve to find the unknown density. Accuracy is very good with this test, to within 0.002 g/ml, when proper care is taken; see Section 5.1.2.

Vapor Transmission. In the absolute sense of the word, nothing is impermeable. As such, liquids within a pipe can diffuse through the pipe wall via vapor diffusion and recondense on the other side. This permeation phenomenon is concentration-driven and is generally assumed to obey Fick's law of diffusion. For thermoplastic materials of the type and thickness of geopipes, the value will generally be extremely small. Nevertheless, if water vapor or solvent vapor is of concern, the procedures outlined in Section 5.1.2 can be followed.

Polymer Identification. The identification of the polymer from which the plastic pipe has been manufactured can be made by experience or can sometimes be made by putting a flame to the sample and noting its behavior during and after it burns (recall Table 1.3). For a much more definitive identification, however, various thermochemical tests are required. The most widely used methods are: thermogravimetric analysis (TGA), differential scanning calorimetry (DSC), including oxidative induction time (OIT), thermomechanical analysis (TMA), infrared spectroscopy (IR), gas chromatography (GC), regular or high pressure liquid chromatography (RLC or HPLC), melt index (MI), molecular weight determination (MW), and gel permeation chromatography (GPC). Halse et al. [2] have recently described these methods and have given valuable insight as to the information obtained in conducting the various tests along with the tests' major advantages and disadvantages (recall Table 1.4 and Section 1.2.2).

7.1.2 Mechanical Properties

As we described in the introductory section, some failures of plastic pipe have occurred due to mechanical stressing. Therefore, it is important to review the various test methods and procedures used to obtain geopipe strength and deformation characteristics.

Concentrated Line Load Test. The most common test for determining the compressive stiffness of the plastic pipes is ASTM D2412 "Determination of External Loading Characteristics by Parallel-Plate Loading." A short (150 mm) length of pipe is placed between parallel steel plates and loaded in compression to a given deflection or until failure. To illustrate how failure occurs, consider Figure 7.3a, which shows the

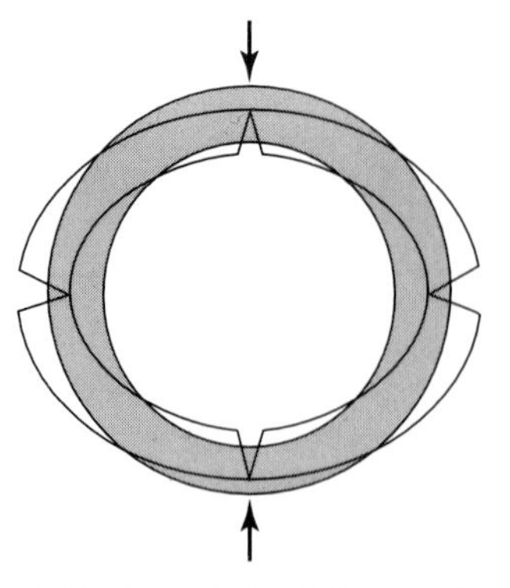

(a) Rigid pipe brittle failure mode

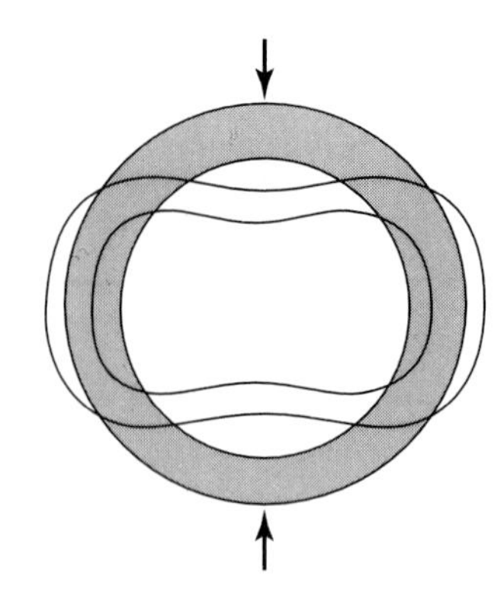

(b) Unsupported flexible pipe buckling failure mode

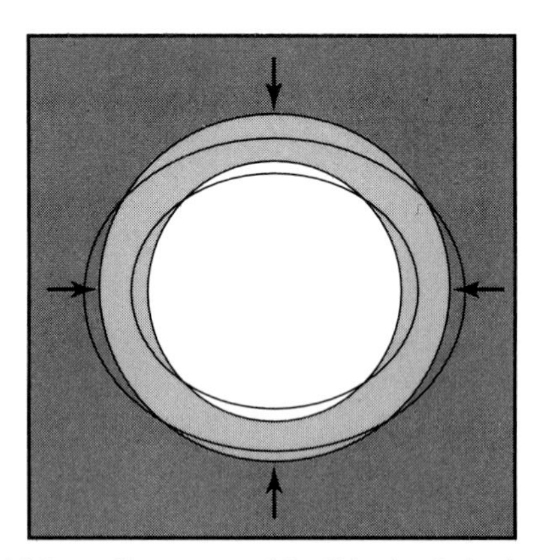

(c) Laterally supported flexible pipe behavior

Figure 7.3 Behavior of rigid-versus-flexible geopipe due to external loads.

failure mode of rigid pipes (e.g., concrete or clay pipes). Note the existence of tension cracks opening on the outside of the pipe, which accompanies contrasting zones of compression on the inside of the pipe. At larger deflection, tension cracks also occur at the top and bottom but now on the inside of the pipe.

Figure 7.3b shows the mode of failure of a flexible pipe (e.g., one made from polymers) in a laterally unconfined state. This is the condition of the pipes when tested in ASTM D2412. Note that the plastic pipe does not crack, but instead deforms and eventually buckles when the deflection becomes excessive. While the entire load-versus-deflection curve is interesting to observe, pipe stiffness is usually calculated at either 5 or 10% deflection. The relevant equations for pipe deflection, pipe stiffness, and stiffness factor calculations according to D2412 are as follows.

$$P = \frac{y}{d}(100) \tag{7.2}$$

where

P = pipe deflection (%),
y = measured change of inside diameter (deflection) (m), and
d = initial inside pipe diameter (m).

$$\text{PS} = \frac{F}{y} \tag{7.3}$$

where

PS = pipe stiffness (kPa),
F = force per unit length of pipe (kN/m), and
y = deflection (m).

$$\text{SF} = 0.149r^3(\text{PS}) \tag{7.4}$$

where

SF = stiffness factor (Pa-m^3), and
r = mid-wall radius (m).

Note that this test is clearly of the index variety because when in service, the pipe will be supported by the backfill soil as shown in Figure 7.3c. The passive pressure exerted by the soil will be seen to be extremely effective and very necessary in the proper functioning of flexible plastic pipes to help support the imposed vertical loads.

The complementary theoretical equivalents to the pipe stiffness terms just mentioned are

$$\text{stiffness factor} \qquad \text{SF} = EI \tag{7.5}$$

$$\text{pipe stiffness} \qquad \text{PS} = 6.71\frac{EI}{r^3} \tag{7.6}$$

$$\text{ring stiffness} \qquad \text{RS} = \frac{EI}{r^3} \quad \left(\text{sometimes } \frac{EI}{D^3}\right) \tag{7.7}$$

where

E = modulus of elasticity (kPa),
I = moment of inertia of pipe cross section per unit length (m^3),
r = mean radius of pipe (m), and
D = mean diameter of pipe (m).

The values of E, I, r, and D for commonly used pipe materials and sizes are found in various pipe manufacturers' literature and technical brochures.

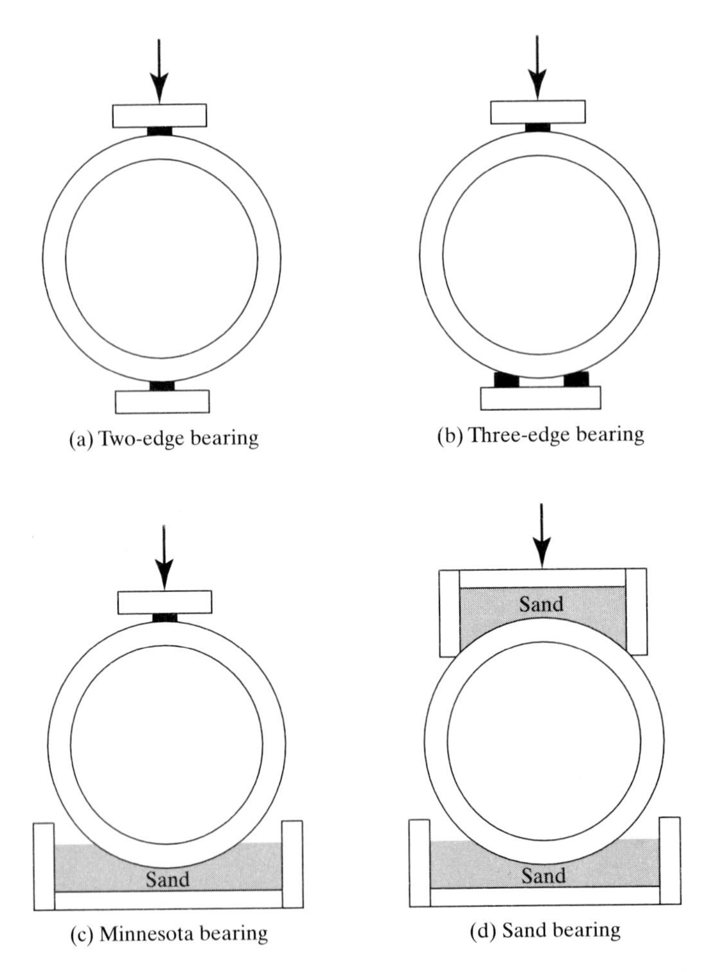

Figure 7.4 Various types of pipe load test configurations and selected results.

Distributed Load Tests. In order to better simulate the effects of soil support beneath the pipe and load spreading above the pipe, several variations of ASTM D2412 have been developed. Figure 7.4a–d illustrates some of these variations, but note that their past use has generally been with rigid pipes and such testing is not common with flexible polymeric pipes. For flexible plastic pipes of the type described in this chapter, the complete lateral confinement of the soil should be considered. Watkins and Reeve [3] have performed tests, as shown in Figure 7.4e. Here the pipe is completely confined in its simulated environment and loaded accordingly. The pressure-versus-deflection response is very different than with nonconfined boundary conditions. Furthermore, there is even a difference in response by virtue of the difference in behavior between the backfill soil at 75 and 85% relative density (see Figure 7.4f). This type of simulated performance test illustrates the importance of uniform backfill around the entire cir-

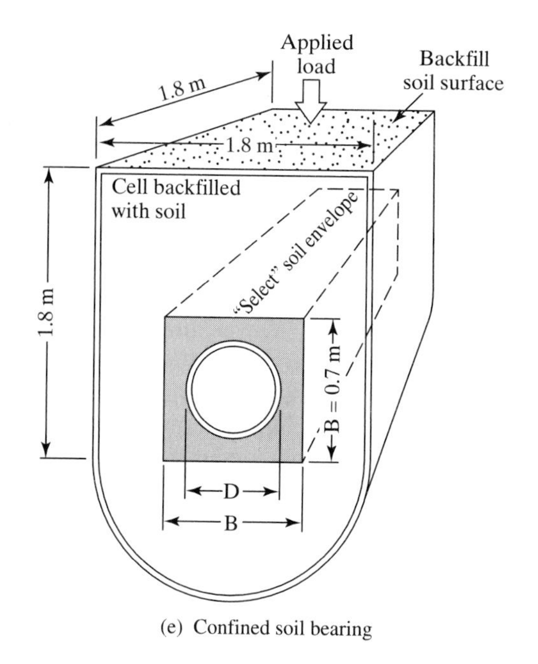

(e) Confined soil bearing

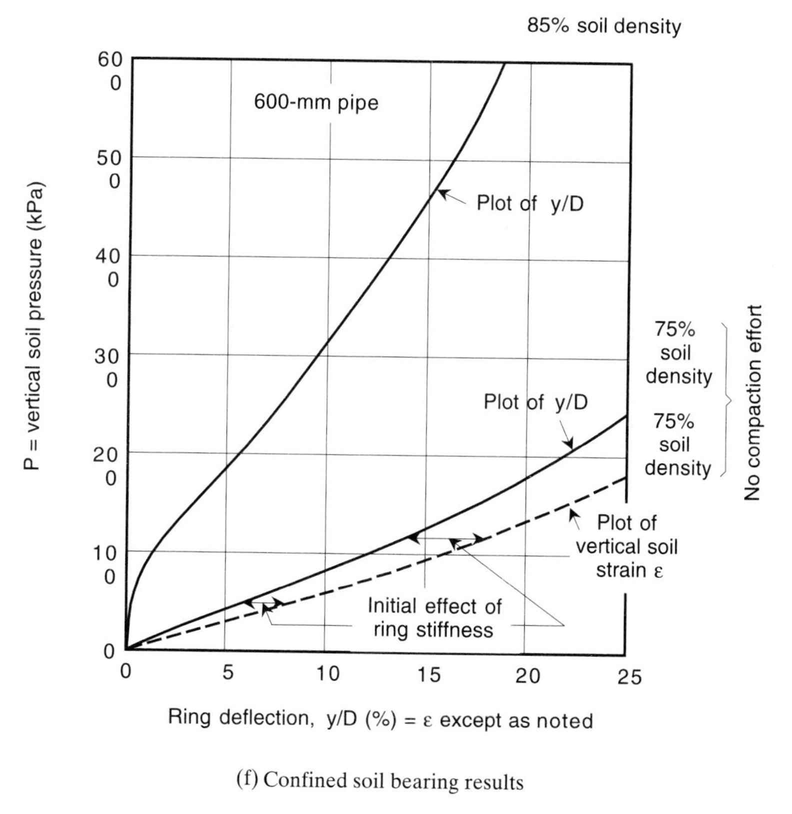

(f) Confined soil bearing results

Figure 7.4 (*continued*) (Parts [e, f] after Watkins and Reeve [3])

cumference of the pipe. Unfortunately, this type of test is not regularly used nor standardized at this time.

Hydrostatic Pressure Test. This test method, ASTM D1598, consists of subjecting samples of thermoplastic pipe or reinforced thermosetting-resin pipe to a constant internal pressure until loss of pressure, rupture, or ballooning of the pipe occurs. The data obtained from the test is subsequently used to establish stress-versus-failure time relationships, from which the hydrostatic design stress can be obtained (recall Figure 5.15).

The test method consists of taking capped sections of pipe that have lengths that are five times the diameter (for pipes less than or equal to 150 mm in diameter) or three times the diameter (for pipes greater than 150 mm in diameter), and immersing them in a water bath at a constant temperature of 23°C, or as otherwise agreed upon. The pipes are filled with water and pressurized according to agreed upon sequences; either constantly increasing the stress rate or maintaining the stress over time in a creep mode are the two major options.

The hoop stress in the pipe is calculated as follows.

$$\frac{2S}{P} = \frac{D}{t} - 1 \tag{7.8}$$

where

S = hydrostatic design stress (MPa),
P = internal pressure rating which includes a factor of safety (MPa),
D = average outside diameter (mm), and
t = minimum wall thickness (mm).

Using thermoplastic pipe materials conforming to ASTM D2513 or D2517 and analyzing the above formula in accordance with D2837 gives the values in Table 7.1 for long-term hydrostatic strength. Note that this test is not used for gravity flow situations but is included to illustrate the capability of plastic pipe in pressurized situations.

Sustained Load (Creep) Resistance. Any one of the mechanical tests just described can be modified to sustain a constant load and thus the pipe can be forced into

TABLE 7.1 LONG-TERM HYDROSTATIC PIPE STRENGTH, EQ. (7.8)

Plastic Pipe Material	Long-Term Hydrostatic Strength (MPa)
PE-Type II, Grade 3 (PE 2306)	8.6
PE-Type III, Grade 3 (PE 3306)	8.6
PVC-Type I, Grade 1 and 2 (PVC-1120, PVC-1120)	27.6
PVC-Type II, Grade 1 (PVC-2110)	13.8
ABS-Type I, Grade 2 (ABS-1210)	13.8
CAB-MH (CAB-MH08)	11.0
CAB-S (CAB-S004)	5.5

a creep mode. While the external load test methods of Figure 7.4a–d are not overly informative due to their nonfield-representative boundary conditions, the uniform pressure test of Figure 7.4e in the creep mode is very informative. Some long-term field tests will be described in Section 7.2.2.

Stress Crack Resistance. Since the plastics used for many types of flexible pipes are semicrystalline (e.g., polyethylene and polypropylene) there is a propensity for a brittle type failure, which is called *environmental stress cracking* or simply *stress cracking*. The tests used to evaluate this phenomenon are the bent strip test covered in Section 5.1.3 (which is not recommended), the NCTL test covered in Section 5.1.3 (which is the preferred performance test method) and the SP-NCTL test covered in Section 5.1.3 (which is the preferred quality control test method). These tests are described in Chapter 5 for geomembranes, where the material comes in a flat sheet that can be used directly. For example, a dumbbell-shaped specimen can be cut directly from the geomembrane sheet for its subsequent testing. For pipe, this type of sampling is not possible. The thickness, curvature, and surface detail (like corrugations) cannot be easily accommodated. As a result it is necessary to melt incoming resin pellets or chips from the manufactured pipe in a compression molding press to fabricate a flat, uniform thickness sheet or *plaque* that, when properly cooled and removed from the press, can be cut into the required test specimen size and shape. The ASTM test protocol for forming these sheets is D1928, "Preparation of Compression-Molded Polyethylene Test Sheets and Test Samples."

Hsuan [4] has performed SP-NCTL tests per ASTM D5397 Appendix on a number of HDPE pipe resins and chips. The results are presented in Table 7.2 for loadings at 10 and 20% of yield stress. The data show that the time for failure to occur in a brittle mode is significantly different for the various materials evaluated. While the significance of this data with respect to stress crack performance in the field is uncertain, the results clearly show that a variety of materials are currently being used for HDPE geopipe.

Fatigue Resistance. For plastic pipe subjected to cyclic loading, fatigue failure of the pipe or of its joints might be of concern. There is no standardized test, but a modification of ASTM D2412, the parallel plate loading test, seems readily adaptable to dynamic loading. Obviously, the frequency, amplitude, wave form, and so on must be mutually agreed upon by the parties involved. It might also be advisable to use a lower containment detail, as shown in Figure 7.5b, to guarantee stability of the test specimen during the test. Optimally, we could suitably modify the confined soil test setup shown in Figure 7.4e.

Impact Resistance. There have been known brittle failures of plastic pipe due to impact loads, resulting in shattered pipe. In addition to the subsequent nonperformance of the pipe, if a geomembrane is in close proximity to the shattered pipe, a puncture could easily occur. The situation of impact failures is greatly aggravated in cold weather. Note that all polymers are subject to such behavior to widely varying degrees.

A number of different test methods can be used to determine impact resistance. For compression molded test specimens of the pipe material we could use an Izod-type

TABLE 7.2 SINGLE POINT-NOTCHED CONSTANT TENSION LOAD (SP-NCTL) TEST DATA ON RESINS USED FOR GEOPIPE

Resin Code	Resin Description	Failure Time at 20% Yield Stress	Failure Time at 10% Yield Stress
Virgin Resins			
1	HDPE-A	1.8	16
2	HDPE-B	7.3	80
3	MDPE-A	3000	—
5	HDPE-C	553	4000*
6	HDPE-D	1.4	14
7	HDPE-E		35
8	HDPE-F		25
9	HDPE-G		38
10	HDPE-H		72
11	HDPE-I		21
12	100% rework HDPE-J		850*
Blended Resins with Post-Consumer Resin (PCR)			
1	HDPE-K with PCR (I)	0.8	3.4
2	HDPE-L with PCR (II)	0.9	6
3	HDPE-M with 25% PCR (III)	3.3	41
4	HDPE-N with PCR (IV)		260*
5	HDPE-O with PCR (V)		1400*
6	HDPE-J with 50% PCR (VI)		960*
7	HDPE-C with 25% PCR (VII)		930*
8	HDPE-C with 50% PCR (VII)		263
9	HDPE-C with 75% PCR (VII)		75
12	HDPE-H with 25% PCR (VII)		55
13	HDPE-H with 50% PCR (VII)		34
14	HDPE-H with 75% PCR (VII)		24
16	100 PCR (VII)		21
17	HDPE-P with 25% PCR (VIII)		27
18	HDPE-P with 50% PCR (VIII)		24
19	HDPE-P with 75% PCR (VIII)		16

*Data obtained by extrapolation.

Source: After Hsuan [4].

falling-pendulum device, in which the specimen is held as a vertical cantilever beam, or a Charpy-type falling-pendulum device, in which the specimen is supported as a horizontal simple beam. These methods are treated in ASTM D256, "Impact Resistance of Plastics and Electrical Insulating Materials."

For plastic pipe in its actual configuration, however, ASTM D2444, "Impact Resistance of Thermoplastic Pipe and Fittings by Means of a Tup (Falling Weight)," is usually used. Here a test specimen is supported on a V-block or on a flat plate with a seating groove in it. The pipe specimen is usually of a length equal to its diameter but not less than 150 mm. Temperature control is important, and the procedure usually calls for the pipe to be maintained at a subfreezing temperature and then rapidly removed from the chamber and tested. A *tup* (that is, a falling weight) of 2.5, 5, 10, or 15 kg is then dropped within a tube of sufficient length to cause impact failure. The tube is approximately 4 m tall and must be aligned so that chattering of the falling tup does not occur, or, at least, is minimized. The tup nose detail is important and the test method allows for three different types: one type is tapered to a 13 mm radius, another is hemispherical with a 50 mm radius, and the third has a protruding tip of 6 mm at its end.

Twenty test specimens are initially required to determine the proper combination of the above variables to cause the pipe to fail and then up to 150 test specimens are required to generate the required data set. With a set of failures and nonfailures, the two percentages are plotted on probability graph paper for the desired value of impact resistance of the pipe.

Abrasion Resistance. The *external* abrasion of geopipes is not considered to be of concern for most situations of static loads on or within stable backfill soils. For exposed pipes or water-submerged pipes, problems could certainly exist but are beyond the scope of this chapter.

The *internal* abrasion of geopipes, however, does occur. Slurry pipelines, used to convey solid particulates like dredged soil materials, are applications where moving particles will cause abrasion. Haas and Smith [5] have performed erosion studies that compare the abrasion performance of several types of plastic pipe with steel and aluminum pipe. The test setup consisted of a closed loop of test pipe with a sand slurry continuously circulated by a pump. Four sets of data were generated, using two silica sand gradations, coarse ($d_{50} = 0.58$ mm) and fine (with $d_{50} = 0.31$ mm) at 40% by weight in a water slurry; each at two velocities, 2.1 and 4.6 m/s. The annual wear rates were measured in terms of loss of thickness and are given in Table 7.3. Overall, the results indicate the following:

- The wear rates vary considerably, from about 0.1 to 4 mm per year under continuous flow of abrasive slurry.
- Polyethylene pipe has good abrasion resistance compared to the other materials tested, including other types of polymers.

The above study indicates that the plastics used for the pipe tested are comparable to or better than metals in resistance to abrasion by sand slurries. PVC, while not covered

TABLE 7.3 WEAR RATE OF PLASTICS AND METALS UNDER ABRASIVE SLURRIES

Pipe Material	Wear Rate (mm/year)			
	Coarse Sand		Fine Sand	
	2.1 m/s	4.6 m/s	2.1 m/s	4.6 m/s
Steel	0.65	1.81	0.04	0.02
Aluminum	1.81	7.48	0.14	0.86
Polyethylene	0.06	0.46	nil	0.06
ABS	0.36	2.07	0.07	0.51
Acrylic	0.99	4.10	0.17	1.42

Source: After Haas and Smith [5].

in the reported tests, has modulus, yield strength, and ultimate strength (the significant parameters in abrasion resistance) in the range of ABS and acrylic. Hence, the abrasion resistance of PVC should show a similar resistance as that of the ABS and acrylic pipe resins tested [5].

Overall, however, abrasion test data should not be interpreted quantitatively without detailed consideration. This study is based on sand slurries circulating at reasonably high velocities. In situations where larger aggregates are transported, abrasion or scour would be expected to be more severe for both metal and plastic pipe.

Regarding a laboratory index test for abrasion we could consider using ASTM D1175. This test method covers six abrasion methods directed at geotextiles placed in a flat configuration (recall Section 2.3.5). Pipe materials could be evaluated by any one of these techniques but only after they were recast into plaques by the compressive molding techniques discussed above.

Connection Tests. As with most geosynthetic materials, the connections between separate geopipe sections can be the weak link of the completed pipe system. The various joining techniques are described in Section 7.5.2. There are no specific seam strength tests of a formalized nature; however, many of the tests described in this section on mechanical testing can be modified to include a joined section. If the pipe is solid wall the adaptation is straightforward, since plastic pipes are often butt-joined in a thermal fusion procedure. If the pipe is profiled, however, joining is usually accomplished by a prefabricated fitting or coupling. In this case a longitudinal tension or large scale three-point bending test might be considered. Future research efforts on pipe connections appears to be warranted. It should be noted that there is an ASTM test for joint watertightness, D 3212, for push-on or mechanical-type joints. In this test either internal pressure or a vacuum on the capped sections of pipe is used with the connections in the center of the test section.

7.1.3 Chemical Properties

Geopipes are often called upon to convey aggressive chemicals or are immersed in chemicals. Both within and without, the chemical resistance of the pipe must be assured. A number of different mechanisms might be considered.

Swelling Resistance. The test for liquid absorption and the monitoring of the amount of swelling of a plastic material is a standard test (ASTM D570) and is covered in Section 5.1.4. The test must be modified for a pipe configuration or be conducted on compression-molded samples of the polymer from which the pipe is manufactured.

Chemical Resistance. Most manufacturers have a good database on the chemical resistance of their pipe materials to commonly conveyed liquids. Heterogeneous liquids, like landfill leachates, however, can be somewhat problematic. As with the geomembrane chemical resistance tests described in Section 5.1.4, sections of the pipe must be incubated, periodically removed, and then tested to evaluate whether changes have occurred when compared to the as-received material. There is no established test procedure for geopipe. Even the incubation of full cross sections of pipe is problematic because large containers must be used to accommodate the required number of pipe sections. For example, if a 150 mm diameter pipe is being evaluated in D2412 concentrated line load tests, the pipe lengths must be 450 mm long and five replicates at 30, 60, 90, and 120 day incubation times would require a container of approximately 600 l. When handling this much leachate, problems are sure to arise and testing costs will certainly rise accordingly.

To alleviate the problem of testing laboratory containers with large amounts of leachate we might consider cutting 90 or 120° sections of the candidate pipe and immersing these in a stacked configuration [6]. The stacking and incubation container would appear as shown in Figure 7.5a. In such a case, as little as 75 l of leachate would be required. The load-versus-deflection test would be conducted in a steel channel section as shown in Figure 7.5b. Note that careful machining of the supporting ends of the pipe is important to alleviate stress concentrations as much as possible. Preliminary data from this type of laterally confined test match remarkably well with those from geopipe tests under full soil confinement. Work is ongoing in this regard.

Ultraviolet Light Resistance. All plastics will suffer from ultraviolet degradation if left exposed to sunlight for long enough. There are a number of outdoor tests

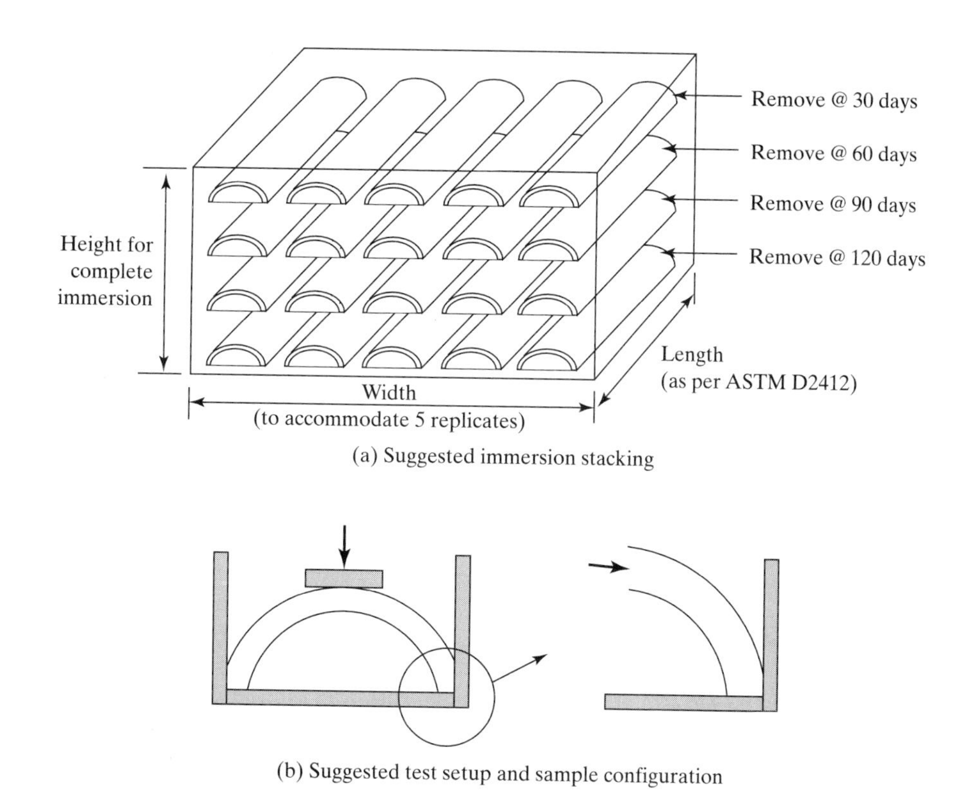

(a) Suggested immersion stacking

(b) Suggested test setup and sample configuration

Figure 7.5 Suggested incubation setup and testing configuration for chemical resistance evaluation of geopipes.

that are available, as well as accelerated laboratory tests. Most notable in this latter group are the following ASTM tests: G23, "Operating Light-Exposure Apparatus (Carbon-Arc Type) With and Without Water for Exposure of Nonmetallic Materials," D4355, "Deterioration of Geotextiles from Exposure to Ultraviolet Light and Water (Xenon-Arc Type Apparatus)" (note that this method is similar to G26, which is also a Xenon-Arc Test Method but uses a prescribed set of exposure conditions), and G53, "Operating Light- and Water-Exposure Apparatus (Fluorescent UV-Condensation Type) for Exposure of Nonmetallic Materials." However, the susceptibility of geopipe to ultraviolet light with respect to other geosynthetics is relatively low. This is based upon the low surface area of geopipes and the usually high wall thickness in comparison to geotextiles, geogrids, and geomembranes. Thus, with timely cover, geopipes (with properly formulated compounding materials including antioxidants and carbon black) should not be of concern as far as ultraviolet degradation is concerned.

7.1.4 Biological Properties

The resistance of plastic pipe to animals and to smaller forms of life (like fungi and bacteria) might be considered. Section 5.1.4 covers these topics; under usual conditions, high molecular weight polymers are not felt to be a source of great concern as to their biological degradation. Furthermore, wall thicknesses are sufficiently great that burrowing animals should not be of general concern.

7.1.5 Thermal Properties

The warm-weather and cold-weather behavior of geopipes can be addressed by simulated laboratory testing. Usually the material in question is incubated and then tested on the basis of the most vulnerable property. Section 5.1.4 covers the ASTM test procedures for the warm and cold behavior of plastics. In general, it will be necessary to compression mold flat test samples and, as such, it is the polymer compound that is being evaluated and not the final structure (recall Section 7.1.2).

As with geomembranes, the coefficient of thermal expansion is important in geopipes. If temperatures are allowed to fluctuate, the pipe will go into either compression (for cold temperatures) or expansion (for warm temperatures) (recall Table 5.10). In the case of compression, tensile stress buildup at the connections may be of concern. For expansion, side-wall pressures and the accommodation of the extra length may be of concern.

7.1.6 Geopipe Specifications

With the variety of material properties that are available to geopipe manufacturers, even within a specific resin type, it is only natural that some standardization be attempted. This is the case with HDPE pipe, which has been put in *cell* classifications in accordance with the resin's density (see Table 7.4). As expected, the properties are interrelated in that higher density produces: a lower melt index, a higher flexural modulus, a higher yield strength, and a higher hydrostatic design-basis strength.

TABLE 7.4 PRIMARY PROPERTIES: CELL CLASSIFICATION LIMITS FOR POLYETHYLENE DRAINAGE PIPE MATERIALS, ASTM D3350

Property	Test Method	Cell Limits 0	1	2	3	4	5	6
Density (mg/l)	D1505	—	0.910 to 0.925	0.926 to 0.940	0.941 to 0.955	> 0.955		
Melt index (g/10 min)	D1238	—	> 1.0	1.0 to 0.4	< 0.4 to 0.15	< 0.15		
				138	276	551	758	
Flexural modulus (MPa)	D790	—	< 138	138 to 276	276 to < 551	551 to < 758	758 to < 1100	> 1100
Tensile strength at yield (MPa)	D638	—	15.1	15.1 to < 17.9	17.9 to < 20.7	20.7 to < 24.1	24.1 to < 27.6	> 27.6
Environmental stress crack resistance	D1693	—						
Test duration, condition			A	B	C			
Test duration (h)			48	24	192			
Failure, max (%)			50	50	20			
Hydrostatic design basis (MPa) @ 23°C	D2837	NPR	5.5	6.9	8.6	11.0		

Materials with melt index less than cell 4 but which have flow rate < 4.0 g/10 min when tested in accordance with D1238, Condition F.

Materials with melt index less than cell 4 but which have flow rate < 0.30 g/10 min when tested in accordance with D1238, but at 310°C with total load of 12,480 g.

NPR = Nonpressure-rated.

Note: AASHTO M294 requires a classification of "33542" for profiled HDPE pipe. This means density is cell 3; melt index is cell 3; modulus is cell 5; strength is cell 4; and ESCR is cell 2.

TABLE 7.5 CLASS REQUIREMENTS FOR RIGID PVC DRAINAGE PIPE COMPOUNDS, ASTM D1784

	Cell Limits								
Property and Unit	0	1	2	3	4	5	6	7	8
Base resin	unspecified	PVC homo-polymer	chlorinated PVC	vinyl copolymer					
Impact strength (Izod), min (J/m of notch)	unspecified	< 35	35	80	270	535	800		
Tensile strength, min (MPa)	unspecified	< 35	35	40	50	55			
Modulus of elasticity in tension, min (MPa)	unspecified	< 1900	1900	2200	2500	2750	3000		
Deflection temperature under load, min 1.82 MPa (°C)	unspecified	< 55	55	60	70	80	90	100	110
Flammability	A	A	A	A	A	A	A	A	A

A = All compounds covered by this specification when tested in accordance with Method D635 shall yield the following results: average extent of burning < 25 mm; and average time of burning of < 10 s.

Also note the seemingly uncertain position regarding stress crack resistance. The issue is controversial. It is generally accepted that ASTM D1693, the bent strip test, is not challenging (and hence unacceptable) in its assessment of stress crack resistance. Furthermore, quantification insofar as the relative ranking of resins is not possible. As a result, the current tendency is to use ASTM D5397, the NCTL test, for an assessment of stress crack resistance. The test is covered in detail in Section 5.1.3 and briefly in Section 7.1.2, with pipe data in Table 7.2. If we are to use this test for a specification the issue under contention is what minimum value to select. The decision will ultimately be made on the basis of forensic evaluations of field-retrieved stress crack failures, and work is ongoing in this direction.

The requirements for PVC resin and its related properties are also available in specification form via ASTM D1784. Table 7.5 presents the minimum values that determine the relevant cell number for use in different applications.

7.2 THEORETICAL CONCEPTS

Before considering design problems, some consideration must be given to selected theoretical concepts. These concepts have to do with the hydraulic considerations of pipes flowing full and pipes that are only partially full. Also in this section are selected concepts and formulas associated with the stresses imposed by soil backfill on the pipe during its service lifetime. For a greater treatment of the topics there are numerous textbooks available (see, for example, [7, 8, 9, 10, 11]).

7.2.1 Hydraulic Issues

For pipelines that are flowing full, the *Hazen-Williams formula* is often used to describe flow behavior. It is empirical and is only used for pipes greater than 50 mm in diameter and velocities less than 3 m/s.

$$V = 0.85CR_H^{0.63}S^{0.54} \tag{7.9}$$

where

V = velocity of flow (m/s),
R_H = hydraulic radius (m),
= flow area/wetted perimeter = $(\pi D^2/4)/\pi D = D/4$,
D = internal diameter of pipe (m),
S = slope or gradient of pipeline (m/m), and
C = a constant which depends on the pipe material and the condition of the inner surfaces (see Table 7.6) (dimensionless).

TABLE 7.6 HAZEN-WILLIAMS COEFFICIENT FOR DIFFERENT TYPES OF PIPE

Pipe Materials	C
Asbestos Cement	140
Brass	130–140
Brick sewer	100
Cast-iron	
New, unlined	130
10 yr. old	107–113
20 yr. old	89–100
30 yr. old	75–90
40 yr. old	64–83
Concrete or concrete lined	
Steel forms	140
Wooden forms	120
Centrifugally spun	135
Copper	130–140
Galvanized iron	120
Glass	140
Lead	130–140
Plastic*	140–150
Steel	
Coal-tar enamel lined	145–150
New unlined	140–150
Riveted	110
Tin	130
Vitrified clay (good condition)	110–140
Wood stave (average condition)	120

*The table does not distinguish between different types of plastic, or between smooth wall, profiled wall, or pipes with perforations.

Source: After Hwang [7].

The above formula can be readily converted to flow rate, Q, by multiplying the velocity by the cross-sectional area A of the pipe.

For pipelines that are either flowing full or flowing partially full, the *Manning equation* is generally used.

$$V = \frac{1}{n} R_H^{0.66} S^{0.5} \tag{7.10}$$

where

V = velocity of flow (m/s),
R_H = hydraulic radius (m),
S = slope or gradient of pipeline (m/m), and
n = coefficient of roughness (see Table 7.7) (dimensionless).

Note that plastic pipe of the type discussed in this chapter, with a *smooth interior,* has a Manning coefficient from 0.009 to 0.010. Plastic pipe with a *profiled or corrugated interior* has a Manning coefficient ranging from 0.018 to 0.025.

Eqs. (7.9) and (7.10) are generally used in the form of charts or nomographs to determine pipe sizes, flow velocity or discharge flow rates (see Figures 7.6 and 7.7). For each chart we include an example from Hwang [7], illustrated on the respective nomographs by heavy lines. Note that both nomographs are for pipes flowing full.

Example 7.1

A 100 m long pipe with $D = 200$ mm and $C = 120$ carries a discharge of 30 l/s. Determine the head loss in the pipe. (See the Hazen-Williams chart in Figure 7.6.)

Solution: Applying the conditions given to the solution chart in Figure 7.6, the energy gradient is obtained.

$$S = 0.0058 \text{ m/m}$$

TABLE 7.7 VALUES OF MANNING ROUGHNESS COEFFICIENT, *N*, FOR REPRESENTATIVE SURFACES

Type of Pipe Surface	Representative *n* value
Lucite, glass, or plastic*	0.010
Wood or finished concrete	0.013
Unfinished concrete, well-laid brickwork, concrete or cast iron pipe	0.015
Riveted or spiral steel pipe	0.017
Smooth, uniform earth channel	0.022
Corrugated flumes, typical canals, river free from large stones and heavy weeds	0.025
Canals and rivers with many stones and weeds	0.035

*The table does not distinguish between different types of plastic, or between smooth wall, profiled wall, or pipes with perforations.

Source: After Fox and McDonald [9].

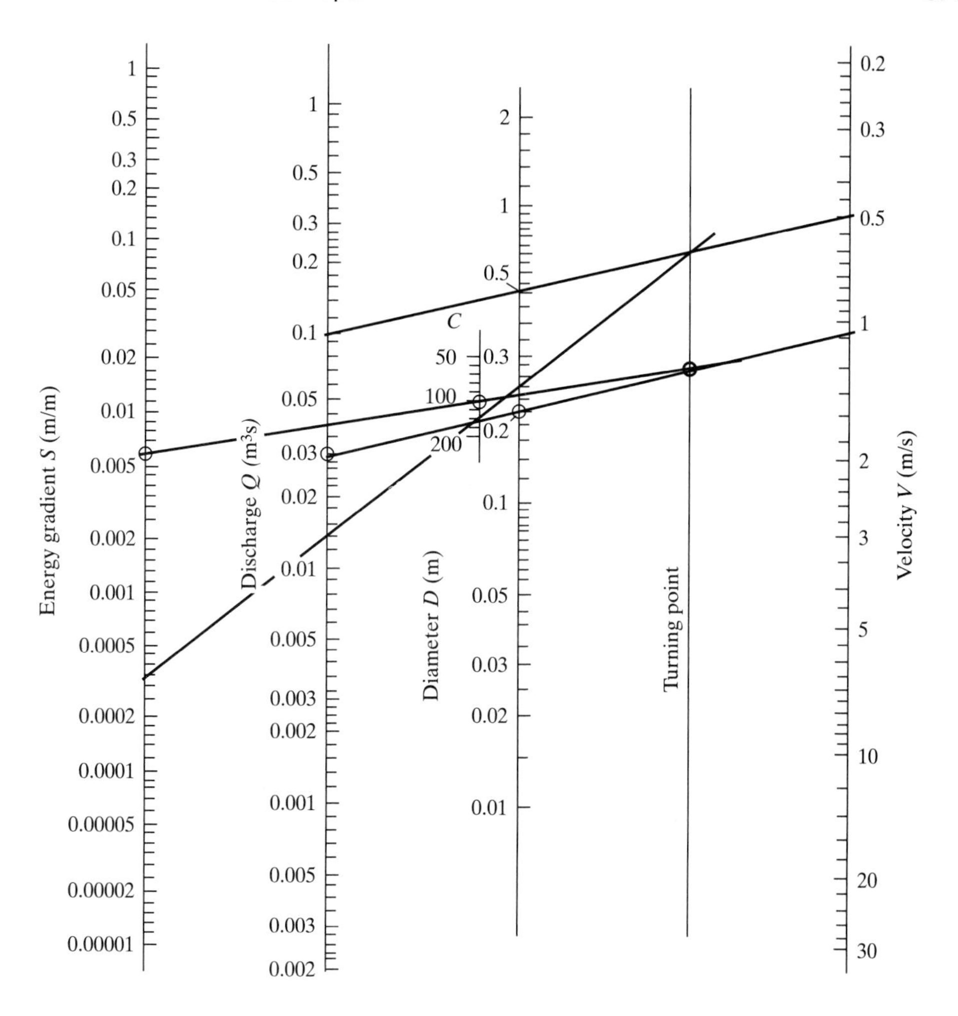

Figure 7.6 Solution nomograph for Hazen-Williams formula.

The energy gradient is defined as the energy loss per unit pipe length or

$$S = \frac{h_f}{L} = \frac{h_f}{100} = 0.0058 \text{ m/m}$$

Hence,

$$h_f = 0.0058 \text{ m/m} \cdot 100 \text{ m} = 0.58 \text{ m}$$

Example 7.2

A horizontal pipe with 100 mm uniform diameter is 200 m long. The Manning roughness coefficient is $n = 0.015$ and the measured drop is 24.6 m in a water column. Determine the discharge. (See Manning chart in Figure 7.7.)

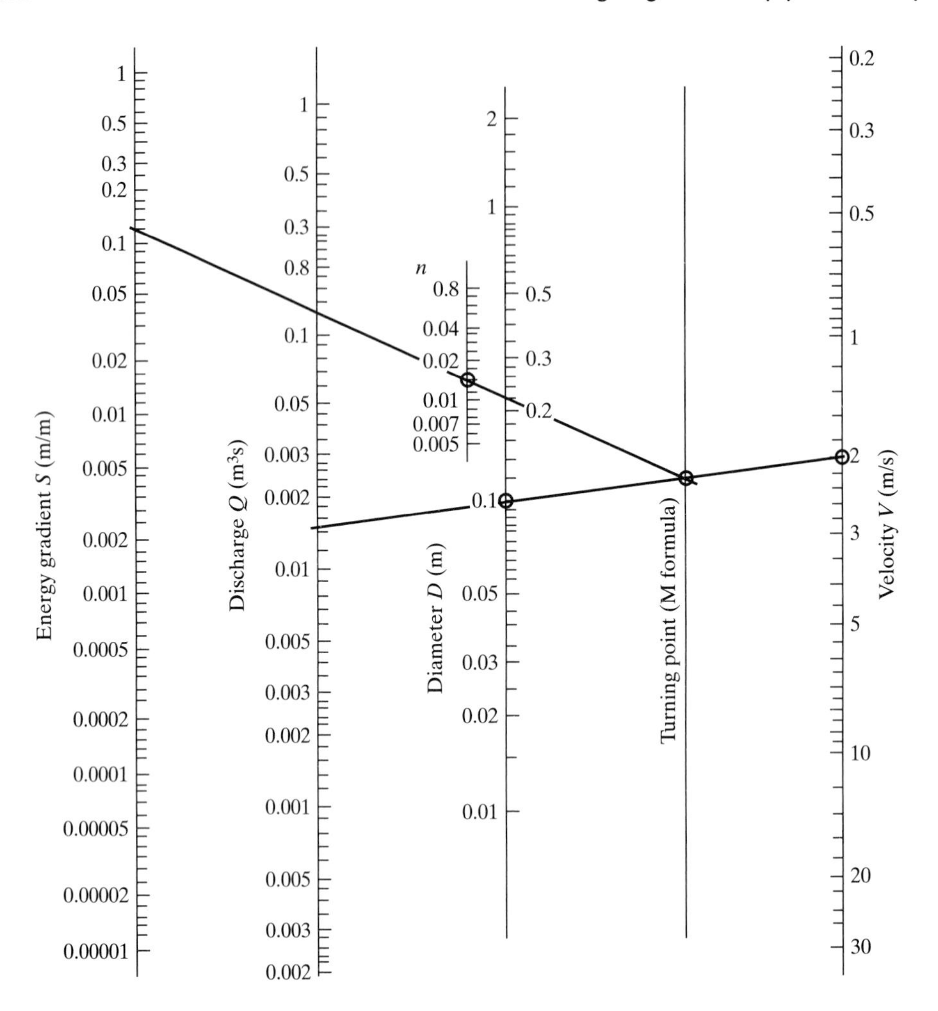

Figure 7.7 Solution nomograph for the Manning formula.

Solution: The energy gradient in the pipe is

$$S = \frac{h_f}{L} = \frac{24.6}{200} = 0.123$$

Applying this value and the conditions given above to the solution chart in Figure 7.7, we obtain the following discharge.

$$Q = 0.015 \text{ m}^3/\text{s}$$

Note that a pipe that is flowing slightly less than full can carry more liquid than it can when it is completely full. This is due to the reduced friction against the inner pipe wall due to the liquid's lesser contact with the wall's surface. However, when the pipe fullness becomes significantly lower, the carrying capacity decreases until it eventually

TABLE 7.8 LIQUID CARRYING CAPACITY OF PARTIALLY FULL PIPELINES

Pipe Fullness (%)	Flow Rate (% of full)	Flow Velocity (% of full)
100	100	100
95	106.3	111
90	107.3	115
80	98	116
70	84	114
60	67	108
50	50	100
40	33	88
30	19	72
20	9	56
10	3	36
0	0	0

Source: After Driscopipe [11].

reaches zero. Table 7.8 gives the resulting values of Q and V for varying pipe fullness percentages.

For many of the examples of gravity drainage using perforated plastic pipe that are discussed in Section 7.3, the partially full condition is the typical case. The modified Manning equations are as follows [11].

$$Q = 1.137AR_H^{0.66}S^{0.5} \tag{7.11}$$

$$V = \frac{Q}{A} \tag{7.12}$$

where

Q = flow rate (m^3/s),
V = velocity of flow (m/s),
A = flow area cross section (m^2),
R_H = hydraulic radius (mm) = [flow area cross section divided by the wetted perimeter], and
S = slope or gradient (m/m).

Another relationship used in the analysis of pipeline flow is the Darcy-Weisbach equation for head loss.

$$h_L = (f)\left(\frac{LV^2}{2gD}\right) \tag{7.13}$$

where

h_L = head loss (m),
f = friction factor (dimensionless),
L = pipe length (m),

D = pipe diameter (m), and
V = flow velocity (m/s).

For laminar flow, Eq. (7.13) becomes

$$h_L = \left(\frac{64}{N_R}\right)\left(\frac{LV^2}{2gD}\right) \tag{7.14}$$

where

$f = 64/N_R$, and
N_R = Reynold's number (dimensionless).

In general, the dimensionless friction factor f in Eq. (7.14) is related to both the pipe roughness and the Reynold's number and is usually the major pipe loss factor. The Colebrook-White expression that follows is sometimes used for its evaluation.

$$\frac{1}{\sqrt{f}} = -2\log\left(\frac{2.51}{N_R\sqrt{f}} + \frac{\varepsilon/D}{3.7}\right) \tag{7.15}$$

where

f = friction factor or resistance coefficient,
d = pipe diameter,
N_R = Reynold's number, and
ε = smoothness factor.

The Reynold's number defines the flow boundary between laminar and turbulent conditions. It is defined as follows:

$$N_R = \frac{Vd\rho}{\mu} = \frac{Vd}{\nu} \tag{7.16}$$

where

N_R = Reynold's number (dimensionless),
V = flow velocity (m/s),
d = pipe diameter (m),
ρ = fluid density (kN-s^2/m^4) = specific weight divided by g = 9.81 m/s^2,
μ = coefficient of viscosity (kN-s/m^2), and
ν = kinematic viscosity (m^2/s) = μ/ρ.

The above rather nonintuitive set of relationships is greatly sorted out by use of a Moody's diagram, as shown in Figure 7.8. In this important graph it is seen that laminar flow conditions exist for $N_R < 2000$. Above 2000, flow becomes more and more unstable, until at higher N_R values it eventually becomes turbulent. These turbulent values are seen to be highly dependent on the pipe's roughness.

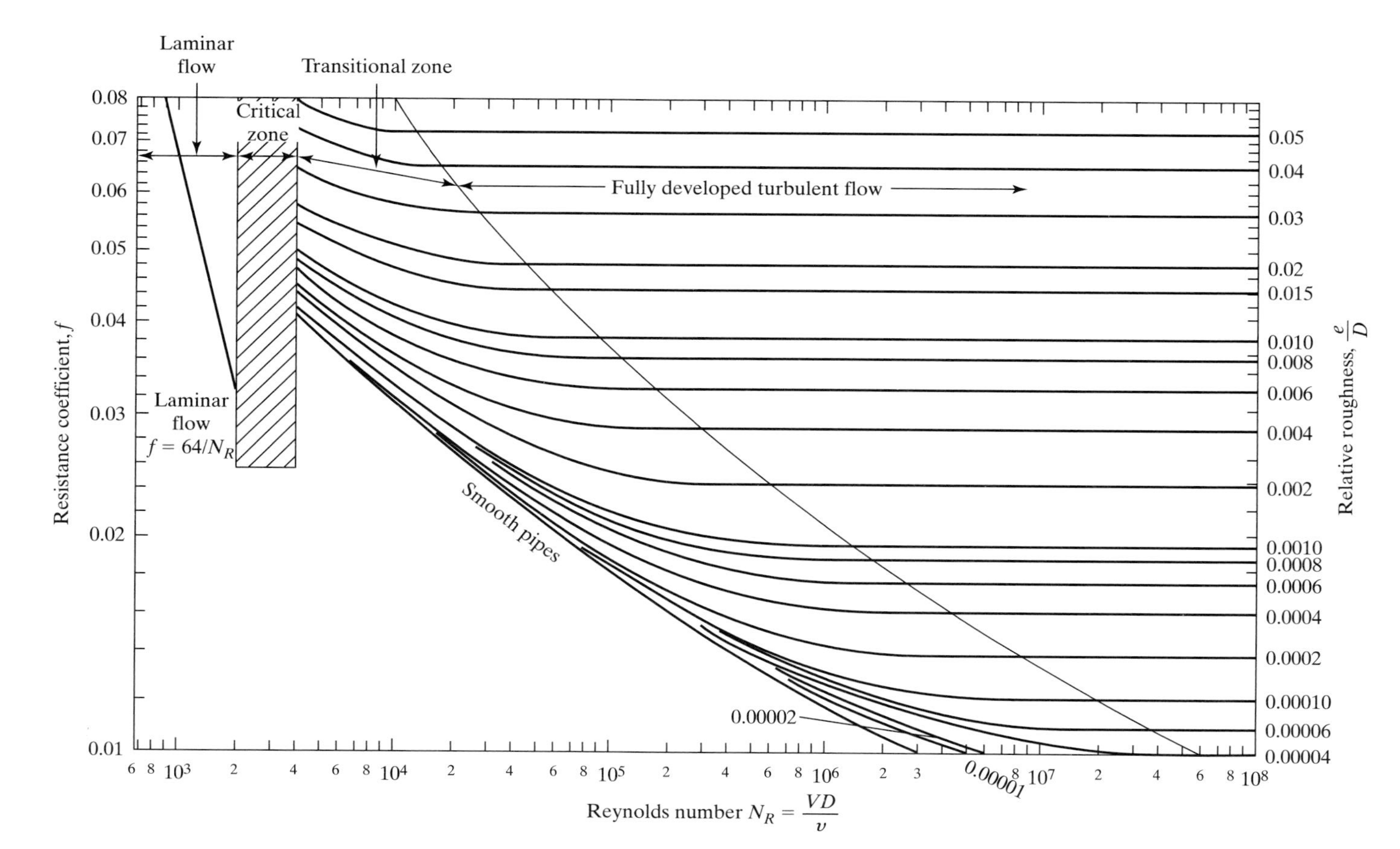

Figure 7.8 Moody chart of resistance to uniform flow in circular pipes. (After Hwang [7])

7.2.2 Deflection Issues

An engineering approach to the quantification of deflection of buried pipelines has been developed by a sequential group of research faculty and students at Iowa State University. Beginning with Marston in the 1920s evaluating rigid conduits (the term used for shallow buried pipes), followed by Spangler in 1950–1970 evaluating flexible conduits, and into the present by Watkins, the group and their colleagues have "written the book" for this type of research [12]. Key issues in the development are the use of arching theory for gravitational force dissipation, the importance of subgrade stability, backfill type, and compaction conditions, and finally the flexibility of the pipe structure itself. Moser [13] presents the following equation, summarizing the Iowa State group's effort for the deflection behavior of flexible (in our case plastic) pipe.

$$\Delta X = \frac{D_L K_b W_c}{(\mathrm{EI}/r^3) + (0.061E')} \cong y \tag{7.17}$$

where

ΔX = horizontal increase in diameter (m),
y = vertical deflection (m),
D_L = deflection lag factor, which varies from 1.0 to 1.5 (dimensionless),
K_b = bedding constant, which varies from 0.83 to 0.110 (dimensionless),
W_c = Marston's prism load per unit length of pipe (kN/m) (note that arching is not taken into account in this formula),
E = modulus of elasticity of the pipe material (kPa),
I = moment of inertia of the pipe wall per unit length (m^3),
EI = bedding stiffness of the pipe ring per unit length (kN-m),
r = mean radius of the pipe (m), and
E' = modulus of soil reaction (kPa).

The last term (E′) has been the subject of intense discussion and research. Howard [14] of the U.S. Bureau of Reclamation has recommended the values given in Table 7.9, which have relatively wide acceptance.

Eq. (7.17) can also be cast in terms of the laboratory plate loading test with the following result. The equation assumes a bedding constant $K_b = 0.2$ and uses the ring stiffness constant (RSC).

$$\frac{y}{D} = \frac{P(0.1L)}{[14.9(\mathrm{RSC})/D + 0.061E']} \tag{7.18}$$

where

y = vertical deflection (m),
D = inside pipe diameter (m),
P = load on pipe (kPa),
L = deflection lag factor (usually 1.0 to 1.5),
RSC = ring stiffness constant (kN/m), and
E' = modulus of soil reaction (kPa).

TABLE 7.9 U.S. BUREAU OF RECLAMATION VALUES OF MODULUS OF SOIL REACTION E' (kPa) FOR BURIED PIPELINES

Class ASTM D-2321	Soil type for pipe bedding material (Unified Classification System[A])	Dumped	Slight < 85% Std. Proctor[C] < 40% Rel. Den.[D]	Moderate 85–95% Std. Proctor 40–70% Rel. Den.	High > 95% Std. Proctor >70% Rel. Den.
I	Crushed rock: manufactured angular, granular material with little or no fines (6 to 38 mm)	7,000	21,000	21,000	21,000
II	Coarse-grained soils with little or no fines: GW, GP, SW, SP[B] containing less than 12 percent fines (max. particle size 38 mm)	NR	7,000	14,000	21,000
III	Coarse-grained soils with fines: GM, GC, SM, SC[B] containing more than 12 percent fines (max. particle size 38 mm)	NR	NR	7,000	14,000
IV(a)	Fine-grained soil (LL < 50): Soils with medium to no plasticity CL, ML, ML-CL, with more than 25 percent coarse-grained particles	NR	NR	7,000[E]	14,000[E]
IV(b)	Fine-grained soils (LL > 50): Soils with high plasticity CH, MH, CH-MH	NR	NR	NR	NR
	Fine-grained soils (LL < 50): Soils with medium to no plasticity CL, ML, ML-CL with less than 25 percent coarse-grained particles				

Organic soils OL, OM, and PT as well as soils containing frozen earth, debris, and large rocks are not recommended for initial backfill; NR = Not recommended for use per ASTM D-2321; LL = Liquid Limit.

[A]ASTM Designation D-2487

[B]Or any borderline soil beginning with some of these symbols (i.e., GM, GC, GC-SC).

[C]Percent Proctor based on laboratory maximum dry density from test standards using about 598,000 joules/m^3 (ASTM D-698)

[D]Relative Density per ASTM D-2049.

[E]Under some circumstances Class IV(a) soils are suitable as primary initial backfill. They are not suitable under heavy dead loads, dynamic loads, or beneath the water table. Compact with moisture content at optimum or slightly dry of optimum. Consult a Geotechnical Engineer before using.

Source: After Howard [14].

The ring stiffness constant (RSC) reflects the sensitivity of the pipe to installation stresses. It is defined in terms of the pipe's deflection resulting from the load applied between parallel plates as per ASTM D2412 (recall Section 7.1.2). As described in ASTM F-894, RSC is the value obtained by dividing the parallel plate load by the resulting deflection (in percent) at 3% deflection. Note that most plastic pipe manufacturers have an empirical formula, along with the necessary tables of their pipe products, for the evaluation of RSC values (e.g., see [15]). Eq. (7.18) also reflects strongly on the type, condition, and placement of backfill both on the sides of the pipe and above it (recall Table 7.9) for values of the modulus of soil reaction E'.

Due to the importance of the above formulation, several full-scale field and large-scale laboratory trials have been published, which give valuable information. Watkins and Reeve [3] have evaluated 375, 450, and 600 mm corrugated plastic pipe under standard H-20 truck loadings to determine the minimum cover necessary to prevent pipe damage and have also performed high pressure large-scale laboratory tests. Regarding the minimum cover tests, their results show the response given in Figure 7.9. Here it can be seen that for a limiting ring deflection of 5% (for this particular pipe) 300 to 375 mm of soil cover is necessary. For the large-scale laboratory tests, the setup and typical data is shown in Figure 7.4e.

Using the finite element computer program "Culvert Analysis and Design" (CANDE), Katona [16] has developed a series of design charts for allowable maximum fill heights. The program has the pipe and surrounding soil in an incremental plane strain formulation. The pipe is modeled with connected beam-column elements and the soil with continuous elements. The assumptions used are all reasonable, with the possible exceptions of a bonded pipe-to-soil interface and linear elastic polyethylene properties. Allowable fill heights for 108 cases are analyzed. The variations are as follows: pipe diameters ranging from 100 to 750 mm; three pipe corrugation areas in each pipe size; good and fair soil backfills; and short-term and long-term pipe properties ($E = 750$ MPa and $\sigma_y = 20$ MPa for a short-term life of 0.05 years, and $E = 150$ MPa

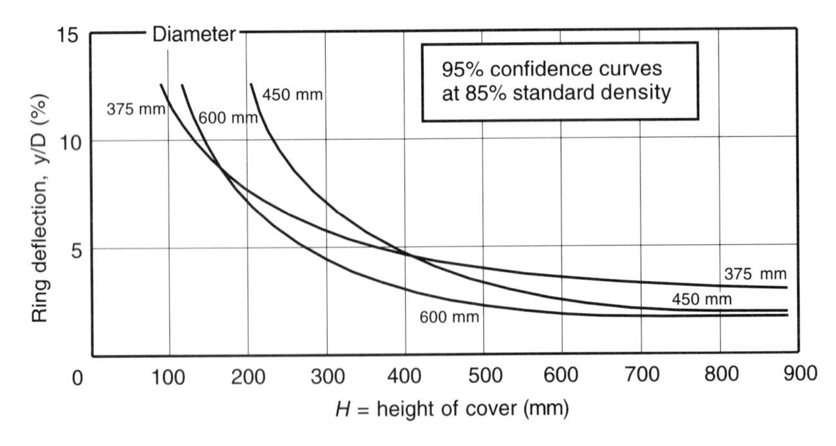

Figure 7.9 Minimum cover values for H-20 loading on HDPE pipe. (After Watkins and Reeve [3])

TABLE 7.10 DESIGN TABLE FOR ALLOWABLE FILL HEIGHT

Pipe properties		Allowable fill heights			
		Good Quality Soil		Fair Quality Soil	
Inside Pipe Diameter (mm)	Corrugated Area (mm^2/m)	S* (m)	L† (m)	S* (m)	L† (m)
100	1.0	13.7	7.8	9.6	3.7
	2.3	26.6	12.6	19.3	7.2
	3.6	38.2	17.3	23.4	10.7
150	1.3	11.9	7.0	8.1	3.1
	2.5	20.9	10.5	15.6	5.5
	3.8	29.1	13.7	20.0	7.9
200	1.5	10.9	6.7	7.2	2.9
	2.8	17.7	9.3	12.9	4.6
	4.1	24.1	11.8	17.9	6.5
250	2.0	11.4	6.9	7.6	3.1
	3.3	16.9	9.1	12.2	4.5
	4.6	22.1	11.0	16.7	5.9
300	2.5	11.7	7.1	7.9	3.2
	3.8	16.3	8.9	11.7	4.4
	5.1	20.7	10.5	15.6	5.6
375	3.8	13.4	7.9	9.4	3.7
	5.1	17.0	9.2	12.4	4.6
	6.3	20.6	10.5	15.4	5.5
450	5.1	14.7	8.4	10.4	3.9
	6.3	17.6	9.6	12.9	4.8
	7.6	20.6	10.6	15.4	5.6
600	6.3	14.6	8.7	10.2	4.0
	7.6	17.0	9.6	12.2	4.6
	8.9	19.4	10.5	14.2	5.2
750	7.6	13.3	8.1	9.2	3.7
	8.9	15.1	8.8	10.7	4.2
	10.2	16.9	9.6	12.3	4.7

*S = Short-term design life (0.05 yr)

†L = Long-term design life (50 yr)

Source: After Katona [16].

and $\sigma_y = 6.2$ MPa for a long-term life of 50 years). Table 7.10 gives Katona's results for these variations.

The values listed in Table 7.10, however, may be somewhat conservative, as indicated by the full-scale field study of Selig and Hashash [17]. A 600 mm corrugated HDPE pipe was placed in a carefully constructed trench and backfilled with a well-graded sand and gravel. It was then backfilled with 30 m of highly compacted soil of 23 kN/m^3 density. Pipe wall strain, circumferential shortening, trench strain, free field soil strain, temperature, soil stress, and shape changes were monitored over a two-year period. (The evaluation is still ongoing.) Some indication of the information gained is

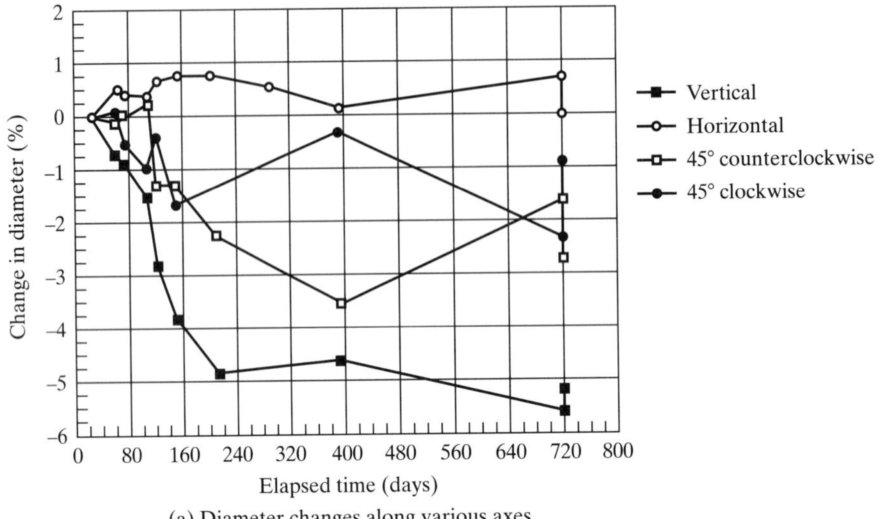

(a) Diameter changes along various axes

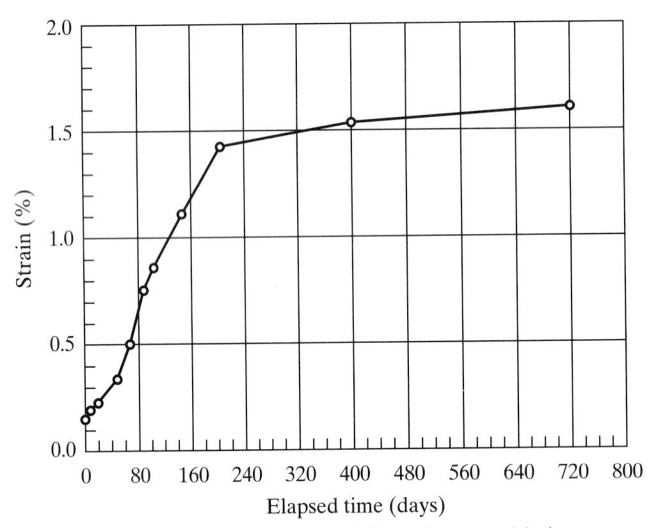

(b) Percent shortening of circumference with time

Figure 7.10 Results of 700 kPa highway loading on a 600 mm diameter profiled pipe. (After Selig and Hashash [17])

given in Figure 7.10a for the change in pipe diameter and Figure 7.10b for the circumferential shortening. All the values are within reason and the performance seems to indicate the proper functioning of the pipe under this very heavy static overburden stress, which is approximately twice the value predicted in Table 7.10. This information seems to suggest that predicted values of overburden stress on polymer pipes (when properly placed and backfilled) may be quite conservative.

Regarding the *internal* stress mobilization from pressurized fluids or gases, there is a tremendous body of knowledge that is available. Pressurized-water and natural-gas transmission lines form the bulk of this information. However, the focus of this chapter is on gravity flow situations.

7.3 DESIGN APPLICATIONS

This section will illustrate the design of various situations that use either profile wall or solid wall plastic pipes. In general, underdrains will require perforated, profile wall pipes and be under gravity-flow conditions. Conversely, landfill pipe risers and manholes will generally be solid wall pipes, also with gravity-flow conditions.

7.3.1 Pavement Underdrains—Perforated Profile Collection Pipes

An underdrain system should be installed parallel to the downward sloping side of all highways, airfields, railroads, parking lots, staging areas, and so on, to collect the water that is transmitted beneath the pavement's surface. That this is necessary has been ably brought out by Cedergren [18], who shows that pavement life is seriously reduced when water remains within the pavement cross section; see Figure 7.11. It is also shown in this figure that the situation of reduced pavement life is greatly aggravated by cold weather conditions.

The water that penetrates through cracks in the pavement's surface flows within the stone base (on a gravitational basis) to the edge of the highway, where it must be collected and transported away from the pavement area (e.g., into a drainage swale,

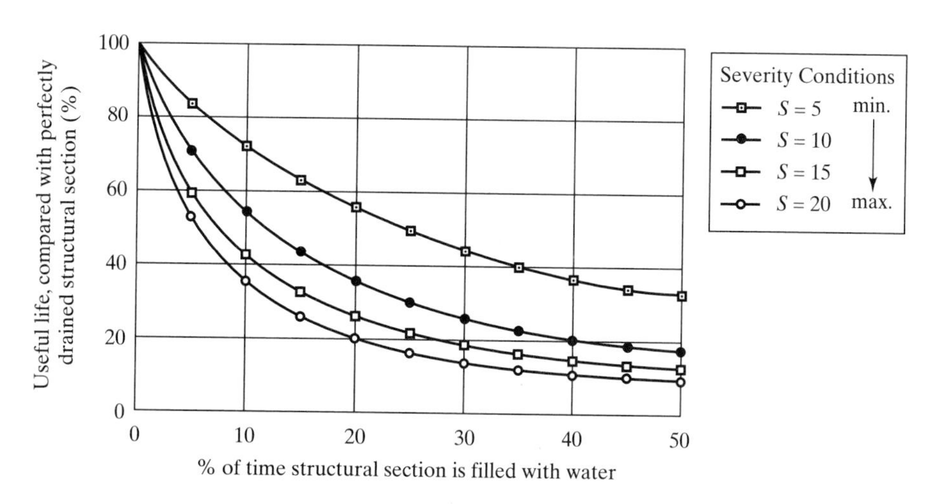

Figure 7.11 Curves showing effect of water in pavement base course on useful life. (After Cedegren [18])

ditch, or stream). This quantity is added to the surface runoff occurring above the pavement and shoulder if the latter is present. A variety of configurations of underdrain systems exist (recall Figure 2.68), but many use a perforated pipe as the central conveyance element. This type of drain is often surrounded by gravel with a sand filter and/or a geotextile filter. This situation will be the focus of the design in Example 7.3.

Example 7.3

Consider a 7.5 m wide pavement in an area with an extremely high rainfall intensity of 150 mm/hr for a 10-year design storm. The runoff coefficient is 0.50. The infiltration coefficient from the pavement is 0.30; thus 80% of the rainfall will reach the underdrain in a relatively short time. (Consider the release factor from an open graded stone base to be 1.0.) What quantity of flow will the underdrain have to convey at the end of 50 m between inlets at an average slope of 0.03, and what size profiled pipe is necessary? (Use a roughness coefficient of 0.010 for smooth interior pipe, i.e., CPP-SP pipe.)

Solution: Using a straightforward approach (see Gupta [10] and Koerner and Hwu [19]), the following formulation is obtained.

$$q = i_R(c_i f_R + c_s)W \tag{7.19}$$

where

q = flow rate per meter of length,
i_R = rainfall intensity,
c_i = pavement infiltration coefficient,
f_R = release factor for the stone base,
c_s = pavement surface runoff coefficient, and
W = pavement width.

$$q = \frac{0.150[0.30(1.0) + 0.50]7.5}{(60)(60)}$$

$$= 0.00025 \text{ m}^2/\text{s}$$

$$= 0.00025(50)$$

$$q = 0.0125 \text{ m}^3/\text{s}$$

From this point either the Hazen-Williams or the Manning chart is used for the proper pipe size. Figure 7.12, based on the Manning formula, has been prepared for $n = 0.010$. Using a slope of 3% and a flow rate of 0.0125 m^3/s gives the required pipe size of 100 mm diameter.

7.3.2 Primary Leachate Collection Systems—Perforated Profile Collection Pipes

Contained within the drainage stone of a leachate collection system (e.g., located above a primary liner system of a landfill) is a pipe-manifold system consisting of interconnected perforated pipes. These pipe systems have made a complete change in the recent past from concrete to plastic pipes. A number of polymers are possible, however, high density polyethylene is widely used due to its good chemical resistance to a wide range of leachates. Furthermore, the pipes are usually of the profiled configuration. The

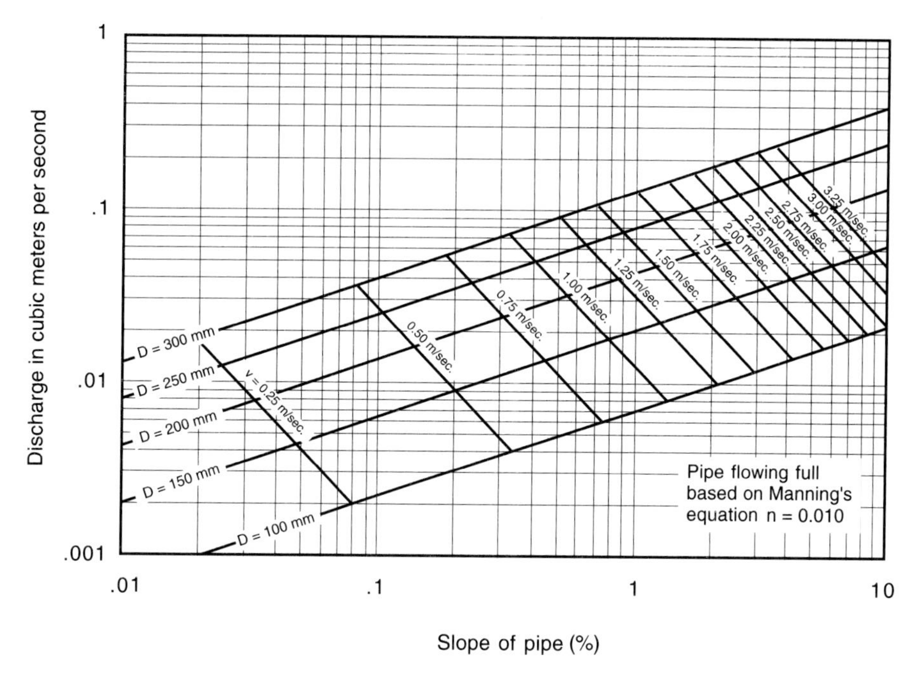

Figure 7.12 Sizing of leachate collection pipe. (After EPA [20])

pipes have to be designed on the basis of both spacing and diameter considerations. In the analysis to follow it will be seen that estimate of the amount of precipitation falling on the site is critical.

Example 7.4

Consider a 3 ha landfill cell ($L = 300$ m and $W = 100$ m) with a tentative primary leachate removal system, the perforated pipe system as shown in Figure 7.13. The cell is uniformly sloped to the sump at 2% for both header and feeder pipes. The landfill is located near Philadelphia, Pennsylvania. Determine (a) the pipe spacing, (b) the pipe size required before waste is placed in the cell; that is, when it is functioning as a dewatering system immediately after its construction, and (c) the pipe size after an initial 4 m of waste is placed in the cell.

Solution: All parts of the solution require an estimate of the rainfall intensity at the specific site. Figure 7.14 gives data for the contiguous U.S. for maximum rainfall in mm per hour over a one-year period [20].

(a) From Figure 7.14, a value of 30 mm/hr is obtained. This is used in the mound equation [21] to determine the pipe spacing.

$$h_c = \frac{S\sqrt{c}}{2}\left[\frac{\tan^2 \alpha}{c} + 1 - \frac{\tan \alpha}{c}\sqrt{\tan^2 \alpha + c}\right] \tag{7.20}$$

where

h_c = maximum height of leachate between adjacent pipes, that is, the allowable height of the *mound* (m),

S = spacing between adjacent perforated pipes (m),

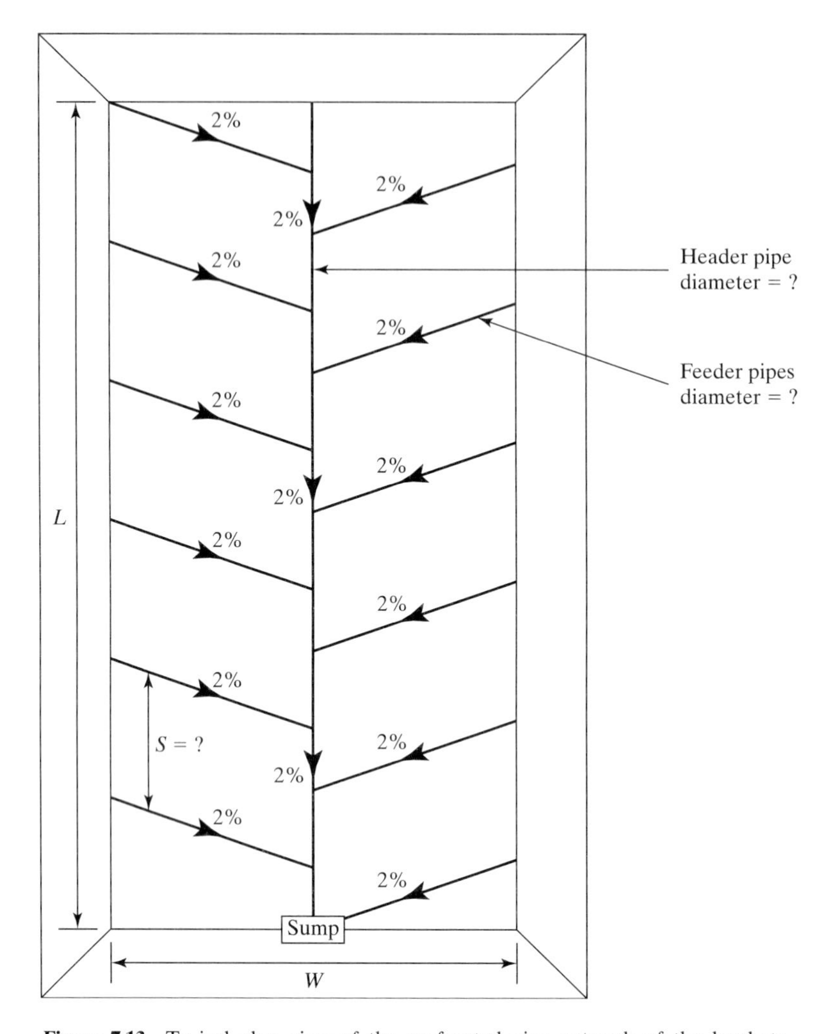

Figure 7.13 Typical plan view of the perforated pipe network of the leachate collection system for a landfill.

α = slope of ground surface between adjacent perforated pipes (dimensionless),
$c = q/k$ (dimensionless),
q = infiltration intensity, which equals the actual rainfall with no waste in cell or requires water-balance calculations with waste in cell (mm/hr), and
k = hydraulic conductivity (permeability) of the drainage stone (mm/hr).

For our problem assume $h_c = 300$ mm, $k = 0.01$ m/s, a 2% slope (which is $\tan \alpha = 0.02$) and solve Eq. (7.20) for the unknown value S.

$$c = \frac{q}{k}$$

$$= \frac{0.030}{(0.01)(60)(60)}$$

Figure 7.14 Precipitation contours in millimeters per hour for one-year storm frequency. (After FHwA [22])

$$c = 8.3 \times 10^{-4}$$

$$0.33 = \frac{S\sqrt{8.33 \times 10^{-4}}}{2}\left[\frac{(0.02)^2}{8.33 \times 10^{-4}} + 1 - \frac{0.02}{8.33 \times 10^{-4}}\sqrt{(0.02)^2 + 8.33 \times 10^{-4}}\right]$$

$S = 32.4$ m use 30 m spacing between feeder collection pipes

(b) Obtain the required pipe diameter from either the Hazen-Williams or Manning formula. Since the pipe for this problem is a smooth interior plastic pipe (use $n = 0.010$) and values for the Manning equation are in the chart of Figure 7.12, this procedure will be used.

For a slope of 2% and a required flow rate of 30 mm/hr over the entire 3 ha site, we get

$$Q = \frac{0.030(100)(300)}{(60)(60)}$$

$$= 0.25 \text{ m}^3/\text{s}$$

From Figure 7.12, this requires a pipe slightly in excess of 300 mm diameter at the sump area. For the feeder pipes, the flow is considerably lower since these pipes must each collect only a portion of the total area. Their size is calculated as follows.

$$Q = \frac{0.030(50)(30)}{(60)(60)}$$

$$= 0.0125 \text{ m}^3/\text{s}$$

From Figure 7.12, the required diameter of the feeder pipes is

$$d = 150 \text{ mm}$$

Note that both of these pipe sizes, 300 and 150 mm, are very large with respect to conventional practice. Since there is no waste in the cell to attenuate the precipitation, the pipe network is acting as a dewatering system that can impound water in excess of the regulatory limits (e.g., in excess of 300 mm of head). The liquid collected is runoff water and is not leachate. Therefore smaller pipe sizes are used. The required flow rate values become much lower when even the first lift of waste is placed.

(c) If a 4 m lift of waste is placed on the collection system some of the precipitation will be captured by the waste and a strong retarding influence will occur. The HELP model, a water-balance computer program developed by Schroeder for the U.S. EPA [23], gives a flow rate for this problem of 0.26 mm/hr (i.e., 115 times less than the solution in part (b), where no waste is in the cell). Clearly, in this case the pipe size will be greatly reduced as shown in the following calculation.

$$Q = \frac{(0.00026)(100)(300)}{(60)(60)}$$

$$= 0.00217 \text{ m}^3/\text{s}$$

This, using Figure 7.12, requires approximately a 50 mm diameter header pipe. The parallel calculation shows that the feeder pipes will be still smaller in diameter. The difference between the two extreme values in parts (b) and (c) of the problem gives rise to an average solution of using 150 to 200 mm diameter pipe for the header pipe, and 100 to 150 mm diameter pipe for the feeder pipes. These are the typical pipe sizes seen in many solid waste landfill cells for leachate removal systems.

7.3.3 Liquid Transmission—Solid-Wall Uniform Pipe with Deflection Calculations

The design of a full-flow buried transmission pipeline requires both pipe sizing and the verification of a limiting deformation. The first part uses the Hazen-Williams formula developed in Section 7.2.1; the second part uses the modified Iowa State equation developed in Section 7.2.2.

Example 7.5

Consider a long gravity transmission PVC pipeline ($C = 150$) with a 0.035% slope and a required discharge of 0.1 m^3/s. What is the required diameter? Consider that this pipe is buried under 6 m of soil weighing 19 kN/m^3 and that the pipe bedding material is Class II with high compaction. What is the total deflection from the soil backfill when added to an assumed installation load deflection?

Solution: Using the Hazen-Williams nomograph in Figure 7.6, the heavy lines indicate that the necessary pipe diameter is 0.5 m.

To enter into the deflection calculation, assume that the pipe is the next largest nominal size, which has the following characteristics for T1-PVC pipe from ASTM F 679.

diameter (inside) = 525 mm
diameter (outside) = 560 mm
wall thickness = 16.0 mm
pipe stiffness = 317 kN/m^2

In the modified Iowa State formula given as Eq. (7.17), assume $D_L = 1.2$, $K_b = 0.2$, and $E' = 21{,}000$ kPa;

$$W_c = H(\gamma)D_0$$

$$= (6.0)(19)(0.560)$$

$$W_c = 63.8 \text{ kN/m}$$

Note that this is the full weight of the soil prism above the diameter of the pipe. There is no reduction due to soil arching via pipe deflection.

$$\Delta X = \frac{D_L K_b W_c}{(EI/r^3) + (0.061E')} \tag{7.17}$$

$$= \frac{(1.2)(0.2)(63.8)}{(317/6.71) + (0.061)(21{,}000)}$$

TABLE 7.11 TENTATIVE DESIGN INSTALLATION DEFLECTIONS FOR HAUNCHED PIPE

Pipe Stiffness (kPa)	Installation Deflection (%)*		
	Less than 85% of Max. Dry Density† or Dumped‡	85 to 95% of Max. Dry Density†	Greater than 95% of Max. Dry Density†
< 275	6+	4	3
275 to 700	4+	3	2
> 700	2+	2	1

*Deflections of unhaunched pipe are significantly larger.

†Maximum dry density determined in accordance with AASHTO T 99.

‡Dumped materials and materials with less than 85% of maximum dry density are not recommended for embedment. Deflection values are provided for information only.

Source: After Chambers et al. [24].

$$= 0.0115 \text{ m}$$

$$\Delta X = 11.5 \text{ mm} \qquad (\cong y, \text{ per ASTM D2412})$$

Therefore,

$$\delta = \frac{y}{D} = \frac{11.5}{525}$$

$$= 0.0219 \times 100$$

$$= 2.2\%$$

The installation deflection is now estimated from the empirical values in Table 7.11. For a pipe stiffness of 317 kN/m^2 and a high compaction density, the installation deflection is 2.0%.

Thus the total deflection is 2.2 to 2.0%, which is 4.2%. This is seen to be acceptable; that is, it is less than 5%, and is therefore an acceptable pipe for the situation.

7.4 DESIGN CRITIQUE

The designs just presented are a selection of geopipe topics that are felt to be of interest to those involved in the geosynthetics area. They are far from being representative of the general use of plastic pipe. Clearly for HDPE pipe, the major use is natural gas transmission pipelines. Equally, the use of PVC pipe in water transmission and plumbing pipelines is widespread. The use of pipe in these types of public utility applications dwarfs the use of plastic pipe in geosynthetics related applications. It is nevertheless a growing area within geosynthetics and it is hoped that these applications gave some insight into the technology.

Perhaps of most current interest is the load-carrying capacity behavior of geopipe in leachate collection systems under large landfills. The theories presented in this chapter were originally developed with significantly lower fills in mind. For large megafills, it appears as though design models incorporating soil arching are appropriate; this comes about from the deflections of the pipe itself as the overlying lifts of waste are imposed. Research and full-scale field testing are ongoing. Of significance is work at Ohio University evaluating a number of leachate collection cross sections (see Sargand et al. [25]). Both short-term and long-term tests are being evaluated using HDPE (smooth and profiled) and PVC pipes.

One additional detail that is extremely difficult to design is the connection of plastic pipe to plastic manholes, such as shown in Figure 7.15. In this case there are many applications with an extremely wide variety of possibilities. One driving force for the use of these plastic manholes is the necessity of providing chemical resistance against various leachates in solid waste landfilling operations. The growing tendency in this regard is to use high density polyethylene for the pipe and manhole system. In so doing, factory prefabrication can be used to great advantage. Many standardized systems are becoming available. The manufacturers of HDPE pipe and fittings should be consulted for information and details.

7.5 CONSTRUCTION METHODS

Subgrade preparation, joining, placement, and backfilling of plastic pipe is relatively simple in comparison to pipes made from other materials. This comes about by virtue of their light weight, relatively long lengths, and ease of joining. As with all geosynthetics,

Figure 7.15 Illustrations of the use of plastic pipe with plastic manholes. (After Driscopipe Corp. and ADS, Inc.)

however, there are proper procedures that must be followed. CQC and CQA are as important with geopipe as they are with any other type of geosynthetic material.

7.5.1 Subgrade Preparation

Plastic pipe is usually placed in a prepared trench or within other prepared subgrade materials. If soil is the subgrade, as it usually is, the compaction should be to 95% of Standard Proctor compaction so as to minimize the deformation of the pipe while it is in service. Note that pipe trenches are often over-excavated so that bedding soil of a cohesionless nature can bring the grade up to the plan elevation (see Figure 7.16a). Granular soil can easily be graded, gives uniform support to the pipe, and readily accommodates the haunching material, which is compacted above it after the pipe is placed. Sufficient trench length should be available such that pipe laying can continue in a uniform manner. Conversely, too much open trench may allow for local instability or sloughing failures and negatively influence pipe placement and/or backfilling. These de-

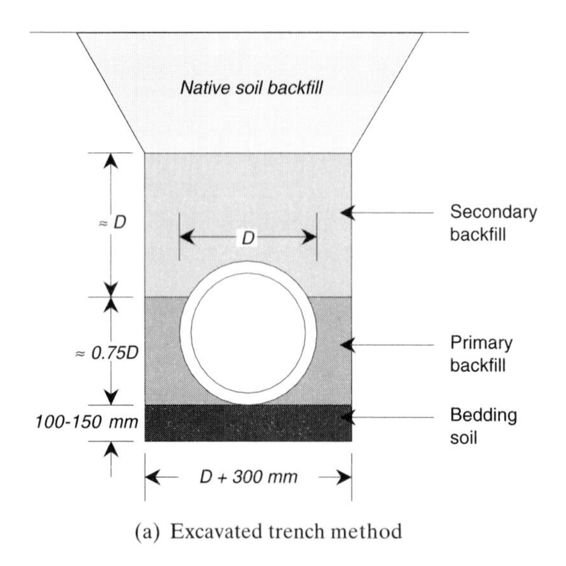

(a) Excavated trench method

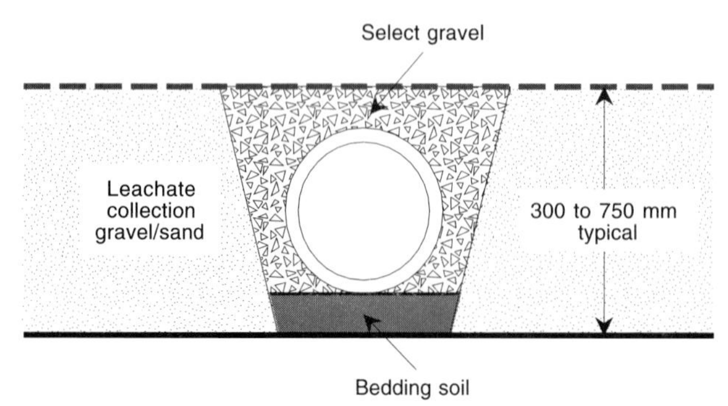

(b) Leachate collection method

Figure 7.16 Pipe placement and backfilling procedures.

cisions depend greatly on the soil type and depth of trench. In some cases, trench support systems may be required. Local conditions and safety considerations will govern each situation.

If the pipe is to be placed directly on a geomembrane (as in the leachate collection system shown in Figure 7.16b), the full-depth drainage stone should be placed before pipe installation. Small excavations of at least the diameter of the pipe are then made, and the pipe is then placed in these shallow excavations. Thus we end up with a trench, albeit a shallow one, even in cases of pipe placement in leachate collection stone.

Note that various organizations have guides or practices relating to underground installation methods for plastic pipe: ASTM D 2774, "Underground Installation of Thermoplastic Pressure Piping;" ASTM D F481, "Standard Practice for Installation of Thermoplastic Pipe and Fittings;" AWWA M23, "PVC Pipe Design and Installation;" PPI TR8, "Installation Procedures for Polyethylene Plastic Pipe;" and PPI TR31, "Underground Installation of Polyolefin Piping."

An approach completely different from the excavation, placement, and backfilling just described can be taken under certain circumstances. Shown in Figure 7.17 is a trenchless pipe placement. It uses a saw-type digging chain followed immediately by a pipe boot into which the pipe is transferred to the bottom of the trench. The soil collapses around the pipe after it leaves the boot and a vibratory sled compactor completes the procedure. This same concept, using a rotary wheel excavator, is being used for geocomposite edge drain installation as will be described in Chapter 8.

7.5.2 Connections

There are essentially three methods used to connect the ends of geopipe together; see Figure 7.18.

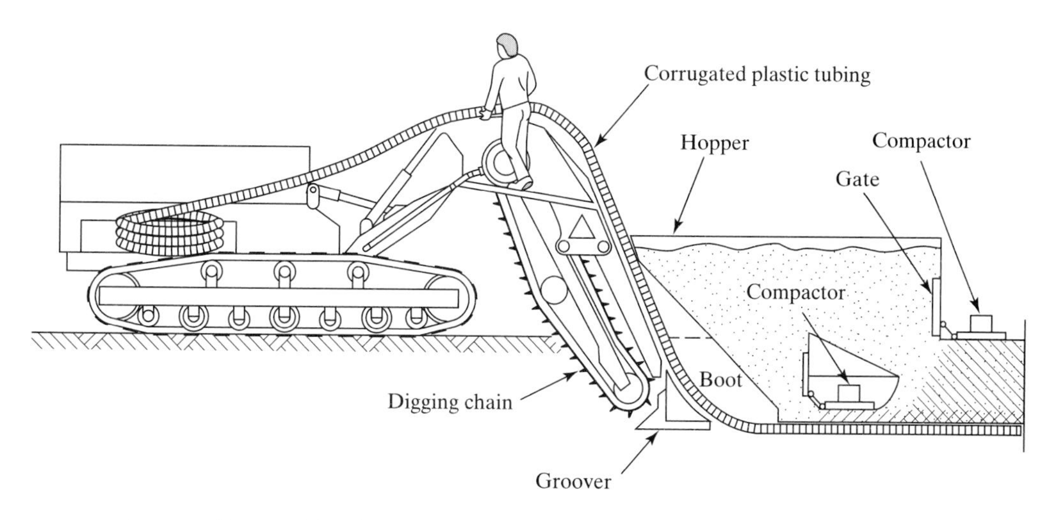

Figure 7.17 Trenchless installation of perforated profiled drainage pipe. (After Chambers et al. [24])

The *butt welding* method is used for thick-walled HDPE pipe (either solid or perforated) and is exactly the same as that used in the natural gas pipe industry; see ASTM D-2657. The ends of the pipe are brought together with a heated plate placed between them. An axial force brings the ends of each pipe against opposite sides of the heat plate. When adequate thermal energy is generated and the pipe ends become viscous, the heat plate is removed and the pipe ends are quickly brought together. Adequate force is applied to the opposing pipes to extrude a slight amount of the molten material out from the seam area. After cooling, the force is released and the seam is completed. This technique is routinely used in natural gas pipeline installation and has been shown to make seams of quality equal to that of the pipe itself. *Electric-socket welding* is similar. In contrast, PVC pipe is usually chemically seamed using a solvent on the pipe ends before pressure-joining them together; see ASTM D 2672.

The *insert* type of plastic pipe connection can only be made if the pipe thickness is adequate to form or machine the pipe ends to accept one another. Lip seals and screw connections are of this type. Various configurations can be made and several patents are available on specialty products. To make a tight connection, gaskets are sometimes used, which reside in slotted seats of the thicker section of the connection. To make a leak-free connection it is possible to extrusion seam the outside separation for small diameter pipe or to extrusion seam the inside for very large diameter pipe. This latter situation could easily arise in connecting HDPE plastic pipe to HDPE plastic manholes, as seen in Figure 7.15.

Sleeve couplings are used to connect the ends of all profiled wall pipe and some solid wall pipe. Each of these couplings must be mated to the type of pipe for which they were designed. It is not an acceptable practice to use couplings made for one style of profiled pipe on a different style.

7.5.3 Placement

The placement of plastic pipe in a prepared trench, after it has been seamed into an essentially continuous length, is very rapid and straightforward. The major consideration that *must* be addressed is temperature. Plastic pipe will expand and contract in direct relationship to the temperature and the pipe material's coefficient of linear expansion/contraction. As seen in Table 5.10, HDPE pipe has the highest value of all the polymers used in the manufacture of geosynthetics. To illustrate these effects numerically, consider Example 7.6.

Example 7.6

Calculate the elongation of 30 m of HDPE pipe in a trench undergoing a 25°C temperature increase, if it is allowed to deflect in an unrestricted manner. Also calculate the compressive stress mobilized if the same pipe is constrained in a fixed position.

Solution: Using a coefficient of expansion of 12×10^{-5}/°C for HDPE, the elongation of the pipe if unconstrained is

$$\delta = \alpha(\Delta T)L \tag{7.21}$$

$$= (12 \times 10^{-5})(25)(30)(1000)$$

$$\delta = 90 \text{ mm}$$

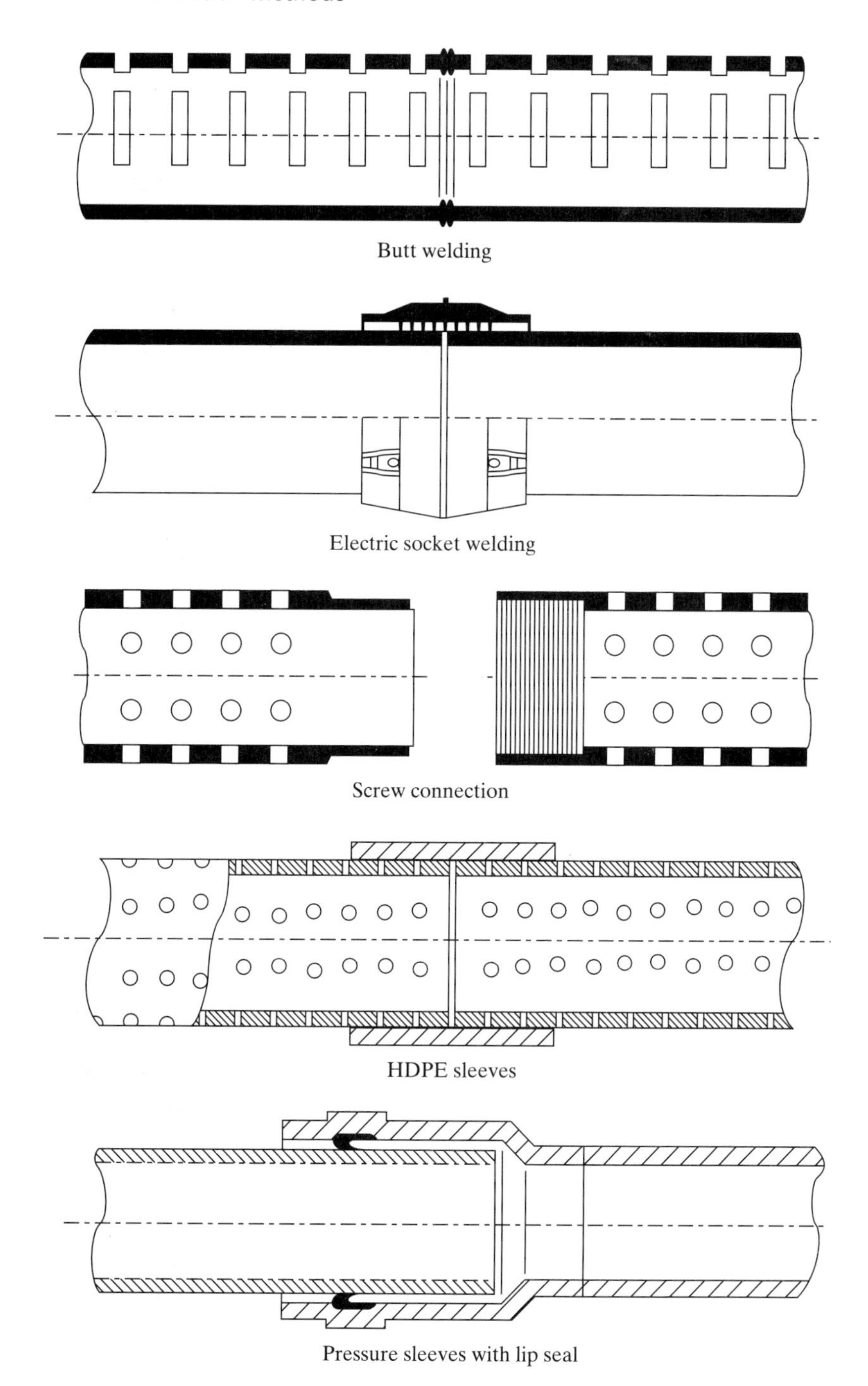

Figure 7.18 Methods of joining geopipe.

If the pipe is restricted (fixed) in position, the compressive stress that is induced is as follows (assuming a modulus of elasticity of HDPE of 760 MPa):

$$\sigma_c = \alpha(\Delta T)E \tag{7.22}$$

$$= (12 \times 10^{-5})(25)(760)$$

$$\sigma_c = 2.3 \text{ MPa} \qquad (\simeq 12\% \text{ of the yield strength of typical HDPE})$$

Obviously, if the temperature decreases after the pipe is laid, the movement will tend to be a contraction and the stress will be tensile. To avoid problems due to thermal effects after backfilling is complete and the pipe is in service, the installation temperature should be within 10°C, and preferably 5°C, of the backfilled service temperature. This certainly could limit the time of day when pipe installation and/or backfilling is performed. The number of variations on this theme is essentially limitless.

7.5.4 Backfilling Operations

The backfilling of plastic pipe is performed in stages, as shown in Figure 7.16a. The primary backfill, which is usually a granular soil with adequate fines to provide good placement stability, is most important and requires special care in its placement in the haunch areas beneath the pipe. If soil is not placed beneath the pipe, voids will result and when loaded the pipe will deform into these voids. Such deformations will always be excessive, thereby jeopardizing the pipe material in a manner not intended in the pipe's design. Conversely, too much compaction (called *slicing* since it is usually done by shovels in a sideways jabbing motion) will lift the pipe off of the subgrade and destroy the elevation control. Also the backfilling should be done in a reasonably symmetric manner so the pipe is not pushed laterally out of alignment. With the proper backfill soil, such as Class I or II material in Table 7.9, the lift thickness can be up to 450 mm. For Class III material, the lift thickness should be a minimum of 200 to 300 mm. Cavalier dumping of backfill soil adjacent to and above the pipe is simply courting disaster. A relatively large stone bearing against the pipe results in a stress concentration of the highest order. Such practices must be not permitted under any circumstances.

The secondary backfill provides support and load transfer to the top of the pipe and support for the subsequent backing operation. The soil type is often one grade class lower than the primary backfill soil; for example, if Class II soil is used as primary backfill then Class III is used for secondary backfill. The thickness of this layer should be at least one-half of the diameter of the pipe itself and preferably equal to the diameter. Backfilling can be mechanized but never on a lift thickness less than 300 mm (recall Figure 7.9).

Following placement of the secondary backfill, natural soil or the soil being used to construct the facility is brought up to final grade in lifts, as per the plans and specifications. This aspect follows standard earthwork construction procedures.

REFERENCES

1. Jeyapalan, J. K., and Jeyapalan, M., *Advances in Pipeline Engineering*, New York: ASCE, 1995.
2. Halse, Y., Wiertz, J., Rigo, J. M., and Cazzuffi, D., "Chemical Identification Methods Used to Characterize Polymeric Geomembranes," in *Geomembranes: Identification and Performance*

Testing, ed A. L. Rollin and J. M. Rigo, RILEM, London: Chapman and Hall, 1990, pp. 316–336.

3. Watkins, R. K., and Reeve, R. C., "Structural Performance of Buried Corrugated Polyethylene Testing," *30th Annual Geology Symp.*, Federal Highway Administration, Washington, DC, 1979.
4. Hsuan, Y. G., "Evaluation of Stress Crack Resistance of Polyethylene Non-Pressure Pipe Resins," Final Report, Plastic Pipe Institute, SPI, Washington, DC, 1997.
5. Haas, D. B., and Smith, L. G., "Erosion Studies," A Report of Canada Ltd., Saskatchewan Research Council, E 57-7, 1975.
6. Goddard, J., "Advanced Drainage Systems, Columbus, OH" (suggested test method for evaluating pipe immersed in leachates), ASTM D35.02 Task Group Leader for Geopipe Immersion at Testing, W. Conshohocken, PA: ASTM.
7. Hwang, H. C., *Fundamentals of Hydraulic Engineering Systems*, Englewood Cliffs, NJ: Prentice-Hall, 1981.
8. Viessam, W., Jr., and Hammer, M. J., *Water Supply and Pollution Control*, 4th ed., New York: Harper and Row, 1985.
9. Fox, R. W., and McDonald, A. T., *Introduction to Fluid Mechanics*, 3d ed., New York: J. Wiley and Sons, 1985.
10. Gupta, R. S., *Hydrology and Hydraulic Systems*, Englewood Cliffs, NJ: Prentice-Hall, 1989.
11. *Driscopipe Systems Design*, Bartlesville, OK: Phillips Driscopipe, Inc.
12. Spangler, M. G., *Soil Engineering*, 2d ed., Scranton, PA: International Textbook Co., 1971.
13. Moser, A. P., *Buried Pipe Design*, New York: McGraw-Hill, 1990.
14. Howard, A. K., "Soil Reaction for Buried Flexible Pipe," *J. Geotechnical Eng. Div., ASCE*, Vol. 103, No. GTI, 1977, pp. 33–43.
15. Spirolite HDPE Pipe Product Data CHM-916, Chevron Chemical Co., Reno, NV, 1989.
16. Katona, M. G., "Allowable Fill Height for Corrugated Polyethylene Pipe," TRB #1191, Transportation Research Board, National Research Council, Washington, DC, 1988.
17. Selig, E. T., and Hashash, N., "Analysis of the Performance of a Buried High Density Polyethylene Pipe," *Structural Performance of Flexible Pipes*, ed S. M. Sargand, G. G. Mitchell, and J. O. Hurd, Rotterdam: A. A. Balkema, 1990, pp. 95–103.
18. Cedergren, H. R., *Drainage of Highway and Airfield Pavements*, New York: John Wiley and Sons, 1974.
19. Koerner, R. M., and Hwu, B-L., "Prefabricated Highway Edge Drains," ASCE/PennDOT Geotechnical Seminar, Hershey, PA, 1990.
20. U.S. EPA, "Lining of Waste Containment and Other Impoundment Facilities," Report No. 600/2-88/052, 1988.
21. Moore, C., "Landfill and Surface Impoundment Performance Evaluation," U.S. EPA, SW-869, Washington, DC, 1983.
22. Federal Highway Administration, "Guidelines for the Design of Subsurface Drainage Systems for Highway Structural Sections," Washington, DC, 1973.
23. Schroeder, P. R., Dizier, T. S., Zappi, P. A., McEnroe, B. M., Sjostrom, J. W., and Peyton, R. L. (1994), "The Hydrologic Evaluation of Landfill Performance (HELP) Model: Engineering Documentation for Version 3," EPA/600/R-94/168b, U.S. Environmental Protection Agency, Risk Reduction Engineering Laboratory, Cincinnati, OH.
24. Chambers, R. E., McGrath, T. J., and Heger, F. J., "Plastic Pipe for Subsurface Drainage of Transportation Facilities," NCHRP Rept. 225, Transportation Research Board, National Research Council, Washington, DC, 1980.
25. Sargand, S. M., Mitchell, G. G., and Hurd, J. O., eds., *Structural Performance of Flexible Pipes*, Rotterdam: A. A. Balkema, 1990, 1993.

PROBLEMS

7.1 Why are profiled pipes not directly measurable in terms of SDR or Schedule?

7.2 How do you find an equivalent SDR of a profiled pipe?

7.3 What are some possible differences between the formulations of a HDPE geomembrane and a HDPE pipe?

7.4 What are some possible differences between the formulations of a PVC geomembrane and a PVC pipe?

7.5 On a normalized graph of stress-versus-deformation, sketch the relative behavior of a plastic pipe, concrete pipe, and steel pipe in an unconfined compression mode, such as ASTM D2412.

7.6 Why is the ultimate deflection of a plastic pipe usually taken as 10%?

7.7 The seaming of solid wall HDPE and PVC pipe is fundamentally different than the seaming of geomembranes.

- **(a)** What are the general seaming methods for solid wall plastic pipe?
- **(b)** What property does pipe have, versus geomembranes, that allows this to be accomplished?
- **(c)** What do you suspect is the general strength requirement for geopipe seams?

7.8 How are profiled HDPE pipes seamed?

7.9 The use of geopipes as leachate collection removal pipes in landfills are of concern because of high vertical stresses and elevated temperatures. Describe how these considerations are handled when using Eq. (7.17) for the calculation of pipe deflection.

7.10 Describe how the modulus of soil reaction (E') in Eq. (7.17) is obtained and discuss its sensitivity.

7.11 Calculate and graph the deflection response for Example 7.5 (Section 7.3.3) for varying values of modulus of soil reaction (E').

7.12 Repeat Problem 7.11, varying the depth at which the pipe is buried.

7.13 Calculate the manifold pipe spacing of a leachate removal system (as shown in Figure 7.13) using Eq. (7.20) for your particular area's rainfall. (*Hint:* Figure 7.14 gives the precipitation contours for the U.S.).

7.14 Recalculate Example 7.4 (Section 7.3.2), using the same parameters except for the permeability of the leachate collection stone. Vary k from 0.0001, 0.001, 0.01 (the example), and 0.1 m/s. Draw a graph of the results on semilog paper.

7.15 In the installation of geopipes, the backfilling beneath the pipe and under the haunches is critical. How do you inspect these two areas of backfilling?

7.16 Why is it necessary to have a minimum amount of soil cover over a geopipe before it can be trafficked by construction equipment? (*Hint:* Recall Figure 7.9).

8 Designing with Geocomposites

8.0 INTRODUCTION

As originally described in Section 1.9, geocomposites consist of various combinations of geotextiles, geogrids, geonets, geomembranes, and/or other materials. In keeping with the general theme of this book, the geocomposites to be discussed will be made from synthetic, or human-made, materials rather than naturally occurring ones. Sometimes, however, it is necessary to include gravel, sand, silt, and/or clay within the composite system. For background information, the reader should review Section 1.9. The general reason for the existence of geocomposites is the higher performance that can often be attained by combining the attributes of two or more materials. Such high performance can be used for any of the basic functions that have already been introduced to the reader: separation, reinforcement, filtration, drainage, and liquid/vapor containment. This chapter focuses on these five functions insofar as they are addressed by geocomposites. Some of the individual sections will be subdivided according to application area due to the large amount of material to be discussed (e.g., drainage geocomposites).

Although the general situation might change with time, the growth rate of geocomposites is currently proceeding faster than any other area discussed thus far. Of course, these hybrid materials have been introduced only recently, but some application areas have completely swung in their direction. Drainage composites having high flow capability are such a situation. Examples are discussed in connection with wick drains, sheet drains, and edge drains. Note that high-performance reinforcement systems can be obtained by different approaches; the role of geocomposites in reinforcement will be emphasized. The area is so dynamic that GCLs were treated in the second edition of the book as liquid containment geocomposites. In the third edition they were treated as a separate geosynthetic in their own right, and have been now expanded still further in Chapter 6 of the fourth edition.

Before beginning the chapter, however, the reader is cautioned that many of these geocomposites serve multiple functions. While the primary function will always be the focus, the proper performance of secondary (and perhaps tertiary) functions are also required. If the design of these additional functions has been covered previously in this book, it will not be repeated here; the appropriate sections will, however, be cross-referenced.

8.1 SEPARATION

When geotextiles are designed as separators between dissimilar (Section 2.5), the mechanical demands on the material are quite low. Many geotextiles between 200 and 300 g/m^2 are adequate to handle general situations (e.g., where a stone base is separated from soil subgrade or different zones are separated within an earth dam). Such values as tensile strength, modulus, burst strength, puncture strength, or tear strength are rarely used fully for the geotextiles discussed. In fact, the recommended installation survivability properties often take precedence over the calculated values. However, in the above situations, the geotextile is usually not used as a ground surface cover by itself; it is not separating the ground surface from the prevailing atmospheric conditions (i.e., wind, rain, snow, etc.). When the design function of geocomposites is separation, the application is in erosion control systems.

Although an extremely porous geotextile could be used for this purpose, specialty geocomposites have been developed for the specific purpose of *erosion control.* The general goal of such systems is to protect the soil from sheet, gully, or rill erosion either indefinitely or until vegetation can establish itself. While water is the predominant medium for erosion (or detachment) and the subsequent transportation of soil particles, wind is also a potential medium as seen in Figure 8.1. Here the interaction of the water or air velocity and the size of soil particles gives rise to the sequence of soil erosion, namely, detachment, transportation, and deposition. Also shown is that water is somewhat more severe in causing erosion than is air.

The International Erosion Control Association (IECA) is an organization focusing on erosion control practices, materials, conferences, publications, and standards. Most of the products dealt with by erosion control specialists use geosynthetic materials in whole or in part; they are shown collectively in Figure 8.2. The relatively large number of erosion control products can be broadly separated into temporary and permanent materials, as will be further described later.

The installation of many of the flexible erosion control products is straightforward and is illustrated in Figure 8.3 for use as erosion protection in a stream channel and on a steep side slope. The products are usually placed on a prepared soil subgrade by pinning them to the soil with U-shaped staples or small ground anchors. *Intimate contact* of the blanket or mat to the soil subgrade is very important, since water flow beneath the material has usually been the cause of poorly functioning and failed systems. In a similar vein, proper installation of the roll edges and ends is important, so that flow does not cause local undermining that can continue under the adjacent rolls in a pro-

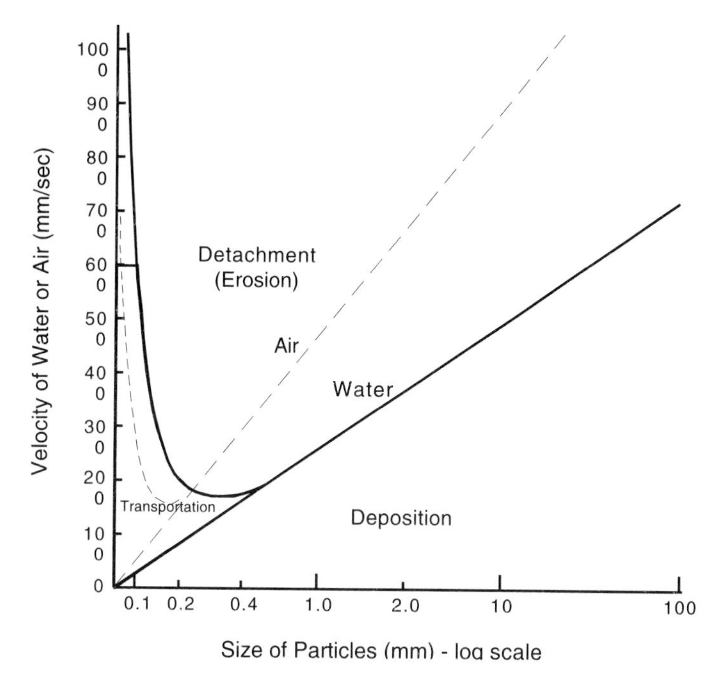

Figure 8.1 Comparison of erosion (or detachment), transportation, and deposition responses due to air and water. (Modified from Garrels [1])

gressive manner. The manufacturers' installation recommendations must be closely followed.

8.1.1 Temporary Erosion and Revegetation Materials

Temporary erosion and revegetation materials (TERMS) consist of materials that are wholly or partly degradable. They provide temporary erosion control and are either disposable after a given period, or only function long enough to facilitate vegetative growth; after the growth is established, the TERM becomes sacrificial. Some of the products are completely biodegradable, while others are only partially so. Theisen [2] groups the following materials (listed in Table 8.1) in the TERM category.

The first two products are self-explanatory, consisting of traditional methods of soil erosion control using straw, hay, or mulch loosely bonded by asphalt or adhesive. Their stability is often quite poor. Geofibers in the form of short pieces of fibers or microgrids can be mixed into soil with machines or rototillers to aid in lay-down and continuity. The fiber or grid inclusions provide for greater stability over straw, hay, or mulch simply broadcast over the ground surface.

ECMNs are biaxially oriented nets manufactured from polypropylene or polyethylene. They do not absorb moisture, nor do they dimensionally change over time. They are lightweight and are stapled to the previously seeded ground using hooked nails or U-shaped pins. This is the practice for many of the rolled sheet products that

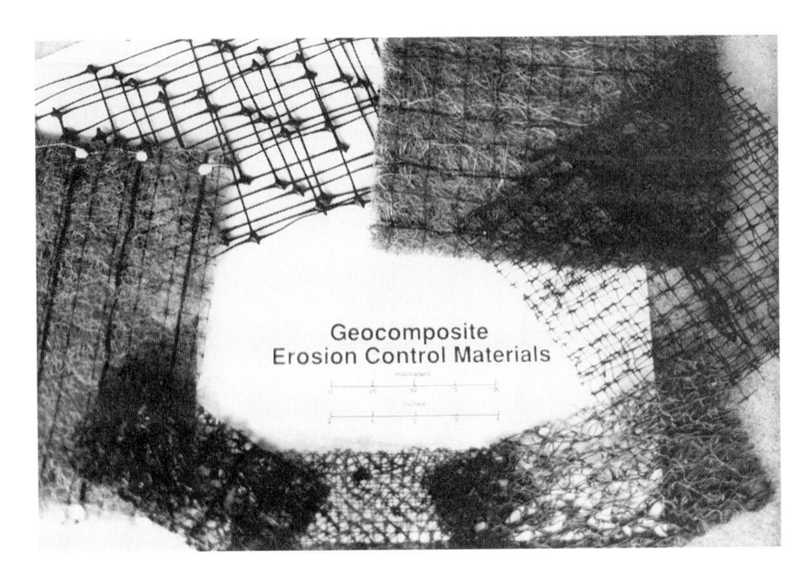

Figure 8.2 Geocomposites used in erosion control.

follow. The stability is obviously greatly improved over the two previously mentioned approaches.

ECBs are also biaxially oriented nets manufactured from polypropylene or polyethylene, but these are now placed on one or both sides of a blanket of straw, excelsior, cotton, coconut, or polymer fibers. The fibers are held to the net by glue, lock stitching, or other threading methods.

TABLE 8.1 GEOSYNTHETIC EROSION CONTROL MATERIALS

	PERMs	
TERMs	Biotechnical-related	Hard armor-related
Straw, hay and hydraulic mulches	UV-stabilized fiber roving systems (FRSs)	Geocellular containment systems (GCSs)
Tackifiers and soil stabilizers	Erosion control revegetation mats (ECRMs)	Fabric formed revetments (FFRs)
Hydraulic mulch geofibers	Turf reinforcement mats (TRMs)	Vegetated concrete block systems
Erosion control meshes and nets (ECMNs)	Discrete length geofibers	Concrete block systems
Erosion control blankets (ECBs)	Vegetated geocellular containment systems (GCSs)	Stone riprap
Fiber roving systems (FRSs)		Gabions

Source: After Theisen [2].

(a) and (b) Erosion control in water runoff channels (compliments of TC Mirafi)

(c) Erosion control of steep side slopes (compliments of Akzo, Inc.)

Figure 8.3 Use of a geocomposite separators in various erosion control applications.

FRSs are continuous strands or yarns, usually of polypropylene, that are fed continuously over the surface that is to be protected. They can be hand-placed or dispersed using compressed air. After placement on the ground surface an emulsified asphalt or other soil stabilizer is used for controlled positioning.

8.1.2 Permanent Erosion and Revegetation Materials—Biotechnical-Related

Within the permanent erosion and revegetation materials (PERMs) are a biotechnical-related group, as shown in Table 8.1. These polymer products furnish erosion control, aid in vegetative growth, and eventually become entangled with the vegetation to provide reinforcement to the root system. As long as the material is shielded from sunlight, via shading and soil cover, it will not degrade (at least within the limits of other polymeric materials). The seed is usually applied after the PERM is placed and is often carried directly in the backfilling soil.

The polymers in FRSs can be stabilized with carbon black and/or chemical stabilizers, so they can be sometimes considered in the PERM category; see Table 8.1. They are described earlier.

ECRMs and TRMs are closely related to one another. The basic difference is that ECRMs are placed on the ground surface with a soil infill, while TRMs are placed on the ground surface with soil filling in and above the material. Thus TRMs can be expected to provide better vegetative entanglement and longer performance. Other subtle differences are that ECRMs are usually of greater density and lower mat thickness. Seeding is generally done prior to installation with ECRMS, but is usually done while backfilling within the structure of TRMs.

Discrete-length geofibers are short pieces of polymer yarns mixed with soil for the purpose of providing a tensile strength component against sudden forces for facilities such as athletic fields, slopes, and so on. GCSs consist of three-dimensional cells of geomembranes or geotextiles that are filled with soil and, when used for erosion control, are vegetated. They are described in Section 8.2.3 from the perspective of their reinforcement capabilities.

8.1.3 Permanent Erosion and Revegetation Materials—Hard Armor-Related

In a separate category of inert materials, we can include a number of PERMs that are essentially hard armor systems (see Table 8.1).

Whenever the infill material is permanent, as with concrete or grout, GCSs can be considered in this category. Clearly FFRs, which are covered in Section 2.10.4, are hard armor materials. They were included in Chapter 2, because the geotextiles in the upper and lower surface hold the key to the installation, but it should be clearly recognized that erosion control is the major feature that is being provided.

Numerous concrete block systems are available for erosion control. Hand-placed interlocking masonry blocks are very popular for low-traffic pavement areas such as carports, driveways, off-street parking, and so on. The voids in the blocks and between them are usually vegetated. Alternatively, the system can be factory-fabricated as a unit, brought to the job site, and placed on prepared soil. The prefabricated blocks are either laid on or bonded to a geotextile substrate. The finished mat can bend and torque by virtue of the blocks being articulated with joints, weaving patterns, or cables. Such systems are generally not vegetated.

Stone riprap can be a very effective erosion control method whereby large rock is placed on a geotextile substrate. A geotextile placed on the proposed soil surface before rock placement serves as a filter and separator, and is described in Section 2.8.5. The stone can vary from small hand-placed pieces to machine-placed pieces of enormous size. Canals and waterfront property are often protected from erosion using stone riprap.

Closely related are gabions, which consist of discrete cells of wire netting filled with hand-placed stone. The wire is usually galvanized steel hexagonal wire mesh, but in some cases it can be a plastic geogrid. Gabions require that a geotextile be placed behind them, acting as a filter and separator for the backfilled soil. The topic is covered in Section 2.8.3.

8.1.4 Theoretical Considerations

While Figure 8.1 shows the general behavior of the soil erosion process, it does little to quantify the variety of complex processes that are involved. Weggel and Rustom [3] have reviewed the state of the art. Beginning with the impact of a raindrop on the soil, a splash mechanism is setup whereby the shear strength of the soil is challenged. Once detachment occurs, surface flow transports the soil particles in a gravitational manner until the hydraulics and topography result in disposition. There are an incredible number of variables involved in the three basic mechanisms of detachment, transportation, and deposition.

The most often-used model for soil loss by erosion is the Universal Soil Loss Equation (USLE), developed by Wischmeier and Smith [4]. The equation is given by

$$E = RK(LS)CP \tag{8.1}$$

where

E = soil loss (tons per square kilometer per year depending upon constants used),
R = rainfall factor (dimensionless),
K = soil erodibility factor (dimensionless),
LS = length of slope or gradient factor (dimensionless),
C = vegetative cover factor (dimensionless), and
P = conservation practice factor (dimensionless).

Charts and tables in [4] describe the various factors involved.

There are many limitations to Eq. (8.1). Among them are that it is not applicable for predicting erosion from the following: gully type runoff, small localized sites, steep slopes, seasonal variations, and short-term water surges. While the equation is useful in a global sense (e.g., to predict large-scale farmland soil loss), it is generally felt to be weak in providing accurate guidance in the quantification of site specific soil erosion. Note that a modified USLE for point-source erosion is also available.

8.1.5 Comparative Behavior

While a theoretical predictive method for soil erosion and its prevention is still an illusive target, there is no lack of laboratory rainfall simulators and wave tanks to study the situation. Weggel and Rustom [3] have surveyed the situation in this regard and have performed laboratory erosion tests on a number of geosynthetic products using a rainfall simulator. Following earlier studies by Ingold and Thompson [5], Rustom and Weggel [6] have made measurements of surface runoff and sediment yield resulting from nominal rainfall intensities of 50, 125, and 200 mm/hr on the upper reaches of a 1 (V) to 2.5 (H) soil slope. The resulting data were analyzed and include the characteristic rising limb of the S-hydrograph, the equilibrium discharge, and the typical recession curve after the cessation of rainfall. Results are presented for bare soil slopes and for soil slopes protected by 12 different natural and geosynthetic erosion control systems. The ability of the various erosion control systems to reduce sediment loss is quantified in Figure 8.4.

In general, all of the products tested demonstrate some ability to reduce soil detachment and subsequent erosion from the test plots. Sediment yield is reduced in all cases to below 60% of the sediment yield from the unprotected soil. Obviously, there are variations in the performance of the various products tested. Some products are more successful in reducing sediment yield than others. Also, some products are more successful in reducing raindrop impact and soil detachment, while others are successful in slowing the rate of overland flow.

Note that the variations in performance may be due to a difference between the manufacturer's intended product application and the way it was actually tested. The products in this study were evaluated on their performance in the absence of vegetation, unfilled with soil, and fastened to bare soil at the upper reaches of a slope where raindrop impact is important and shear stresses imposed by overland flow are somewhat less important. Some products tested are intended for use with vegetation and to provide interlocking for plant root systems. A true evaluation of these products would involve testing each of them under the conditions for which it is specifically intended. Nevertheless, all of the products tested will probably experience the conditions under which they were tested in this study at some time immediately following installation. That is, they are filled with soil before vegetation is established and invariably they will experience rainfall. Clearly, a major research and development thrust should be focused on such laboratory testing programs.

Until a product-specific database along with standardized testing becomes available, broad generalized approaches toward controlling soil erosion behavior are ongoing. Rather than focusing on specific products, the generalized classes of materials listed in Table 8.1 are being tested. Figure 8.5 presents long-term design velocities versus storm flow duration. Note that allowable flow velocities decrease with flow duration. Manufacturers of both natural and geosynthetic erosion control products often present the erosion resistance of their materials in terms of maximum allowable flow velocity. Though unstated, these flow limits are typically for very short durations (minutes rather than hours) and do not reflect the potential for severe erosion damage that results from moderate flow velocities over a period of several hours. For further details refer to the discussion in Theisen and Carroll [7].

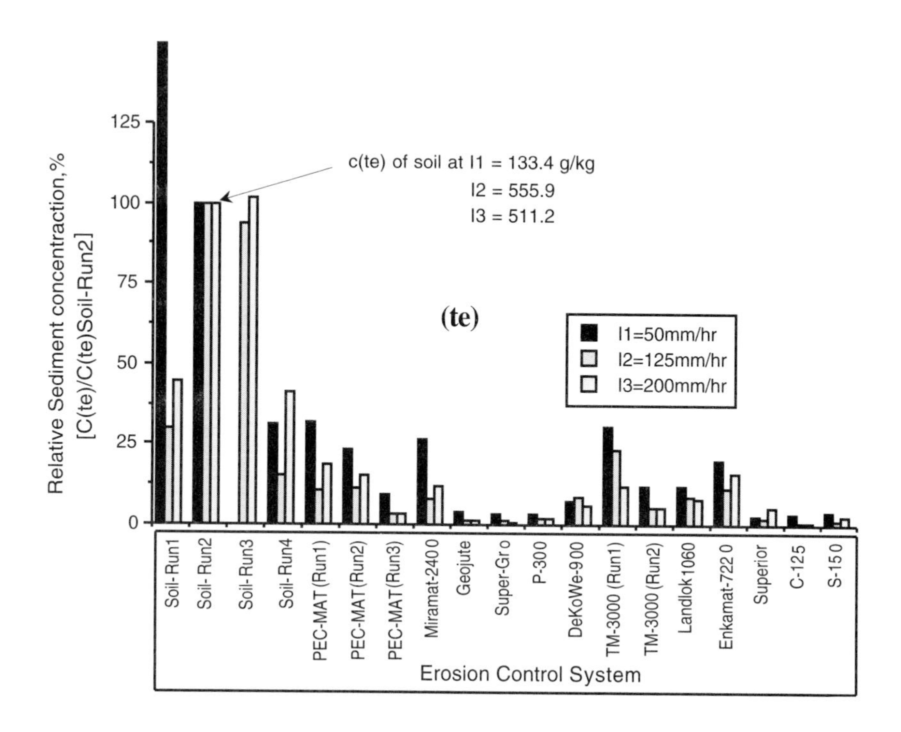

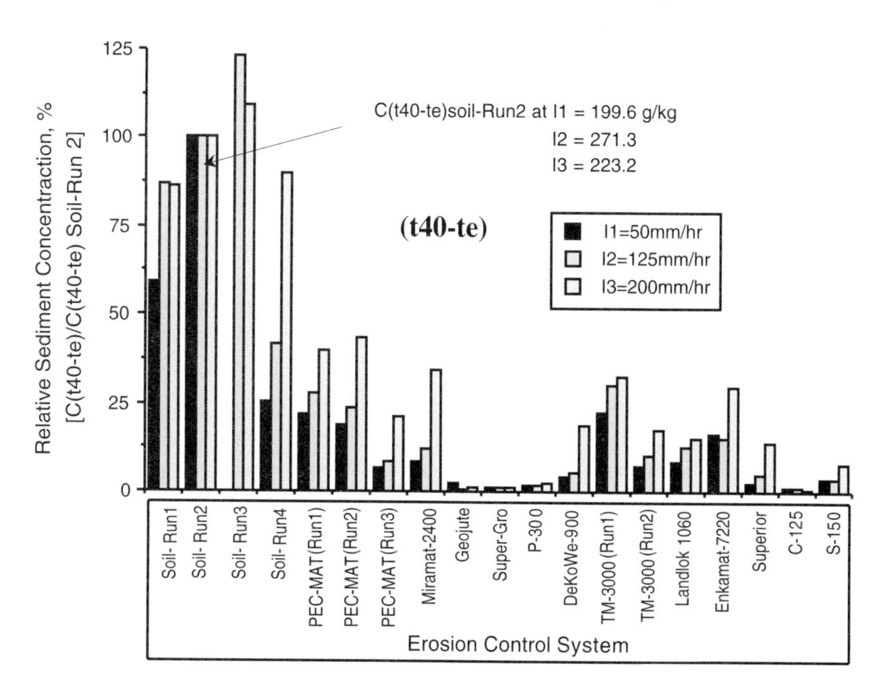

Figure 8.4 Laboratory soil erosion test results using various erosion control products. (After Rustom and Weggel [6])

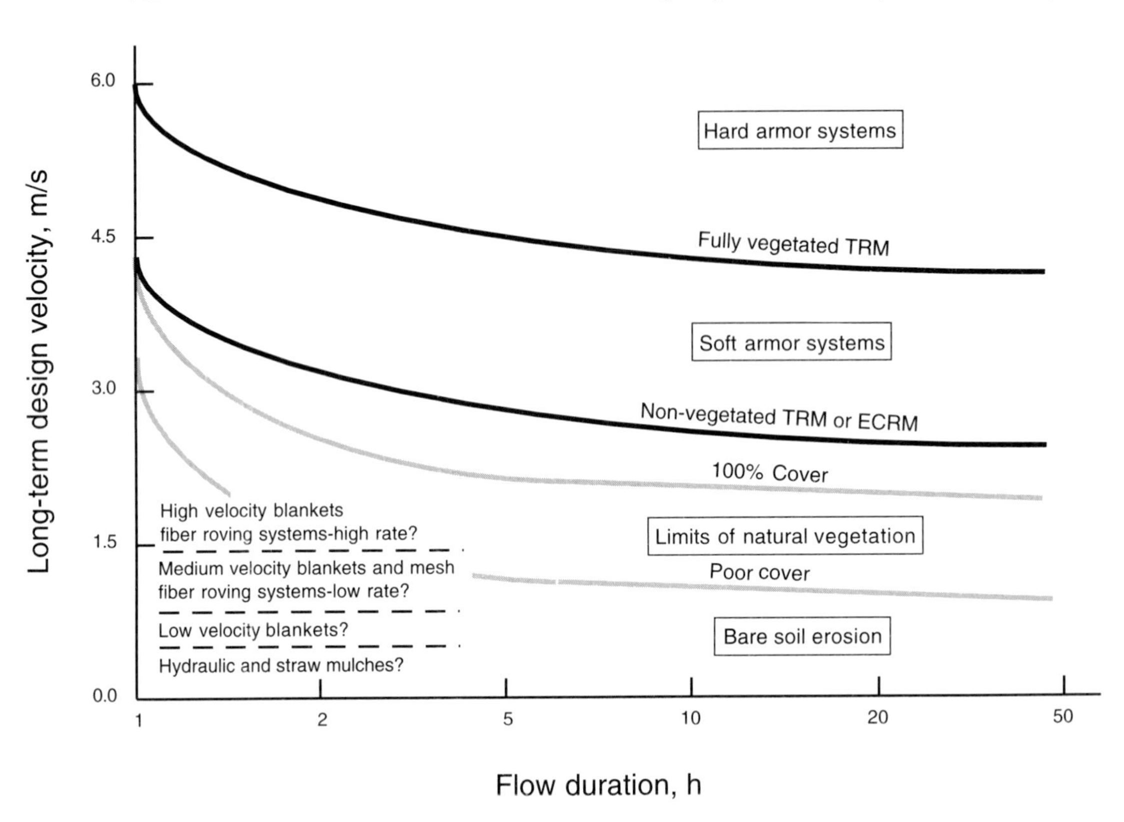

Figure 8.5 Recommended maximum design velocities for various classes of erosion control materials. (After Theisen [2])

8.2 REINFORCEMENT

While conventional geotextiles can be made very strong—woven multifilament fabrics have been made and used with tensile strengths up to 500 kN/m—the use of polymeric fibers with other materials can make the result synergistically stronger, and sometimes for less cost, too. Furthermore, configurations completely different from the fabric-like systems we've discussed can sometimes be created. The focus in this section is the use and potential synergism between two (or more) materials in reinforced geocomposites: reinforced geotextiles, reinforced geomembranes, reinforced soil, or other reinforced construction materials, such as concrete or bitumen. As will be seen, many innovative systems have been developed and are currently available.

8.2.1 Reinforced Geotextile Composites

Geotextiles have been reinforced by other polymers, by fiberglass, and by steel. Each will be described in this section.

Polymers. Many possibilities exist for making fibers from two polymers or fabrics from two fibers. One of the early bicomponent fiber types, made into a nonwoven heat bonded geotextile, was of the former type. It had a polyester core surrounded by a polypropylene sheath. The outer polypropylene sheath was bonded to crossover fibers at the intersections. This was nicely designed, since the melting temperature of polypropylene is somewhat lower than polyester. More common, however, is the use of bundled high-strength fibers protected by an outer covering. Some of these systems (called ParaProducts by ICI Fibres, Ltd., U.K.) consist of parallel high tenacity polyester or polyaramid (nylon) fibers encased in a polyolefine sheath. The polyolefine is either polypropylene or polyethylene. The core, however, is what gives the material its high strength. These materials are available in a number of different forms trademarked under the names: ParaStrips, ParaTies, ParaGrids, ParaLinks, and ParaWebs; the last having wide width tensile strengths up to 100 kN/m.

Fiberglass. As a synthetic material, fiberglass represents a great potential resource for geotechnical engineering reinforcement applications. Fiberglass has excellent mechanical properties, including high tensile strength, high modulus, and high creep resistance. Conversely, when buried in soil fiberglass can experience corrosion or pitting with a related strength loss (unless specially formulated), and its resistance to abrasion can be low. Nevertheless, many uses of fiberglass have been successful. For reinforcement, fiberglass certainly can be used in conjunction with polymeric fibers or as a product by itself. Some possible manufacturing processes for accomplishing a fiberglass fiber inclusion in a geotextile are weft- or warp-insertion knit fabrics, and triaxial, diagonal, or bias woven fabrics. Most systems of this type are still in the development stage. Fiberglass geogrids, however, are being manufactured and installed on a regular basis (recall Chapter 3).

Steel. In a manner similar to that just described, steel strand or cable can be used with different polymeric fabrics as the host material. DeGroot [8] describes a system called Geo-Fleximat, developed by Bekaert in Belgium. This product is a woven network of steelcord and geotextile with tensile strengths up to 3000 kN/m. Such high-strength materials are used as direct road support, sometimes being placed directly beneath the asphalt surfacing where strains tend to be a maximum. Field results show that the number of load repetitions can be raised from 3 to 10 times over nonreinforced sections.

Kenter et al. [9] report on a steel reinforced polypropylene woven geotextile with strength up 2000 kN/m called Mommoth-Mat. Developed by Robusta in the Netherlands, this material acts as a direct unpaved road support without aggregate. This type of "instant road," however, when placed on soft subgrade soils, must act as its own anchoring system. For the prevention of continuing rutting, Mommoth-Mat has prestressing rods or springwire bars in the cross machine direction. Two types are available, depending on the stability of the subsoil.

No details are available on the joining methods for either Geo-Fleximat or Mommoth-Mat. It is a potential problem, since load transfer across transverse joints is difficult with these extremely high strength geocomposites. The longitudinal joint would be even a larger potential problem if the width of the mat were smaller than the width of the roadway. Sewing is simply not possible; for example, Figure 2.12 shows that strength efficiencies become unacceptably low long before the strengths of these types of geocomposites are realized. This leaves resin bonding, perhaps in conjunction with mechanical joining, as the only possibility. A major research effort seems to be warranted in this regard. The other option might be to design the system to avoid connections altogether. The next case history describes how this could be accomplished.

A spectacular application in which steel reinforced woven geotextiles have been used to great advantage was as the support system for sea-bottom mattresses used to support large concrete piers in The Netherlands. These mattresses consist of three layers of filter soils, each 120 mm thick. Each mattress measures 42 m $\times$ 200 m and weighs 5500 tonnes. The high strength fabric supports the mattress as it is being constructed, rolled onto a handling drum, and transported onto a seagoing vessel. The mattresses are deployed from the roll directly onto the ocean floor in water up to 35 m deep. This particular steel reinforced fabric has steel cable running in the length direction and is capable of developing a 790 kN/m tensile strength. The manufacturer was Robusta, which worked with the consortium of companies that constructed the Eastern Scheldt Storm Surge Barrier. For this project a special on-site factory was constructed for the fabrication of the mattress. Its details will be described later in Section 8.3 on geocomposite filtration. A major point regarding the high strength fabric support material is that seams were not necessary. The mats were constructed in one huge continuous roll!

8.2.2 Reinforced Geomembrane Composites

While the primary function of a geomembrane is indeed as liquid or vapor containment, it still can be subjected to tensile stresses and, as such, must be capable of adequate performance in this regard. The scrim reinforced multiply materials described in Chapter 5 are of this type. Also described in Chapter 5 are spread coated geomembranes, in which the polymer is applied directly to a geotextile substrate. Here the geotextile vastly changes the tensile performance of the composite and is certainly a reinforcement component. Not described in Chapter 5 are geomembrane/geonet composites and geomembrane/geogrid composites. These variations will be described in Section 8.5 on moisture barrier geocomposite systems.

8.2.3 Reinforced Soil Composites

By suitably mixing soil and polymer element(s), a reinforced soil composite results. These interesting systems are described in this section.

Fibers and Meshes. Fiber reinforcement has long been used to enhance the brittle nature of cementatious materials, so it should come as no surprise that similar attempts should be made with polymer fibers in soil. Most work has been done with cohesionless sands and gravels, but cohesive silts and clays might benefit as well. Usually, the fibers [10] are 25 to 100 mm in length; meshes, or microgrids [11], are of a similar size. The composite material must be uniformly mixed as a first step and then placed and compacted in layers or sections where desired. Based on laboratory tests, McGown et al. [12] have found that mesh elements in 0.18% weight proportion resulted in an apparent cohesion of 50 kPa for a granular soil. What the optimal behavior is for different soils, different fibers or meshes, different sizes and percentages of fibers or meshes, and so on, all awaits additional research.

Continuous Fibers. Laflaive [13] has pioneered the application of mixing continuous polyester threads with granular soil to steepen and/or stabilize embankments and slopes. The technique, called Texsol, uses a specially designed machine capable of dispensing 23 m^3/hr of soil mixed with fibers coming from 40 bobbins, resulting in a weight percent of 0.1 to 0.2%. The finished fiber reinforced soil has fascinating properties. The system has been used in France where highway slopes of 60° have been constructed and have remained stable. Large field trials with enormous surcharges have failed to destroy the thread-reinforced soil mass. Laboratory studies on continuous fiber reinforced granular soils have resulted in apparent cohesion values in excess of 100 kPa [14]. The use of the technique in the widening of highways or railroads that are in cut areas is quite attractive.

Three-Dimensional Geocells. Rather than rely on friction, arching, and entanglements of fiber or mesh for improved soil performance, geosynthetics can be manufactured so that they physically confine the soil. Such confinement is known to vastly improve granular soil shear strength, as any triaxial shear test will substantiate. Furthermore, the increased shear strength due to confinement results in excellent bearing capacity.

The U.S. Army Corps of Engineers [15] in Vicksburg, Mississippi, has experimented with a number of confining systems, from short pieces of sand filled plastic pipes standing on end to cubic confinement cells made from slotted aluminum sheets to prefabricated polymeric systems called sand grids or *geocells.* Geocells are typically made from HDPE strips 100 mm wide and approximately 1.2 mm thick. They are ultrasonically welded along their 100-mm width at approximately 300 mm intervals and are shipped to the job site in a collapsed configuration (see Figure 8.6). At the job site they are placed directly on the subsoil's surface and propped open in an accordian-like fashion with an external stretcher assembly. This section expands to a 5 by 10 m area of hundreds of individual cells, each approximately 250 mm in diameter. They are then filled with sand and compacted using a vibratory hand-operated plate compactor. The final step is spraying on an emulsified asphalt (approximately 60% asphalt in a 40% water suspension) at the rate of approximately 5 l/m^2. The water drains through the sand, leaving the asphalt globules in the upper portion of the sand, thereby forming a temporary wearing surface. Expanded, the system appears as shown in Figure 8.6. There are a number of manufacturers that make different products within this geocell category. For example, Tenweb by Texax Corp., TerraCell by Webtec Inc., and Geoweb by Presto Products, Inc. all use high density polyethylene for the cell material, while Armater by Akzo Inc. and Nidaplast by Induplast, Inc., use geotextiles for the cell materials. The manufacturers should be consulted for the various material and geometric properties and for the latest styles that are available. Using the system described above, tests have been conducted that have supported tandem-axle truck loads of 230 kN for 10,000 passes with only slight rutting. Without the system the same trucks become bogged down in deep ruts after only 10 passes.

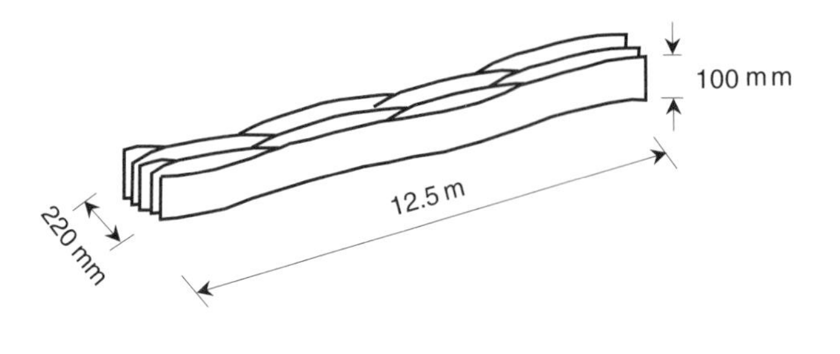

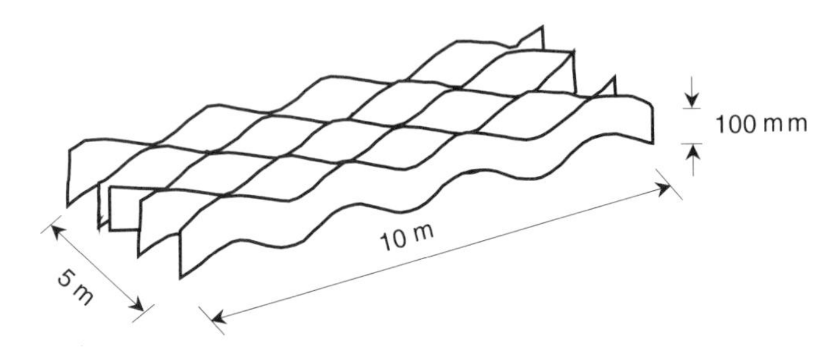

Figure 8.6 Sketches and photograph of three-dimensional cell for soil stabilization. (After Tenax Corp.)

In terms of design, such systems are quite complex to assess. If we adapt the conventional plastic limit equilibrium mechanism, as used in statically loaded shallow foundation bearing capacity (see Figure 8.7a), its failure mode is interrupted by the vertically deployed strips. For such a failure to occur, the sand in a particular cell must overcome the side friction, and punch out of it, thereby loading the sand beneath the level of the mattress (see Figure 8.7b). This, in turn, fails in bearing capacity, but now

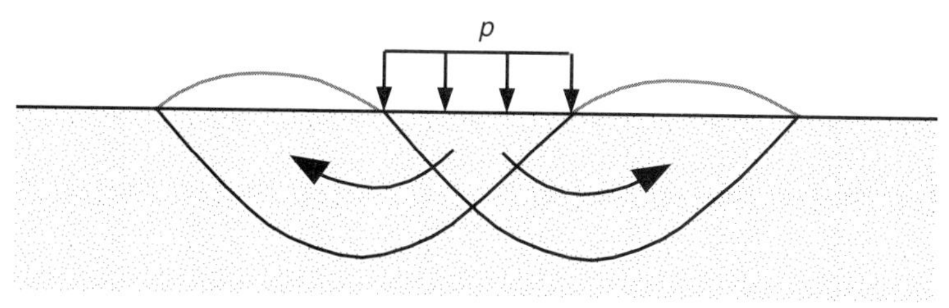

(a) Without mattress

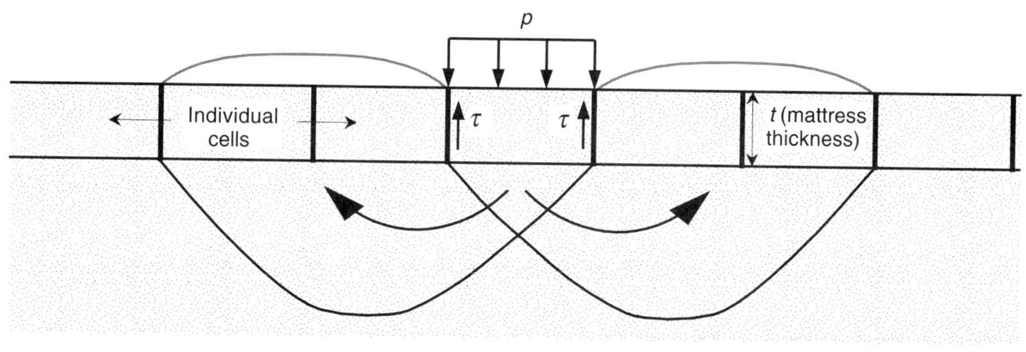

(b) With mattress

Figure 8.7 Bearing capacity failure mechanisms of sand without and with a geocell confinement system.

with the positive effects of a surcharge loading and higher density conditions. The relevant equations are as follows, illustrated by Example 8.1.

Without mattress:

$$p = cN_c\zeta_c + qN_q\zeta_q + 0.5\gamma BN_\gamma\zeta_\gamma \tag{8.2}$$

With mattress:

$$p = 2\tau + cN_c\zeta_c + qN_q\zeta_q + 0.5\gamma BN_\gamma\zeta_\gamma \tag{8.3}$$

where

p = maximum bearing capacity load (≅ tire inflation pressure of vehicles driving on the system if this is the application);
c = cohesion (equal to zero when considering granular soil such as sand);
q = surcharge load (= $\gamma_q D_q$), in which
γ_q = unit weight of soil within geocell, and
D_q = depth of geocell;
B = width of applied pressure system;
γ = unit weight of soil in failure zone

N_c, N_q, N_γ = bearing capacity factors, which are all functions of ϕ (where ϕ = the angle of shearing resistance (friction angle) of soil; see any geotechnical engineering text);

$\zeta_c, \zeta_q, \zeta_\gamma$ = shape factors used to account for differences from the plane strain assumption of the original theory; and

τ = shear strength between geocell wall and soil contained within it; note that $\tau = \sigma_h \tan \delta$ (for granular soils), in which

σ_h = average horizontal force within the geocell ($\cong pK_a$),

p = applied vertical pressure,

K_a = coefficient of active earth pressure. Note $K_a = \tan^2 (45 - \phi/2)$, for Rankine theory, and

δ = angle of shearing resistance (friction angle) between soil and the cell wall material ($\cong$ 15 to 20° between sand and HDPE, $\cong$ 25 to 35° between sand and the nonwoven geotextile).

Example 8.1

Compare the ultimate bearing capacity of a sand soil (a) without and (b) with a geocell 200 mm thick under the conditions given below:

(a) Without geocell

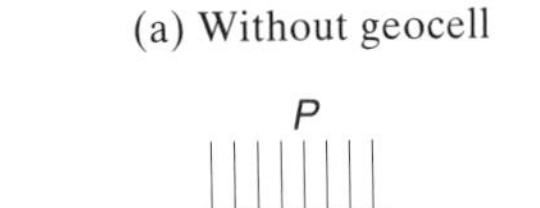

γ = 15kN/m^3
ϕ = 20 deg.
c = 0

(b) With geocell

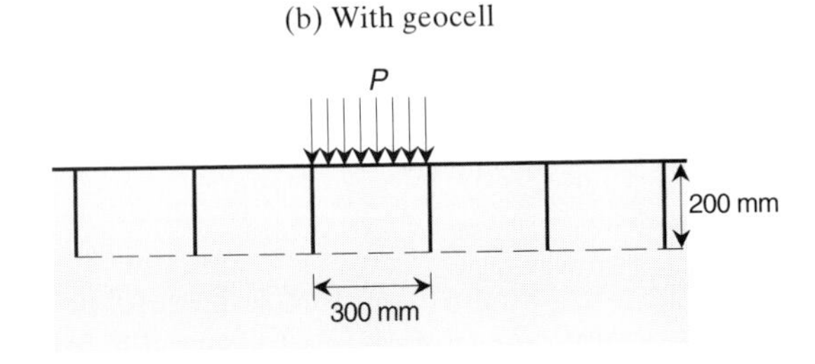

γ = 16kN/m^3
ϕ = 27 deg.
c = 0
σ_h = 20kPa (est.)
δ = 18 deg.

Solution

(a) Without a geocell

$$p = cN_c\zeta_c + qN_q\zeta_q + 0.5\gamma BN_\gamma\zeta_\gamma$$

Since $c = 0$ and $q = 0$;

$$p = 0 + 0 + (0.5)(15)(0.30)(5.39)(0.60):]$$

$$= 7.3 \text{ kPa}$$

(b) With a geocell, only $c = 0$.

$$p = 2\tau + cN_c\zeta_c + qN_q\zeta_q + 0.5\gamma BN_\gamma\zeta_\gamma$$

$$= 2(20)\tan 18^\circ + 0 + (0.2)(16)(13.2)(1.51) + 0.5(16)(0.30)(14.47)(0.60)$$

$$= 13.0 + 0 + 63.8 + 20.8$$

$$= 97.6 \text{ kPa} \quad \text{which is 13 times greater than without the geocell}$$

To the description of geocells, the design method, and the various types of geocell products that are available, we can add the following.

- The improvement shown using geocells is very pronounced.
- The use of thicker (i.e., greater depth) geocells will give proportionately greater improvement (see below).
- The use of a geotextile cell wall material with higher δ values than that used in the above analysis will give a proportionate improvement.
- With an increased densification of the soil infill, the improvement can be exponential.
- The dynamic effects of sand working under the mattress and gradually lifting it up out of position due to moving vehicles has not been considered. It is relevant, however, since it is a possible mode of failure.
- The solution given is for static bearing capacity; thus it is also suited for such problems as building foundations, embankment loads, earth dams, and retaining walls.
- Neither the foundation conditions nor the backfill types have to be cohesionless soils, as illustrated. Cohesive soils can be used in both situations and easily accounted for in the analysis.
- Geocell mattresses have been successfully used in slope stability situations but the analysis requires some major assumptions. Work appears warranted in this regard.

Three-Dimensional Mattresses. A deeper, more rigid mattress can be developed using a three-dimensional geosynthetic structure consisting, for example, of gravel-filled geogrid cells (recall Figures 3.20 and 3.21). These cells are typically 1.0 m deep and can be either square or triangular in plan view. They are joined together by an interlocking knuckle joint with a steel or plastic rod threaded through the intersection forming the coupling. This is called a *bodkin* joint. Both Tensar and Tenax geogrids (recall Chapter 3) can be joined in this manner. Other geogrids mentioned in Chapter 3 must be joined by hog rings or other mechanical fasteners. The filling sequence is important, and John [16] suggests the following:

1. Fill the first two rows of cells to half height
2. Fill the first row to full height
3. Fill the third row to half height
4. Fill the second row to full height
5. Advance by repeating the sequence of half-height and full-height filling in Steps 1–4

Edgar [17] reports on a three-dimensional geogrid mattress that somewhat parallels the geocells described above. The soil-filled geogrid mattress was constructed as shown in Figure 3.20 over soft fine grained soils. On top of it a 32 m high embankment was successfully placed. It was felt that the nonreinforced slip plane was forced to pass vertically through the mattress and therefore deeper into the stiffer layers of the underlying subsoils. This improved the stability to the point where the mode of failure was probably changed from a circular arc to a less critical plastic failure of the soft clay. The experience was considered to be a successful (and economical) one and parallels the mattress support system for a landfill in Germany shown in Figure 3.21.

8.2.4 Reinforced Concrete Composites

Historically, fibers have been used to reinforce many different types of building materials. Some classic uses are straw in bricks, animal hair in plaster, and asbestos in cement. More recently, however, fiber reinforced concrete has concentrated on steel, glass, and plastic fibers being placed in mortar, concrete, gunite, and shotcrete to improve their mechanical characteristics, particularly those of tensile, flexural, and impact strength. An overview of these fibers is given in Table 8.2. The steel fibers are either round (made from wire) or rectangular (made from shearing sheets or from flattened wire) with lengths of 6 to 75 mm. Aspect ratios (length to thickness) range from 30 to 150. Smooth, crimped, and deformed fibers have also been used.

Typical glass fibers have diameters ranging from 0.005 to 0.15 mm but are usually bonded together to form diameters of 0.013 to 1.3 mm. Lengths are generally from 13 to 50 mm.

Typical plastic fibers that have been used are nylon, polypropylene, polyethylene, polyester, and rayon. Fiber diameters 0.020 to 0.38 mm have been used; lengths of 13 to 50 mm being customary.

In general, the addition of fibers to cementatious materials results in the following improvements: greater resistance to cracking; holding cracked sections together;

TABLE 8.2 TYPICAL PROPERTIES OF FIBERS

Type of Fiber	Specific Gravity	Tensile Strength (kPa)	Elastic Modulus (kPa)	Ultimate Elongation (%)
Acrylic	1.1	2 to 4 × 10^5	2 × 10^6	25 to 45
Asbestos	3.2	5 to 9 × 10^5	80 to 140 × 10^6	~0.6
Cotton	1.5	4 to 7 × 10^5	5 × 10^6	3 to 10
Glass	2.5	10 to 40 × 10^5	70 × 10^6	1.5 to 3.5
Nylon (high tenacity)	1.1	7 to 8 × 10^5	4 × 10^6	16 to 20
Polyester (high tenacity)	1.4	7 to 9 × 10^5	8 × 10^6	11 to 13
Polyethylene	0.95	~7 × 10^5	0.14 to 0.41 × 10^6	~10
Polypropylene	0.90	5 to 7 × 10^5	3 × 10^6	~25
Rayon (high tenacity)	1.5	4 to 6 × 10^5	7 × 10^6	10 to 25
Rock wool (Scandinavian)	2.7	5 to 7 × 10^5	70 to 110 × 10^6	~0.6
Steel	7.8	3 to 40 × 10^3	200 × 10^6	0.5 to 3.5

Source: After Batson [18].

greater resistance to thermal changes, particularly shrinkage; thinner design sections; less maintenance, and longer life.

There is little standardization available regarding the amount of fibers to add to cementatious materials. The criterion is often dictated by how much fiber can be added before the mix becomes unworkable. This depends not only on the volume, aspect ratio, type, and kind of fiber but also on the aggregate size, amount of sand, and amount of cement. Hoff [19] recommends 9.5 mm maximum aggregate size, using more sand than normally required, aggregate ratios between 50:50 and 70:30, and high cement contents up to 5.0 kN/m^3.

8.2.5 Reinforced Bitumen Composites

Including polymeric geogrids with bituminous pavement base courses with the idea of increasing the lateral modulus has been attempted. A number of the geogrids mentioned in Chapter 3 have been used. The attempts have only been of marginal success for two reasons: the softening (and sometimes the partial loss of prestress) of the polymers under the rather high temperature of the bituminous material as it is placed, and placement difficulties where the paving machine tends to pick up the geogrid and distort or tear it.

Rather than using geogrids, an alternative is to use discrete fibers generally made from polypropylene, although polyester is also a possibility. The fibers are usually #4 denier, approximately 10 to 12 mm in length with tensile strength of 15 g or higher. The melting point must be at least 160°C. The fibers (approximately 3 kg/ton) have been used in many applications where measurable benefits have arisen, most importantly, that crack formation is delayed for an additional one to four years over nonreinforced mixes, and that the tendency of creep is also delayed, leading to less rutting of the pavement's surface. Jenq [20], among others, suggests that rigorous field testing be pursued for further quantification.

8.3 FILTRATION

When a geosynthetic is asked to perform multiple functions, a geocomposite can be formed wherein each function is addressed using a single material. When an ample market is available, or envisioned, such products are quick to appear. An example is in the railroad industry, where two material properties of geotextile separators are critically important: high abrasion resistance on the upper side against ballast stone, and filtration of very fine particles on the lower side against the subsoil (recall Figure 2.78). To accomplish this with a single geotextile, for example, is quite difficult. The high abrasion resistance can be achieved nicely with a resin dipped, force-air-dried geotextile, but this process leaves quite large voids in the fabric, where loss of soil would likely be a problem. The problem is overcome by attaching (e.g., by needling) the abrasion resistant geotextile on top with a tight nonwoven geotextile on the underside, to allow for the desired result.

Another example is that of a geotextile required above a drainage geonet and below the clay layer of a composite primary liner in a landfill (recall Figure 5.41f). In order for the clay not to extrude into the geonet, we require the use of a nonwoven geotextile. Yet such geotextiles, particularly if needle punched, have relatively low modulus values and intrude into the geonet's core space, markedly reducing flow capacity. One possible solution to this situation is the use of a scrim reinforced nonwoven needle punched geotextile. The resulting geocomposite filter serves both the filtration and the reinforcement functions. Potentially any number of such combinations are possible.

By far, the most elaborate filtration application using geosynthetics in connection with natural soils has been by the Dutch in connection with the Eastern Scheldt Storm Surge Barrier. (A series of papers from the Second International Conference on Geotextiles provides many details on this fascinating project [21, 22, 23, 24]). To form a gate-type dam across the delta of the Scheldt River, 66 prefabricated piers and 63 sliding gates fitting between the piers have been constructed. The gates are open during normal flow and closed only when high seas are in the North Sea. To ensure the stability of the sandy soil beneath the piers (which are 35 to 45 m high and weigh 18,000 kg each; see Figure 8.8a), a very elaborate filter mattress consisting of geotextiles, wire-mesh containment baskets, and natural soils was developed. Each bottom mattress, measuring 200 m $\times$ 42 m and 0.36 m thick, consists of three layers (see Figure 8.8b): a lower layer of 110 mm of sand (0.3 to 2.0 mm), a middle layer of 110 mm of sand-gravel (2.0 to 8.0 mm), and an upper layer of 140 mm of gravel (8.0 to 40 mm). The lower support geotextile is a woven multifilament geotextile reinforced with steel cables described previously in Section 8.2.1. Its function is reinforcement, since it supports the entire 5,000,000 kg mattress during its deployment. Above this, and separating the three soil layers, is a lightweight nonwoven heat bonded polypropylene geotextile used as a separator between the different-sized soils. Within each layer of filter soil, partitions are created by using steel wire baskets lined with geotextiles to prevent the soil from shifting during construction. The uppermost fabric is a 400 g/m^2 knit polypropylene geotextile. To tie the mat together vertically, 6-mm-diameter steel pins are inserted and secured with the snap-lock grommets on the top and bottom of the mattress.

This entire mattress is placed on a vibratory compacted seafloor in water up to 30 m deep at 90° to the axis of the dam itself. Each mattress is centered under each concrete pier, the piers being placed at 45 m spacings; thus a 3 m space is between each mattress. It was felt that the stiffness of the mat was too great for a continuous system, which therefore necessitated the gap. This gap was filled in later with larger stones, forming, in effect, another filter.

As shown in Figure 8.8a, a second and smaller mattress measuring 60 m $\times$ 31 m $\times$ 0.36 m thick was placed above the lower one. This upper mat is assembled in exactly the same way as the lower mat except that it is composed of three layers of uniform-size 8.0 to 40 mm gravel. The purpose of this upper mat is to distribute the weight of the concrete pier sitting directly on top of it and to protect the lower mat. The remainder of the concrete pier is protected with large armor stone.

Indeed, this project is among the greatest achievements of any type of heavy construction and is one that uses geocomposites as filter mattresses as a key element in its successful construction and anticipated durability, which is estimated to be 200 years.

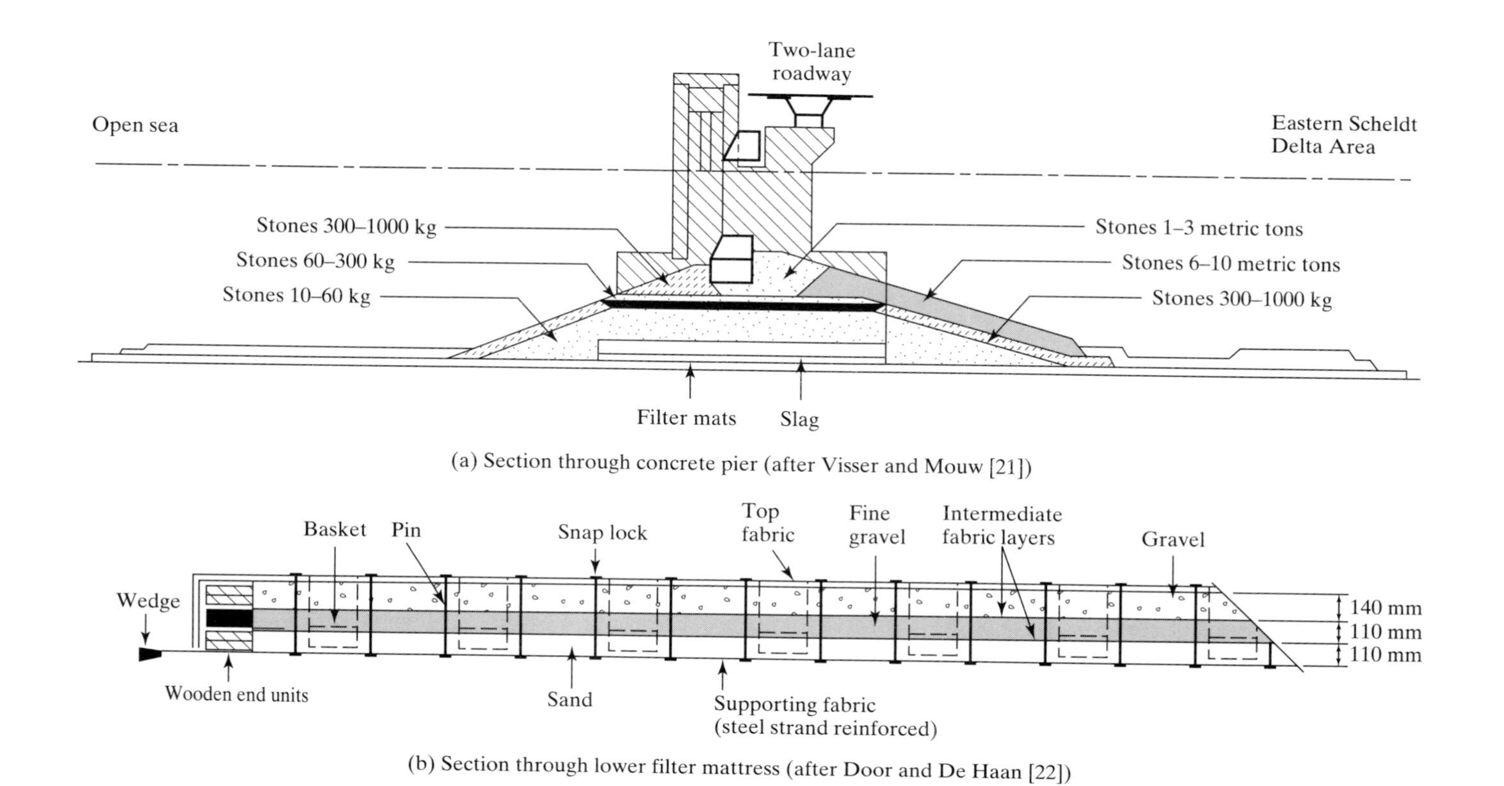

Figure 8.8 Cross sections of the Eastern Scheldt Storm Surge Barrier in Holland.

8.4 DRAINAGE

Geocomposites used as drainage media have completely taken over certain geotechnical application areas in an amazingly short time. Wick drains, used instead of sand drains, and sheet drains behind retaining walls are such areas. Prefabricated highway edge drains are also being used to a great extent. Each of the above topics will be discussed.

All of the drainage products to be discussed consist of an encapsulated polymeric core for accommodating the anticipated in-plane flow. This is necessary since the typical geotextiles described in Chapter 2 are generally not thick enough to handle the required flow rates. For example, the different types of geotextiles have been evaluated by Gerry and Raymond [25], who give typical values of transmissivity. However, as shown in Examples 2.25–2.28 (Section 2.9), some drainage situations result in flow rates that are too great to be handled by any conventional geotextile. Thus there exists a need for specialty products capable of handling significantly higher flow rates.

As a reference plane for many of the problems to follow, it is customary to see drainage designs using 300 mm of clean sand soil. For a typical permeability of 0.001 m/s at a hydraulic gradient of 1.0, this is equivalent to the following Darcian flow rate, which can be modified into a flow rate per unit width or q/W value.

$$q = kiA \tag{8.4}$$

$$q = ki(W \times t)$$

$$\frac{q}{W} = (k)(i)(t) \tag{8.5}$$

where

q = in plane flow rate,
k = hydraulic conductivity (permeability),
i = hydraulic gradient,
A = cross-sectional area of flow,
W = width of flow, and
t = thickness.

$$\frac{q}{W} = 0.001(1.0)(0.3)$$

$$\frac{q}{W} = 3 \times 10^{-4}\ \text{m}^2/\text{s}$$

In comparison to even the thickest geotextile, this value is two orders of magnitude higher, hence the need for specialized geocomposite drainage materials.

8.4.1 Wick (Prefabricated Vertical) Drains

Generally, geosynthetic materials made from polymers (polypropylene, polyester, polyethylene, etc.) *do not wick.* These polymers by themselves are quite hydrophobic—they actually repel water. If the fabric pore structure is full or partially saturated with water, it will exist in the voids and wait for some external source like gravity or pressure to initiate its movement. So why do most people refer to the subject of this section as *wick drains*? The answer is probably that when these geocomposites are vertically inserted in the ground with their ends protruding at the ground surface and with water being forced out of them under pressure, they resemble a set of giant wicks. The mechanism of flow, however, is *not* by wicking, as with a candle wick, so it must be clearly explained. A far better term for these materials is *prefabricated vertical drains* or PVDs, which is a commonly used term in Europe and Asia. It is also, however, a somewhat limiting term since the drains are often used at an angle other than vertical, and sometimes they are even used horizontally.

The method of rapid consolidation of saturated fine grained soils (silts, clays, and their mixtures) has been actively pursued using sand drains since the 1930s. The practice involves the placement of vertical columns of sand (usually 200 to 600 mm in diameter) at spacings of 1.5- to 6.0-m centers throughout the subsurface to be dewatered. Their lengths are site-specific but usually extend to the bottom of the soft layer(s) involved. Once installed, a surcharge load is placed on the ground surface so as to mobilize excess pore water pressure in the water in the soil voids. This surcharge load is placed in incremental lifts, and with each increment (and the simultaneous increase in excess pore water pressure in the underlying soil) drainage occurs via the installed sand drains. The water takes the shortest drainage path—which is horizontally—to the sand drain, at which point flow becomes vertical and very rapid since the sand is many orders of magnitude greater in permeability than the fine grained soil being consolidated. Critical in this method of consolidating soils rapidly (it does little insofar as the amount or magnitude of settlement is concerned) is the rate at which surcharge fill is added. Surcharge fill placement is controlled by the effective stress equation

$$\bar{\sigma} = \sigma - u_w \tag{8.6}$$

where

$\bar{\sigma}$ = effective (or intergranular) stress,
σ = total stress, and
u_w = excess pore water pressure (*excess,* since it is higher than normal hydrostatic conditions).

The increase in the excess pore water pressure increment (via the surcharge) should never be higher than the increase in the total stress increment. Thus the effective stress (directly related to the soil's shear strength) is never decreased, and as surcharge loading proceeds it should actually increase. Surcharge loads are usually earth fills, but have been accomplished using a geomembrane-contained water loading.

With this brief background of the concept (there is a wealth of information available on the technique [26]), and recognizing that millions of sand drains have been installed, there are nonetheless the following shortcomings:

- There is a distinct possibility that the sand drain may not be continuous if the installation mandrel is withdrawn too fast or if insufficient sand is in the mandrel during the withdrawal.
- There is a a vulnerability to shear failure of the foundation soil as surcharge load is being placed. Since the small diameter sand drains offer essentially no shear resistance, the technique is limited to very slow placement of the surcharge. Often this is as little as 2.0 kPa per day, which is only 100 mm per day, and even then instrumentation of the site is mandatory [26].
- A relatively large crane is needed to install the sand drains. This, in turn, requires a substantial soil layer, typically 1 to 2 m thick, to be placed over the site to begin the project, for the purpose of adequate bearing capacity.
- The sand for this soil layer and for the sand drains themselves may be difficult and costly to obtain.
- The material for the surcharge (which should be somewhat greater in its final ground surface pressure than the contact pressure of the proposed structure) might be difficult to obtain. Surcharge fill heights of up to 10 m are not uncommon.
- Most of this surcharge fill must be removed once the site is consolidated, which is sometimes difficult to do and often expensive.

By contrasting sand drains to the alternative of geocomposite wick drains (see Figure 8.9), a number of interesting features are revealed. The wick drains, usually con-

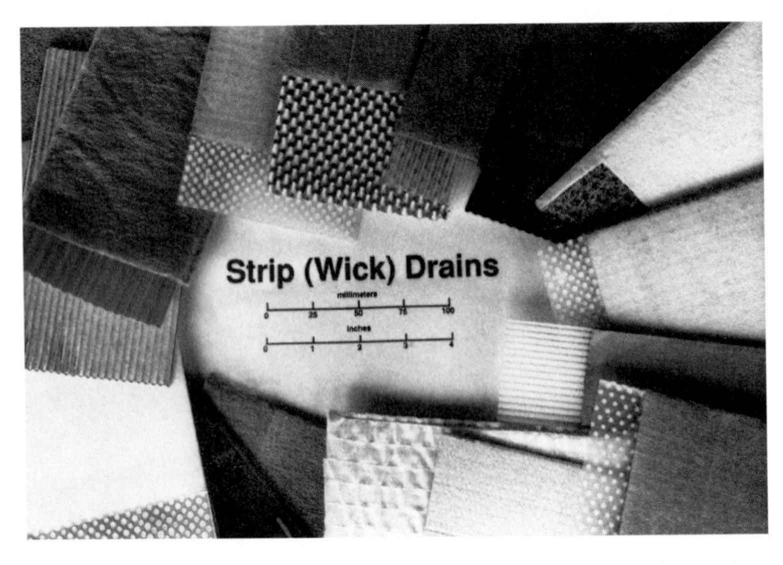

Figure 8.9 Various commercially available geocomposite wick drains (in previous editions these were called strip drains).

sisting of plastic fluted or nubbed cores that are surrounded by a geotextile filter, have considerable tensile strength. Typically, the breaking strength of a 100-mm-wide wick drain is 5 to 15 kN. When threaded throughout a site on centers of 1 to 2 m, they offer a sizable reinforcing effect. Although the effect has not been quantified in a three-dimensional analysis, the equivalent plane strain force is sizable. Furthermore, wick drains do not require any sand to transmit flow. The photographs of Figure 8.10 illustrate the installation process. The wick drain is threaded into a hollow lance, which is pushed (or driven) through the soil layer to be consolidated. The lance is withdrawn leaving the wick drain within the soil layer, which collapses around it. At the ground surface the ends of the wick drains (typically at 1 to 2 m spacing) are interconnected by a granular soil drainage layer or a geocomposite sheet drain layer. There are a number

Figure 8.10 Installation rig and associated details for the installation of wick (prefabricated vertical) drains. (compliments of Akzo Nobel Geosynthetics Co.)

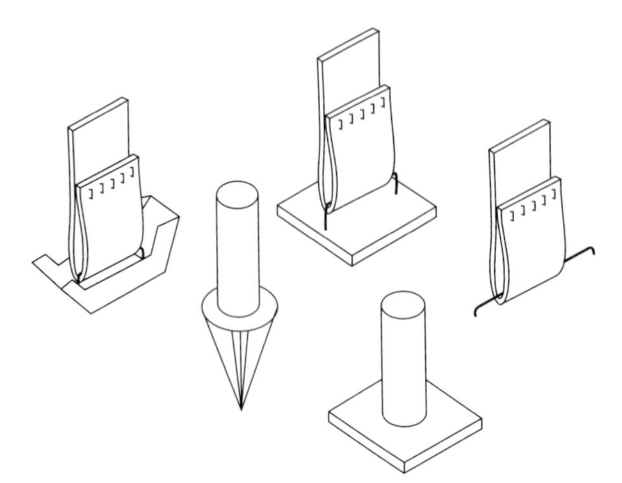

Figure 8.11 Different base plates to maintain wick drains at the base of the installation lance. (Compliments of M. Gambin).

of commercially available wick drain manufacturers and installation contractors who readily provide information on the current products, styles, properties, and estimated costs.

For installation, wick drains arrive at the site in rolls and are placed on the installation rig in dispensers like a huge roll of toilet paper. The end is threaded down inside a hollow steel lance, which must be as long as the depth to which the strip drains are to be installed. At the bottom of the lance, the wick drain is folded around a steel bar or other type of base plate (see Figure 8.11). The purpose of the base plate is to keep the wick drain down at the bottom of the lance and at the same time to keep the soft soil through which it will be placed out of the lower portion of the lance so that the drain properly releases when the lance is withdrawn. The entire assembly (lance, base plate, and wick drain) is now pressed into the ground to the desired depth. If a hard crust of soil or a high strength geotextile or geogrid is at the original ground surface, it must be pre-augered or suitably pierced beforehand. When it reaches the desired depth, the lance is withdrawn, leaving the base plate and wick drain behind. The rig moves and the process is repeated at the next location. It is a very rapid construction cycle (approximately 1 min), requiring no other materials than the wick drains and base plates.

Concerning the design method for determining wick drain spacings, the initial focal point is on the time for consolidation of the subsoil to occur. Generally, the time for 90% consolidation (t_{90}) is desired, but other values might also be of interest. Two approaches to such a design are possible. The first is an equivalent sand drain approach that uses the wick drain to estimate an equivalent sand drain diameter and then proceeds with design in the manner of sand drains. This is done by taking the actual cross-sectional area of the candidate wick drain and making it into an open void circle. This open void circle is then increased using the estimated porosity of sand to obtain the equivalent sand drain diameter.

Example 8.2

What is the equivalent sand drain diameter of a wick drain measuring 96 mm wide × 2.9 mm thick that is 92% void in its cross section? Use an estimated sand porosity of 0.3 for typical sand in a sand drain.

Solution:

$$\text{Total area of wick drain} = 96 \times 2.9 = 279 \text{ mm}^2$$

$$\text{Void area of wick drain} = 279 \times 0.92 = 257 \text{ mm}^2$$

The equivalent void circle diameter is

$$d_v = \sqrt{\frac{(4)(257)}{\pi}}$$

$$= 18.1 \text{ mm}$$

The equivalent sand drain diameter is

$$d_{sd} = \sqrt{\frac{d_v^2}{0.3}} = \sqrt{\frac{18.1^2}{0.3}}$$

$$= 33 \text{ mm}$$

Note that equivalent sand drain diameters for the various commercially available wick drains vary from 30 to 50 mm. Design for spacing versus time for consolidation now proceeds as per standard radial consolidation theory.

The second approach to wick drain design is more straightforward than the preceding approach and is the preferable one. As developed by Hansbo [27], the time for consolidation is given by the following equations, which are illustrated in Example 8.3.

$$t = \frac{D^2}{8c_h}\left[\frac{\ln(D/d)}{1-(d/D)^2} - \frac{3-(d/D)^2}{4}\right]\ln\frac{1}{1-\overline{U}} \tag{8.7}$$

This can be simplified, since d/D is small, to

$$t = \frac{D^2}{8c_h}\left(\ln\frac{D}{d} - 0.75\right)\ln\frac{1}{1-\overline{U}} \tag{8.8}$$

where

t = time for consolidation,
c_h = coefficient of consolidation of soil for horizontal flow,
d = equivalent diameter of strip drain ($\cong$ circumference/π),
D = sphere of influence of the strip drain (for a triangular pattern use 1.05 × spacing; for a square pattern use 1.13 × spacing), and
$\overline{U}$ = average degree of consolidation.

Example 8.3

Calculate the times required for 50, 70, and 90% consolidation of a saturated clayey silt soil using wick drains at various triangular spacings. The wick drains measure 100 × 4 mm and the soil has a $c_h = 6.5 \times 10^{-6}$ m²/min.

Solution: In the simplified formula above for d

$$d = \frac{100 + 100 + 4 + 4}{\pi}$$

$$= 66.2 \text{ mm}$$

so using Eq. (8.8),

$$t = \frac{D^2}{8(6.5 \times 10^{-6})}\left(\ln\frac{D}{0.0662} - 0.75\right)\ln\frac{1}{1-\bar{U}}$$

which results in the following table for consolidation times.

Theoretical Wick Drain Spacing, D (m)	Targeted Percent Consolidation ($\bar{U}$)		
	50%	70%	90%
2.1	159,000 (110)	276,000 (192)	529,000 (367)
1.8	110,000 (77)	192,000 (133)	366,000 (254)
1.5	71,000 (49)	123,000 (86)	236,000 (164)
1.2	41,000 (29)	72,000 (50)	137,000 (95)
0.9	20,000 (14)	35,000 (24)	67,000 (46)
0.6	7,000 (4.8)	12,000 (8.4)	23,000 (16)
0.3	910 (0.6)	1,590 (1.1)	3,030 (2.1)

Consolidation times are in minutes; the equivalent number of days are in parentheses.

These values are now plotted for the required design curves. Note that the D spacings must be decreased by 1.05 using a triangular strip drain pattern. When compared to the results using the equivalent sand drain method, these values are seen to agree very closely. Note the agreement in the 90% consolidation curves.

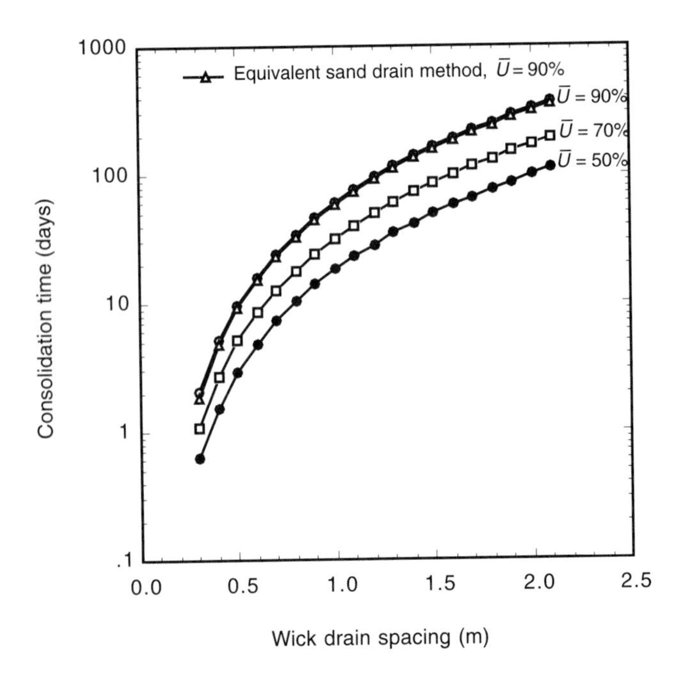

In summary, it is felt by the author that wick drains offer so many advantages over sand drains that wick drains should be used exclusively in the future. Strongly in their favor are the following items:

- The analytic procedure is available and straightforward in its use.
- Tensile strength is definitely afforded to the soft soil by the installation of the wick drains. It is, however, a difficult three-dimensional problem to quantitatively assess.
- There is only nominal resistance to the flow of water once it enters the wick drain (unless it kinks and limits the flow).
- Construction equipment is generally small, imparting low ground contact pressures on the soft soils.
- Installation is simple, straightforward, fast, and economic.

Additional research into wick drains is needed in the areas of the effects of soil smear on the geotextile filter and the effects of the kinking of the wick drain core. Soil smear stems from the distortion of the soil due to installation, withdrawal, and collapse of the in situ soil on the wick drain. Its effect is mainly on the horizontal coefficient of consolidation (c_h) and it is yet to be understood [28].

Kinking refers to the necessary shortening of the wick drain during the consolidation process. In some wick drains the tendency might be to fold into a tight S-shape, that is, to kink, thereby restricting or even cutting off flow. The subject of kinking on actual flow rates has been evaluated by Suits et al. [29] and Lawrence and Koerner [30]. The flow rate evaluation device is shown in Figure 8.12. Here flow rate in a wick drain in a standard (unkinked) condition can be evaluated and then compared to a kinked, or crimped, condition to determine what reduction is occurring. The results from two different kinking devices (one rounded and one pointed) at different applied pressures are shown in Table 8.3. Here the types of wick drains evaluated are classified according to their perceived stiffness as rigid, semiflexible, or flexible. The data indicate, however, that such apparent stiffness is not a particularly good indicator of flow rate behavior in a kinked state. Indeed, if kinking is an anticipated problem, the candidate wick drain(s) should be experimentally assessed.

The importance of knowing the flow rate in a kinked state is in allowing the required flow to pass through the wick drain at all times during its operational lifetime. This is the second part of design using wick drains (the first is to determine the spacing for a required degree of consolidation). Here the required flow rate (q_{reqd}) has been investigated by a number of researchers; see Table 8.4. A value from this table must now be compared to an allowable flow rate (q_{allow}) for the conventional design formulation used throughout this book.

$$FS = \frac{q_{\text{allow}}}{q_{\text{reqd}}} \tag{8.9}$$

where

FS = (global) factor of safety,
q_{allow} = allowable flow rate for the candidate wick drain, and
q_{reqd} = required flow rate from Table 8.4.

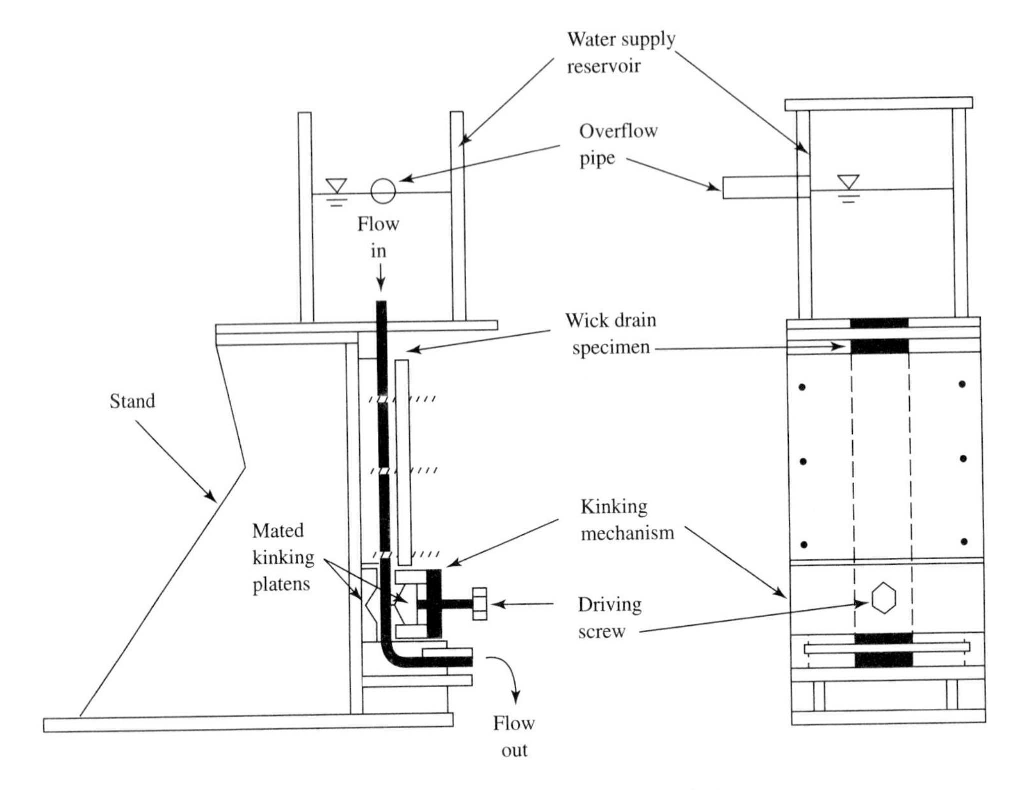

Figure 8.12 Side and front views of laboratory kinking test device.

For the allowable flow rate q_{allow} the ultimate flow rate from a ASTM D4716 test method should be obtained (recall Section 4.1.3). Typical values of ultimate flow rate at a hydraulic gradient of 1.0 under 200 kPa normal stress vary from 2.5 to 5.0 l/min for a 100 mm wide wick drain. This value must then be reduced on the basis of site specific reduction factors,

$$q_{allow} = q_{ult}\left[\frac{1}{RF_{IN} \times RF_{CR} \times RF_{CC} \times RF_{BC}}\right] \tag{8.10}$$

where

q_{allow} = allowable flow rate to be used in design,
q_{ult} = ultimate flow rate (as determined from ASTM D4716) for short-term tests,
RF_{IN} = reduction factor for elastic deformation of the adjacent geotextile intruding into the drainage core space,
RF_{CR} = reduction factor for creep deformation of the drainage core itself and/or intrusion of the adjacent geotextile into the drainage core space,
RF_{CC} = reduction factor for chemical clogging and/or precipitation of chemicals onto the geotextile or within the drainage core space, and

TABLE 8.3 FLOW RATE (DISCHARGE) TEST RESULTS FOR WICK DRAINS UNDER STANDARD AND KINKED CONDITIONS

		90° Wedge						13-mm Diameter Cylinder					
	Standard (Initial)	Seating Load (≅0.6 kN)		Moderate Load (3 kN)		Heavy Load (12 kN)		Seating Load (≅0.6 kN)		Moderate Load (3 kN)		Heavy Load (12 kN)	
Product	Flow Rate (lpm)	Flow Rate (lpm)	% Decrease	Flow Rate (lpm)	% Decrease	Flow Rate (lpm)	% Decrease	Flow Rate (lpm)	% Decrease	Flow Rate (lpm)	% Decrease	Flow Rate (lpm)	% Decrease
Rigid Types													
Bando	7.6	6.4	16	5.4	29	0.3	97	5.7	25	0.6	92	0.2	98
Castle Board	4.3	1.6	62	1.4	67	0.9	78	3.8	11	2.9	33	1.1	75
Desol	3.3	0.9	72	0.6	81	0.2	95	1.6	50	0.1	97	negl.	100
Semi-Flexible Types													
Aliwick	7.9	6.9	13	6.7	15	3.6	54	7.5	5	7.4	7	2.1	73
Ameridrain	6.3	5.1	20	4.5	28	0.3	95	6.1	4	5.8	8	1.8	71
Mebra	10.2	8.2	20	6.7	34	1.3	87	6.1	40	3.1	70	1.8	82
Vinylex	10.2	6.4	37	6.1	41	5.1	50	8.4	18	7.3	29	1.5	86
Flexible Types													
Alidrain	8.5	7.7	9	4.4	48	1.8	79	5.5	35	4.1	52	0.1	99
Alidrain "S"-F	5.1	4.0	21	3.9	23	3.4	32	4.7	7	4.4	13	3.1	38
Alidrain "S"-B	5.1	3.1	40	2.6	48	2.3	54	4.4	13	4.3	15	2.3	54
Colbond	3.8	2.8	25	2.4	36	0.6	85	3.4	9	2.2	43	0.2	96
Hitek Flodrain	12.8	6.5	49	2.2	83	0.1	99	10.2	20	2.9	78	0.1	99

Source: After Lawrence and Koerner [30].

TABLE 8.4 ESTIMATES OF REQUIRED FLOW RATES (DISCHARGE) FOR WICK DRAINS

Source	Required Flow Rate* (l/min)	Normal Stress (kPa)	Hydraulic Gradient
den Hoedt [31]	0.17	—	—
Kremer et al. [32]	0.30	100	0.62
Kremer et al. [32]	0.09†	—	—
Kremer et al. [33]	1.51	15	1.00
Holtz et al. [34]	0.23	400	—
Koerner et al. [35]	0.10	in situ	1.00
Rixner et al. [28]	0.19	in situ	1.00
Holtz and Christopher [36]	0.95	in situ	1.00
Bergado et al. [37]	0.76	in situ	1.00
	0.47†	in situ	1.00

*Note, in the literature many authors use the unit of m^3/yr for the required rate. 1.0 l/min = 526 m^3/yr

†In flattened S-configuration, i.e., in deformed state.

RF_{BC} = reduction factor for biological clogging of the geotextile or within the drainage core space.

A guide for typical values in Eq. (8.10) is presented as Table 8.5 (compare this with Table 4.2 for geonets). Note, however, that wick drains are temporary construction expedients, thus the chemical and biological clogging potential is probably quite low. Creep is dependent on the time the strip drains are required and the normal stress arising from the depth within the soil to be consolidated. For intrusion RF_{IN}, ASTM D4716 can be evaluated with soil above and below the wick drains. In this case, the intrusion reduction factor would be included as a value of unity. Now, having an in situ modified value of q_{allow}, a traditional design-by-function can be performed. See Example 8.4.

TABLE 8.5 RECOMMENDED REDUCTION FACTORS FOR EQ. (8.10) TO DETERMINE ALLOWABLE FLOW RATE OF DRAINAGE GEOCOMPOSITES (WICK DRAINS, SHEET DRAINS AND EDGE DRAINS)

Application Area	RF_{IN}	RF_{CR}^{*}	RF_{CC}	RF_{BC}
Sport fields	1.0 to 1.2	1.0 to 1.2	1.0 to 1.2	1.1 to 1.3
Capillary breaks	1.1 to 1.3	1.0 to 1.2	1.1 to 1.5	1.1 to 1.3
Roof and plaza decks	1.2 to 1.4	1.0 to 1.2	1.0 to 1.2	1.1 to 1.3
Retaining walls, seeping rock and soil slopes	1.3 to 1.5	1.2 to 1.4	1.1 to 1.5	1.0 to 1.5
Drainage blankets	1.3 to 1.5	1.2 to 1.4	1.0 to 1.2	1.0 to 1.2
Surface water drains for landfill caps	1.3 to 1.5	1.2 to 1.4	1.0 to 1.2	1.2 to 1.5
Secondary leachate collection (landfill)	1.5 to 2.0	1.4 to 2.0	1.5 to 2.0	1.5 to 2.0
Primary leachate collection (landfill)	1.5 to 2.0	1.4 to 2.0	1.5 to 2.0	1.5 to 10
Wick drains†	1.5 to 2.5	1.0 to 2.5	1.0 to 1.2	1.0 to 1.2
Highway edge drains	1.2 to 1.8	1.5 to 3.0	1.1 to 5.0	1.0 to 1.2

*These values assume that the ultimate value was obtained using an applied normal pressure of approximately 1.5 times the field anticipated maximum value. If not, the values must be increased.

†An additional term for kinking should be included, where RF_{KG} = 1.0 to 4.0.

Example 8.4

What is the flow rate factor of safety for a wick drain in a stratified soil profile requiring a flow rate of 0.10 l/min? Note from Table 8.4, however, that there is quite a difference of opinion as to the proper value of the required flow rate. The laboratory measured value of the candidate wick drain using solid plates as per ASTM D4716 is 2.8 l/min.

Solution: Using Eq. (8.10) but now adding an additional term for kinking (RF_{KG}), and the values from Table 8.5, we have

$$q_{\text{allow}} = q_{\text{ult}}\left[\frac{1}{RF_{IN} \times RF_{CR} \times RF_{CC} \times RF_{BC} \times RF_{KG}}\right]$$

$$= 2.8\left[\frac{1}{2.0 \times 1.5 \times 1.0 \times 1.0 \times 1.5}\right]$$

$$= 2.8\left(\frac{1}{4.5}\right)$$

$$= 0.62$$

Using Eq. (8.9), we have

$$FS = \frac{q_{\text{allow}}}{q_{\text{reqd}}}$$

$$= \frac{0.62}{0.10}$$

$$FS = 6.2 \qquad \text{which is acceptable}$$

and would be so even if a significantly higher required flow rate had been used.

Additional details on wick drains (prefabricated vertical drains) are found in Holtz et al. [34].

8.4.2 Sheet Drains

In a number of application areas involving sheet drainage, the requirements are higher than those indicated in Table 2.7 for geotextiles. Some of these areas are: behind retaining walls, against fractured rock slopes that are seeping, against soil slopes that are seeping, beneath athletic fields, beneath geomembrane liners, beneath floor slabs, beneath plaza decks, beneath surcharge fills, as capillary breaks, as vertical drainage inceptors, and as horizontal drainage inceptors. To meet the need, there are a number of commercially available geocomposite sheet drainage systems available. Some of these products are illustrated in Figure 8.13. We will refer to them collectively as *sheet drains,* although the reader should recognize that other names for these products are also used, such as *geomats* and *geospacers.* All have high flow capacities in their as-manufactured state, but vary greatly in their normal compression behavior (see Figure 8.14). It is, of course, the behavior in the compressed state that will dictate the amount of flow available for a given situation.

Figure 8.13 Various commercially available geocomposite sheet drains.

Most manufacturers have quantified their products for the actual flow rates under load and at various flow gradients. Figure 8.15 presents the typical type of information: illustrating a stiff core behavior in Figure 8.15a, versus a flexible core behavior in Figure 8.15b, versus two types of geonets in Figures 8.15c, d. Note, however, that these are index tests conducted between rigid platens and do not account for intrusion. To gain a better perspective of how these various geocomposite drains compare to one another, Figure 8.16 has been prepared. Here, seven different geocomposite drains are compared to one another (including geonets), and to two relatively thick nonwoven needle

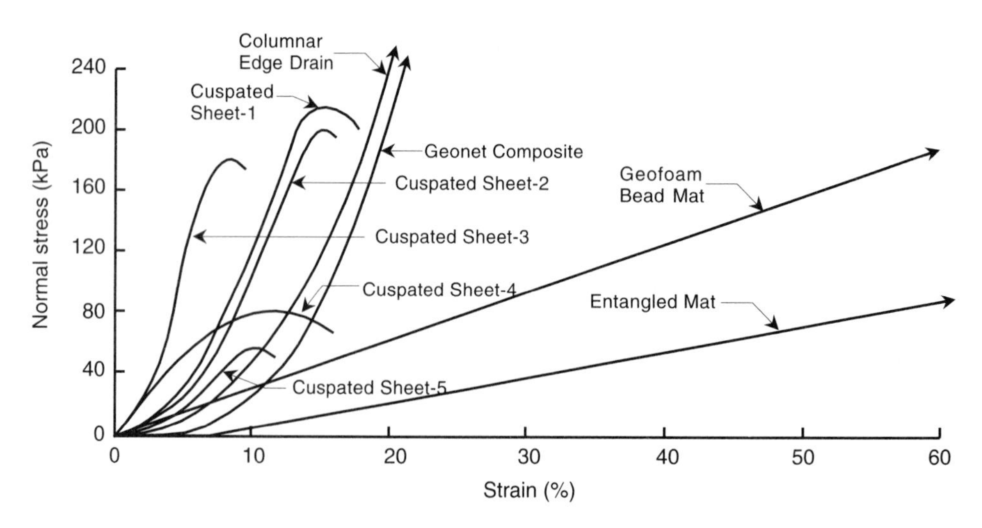

Figure 8.14 Compressibility behavior of selected geocomposite sheet drain materials.

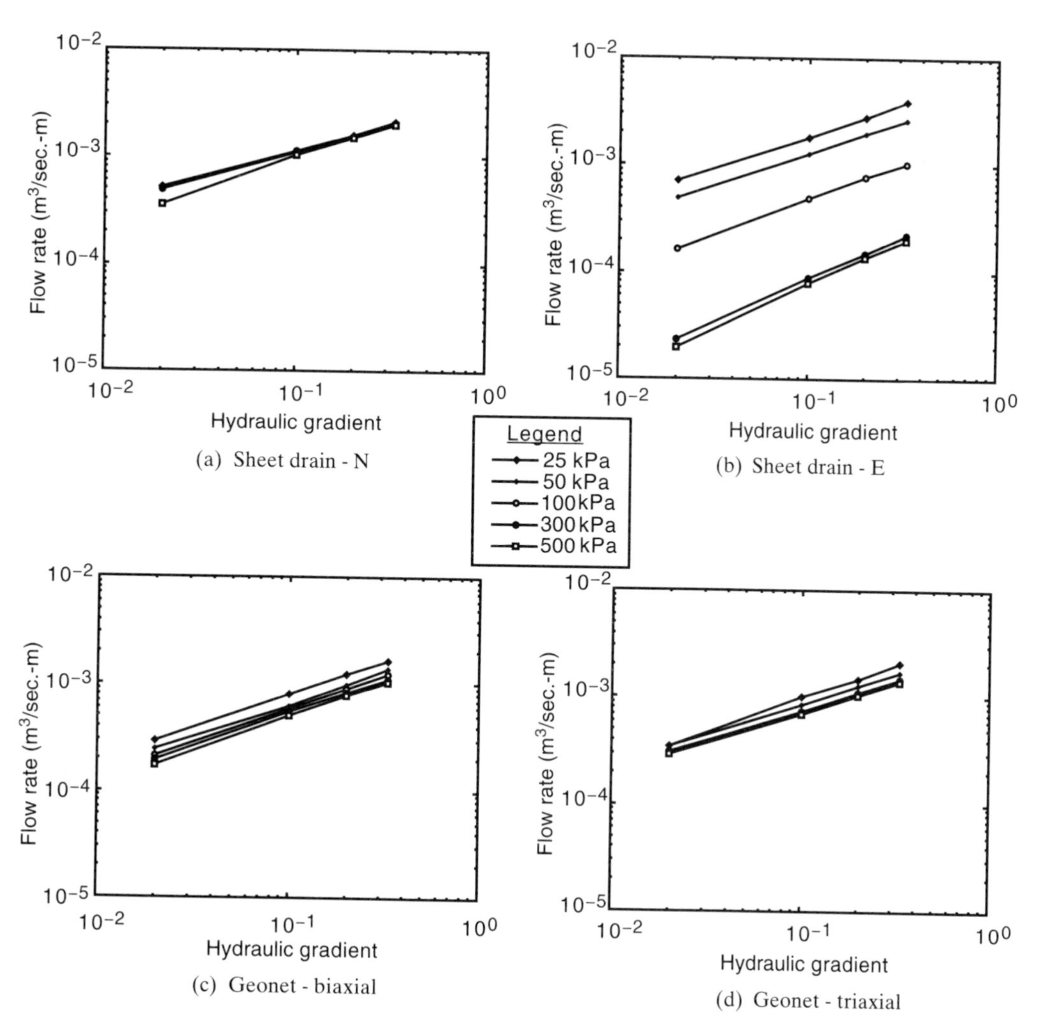

Figure 8.15 Index flow rate behavior of selected sheet drains compared to different types of geonets.

punched geotextiles. The upper figure has been generated by holding normal stress constant and varying the hydraulic gradient. The lower figure has been generated by holding hydraulic gradient constant and varying the normal stress. From these curves the following trends are observed.

- All geocomposite drains are significantly greater in their in-plane flow capability than even very thick geotextiles.
- The biaxial geonet composite (with geotextiles on the surfaces) is greater in flow rate than a relatively thick 1500 g/m^2 geotextile.
- The triaxial geonet composite is higher in its flow rate than the biaxial geonet composite due mainly to lower intrusion.

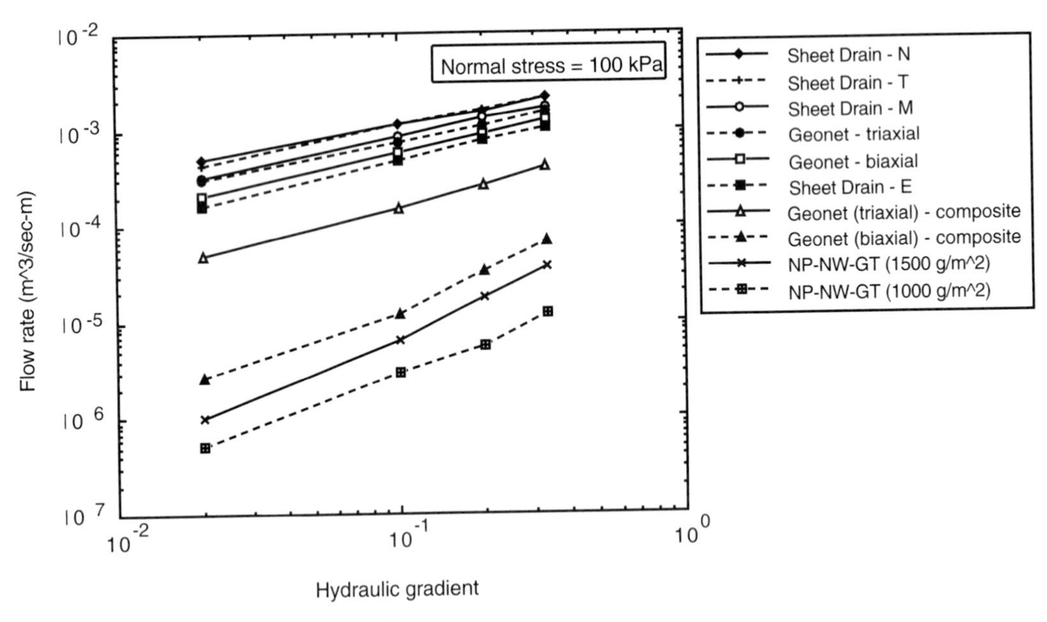

(a) Variation of hydrolic gradient

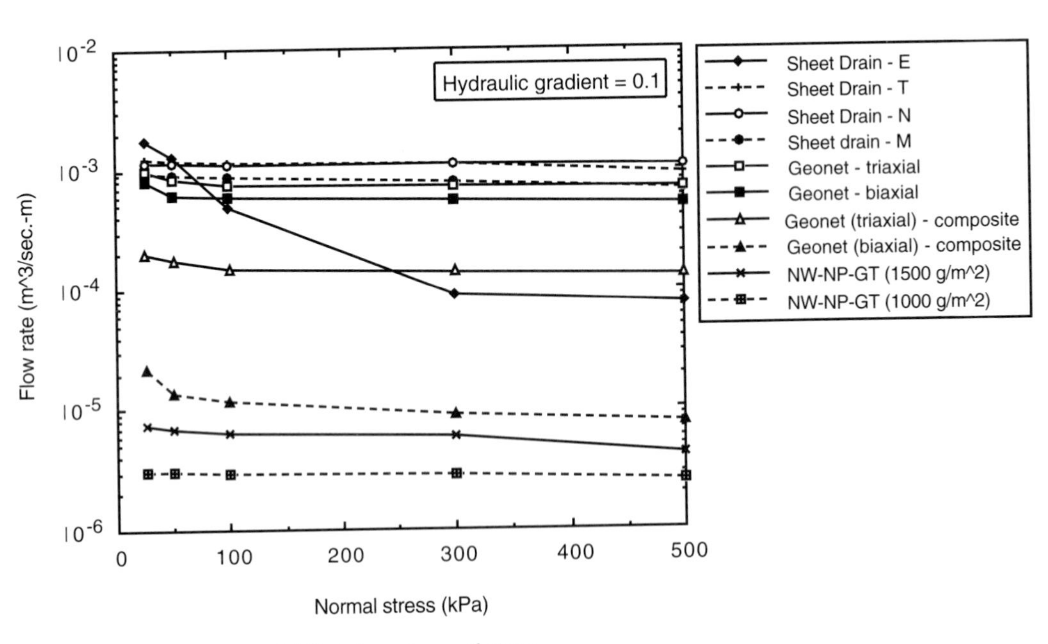

(b) Variation of normal stress

Figure 8.16 Index flow rate behavior of various sheet drains compared to geotextiles and geonets.

- The stiff cuspated and columnar drainage geocomposites offer the highest flow-rate capabilities of the products evaluated.

Note that the design procedure for geocomposite drains is exactly like that described in Chapter 4 for geonets. Geonets and geocomposite sheet drains are indeed competing geosynthetics in many application areas. Example 8.5 illustrates the use of sheet drains.

Example 8.5

Recalculate the retaining wall drainage problem originally given in Example 2.26 (Chapter 2) for geotextiles (where FS = 0.0062) and then again in Example 4.5 (Chapter 4) for geonets (where FS = 1.48), this time for the geocomposite whose response is given in Figure 8.15a. Recall that the required flow rate (which is equal to the transmissivity since the hydraulic gradient is 1.0) is 0.024 m³/min-m.

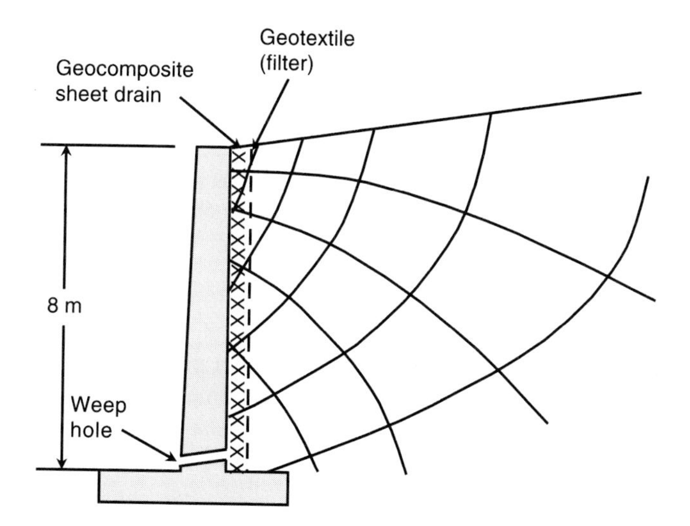

Solution: From the ultimate flow rate, an allowable value of the flow rate is obtained, which is then compared to the required (design) value for the global factor of safety.

(a) From Figure 8.15a at the maximum applied lateral pressure of

$$\sigma_n \cong 0.5(8.0)(18)$$

$$= 72 \text{ kPa}$$

and extrapolating to a hydraulic gradient of 1.0, the ultimate flow rate is

$$q_{\text{ult}} = 3.6 \times 10^{-3} \text{ m}^3\text{/sec-m}$$

$$q_{\text{ult}} = 0.216 \text{ m}^3\text{/min-m}$$

(b) Using Eq. (8.10) and Table 8.5 to obtain q_{allow} (note that the values taken from Table 8.5 are site-specific and up to the design engineer).

$$q_{\text{allow}} = q_{\text{ult}}\left[\frac{1}{\text{RF}_{IN} \times \text{RF}_{CR} \times \text{RF}_{CC} \times \text{RF}_{BC}}\right]$$

$$= 0.216\left[\frac{1}{1.4 \times 1.3 \times 1.3 \times 1.25}\right]$$

$$= 0.216\left(\frac{1}{2.96}\right)$$

$$q_{\text{allow}} = 0.073 \text{ m}^3/\text{min-m}$$

(c) The final comparison can now be made using Eq. (8.9).

$$\text{FS} = \frac{q_{\text{allow}}}{q_{\text{reqd}}}$$

$$= \frac{0.073}{0.024}$$

$$\text{FS} = 3.0 \qquad \text{which is acceptable}$$

Note that this is twice as great as the geonet solution (FS = 1.48) of Example 4.5 and almost 500 times greater than the geotextile solution (FS = 0.0062) of Example 2.26.

To be sure, there is a need for high flow rate sheet drain geocomposites, and the need is being satisfied by the currently available sheet drain products. There are, however, some unanswered questions that will probably lead to variations of the existing products and to the development of new ones as well.

- As seen in Figure 8.16, the core material drainage response varies significantly from product to product. The entire database must be quantified and greatly expanded.
- The creep of the system is of concern, particularly with respect to the breakdown stress of the cuspated cores for permanent installations and at high normal stress levels.
- The strength and creep of the geotextile filter are important in that the filter must span between protrusions of the core material. Flow rates will rapidly diminish if it excessively intrudes or plastically deforms into the flow channels of the core material, and will shut off completely if it fails.
- The geotextile covering must be designed with respect to the flow passing through it and the soil to be retained since it is indeed a filter in every sense of the word. The designs offered in Section 2.8 must be considered in this regard.
- The connection of the sheet drain to the interceptor drain must be such that flow is uniform and continuous. It involves both the drainage core and the geotextile filter.

- In some extreme cases, high or low temperatures may adversely affect the systems. High temperatures may lead to creep, while low temperatures can cause icing problems that block upstream flow. No known information is available in this regard.
- Claims about the insulation value of some of the systems have been noted, particularly when they are used adjacent to earth-sheltered structures. Although possible, the effect of moisture in the system versus its insulation value needs to be properly quantified.
- If a vapor barrier is required when placing the system next to an earth-sheltered structure, this can possibly be achieved by using a thin geomembrane on the side of the core facing the structure.
- Traditional vapor barriers on walls made from bitumen must be protected against some of the geocomposite sheet drain products that have hollow recesses on the side facing the structure. These recesses can indent the waterproofing, rendering it thinner than desired. A geomembrane on the back side of the sheet drain avoids the potential problem.
- The allowable flow rate to be used in a functional factor of safety design must come from a laboratory test replicating the actual in situ conditions. If not possible or practical (which is often the case), a modification of the laboratory determined value using reduction factors must be made. Eq. (8.10) and Table 8.5 are recommended.

8.4.3 Highway Edge Drains

Highway performance and lifetime are both directly related to the drainage capability of the stone base course beneath the pavement. This is clearly illustrated in Chapter 2, dealing with geotextiles as separators for the purpose of preserving stone base drainage, as well as for geotextile filters protecting the perforated pipe underdrain system at the edge of the pavement. But why do we need a perforated pipe, embedded in drainage stone, and further enclosed with a geotextile? Why not use a prefabricated drainage geocomposite serving as a complete edge drain by itself? Indeed such systems are now available under the category of geocomposite *highway edge drains*; see Figure 8.17.

The concept of prefabricated highway edge drains was pioneered by the Monsanto Co. and brought into a rational design context by Dempsey [38]. Of particular note is that the flow mechanism within highway edge drains is very different than the wick drains and sheet drains described in Sections 8.4.1 and 8.4.2. Figure 8.18 indicates that flow comes mainly from the stone base, through the geotextile filter, and then drops into the bottom portion of the vertically deployed core. Additional flow comes from the surface, where the shoulder often pulls away from the more stationary highway pavement. Generally no flow (or certainly very little) will come from the subsoil beneath the stone base or from the soil beneath the shoulder. Note that conceptually only the region of the geocomposite edge drain beneath the bottom of the stone base is carrying flow. The rest of the edge drain above this flow zone is just acting as an accumulator for the incoming water. The flow region then conveys the gathered water

Figure 8.17 Various commercially available geocomposite highway edge drains.

parallel to the highway to an appropriate outlet. Outlets are required at intervals of 50 to 150 m, depending upon the highway grade and hydrologic conditions. There are numerous types of product-specific prefabricated fittings that fit directly into the end of the drain and exit the accumulated water at 45, 60, or 90° away from the pavement section.

Regarding design, the basic flow rate equation will again apply.

$$FS = \frac{q_{\text{allow}}}{q_{\text{reqd}}} \tag{8.9}$$

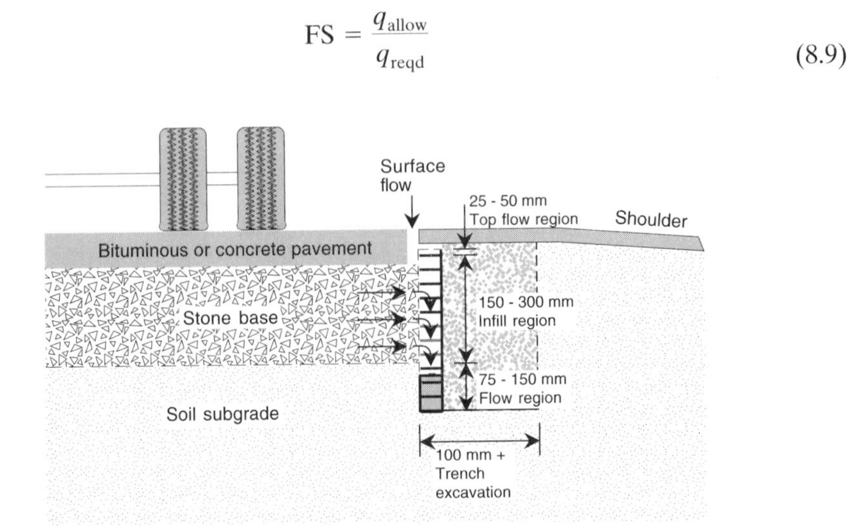

Figure 8.18 Flow path and configuration using prefabricated highway edge drains.

where

$$\text{FS} = \text{factor of safety},$$
$$q_{\text{allow}} = \text{allowable flow rate from a laboratory test (to be described), and}$$
$$q_{\text{reqd}} = \text{required (or design) flow rate.}$$

The laboratory-obtained flow rate should be determined by a flow box of the type shown in Figure 8.19. Here the simulation models the actual flow regime quite nicely. The major limitation of the test device is its complexity and its length (Dempsey [38] uses as 6.1 m). Laboratory-wise it is limited by practical considerations. Other nonrepresentative conditions such as solid side walls, short-term conditions, and the effects of liquids other than tap water can be handled according to Eq. (8.10) and Table 8.5. Thus a q_{ult} value obtained from this type of test can be modified to a q_{allow} value for design.

The required flow rate q_{reqd} has been measured by Dempsey [38] at a number of sites with precipitation rates from 20 to 35 mm/hr., giving values ranging from 760 to 1900 l/hr. Note, however, that this is a very site-specific value depending upon the following: the type of pavement surface, the condition and age of pavement, the type of stone base, the thickness of stone base, the fouling (contamination) of stone base, the edge joint condition, the rainfall and snow melt, the temperature, the shoulder type, the system gradient, the outlet spacing, the outlet type, the normal stress, and the response initiation time. Example 8.6 illustrates the design methodology.

Example 8.6

The maximum anticipated flow rate to a highway edge drain is 1150 l/hr. What is the factor of safety using a geocomposite edge drain whose laboratory test value is 4000 l/hr?

Solution: First calculate an allowable flow rate using Eq. (8.10) along with the values of Table 8.5 tuned to the site-specific conditions.

$$q_{\text{allow}} = q_{\text{ult}}\left[\frac{1}{\text{RF}_{IN} \times \text{RF}_{CR} \times \text{RF}_{CC} \times \text{RF}_{BC}}\right]$$

$$q_{\text{allow}} = 4000\left[\frac{1}{1.3 \times 1.5 \times 1.2 \times 1.1}\right]$$

$$= 4000\left(\frac{1}{2.57}\right)$$

$$= 1560 \text{ l/hr}$$

Then calculate the flow-rate factor of safety in the conventional manner per Eq. (8.9).

$$\text{FS} = \frac{q_{\text{allow}}}{q_{\text{reqd}}}$$

$$= \frac{1560}{1150}$$

$$= 1.4 \quad \text{which is low, but acceptable for this noncritical situation}$$

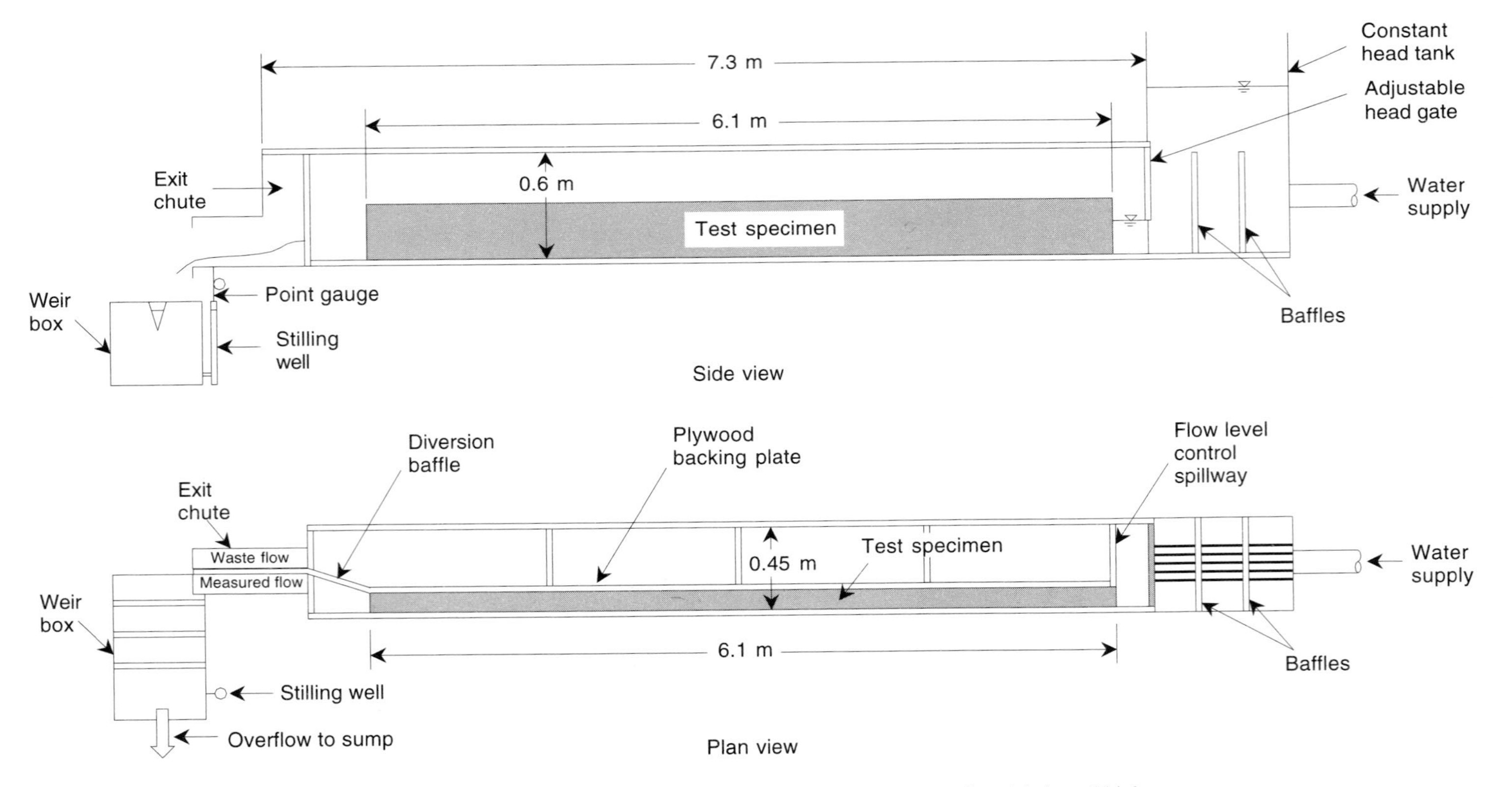

Figure 8.19 Side and plan view of flow box for evaluating flow rate of prefabricated highway edge drains. (After Dempsey [38])

A secondary consideration in the design methodology is the issue of the compressive strength of the drainage core. According to Koerner and Hwu [39] the maximum loading that a highway edge drain will experience is a parked truck on the shoulder of the highway. Using this type of loading and a Boussinesq analysis, the maximum normal stress is approximately 140 kPa at a depth of approximately 18 mm. A number of less tangible issues must now be resolved to obtain a required value, for example: the effect of dynamic stresses, the effect of overweight vehicles, the effect of creep stresses (e.g., a broken or abandoned truck), and the effect of inclined loads due to friction. Thus the calculated value should be increased. Koerner and Hwu [39] use a factor of 3.0, while some highway engineers use 2.0. Thus the required strength should probably be between 280 and 420 kPa.

Concerning the field performance of all types of highway edge drains, Koerner et al. [40] have exhumed 91 sites across the USA, including 41 geocomposite edge drains of the type described in this section. The cores were seen to perform quite well. The only failures, per se, were due to the use of a compressible product that was never intended for this application and a geotextile that was punctured by the core protrusion, due to its inadequate strength.

Problems were encountered, however, with the geotextile filter covering the drainage cores. In the above total of 41 exhumed geocomposite edge drains, there were 8 sites that allowed excessive amounts of soil to pass the geotextile and reside in the core. It was found that the soil that passed through the geotextile was always less than the #100 sieve in its particle size characteristics. Since the geotextiles used on these products were #40 to #70 AOS sieve size, it was felt that the failures were construction related, in that intimate contact of the upstream soil to the geotextile did not occur. Figure 8.20(a) illustrates the situation. The situation usually occurs with machine excavated trenches, which sometime create overexcavation beneath the pavement slab, particularly when coarse gravel or boulders are present. The suggested remedy [40] shown in Figure 8.20(b) is to move the edge drain to the shoulder side of the small 100 mm wide trench and then backfill with sand that is slurried into the open space. The water should carry the sand beneath the slab to fill any holes that may be present. In this situation the geotextile is designed as a filter for the sand, which is readily accomplished using the principles described in Section 2.8.

Clearly, geocomposite highway edge drains are the future in both retrofitted highway pavements and also in new construction. Their installed costs are extremely low due to the automated method of installation [40]. With adequate design of the geotextile filters [41], either through use of upstream sand or proper design to handle local situations, it is felt that these products can be used with confidence.

8.5 CONTAINMENT (LIQUID/VAPOR BARRIERS)

There are a wide variety of geosynthetic combinations that can be used as moisture barriers either to keep liquids within an area or out of an area. Already discussed are membrane encapsulated soil layers (MESLs) in connection with paved or unpaved road construction (see Section 2.6.2) and asphalt saturated geotextiles for use in the prevention of bituminous pavement crack reflection problems (see Section 2.10.2).

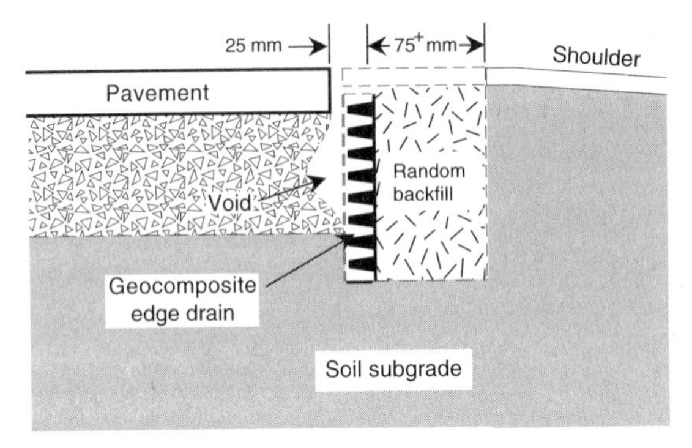

(a) Occurence of large void(s) beneath pavement slab preventing intimate contact of geotextile to upstream stone base and subgrade soil.

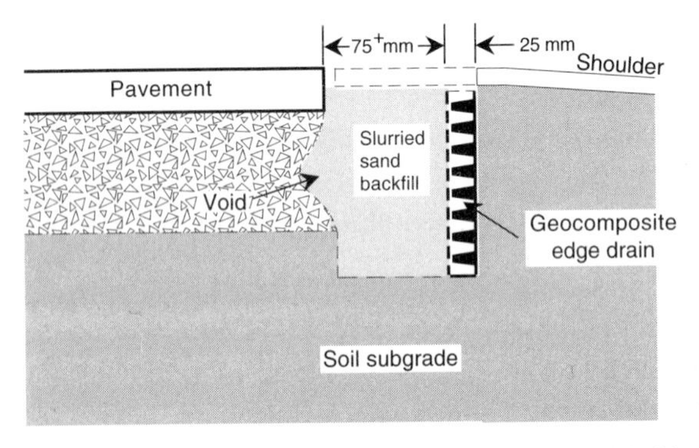

(b) Suggested remedy for backfilling large voids via slurried sand with geocomposite edge drain moved to shoulder side of trench.

Figure 8.20 Intimate contact issue and the avoidance of upstream voids using geocomposite highway edge drains. ([40, 41])

Both of these situations utilize composite behavior, but the discussions pertaining to them fell more appropriately in Chapter 2 than in this section. This does not imply that this section has a lower priority. Indeed, there is so much innovation occurring in the development of geocomposite moisture barriers that we scarcely know where to begin (see, for example, Figure 8.21). Thus without any particular order, a description will be made of various products fitting into this category.

Prefabricated *geotextile/bitumen* products are available wherein the two-ply systems are discrete, yet act together in composite form. The geotextile gives tensile strength, while the bitumen provides the moisture barrier. One particular variant of this class is the use of woven fiberglass fabric as the geotextile. This high strength, high modulus, low creep material is best used in the prevention of reflective cracking in distressed pavements. Rather than covering the entire roadway surface, as described in Section 2.10.2, however, these materials are meant for application directly spanning the

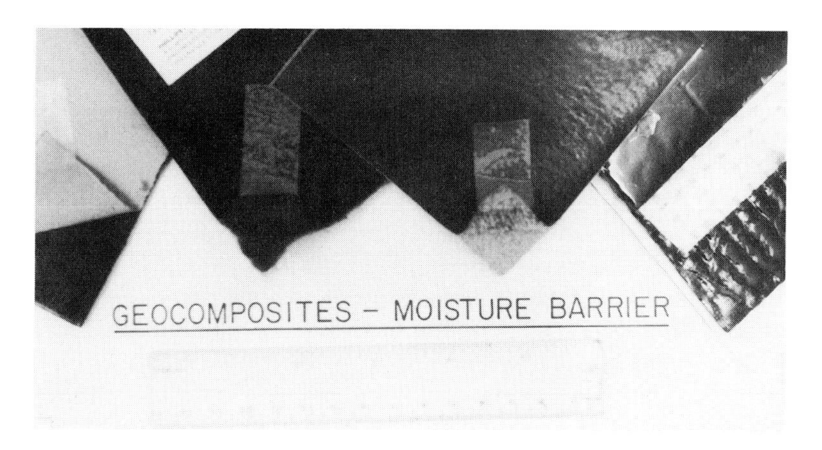

Figure 8.21 Various types of geocomposite moisture barriers.

cracks. The materials are laid down in 450 mm wide strips and undoubtedly provide a reinforcement function as well as a moisture barrier function.

Another product in this category, but now serving as a containment material, consists of the following nine layers: strip-film on top; slate, sand, or geotextile; bitumen; polyester film; bitumen; reinforcement geotextile; bitumen; sand or nonwoven geotextile; strip-film on bottom. The product, produced by Akzo, Inc. and called Hypofors, is supplied in large rolls placed directly on a prepared subgrade for landfill or reservoir containment liners. The seams are made using a hot bitumen mix poured directly on the overlapped joint. In this sense the product is actually a geomembrane and could have been included in Chapter 5; it is not because usual practice uses polymer-base (noncomposite) liners almost exclusively. However, in some countries (e.g., France and The Netherlands) bituminous liners and bituminous composite liners are seeing some use for lining of reservoirs, surface impoundments, and landfills.

Recognizing that geotextile underliners beneath geomembranes serve a number of valuable purposes, several one-piece *geotextile-geomembrane* geocomposites are available. These two-ply laminates are generally meant to serve as conventional geomembranes but can be placed much more efficiently than in two separate steps. Conceivably, any combination of geomembrane and/or geotextile could be developed in this manner. An area that promises to see growing use is a *geotextile-geomembrane-geotextile* composite used as a secondary containment system into which are placed underground gasoline storage tanks. Leaks from such tanks pose a serious threat to aquifers unless suitably contained.

The idea of encapsulating a thin metallic sheet between geomembranes has as its goal the complete barrier to the permeation of HC, CHC, and CFC organic liquids and vapors. Made commercially available by Agru Inc. in Austria and the U.S. are composite HDPE/aluminum/HDPE products that are 2.0 to 3.0 mm thick and supplied in rolls in a standard format. Complete impermeability to such organics as trichlorethylene, chloroform, tetrachlorethylene, toluene, tetrachlormethane, xylene, chlorobenzene, octave, and so on, is possible with such liner systems. The bonding of the layers is very strong and seaming is afforded by the HDPE covering layers using conventional thermal fusion methods.

Using this same concept, radioactive wastes could be contained by encapsulating a lead sheet between two HDPE outer layers. Lead is, of course, known to be resistant to the permeation of radioactive vapors and liquids.

Still further, Agru Inc. manufactures a geomembrane with uniformly spaced protrusions, or nubs, rising out from its surface. Such protrusions can be any height up to 25 mm and placed at any spacing or pattern. With a geotextile as a filter lying on top of the nubs, we have a combined liquid barrier and drainage system in the form of an integral geonet-geomembrane composite. With a geomembrane covering the top of the nubs, we have a double liner system with leak detection capability.

A parallel development, now with a high strength (in the *supertuff* class of polymers) geogrid manufactured with the geomembrane, is also in the prototype stage. Indeed, the era in which completely prefabricated, reinforced filter-drainage-barrier systems are mass-produced in the factory, with superb quality control, is about to begin. Truly a thrilling development is occurring.

8.6 GEOFOAM

Geofoam is the term being used more and more when referring to expanded polystyrene (EPS) or extruded polystyrene (XPS) used in below-grade construction applications. While it is sometimes used with a geotextile filter or reinforced with another material, it is somewhat of a stretch to include the topic in this chapter on geocomposites. We will consider it here because some exposure is clearly warranted, yet (at this time) not to the point of a complete chapter to itself. The primary function varies considerably; in fact, the usual geofoam applications of compressible inclusion, lightweight fill, and thermal insulation are difficult to categorize in the context of the five primary functions emphasized throughout this book.

Horvath [42] and Negussey [43] describes the manufacturing process for EPS (generally resulting in large blocks or panels but also in a spray-on format) and some typical properties.

- Density is in the range of 15 to 30 kg/m^3 (about 1/100th the weight of soil)
- Compression behavior is linear elastic up to 1 to 2% strain and then is plastic out to 10 to 20% strain (see Figure 8.22)
- The modulus in this range is approximately

$$E = 0.479\rho - 2.875 \tag{8.11}$$

where

E = initial tangent modulus (MPa), and
ρ = density (kg/m^3).

- Yield occurs over a range of 5 to 10% strain
- Post yield behavior exhibits a work-hardening response

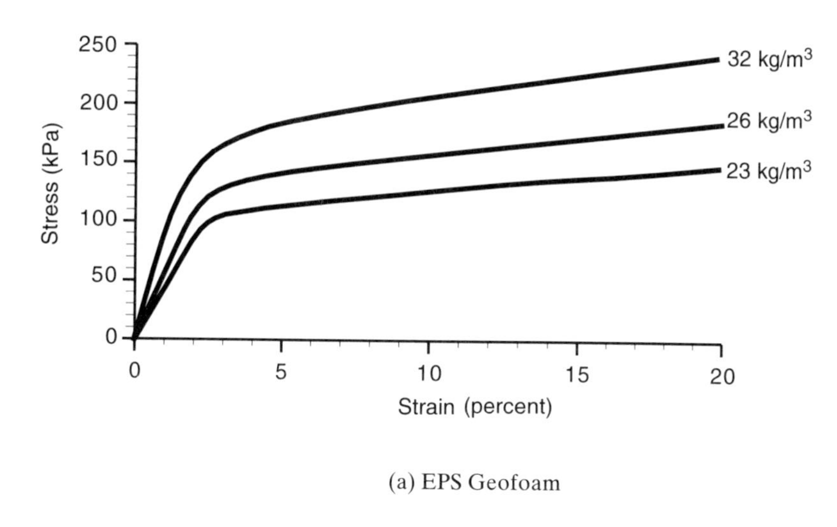

(a) EPS Geofoam

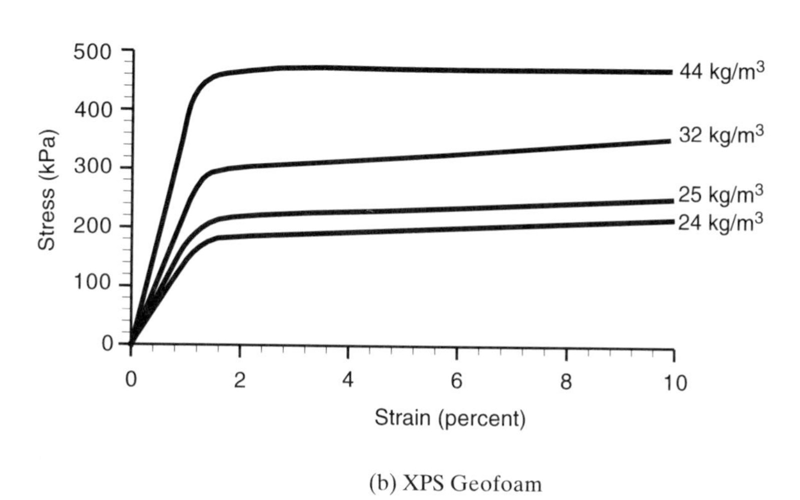

(b) XPS Geofoam

Figure 8.22 Unconfined compression behavior of geofoam and the effect of density. (After Negussey [43])

- Compression-creep and stress-relaxation behaviors are deterministic
- Thermal conductivity is in the range of 25 to 40 mW/m-K when the geofoam is dry

A number of interesting case histories illustrate the use of EPS geofoam. Partos and Kazaniwsky [44] used geofoam blocks to balance earth pressures on each side of an underground structure. Horvath [45] also describes such applications, and adds tensile reinforcement, which could possibly reduce the lateral pressure from active conditions to essentially zero. Stulgis et al. [46] use geofoam for its insulation value in keeping a geocomposite sheet drain behind a retaining wall from freezing. Horvath [47] adds to the use of geofoam as a compressible inclusion by posing its use as a damping medium for vibrations (trains, subways, highways). Potentially, even seismic waves may be attenuated.

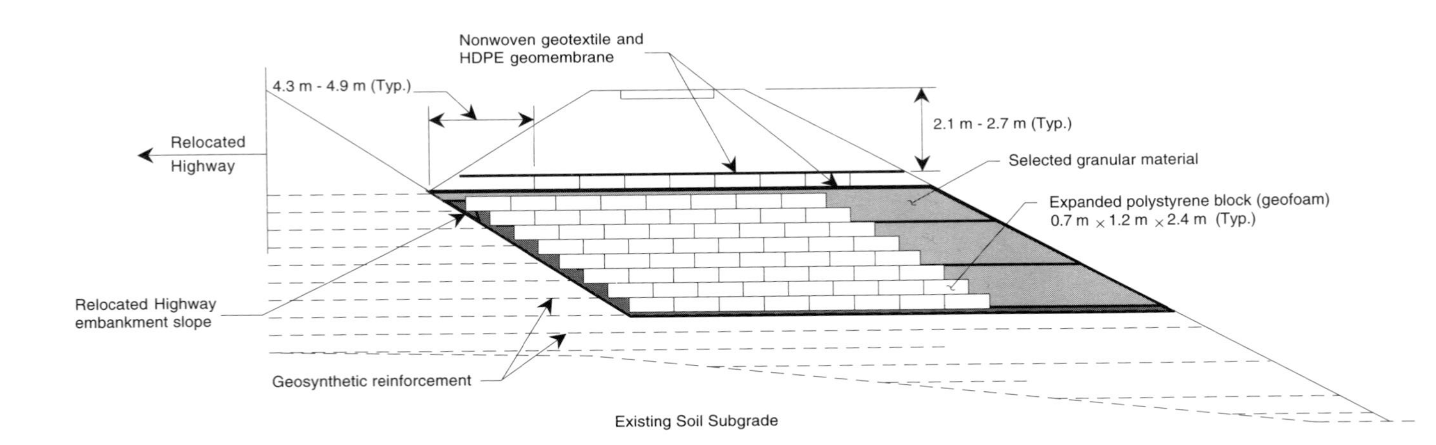

Figure 8.23 Cross section of 21 m high reinforced highway embankment over soft foundation soil using geofoam as a lightweight fill. (After Mimura and Kimura [49])

Geofoam use as a lightweight fill has been exploited in Scandinavia for many years. In Norway, as many as 100 highway projects have used geofoam. The typical situation is the construction of highway fills over frost sensitive soils. Its thermal insulation properties are generally referenced in this regard. However, soft soils of any type can be built on using geofoam blocks. Frydenlund and Aaboe [48] report on a number of applications, and Mimura and Kimura [49] illustrate a case history of geosynthetic reinforcement inclusions. The latter situation is particularly interesting, as shown in Figure 8.23. Note that this case history used almost every type of geosynthetic material that is available.

The environmental uses of geofoam have been as a spray for daily landfill cover (Gasper [50]), to create a floating pond cover (Borgaard and Anderson [51]), to prevent frost damage to compacted clay liners, and to assist in methane and radon gas venting (White [52]).

Clearly, geofoam is a growing aspect of geosynthetic engineering and only time will tell how great its impact will be. Perhaps in future editions, geofoam will be a stand-alone chapter. This seems to be indicated by the outflow of publications; for example, a separate conference proceedings is available (Horvath [53]).

8.7 CONCLUSION

The word *innovation* best summarizes the heart of this chapter. By knowing the strongest and weakest points of two materials, they can sometimes be used together to emphasize the strongest points of both of them. When done cleverly, even a synergistic effect can be the result, where the combined effect is better than the sum of performances of the separate materials. These performance thoughts, however, must be weighed in light of the economics of the production of the geocomposites.

It is of little surprise that manufacturers are taking the lead in the area of geocomposites. New products spring up regularly, leaving the designer to play catch up in assessing these new products. The phrase *or equal,* commonly used in specification writing, is completely inappropriate in this area (as it is with most areas involving geosynthetics). Certainly new products will appear and old ones will disappear in the future. It is up to the designer to understand the application, evaluate the function or functions, and compare these requirements to that of the candidate geocomposite's properties. The resulting factor of safety is then assessed in light of the particular application. The process is not different in any way from other engineering materials using *design-by-function*—the byword of this entire book.

REFERENCES

1. Garrels, R. M., *A Textbook of Geology,* New York: Harper Bros., 1951.
2. Theisen, M. S., "The Role of Geosynthetics in Erosion and Sediment Control: An Overview," *J. Geotextiles and Geomembranes,* Vol. 11, Nos. 4–6, 1992, pp. 199–214.
3. Weggel, J. R., and Rustom, R., "Soil Erosion by Rainfall and Runoff—State of the Art," *J. Geotextiles and Geomembranes,* Vol. 11, Nos. 4–6, 1992, pp. 215–236.

4. Wischmeier, W. H., and Smith, D. D., "A Universal Soil-Loss Equation to Guide Conservation Farm Planning," *Proc. 7th Intl. Conf. on Soil Science,* Washington, DC: Soil Science Society of America, 1960.
5. Ingold, T. S., and Thompson, J. C., "The Comparative Performance of Some Geosynthetic and Geochemical Erosion Control Systems," *Proc. Environmental Geotechnology Conf.,* ed. H. Y. Fang, Bethlehem, PA: Envo Publ. Co., 1986, pp. 413–424.
6. Rustom, R., and Weggel, J. R., "A Study of Erosion Control Systems: Experimental Results," *Proc. Intl. Erosion Control Assoc.,* Steamboat Springs, CO, 1993, pp. 239–251.
7. Theisen, M. S., and Carroll, R. G., Jr., "Turf Reinforcement—The Soft Armor Alternative," *Proc. 21st Conf. Prof. Intl. Erosion Control Assoc.,* Steamboat Springs, CO, 1990, pp. 255–270.
8. DeGroot, M. T., "Woven Steelcord Networks as Reinforcement of Asphalt Roads," *Proc. 3d Intl. Conf. on Geotextiles,* Vienna: Austrian Society of Engineers and Architects, 1986, pp. 113–118.
9. Kenter, C. J., DeGroot, M. T., and Dunnewind, H. J., "An Instant Road of Steel Reinforced Geotextile," *Proc. 3d Intl. Conf. on Geotextiles,* Vienna: Austrian Society of Engineers and Architects, 1986, pp. 67–70.
10. Hoare, D. J., "Laboratory Study of Granular Soils Reinforced with Randomly Oriented Discrete Fibers," *Proc. Intl. Conf. on Soil Reinforcement,* Paris: International Society of Soil Mechanics and Foundation Engineering, French Chapter, 1979, pp. 47–52.
11. Mercer, F. B., Andrawes, K.Z., McGown, A., and Hytiris, N., "A New Method of Soil Stabilization," in *Polymer Grid Reinforcement,* London: Thomas Telford, 1985, pp. 244–249.
12. McGown, A., Andrawes, K. Z., Hytiris, N., and Mercer, F. B., "Soil Strengthening Using Random Distributed Mesh Elements," *Proc. 11th ISSMFE Conf.,* Vancouver, Canada: BiTech Publications Ltd., 1985, pp. 1735–1738.
13. Leflaive, E., "The Reinforcement of Granular Materials with Continuous Fibers," *Proc. 2d Intl. Conf. on Geotextiles,* St. Paul, MN: IFAI, 1982, pp. 721–726.
14. Leflaive, E., and Liausu, Ph., "The Reinforcement of Soils by Continuous Threads," *Proc. 3d Intl. Conf. on Geotextiles,* Vienna: Austrian Society of Engineers and Architects, 1986, pp. 1159–1162.
15. "WES Developing Sand-Grid Confinement System," *Army Res. Dev. Acquisition Magazine,* July–Aug., 1981, p. 7.
16. John, N. W. M., *Geotextiles*, Glasgow: Blackie Publ. Co. Ltd., 1987.
17. Edgar, S., "The Use of High Tensile Polymer Grid Mattress on the Musselburgh and Portobello Bypass," in *Polymer Grid Reinforcement,* London: Thomas Telford, 1985, pp. 103–111.
18. Batson, G. B., "Introduction to Fibrous Concrete," *Proc. CERL Fibrous Concrete Conf.,* Champaign, IL: University of Illinois Press, 1972, pp. 1–25.
19. Hoff, G. C., *Research and Development of Fiber Reinforced Concrete in North America,* Misc. Paper No. C-74-3, U.S. Army Waterways Experiment Station, Vicksburg, MS, 1974.
20. Jenq, Y.-S., "Peformance Evalution of Fiber Reinforced Asphalt Concrete," FHWA/OH/94/018, 1994, Washington, DC.
21. Visser, T., and Mouw, K. A. G., "The Development and Application of Geotextiles on the Oosterschelde Project," *Proc. 2d Intl. Conf. Geotextiles,* St. Paul, MN: IFAI, 1982, pp. 265–270.
22. Door, H. C., and DeHaan, D. W., "The Oosterschelde Filter Mattress and Gravel Bag," *Proc. 2d Intl. Conf. Geotextiles,* St. Paul, MN: IFAI, 1982, pp. 271–276.
23. Van Harten, K., "Analysis and Experimental Testing of Load Distribution in the Foundation Mattress," *Proc. 2d Intl. Conf. Geotextiles,* St. Paul, MN: IFAI, 1982, pp. 277–282.
24. Wisse, J. D. M., and Birkenfeld, S., "The Long-Term Thermo-Oxidative Stability of Polypropylene Geotextiles in the Oosterschelde Project," *Proc. 2d Intl. Conf. Geotextiles,* St. Paul, MN: IFAI, 1982, pp. 283–286.

25. Gerry, B. S., and Raymond, G. P., "The In-Plane Permeability of Geotextiles," *Geotech. Testing J.,* ASTM, Vol. 6, No. 4, 1983, pp. 181–189.
26. *Study of Deep Soil Stabilization by Vertical Sand Drains,* Bureau of Yards and Docks, U.S. Navy, Rep. No. 88812, 1958.
27. Hansbo, S., "Consolidation of Clay by Band Shaped Perforated Drains," *Ground Eng.,* 1979, pp. 16–25.
28. Rixner, J. J., Kraemer, S. R., and Smith, A. D., *Prefabricated Vertical Drains,* Report No. FHwA/RD-86/168, Washington, DC, 1986.
29. Suits, L. D., Gemme, R. L., and Masi, J. J., "The Effectiveness of Prefabricated Drains in the Laboratory Consolidation of Remolded Soils," *ASTM Symp. Consolidation of Soils: Laboratory Testing,* W. Conshohocken, PA: ASTM, 1985, pp. 114–126.
30. Lawrence, C. A., and Koerner, R. M., "Flow Behavior of Kinked Strip Drains," *Proc. Sym., Geosynthetics for Soil Improvement,"* New York: ASCE, 1988, pp. 22–39.
31. den Hoedt, G., "Laboratory Testing of Vertical Drains," *Proc. 10th ISSMFE,* Stockholm: Swedish Society of Civil Engineers, 1981, pp. 627–630.
32. Kremer, R. et al., "Quality Standards for Vertical Drains," *Proc. 2d Intl. Conf. Geotextiles,* St. Paul, MN: IFAI, 1982, pp. 319–324.
33. Kremer, R. et al., "The Quality of Vertical Drainage," *Proc. 8th European Conf. SSMFE,* Helsinki: Finnish Society of Civil Engineers, 1983, pp. 721–726.
34. Holtz, R. D., Jamiolkowski, M., Lancelotta, R., and Pedroni, S., *Prefabricated Vertical Drains,* Newton, MA: Butterworth/Heinemann, 1990.
35. Koerner, R. M., Fowler, J., and Lawrence, C. A., "Soft Soil Stabilization Study for Wilmington Harbor South Dredge Materials Disposal Area," U. S. Army Corps of Engineers, WES, Misc. Paper GL-86-38, Vicksburg, MS, 1986.
36. Holtz, R. D., and Christopher, B. R., "Characteristics of Prefabricated Vertical Drains for Accelerating Consolidation," *Proc. 9th European Conf. SSMFE,* Dublin: Irish Society of Civil Engineers, 1987, pgs 453–466.
37. Bergado, D. T., Manivannan, R., and Balasubramanian, A. S., "Proposed Criteria for Discharge Capacity of Prefabricated Vertical Drains," *J. Geotextiles and Geomembranes,* Vol. 14, No. 9, 1996, pp. 481–506.
38. Dempsey, B. J., "Core Flow Capacity Requirements of Geocomposite Fin-Drain Materials Utilized in Pavement Subdrainage," 67th Annual Transportation Research Board Meeting, Washington, DC, 1988.
39. Koerner, R. M., and Hwu, B.-L., "Prefabricated Highway Edge Drains," *Transportation Research Record* No. 1329, Transportation Research Board, Washington, DC, 1991, pp. 14–20.
40. Koerner, G. R., Koerner, R. M., and Wilson-Fahmy, R. F., "Field Performance of Geosynthetic Highway Drainage Systems," *Recent Developments in Geotextile Filters and Prefabricated Drainage Geocomposites, ASTM STP 1281,* ed Shobha K. Bhatia and L. David Suits, W. Conshohocken, PA: ASTM, 1996, pp. 165–181.
41. Wilson-Fahmy, R. F., Koerner, G. R., and Koerner, R. M., "Geotextile Filter Design Critique," *Recent Developments in Geotextile Filters and Prefabricated Drainage Geocomposites, ASTM STP 1281,* ed Shobha K. Bhatia and L. David Suits, W. Conshohocken, PA: ASTM, 1996, pp. 132–161.
42. Horvath, J. S., "Expanded Polystyrene (EPS) Geofoam: An Introduction to Material Behavior," *J. of Geotextiles and Geomembranes,* Vol. 13, No. 4, 1994, pp. 263–280.
43. Negussey, D., "Properties and Applications of Geofoam," Society of Plastics Engineers, Washington, DC, 1997.
44. Partos, A. M., and Kazaniwksy, P. M., "Geoboard Reduces Lateral Earth Pressures," *Proc. Geosynthetics '87,* St. Paul, MN: IFAI, 1987, pp. 409–424.

45. Horvath, J. S., "Using Geosynthetics to Reduce Earth Loads on Rigid," *Proc. Geosynthetics 1991,* St. Paul, MN: IFAI, 1991, pp. 409–424.
46. Stulgis, R. P., Dykstra, T. A., Telgener, R. J., and Oosterbaan, M. D., "Design and Construction of a Permanent Soil Nail Wall," *Proc. Reinforced Retaining Walls*, Denver, CO: University of Colorado Press, 1997, pp. 1–13.
47. Horvath, J. S., "EPS as a Vibration Damper and Drainage Product," *Geosynthetics World*, Vol. 4, No. 2, 1994, p. 11.
48. Frydenlund, T. E., and Aaboe, R., "Expanded Polystyrene—A Superlight Fill Material," *International Geotechnical Symposium on Theory and Practice of Earth Reinforcement,* Rotterdam: A. A. Balkema, 1988, pp. 383–388.
49. Mimura, C. S., and Kimura, S. A., "A Lightweight Solution," *Proc. Geosynthetics '95 Conf.*, St. Paul, MN: IFAI, 1995, pp. 39–52.
50. Gasper, A. J., "Stabilized Foam as Landfill Daily Cover," *Proc. Municipal Solid Waste Management: Solutions for the 90's,* Washington, DC: U.S. EPA, 1990, pp. 1113–1121.
51. Bogaard, D., and Anderson, R., "Geosynthetics Combine to Create an Efficient Floating Insulated Pond Cover," *Geotechnical Fabrics Rpt.,* Vol. 12, No. 6, 1994, pp. 26–27.
52. White, R., "EPS Used to Assist in Methane and Radon Gas Venting," *Geosynthetics World,* Vol. 5, No. 2, 1995, p. 12.
53. Horvath, J. S., ed., *Geofoam Geosynthetic,* Scarsdale, NY: Horvath Engineering, P.C., Scarsdale, NY, 1995.

PROBLEMS

8.1. List some of the various joining methods that you could suggest to make a laminated geocomposite from different geosynthetic materials.

8.2. Describe or illustrate the differences between sheet, gully, and rill erosion mechanisms.

8.3. Using the Unified Soil Classification for typical soil particle sizes (Figure 8.1), what velocities are required for the erosion/transportation of gravels, sands, silts, and clays. Answer the problem for both water and air.

8.4. What test methods and properties should be included in a generic specification for an erosion control geocomposite?

8.5. When used on exposed slopes, synthetic materials eventually degrade. Why does this occur, how it is usually minimized by geosynthetic manufacturers, and how is it prevented when using a PERM?

8.6. Erosion control is very much related to the intimate contact of the system to the soil that it is to protect. How is this statement evidenced by the data of Figure 8.4?

8.7. In Section 8.2.1 it is mentioned that Mommoth-Mat could be used for the direct support of vehicles without any soil or paved surface covering. In such instances, what test methods and properties should be included in a generic specification?

8.8. What shear strength properties of a sand are modified by the inclusion of fibers per Section 8.2.3? What differences do you envision between short fibers and continuous fibers?

8.9. Determine the ultimate bearing capacity of a cohesionless sand using a geocell reinforcement, as described in Example 8.1 as a function of the depth of the mattress. Vary the depth using 50, 100, 200 (the example), and 300 mm, and plot the response curve. Use the following soil properties: $\gamma = 17$ kN/m^3, $\phi = 27°$, and $\sigma_h = 25$ kPa (avg.).

8.10. Three-dimensional geogrid mattresses filled with gravel are mentioned in Section 8.2.3. When placed on soft foundation soil, how do these mattresses provide stability, that is, what are the design elements that are improved? [*Hint:* Revisit Chapter 3 for information in this regard.]

8.11. What are the advantages and disadvantages of mixing polymeric fibers into the following materials:

(a) Portland cement concrete, per Section 8.2.4

(b) Bituminous concrete, per Section 8.2.5

8.12. As with graded soil filters, geotextiles can be used in a similar composite form, per Section 8.3. If a cohesionless silt (ML classification) were to be placed adjacent to a poorly graded gravel (GP classification), how could a composite geotextile filter be fabricated to accommodate these very dissimilar soils? (Use assumed values for your opening size design.)

8.13. Concerning the wick drains described in Section 8.4.1, answer the following questions:

(a) Do these geocomposites actually *wick* the water out of the soil as a wick in a candle brings wax up to the flame?

(b) Does the installation of wick drains affect the amount of settlement or the rate of settlement?

(c) What function does the geotextile covering the plastic core play?

(d) What is the role of the core?

(e) How does the expelled water exit the wick drain (top and/or bottom) so that excess pore water pressures do not build up under the surcharge?

8.14. Describe and illustrate what is meant by wick drain *smear,* as probably occurs during installation.

8.15. What is meant by *well resistance*, as might occur in the performance of sand drains versus wick drains?

8.16. What is the time for 95% consolidation of a proposed building site using wick drains measuring 92 × 10 mm in cross-sectional area to be placed on a square pattern of 1.25 m centers in a saturated silt having a horizontal coefficient of consolidation of 12 mm²/min?

8.17. Regarding the Hansbo analysis for wick drain spacing:

(a) What spacing is required for wick drains of 100 mm × 6 mm placed on a triangular pattern, if the required time for 50% consolidation is 90 days in a saturated silty clay with a horizontal coefficient of consolidation of 0.65 mm²/min?

(b) Check the spacing for part (a) if the time for 90% consolidation is 300 days.

8.18. Regarding traditional consolidation soil testing:

(a) Describe how you measure the vertical coefficient of consolidation (c_v) of an in situ soil.

(b) On a conceptual basis, how could the value in part (a) be related to the horizontal coefficient of consolidation (c_h)?

(c) How could a soil sample be oriented to obtain c_h directly?

8.19. Calculate the time for 90% consolidation settlement for a silty clay soil with c_h = 4.5 mm²/min using wick drains measuring 100 mm × 3.8 mm on triangular spacings of 4.0, 3.0, 2.0, and 1.0 m and plot the results on a semilog graph as shown in Example 8.3 (Section 8.4.1).

8.20. Repeat Problem 8.19 for 70% and 50% consolidation, and compare the three response curves.

8.21. Given a measured flow rate between solid end platens of a wick drain in the laboratory ASTM D4716 test of 45 l/min-m at site-specific pressure and a hydraulic gradient of 1.0:

(a) Reduce this value according to Table 8.5 to an allowable value (use average values).

(b) Calculate the flow rate factors of safety using the required flow rates from Table 8.4 following the Koerner value of 0.10 l/min. and then the Holtz/Christopher value of 0.95 l/min.

(c) The calculation in part (b) is based on a typical length of wick drain. What are the implications of extremely long wick drains?

8.22. Given the measured flow rate of a sheet drain geocomposite of 95 l/min-m when tested between solid platens in a planar flow test, what would be the allowable flow rate using Table 8.5 if used in

(a) sport-field drainage?

(b) a landfill cap?

(c) a highway edge drain?

8.23. Using sheet drain E, shown in Figure 8.15b, check to see if it could handle the flow in Example 8.5 (Section 8.4.2) for the concrete cantilever retaining wall drainage.

8.24. Determine the flow rate that can be handled by sheet drain N, shown in Figure 8.15a, for the following situation of drainage of a roof garden. Repeat the problem, but now use sheet drain E from Figure 8.15b.

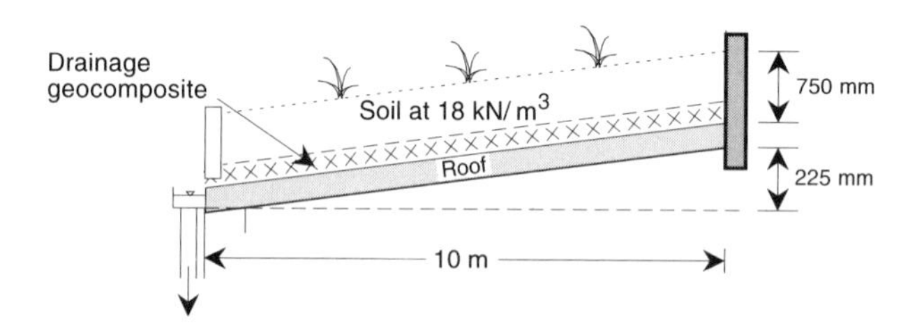

8.25. Of all the geosynthetic drainage materials currently existing, the highway edge drains presented in Section 8.4.3 have the highest flow rates. Their thicknesses range from 12 to 38 mm. Estimate where their response curve falls on Figures 8.16a,b. What are the range of flow rates for these products?

8.26. For geocomposite highway edge drains, why is the test method illustrated in Figure 8.19 preferred over the ASTM D4716 test method? If the only laboratory test data available is from an ASTM D4716 test, how do you modify the value to be representative (i.e. what additional reduction factor should be used)?

8.27. For the geocomposite barriers in Section 8.5:

(a) What are some reasons for attaching a geotextile to the upper side of a geomembrane used in a landfill cover?

(b) What are some reasons for attaching a geotextile to the underside of a geomembrane that is used as a single moisture barrier for a building foundation?

(c) In such cases, why are the seams difficult to make?

8.28. For geocomposite barriers per Section 8.5 an aluminum sheet (as a hydrocarbon barrier) and a lead sheet (as a radioactive barrier) were mentioned.

(a) How is the HDPE bonded to the metallic sheet?

(b) How are seams made?

(c) What installation difficulties do you envision?

8.29. In Section 8.7, it was mentioned that the phrase "or equal" in specifications is inappropriate. Why is this the case?

Index

A

B

C

D

E

F

G

H

I

Q

R

S

V

W

X

Y